Phage Therapy

Current Research and Applications

Edited by

Jan Borysowski

Department of Clinical Immunology
The Medical University of Warsaw
Warsaw
Poland

Ryszard Międzybrodzki and Andrzej Górski

Department of Clinical Immunology
The Medical University of Warsaw
Warsaw
and
Ludwik Hirszfeld Institute of Immunology and Experimental Therapy
Polish Academy of Sciences
Wrocław
Poland

Caister Academic Press

Caister Academic Press
Norfolk, UK

www.caister.com

British Library Cataloguing-in-Publication Data
A catalogue record for this book is available from the British Library

ISBN (hardback): 978-1-908230-40-9
ISBN (ebook): 978-1-908230-74-4

Cover design adapted from various images. Top-left image: *Enterococcus faecalis* EF1 phage (courtesy of A. Gozdek and D. Izak, Institute of Biochemistry and Biophysics, Polish Academy of Sciences, Warsaw, Poland). Top-right image: *Stenotrophomonas maltophilia* Steno11/8c phage taken with electron microscope (courtesy of J. Kassner, University of Wrocław, and B. Weber-Dąbrowska). Bottom-left image: plaques of *Enterococcus faecalis* Ent23 phage (courtesy of R. Międzybrodzki). Bottom-right image: *Enterococcus faecalis* EF56 phage (courtesy of A. Gozdek and D. Izak, Institute of Biochemistry and Biophysics, Polish Academy of Sciences, Warsaw, Poland). The top-left and bottom-right images were taken in the Laboratory of Electron Microscopy at the Nencki Institute of Experimental Biology, Warsaw, Poland, using JEM 1400 (JEOL Co., Japan, 2008) transmission electron microscope.

Contents

Contributors

Stephen T. Abedon
Department of Microbiology
The Ohio State University
Columbus, OH
USA

abedon.1@osu.edu

Hans-Wolfgang Ackermann
Department of Microbiology
Laval University
Quebec, QC
Canada

ackermann@mcb.ulaval.ca

Revaz Adamia[†]
Eliava Institute of Bacteriophages, Microbiology and Virology
Tbilisi
Georgia

radamia@gmail.com

Zemphira Alavidze
Eliava Institute of Bacteriophages, Microbiology and Virology
Tbilisi
Georgia

zemal@caucasus.net

Joana Azeredo
Institute for Biotechnology and Bioengineering
University of Minho
Braga
Portugal

jazeredo@deb.uminho.pt

Botond Balogh
Nichino America Inc.
Apollo Beach, FL
USA

bbalogh@nichino.net

Itai Benhar
Department of Molecular Microbiology and Biotechnology
Tel-Aviv University
Ramat Aviv
Israel

benhar@post.tau.ac.il

Udo Bläsi
Department of Microbiology, Immunobiology and Genetics
University of Vienna
Vienna
Austria

udo.blaesi@univie.ac.at

Jan Borysowski
Department of Clinical Immunology
The Medical University of Warsaw
Warsaw
Poland

jborysowski@interia.pl

Carl Ceulemans
Department of Philosophy
Royal Military Academy
Brussels
Belgium

carl.ceulemans@rma.ac.be

Aidan Coffey
Department of Biological Sciences
Cork Institute of Technology
Bishopstown
Ireland

aidan.coffey@cit.ie

Anneleen Cornelissen
Laboratory of Gene Technology
Katholieke Universiteit Leuven
Leuven
Belgium

anneleencornelissen@hotmail.com

Krystyna Dąbrowska
Ludwik Hirszfeld Institute of Immunology and Experimental Therapy
Polish Academy of Sciences
Wrocław
Poland

dabrok@iitd.pan.wroc.pl

Michał Dadlez
Institute of Biochemistry and Biophysics
Polish Academy of Sciences
Warsaw
Poland

michald@ibb.waw.pl

Masanori Daibata
Department of Microbiology and Infection
Kochi University Medical School
Kochi
Japan

daibatam@kochi-u.ac.jp

Daniel De Vos
Laboratory for Molecular and Cellular Technology
Burn Wound Centre
Queen Astrid Military Hospital
Brussels
Belgium

daniel.devos2@mil.be

Grzegorz Figura
Ludwik Hirszfeld Institute of Immunology and Experimental Therapy
Polish Academy of Sciences
Wrocław
Poland

grzegorz.figura@iitd.pan.wroc.pl

Vincent A. Fischetti
Laboratory of Bacterial Pathogenesis
Rockefeller University
New York, NY
USA

vaf@rockefeller.edu

Urszula Gągała
Autonomous Department of Microbial Biology
Warsaw University of Life Sciences
Warsaw
Poland

ula.gagala@gmail.com

Marina Goderdzishvili
Eliava Institute of Bacteriophages, Microbiology and Virology
Tbilisi
Georgia

mgoderdzishvili@gtu.ge

Andrzej Górski
Ludwik Hirszfeld Institute of Immunology and Experimental Therapy
Polish Academy of Sciences
Wrocław;
Department of Clinical Immunology
The Medical University of Warsaw
Warsaw
Poland

agorski@ikp.pl

Monika S. Hejnowicz
Institute of Biochemistry and Biophysics
Polish Academy of Sciences
Warsaw
Poland

hejmonika@ibb.waw.pl

Geraldine R. Huff
USDA–ARS Poultry Production and Product Safety Research Unit
University of Arkansas
Fayetteville, AR
USA

grhuff@uark.edu

William E. Huff
USDA–ARS Poultry Production and Product Safety Research Unit
University of Arkansas
Fayetteville, AR
USA

huff@uark.edu

Isabelle Huys
Centre for Intellectual Property Rights and Centre for Pharmaceutical Care and Pharmacoeconomy
Katholieke Universiteit Leuven
Leuven
Belgium

isabelle.huys@pharm.kuleuven.be

Jeffrey B. Jones
Plant Pathology Department
University of Florida
Gainesville, FL
USA

jbjones@ufl.edu

David Kelly
Department of Biological Sciences
Cork Institute of Technology
Bishopstown
Ireland

davidkelly_24@yahoo.co.uk

Marlena Klak
Ludwik Hirszfeld Institute of Immunology and Experimental Therapy
Polish Academy of Sciences
Wrocław
Poland

mklak@iitd.pan.wroc.pl

Mzia Kutateladze
Eliava Institute of Bacteriophages, Microbiology and Virology
Tbilisi
Georgia

kutateladze@pha.ge

Elizabeth Kutter
The Evergreen Phage Biology Laboratory
Evergreen State College
Olympia, WA
USA

kutterb@evergreen.edu

Rob Lavigne
Laboratory of Gene Technology
Katholieke Universiteit Leuven
Leuven
Belgium

rob.lavigne@biw.kuleuven.be

Małgorzata Łobocka
Institute of Biochemistry and Biophysics
Polish Academy of Sciences;
Autonomous Department of Microbial Biology
Warsaw University of Life Sciences
Warsaw
Poland

lobocka@ibb.waw.pl
malgorzata_lobocka@sggw.pl

Olivia McAuliffe
Teagasc Food Research Centre
Fermoy
Ireland

olivia.mcauliffe@teagasc.ie

Salim Manoharadas
Department of Microbiology, Immunobiology and Genetics
University of Vienna
Vienna
Austria

salimmdas@gmail.com

Shigenobu Matsuzaki
Department of Microbiology and Infection
Kochi University Medical School
Kochi
Japan

matuzaki@kochi-u.ac.jp

Ryszard Międzybrodzki
Ludwik Hirszfeld Institute of Immunology and Experimental Therapy
Polish Academy of Sciences
Wrocław;
Department of Clinical Immunology
The Medical University of Warsaw
Warsaw
Poland

mbrodzki@iitd.pan.wroc.pl

Paulina Miernikiewicz
Ludwik Hirszfeld Institute of Immunology and Experimental Therapy
Polish Academy of Sciences
Wrocław
Poland

pola@iitd.pan.wroc.pl

Sylvain Moineau
Groupe de recherche en écologie buccale (GREB)
Faculté de médecine dentaire;
Felix d'Hérelle Reference Center for Bacterial Viruses;
Département de biochimie, de microbiologie et de bio-informatique
Université Laval
Québec city, QC
Canada

sylvain.moineau@bcm.ulavl.ca

Aleksa Obradović
Plant Pathology Department
University of Belgrade
Belgrade-Zemun
Serbia

aleksao@agrif.bg.ac.rs

Jim O'Mahony
Department of Biological Sciences
Cork Institute of Technology
Bishopstown
Ireland

jim.omahony@cit.ie

Jean-Paul Pirnay
Laboratory for Molecular and Cellular Technology
Burn Wound Centre
Queen Astrid Military Hospital
Brussels
Belgium

jean-paul.pirnay@mil.be

R. Paul Ross
Teagasc Food Research Centre
Fermoy
Ireland

paul.ross@teagasc.ie

Sanna Sillankorva
Institute for Biotechnology and Bioengineering
University of Minho
Braga
Portugal

s.sillankorva@deb.uminho.pt

Iyo Takemura-Uchiyama
Department of Microbiology and Infection
Kochi University Medical School
Kochi
Japan

jm-itakemura@kochi-u.ac.jp

Jumpei Uchiyama
Department of Microbiology and Infection
Kochi University Medical School
Kochi
Japan

jumpeiu@ kochi-u.ac.jp

Lilach Vaks
Department of Molecular Microbiology and Biotechnology
Tel-Aviv University
Ramat Aviv
Israel

linais07@gmail.com

Gilbert Verbeken
Laboratory for Molecular and Cellular Technology
Burn Wound Centre
Queen Astrid Military Hospital
Brussels
Belgium

gilbert.verbeken@mil.be

Beata Weber-Dąbrowska
Laboratory of Bacteriophages
Ludwik Hirszfeld Institute of Immunology and Experimental Therapy
Polish Academy of Sciences
Wrocław
Poland

weber@iitd.pan.wroc.pl

Grzegorz Węgrzyn
Department of Molecular Biology
University of Gdańsk
Gdańsk
Poland

wegrzyn@biotech.ug.gda.pl

Current books of interest

Microarrays: Current Technology, Innovations and Applications	2014
Metagenomics of the Microbial Nitrogen Cycle: Theory, Methods and Applications	2014
Pathogenic *Neisseria*: Genomics, Molecular Biology and Disease Intervention	2014
Proteomics: Targeted Technology, Innovations and Applications	2014
Biofuels: From Microbes to Molecules	2014
Human Pathogenic Fungi: Molecular Biology and Pathogenic Mechanisms	2014
Applied RNAi: From Fundamental Research to Therapeutic Applications	2014
Halophiles: Genetics and Genomes	2014
Molecular Diagnostics: Current Research and Applications	2014
Bioinformatics and Data Analysis in Microbiology	2014
The Cell Biology of Cyanobacteria	2014
Pathogenic *Escherichia coli*: Molecular and Cellular Microbiology	2014
Campylobacter Ecology and Evolution	2014
Burkholderia: From Genomes to Function	2014
Myxobacteria: Genomics, Cellular and Molecular Biology	2014
Next-generation Sequencing: Current Technologies and Applications	2014
Omics in Soil Science	2014
Applications of Molecular Microbiological Methods	2014
Mollicutes: Molecular Biology and Pathogenesis	2014
Genome Analysis: Current Procedures and Applications	2014
Bacterial Membranes: Structural and Molecular Biology	2014
Bacterial Toxins: Genetics, Cellular Biology and Practical Applications	2013
Cold-Adapted Microorganisms	2013
Fusarium: Genomics, Molecular and Cellular Biology	2013
Prions: Current Progress in Advanced Research	2013
RNA Editing: Current Research and Future Trends	2013
Real-Time PCR: Advanced Technologies and Applications	2013
Microbial Efflux Pumps: Current Research	2013
Cytomegaloviruses: From Molecular Pathogenesis to Intervention	2013
Oral Microbial Ecology: Current Research and New Perspectives	2013
Bionanotechnology: Biological Self-assembly and its Applications	2013
Real-Time PCR in Food Science: Current Technology and Applications	2013
Bacterial Gene Regulation and Transcriptional Networks	2013
Bioremediation of Mercury: Current Research and Industrial Applications	2013

Full details at www.caister.com

Foreword

This book represents a number of impressive efforts aimed at developing phage applications for the service of the mankind. In many ways it is ironic that we are still struggling to find acceptance for such phage applications, as their capacity to serve as antibacterial agents was discovered over three decades before the discovery of the widely employed antibiotics. Even now, with the failure of antibiotics to treat numerous bacterial infectious agents (including three that the Centers for Disease Control has recently classified at the highest threat level: *Clostridium difficile, Neisseria gonorrhoeae,* and carbapenem-resistant *Enterobacteriaceae*), applications of phage therapy that could alleviate these problems are still not clinically available.

Some barriers to clinical applications of phage therapy arose shortly after the initial discovery of phage. One of the problems has been the lack of carefully designed clinical studies to determine the efficacy of phage therapy. This problem was well illustrated soon after the discovery of phage in the prescient 1924 fictional book, Arrowsmith, written by Sinclair Lewis and his silent collaborator, the microbiologist, Paul de Kruif (Lewis, 1924). In this story, Arrowsmith, a research physician, attempts to use phage in a carefully designed clinical study to treat a plague epidemic on a fictional island (St. Hubert) in the Virgin Islands. However, when Arrowsmith's wife dies of the plague on the island he decides to treat everyone, destroying the information that could have been provided by the non-treated control population. Now, nine decades later, we are still struggling with a scarcity of required controlled clinical trials of phage preparations. The current difficulty is that such clinical studies are relatively expensive and the pharmaceutical industry is reluctant to invest in any antibacterial therapies. This hesitancy is due to the relatively low rate of economic return of such therapies, especially when they are compared with the remuneration associated with therapies developed for chronic diseases, such as heart disease and cancer. As noted in this book, this is particularly unfortunate situation, given the clear need for development, expedited regulatory review, and clinical clearance for phage therapy.

Another important problem for clinical phage therapy is unfavourable pharmacokinetics of phage preparations. In particular, as shown by our group as early as 1973, a major factor which determines short half-life of phages in blood is rapid clearance of phage particles by cells of the reticulo-endothelial system of the spleen and the liver (Geier *et al.*, 1973). To address this problem, in another study (Merril *et al.*, 1996) we developed a serial-passage technique which enabled us to obtain phage particles with substantially prolonged half-life; predictably, such particles also displayed higher antibacterial activity *in vivo* compared to parental phages. This is but one example of present efforts to overcome main obstacles to further development of clinical phage therapy in Western medicine (Merril *et al.*, 2003).

In addition to the need for new antibacterial agents to treat antibiotic resistant infections, we just are beginning to realize the importance of maintaining a diverse natural microbiome, which is often adversely affected by the use of the broad-spectrum antibiotics used to treat infectious diseases. The employment of highly specific phage strains as therapeutic anti-bacterial agents would be less disruptive to the microbiome than the broad spectrum antibiotics and they should be preferred once they are further developed. Also as addressed in this book, it will be important to improve methods for rapidly determining which phage strains to use in each specific clinical infection, to fully realize the capacities of phage therapy.

Given the current economic environment, it may be necessary for clinicians and researchers who are attempting to develop phage therapy to explore the alternate means to support their efforts. One such mechanism might be to develop self-governing organizations that are capable of cooperatively pooling resources, as described in numerous publications by the 2009 Economics Nobel Laureate, Elinor Ostrom, without the need for additional industrial or governmental grants.

These are some of the issues that the regulatory agencies, such as the FDA will have to address, by carefully balancing "risk benefit ratios". The FDA has faced such a problem in the past, as illustrated by their decision to permit the continued use of vaccines following the discovery that many vaccines were contaminated by phage (Merril *et al.*, 1971; Merril, 1975). At that time the FDA decided that the risk of interrupting the vaccine program was outweighed by the risks posed by phage contamination, and an Executive Order was issued to permit the use of phage-contaminated vaccines. A number of the authors in this book have addressed similar concerns and while we cannot be certain that the right decision will always be made, it should be apparent that we must continue to gather experimental data and vigorously explore the potentials for phage therapy. So far, the therapy has proven to be safe and relatively free of side effects. Importantly, in 2009 FDA broadened the rules of "expanded access" to investigational drugs which may provide seriously ill patients with a lifeline where none had existed before (Federal Register 2009, 74, No 155) and similar approach is being developed in Europe by EMA ("compassionate use"). Those actions should pave the way towards the broader application of phage therapy in the coming years.

Carl R. Merril
National Institutes of Health, USA

References

Geier, M.R., Trigg, M.E., and Merril, C.R. (1973). Fate of bacteriophage lambda in non-immune germ-free mice. Nature *246*, 221–223.

Lewis, S. (1924). Arrowsmith (The Designer Publishing Company, Inc., New York).

Merril, C.R., Friedman, T.B., Attallah, A., Geier, M.R., Krell, K., and Yarkin, R. (1972). Isolation of bacteriophages from commercial sera. In Vitro *8*, 91–93.

Merril, C.R. (1975). Phage in Human Vaccines. Science *188*, 8.

Merril, C.R., Biswas, B., Carlton, R., Jensen, N.C., Creed, G.J., Zullo, S., and Adhya, S. (1996). Long-circulating bacteriophage as antibacterial agents. Proc. Natl. Acad. Sci. U.S.A. *93*, 3188–3192.

Merril, C.R., Scholl, D., and Adhya, S.L. (2003). The prospect for bacteriophage therapy in Western medicine. Nat. Rev. Drug Discov. *2*, 489–497.

Introduction

Jan Borysowski, Ryszard Międzybrodzki and Andrzej Górski

We live in the era marred by the increasing drama of antibiotic resistance. According to the World Economic Forum the greatest risk to human health comes in the form of antibiotic – resistant bacteria. "We live in a bacterial world where we will never be able to stay ahead of the mutation curve. A test of our resilience is how far behind the curve we will allow ourselves to fall" (Spellberg, 2013). The World Health Organization's Director General warned that the world risked entering "post antibiotic" age unless new antibacterial drugs become available quickly, a threat aggravated by a worrying trend of the emergence of bacterial strains resistant to vaccines which undermines the hope that this approach may offer a reliable weapon against resistant bacteria (Hunter, 2012). Furthermore, the UK chief medical officer believes that antimicrobial resistance is a threat as grave as climate change (Torjesen, 2013). Worse still, mounting antibiotic-resistance of bacteria is accompanied by inadequate investment of pharmaceutical industry in search for new antibiotics due to their relatively low profitability, especially as compared with drugs used in the treatment of chronic diseases. Of particular concern is a shortage of new classes of antibiotics which are essential to combat multidrug-resistant bacteria. This has created an urgent need to develop non-antibiotic antibacterial agents (Stanton, 2013). Currently the most promising non-antibiotic antibacterials are virulent bacteriophages (phages), viruses that specifically kill bacterial cells. In view of a combination of several features, especially a new mode of action, a capacity to kill bacteria regardless of their antibiotic resistance, a narrow antibacterial range, and an ability to replicate in bacterial cells, virulent bacteriophages are unique antibacterial agents and their very high bactericidal potential makes them the major therapeutic modality in phage therapy (Górski *et al.*, 2009). Phages were first administered to humans for therapeutic purposes in 1919. Since then, many thousands of patients with bacterial infections have been treated with bacteriophages, as reflected by the vast literature on phage therapy (Sulakvelidze and Kutter, 2005). Unfortunately, the majority of earlier studies which showed high efficacy and safety of bacteriophages have serious methodological shortcomings and do not meet current rigorous standards for clinical trials. However, the first small randomized controlled trials of phage therapy conducted in the last decade seem to preliminarily confirm the safety and efficacy of phages (Rhoads *et al.*, 2009; Wright *et al.*, 2009; Sarker *et al.*, 2012). Very encouraging results were also obtained in many high-quality studies performed on animal models of clinically relevant infections (Matsuzaki *et al.*, 2003; Capparelli *et al.*, 2010). Thus at the current state of research therapy using virulent phages appears to be a realistic means of treatment of bacterial infections including those caused by antibiotic-resistant bacteria (Międzybrodzki *et al.*, 2012).

The main objective of this book is to present virulent phages as antibacterial agents and their main therapeutic applications. In addition, the book discusses other phage-based and phage-derived antibacterials including lysins (lytic enzymes of bacteriophage origin), genetically-engineered phages, and filamentous phages used as delivery vehicles for other antimicrobial compounds; while the safety and the efficacy of these products have not been evaluated in clinical trials as yet, they also hold promise for future therapeutic use, as discussed in relevant chapters.

The book is composed of two major parts. Part I which includes Chapters 1–5 provides comprehensive characteristics of virulent bacteriophages as antibacterial agents, their interactions with bacterial cells, as well as the effects of phages on eukaryotic cells which can occur following the administration of phage preparations to patients. This part of the book starts with general characteristics of phages including but not limited to their structure and the interactions between phages and bacterial cells (Chapter 1). Chapter 2 discusses main stages of the complex process of developing bacteriophage preparations for therapeutic use. These stages include isolation of phages from environmental samples, preliminary grouping of phage isolates based on electron microscopy and analysis of phage DNA and proteins, selection of therapeutic phages, and purification of phage lysates. Included in this chapter are also insightful considerations about selection of correct bacterial strains for therapeutic phage propagation, which may involve modification of propagation host to prevent contamination of phage preparations with other phages, plasmids, or bacterial toxins. Chapter 3 is an interesting attempt to characterize phages as a unique "class" of drugs, as well as to formulate basic principles of phage therapy using terms of classic pharmacology. It includes a discussion of different issues relevant to both the pharmacokinetics and pharmacodynamics of phage preparations. Chapter 4 focuses on different mechanisms of bacterial resistance to phages, which is one of the major factors that may account for failures of phage therapy. In addition, this chapter discusses strategies employed by phages to overcome bacterial resistance. Chapter 5 presents non-bactericidal effects of phages. While phages can infect only prokaryotic cells, they can also interact with some populations of eukaryotic cells, including immune cells. Interestingly, there are also some data to show that bacteriophages may interfere with infections by pathogenic viruses. We believe that knowledge of these generally less known effects of phages, especially their immunomodulatory activity and immunogenicity is necessary for rational use of phage therapy.

Part II of the book is devoted to applications of bacteriophages and enzymes of bacteriophage origin to combat bacteria. This part starts with an overview of different phage-based and phage-derived antibacterials, including virulent phages, phage lysins, genetically-engineered phages, and filamentous phages used as delivery vehicles, along with their possible applications (Chapter 6). Main applications of virulent phages are discussed comprehensively in Chapters 7–12. One of these applications is the use of phages to eliminate plant pathogens (Chapter 7). Although several studies have shown potential of phages to control different plant pathogens both in the rhizosphere and the phyllosphere, a number of factors, both environmental and bacteria-related can contribute to inadequate efficacy of phage preparations. The main objectives of Chapter 7 are to discuss main factors that determine the outcome of phage treatment of plant diseases as well as methods to improve phage efficacy. Phage therapy can be used also to prevent and treat bacterial infections in animal production (Chapter 8). Based upon a comprehensive discussion of the results of experimental studies of the efficacy of the use of bacteriophages in different production animals, the authors indicate possible applications of phage therapy in commercial cattle, poultry, swine, and aquaculture systems. These applications of phage therapy gain special importance in view of current tendency to restrict the use of antibiotics in production animals to curb the development of antibiotic resistance both in the US and the European Union. Chapter 9 documents major advances in the use of phages as biocontrol agents against foodborne pathogens. In order to prevent foodborne infections, bacteriophages can be added directly to foods (at all stages of food production) or can be applied onto food contact surfaces; many studies have shown relatively high efficacy of phages in controlling the growth of major foodborne pathogens especially *Listeria monocytogenes*, *Escherichia coli*, *Salmonella enterica*, *Campylobacter jejuni*, and *Staphylococcus aureus* in a variety of foods. Recent clearance by FDA of four bacteriophage preparations for food applications indicates that bacteriophages are gradually gaining acceptance as novel biocontrol agents. Chapter 11 discusses the most important and at the same time the most controversial application of phage therapy – the treatment of bacterial infections in humans. This chapter focuses on recent randomized controlled trials of phage preparations and clinical experiences of the only two major worldwide centres of phage therapy, Phage Therapy Unit at the Ludwik Hirszfeld Institute of Immunology and Experimental Therapy, Wrocław, Poland, and the Eliava Institute, Tbilisi, Georgia. This chapter is complemented by a discussion of the results of phage therapy performed on preclinical (largely murine) models of clinically relevant infections (Chapter 10), and of main regulatory paths that may enable phage preparations to be officially approved for clinical use, as well as intellectual property rights relevant to the development of phage preparations (Chapter 12). The latter chapter is particularly important due to unique (as compared to antibiotics and other antibacterial agents) biological nature of bacteriophages which results in phage therapy not being covered by any existing regulatory pathway. A lack of suitable regulatory framework is considered one of the most important hurdles to further development of clinical phage therapy.

The last four chapters of the book (Chapters 13–16) present modified bacteriophages and enzymes of bacteriophage origin as antibacterial agents. Chapter 13 focuses on the use of phages to control the growth of biofilms. While the majority of experimental studies discussed in this chapter involve the use of unmodified virulent bacteriophages to control the growth of biofilms *in vitro*, there are also reports on successful use of genetically-engineered phages and phage lysins to reduce biofilms. Outlined in this chapter are also prospects for clinical use of phages to control biofilms, as well as specific bacteriophage preparations which could be used in the treatment of infections involving biofilms. Chapter 14 presents general characteristics and potential prophylactic and therapeutic applications of lysins. A unique combination of several features, including a new (as compared to antibiotics) mode of action based on peptidoglycan cleavage, narrow antibacterial range, and a low probability of the development of bacterial resistance makes lysins novel antibacterial agents which are a major alternative to virulent phages. While no clinical trials of the safety or efficacy of lysins have been completed so far, very promising results of *in vitro* experiments and studies performed on animals models of infections indicate high potential of lysins as antibacterials. Chapter 15 presents different possibilities for genetic engineering of phages to modify their basic features as antibacterial agents; as discussed in this chapter, genetically modified bacteriophages could find diverse applications as antibacterial agents (e.g. augmentation of antibacterial effects of antibiotics or reduction of biofilms). Chapter 16 presents an interesting application of filamentous (and to a lesser extent of tailed) phage particles as delivery vehicles for other antibacterial agents such as antibiotics or compounds used in photodynamic therapy. Discussed in the chapter are both methods for conjugating antimicrobials to phage particles and the results of *in vitro* studies to evaluate the efficacy of this approach.

This book has its roots in Poland, a country with a special history and tradition of phage therapy. The first published reports of the therapy in Poland date back to the 1920s. For about 30 years phage therapy in Poland has been performed and supervised by the Ludwik Hirszfeld Institute of Immunology and Experimental Therapy of the Polish Academy of Sciences in Wrocław. The editors of the book have been engaged in research into bacteriophages and phage therapy of patients thus having practical expertise with the subjects addressed in the book, which we wish to dedicate to the founder of the Institute and one of the pioneers of phage therapy who believed that "a conflict between the invisible and the imperceptible may one day be used for protection of human health and life" (Hirszfeld, 1948). We hope that publication of this book will serve this purpose.

References

Capparelli, R., Nocerino, N., Iannaccone, M., Ercolini, D., Parlato, M., Chiara, M., and Iannelli, D. (2010). Bacteriophage therapy of *Salmonella enterica*: a fresh appraisal of bacteriophage therapy. J. Infect. Dis. *201*, 52–61.

Górski, A., Międzybrodzki, R., Borysowski, J., Weber-Dabrowska, B., Lobocka, M., Fortuna, W., Letkiewicz, S., Zimecki, M., and Filby, G. (2009). Bacteriophage therapy for the treatment of infections. Curr. Opin. Investig. Drugs *10*, 766–774.

Hirszfeld, L. (1948). The battle of the invisible with the imperceptible. Wroc. Tow. Nauk. *3*, 3–18.

Hunter, P. (2012). Where next for antibiotics? The immune system and the nature of pathogenicity are providing vital clues in the fight against antibiotic-resistant bacteria. EMBO Reports *13*, 680–683.

Matsuzaki, S., Yasuda, M., Nishikawa, H., Kuroda, M., Ujihara, T., Shuin, T., Shen, Y., Jin, Z., Fujimoto, S., Nasimuzzaman, M.D., *et al.* (2003). Experimental protection of mice against lethal *Staphylococcus aureus* infection by novel bacteriophage φMR11. J. Infect. Dis. *187*, 613–624.

Międzybrodzki, R., Borysowski, J., Weber-Dąbrowska, B., Fortuna, W., Letkiewicz, S., Szufnarowski, K., Pawełczyk, Z., Rogóż, P., Kłak, M., Wojtasik, E., *et al.* (2012). Clinical aspects of phage therapy. Adv. Virus Res. *83*, 73–121.

Rhoads, D.D., Wolcott, R.D., Kuskowski, M.A., Wolcott, B.M., Ward, L.S., and Sulakvelidze, A. (2009). Bacteriophage therapy of venous leg ulcers in humans: results of a phase I safety trial. J. Wound Care *18*, 237–238, 240–243.

Sarker, S.A., McCallin, S., Barretto, C., Berger, B., Pittet, A.C., Sultana, S., Krause, L., Huq, S., Bibiloni, R., Bruttin, A., *et al.* (2012). Oral T4-like phage cocktail application to healthy adult volunteers from Bangladesh. Virology *434*, 222–232.

Spellberg, B. (2013). The future of antibiotics and resistance. New Eng. J. Med. *368*, 299–302.

Stanton, T.B. (2013). A call for antibiotic alternatives research. Trends Microbiol. *21*, 111–113.

Sulakvelidze, A., and Kutter, E. (2005). Bacteriophage therapy in humans. In Bacteriophages: Biology and Applications, Kutter, E., and Sulakvelidze, A., eds (CRC Press, Boca Raton, Florida), pp. 381–436.

Torjesen, I. (2013). Antimicrobial resistance presents an "apocalyptic" threat similar to that of climate change, CMO warns. Brit. Med. J. *346*, f1597.

Wright, A., Hawkins, C.H., Anggård, E.E., and Harper, D.R. (2009). A controlled clinical trial of a therapeutic bacteriophage preparation in chronic otitis due to antibiotic-resistant *Pseudomonas aeruginosa*; a preliminary report of efficacy. Clin. Otolaryngol. *34*, 349–357.

Part I

Characteristics of Phages as Antibacterial Agents

General Characteristics of Bacteriophages

1

Hans-Wolfgang Ackermann and Grzegorz Węgrzyn

Abstract

Bacteriophages are the most numerous biological entities in the biosphere and the largest virus group known. Tailed phages constitute approximately 96% of known phages. Bacteriophage research is an extremely dynamic branch of microbiology. This review starts with the discovery of bacteriophages and covers basic phage techniques, ecology and incidence, aspects of phage physiology and replication, host range, classification, morphology, physico-chemical properties, genomics, and selected practical aspects of phage research. It concludes with a section on electron microscopy for rapid phage identification.

Introduction

Bacteriophages or 'phages' are viruses of bacteria. 'Bacteriophage,' derived from the Greek word φαγειν, to eat, literally means 'eater of bacteria.' The term excludes archaeal viruses, although archaea were formerly considered as bacteria and some of their viruses have the morphology of typical tailed phages of the *Myoviridae* and *Siphoviridae* families. These viruses all occur in the Crenarchaeota phylum of Archaea (Ackermann, 2007; Prangishvili, 2006; Abedon and Murray, 2013) and their existence suggests very ancient links between bacteria and archaea. A 'bacteriophage' is a viable virus that infects bacteria. This excludes defective phages, e.g. particulate bacteriocins, the 'killer particles' of *Bacillus*, gene transfer elements, and the many morphologically defective entities that are regularly seen in bacterial lysates. Particulate bacteriocins are phage-like particles able to kill bacteria, harboured and produced by lysogenic bacteria, but unable to reproduce on their own. They may be morphologically defective and consist only of a phage tail. Killer-particles, exemplified by *Bacillus subtilis* phage PBSX, are defective lysogenic myoviruses with very small heads and long thick tails. The heads contain bacterial DNA. The particles kill bacteria and are unable to replicate. 'GTA' or 'gene transfer elements' (Solioz *et al.*, 1977; Lang *et al.*, 2012) are phage-like particles that transfer only host DNA (see 'Transduction and conversion', below). Similarly, plasmids and episomes are not phages, but extrachromosomal genetic elements that may be parts of the life cycle of certain phages. Plasmids consist typically of circular dsDNA and inhabit the cytoplasm of bacteria and some eukaryotes. The term 'episome' designates both plasmids and prophages (see 'Lysogeny', below) that can integrate into the host chromosome and replicate with it.

By 2012, nearly 6300 prokaryote viruses, of which only 88 were archaeal viruses, had been examined in the electron microscope (Ackermann and Prangishvili, 2012). Phages thus appear, at least on paper, as the largest viral group in existence. New phages are described at the rate of about 150 per year. This corresponds to an enormous number of publications. The phage bibliography of Raettig (1967) lists 11,405 publications for the years 1917 to 1965. The bibliography of H.-W.A., now published on the internet (www.phage.ulaval.ca), comprises over 33,000 phage publications for 1965–2010 and is certainly incomplete. The total volume of phage literature is probably about 45,000 publications. For brevity, this chapter relies heavily on two excellent, very complete general books on bacteriophages that cover most topics treated here (Kutter and Sulakvelidze, 2005; Calendar, 2006).

Discovery

Phages were discovered twice at the beginning of the 20th century (Summers, 2012). Their discovery was possible because bacteriological techniques and instrumentation were already well developed and viruses of plants and animals were known, namely the tobacco mosaic virus, the Rous sarcoma virus, and the virus of the foot-and-mouth disease.

In London, Frederick William Twort, a medical doctor, observed that cultures of a 'micrococcus' from vaccine lymph became glassy and transparent. This was caused by an agent that passed through bacterial filters, could be transmitted to other cultures, and was heat-sensitive. It was said to be a bacterium, an amoeba, an enzyme, a virus, or a living protoplasm. Twort published his observation in 1915 (Twort, 1915), but it apparently went unnoticed. He abandoned phage research and devoted the rest of his scientific life to the cultivation of animal viruses on inert media. It seems that Twort's agent and its host, now identified as a *Staphylococcus hyicus*, are still in existence (Klumpp *et al.*, 2010). Twort is said to have deposited the phage in the Pasteur Institute of Paris in 1948. A *Staphylococcus* phage labelled 'Twort' was indeed held in the collection of the institute and is now in the Félix d'Hérelle collection in Quebec (Ackermann *et al.*, 2004) (Fig. 1.1). Félix d'Hérelle was a French Canadian from Montreal and a self-styled bacteriologist. In 1915, he investigated a dysentery outbreak near Paris. He observed that stool filtrates from patients lysed cultures of Shiga bacilli and that the appearance of the lytic principle correlated with the onset of convalescence. He published his findings in 1917, coining the term 'bacteriophage' and clearly stating that his agent was an invisible microbe and an obligate parasite of living bacteria (d'Hérelle, 1917). From then on, d'Hérelle devoted his life to bacteriophages and their therapeutic use in humans and animals. He worked in Indochina, Assam, Egypt, the USA, France, and Georgia, where he and G. Eliava founded a bacteriophage institute in Tbilisi. During his peripatetic life, d'Hérelle published over 120 books and articles on bacteriophages, ceaselessly advocating their use for the treatment of human infections (Ackermann *et al.*, 1982; Summers, 1999).

We do not know the original phage of d'Hérelle. His publications suggest that it was a mixture of phages that included a T4-like virus. He maintained that there was only one bacteriophage with many races, the '*Bacteriophagum intestinale*' (d'Hérelle, 1918), and was apparently the first to postulate the concept of intracellular multiplication of viruses (d'Hérelle, 1921). D'Hérelle was

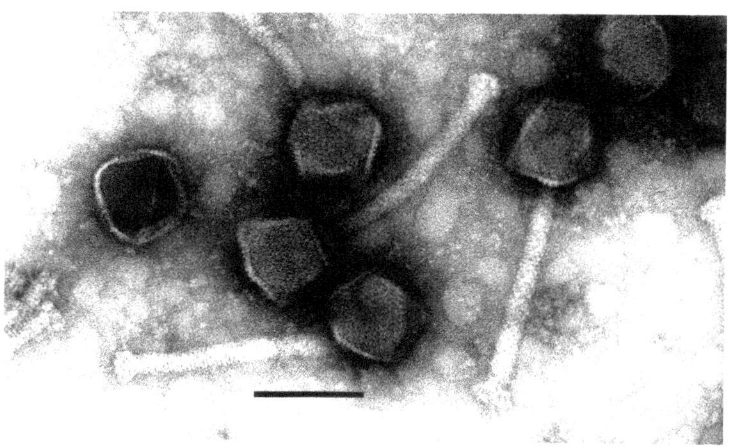

Figure 1.1 *Staphylococcus* phage Twort, uranyl acetate, final magnification x 297,000. The bar indicates 100 nm.

never awarded the Nobel prize, probably because of an ongoing dispute on the primacy of phage discovery and the nature of phages. It was contended that phages were not viruses, but enzymes, cell fragments, products of bacterial autolysis, or filterable forms of bacteria. The controversy was only settled by the advent of the electron microscope and the publication of the first phage micrographs in 1940 (Pfankuch and Kausche, 1940; Ruska, 1940).

Basic phage techniques (Carlson, 2005)

Phages are generally isolated by enrichment from the environment or by induction of lysogenic bacteria. Many phage techniques were devised by d'Hérelle himself. The basis of all phage work is that phages lyse bacterial cultures. On agar dishes with bacterial lawns, phages produce clear areas. If phage lysates are sufficiently diluted, they produce small lysed areas or holes of 0.1–3 mm, called 'plaques' (Abedon and Yin, 2009). Each plaque represents the progeny of a single phage. In liquid media, phages may (but often do not) destroy the whole bacterial culture. The subject has been dealt with many times and we do not wish to elaborate.

Phages are propagated in three ways: (a) as broth cultures inoculated with phages and bacteria, (b) on agar surfaces covered with a bacterial lawn, and (c) in agar double layers consisting of normal agar covered with a mixture of soft agar of about 0.7%, phages, and bacteria. The latter technique is quite cumbersome, time-consuming, error-ridden, and prone to contamination. Phages on agar are harvested by scraping or extracted by broth. Phage-containing lysates are then filtered, best through membrane filters of 0.45 μm pore size. If many bacteria are present, the lysate should be clarified by centrifugation. Phages are then titrated and purified by three successive isolations of single plaques. Sterile phage lysates are easily stored at +4°C or in a deep freezer, and may be kept for years. Their longevity depends on the phage and the preservation technique used (Ackermann *et al.*, 2004). Phage lysates may also be sterilized with chloroform, but this is dangerous as about one third of phages, whether tailed or not, are chloroform sensitive.

For phage isolation from natural substrates, samples must be liquid. Solid material, e.g. soil or faeces, is homogenized in broth or buffer. The sample is then mixed in a flask with an equal volume (e.g. 20 ml) of liquid growth medium and a suspension of bacteria (e.g. 0.5 ml) in logarithmic growth and incubated. It is then filtered and titrated. Lysogenic phages are produced by bacteria, either spontaneously or by induction (Raya and Hébert, 2009). This is generally done by application of UV light for 1 min or mitomycin C (0.5 ml per ml for 10 min contact time). Bacteria treated with chemicals must be transferred into fresh medium for phage production. A wide variety of agents is available for induction. Countless variations of these basic techniques have been devised and are described elsewhere (Carlson, 2005).

Habitats and ecology

Phages occur in enormous, even astronomical numbers almost everywhere, especially in the oceans (Weinbauer, 2004; Hurwitz and Sullivan, 2013; Matteson *et al.*, 2013; Zhao *et al.*, 2013). From counts of marine phages, the total number of phages in the biosphere has been estimated at over 10^{30} to 10^{32} particles (Brüssow and Hendrix, 2002; Breitbart and Rohwer, 2005; Suttle, 2005; Hatfull and Hendrix, 2011). An extrapolation from metagenomic data suggests that less than 0.0002% of the global phage metagenome has been sampled and that 100 million phage genomes exist in nature (Rohwer, 2003). Phages occur abundantly in nature and in every conceivable bacterial habitat, for example in:

- water: seawater, freshwater, sewage, volcanic hot springs, salt ponds, hypersaline lagoons;
- soil, mostly from the rhizosphere, but even in subsoil from the Sahara;
- air, as droplets or dust in the air;
- plants: mainly on surfaces, but also in nodules of leguminous plants and on crown gall tumours;
- animals: body fluids and excreta, body cavities including the rumen and crop of birds;
- food: a wide variety of dairy products, meat and fish, oriental fermented food (see Ackermann

and DuBow, 1987; Weinbauer, 2004; Srinivasiah *et al.*, 2008; Letarov and Kulikov, 2009; Mateson *et al.*, 2013; Zhao *et al.*, 2013).

The most important habitat of phages, or rather prophages, seems to be the lysogenic bacterium (Canchaya *et al.*, 2003; Bobay *et al.*, 2013). As lysogeny is extremely widespread (*P. aeruginosa* is even 100% lysogenic), this is a handy way to preserve phage genomes over enormous timespans; for example, there is a giant *Bacillus* phage with a contractile tail that may antedate the separation of bacilli and clostridia 2 billion years ago (Ackermann *et al.*, 1995). Phages may be detected by isolation and the spot test, but this technique works with phages of cultivable bacteria only. On the other hand, phages once isolated can be studied in detail. Alternatively, phages can be detected and counted by transmission electron microscopy (TEM), epifluorescence microscopy, flow cytometry, and metagenomics. Each technique has its problems.

- **TEM** provides family and sometimes even species identification. It depends very much on the qualification of the observer and is expensive. A technique for phage counts in water has been devised and is a basis for current estimates of phage frequency in the oceans (Bratbak and Heldal, 1993; Ackermann and Heldal, 2010). It relies on positive staining by uranyl acetate and this hinders or precludes a proper identification of phage particles.
- **Epifluorescence microscopy and flow cytometry** rely on fluorescent stains, e.g. SYBR Green. Both techniques are fast and inexpensive (Brussaard, 2009; Ortmann and Suttle, 2009; Zemb *et al.*, 2013). Unfortunately, they seem to be fraught with false positive results. H.-W.A. observed strongly fluorescent preparations which contained no phages, but instead large numbers of small bacterial debris. Indeed, bacterial DNA adsorbed to bacterial debris could give a positive fluorescence and simulate phage DNA. Recently, confocal laser scanning microscopy was proposed as an alternative to conventional epifluorescence microscopy to enumerate virus-like particles in aquatic systems (Peduzzi *et al.*, 2013).
- In **metagenomics**, the total viral component from a particular environment is collected, amplified, and sequenced for the purpose of identifying genomes or genes without resorting to propagation (Fancello *et al.*, 2012; Reyes *et al.*, 2012; Dinsdale *et al.*, 2013). Since about 95% of bacteria are non-cultivable, their phages cannot be isolated. The technique has essentially been applied to seawater and marine sediments (Breitbart *et al.*, 2004; Breitbart and Rohwer, 2005; Angly *et al.*, 2006), but is also applicable to horse faeces or fermented Asian food. Moreover, metagenomic approaches have been used to study phage population in the human gut (Reyes *et al.*, 2012); such studies can provide novel insights into the pathogenesis of inflammatory bowel disease (Wagner *et al.*, 2013) and in the future possibly other diseases. However, metagenomics suffers from a severe limitation, i.e. it depends totally on virus descriptions in the literature and known sequences in databases (Clokie *et al.*, 2011).

The goal of establishing a phage biogeography (Thurber, 2009) remains remote and elusive. One reason is that ecological studies are essentially limited to the oceans and a few developed countries, while we know next to nothing on the phage flora in such vast areas as all of China, South America, and sub-Saharan Africa. The other reasons are that fluorescence microscopy identifies total phage numbers and not individual phages and that metagenomics may identify, for example, the major capsid gene of phage T4, but this may belong to any phage with T4 affinities. TEM could help here, but most studies are carried out with poor electron microscopes. On the other hand, phage identification by TEM is now far advanced. In a pilot study, a dozen phage genera or species could be visually identified in Nigerian sewage (Koko *et al.*, 2011). It seems that phages of mesophilic bacteria have a global distribution and acidophilic, thermophilic or halophilic phages occur in special, limited habitats such as volcanic hotsprings, that may be located, for example, in regions as far apart as Kamchatka and the Yellow Stone Park in the USA (Yu *et al.*, 2006).

Phage life

This subject has been covered elsewhere in comprehensive reviews (Ackermann and DuBow, 1987; Guttman *et al.*, 2005; Kutter *et al.*, 2005). In addition, the reader is referred to an excellent recent review of bacteriophage replication (Weigel and Seitz, 2006). We describe here the general course of phage replication.

Life cycles

The lytic cycle

The **lytic cycle**, also called productive, vegetative or virulent, results in the production of new phages and the death of the bacterial host. The first stage is a period of adsorption to specific receptors located on the cell wall, flagella, pili, or capsules. The phage then infects the bacterium. In most phage groups, only the nucleic acid enters the host while the capsid or shell remains outside. In filamentous phages of the *Inovirus* genus and the icosahedral *Cystoviridae*, the virus penetrates the cell wall, but not the cytoplasmic membrane. The next stage is that of phage synthesis and assembly, also called *eclipse* because phages have temporarily disappeared from the lysate. Typically, bacterial syntheses are shut down, modified and directed towards phage production. Phage DNA is transcribed into mRNA, but the RNA of leviviruses (see 'Replication', below) acts itself as mRNA and needs no transcription. Nucleic acid and proteins are synthetized separately and often in different parts of the cell. In relatively complex tailed phages such as T4, the protein shell is assembled in three different lines for the capsid, the tail, and tail fibres, respectively. In most tailed phages, the DNA is cut to size from long filaments or 'concatemers' (see 'Circle-to-circle replication', below). In all tailed phages, the progeny DNA enters a preformed capsid. Phage assembly results in novel phages. The length of the latent period depends on the nature and nutritional state of the host. It varies between 20 min for enterobacteria and bacilli and 30–50 h for some cyanobacteria. The final step in the lytic cycle is the release period. In most phages, the infected cell bursts and releases novel phages and bacterial debris of all kinds. In addition to novel infectious viruses, phage assembly often results in freak, monster, or incomplete particles. The number of novel phages per bacterial cell is called the burst size and this can vary enormously, but is often around 50–100.

In two varieties of the lytic cycle, the host bacteria are not immediately killed. The filamentous inoviruses are slowly extruded from their host, which may survive for up to 300 generations. This is called 'steady-state infection' or 'carrier state' and should be distinguished from pseudolysogeny (see 'Filamentous phages', below). The enveloped plasmaviruses (see 'Classification', below) are released by budding and acquire their envelope in the process.

The lysogenic or temperate cycle

In the lysogenic or temperate cycle, phage DNA infects a bacterial cell and either starts a lytic cycle with phage synthesis or enters a state of equilibrium, in which it becomes part of the bacterial cell. The phage genome becomes circular and latent, persisting in an integrated form within the bacterial genome or, independently, in the cytoplasm as a circular 'plasmid' (Campbell, 2006). A latent phage genome is called a 'prophage.' Integration is mediated by enzymes called 'integrases.' The prophage of coliphage P1 is always a plasmid. Coliphage λ persists as an integrated prophage, although specific mutants form plasmids (Węgrzyn *et al.*, 2012). The prophage replicates more or less synchronously with its bacterial host and may persist indefinitely, until the link between phage and host breaks down, either spontaneously or under the action of physical and chemical agents, notably antitumour agents, carcinogens, and mutagens. This is called 'induction' and results in the production of novel phages (see 'Basic phage techniques').

Pseudolysogeny

Pseudolysogeny is a condition where a bacteriophage is maintained in a mixture of phage-resistant and non-infected, phage-sensititive bacteria. The pseudo-lysogenic state is unstable and may, by selection of survivors, lead to phenotypic changes in the host bacterium (e.g. pigment or toxin production) (Abedon, 2009; Łoś and Węgrzyn, 2012).

Table 1.1 Phage hosts by bacterial phylum

Phylum	Example
Actinobacteria	*Corynebacterium*, *Mycobacterium*, *Streptomyces*
Bacteroidetes	*Bacteroides*, *Cytophaga*
Chlamydiae	*Chlamydia*
Cyanobacteria	*Prochlorococcus*, *Synechococcus*
Deinococcus–Thermus	*Thermus*
Firmicutes	*Bacillus*, *Lactobacillus*, *Staphylococcus*, *Streptococcus*
Clostridia	*Clostridium*
Mollicutes	*Mycoplasma*
Fusobacteria	*Fusobacterium*
Proteobacteria	Divisions α, β, γ, δ, ε *Aeromonas*, *Brucella*, enterobacteria, *Pseudomonas*, *Rhizobium*
Spirochetes	*Borrelia*, *Treponema*

Replication

Phages have several fundamentally different types of genomes (Fig. 1.2) which replicate in different ways. Some examples of replication strategy and mechanisms are given here. Efficient genome replication is absolutely crucial for effective propagation of phages. This propagation is, in turn, fundamental in phage therapy, as only highly effective production of progeny phages can ensure a quick destruction of pathogenic bacteria during an infection of a macroorganism.

RNA genomes

The simplest strategy is that of bacteriophages with RNA genomes. They encode their own replicases, which catalyse the synthesis of daughter genomic molecules (Hosoda *et al.*, 2007). In the case of ssRNA genomes, such a synthesis involves the production of a complementary RNA nucleotide strand. The phage genome acts as its own messenger RNA.

Filamentous phages

The genomes of filamentous phages or inoviruses, such as coliphages fd, f1 and M13, are circular ssDNA molecules (Rakonjac *et al.*, 2011). Their replication starts with the synthesis of complementary or RF (replication form) dsDNA. Once a circular dsDNA molecule is formed and supercoiled by host gyrase, a phage-encoded protein cuts one of the strands in a specific site. This is the first stage of rounds of unidirectional rolling-circle (also called σ) replication. This replication proceeds with the displacement of the ssDNA strand, which after one round of such a replication is sealed, forming a copy of the genome that can be used for packaging into a progeny capsid.

Coliphage P2

A similar rolling-circle mechanism occurs in coliphage P2, a myovirus (Odegrip *et al.*, 2000). Although this phage has a linear dsDNA genome and synthesis of a complementary strand is not required, one strand is nicked after circularization and supercoiling. This promotes unidirectional rolling-circle replication. The difference from fd-like genomes is that a complementary strand of the P2 genome is synthesized on the newly produced daughter DNA strand.

Circle-to-circle replication

Bacteriophages with circular dsDNA, or whose linear dsDNA is circularized shortly after injection into host cells, often employ the circle-to-circle (also called θ) mode of DNA replication. A classic example is bacteriophage λ (Węgrzyn *et al.*, 2012). To initiate θ replication, it is necessary to recognize a specific region of the phage genome, called *origin* of replication (*ori*). Such bacteriophages encode replication initiator proteins, which bind specifically to the *ori* region and recruit a helicase to this site. This allows for the start of a bi-directional θ replication that produces daughter copies of phage genomes. However, if the phage genome is linear, it is necessary to convert the DNA circles into linear molecules. Generally, a switch from θ to rolling-circle (σ) replication occurs, which may be initiated by various mechanisms. One possibility is nicking one DNA strand, similarly to the mechanism described above for the P2 phage. Another variety of the switch is a recombination-mediated initiation of σ replication, where a recombination intermediate serves as an untypical primer in DNA synthesis. Finally, if unidirectional, rather than bidirectional, θ replication starts, it can be switched after one round to σ replication by displacement of the 5′-end of the newly synthesized DNA strand to its growing 3′-end.

Coliphage T7

Another way of replicating linear dsDNA phage genomes is to use of RNA polymerase-produced transcripts as primers for DNA synthesis. This mechanism is employed, among others, by coliphage T7 (Lee and Richardson, 2011). DNA polymerase can extend the transcript and synthesize a new DNA strand.

Coliphage T4

In coliphage T4, the first steps in replication are similar to those in phage T7 (Mueser *et al.*, 2010). Both processes depend on RNA polymerase-mediated transcription. Later stages of T4 DNA replication also depend on recombination. As said below, the phage T4 genome has terminal redundancies allowing for circular permutations. This means that each T4 DNA molecule has the same set of genes, although various segments of the genome are repeated in various DNA molecules. Such a genome structure allows for efficient recombination between novel T4 genomes, as the redundant sequences, when being replicated, can easily act as invading strands at corresponding sites of the phage genome. Extension of the invading strands by phage-encoded DNA polymerase results in formation of the specific replisome, which is very efficient in replication of phage DNA. However, this kind of replication implies that there is no specific starting point of the late T4 DNA synthesis mode. To avoid a scarcity in newly formed DNA molecules, concatemers or large, branched molecules of T4 DNA, encompassing DNA of the length of several T4 genomes, are formed. They are then packaged into capsids in such a way that the length of the packaged DNA (about 173 kb) exceeds that of the complete T4 genome (168.8 kb). This 'excess' of phage DNA serves as the terminally redundant part of the genome.

Bacillus subtilis phage φ29

A different strategy for replicating a linear dsDNA genome is to start replication at the very end of each DNA strand. *Bacillus subtilis* phage φ29, a podovirus, encodes a terminal protein that is covalently bound to the 5′-terminal nucleotides (Salas *et al.*, 2008; Tone *et al.*, 2012). It is responsible for the priming reaction, resulting in addition of a few first nucleotides of the new DNA strand. To avoid shortening of the phage genome after each replication event, a sliding-back process occurs, in which the terminal protein with attached nucleotides moves back along the template strand. In this way, the replication can actually start at the very end of the linear genome. This movement is possible because there are a few nucleotide residues repeated at each end of the DNA molecule, so the bases of two strands are still complementary despite the sliding-back of the nucleoprotein structure.

Transduction and conversion

Transduction is DNA transfer by viruses and is normally rare. In generalized transduction, any fragment of bacterial DNA can be loaded into a phage capsid and transferred into a novel bacterium (Thierauf *et al.*, 2009). This is a normal event and can be achieved by lytic and lysogenic phages. The majority of the phage population is normal and gives rise to infectious progeny. Gene transfer agents or GTA, by contrast, are phage-like entities that carry and transfer bacterial DNA only (Solioz and Marrs, 1977). Insofar as known, they occur in α-proteobacteria (*Rhodobacter, Rhodopseudomonas, Roseobacter*) and appear as typical siphoviruses. Specialized transduction is the domain of lysogenic phages that, like λ, can integrate into the bacterial genome. If, after induction, the phage DNA is not properly excised, bacterial genes close to the prophage site may be excised and packaged into progeny phages. The resulting particle carries bacterial and phage genes and may be non-viable.

Conversion is a frequent event and affects a whole bacterial population. It is carried out by normal lysogenic phages that carry certain bacterial genes as part of their genetic make-up and transmit these same genes to a whole bacterial population, which they 'convert' by integration into their novel hosts. Such phages may confer new properties to bacterial cells, for example, new antigens, antibiotic resistance, haemolysins and production of diphtheria, botulinus, cholera or Shiga toxins (Boyd, 2005; Wagner and Waldor,

2006; Loś *et al.*, 2011). Converting phages play a major role in bacterial pathogenesis and are evidently inappropriate for phage therapy.

Host range

General

Phages infect widely different bacteria: aerobes and anaerobes, Gram-positives and Gram-negatives, aerobes and anaerobes, budding, sheathed, ramified, capsulated, stalked, sporulating or non-sporulating bacteria, spirochetes, and cyanobacteria. The bacterial hosts belong to 11 bacterial phyla (Table 1.1) and some 150 genera, enterobacteria being considered as a single 'genus' because of their close relationships. Most hosts belong to the phyla Actinobacteria, Firmicutes, and Proteobacteria. This reflects the availability of media and the amount of work invested. No phages have been found in the phyla Green-sulfur-bacteria, Planctomycetes or Aquificales.

Polyvalence and specificity

Individual phage host ranges depend on the bacterium and its properties and defences against phage attack. The first and best-known line of defence is the availability of receptors on the bacterial surface. They include specific somatic or capsular antigens, flagella, or pili. One example of the former is the Vi antigen of *Salmonella enterica* serovar Typhi. It was formerly thought to be associated with virulence (hence the name 'Vi') and is the receptor for a series of specific phages. One of them, ViII, is the basis of a typing scheme and thus epidemiologically important. The role of pili is evident in RNA phages of the *Leviviridae* family. If the pili are not formed, the corresponding RNA phage is not adsorbed, cannot be propagated and is for all purposes dead. Many other factors affect phage host ranges, for example inhibition of DNA injection, superinfection immunity by resident prophages, restriction endonucleases destroying infecting phage DNA, or CRISPR (clustered regularly interspaced palindromic repeat, a bacterial immune system that cleaves bacteriophage and plasmid DNA (Barrangou *et al.*, 2007)).

Many phages are polyvalent and infect members of several bacterial genera with phylogenetic relationships. This has important consequences for phage therapy. Polyvalent phages are exceedingly frequent in enterobacteria and, apparently, care little for bacteriological classification. As an example, they will not, as a rule, differentiate between *Escherichia, Shigella* or *Klebsiella* bacteria, but will respect to some extent bacteriological divisions between enterobacteria tribes. Thus, a coliphage will generally not infect a *Proteus* or a *Salmonella*, but this rule is not absolute. In other instances, host range depends on plasmids specifying certain pili. For example, tectivirus phage PRD1 will infect enterics, acinetobacters or vibrios provided that they have the necessary pilus antigens for adsorption. Polyvalent phages seem to be frequent in actinomycetes; e.g. they were frequently reported in *Streptomyces* and *Verticillium*. This may reflect outdated bacteriological classifications. On the other end of the spectrum, there are phages specific to particular antigens, strains, bacterial species, or genera. The general rule is that, except in enterobacteria, phages are generally specific to the genus of their host bacteria and will lyse or attack only bacteria belonging to this genus.

Classification

General

Modern virus classification started in 1962 when the physico-chemical properties of viruses and their nucleic acid were used to define virus families (Lwoff *et al.*, 1962). Phage classification started in 1967. In a seminal paper (Bradley, 1967), six basic phage types were defined by gross morphology and type of nucleic acid. Three were tailed and had dsDNA and contractile, long, non-contractile, or short tails. In addition, there were filamentous phages with dsDNA and small isometric phages with ssDNA or RNA. The first report of the future ICTV or International Committee on Taxonomy of Viruses (Wildy, 1971) included six phage 'genera' and their descriptions: T4 and relatives, λ, lipid-containing phage PM2, φX174-like phages, filamentous phages, and the 'ribophage group.' This was followed by several reports published at irregular intervals. The Ninth Report includes six orders, 87 families, 19 subfamilies, and 348 genera (Anonymous, 2011; King *et al.*, 2011). The ICTV uses every available virus property, but for practical reasons, the most important properties for family classification are virion nucleic acid and physicochemical properties, increasingly completed by genomic data. The ICTV does not classify genes, proteins, or prophages.

Phage orders and families

Phages have double-stranded or single-stranded DNA or RNA. Virions are tailed or isometric, filamentous, or pleomorphic (Fig. 1.2). Two types contain lipids or have lipid envelopes. Their general properties have been reviewed many times and may be found elsewhere (Ackermann, 2005, 2006; Ackermann and DuBow, 1987; Fauquet *et al.*, 2005). Currently, phages have one order, the *Caudovirales*, and 10 families (Table 1.2). Dimensions and important physiological and physicochemical properties of phage families are given in Table 1.3. The number of phage genera and species is in constant expansion and can only be guessed.

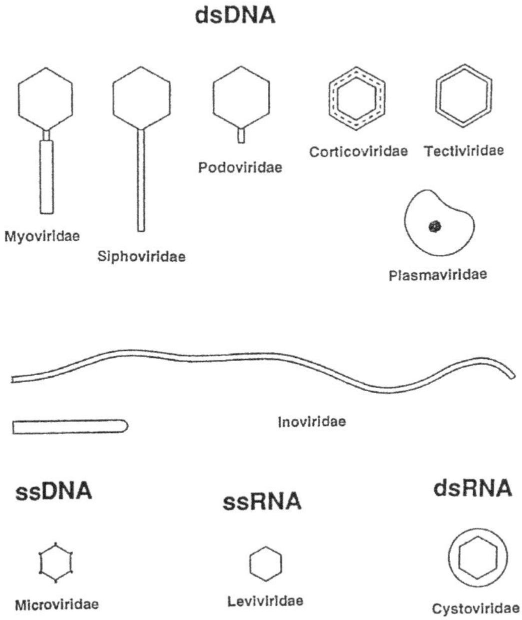

Figure 1.2 Schematic representation of phage families.

- **Tailed phages** form the order *Caudovirales* and fall into three very large families, namely the *Myoviridae* with contractile tails (25%), the *Siphoviridae* with long, non-contractile tails (61%), and the *Podoviridae* with short tails (14%) (Ackermann, 2007). These families are clearly related by evolution and share lifestyles and sometimes specific proteins and genes. They represent over 96% of phages. Virions consist of a protein shell, contain a single molecule of linear dsDNA, and are lytic or lysogenic. They adsorb to bacteria by their tails and are liberated by burst of the infected cell.
- **Isometric, filamentous, and pleomorphic phages** fall into seven families. They differ profoundly by their properties and are clearly polyphyletic. Families are small, sometimes having one member only, and have narrow host ranges.

Isometric phages are icosahedra. The *Microviridae* (ssDNA) include the well-known coliphage φX174, characterized by conspicuous capsomers. The only certain member of the *Corticoviridae* (dsDNA) was isolated from seawater and has a complex coat with alternating layers of protein and lipid. The *Tectiviridae* (dsDNA) have a protein shell which surrounds a lipid vesicle. For DNA ejection, the vesicle transforms itself into a tail-like tube. *Leviviridae* (ssRNA) are often plasmid-specific and adsorb to sex pili. The *Cystoviridae*, again a very small family, consist of a lipidic envelope, a capsid, and three molecules of dsRNA. They have been found in *Pseudomonas* only. The envelope is synthetized within the bacterial host and not acquired by budding. Filamentous phages form the *Inoviridae* family (ssDNA). The family is clearly heterogeneous and includes the genus *Inovirus*, consisting of long flexible filaments, and the genus *Plectrovirus*, consisting of short rods. *Inovirus* phages are not liberated by cell burst, but are synthetized at the plasma membrane and then slowly extruded. Little is known about plectroviruses, except that they are specific to mycoplasmas. The pleomorphic *Plasmaviridae* (dsDNS) are another small group of mycoplasma viruses. They have no visible capsid and apparently consist of nothing more that a ball of DNA and an envelope acquired by budding from the host.

Table 1.2 General properties of bacteriophage families

Shape	Tailed	Polyhedral					Filamentous	Pleomorphic
Order or family	*Caudovirales*	*Microviridae*	*Corticoviridae*	*Tectiviridae*	*Leviviridae*	*Cystoviridae*	*Inoviridae*	*Plasmaviridae*
Nucleic acid	DNA	DNA	DNA	DNA	RNA	RNA	DNA	DNA
Strands	2	1	2	2	1	2	1	2
Linear or circular	L	C	C	L	L, S	L	C	C
Envelope	–	–	–	–	–	+	–	+
Lipids	–	–	+	+	–	+	–	+
Particulars			Complex capsid	Pseudo-tail				No capsid
Virulent or temperate	V, T	V	V	V	V	V	Carrier, T	Carrier
Exit from cell	Burst	Burst	Burst	Burst	Burst	Burst	Extrusion	Budding
Members	4245	38	3?	19	38	3	66	5
Example	T4, λ, T7	ϕX174	PM2	PRD1	MS2	ϕ6	M13	L2

C, circular; L, linear; S, segmented; T, temperate; V, virulent; 1, single; 2, double; +, present; –, absent. Carrier, steady-state infection. Members: number of phages examined by electron microscopy, excluding phage-like bacteriocins and known defective phages (modified from Ackermann, 2007).

Table 1.3 Dimensions and physico-chemical properties of phage families

	Caudovirales								
Order or family	Average	Range	*Microviridae*	*Corticoviridae*	*Tectiviridae*	*Leviviridae*	*Cystoviridae*	*Inoviridae***	*Plasmaviridae*
Capsid (nm)	68*	38–160	27	60	63	23	75–80	760–1950 x 7	80
Tail length (nm)	153	3–820							
Particle weight ($M_r \times 10^{-6}$)	100	29–470	6–7	49	70	3.6–4.2	99	12–34	
Buoyant density (g/ml, CsCl)	1.49	1.4–1.54	3.6–4.2	1.28	1.29	1.46	1.27	1.3	
Lipids (%)	–	–	–	13	15	–	20	–	11
NA (%)	46	30–62	26	14.3	14	30	10	60	
NA mass (kb)	79	17–498	4.4.–6.1	9.0	15–16	3.5–4.3	13.4	5.8–7.3	12
NA (G+C, %)	48	27–72	44	43–44	51	51	56	40–60	32

kb, kilobase; Mr, molecular mass; NA, nucleic acid. Data are updated from Ackermann, 2005; *Isometric capsids; ***Inovirus* genus only; –, absent.

Subfamilies, genera, species

While isometric, filamentous, and pleomorphic phages are generally easy to classify, this is not the case for tailed phages. Several tailed phage genera had already been established in previous ICTV reports, but large-scale genome sequencing and the availability of numerous completely sequenced phage genomes changed the situation. The sequenced genomes of 102 *Myoviridae* and 55 *Podoviridae* were compared by the CoreGenes and CoreExtractor programs (Lavigne *et al.*, 2008, 2009) in order to define genera by the number of shared homologous/orthologous proteins. In a general way, ICTV phage families were confirmed and often subdivided into subfamilies and genera. In addition, large numbers of new genera were individualized. Results show a close correlation of morphology and genomics and confirm the enormous diversity of tailed phages that had already been suggested by metagenomics. Subfamilies are generally heterogeneous and group phages active on several host genera. For example, the large *Spounavirinae* or SPO1 group comprises phages of *Bacillus*, *Brochothrix*, *Enterococcus*, *Lactobacillus*, *Listeria*, *Staphylococcus* and *Streptococcus* bacteria (Klumpp *et al.*, 2010). Unfortunately, there is no certain way to define a species and this is left to the intuition of individual taxonomists. The ICTV has adopted the 'polythetic species definition,' meaning that a virus species is a class of individuals with common properties that may or not be present in all members, constitute a replicating lineage and share a particular biotic niche (Van Regenmortel, 1990). This definition is of little practical help.

More on tailed phages

Tailed phages include some of the most complex of all viruses and are evidently highly evolved. Particles vary enormously in dimensions, physico-chemical, genomic, and biological properties. Heads are icosahedra or derivatives thereof and built of capsomers. Owing to the presence of surface proteins, they appear generally smooth. Tails are of helical symmetry and built of rings of subunits. Their complexity is exemplified by coliphage T4. Its head is a prolate icosahedron and its tail consists of a neck, a collar with appendages, a contractile sheath with rows of subunits, a tail core, a base plate, and six long tail fibres. Tailed phage genomes are generally constructed of 'modules' of related genes and often show evidence of horizontal evolution by genetic exchange.

Sequencing has revealed that certain phage components, such as gp23, the major capsid protein of phage T4, occur in a wide variety of morphologically unrelated myoviruses (e.g. marine cyanophages, various phages of enterics, campylobacters, *Rhodothermus*, and *Stenotrophomonas* (Petrov *et al.*, 2010). There is no satisfactory explanation for this fact.

Genomics

Genome size and structure

Phage genomes are single-stranded or double-stranded DNA or RNA, may be linear or circular, and vary enormously in size (Tables 1.2 and 1.3). In June 2013 the EMBL-EBI and GenBank (NCBI) databases listed 1289 and 936 complete bacteriophage genomes, respectively. These numbers are surprisingly low in the light of relatively small sizes of phage genomes. Tailed phage genomes will be discussed in some detail because of their numbers and complexity.

Tailed phages carry a single molecule of linear dsDNA genome in their capsid. Shortly after infection, this molecule is 'circularized' or becomes a circle; thus, the linear form exists in the infected cell only for an extremely short period. Their genomes also contain short single-stranded (ss) segments at both ends, called 'cohesive ends.' These ssDNA parts complement one another, facilitating circularization of the molecule after its entry into the host cell. Another interesting feature of some dsDNA phage genomes is the terminal redundancy of sequences. For example, in coliphage T4, a myovirus, virtually every single genome is different from any other genome, though they all contain exactly the same genes. This is possible because of the terminal redundancy; namely, a part of the genome is repeated at each end of the DNA molecule, but in every single genome, a different part of the genetic material can be duplicated. This circular permutation facilitates specific processes of genome replication and packaging into newly formed capsids of progeny phages.

The size of phage genomes is extremely variable. The largest bacteriophage genome reported is the 497,513-bp-long dsDNA of *Bacillus megaterium* phage G, a myovirus. Unfortunately, the nucleotide sequence of this genome, though deposited in the NCBI database, has not been published. Another very large phage genome belongs to *Pseudomonas* phage 201φ2-1. This genome is a 316,674-bp-long dsDNA. It contains 462 genes, including a gene with an intron, and a tRNA-encoding gene. Interestingly, the number of genes present in this genome is higher than the estimated number of genes required to build a theoretical minimal genome of the putative simplest bacterial cell (Henry *et al.*, 2010; Jewett and Forster, 2010; Wegrzyn, 2001). Some phages encode even more tRNA molecules. For example, *Mycobacterium* phage Myrna, a myovirus whose circularly permuted DNA genome is 164,602 bp long, encodes 41 different tRNAs. Another interesting genome belongs to the T4-like *Vibrio* phage KVP40. It consists of 244,834 bp of linear dsDNA and contains 381 protein-encoding genes, 29 genes for tRNAs, and five 'pseudogenes'. The presence of pseudogenes in a phage genome may be deemed surprising in the light of a general tendency of viral genomes to minimize their sizes and to compact genes.

The smallest known phage genome is that of coliphage BZ13, a levivirus. It is a 3412-bp-long ssRNA genome and has four genes only, coding for the assembly/maturation protein, capsid protein, replicase and lysis protein, respectively. Similar sets of genes occur in other leviviruses, for example *Caulobacter* phage φCb5. Its genome is a 3762-bp-long RNA molecule encoding four proteins with functions analogous to those of phage BZ13. This suggests that for autonomous propagation in a bacterial cell, a phage requires at least a replicase, a coat protein, a maturation protein and a lysis protein. An even smaller genome is attributed to *Leuconostoc* 'phage' L5. This entity is said to have a circular dsDNA genome of 2435 bp, but this was never published. Both DNA size and circular nature suggest that it is a plasmid. It is also possible that the L5 genome was fragmented during isolation and that its sequence represents only a part of the L5 genome.

Gene functions

Interestingly, even in phages with genomes 50 or 100 times larger than those of BZ13 or φCb5, the number of groups of genes coding for particular functions is not dramatically higher than that observed in small-genome bacteriophages. The major difference is that in the latter phages, particular groups of genes are represented by single open reading frames (ORF), sometimes even overlapping one another, while in large-genome phages there are several or more phage-encoded proteins for particular biological functions or particular structures. Therefore, instead of a single coat protein as in BZ13 or φCb5, large phages code for many proteins forming complicated structures of capsids and tails. However, if we consider the formation of the capsid as a function, then single coat proteins of BZ13 or φCb5 and multiprotein structures of large virions are analogues. Similarly, all phage genomes code for (a) proteins ensuring the replication of their genomes (either replicases or proteins recruiting host replication proteins to the bacteriophage *origin* sequences, or both), (b) the assembly of newly formed progeny virions, and (c) lysis of host cells or liberation of progeny virions from bacterial cells by extrusion (as in filamentous phages). Generally, what is absent in small bacteriophage genomes and present in large phage genomes is a group of genes coding for regulatory proteins or for proteins and tRNAs which can replace host functions under specific conditions of bacterial cell growth. Prominent among regulatory proteins are transcription activators and repressors, as well as other regulators of gene expression. Enzymes involved in genetic recombination are common too. Many phages also encode proteins interfering with host gene expression or host cell metabolism. Alleviation of host-encoded systems directed to protect the cell against foreign nucleic acids is another function of a group of products of bacteriophage genes. As an example of phage genomes coding for multiple functions, a scheme of the bacteriophage λ genome is shown in Fig. 1.3. This bacteriophage is a medium-size virus, with a genome 48,502 bp of dsDNA, and a workhorse for molecular biology. It may serve as a paradigm of phage genome structure, function, and grouping of genes in functional regions.

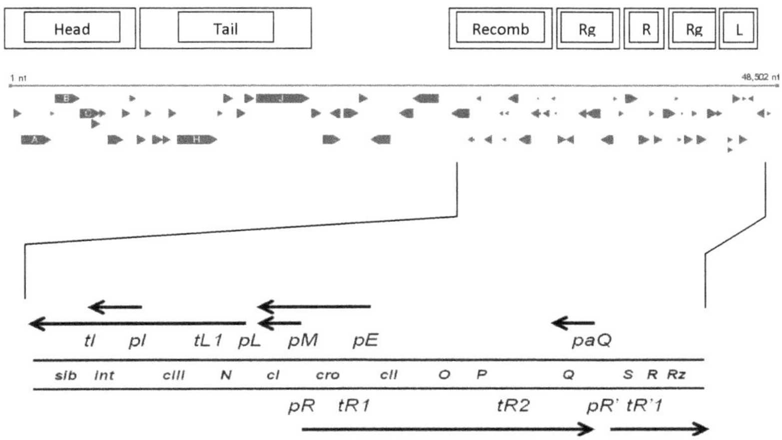

Figure 1.3 Map of the bacteriophage λ genome, showing the position of ORFs (open reading frames) the direction of transcription of particular genes (middle panel), and clusters of genes coding for proteins with specific functions (indicated above). Functions of gene products are abbreviated as follows: Head, assembly of the virion head; Tail, assembly of the virion tail; Recomb, DNA recombination; Rg, regulation of gene expression; R, phage DNA replication; L, lysis of host cell. A map of the right arm of the genome, not drawn to scale, indicates major transcription promoters (*p*) and transcription terminators (*t*) is shown below. Arrows indicate the direction of transcription. The whole genome sequence of phage λ may be found at (http://www.ncbi.nlm.nih.gov/genomes/genlist.cgi?taxid=10239&type=6&name=Phages) in the NCBI database. A colour version of this figure is available in the plate section at the back of the book.

Many phages (including λ) have genomes organized into modules, or groups of genes with related functions (Fig. 1.3), meaning that genes coding for particular functions are grouped in certain genome regions. In the paradigmatic genome of bacteriophage λ, there are clusters of genes coding for: (i) capsid proteins (genes of proteins forming the head and tail of the virion are also located in two separate subregions of the genome), (ii) recombination proteins, (iii) regulators of gene expression, (iv) replication factors, (v) lysis proteins. Such a structure of bacteriophage genetic material allows for the exchange of large parts of genomes if two different phages infect a single bacterial cell at the same time. Perhaps such events led to the currently observed mosaicism of many bacteriophage genomes, namely the presence of fragments of various phage genomes in the DNA of other phages (Hatfull, 2008).

Transcription

The huge variability of phages is reflected in different mechanisms for production of a single transcript or in complicated mechanisms for sequential activation and silencing of particular promoters, transcription antitermination, regulation, and control of transcript stability. Phages thus have many transcription strategies. Bacteriophages played a crucial role in the study of gene transcription. Although about 50 years ago it was known that both DNA strands can be used *in vitro* as templates for RNA synthesis, early studies on *Bacillus* phage α and coliphage φX174 suggested that *in vivo* only one DNA strand may be coding (Hayashi *et al.*, 1963; Tocchini-Valentini *et al.*, 1963). However, a classic study by Taylor *et al.* (1967) ascertained that both strands of bacteriophage λ DNA contain regions which can be used as templates for RNA synthesis *in vivo*. We know today that results obtained with phages α and φX174 were specific cases, that the general rule is that coding sequences can be located on either DNA strand, and that various genes can be transcribed in different directions. In fact, the same DNA region of a bacteriophage genome may contain fragments transcribed in opposite directions, although both transcripts are functional. This is again exemplified in the bacteriophage λ genome (Fig. 1.3).

Practical aspects of phage research

Phages may be both helpful and harmful and innumerable applications have been devised to put them to work or to tame them. They are summarized below. The subject has been reviewed many times and the reader is referred to these publications. Phage therapy and biocontrol of bacteria by phages will be discussed outside this chapter.

Phages for detection and diagnosis (Ackermann and DuBow, 1987; Rees and Loessner, 2005; Singh *et al.*, 2012)

The use of phages for the identification of bacteria was proposed as early as 1925. The idea was excellent, but hampered by the fact that phages never lyse all strains of a species and only these. Nevertheless, a few phages continue to be in use, for example phage O1 for the identification of *S. enterica* serovar Typhi and phage γ for that of *B. anthracis*. Between 1940 and 1965, over 150 typing sets were developed for a large number of bacteria (Kasatiya and Nicolle, 1978), including *Staphylococcus aureus* and *S. enterica* serovars Paratyphi B and Typhi. Bacteria were identified by their responses against a series of phages and divided into 'phage types.' The aim was to detect sources of infection, especially in hospitals. Phage typing has now been largely superseded by other techniques and is performed in specialized institutions only. A very important type of diagnostic phages is represented by recombinant reporter phages. A bacterial luciferase gene (*lux*) is inserted into a phage genome. The modified phage is applied to target bacteria. Infected cells become luminescent and may be detected by the emission of light. Another detection method is the phage amplification assay or 'titre increase method'. Phages are allowed to infect bacteria and titrated after completion of their cycle. The method has been refined by adding a virucide to kill remaining external phages. Apart from whole virions, also phage receptor binding proteins have been used to detect bacteria (Singh *et al.* 2012).

Phages as pollution indicators (Goyal *et al.*, 1987; Gerba, 2005; Santiago-Rodriguez *et al.*, 2013)

The aim is to detect faecal pollution by humans and animals. Four groups of phages are used as indicators of sewage pollution: somatic coliphages, F-specific RNA phages (*Leviviridae*), enterococcal phages, and phages of *Bacteroides fragilis*. 'Somatic coliphages' include any tailed phage and microvirus φX174. The RNA phages are pilus specific. *B. fragilis* phages are siphoviruses. All four types of phages occur in human faeces. The material is plated on selected indicator bacteria. Lysis indicates that indicator phages are present, but their specificity in indicating human faecal pollution is still under discussion. RNA phages seem to be useful indicators of human enteric viruses in water and shellfish.

Phages in genetic manipulations (Westwater and Schofield, 2005; Murray, 2006)

Phages may be used as vectors and cloning vehicles. The most important phages for this purpose are filamentous coliphages, notably M13, λ, and P1. Filamentous phages (M13) have had a major role in DNA sequencing by the Sanger method. The capacity of filamentous phages can be improved by the creation of 'phagemids,' a combination of phages and plasmids. Donor and vector DNA are cleaved with a restriction endonuclease and the fragments are ligated, producing hybrid molecules called 'chimeras.' Lambda and P1 have a much larger capacity than filamentous phages. Lambda has been improved by the introduction of promoters from coliphages T3 and T7. Promoters from λ itself can be used in expression plasmids. P1-derived vectors are suitable for transfer of relatively large amounts of DNA into Gram-negative bacteria.

Phage display (Lindquist, 2006; Matsubara, 2009; Tohidkia *et al.*, 2012)

A peptide or a protein is fused to a phage gene and expressed on the phage surface. Phage display started with filamentous phages (fd, M13) or inovirus-derived phagemids, but now has been extended to coliphages λ, T4, and T7. When a large number of heterologous DNA sequences are inserted into a phage genome, the corresponding proteins are fused on the coat protein. By creating large populations of phages, each expressing a different peptide or protein, phage 'libraries' can be obtained for isolation of interesting peptides or proteins by enrichment, called 'biopanning'.

Nuisance phages (Ackermann and DuBow, 1987; Moineau and Lévesque, 2005; Millen *et al.*, 2012)

The fermentation industry has been, and still is, plagued by harmful phages. As early as 1935, it was noted that phages interfered with cheese production, destroying valuable starter cultures and causing faulty fermentations and economic losses. Phages also destroyed clostridia in the now abandoned acetone–butanol fermentation and affected the production of a great variety of bacterial products, e.g. amylase, glutamic and gluconic acid, or antibiotics produced by bacilli and streptomycetes. They also made themselves undesirable in the production of fermented oriental food and even sake and wine. The most persistent and indeed universal problem is phage infestation in the dairy industry. It is much attributable to the use of a few standardized starter monocultures in all industrial countries with a large dairy industry. The principal bacteria affected, called lactic acid bacteria, belong to the genera *Lactobacillus*, *Lactococcus*, *Leuconostoc* and *Streptococcus thermophilus*. A considerable research effort has been made to trace and combat phage infestation; for example, over 1200 phages of lactic acid bacteria have been identified by electron microscopy (Ackermann, 2011).

Miscellaneous applications (Ackermann and DuBow, 1987).

This variegated and colourful group includes:

- tracers of water movements in the ground and of airborne infections;
- testing of air filters, aerosol samplers, antivirals, condoms, and disinfectants;
- detection of carcinogens and antitumour agents;
- decontamination of vaccine preparations and cultures of *Entamoeba histolytica*, *Paramecium* and *Leptospira*;
- isolation of rare actinomycetes (Kurtböke, 2009).

Electron microscopy for rapid diagnosis (Ackermann and Heldal, 2009; Ackermann, 2009)

Phage electron microscopy (EM) is nearly always transmission electron microscopy (TEM) of negatively stained viruses (Brenner and Horne, 1959). Cryo-electron microscopy is the domain of specialized laboratories; sectioning and shadowing are seldom used. TEM is perhaps the fastest technique in virology. In a general way, it permits instant identification of phage families and, often, of genera and species as well. Very often, virus identification by TEM takes as little as 30 s. In other cases, the virus must be photographed and measured for identification. Species diagnosis depends critically on the know-how of the observer. The observing scientist must be aware of staining artefacts and sources of errors (bacterial debris and slime, incomplete and adventitious phages, and malformations such as abnormally long tails). EM is particularly valuable in the contexts of phage therapy and biocontrol of pathogenic microorganisms. It provides a handy way of identifying phages and checking the purity and viability of phage preparations. It is foreseeable that every therapeutic phage or phage cocktail, like any pharmaceutical specialty, must carry a flyer with information on the virion (host, morphology, species or genus), titre, and shelf-life. Further, EM is likely to pinpoint interesting novel phages. For example, all but one of the *Staphylococcus* phages in the literature considered for therapy seem to belong to the Twort species. Any novel phage of this morphology is therefore interesting and worth investigating. Conversely, by comparison with known phages, EM is well able to pinpoint phages likely to be temperate.

Purification and staining

Examinations should be carried out by the investigating scientist and not be farmed out to some foreign institution. Phages should be concentrated and purified before examination to save time and energy. Crude, non-purified lysates should never be examined. Purification is best done by centrifugation and two or more subsequent washings in buffer or even tap water. Phages are then stained with phosphotungstate (0.5–2%) or uranyl acetate (1–4%). Both stains have been in use for over 50 years and are very satisfactory. Other uranyl salts and ammonium molybdate may be used too. Staining is near instantaneous and no fixation is necessary. Uranyl acetate produces both negative and positive staining. The latter is due to the affinity of uranyl acetate to dsDNA. Positively stained heads are invariably deep black and shrunken and their dimensions are perfectly useless for virus identification.

Dimensions

As a rule, the specifications of manufacturers on magnification are unreliable. They must be controlled by means of test specimens. Unfortunately, only beef liver catalase crystals are acceptable for high magnification. Diffraction grating replicas or latex spheres are inappropriate. Phage scientists may also use T4 phage tails (length 114 nm). Magnification correction of conventional manual microscopes is very easy and achieved in the dark room. In digital microscopes, it is suggested to calculate correction factors from photographs of test specimens.

Outlook

In our opinion, the most interesting topics for future research are genomics, phage ecology, and the object of this book, phage therapy. Genomics should be combined with electron microscopy to maximize conclusions. Now that software for the detection and comparison of core genes exists, it should be applied to siphoviruses. This will lead to the definition of natural groups in siphoviruses and thus improve their classification. It also appears rewarding to reannotate the first phage genomes sequenced in the 1990s by comparing them to the very considerable number of data now in databases. The latter, however, should be reappraised because they contain enormous amounts of unpublished data, that may never see peer review and print. Much progress is to be made in phage ecology. For example, it appears necessary and urgent to assemble an atlas of cyanophages with high-quality micrographs. Although many cyanophages have been depicted in the literature, their micrographs are very often poor and no identification is possible. Likewise, it appears desirable to extend ecological studies to tropical Africa, South America, Siberia, and China. Ecological studies should focus on precise virus identification by EM.

Web resources

Basic information on nearly all known bacteriophage sequences can be found in the two largest genomic databases, EMBL-EBI and NCBI. Descriptions of ICTV-recognized phage taxa and micrographs are available at the Universal Virus Database, version 4 (Columbia University, New York, ed. C. Büchen-Osmond). A rewrite of the ICTVdb is under way. A very useful list of Net resources for phage workers is available (Kropinski, 2009). This paper also lists companies involved in phage therapy and research and indicates phage collections.

- EBI/EMBL: http://www.ebi.ac.uk/genomes/phage.html
- NCBI:
  - Phage genomes: http://www.ncbi.nlm.nih.gov/genomes/genlist.cgi?taxid=10239&type=6&name=Phages
  - Taxonomy browser: http://www.ncbi.nlm.nih.gov/genomes/Taxonomy/taxonomy-home.html
- ICTV: www.ictv.org/
  - See for example 02. Caudovirales; 02.043, Myoviridae; 02.045 Podoviridae; 02.066 Siphoviridae

References

Abedon, S.T. (2009). Disambiguating bacteriophage pseudolysogeny: an historical analysis of lysogeny, pseudolysogeny, and the phage carrier state. In Contemporary Trends in Bacteriophage Research, H.T. Adams, ed. (Hauppauge, NY, Nova Science Publishers), pp. 285–307.

Abedon, S.T., and Murray, K.L. (2013). Archaeal viruses, not archaeal phages: an archaeological dig. Archaea *2013*, 251245.

Abedon, S.T., and Yin, J. (2009). Bacteriophage plaques: theory and analysis. Methods Mol. Biol. *501*, 61–74.

Ackermann, H.-W. (2005). Bacteriophage classification. In Bacteriophages: Biology and Applications, E. Kutter and A. Sulakvelidze, eds. (Boca Raton, FL, CRC Press), pp. 169–187.

Ackermann, H.-W. (2006). Classification of bacteriophages. In The Bacteriophages, 2nd edn, R. Calendar, ed. (New York, Oxford University Press), pp. 8–16.

Ackermann, H.-W. (2009). Basic phage electron microscopy. In Bacteriophages. Methods and Protocols, Vol. I, Isolation, Characterization, and Interactions, M.R.J. Clokie and A.M. Kropinski, eds. (Clifton, NJ, Humana Press), pp. 113–126.

Ackermann, H.-W. (2011). Phages of lactic acid bacteria: discovery and classification. In Bacteriophages in Dairy Processing, A. Quiberoni and J.A. Reinheimer, eds. (Hauppauge, NY, Nova Science Publishers), in print.

Ackermann, H.-W., and DuBow, M.S. (1987). Viruses of Prokaryotes, Vol. I, General Properties of Bacteriophages (Boca Raton, FL, CRC Press), pp. 13–28, 33–47, 49–101, 143–172.

Ackermann, H.-W., and Heldal, M. (2009). Electron microscopy of aquatic viruses. In Manual of Aquatic Virus Ecology, C.A. Suttle, S. Wilhelm, and M. Weinbauer, eds. (Waco, TX, American Society for Limnology and Oceanography), pp. 182–192.

Ackermann, H.-W., and Prangishvili, D. (2012). Prokaryote viruses studied by electron microscopy. Arch. Virol. *157*, 1843–1849.

Ackermann, H.-W., Martin, M., Vieu, J.-F., and Nicolle, P. (1982). Félix d'Hérelle: his life and work and the foundation of a bacteriophage reference center. ASM News *48*, 346–348.

Ackermann, H.-W., Yoshino, S., and Ogata, S. (1995). A Bacillus phage that is a living fossil. Can. J. Microbiol. *41*, 294–297.

Ackermann, H.-W., Tremblay, D., and Moineau, S. (2004). Long-term bacteriophage preservation. World Fed. Culture Coll. Newslett. *38*, 35–40.

Angly, F.E., Felts, B., Breitbart, M., Salamon, P., Edwards, R.A., Carlson, C., Chan, A.M., Haynes, M., Kelley, S., Liu, H., Mahaffy, J.M., Mueller, J.E., Nulton, J., Olson, R., Parsons, R., Rayhawk, S., Suttle, C.A., and Rohwer, F. (2006). The marine viromes of four oceanic regions. PloS Biol. *4*, 2121–2131.

Anonymous (2011). Virus classification. Wikipedia, http://en.wikipedia.org/wiki/Virus_classification. Accessed on 15 March 2011.

Barrangou, R., Fremaux, C., Deveau, H., Richards, M., Boyaval, P., Moineau, S., Romero, D.A., and Horvath, P. (2007). CRISPR provides acquired resistance against viruses in prokaryotes. Science *315*, 1709–1712.

Bobay, L.M., Rocha, E.P., and Touchon, M. (2013). The adaptation of temperate bacteriophages to their host genomes. Mol. Biol. Evol. *30*, 737–751.

Boyd, E.F. (2005). Bacteriophages and bacterial virulence. In Bacteriophages: Biology and Applications, E. Kutter and A. Sulakvelidze, eds. (Boca Raton, FL, CRC Press), pp. 223–265.

Bradley, D.E. (1967). Ultrastructure of bacteriophages and bacteriocins. Bacteriol. Rev. *31*, 230–314.

Bratbak, G., and Heldal, M. (1993). Total count of viruses in aquatic environments. In Handbook of Methods in Aquatic Microbial Ecology, P.F. Kemp, B.F. Sherr, E.B. Sherr, and J.J. Cole, eds. (Boca Raton, FL, Lewis Publishers), pp. 135–138.

Breitbart, M., and Rohwer, F. (2005). Here a virus, there a virus, everywhere the same virus? Trends Microbiol. *13*, 278–284.

Breitbart, M., Felts, B., Kelley, S., Mahaffy, J.M., Nulton, J., Salamon, P., and Rohwer, F. (2004). Diversity and population structure of a near-shore marine-sediment viral community. Proc. Royal Soc. Lond. B *271*, 565–574.

Brenner, S., and Horne, R.W. (1959). A negative staining method for high resolution electron microscopy of viruses. Biochim. Biophys. Acta *34*, 103–110.

Brussaard, C.P.D. (2009). Enumeration of bacteriophages using flow cytometry. In Bacteriophages. Methods and Protocols, Vol. I, Isolation, Characterization, and Interactions, M.R.J. Clokie and A.M. Kropinski, eds. (Clifton, NJ, Humana Press), pp. 97–111.

Brüssow, H., and Hendrix, R.W. (2002). Phage genomics: small is beautiful. Cell *108*, 13–16.

Calendar, E., ed. (2006). The Bacteriophages, 2nd edn (New York, Oxford University Press).

Campbell, A. (2006). General aspects of lysogeny. In The Bacteriophages, R. Calendar, ed. (New York, Oxford University Press), pp. 66–73.

Canchaya, C., Proux, C., Fournous, G., Bruttin, A., and Brüssow, H. (2003). Prophage genomics. Microbiol. Mol. Biol. Rev. *67*, 238–276.

Carlson, K. (2005). Working with bacteriophages: common techniques and methodological apporaches. In Bacteriophages: Biology and Applications, E. Kutter and A. Sulakvelidze, eds. (Boca Raton, FL, CRC Press), pp. 437–494.

Clokie, M.R.J., Millard, A.D., Letarov, A.V., and Heaphy, S. (2011). Phages in nature. Bacteriophage *1*, 31–45.

D'Hérelle, F. (1917). Sur un microbe invisible antagoniste des bacilles dysentériques. C.R. Hebd. Seances Acad. Sci. D *165*, 373–375.

D'Hérelle, F. (1918). Technique de la recherche du microbe filtrant bacteriophage (Bacteriophagum intestinale). C.R. Seances Soc. Biol. Filiales *81*, 1160–1162.

D'Hérelle, F. (1921). Le bactériophage: son comportement dans l'immunité (Paris, Masson), pp. 112 and 374.

Dinsdale, E.A., Edwards, R.A., Bailey, B.A., Tuba, I., Akhter, S., McNair, K., Schmieder, R., Apkarian, N., Creek, M., Guan, E., *et al.* (2013). Multivariate analysis of functional metagenomes. Front. Genet. *4*, 41.

Fancello, L., Raoult, D., and Desnues, C. (2012). Computational tools for viral metagenomics and their application in clinical research. Virology *434*, 162–174.

Fauquet, C.M., Mayo, M.A., Maniloff, J., Desselberger, U., and Ball, J.A., eds. (2005). Virus Taxonomy. VIIIth Report of the International Committee on Taxonomy of Viruses (Elsevier Academic Press), pp. 43–94, 279–299, 443–446, 741–749.

Gerba, C.P. (2006). Bacteriophage as pollution indicator. In The Bacteriophages, 2nd edn, R. Calendar, ed. (New York, Oxford University Press), pp. 695–701.

Goyal, S.M., Gerba, C.P., and Bitton, G., eds. (1987). Phage Ecology (New York, John Wiley & Sons).

Guttman, B., Raya, P., and Kutter, E. (2005). Basic phage biology. In Bacteriophages: Biology and Applications, E. Kutter and A. Sulakvelidze, eds. (Boca Raton, FL, CRC Press), pp. 29–66.

Hatfull, G.F. (2008). Bacteriophage genomics. Curr. Opin. Microbiol. *11*, 447–453.

Hatfull, G.F., and Hendrix, R.W. (2011). Bacteriophages and their genomes. Curr. Opin. Virol. *1*, 298–303.

Hayashi, M., Hayashi, M.N., and Spiegelman, S. (1963). Restriction of *in vivo* genetic transcription to one of the complementary strands of DNA. Proc. Natl. Acad. Sci. U.S.A. *50*, 664–672.

Henry, C., Overbeek, R., and Stevens, R.L. (2010). Building the blueprint of life. Biotechnol. J. *5*, 695–704.

Hosoda, K., Matsuura, T., Kita, H., Ichihashi, N., Tsukada, K., and Yomo, T. (2007). Kinetic analysis of the entire RNA amplification process by Qβ replicase. J. Biol. Chem. *282*, 15516–15527.

Hurwitz, B.L., and Sullivan, M.B. (2013). The Pacific Ocean virome (POV): a marine viral metagenomic dataset and associated protein clusters for quantitative viral ecology. PLoS One *8*, e57355.

Jewett, M.C., and Forster, A.C. (2010). Update on designing and building minimal cells. Curr. Opin. Biotechnol. *21*, 697–703.

Kasatiya, S.S., and Nicolle, P. (1978). Phage typing. In CRC Handbook of Microbiology, Vol. II, A.I. Laskin and H.A. Lechevalier, eds. (Boca Raton, FL, CRC Press), pp. 699–715.

King, A.M.Q., Lefkowitz, E., Adams, M.J., and Carstens, E.B., eds. (2011). Virus Toxonomy. Ninth Report of the International Committee on Taxonomy of Viruses (London and San Diego, Academic Press), p. 1791.

Klumpp, J., Lavigne, R., Loessner, M.J., and Ackermann, H.-W. (2010). The SPO1-related bacteriophages. Arch. Virol. *155*, 1547–1561.

Koko, A.S., Ackermann, H.-W., Taiwo, M., and Omilabu, S.A. (2011). Nigerian phages: the first bacteriophages from Tropical Africa. African J. Microbiol. Res. *5*, 2207–2210.

Kropinski, A.M. (2009). Internet resources of interest to bacteriophage workers. In Bacteriophages, Methods and Protocols, Vol. 2, Molecular and Applied Aspects, M.R.J. Clokie and A.M. Kropinski, eds. (Clifton, NJ, Humana Press), pp. 365–370.

Kurtböke, I. (2009). Use of phage-battery to isolate industrially rare actinomycetes. In Contemporary Trends in Bacteriophage Research, H.T. Adams, ed. (Hauppauge, NY, Nova Science Publishers), pp. 119–149.

Kutter, E., and Sulakvelidze, A., eds. (2005). Bacteriophages: Biology and Applications (Boca Raton, FL, CRC Press).

Kutter, E., Raya, R., and Carlson, K. (2005). Molecular mechanisms of phage infection. In Bacteriophages: Biology and Applications, E. Kutter and A. Sulakvelidze, eds. (Boca Raton, FL, CRC Press), pp. 165–222.

Lang, A.S., Zhaxybayeva, O., and Beatty, J.T. (2012). Gene transfer agents: phage-like elements of genetic exchange. Nat. Rev. Microbiol. *10*, 472–482.

Lavigne, R., Seto, D., Mahadevan, P., Ackermann, H.-W., and Kropinski, A.M. (2008). Unifying classical and molecular taxonomic classification: analysis of the Podoviridae using BLASTP-based tools. Res. Microbiol. *159*, 406–414.

Lavigne, R., Darius, P., Summer, E.J., Seto, D., Mahadevan, P., Nilsson, A.S., Ackermann, H.-W., and Kropinski, A.M. (2009). Classification of Myoviridae bacteriophages using protein sequence similarity. BMC Microbiol. *9*, 224.

Lee, S.J., and Richardson, C.C. (2011). Choreography of bacteriophage T7 DNA replication. Curr. Opin. Chem. Biol. *15*, 580–586.

Letarov, A., and Kulikov, E. (2009). The bacteriophages in human- and animal body-associated microbial communities. J. Appl. Microbiol. *107*, 1–13.

Lindqvist, B.H. (2006). Phage in display. In The Bacteriophages, 2nd edn, R. Calendar, ed. (New York, Oxford University Press), pp. 686–694.

Lwoff, A., Horne, R.W., and Tournier, P. (1962). A system of viruses. Cold Spring Harbor Symp. Quant. Biol. *27*, 51–62.

Loś, J.M., Loś, M., and Węgrzyn, G. (2011). Bacteriophages carrying Shiga toxin genes: genomic variations, detection and potential treatment of pathogenic bacteria. Future Microbiol. *6*, 909–924.

Łoś, M., and Węgrzyn, G. (2012). Pseudolysogeny. Adv Virus Res. *82*, 339–349.

Matsubara, T. (2009). Usage of bacteriophage in display technology: an effective identification of target-binding peptide/protein from large repertoires of sequences. In Contemporary Trends in Bacteriophage Research, H.T. Adams, ed. (Hauppauge, NY, Nova Science Publishers), pp. 407–419.

Matteson, A.R., Rowe, J.M., Ponsero, A.J., Pimentel, T.M., Boyd, P.W., and Wilhelm, S.W. (2013). High abundances of cyanomyoviruses in marine ecosystems demonstrate ecological relevance. FEMS Microbiol. Ecol. *84*, 223–234.

Millen, A.M., Horvath, P., Boyaval, P., and Romero, D.A. (2012). Mobile CRISPR/Cas-mediated bacteriophage resistance in Lactococcus lactis. PLoS One *7*, e51663.

Moineau, S., and Lévesque, C. (2005). Control of bacteriophages in industrial fermentations. In Bacteriophages: Biology and Applications, E. Kutter and A. Sulakvelidze, eds. (Boca Raton, FL, CRC Press), pp. 285–295.

Mueser, T.C., Hinerman, J.M., Devos, J.M., Boyer, R.A., and Williams, K.J. (2010). Structural analysis of bacteriophage T4 DNA replication: a review in the Virology Journal series on bacteriophage T4 and its relatives. Virol. J. *7*, 359.

Murray, N.E. (2006). Phage-based expression systems. In The Bacteriophages, 2nd edn, R. Calendar, ed. (New York, Oxford University Press), pp 674–685.

Odegrip, R., Schoen, S., Haggård-Ljungquist, E., Park, K., and Chattoraj, D.K. (2000). The interaction of bacteriophage P2 B protein with *Escherichia coli* DnaB helicase. J. Virol. *74*, 4057–4063.

Ortmann, A.C., and Suttle, C.A. (2009). Determination of virus abundance by epifluorescence microscopy. In Bacteriophages. Methods and Protocols, Vol. I, Isolation, Characterization, and Interactions, M.R.J. Clokie and A.M. Kropinski, eds. (Clifton, NJ, Humana Press), pp. 87–95.

Peduzzi, P., Agis, M., and Luef, B. (2013). Evaluation of confocal laser scanning microscopy for enumeration of virus-like particles in aquatic systems. Environ Monit. Assess. *185*, 5411–5448.

Petrov, V.M., Ratnayaka, S., Nolan, J.M., Miller, E.S., and Karam, J.D. (2010). Genomes of the T4-related bacteriophages as windows on microbial genome evolution. Virol. J. *7*, 292.

Pfankuch, E., and Kausche, G.A. (1940). Isolierung und übermikroskopische Abbildung eines Bakteriophagen. Naturwissenschaften *28*, 46.

Prangishvili, D., Forterre, P., and Garrett, R.A. (2006). Viruses of the Archaea: a unifying view. Nature Rev. Microbiol. *4*, 837–848.

Raettig, H. (1958). Bakteriophagie 1917 bis 1956. Vol. I. Autorenregister (Stuttgart, Gustav Fischer).

Raettig, H. (1967). Bakteriophagie 1957–1965. Vol. II. Bibliographie (Stuttgart, Gustav Fischer).

Rakonjac, J., Bennett, N.J., Spagnuolo, J., Gagic, D., and Russel, M. (2011). Filamentous bacteriophage: biology, phage display and nanotechnology applications. Curr. Issues Mol. Biol. *13*, 51–76.

Raya, R.R., and Hébert, E.M. (2009). Isolation of phages via induction of lysogens. In The Bacteriophages. Methods and Protocols, Vol. I, Isolation, Characterization, and Interactions, M.R.J. Clokie and A.M. Kropinski, eds. (Clifton, NJ, Humana Press), pp. 23–32.

Rees, C.E.D., and Loessner, M.J. (2005). Phage for the detection of pathogenic bacteria. In Bacteriophages: Biology and Applications, E. Kutter and A. Sulakvelidze, eds. (Boca Raton, FL, CRC Press), pp. 267–284.

Reyes, A., Semenkovich, N.P., Whiteson, K., Rohwer, F., and Gordon, J.I. (2012). Going viral: next-generation sequencing applied to phage populations in the human gut. Nat. Rev. Microbiol. *10*, 607–617.

Rohwer, F. (2003). Global phage diversity. Cell *113*, 141.

Ruska, H. (1940). Über die Sichtbarmachung der bakteriophagen Lyse im Übermikroskop. Naturwissenschaften *28*, 45–46.

Salas, M., Blanco, L., Lázaro, J.M., and de Vega, M. (2008). The bacteriophage phi29 DNA polymerase. IUBMB Life. *60*, 82–85.

Santiago-Rodriguez, T.M., Marcos, P., Monteiro, S., Urdaneta, M., Santos, R., and Toranzos, G.A. (2013). Evaluation of Enterococcus-infecting phages as indices of fecal pollution. J. Water Health *11*, 51–63.

Singh, A., Arutyunov, D., Szymanski, C.M., and Evoy, S. (2012). Bacteriophage based probes for pathogen detection. Analyst *137*, 3405–3421.

Srinivasiah, S., Bhavsar, J., Thapar, K., Liles, M., Schoenfeld, T., and Wommack, K.E. (2008). Phages across the biosphere: contrasts of viruses in soil and aquatic environments. Res. Microbiol. *159*, 349–357.

Solioz, M., and Marrs, B. (1977). The gene transfer agent of Rhodopseudomonas capsulata – purification and characterization of its nucleic acid. Arch. Biochem. Biophys. *181*, 300–307.

Summers, W.C. (1999). Félix d'Hérelle and the origins of molecular biology (New Haven, CT, Yale University Press), pp. 60–81.

Summers, W.C. (2012). The strange history of phage therapy. Bacteriophage *2*, 130–133.

Suttle, C.A. (2005). Viruses in the sea. Nature *437*, 356–361.

Taylor, K., Hradecna, Z., and Szybalski, W. (1967). Asymmetric distribution of the transcribing regions on the complementary strands of coliphage lambda DNA. Proc. Natl. Acad. Sci. U.S.A. *57*, 1618–1625.

Thierauf, A., Perez, G., and Maloy, A.S. (2009). Generalized transduction. Methods Mol. Biol. *501*, 67–86.

Thurber, R.V. (2009). Current insights into phage biodiversity and biogeography. Curr. Opin. Microbiol. *12*, 582–587.

Tocchini-Valentini, G.P., Stodolsky, M., Aurisicchio, A., Sarnat, M., Graziosi, F., Weiss, S.B., and Geiduschek, E.P. (1963). On the asymmetry of RNA synthesis in vivo. Proc. Natl. Acad. Sci. U.S.A. *50*, 935–942.

Tohidkia, M.R., Barar, J., Asadi, F., and Omidi, Y. (2012). Molecular considerations for development of phage antibody libraries. J. Drug Target. *20*, 195–208.

Tone, T., Takeuchi, A., and Makino, O. (2012). Single-stranded DNA binding protein Gp5 of *Bacillus subtilis* phage Φ29 is required for viral DNA replication in growth-temperature dependent fashion. Biosci. Biotechnol. Biochem. *76*, 2351–2353.

Twort, F.W. (1915). An investigation on the nature of ultra-microscopic viruses. Lancet *186*, 1241–1243.

Van Regenmortel, M.H.V. (1990). Virus species: a much overlooked but essential concept in virus classification. Intervirology 32, 241–254.

Wagner, P.L., and Waldor, M.K. (2006). Bacteriophages in bacterial pathogenesis. In The Bacteriophages, 2nd edn, R. Calendar, ed. (New York, Oxford University Press), pp. 710–724.

Wagner, J., Maksimovic, J., Farries, G., Sim, W.H., Bishop, R.F., Cameron, D.J., Catto-Smith, A.G., and Kirkwood, C.D. (2013). Bacteriophages in gut samples from pediatric Crohn's disease patients: metagenomic analysis using 454 pyrosequencing. Inflamm. Bowel Dis. *19*, 1598–1608.

Westwater, C., and Schofield, D.A. (2005). Phage as vectors and targeted vehicles. In Bacteriophages: Biology and Applications, E. Kutter and A. Sulakvelidze, eds. (Boca Raton, FL, CRC Press), pp. 265–297.

Wegrzyn, G. (2001). The minimal genome paradox. J. Appl. Genet. *42*, 385–392.

Węgrzyn, G., Licznerska, K., and Węgrzyn, A. (2012). Phage λ – new insights into regulatory circuits. Adv. Virus Res. *82*, 155–78.

Weigel C., and Seitz, H. (2006). Bacteriophage replication modules. FEMS Microbiol. Rev. *30*, 321–381.

Weinbauer, M.G. (2004). Ecology of prokaryotic viruses. FEMS Microbiol. Rev. *28*, 127–181.

Wildy, P. (1971). Classification and nomenclature of viruses. First Report of the International Committee on Nomenclature of Viruses (Basel, S. Karger).

Yu, M.X., Slater, M.R., and Ackermann, H.-W. (2006). Isolation and characterization of Thermus bacteriophages. Arch. Virol. *151*, 663–679.

Zemb, O., Manefield, M., Thomas, F., and Jacquet, S. (2013). Phage adsorption to bacteria in the light of the electrostatics: a case study using *E. coli*, T2 and flow cytometry. J. Virol. Methods *189*, 283–289.

Zhao, Y., Temperton, B., Thrash, J.C., Schwalbach, M.S., Vergin, K.L., Landry, Z.C., Ellisman, M., Deerinck, T., Sullivan, M.B., and Giovannoni, S.J. (2013). Abundant SAR11 viruses in the ocean. Nature *494*, 357–360.

The First Step to Bacteriophage Therapy: How to Choose the Correct Phage

2

Małgorzata Łobocka, Monika S. Hejnowicz, Urszula Gągała, Beata Weber-Dąbrowska, Grzegorz Węgrzyn and Michał Dadlez

Abstract

Bacteriophages are viruses that can kill bacteria but are harmless to eukaryotic cells. In natural environments they have a dominant role in controlling bacterial populations. Thus, in the era of the emergence and spread of multidrug resistant bacterial pathogens they are more and more often seen as promising antibacterial agents that could be an alternative to antibiotics. Bacteriophages' main advantages as therapeutics are their ability to target bacteria of certain strains or species, without any harmful effect on the rest of the bacterial microflora, as well as their self-limiting propagation which is controlled by the availability of a sensitive host. Moreover, bacterial antibiotic resistance is not a barrier for phage infection. Only a limited number of phages from environmental isolates meet the criteria that are expected for therapeutic phages. Here we describe the most important of these criteria and provide a guide for selecting potential therapeutic phages for further studies in animal models. The ability of many phages to remain in a bacterium in the form of a prophage and increase its adaptive potential, as well as to participate in the horizontal gene transfer between bacterial cells, a priori precludes their use in therapy due to safety concerns. Factors that matter in the prediction of the remaining phages' therapeutic efficacy include host range and killing potential, adsorption kinetics and propagation efficiency, stability during storage and under 'natural conditions', the ability to penetrate encapsulated cells or biofilms, easiness of purification. The choice or modification of phage propagation host, which cannot be a source of contaminating phages, plasmids and toxins, appears nearly as important as the selection of a proper therapeutic phage.

Introduction: therapeutic phages – what requirements should they meet?

Bacteriophages can be found everywhere where bacteria live. They appear to be the most abundant life forms, with the estimated global number of about 10^{31} phage particles, and ~100 million of phage species (Rohwer, 2003; Wommack and Colwell, 2000). Only some phages infect clinically relevant bacteria or plant bacterial pathogens, and thus, only some of them could be interesting for potential application in phage therapy. However, the infective potential of each phage and its host range as well as the requirements that have to be met by a phage as a therapeutic result in a serious narrowing of the number of phages that could find a practical use (Clark and March, 2006; Gill and Hyman, 2010).

The main limitation is dictated by phage propagation strategy. While all phages are obligatory parasites of bacterial cells, only some of them, which are called obligately lytic, inevitably kill the infected bacterium. Opposite to them, the so called temperate phages can either multiply lytically in a bacterium causing its lysis or lysogenize the bacterium. Lysogens contain only temperate phage DNA, in the form of a prophage, which can replicate in a cell cycle and is inherited by daughter cells from the mother. Temperate phages encode numerous proteins, virulence factors among them, which give a selective advantage to prophage carrying cells over cells that lack the prophage, and in this way prevent the overgrowth of bacterial populations by the latter (Brüssow *et al.*, 2004, Wang *et al.*, 2010). They could potentially worsen the outcome of infection by lysogenization of infecting bacteria and increasing their virulence. Additionally, spontaneous activation of prophages in lysogens could enhance biofilm formation by promoting DNA release (Carrolo *et al.*, 2010; Godeke *et al.*, 2011). Thus, temperate phages should be excluded from a therapeutic use, at least as long as they can be replaced by obligately lytic phages (Brüssow, 2005; Kropinski, 2006; Borysowski *et al.*, 2011).

Although the lifestyle of obligately lytic phages predisposes them for therapeutic applications, there are additional factors that limit the use of certain representatives. They may result from instability, weak infective potential under *in vivo* conditions, limited host range and an inability to infect encapsulated bacterial cells or to penetrate biofilms, the capacity to transfer the genetic material from one bacterium to the other, etc. Bacterial phage resistance mechanisms and immune system responses of higher organisms to phages are separate concerns. Certain processes that underlie these interactions are just beginning to be understood. However, with the progress of studies on bacteriophages one can be more and more successful with the selection of phages that will be both effective therapeutically and safe.

Environmental sources of therapeutic phages

As obligate bacterial parasites, bacteriophages are present in all environmental niches where bacteria live – from deep oceanic water through the sands of the Sahara to hot springs. They can be found in soil as well as in city sewage, on plants, and human and animal bodies (see, for example, Adriaenssens *et al.*, 2012a; Ashelford *et al.*, 2003; Górski and Weber-Dąbrowska, 2005; Sawstrom *et al.*, 2008; Srinivasiah *et al.*, 2008; Breitbart, 2012; Zhong *et al.*, 2010).

The variability of phages exceeds that of other organisms. Despite their abundance, phages are underrepresented in biological strain collections. Over 6000 isolates have been morphologically characterized (Ackermann, 2007; Ackermann and Prangishvili, 2012). Phages that infect bacteria were grouped into 10 families whose representatives differ in sizes and shapes, the type and topology of virion nucleic acid (ssRNA, dsRNA, ssDNA, ds circular DNA, ds linear DNA), the presence or a lack of a lipid envelope, and virion complexity (Ackermann, 2009a; Krupovic *et al.*, 2011). Phages of three families, *Myoviridae*, *Siphoviridae* and *Podoviridae*, which have virions that consist of a head and a tail, and thus are called tailed phages, predominate (over 96% of phage isolates). They form a separate order, *Caudovirales*, and have been estimated to evolve before the separation of bacteria and archaea. They have the most variable genomes which differ in size from 12 to 497 kb. The tailed phages predominate in therapeutic phage collections, perhaps due to their coding potential that is larger than that of other phages. However, certain small RNA phages of *Cystoviridae* family also found a therapeutic use, e.g. as components of the phage cocktail AgriPhage against *Xanthomonas campestris* pv. *vesicatoria*, or *Pseudomonas syringae* pv. *tomato* (http://www.omnilytics.com/products/agriphage/agriphage4.html; Jones *et al.*, 2007).

The first note of an unknown agent with bacteriolytic activity was published in 1896 (Hankin, 1896; reviewed by Chanishvili, 2012). Its author, Hankin, isolated this agent from the waters of the Ganges River in India. Twenty years later Frederick Twort and Félix d'Hérelle rediscovered similar agents, reported them in more detail and pointed out the potential usefulness of their antibacterial activity for the treatment of diseases (d'Hérelle, 1917; Twort, 1915). It was d'Hérelle who applied the name bacteriophage for an agent which causes transmissible lysis or the dissolution of bacteria. Since then different phages that were isolated from various sources have been described. In the majority of cases the truly obligately lytic nature of these phages could not have been easily confirmed. However, they showed the lytic activity for bacteria that were used as targets in their isolation. For instance, Dubos *et al.* (1943) isolated a *Shigella* bacteriophage from a New York City sewage sample using a smooth strain of *Shigella dysenteriae* 2308 as the host. He simply refreshed the overnight culture of the *Shigella* strain by dilution and 3 h incubation at 37°C and inoculated it with the sewage sample. After 4 h incubation at 37°C, the culture was filtered through a Selas porcelain candle filter. The isolated phage was used in mice in an experimental intracerebral injection of *Shigella dysenteriae*. The authors showed that the bacteriophage not only can multiply in the brain (at the site of infection and by crossing the blood–brain barrier), but also may protect animals against a new infection. D'Hérelle himself recommended isolation of phages either from wastewater or from specimens from patients recovering from an infection caused by a given bacterium (d'Hérelle, 1938). He claimed the latter to be a source of the most effective therapeutic phages.

Numerous phages that were isolated from animal and human materials as well as from sewage samples were successfully used in experimental therapy of bacterial infections in calves, mice and chickens (Smith *et al.*, 1987; Soothill, 1992; Barrow, 1998; Biswas *et al.*, 2002; Atterbury *et al.*, 2003; Hwang *et al.*, 2009; Oliveira *et al.*, 2009). The virulent R bacteriophage from sewage with lytic activity against *Escherichia coli* K1 bacteremic strain was used by Barrow *et al.* (1998) to prevent septicaemia and meningitis-like infection in chickens and mice. It was demonstrated that phage R was able to multiply (at both 37 and 44°C) in chicken brain, blood as well as in mice tissues. Mankiewicz and Liivak (1967) isolated mycobacteriophages from patients with tuberculosis and sarcoidosis by successive passages of the patient's serum with two mycobacterium indicator strains. The phages that were isolated appeared to have a wide host range among pathogenic and saprophytic mycobacteria.

In the studies of Gantzer *et al.* (2002), Bachrach *et al.* (2003), Hitch *et al.* (2004), Chibani-Chennoufi *et al.* (2004) and Łusiak-Szelachowska *et al.* (2008), potential therapeutic phages against *Bacteroides fragilis*, *Enterococcus faecalis*, *Proteus mirabilis* and *Escherichia coli* were isolated from human stools, saliva and oral cavity with the standard procedure.

Wommack and Colwell (2000) provided an overview concerning viruses in aquatic ecosystems. Phages isolated from aquatic samples often appeared to have much broader host specificity than phages from other samples. For instance phages infecting *Sphaerotilus natans* that were isolated in that study were able to lyse *Pseudomonas aeruginosa* strains as efficiently as the original host. Although such phages may seem superior for therapeutic applications, they are not as effective therapeutically, as phages which propagate in representatives of only one species (d'Hérelle, 1938). Thus, it is advisable to select, from among others, phages that can infect and propagate in the greatest possible number of strains within a given pathogenic species, except when a phage against a specific strain is required.

The success rate in initial attempts to isolate a phage against a given pathogen varies, dependent on the bacterium, the source of a phage and a procedure that is applied for particular samples. Typically, it is easier to isolate phages against common environmental bacteria that are opportunistic pathogens than against specialized pathogens that inhabit a specific environmental niche. The former are usually abundant in water or soil (see, for example, Ashelford *et al.*, 2003; Knezevic *et al.*, 2009; Ogunseitan *et al.*, 1992). The latter can most often be found in sites of infection, in faeces or body fluids from infected individuals or in raw sewage, which appears to contain the most diverse viral biome examined thus far (Callaway *et al.*, 2010; Cantalupo *et al.*, 2011; Kwiatek *et al.*, 2012). Certain insects can also serve as a source of such phages. For instance, an obligately lytic phage that was able to infect *Staphylococcus aureus* was isolated from an extract of house-flies (Shope, 1927; Rippon, 1956). Even narrowly specialized bacterial pathogens, such as *Helicobacter pylori* or *Borrelia burgdorferi* were described to spontaneously release phages (Schmid *et al.*, 1990; Heintschel *et al.*, 1993; Eggers and Samuels, 1999; Luo *et al.*, 2012; Wan *et al.*, 2011). However, all of these phages are either confirmed to be temperate or their genomes have not been sequenced. Thus, obligately lytic phages that infect these pathogens await identification.

A brief overview of phage isolation methods

The possibility to visualize bacterial host lysis caused by bacteriophages, in the form of plaques on lawns of bacterial cells (Mazzocco *et al.*, 2009a,b; Kropinski *et al.*, 2009b), makes the detection and subsequent isolation of most of the environmental phages against cultivable hosts relatively simple, if only they are present (see, for example, Bryner *et al.*, 1973; Yordpratum *et al.*, 2011). Certain compounds, such as 2,3,5-triphenyltetrazolium chloride (TTC), ferric ammonium citrate and sodium thiosulfate have been described to enhance the contrast of bacteriophage plaques on bacterial lawns (McLaughlin and Balaa, 2006). Their use can facilitate the detection of plaques that are too small or too turbid to be easily visible. Another visualization method of very small or even undetectable plaques is the use of sublethal concentrations of antibiotics (Comeau *et al.*, 2007; Łoś *et al.*, 2008). It appears that slowing down the bacterial growth, with a simultaneous provision of nutrients from rich medium still supports the propagation of phages, provides conditions for larger burst size and at the same time prevents an overgrowth of uninfected bacteria, thus facilitating formation of a well-visible plaque.

The metabolic state of bacterial cells influences their susceptibility to phage infection, phage latent period and burst size and hence the success of new phage isolation attempts (e.g. Cerca *et al.*, 2007; Golec *et al.*, 2011b; Kutter *et al.*, 1994; Moreno, 1977; Robb *et al.*, 1978; Schrader *et al.*, 1997; Voroshilova and Kozakova, 1983). Most often, but not always, exponentially growing cells are the most susceptible and can support the fastest and the most efficient phage production. Thus, Wommack and Colwell (2000) recommend using bacteria from an exponentially growing culture for phage isolation.

Van Twest and Kropinski (2009), and Serwer *et al.* (2009) summarized the standard methods of bacteriophage isolation and concentration from environmental samples (water, soil) and sewage, as well as a unique procedure for giant bacteriophage isolation. The general practice is to use the so-called enrichment procedure with more than one potential phage host in a cultivation flask, which is inoculated with an original source of phages (typically a filtrate of water, sewage, soil suspension etc.; Carvalho *et al.*, 2010; van Twest and Kropinski, 2009). Based on early observations, d'Hérelle (1938) recommended using at least 50 isolates of different strains for each target bacterial species. However, in the authors' experience a mixture of a few well differentiated species representatives usually suffices for the isolation of polyvalent phages, which infect a wide range of strains of a given bacterial species, assuming they are present in the tested sample. By mixing *S. natans* and *P. aeruginosa* or *S. natans* and *E. coli* in the enrichment culture, Jensen *et al.* (1998), demonstrated the usefulness of such a procedure even for the isolation of phages that could infect representatives of different bacterial genera.

By estimation, less than 1% of the bacteriophage species present in environmental samples can be detected by plaque assays with cultivable hosts (Ashelford *et al.*, 2003; Williamson *et al.*, 2003). Moreover, they are amplified in enrichment cultures with different efficiency illustrating the problems with standard enrichment procedures that are commonly used for phage isolation. One of those problems is the limitation of phage biodiversity in original samples by promoting the development of those phages that can propagate most efficiently under given experimental conditions (Hendrix, 2002). Another one is the representation of a progeny of each single phage from the original source by multiple plaques on a detection plate, which results in a repeated isolation of the same phages. To avoid these problems Serwer *et al.* (2004) applied a procedure which involves the placement of an original phage source (e.g. soil) directly on an agar plate, overlaying it with a bacterial culture in a molten, dilute agarose gel, waiting for the agarose solidification and incubation of such plates in an appropriate temperature until the appearance of plaques. Four new bacteriophages were isolated by using this approach. The poor growth of two of them makes them unlikely to be isolated by a standard enrichment procedure. However, one cannot estimate how many well propagating phages were lost by omitting the enrichment step. Thus, omission of the enrichment procedure may not be a recommended strategy to isolate potential therapeutic phages, which are expected to be well propagating ones. A solution may be a consensual approach, in which the enrichment procedure is shortened substantially to minimize its adverse effects. It has been successfully used at the Ludwik Hirszfeld Institute of Immunology and

Experimental Therapy (IIET) to isolate over 300 different phages from samples such as crude and purified city sewage, inland and sea waters as well as biological material (stool, sputum, urine), and is briefly summarized below.

Typically, liquid samples that can contain phages are passed through 0.22-μm low protein binding membrane filters and mixed with nutrient broth in a 1:1 ratio. To 10 ml of the sample 0.1 ml of bacterial culture is added, the sample is mixed and incubated at 37°C for 1 h. After that 0.2 ml of each sample is distributed with a T-shaped spreader on the surface of the Mueller–Hinton agar plate. Every stage is repeated twice. The samples are incubated for about 18 h at 37°C until plaques appear. If plaques merge with one another, appropriate dilutions are made. Plaques of similar size (observation on two plates) are transferred to a test tube with 5 ml of the nutrient broth and one drop of bacterial culture (a control contains only nutrient broth and bacteria). The sample is subsequently mixed gently, incubated at 37°C for 18 h, and centrifuged for 30 min at ~2000 × g. The bacterial pellet is resuspended in a small portion of nutrient broth and spread on the surface of a nutrient agar plate. Plates are incubated at 37°C (for the time that depends on the growth rate of bacteria). To obtain single phage clones, plaques are reisolated from freshly infected bacteria five more times. After centrifugation, the supernatant is filtered using membrane filters (0.22 μm) and the phage titre is determined.

Phage lysates are prepared in peptone water or LB (Luria broth) by the standard method with some modifications. Fresh, exponentially growing cultures of bacterial strains (3–5 h, 37°C) are incubated with phages at 37°C till complete lysis occurs (3–6 h), stored at 4°C overnight, filtered (0.22 μm) and tested for phage presence by the double agar method according to Adams's procedure (Adams, 1959).

The tetrazolium colorimetric testing of phage lysis in wells of a microtitre plate, with some modifications, can be used alternatively to a double agar layer method for phage isolation from sewage and environmental samples (McLaughlin, 2007). The application of tetrazolium is based on the enzymatic reduction of TTC to red-coloured triphenylformazan by live bacterial cells. The final quantity of formazan corresponds to the number of viable bacterial cells. In a standard assay, 96-well microplates are loaded with 100 μl of bacterial culture (10^6 to 10^8 CFU/ml), 100 μl of the tested sample and 50 μl of tetrazolium (6 mg/ml), all suspended in nutrient broth. The tetrazolium solution is added in the final stage of incubation. Several dilutions of bacterial culture are tested in parallel. Typically, a 6-hour incubation period of samples at 37°C with 10^6 CFU/ml of host strain gives the complete lysis, or single plaques are confirmed on agar plates. For phages that are newly isolated from environmental waters an 18-hour incubation period at 25°C is usually the most suitable. Whereas the control samples with bacteria and tetrazolium alone become red in this assay, the change of colour in samples containing phages that infect bacteria in a given culture is weaker or cannot be observed at all. The advantage of the colorimetric method is its suitability to detect a low number of phage particles (<10 P/ml) and to test a large number of environmental samples in one experiment, which is important for the progress in fast isolation of new candidate phages for therapeutic or biocontrol purposes.

The abundance of phages in certain environments, such as sewage, makes the isolation of a phage against any bacterium of this environment likely, even when starting from a few hundred millilitre sample as a phage source. Opposite to that, chances of isolating a given phage from e.g. oligotrophic waters are poor, unless phage concentration methods are applied as a preliminary step. The removal of cellular organisms as well as organic and inorganic particulate material is typically the starting point of all phage isolation procedures. It is routinely done by filtering phage suspension through low protein affinity filters of diminishing pore size, where 0.22 μm is the pore size of the final cut-off filter. The impact filtration of outcome phage suspensions through low protein affinity, < 20 μm pore size filters, to concentrate phages is limited to at most a few litre samples of oligotrophic waters, due to low filtration speed and filter clogging problems (Steward and Culley, 2010). Polyethylene glycol (PEG) precipitation or centrifugation of phage suspensions have similar volume limits. Thus, high throughput methods have been introduced to concentrate phages from water with low suspended solids content (for a summary see John *et al.*, 2011). One of them is tangential flow filtration (TFF), in which the stream of sample water passes parallel to the ultrafiltration filter membrane face. One portion of sample water (permeate) passes through the membrane leaving viral particles behind, while the rest (retenate) recirculates back to the feed tank. TFF allows for phage concentration from hundreds of litres of samples (Olszewski *et al.*, 2005; Wommack *et al.*, 2000). However, it is technically tricky and requires expensive equipment. Moreover, virus recovery efficiencies vary highly. To overcome these problems John *et al.* (2011) explored and optimized a flocculation, filtration and resuspension method (FFR), using $FeCl_3$ (1 g of Fe per litre of sea water) as a flocculent. Flocculate can be collected by filtration through the Whatman polycarbonate membrane filter (0.8 mm pore size), and resuspended in Tris-HCl buffer (pH 6) containing oxalic acid and supplemented with Mg_2EDTA to chelate Fe(III) in solution and to prevent re-precipitation. The FFR method is superior over the TFF method when it comes to costs, the equipment needs and the reproducibility of recovered virus fraction (over 90% versus 2–98%). The fraction of infective phage recovery when using this method is also satisfactory. It varies on the average between 47–73%, but can be as low as 13% for certain phages. Differences between TFF and FFR may concern the efficiency of recovery of particular taxa representatives (Hurwitz *et al.*, 2013). Podoviruses and non-tailed viruses were underrepresented in TFF concentrated samples, perhaps due to their leakage through 100-kDa filters. Presumably, the use of 50- or 30-kDa filters could overcome this problem.

The selection of phages that infect encapsulated bacterial cells

To infect a bacterium the phage genome has to penetrate through all that surrounds a cytoplasm, including a cell membrane(s) and cell wall as well as additional protective layers. Main surface components or exudates of most bacteria are high molecular-weight polysaccharides, which are astonishingly diversified. They have a wide range of biological functions, forming, for example, cell-bound capsular polysaccharides, O-antigen components of lipopolysaccharides, certain S-layer glycoprotein glycans as well as unbound slime and main biofilm matrix components (Bazaka *et al.*, 2011; Flemming and Windenger, 2010; Mann and Wozniak, 2012; Ristl *et al.*, 2011). Certain free exopolysaccharides can remain in close proximity to the cell surface despite the lack of a detectable membrane anchor and can form a separable envelope, which encloses a cell.

Exopolysaccharide capsules, slime and biofilm matrix form a barrier that can protect bacterial cells from desiccation, access of drugs and disinfectants, immune system components as well as from the infection by certain phages (reviewed by Roberts, 1995; Scholl *et al.*, 2005; Taylor and Roberts, 2005). They also facilitate bacterial cell adhesion to each other and to biotic and abiotic surfaces, enabling biofilm formation (reviewed by Bazaka *et al.*, 2011). Thus, surface-bound or unbound exopolysaccharides are often determinants of bacterial virulence (Roberts, 1995). For instance, a majority of clinical coagulase-negative isolates of staphylococci (CNS) produce slime (Beard-Pegler *et al.*, 1989; Christensen *et al.*, 1983; Hall *et al.*, 1987; Etienne *et al.*, 1988). In studies of methicillin-resistant *S. aureus* strains (MRSA) from surgical or immunocompromised patients the incidence of slime formation was the highest in the case of MRSA associated with fatal infections (Takahashi *et al.*, 1997). In the case of group A *Streptococcus* (GAS) the production of a hyaluronic acid capsule is one of a few factors indispensable *in vivo* for the switching of phenotype in the superinvasive mutants of GAS M1T1 serotype to a hypervirulent state that allows dissemination into the bloodstream (Cole *et al.*, 2010). The high-molecular-weight polysaccharides with O-specificity are present in extracellular slime produced by *P. aeruginosa* strains isolated from the immunocompromised hosts (reviewed by Knirel, 1990). The extracellular antigenic polysaccharide of another type (bacterial alginate) is expressed by mucoid *P. aeruginosa* strains isolated from patients with cystic fibrosis. Also in the case of *Klebsiella pneumoniae*, hypermucoviscosity of cells correlated with a high serum resistance was more prevalent in invasive strains causing primary liver abscess and septic metastatic complications (Fang *et al.*, 2004).

Several examples of the roles of capsular and extracellular polysaccharides in the protection of bacterial cells from phages have been described. In early studies, Danz and Schultz (1949) isolated a mucoid mutant of *P. aeruginosa*. The mutant lost the susceptibility to phages that could specifically infect the parental strain. In the case of streptococci, the presence of hyaluronic acid capsule conferred resistance to phages on group A *Streptococcus* strain (Maxted, 1952). Cell treatment with hyaluronidase restored their susceptibility. In studies of over 300 CNS from healthy donors or blood cultures, slime-producing CNS appeared to be resistant to one or more antibiotics and, in most cases, non-typeable with phages (Boussard *et al.*, 1993). In the studies of Christensen *et al.* (1983) only a slightly higher proportion of phage untypable strains produced slime. However, of 10 slime-producing *S. epidermidis* strains from endocarditis cases tested by Etienne *et al.* (1988), nine were non-typable by phages. Heavy slime exudates surrounding *E. coli* cells grown in media containing 1% sodium dodecybenzene sulfonate blocked adsorption of coliphages to susceptible strains (Koransky and Anderson, 1973).

To overcome these barriers some phages contain exopolysaccharide depolymerases in their tails. They can adsorb to capsular polysaccharides and catalyse their degradation. The exopolysaccharide depolymerases act to disperse the viscous interfering polysaccharide of capsules or slime layers to enable the phage access to underlying receptor sites (reviewed by Scholl and Merril, 2005). A hallmark of phages whose virions contain proteins of exopolysaccharide depolymerase activity is the production of plaques which are surrounded by halos – turbid zones of clearing that are more transparent than the rest of the exopolysaccharide-producing bacterial cell lawn. They are formed as a result of the reduction or disappearance of bacterial polysaccharide capsules. The concentration of viable bacteria remains unchanged within the halo (Shaburova *et al.*, 2009).

Phages have been described that use bacterial slime or capsule components as receptors for adsorption, and cannot infect mutants of capsular strains that do not have the receptor molecules in their capsular polysaccharides or are unable to produce EPS capsules (see, for example, Reese *et al.*, 1974; Ayers *et al.*, 1979; Cornelissen *et al.*, 2011). Purified slime from *P. aeruginosa* strain BI inactivated phage 2 specific for this strain indicating that it possesses receptor-like properties (Bartell *et al.*, 1971). Consistently, the slime of mutant that did not contain the phage depolymerase substrate could not inactivate phage 2 (Castillo and Bartell, 1974).

Certain phages that contain tail-spike-associated exopolysaccharide depolymerases, e.g. *Pseudomonas putida* phage φ15, were shown to have biofilm-degrading activity (Cornelissen *et al.*, 2011). A summary of the recent literature concerning the application of phages against biofilm-forming bacteria indicates that the most important obstacles to biofilm control by antimicrobial agents, such as antimicrobial tolerance, presence of the biofilm extracellular matrix and the effect of biofilm age might be overcome by phages (Donlan, 2009). Thus, phages that form plaques with a halo zone are especially precious isolates for potential therapeutic applications.

Preliminary grouping of phage isolates

Typically, a search for phages against a given pathogen provides a starting collection containing a few to a few dozen of isolates. Their differentiation based on plaque morphology may be possible only for the limited subset of collection isolates. Similar phages may form different plaques, and morphologies of plaques that are formed by various phages are not always different (see, for example, Ambrozic *et al.*, 2001; Krzywy *et al.*, 1976; reviewed by Gallet *et al.*, 2011). Thus, it is advisable to check similarities between the starting collection phages by electron microscopy, if possible, and by molecular typing.

Electron microscopy as a basic phage classification tool

Phages with similar properties that infect related hosts are often morphologically similar. Moreover, phage capsid diameter limits the capsid capacity and allows for the crude estimation of virion DNA size. Thus, electron microscopy is a recommended tool for the preliminary characterization of new phage isolates (Ackermann, 2009a). Not only does it permit one to classify a phage to one of the morphologically distinguishable families, but also allows for its inclusion to a group of phages of similar size and morphology within a family. The only limit of this method is the access to a transmission electron microscope and experience in the preparation and staining of phages. Readers, who are interested in these technique can find their details in excellent expert guides to phage electron microscopy that have been published recently (Ackermann, 2009b, 2012; Ackermann and Tiekotter, 2012; Tiekotter and Ackermann, 2009). Complementary or more detailed methods of phage preliminary characteristic that are based on the DNA or protein analysis are described below.

DNA analysis-based methods of phage grouping

Conventional DNA-based methods of phage characterization that do not require prior sequence knowledge have been described in detail by Carlson (2005). They require microgram quantities of purified phage DNA, and thus, as quite tedious, can be applied rather for testing relatively small collections of phage isolates. Phages of such collections can be grouped based on the size of their virion DNA, by comparison of their DNA mobility patterns in agarose gels in pulsed-field gel electrophoresis (PFGE). The DNA of phages within each genome size group can be further differentiated based on digestion profiles with restriction endonucleases, so-called RFLP (restriction fragment length polymorphism). The optimal set of enzymes for digestions has to be selected either based on an *in silico* analysis of genomes of known phages that infect bacteria of the same or related species as those infected by the tested phage isolates, or empirically. The latter requires more DNA as not every enzyme selected may appear to be the right one. Additionally, certain phages, even those of large genomes, avoid certain restriction sites in their DNA to avoid the attack of host restriction endonucleases at infection (e.g. Łobocka *et al.*, 2012; Sharp, 1986; Uchiyama *et al.*, 2012). Virion DNA molecules of others, for the same purpose, contain modified bases and thus are either cleaved with low efficiency by certain restriction endonucleases or are even resistant to some of them (Huang *et al.*, 1982). Hence it is often wise to use of a battery of restriction endonucleases.

PCR-based genomic fingerprinting methods allow one to overcome the aforementioned problems. They do not require prior DNA sequence knowledge. It usually suffices to take a trace amount of phage suspension or even to scrape plaque material on the tip of a toothpick or a pipette tip, or transfer the plaque to a nitrocellulose or nylon membrane disc to have enough DNA for further analysis (Gussow and Clackson, 1989; Wang and Cutler, 1992). For instance, to increase the differentiating potential and sensitivity of RFLP analysis, Merabishvili *et al.* (2007) used fluorescent RFLP (fRFLP) with restriction enzymes that recognize and cleave specific 4-bp sequences (HhaI and MseI), and leave single stranded overhangs at ends of cleavage products. Ligation of the restriction fragments to short double-stranded DNA adapters containing the complementary single-stranded overhangs, in the combined restriction-ligation reaction, allowed them to be used as templates for PCR amplification with fluorescently labelled primers, specific for the adapter sequences. Following the separation of fluorescently labelled PCR products by a capillary electrophoresis, a unique fingerprint pattern could be obtained. It could be then digitized and used as an entry in a database. Basically, fRFLP does not differ from modified versions of conventional amplified restriction fragment length polymorphism (AFLP) fingerprint analysis, which is based on the same principles – selective amplification of genomic restriction fragments by PCR to differentiate source organisms' DNA at the subgeneric level (Vos *et al.*, 1995; Roman *et al.*, 1999; Ticknor *et al.*, 2001; Lazzaro *et al.*, 2002). In a dendrogram that was constructed based on the fRFLP patterns for 69 tailed phages' DNA, genomically and/or serologically related phages clustered together, which proves the suitability of this method for the prior testing of relationships between newly isolated and known phages as well as among newly isolated phages (Merabishvili *et al.*, 2007).

Recently, Gutierrez *et al.* (2011) developed a

randomly amplified polymorphic DNA (RAPD)-PCR-based technique to obtain unique and reproducible band patterns from 26 different bacteriophages infecting *Staphylococcus epidermidis, S. aureus, Lactococcus lactis, E. coli, Streptococcus thermophilus, Bacillus subtilis* and *Lactobacillus casei* strains, using phage suspensions as a source of DNA and 10-mer primers. Earlier, other authors used a similar approach to estimate the genetic diversity of newly isolated phages infecting strains of single bacterial species: e.g. enterotoxigenic *E. coli, K. pneumoniae, P. aeruginosa* and *Vibrio harvei* (Jothikumar *et al.*, 2000; Kumari *et al.*, 2010; Li *et al.*, 2010; Shivu *et al.*, 2007, respectively). To increase the number and diversity of amplified genomic fragments for relationship analysis by genomic fingerprinting, Comeau *et al.* (2004, 2006) used degenerate-primed RAPD analysis (DP-RAPD) to diversify marine viruses, phages among them, and to diversify vibriophages from coastal environments. In the recently optimized versions of phage DNA RAPD-PCR analysis the titres of phage suspensions exceeding 10^9 PFU/ml appeared to be high enough to provide reproducible results (Gutierrez *et al.*, 2011). Phage suspensions of lower titres may be easily concentrated by centrifugation for a few hours in a microcentrifuge and resuspension of phage pellet in an appropriate amount of e.g. PCR reaction buffer (authors' unpublished results).

An alternative approach which is DNA-independent and often allows for preliminary grouping of the newly isolated phages is proteomic analysis of the most abundant phage proteins, which is described below.

Protein analysis-based methods of phage grouping

Amino acid sequences of most if not all major virion proteins, which form a head, tail and baseplate are conserved between closely related phages (Proux *et al.*, 2002; Lavigne *et al.*, 2008, 2009b). Thus, the complete or partial amino acid sequences of most abundant virion proteins of an uncharacterized phage may suffice to classify this phage to a known species, genus or subfamily. Proteomic studies appear to be the simplest and the least expensive way to gain this information. The most abundant virion proteins can be easily identified following SDS-PAGE separation of protein virion components, due to their large number of copies.

The method of virion purification is critical for reliable results of virion protein analysis. A standard PEG/NaCl precipitation of phages from DNase and RNase pretreated cell lysates cannot be used for this purpose, without further purification steps (Boulanger, 2009), due to the contamination of the precipitate with bacterial cell debris, lipoproteins and endotoxins (in the case of Gram-negative phage hosts). Only cell debris can be easily removed from the resuspended pellet by low speed centrifugation. Lipoproteins, which are a major fraction of contaminants, are typically so abundant, that they can be erroneously taken for major phage proteins on SDS-PAGE gels (our unpublished results).

Thus, for the purpose of preliminary grouping of numerous phage isolates, we purify virions from cell proteins in the lysate by simple centrifugation in a microcentrifuge for 2–5 h and by washing the pellet twice by resuspension in 0.1 M ammonium acetate and centrifugation under the same conditions again. In our hands, this simple procedure usually suffices to remove, from virion suspensions, major fractions of cellular proteins. SDS-PAGE of virions purified by this method and denatured in a standard SDS-protein denaturation buffer by boiling (Sambrook *et al.*, 1989) allows for easy identification of the most abundant virion components (authors' unpublished results). To prevent a reassembly of certain virion structures, samples should be loaded on a gel immediately after boiling. Sometimes, the increase of SDS concentration in the denaturing buffer helps to overcome this problem.

Certain authors recommend to prepare DNA-less virion particles (phage 'ghosts') prior to virion protein separation by SDS-PAGE, to avoid the disruption of protein migration in a gel by phage DNA (Boulanger, 2009). However, if only a preliminary phage grouping is a goal, we found this unnecessary in the case of the majority of phages. Further virion purification steps, such as single or double CsCl gradient centrifugation or HPLC of lysates on monolithic columns, are not necessary at this stage either. They are recommended in further stages of phage analysis for the preparation of highly pure phage suspensions and will be described further in this chapter.

Major phage proteins isolated from distinct SDS-PAGE gel bands can be analysed by using automated Edman degradation or mass spectrometry (MS) (Boulanger, 2009; Deutzmann, 2004; Lavigne *et al.*, 2009a; Maxwell and Frappier, 2007). The Edman degradation method is precise and suitable for proteins that have homologues in databases as well as for new proteins, but has several disadvantages. It only provides amino acid sequences of N-terminal protein fragments (under the condition that the N-terminus is not modified) and typically requires microgram quantities of a protein to be analysed. Additionally, proteins that are to be analysed by this method need to be electrotransferred to the PFGE membrane or purified by HPLC prior to their further processing. This adds to the entire procedure additional steps that may need to be optimized for individual proteins. Thus, in recent years, virion protein analysis by mass spectrometry (MS) has found a more common use. It requires trace amounts of proteins (~100 fmol or less) and is mainly based on two techniques of transfer of peptides to the gas phase: matrix-assisted laser desorption ionization (MALDI) or electrospray (ESI) followed by the analysis of the fragmentation patterns (MS/MS), which allow peptide sequences to be determined. ESI-based MS analysis is often coupled on-line with liquid chromatography separation (LC) (see, for example, Maxwell and Frappier, 2007). Readers who are interested in a more detailed description of both approaches for identification of microbial proteins can find it in a recent review (Jordana-Lluch *et al.*, 2012). In both cases the output data are generated from mass spectra of peptides that were obtained by in-solution fragmentation (typically proteolytic cleavage by trypsin) of source protein(s) supplemented by gas-phase MS/MS fragmentation spectra. Peptides of known proteins that provide the best match to the experimental data peptides are identified by reference to databases of proteomic or genomic information. A serious limitation of these methods is the necessity to obtain a statistically significant match between the tested peptide and a peptide from database protein for the tested peptide to be identified. Thus, they depend on the availability of proteins that contain peptides identical to the tested peptide in publicly available proteomic or genomic repositories. Despite these limitations they allow one to distinguish new phages from close relatives of known phages quickly and may appear helpful in a preliminary selection of potential therapeutic phages from among new phage isolates. Although thus far they have been mostly used to supplement a characteristic of phages of known genomic sequence (see, for example, Lingohr *et al.*, 2008; Pickard *et al.*, 2008), for instance Eyer *et al.* (2007) used MS to identify virion structural proteins of *S. aureus* phage 812 prior to the genome sequencing of this phage.

To overcome the aforementioned 'perfect match' limitations, Ceyssens *et al.* (2009) used a modified MS strategy [optimized matrix-assisted laser desorption/ionization (MALDI) postsource decay (PSD)] supported by manual calculations to obtain *de novo* peptide sequences (7–18 amino acid residues) representing the most abundant proteins of 15 newly isolated *Pseudomonas* phages. Homology searches of these peptide sequences against proteins of phage origin in databases using publicly available search engines MS-BLAST (Shevchenko *et al.*, 2001), MS-homology (Protein-Prospector, University of San Francisco, CA; http://prospector.ucsf.edu), and BLASTP (Altschul *et al.*, 1990) led to the identification of similar phage proteins and proper classification of all phages analysed, as it was confirmed by genomic and electron microscopy analysis.

An additional advantage of MS-based methods of phage structural proteome analysis is the possibility of using whole virions in place of separated virion proteins as a source of proteins for tryptic digestion and peptide analysis (Lavigne *et al.*, 2006). In this case peptides of the most abundant virion proteins can also be easily identified, as they predominate. Virions for this type of analysis can be purified by CsCl gradient centrifugation or HPLC, or even by simple centrifugation and washing in 0.1 M ammonium acetate (authors' unpublished results). This so-called 'whole-phage shotgun analysis' (WSA) approach was successfully used as a method supporting the analysis of SDS-PAGE separated virion proteins in studies of *Pseudomonas* phage φKMV and certain other phages (Lavigne *et al.*, 2006, 2009a). An additional advantage of the WSA approach is the identification of proteins that do not form distinguished bands in SDS-PAGE gels. However, the disadvantage is that one cannot correlate proteins with bands on a gel.

Proteomic data obtained by any of the aforementioned methods can be useful in a design of degenerate primers to test phage collections for the presence of close relatives of a given phage without a prior DNA sequence knowledge (Clokie, 2009).

Pre-selection of potential therapeutic phages

Even a large collection of phages isolated from different samples does not give any guarantee that there will be an appropriate therapeutic phage among them. Several tests have to be performed to ensure the selection of the right phage.

Reasons and methods to avoid temperate phages

Elimination of temperate phages from the collections of newly obtained isolates is one of the most important early tasks in the selection of phages for a therapeutic use, as temperate phages are estimated to comprise about 50% of environmental isolates (Ackermann, 2005). Lysogeny is the predominant phage–bacteria interaction, especially among certain pathogens and opportunistic pathogens, and in certain soil environments (Brüssow *et al.*, 2004; Ghosh *et al.*, 2008; Mercanti *et al.*, 2011; Paul, 2008; Williamson *et al.*, 2007). As many as 760 potentially functional prophages were identified by bioinformatic analysis in 199 bacterial genomes (Lima-Mendez *et al.*, 2011). Phage DNA in lysogens (a prophage) is usually integrated into a bacterial chromosome (Ackermann, 2003; Guttman *et al.*, 2005). In certain cases the prophage is maintained as a plasmid (Lehnherr, 2005; Ravin, 2011).

In lysogens, the phage repressor protein prevents the transcription of prophage lytic genes and in this way inhibits lytic development. It also causes lysogens to be immune to the superinfection by similar phages. The prophage functions as an integral part of a cell as long as there are enough active phage repressor and accessory regulators to maintain lysogeny. This can last for uncounted generations. A genomic analysis of some phages indicates that the prophage stage has dominated in their evolutionary history (Łobocka *et al.*, 2004). Gaining new prophages and a concomitant loss of integrated prophages or their fragments from bacterial chromosomes are main forces that drive the evolution of numerous bacterial species (Brüssow, 2007; Cortez *et al.*, 2009; Thomson *et al.*, 2004; Weinbauer and Rassoulzadegan, 2004). The plethora of bacterial adaptive functions that are encoded by prophages is astonishing (reviewed by McShan and Feretti, 2007; Ronning *et al.*, 2010). Prophages of clinically relevant strains as well as animal and plant pathogens encode numerous virulence factors (Brüssow *et al.*, 2004; Casas and Maloy, 2011; Tinsley *et al.*, 2006; de Mello Varani *et al.*, 2008; Boyd, 2012). Many of them are essential for bacterial pathogenicity. Cases of drug resistance determinants in temperate phage genomes, albeit rare, have also been described (Hens *et al.*, 2006). The aforementioned genetic load is as dangerous by itself as the ability of its exchange with other phages, plasmids or bacterial chromosomes and by that means extending the borders of the virulence genes' spread far more than could be achieved just by lysogeny (Canchaya *et al.*, 2003; Lima-Mendez *et al.*, 2011). The frequency of such exchanges is facilitated by phage recombinases and recombination enhancing functions (reviewed by Szczepańska, 2009). Although not all temperate phages contain homologues of known virulence-associated genes, all can potentially participate in their horizontal transfer, as implied by the mosaic structures of their genomes as well as homologues of plasmid and bacterial genes (Casjens, 2003; Dinsdale *et al.*, 2008; Hatfull, 2008; Hatfull and Hendrix, 2011; Kwan *et al.*, 2005; Romero *et al.*, 2009). Thus, it is important not only to avoid using temperate phages in therapy, but also to avoid contaminations of obligately lytic, therapeutic phage preparations with temperate phages.

A commonly accepted criterion to distinguish obligately lytic and temperate phages is the ability to form clear plaques by the former (Guttman *et al.*, 2005). In most cases plaque clarity is a good indicator of phage propagation strategy.

However, the preliminary classification of a phage as obligately lytic or temperate just on the basis of plaques observation should be treated with caution. Certain obligately lytic phages that form clear plaques on layers of cells of certain strains, form plaques that are more or less overgrown by resistant or infection escaping bacteria on cell layers of certain other strains, and make the impression of being turbid (e.g. Higgins and Lechevalier, 1969; Ulatowska *et al.*, unpublished results). Others can form plaques of hazy appearance or non-uniform clarity, e.g. clear with turbid centres, which cannot be unambiguously classified either as turbid or clear to begin with (e.g. Farrar *et al.*, 2007; Pedulla *et al.*, 2003). Certain temperate phages form lightly turbid plaques, which may cause their unambiguous classification difficult without additional studies (e.g. Pham *et al.*, 2007). Additionally, bacterial medium components and postinfection incubation conditions may influence the appearance of plaques (Carlson, 2005; Hendry and Fitz-James, 1974; Lockington *et al.*, 1988; Mullan, 1979). Finally, clear plaques can also be formed by mutants of temperate phages (Hranueli *et al.*, 1979; Nakamura *et al.*, 1975; Scott, 1970; Welker and Campbell, 1965). Mutations that cause such phenotype may be located in many different regions of phage DNA as well as in the host genome, reflecting the complexity of factors that influence the lysis/lysogeny decision in a freshly infected cell or contribute to the maintenance of lysogeny (for reviews on lysogeny regulating factors in model phages see, for example, Court *et al.*, 2007; Oppenheim *et al.*, 2005; Yarmolinsky, 2004). Most often, they are located in the gene for the phage repressor protein, in the repressor binding site(s) in phage DNA or in any phage or host gene whose product is involved in the regulation of repressor–DNA interaction or repressor stability (Carlson and Little, 1993; Grodzicker *et al.*, 1972; Lamont *et al.*, 1988; Ljungquist *et al.*, 1984; Murray *et al.*, 1977; Prell *et al.*, 1984; Shi and Friedman, 2001; Tokuno and Gough, 1977; Walker *et al.*, 1973, Wilson *et al.*, 1995). Some temperate phages or their mutants encode a temperature-sensitive repressor protein (Bukhari and Curtin, 1974; D'Ari, 1977; Dhillon and Dhillon, 1976; Grkovic *et al.*, 1995; Vogel *et al.*, 1991; authors' unpublished results). Thus, testing plaques that are formed at 37°C, a standard temperature to study bacterial pathogens of warm-blooded organisms, may lead to an erroneous phage classification. Additionally, certain plasmids may interfere with the establishment of lysogeny, as it was shown for *Bacillus* plasmid pCM194 and phage SPO2 (Marrero and Lovett, 1982). As a consequence, SPO2, which is temperate by its nature, forms almost clear plaques on a layer of pCM194 containing *Bacillus* cells. Examples of phages that are apparently non-temperate derivatives of temperate parents, and despite forming clear plaques have several genes characteristic of temperate phages, have also been described (Pena *et al.*, 1998).

If colony-forming bacteria can be recovered from a turbid plaque, then one can perform simple assays with the tested phage to verify whether they are lysogens. Lysogens may display various prophage-associated phenotypes. Those that are associated with immunity to superinfection (e.g. λ lysogens; reviewed by Court *et al.*, 2007) or, less commonly, with resistance to superinfection (e.g. *P. aeruginosa* phage D3 lysogens; Kuzio and Kropinski, 1983) are the easiest to detect. Simply, the tested phage will not plaque on its lysogen at all or will occasionally plaque only at much higher MOIs than on the wild-type strain. Additionally, the phage should be recoverable from the lysogen spontaneously or upon mitomycin C (or UV) induction, and used to test the ability to form plaques on a lawn of wild-type strain cells. Only prophages that have low frequencies of spontaneous induction and/or cannot be induced by either mitomycin C or UV treatment do not qualify for this test.

A more up to date solution to minimize the misclassification of phages, prior to sequencing of their DNA, may appear to be molecular typing. The increasing number of sequence data in databases allows one for a more and more detailed comparative analysis of genomic sequences of temperate and lytic phages that infect bacteria of certain species and genera. Based on such analysis conserved regions of phage serotype-specific sequences and sequences specific for particular subtypes of integrase genes of staphylococcal temperate phages were selected and used to design specific primers for their multiplex PCR amplification (Pantucek *et al.*, 2004; Kahankova *et al.*, 2010). It is thus possible to detect sequences specific for at least the most common staphylococcal temperate phages by diagnostic multiplex PCR.

A limitation in the wider use of this method is the scarcity of sequences of temperate phages that infect bacteria of certain clinically relevant species or genera as well as the diversity of prophage specific DNA markers, such as e.g. integrase encoding genes (Balding *et al.*, 2005). So far, the comparison of temperate phage genomes, which resulted in the design of a sufficient number of primer pairs to detect a majority of species-specific temperate phages, was performed in the case of *S. aureus* phages and enterobacterial phages (Balding *et al.*, 2005; Goerke *et al.*, 2009; Kahankova *et al.*, 2010; Pantucek *et al.*, 2004). Additionally, in the case of bacteria whose intraspecies diversity is generated mostly by different prophage integrations, such as *Salmonella enterica* (Thomson *et al.*, 2004), numerous primer pairs to prophage sequences have been designed for strain differentiation (see, for example, Cooke *et al.*, 2007; Drahovska *et al.*, 2007; Lan *et al.*, 2007; Ross and Heuzenroeder, 2005; Rychlik *et al.*, 2008). Thirty prophage-specific primer pairs that have been selected from among them for a recently developed *S. enterica* subtyping scheme (Fang *et al.*, 2012) can also be used to detect free *S. enterica* temperate phages using just two multiplex PCR runs with different sets of primer pairs. The increasing number of bacterial genomic sequences that are deposited in databases as well as the development of bioinformatic methods to identify prophages in bacterial genomes makes the prophage comparative analysis and the design of appropriate primer pairs possible even in the case of bacterial species whose phages are underrepresented in bacteriophage collections (Bose and Barber, 2006; Fouts, 2006; Lima-Mendez *et al.*, 2008, 2011).

Intuitively, positive results of PCR with primers specific for conserved regions of temperate phages should allow excluding such phages from further testing for therapeutic purposes. However, this is usually not the case. For instance, in the majority of cases *Staphylococcus* isolates, including strains that are used for phage propagation, are lysogens (Jin *et al.*, 2003; Pantucek *et al.*, 2004; van Wamel *et al.*, 2006; Goerke *et al.*, 2009; our unpublished results). Thus, prophage DNA will always contaminate DNA samples taken for PCR screening directly from plaques that are formed by tested phages on cell layers of such strains, leading to positive results even in the case of plaques of lytic phages (Ulatowska *et al.*, unpublished results). Testing DNA of phage precipitates that were prepared from DNase and RNase pretreated lysates does not provide a solution either. Phages that were released following even infrequent spontaneous induction of lysogens contaminate lysates of such strains. The frequencies of spontaneous excisions of single prophages that were observed e.g. in exponentially growing cells of an *E. coli* polylysogen were as high as 10^{-4}–10^{-5} per cell (Wang *et al.*, 2010). Hence, populations of lysogens are typically contaminated with free temperate phages. Although they are rarely found in numbers exceeding 10^4 phage particles per ml, whereas particles of the propagated phages usually number 10^8 to 10^9 (Rippon, 1956), they typically escape EM detection but their quantities are sufficient enough to provide a template for the DNA amplification (our unpublished results). Thus, PCR with DNA of a phage-propagating strain, as a template, should be always used as a control. If PCR amplification products of the same mobility are formed with both phage DNA and phage propagating strain DNA, their source is nearly always the contaminating phage (author's unpublished results).

Reasons and methods to avoid transducing phages

Some phages can accidentally package bacterial DNA into phage heads and inject such DNA to a new cell at infection, in a process that is known as transduction (Kokjohn, 1989; Sternberg and Maurer, 1991; Masters, 1996, 2004). This way they promote genetic exchange between bacteria and increase the risk of formation of recombinants of increased pathogenicity (Brabban *et al.*, 2005). Transducing phages can inject their DNA load even into bacteria that are unrelated to the phage host if only the latter contain appropriate receptors on the surface of their cells (Kaiser and Dworkin, 1975). Transduction has been shown to occur commonly in natural environments, including organisms of warm-blooded animals, and to contribute to the fitness and adaptability of bacterial populations significantly (reviewed by Miller, 2001; and Wommack and Colwell, 2000). Thus, it is generally accepted that transducing phages should not be used for phage therapy (Gill and Hyman, 2010).

Specialized transduction of chromosomal fragments from a direct neighbourhood of integrated prophage DNA is typical of temperate phages (Kwoh and Kemper, 1978). They 'hijack' such fragments as a result of either an imperfect or developmentally programmed excision of an integrated prophage upon induction of lytic development and carry them in virions as extensions of their own DNA. Generalized transduction in which random DNA fragments from a host can be carried in defective phage virions in place of phage DNA and injected into a cell at infection can be mediated by certain temperate phages and certain obligately lytic phages as well (Jensen et al., 1998; Masters, 2004; Thierauf et al., 2009; Thiyagarajan et al., 2011; Waddell et al., 2009). Such phage-mediated DNA exchange between bacteria can occur with a frequency of 1 per 10^{-3}–10^{-4} PFU, as shown by culture-independent methods (Kenzaka et al., 2007). Numerous cases of plasmid transfer by transducing phages have also been described (see, for example, Coren et al., 1995; Hertwig et al., 1999; Jiang and Paul, 1998; Luchansky et al., 1989; Mann and Slauch, 1997; Mercenier et al., 1988; Ruhfel et al., 1984; Saye et al., 1987; Stepanov et al., 1989; Takahashi and Saito, 1982). In experiments with staphylococcal phages φ80 and φJB the frequency of plasmid carrying phage particles was determined to be as high as 1 per 1700 (Varga et al., 2012). Additionally, transduction was shown to contribute to a spread of a plasmid carrying an antiseptic resistance determinant *qacB* among clinically relevant *S. aureus* strains (Nakanami et al., 2007) and to the intergeneric plasmid transfer between *S. thermophilus* and *L. lactis* cells (Ammann et al., 2008).

The ability to form generalized transducing particles appears to depend on the mechanisms of phage DNA packaging. In general, phages described as able to perform generalized transduction can initiate packaging of their DNA into phage heads from fragments of a non-specific sequence or sequences of low specificity (Casjens and Gilcrease, 2009; Casjens et al., 2004a, 2005). Model phages, such as P1, P22 and Mu are the best known examples (Bächi and Arber, 1977; Hanks et al., 1988; Howe, 1973; Tye et al., 1974; Sternber and Meurer, 1991; for a recent summary of P1 packaging see: Łobocka et al., 2004). Phages that require specific sequences to initiate packaging of their DNA to every phage head are unable to form generalized transducing particles (e.g. Chung et al., 1990a,b; Klumpp et al., 2008; Ziermann and Calendar, 1990). The latter could transduce host or plasmid DNA, which in the case of temperate phages is not derived from the neighbourhood of integrated prophage, only when packaging of such DNA to phage heads could be initiated from cloned junctions of phage DNA ends or from recombinant phage DNA oligomers containing host DNA (Liebeschuetz and Ritchie, 1986; Miva and Matsubara, 1983; Morino et al., 1985; Morino and Takahashi, 1996; Umene et al., 1979; Ziermann and Calendar, 1990).

An additional factor that limits the formation of transducing particles is the degradation of an infected bacterium's DNA during phage development, which is believed to be characteristic of most of the obligately lytic phages (Waddell et al., 2009). Phages which use products of chromosomal DNA degradation as building blocks to synthesize their own DNA typically do not transfer bacterial DNA efficiently (reviewed by Masters, 2004). Simply, the latter is either absent from cells or present in negligible concentration at late stages of phage infection when packaging takes place. Phage T4 is an example. Although T4 DNA packaging can start from sites of unspecific sequence, wild-type T4 is not a transducing phage. However, a T4 mutant, which lost the capacity to degrade host DNA as a result of mutations, could transduce random fragments of bacterial DNA similar in length to T4 virion DNA (Wilson et al., 1979; Young et al., 1982; Young and Edlin, 1983). In the case of phage T1, which can initiate DNA packaging from an unspecific sequence but derives most nucleotides for its DNA synthesis from degraded host DNA, rare transducing particles were detected in a specifically developed assay system (Drexler, 1970). They appeared to be formed at early packaging events, conceivably when cells still contained some undegraded host DNA (Kylberg *et al.*, 1975). Consistently, T1 mutants defective in host DNA degradation appeared to be able to transduce host DNA with a relatively high frequency (10^{-3}–10^{-4} per single infected cell) (Roberts and Drexler, 1981a,b).

Typically, either the high DNA sequence specificity of phage packaging enzymes or the degradation of infected cell DNA is sufficient to seriously limit the formation of phage transducing particles. Consequently, wild-type φ29 and T4 phages have been described as unable to transfer bacterial DNA (Masters, 2004)

Some phages, such as T7, SPO1 and staphylococcal Twort-like phages, require specific nucleotide sequences for their DNA packaging and also degrade DNA of the infected cell (Chung and Hincle, 1990a,b; Krüger and Schroeder, 1981; see Stewart *et al.*, 2009 for SPO1 development review, and Łobocka *et al.*, 2012, for the staphylococcal Twort-like phage properties summary). They could be considered as warranted 'non-transducers', unless they acquire a mutation that will decrease the packaging specificity or the ability to degrade host DNA (Hashimoto and Fujisawa, 1988).

General transducing phages can be identified relatively simply by testing their ability to transfer an antibiotic resistance marker from a resistant strain to a susceptible strain (Waddell *et al.*, 2009) or by testing the presence of bacterial DNA in phage progeny virions that were purified from external remnants of bacterial DNA (Sander and Schmieger, 2001). This could be done by PCR with primers specific for any conserved gene of a given bacterial species or genus, or even with general primers for bacterial 16S rDNA, using bacterial DNA as a positive control (Ghosh *et al.*, 2008). In experiments with single genetic markers, transduction frequencies were shown to differ from as high as 10^{-5} to as low as 10^{-9} per bacteriophage plaque forming unit (PFU) or single infected cell (SIC) (Masters, 2004). However, the number of transducing particles among virions appears to be 2 to 5 orders of magnitude higher, as indicated by the fractions of cells freshly infected by different phages and containing phage-introduced bacterial DNA (Kenzaka *et al.*, 2007). Thus, PCR-based methods are sensitive enough to detect their presence. Positive results of PCR justify the elimination of a given phage as a candidate for phage therapy, if only one can be sure that there is no contamination of tested phage lysate with temperate phages, as the latter may potentially be transducing phages.

Additional hints as to the ability of a given phage to participate in generalized transduction may come from the results of an analysis of virion DNA ends. In phages of low packaging specificity DNA ends vary between virions. However, if such phages derive substrates for their DNA synthesis from the degradation of host DNA, they typically do not transfer bacterial DNA by transduction or the number of transducing particles they accidentally form is negligible (as, for example, in the case of T4 phage).

Phage genome analysis as a tool to choose potential therapeutic phages

In the era of high-throughput and relatively inexpensive DNA sequencing technologies every potential therapeutic phage should be characterized at the level of nucleotide sequence of its genome to prove its truly obligately lytic nature (if possible) and to estimate its safety, especially the risk of participation in horizontal transfer of bacterial, plasmid and temperate phages' genes. An overview of the step by step bioinformatic analysis of phage genomic sequences as well as Internet resources that provide access to appropriate tools have been described in detail elsewhere [Belcaid *et al.*, 2010; Borodovsky and Lomsadze, 2011; Cresawn *et al.*, 2011; Henn *et al.*, 2010; Kropinski, 2009a; Kropinski *et al.*, 2009a; Lavigne *et al.*, 2004, 2009c; see, for example, Miller *et al.* (2003a,b) and Łobocka *et al.* (2004) for examples of single phage genome analysis]. Comparative analysis of most of the currently available phage genomic sequences allowed one to cluster representatives of *Myoviridae* and *Podoviridae* family of tailed phages into related subfamilies, genera and species, based on their phylogenetic relationships (Lavigne *et al.*, 2008, 2009b). Known phage sequences have been collected in a separate database, ACLAME, together with other elements of bacterial mobilome (Leplae *et al.*, 2010). Approaches have been undertaken to cluster phage proteins based on their phylogenetic relationship and to summarize global phage genomics discoveries (Hatfull and Hendrix, 2011; Krupovic *et al.*, 2011; Lima-Mendez *et al.*, 2011; Rohwer and Edwards, 2002). All of these efforts provide more and more support for unprecedented tailed phage diversity and the mosaic structure of phage genomes, and open a way to a better understanding of phage biology. They also contribute to the development of methods that facilitate the prediction of phage taxonomic affiliation, lifestyle and other properties based on *in silico* genome sequence analysis.

Temperate phages can be identified by their DNA sequence analysis, based on at least a few features: (i) the presence in their genomes of genes that encode prophage integration-mediating enzymes, or plasmid-prophage maintenance functions, (ii) the organization of genes in their genomes, (iii) sequence similarity of their genomes to the genomes of known temperate phages, (iv) presence of genes that encode bacterial adaptive functions in their genomes.

Enzymes that mediate prophage integration to bacterial chromosomes belong to two categories of site-specific recombinases. The first and the most common one include bacteriophage integrases (reviewed by Groth and Calos, 2004). They are a hallmark of phages, whose prophages integrate into specific attachment sites in their bacterial host chromosomes (*attB*) and mediate unidirectional site-specific recombination between bacterial *attB* and the phage attachment site, *attP*. Integrases belong to either of the two enzyme families: serine-integrases and tyrosine-integrases, which differ in structure and the mechanism of catalytic action. The crystal structures of representatives of each of these families have been solved (Biswas *et al.*, 2005; Nunes-Duby *et al.*, 1998; Smith and Thorpe, 2002; Yuan *et al.*, 2008). Thus, integrase-encoding genes can be identified readily in the genomes of newly sequenced phages by a comparison of amino acid sequences of predicted phage proteins with proteins in databases.

The second category of prophage integration-mediating recombinases is transposases. They are characteristic of transposable phages, whose prophages integrate into random regions of a host chromosome by transposition mechanism and are thus known as mutator phages (for a recent review see Lima-Mendez *et al.*, 2011). In addition to a model transposable phage Mu, which infects *E. coli*, only a few confirmed transposable phages have been described so far (Fogg *et al.*, 2011; for a summary of earlier data see Lima-Mendez *et al.*, 2011). They all infect Gram-negative bacterial hosts and belong to *Myoviridae* or *Siphoviridae* families. However, the classification of a new phage isolate to the mutator phages based just on the presence of transposase-encoding genes in its genome should be treated with caution. Transposase encoding genes can be as well a part of transposable elements that are carried by certain non-mutator phages (see, for example, Hatfull, 2012). Thus, the neighbourhood of each transposase-encoding gene should be carefully analysed to differentiate between these two possibilities (Lima-Mendez *et al.*, 2011).

Prophages of certain temperate phages are maintained as plasmids in their lysogens. Examples of circular and linear plasmid-prophages have been described (e.g. Casjens *et al.*, 2004b; Eggers *et al.*, 2000; Łobocka *et al.*, 2004; Ravin, 2011; Zabala *et al.*, 2009). All of them are low-copy-number plasmids, and in addition to genes that encode phage lytic functions, they contain homologues of certain plasmid maintenance genes. The latter ensure replication, resolution of plasmid multimers to monomers, proper segregation (partition) of plasmid copies to daughter cells at cell division and, at least in some cases, post-segregational killing of plasmid free cells. Thus, they can provide a basis for the identification of plasmid-prophage lifestyle. However, not all genes that encode these functions can be easily identified. For instance, in the case of N15 the same protein is responsible for the replication of plasmid and phage DNA (Łobocka *et al.*, 1996). Plasmid or plasmid-prophage-encoded resolvases belong to the same families of site-specific recombinases as tyrosine or serine integrases (Janniere *et al.*, 1996; Nunes-Duby *et al.*, 1998; Thomson *et al.*, 2009; Vetcher *et al.*, 2006), which hampers their unambiguous identification based just on protein homology searches (see, for example, Askora *et al.*, 2011). Additionally, while P1 and N15, the best known temperate phages that lysogenize as a circular or linear plasmid, respectively, encode all the aforementioned functions, homologues of only some genes that encode plasmid maintenance functions have been identified so far, e.g. in a circular plasmid prophage LE1 of spirochete *Leptospira biflexa* or in c-st plasmid prophage of *Clostridium botulinum* (Bourhy *et al.*, 2005; Sakaguchi *et al.*, 2005). The remaining genes either await identification or are absent from the genomes of these phages. The latter seems likely in the case of c-st bacteriophage, whose prophage is frequently lost from lysogens. Thus, identification of phages that lysogenize as plasmids just on the basis of homologues of plasmid maintenance genes in their genomes should be supported by additional experimental or sequence analysis data.

The increasing number of complete phage genomic sequences in databases accelerates the development of bioinformatic-based methods that enable prediction of the phage taxon and lifestyle solely on the basis of *in silico* sequence analysis. A recent comparative analysis of different shared, evolutionary conserved protein modules (ECMs), encoded by the genomes of 457 phages and 760 prophages extracted from bacterial genomes reveals certain phage-type specific, phage host specific or lifestyle-associated features (Lima-Mendez *et al.*, 2011). However, among 94 phage genome modules, which have been analysed in this study, only five are significantly associated with the obligately virulent lifestyle. Thirteen are significantly associated with temperate phages. It is worth mentioning that a module which has the strongest association with temperate phages includes integrases and repressor protein families. They are hallmarks of those temperate phages, that integrate their genomes into specific attachment sites in host DNA (*attP*). Two modules include transposases and transposase activator proteins – hallmarks of transposable phages, all of which are temperate. Although neither of these modules is specific for all obligately virulent or temperate phages, conceivably the refinement of the aforementioned or similar methodologies will soon enable the reliable automated prediction of phage lifestyles based solely on *in silico* genome sequence analysis. Results of first attempts to implement such methods have been published recently and are briefly summarized below.

McNair *et al.* (2012) developed a Phage Classification Tool Set (PHACTS), which utilizes a novel similarity algorithm and complementary classification methods to predict the lifestyle of a given phage on the basis of its comparative sequence analysis. The similarity algorithm creates a training set of data from a subset of 227 phages of known lifestyles and known sequence. By excluding certain phages of known lifestyles from the training set, PHACTS predictions for these phages were shown to have a 99% precision rate. Although in the case of phages that have no similarity to phages of known lifestyles the predictions are less certain, their accuracy will be increasing with the supplementation of training data set with new sequences of confirmed temperate or obligately virulent phages.

Deschavanne *et al.* (2010) used tetranucleotide frequencies as genomic signatures and observed that a host–phage distance measured by genomic signatures is much smaller in the case of temperate phages than in the case of obligately lytic phages. Consistently, the genomic signature distance of a phage and its host seems to be a sufficient parameter by itself to predict the lifestyle of a given phage. Although the use of this method is limited by the availability of phage and host genomic sequences, this limitation does not apply to phages of the most important clinically relevant bacterial species, as their genomic sequences are mostly known. An additional advantage of this approach is the possibility to use raw phage sequences as input data in the analysis, without their prior gene content analysis, which should allow for a fast elimination of potentially temperate phages from the collections of sequenced isolates.

Hallmarks of temperate phages that infect bacterial pathogens are genes encoding pathogenicity-related traits. In general, the identification of such genes should exclude a phage for the studies on its applicability in phage therapy. However, some obligately lytic phages of known therapeutic potential contain single homologues of bacterial pathogenicity associated genes, which seem to be involved in phage propagation functions (Łobocka *et al.*, 2012). Thus, the presence of genes of this kind in the phage's genome cannot be a sole basis of the elimination of this phage from further testing.

The success of a therapeutic phage in eradication of its host bacteria in an infected organism depends on its growth rate (see, for example, Bull et al., 2012). This in turn is directly influenced by the degree of conformity between metabolic functions that are required for the phage development and those that are available in a given host. One of the critical processes that limits the phage propagation rate is translation of phage mRNAs in a given host – a process influenced by the conformity between the codon usage frequencies in phage genes, and the accessibility of particular isoacceptor tRNAs in a host. Although certain phages encode a subset of their own tRNAs (reviewed by Dreher, 2010), the accessibility of the remaining tRNAs can limit synthesis of their proteins. Based on this dependency, Sau *et al.* (2005) calculated the codon usage preferences in 920 genes of 16 staphylococcal phages and predicted that genes of three lytic phages (K, 44AHJD and 68) can be expressed especially rapidly by host's translational machinery. Hence these phages should propagate rapidly and be useful for therapeutic purposes. Indeed, phages of genomes highly homologous to K and 44AHJD appeared to be effective in curing staphylococcal infections in a nematode infection model (our unpublished results). Yet another close relative of phage K is an active component of a biodegradable polymer PhageBioDerm, which is successfully used to treat staphylococcal wound infections in humans (reviewed by Chanishvili, 2012). Taken together, this could imply that a comparative analysis of codon usage frequencies in the genes of phages and their possible hosts may be helpful in choosing the optimal host for a given phage propagation. However, results of other studies indicate that the use of this implication is limited to highly expressed phage genes encoding major structural proteins (Lucks *et al.*, 2008). The average synonymous codon usage in certain phages that propagate efficiently in their hosts, e.g. *E. coli* phages T4 and T7, and *P. aeruginosa* phage φKZ differ significantly from that of their hosts. In the case of T4 and T7 this difference was attributed to constraints resulting from the use by these phages of their own replicative DNA polymerases (Kunisawa *et al.*, 1998). In the case of φKZ it was proposed to be a consequence of evolutionary economization of phage development costs by a shift towards the incorporation of predominantly smaller amino acids (from among those of similar properties) into phage proteins (Sau *et al.*, 2005).

Perspectives for the prediction of phage therapeutic efficacy

The successful elimination of a bacterial pathogen by a therapeutic phage can only occur when the phage-mediated bacterial cell killing at the infection site is faster than the rate of the bacterial cell number increase resulting from divisions (reviewed by Levin and Bull, 2004). Starting phage and bacterial population densities, phage and bacteria properties as well as phage–bacterium interactions and environmental conditions contribute to whether such a situation will take place.

The number of phages produced per time unit, which determine the so called phage fitness, can be assayed at laboratory conditions, in liquid cultures of bacteria infected with a given phage (for the outline of methods see Hyman and Abedon, 2009). The exact determination to what extent results of these assays can be extrapolated to *in vivo* conditions, in which bacteria are unlikely to grow exponentially and in which local factors may influence phage stability and phage–bacteria interactions, requires individual studies with each phage. The results of Weld *et al.* (2004) indicate that phage growth in cells in standard laboratory cultures is a poor indicator of phage growth *in vivo*. Consequently, phages of similar fitness and bacteria killing capacity under conditions of standard laboratory assays may vary significantly in therapeutic efficacy (see, for example, Bull *et al.*, 2002, 2012). Of two groups of phages that were used to treat mice infected with *E. coli* O18:K1:H7 and had a similar fitness when grown in their host cells in a standard laboratory medium, phages that had a better fitness when grown in cells of the same host in serum appeared to be therapeutically superior (Bull *et al.*, 2010). Thus, it is recommended to select candidate phages for a therapeutic use from among phages of high *in vitro* fitness, but add to their fitness assays the assays at conditions that mimic as closely as possible the conditions at the therapy target sites, especially with respect to medium content, pH, temperature, level of aeration and phage as well as host densities. Such phages are more likely to propagate efficiently also under *in vivo* conditions.

Phage fitness depends on the phage burst size (the average number of progeny phages that are released from an infected cell) and phage generation time (the time from the start of phage adsorption to the release of progeny phages). The latter is a sum of periods that are needed for phage adsorption, virion attachment, DNA transfer to a host cell, and the time from phage infection to the release of progeny phages (the latent period). The contribution of each of these parameters to phage fitness varies, dependent on a phage, its host, the densities of their populations and environmental conditions. Within certain time frames there is a positive linear relationship between the burst size and latent period (Wang, 2006). An artificially prolonged latent period (also called lysis time) caused an increase in the burst size, indicating that host lysis is not prompted by the lack of available substrates for the increase in the number of progeny phages (Wang *et al.*, 1996). However, phage mutants with an intermediate latent period had the highest fitness, indicating the existence of an optimal latent period (Wang, 2006). According to Abedon *et al.* (2001) the optimal latent period can be defined as the one which allows maximizing the increase in the population of a given phage in the presence of a given number of a given host cells in a given environment. The length of latent period can be precisely determined in 'one-step growth experiments', provided that starting host cell densities are sufficiently high (Delbrück, 1940; Hyman and Abedon, 2009) Opposite to expectations, the impact of latent period on phage fitness appears to be much smaller than the impact of the rate of phage adsorption to host cells (Shao and Wang, 2008). Moreover, the optimal values of both parameters that allow maximizing phage fitness are correlated with each other. A mutant of λ phage with a high adsorption rate had a shorter optimal latent period than mutants with a low adsorption rate.

Adsorption kinetics of phages to their host bacteria is influenced by several factors, including cell growth phase, metabolic state and density, medium components, aeration, temperature, density of receptors for a phage (Delbrück, 1940, summarized by Kropinski, 2009b). It can be measured by incubating phage suspension of a given PFU/ml with bacterial cells, removing incubated mixture samples at minute time points, filtering them through 0.22 μm filters to recover unadsorbed phages in a filtrate, and measuring the filtrate phage titre (see, for example, Golec *et al.*, 2011a). The subtraction of the number of phages in the filtrate from the starting number of phages will indicate the number of adsorbed phages. In the case of the majority of phages which are chloroform-resistant, the entire procedure can be simplified by the addition of chloroform to bacteria, before mixing them with the phage suspension. This allows one to eliminate the filtration step, as chloroform kills the infected bacteria, which will otherwise form infective centres and contribute to the number of PFU (Kropinski, 2009b). It is important to narrow the assay time to less than 10 min, as in the case of short latent period phages the release of progeny phages from infected cells will interfere with the assay results. The adsorption of most but not all phages that have been proven to be effective therapeutically is relatively fast (over 70% or more in 10 min), when assayed with bacterial cultures in standard laboratory media (Golec *et al.*, 2011a; Kwiatek *et al.*, 2012).

Certain factors in an environment in which a given phage is expected to act therapeutically can substantially interfere with the attachment of phage to its host bacterium. For instance, bovine whey proteins inhibited the attachment of phage K – a candidate for mastitis therapy – to *S. aureus* cells, by adsorption to these cells and blocking the access of the phage to phage-specific receptors (Gill *et al.*, 2006). The extent of inhibition varied for whey from different cows' milk. Normal human and rabbit serum inhibited the adsorption of phage 80 to cells of various *S. aureus* strains that are hosts of this phage (Nordstrom *et al.*, 1974). This effect was attributed to the serum IgM and IgG fractions. These few examples show that factors which are present at the expected site of therapeutic phage action can strongly influence the outcome of phage therapy. Thus, if possible, studies on their influence on phage adsorption, infection and development should be routinely included in a standard set of tests with each potential therapeutic phage.

Intuitively, to be effective as therapeutic agents in a target site, phages have to be relatively stable in a target site environment. They have to have access to their host bacteria and must adsorb to phage receptors at bacterial cells quickly and efficiently, in order to minimize the time of their susceptibility to inactivating factors, such as e.g. intestinal content, serum or other body fluid components. The phage generation time has to be relatively short, possibly shorter than the cell generation time. Furthermore, each cell infection should lead to phage multiplication and cell killing. The number of progeny phages released from each cell (burst size or phage fecundity) should be relatively high, to increase the probability that new phages encounter yet uninfected cells. Examples of successful phage therapy trials in animals and in humans prove that the aforementioned phage fitness parameters at certain therapy target sites can achieve values that will lead to the decrease of phage host densities at least to levels at which the pathogenic bacteria could be controlled by the immune system (see, for example, Międzybrodzki *et al.*, 2012; Saussereau and Debarbieux, 2012; Sulakvelidze and Barrow, 2005).

To adsorb to its host bacterium a phage has to establish physical contact with cells. Thus, the time of 'search' for a host cell by a newly released progeny phage contributes to the time of infection cycle (the time from the infection of a cell by a phage to the infection of a cell by the progeny of this phage). However, a phage population can only increase initially, when the density of susceptible bacteria is sufficiently high. When it is low (~10^4 cells/ml or less), there is no measurable phage population increase, even at 100-fold initial excess of phages as compared to their host cells (Wiggins and Alexander, 1985; Zuber *et al.*, 2008). This limiting bacterial concentration for a particular phage-host system in a given environment has been defined as a 'proliferation threshold' or 'replication threshold' (Payne and Jansen, 2000). In the case of stable phages, in environments of no phage inactivating factors, phage population increase can start, albeit with a delay, when bacteria reach a density above the threshold value (Wiggins and Alexander, 1985; Woody and Cliver, 1997; Zuber *et al.*, 2008). In environments that contain phage destructive factors, phages can be inactivated before they are able to infect their host, and thus will not be effective therapeutically, unless fresh portions of phage inoculum can be constantly provided to compensate for the decayed phages. Thus, phage stability under conditions resembling those at the expected site of phage therapeutic action is of crucial importance for the therapeutic effectiveness of a given phage.

A bacterial cell infection with an ideal therapeutic phage should be characterized by the so called 'single hit' kinetics, in which a single molecule of a therapeutic agent (a phage) kills its target upon binding (Bull and Regoes, 2006). Optimally, each infection should lead to the production of phage progeny to ensure phage multiplication at the infection site as long as there are living bacterial hosts around. In practice, the bacterium killing by a bacteriophage does not necessarily signify the productive phage infection. Additionally, certain phage, host and environment properties act to decrease the infection and killing efficiency and to prevent phage spread. Variations can concern the killing efficiencies of different susceptible strains' cells within a species. Thus, the low killing efficiency of one strain cells may not necessarily implicate the similarly low killing efficiency of other strains' cells. However, the closer this efficiency is to the ideal efficiency under laboratory conditions, the better the chances that it will also be satisfactory during therapy, provided the lack of interfering factors.

The killing efficiency of a given host by its specific phage cannot be measured by a standard plaque assay. The formation of a visible plaque requires the production of at least 10–15 progeny phages in each infected cell, and thus a low burst size of a given phage can make the interpretation of plaque assay results problematic (Carlson, 2005). Additionally, the completion of phage development in a bacterium and the release of progeny phages, which is essential for plaque formation, may be not required for phage bactericidal activity. T4 phage ghosts deprived of DNA are still capable of killing bacteria. The cessation of ongoing phage development in a bacterium at the cost of sacrificing cell viability, known as abortive infection, will lead to cell death despite the lack of production and release of mature phage progeny (for a detailed review see Lebrie *et al.*, 2010, and Chapter 4). Abortive infection systems (Abi) are often encoded by plasmids or prophages. Hence, the ratios of so-called phage killing titre/ml to the number of PFU/ml for a given phage can vary substantially, when assayed with different bacterial strains of the same species.

The simplest way to measure the killing-titre is to infect bacteria with phage at MOI 1 or 4–5, leaving enough time for efficient phage adsorption, and then plate bacterial cells on a solid medium at different dilutions to calculate the fraction of survivors among phage-treated cells as compared to the number of bacteria that were not treated with a phage (Carlson, 2005; authors' unpublished results). The subtraction of the number of survivors from the total number of bacteria in a given volume of phage–bacteria mixture reflects the number of phage killing particles. The survivors include cells that escaped the infection due to actions of various phage resistance mechanisms that do not lead to cell death as well as cells that did not encounter and bound the phage. The number of the latter depends, for example, on whether the infected cells do or do not adsorb additional phages. In the first case the maximum number of phages needed to infect over 99% of cells in a population is equal to the number of cells, in the absence of factors that inactivate a phage or interfere with the phage adsorption (thus, 1 is the optimal MOI). In the second case, it has been estimated to be about five per cell (Bull and Regoes, 2006). The ratio of the number of PFU per ml to the number of killing phage particles per ml in the phage suspension reflects the phage productive infection efficiency. Phages of high productive infection efficiency are suitable for active phage therapy, which relies on the assumption of phage replication at a therapy target site (Payne *et al.*, 2000; Abedon and Thomas-Abedon, 2010). They can be used in therapy at doses which are substantially lower than the number of their host bacteria.

Although the infection productivity is necessary for phage multiplication and hence for the prolonged effects of phage application, phages that kill bacteria but do not lyse them to release their progeny may be used in so called passive phage therapy, provided that they can be delivered in a form of high-titre suspensions to their sites of action. The superiority of such phages over propagating ones in certain situations may result from the lack of bacterial cell components (including endotoxins) that are released from lysed cells in the course of therapy (Hagens *et al.*, 2004). However, their inability to multiply and spread limits their possible applications.

Purification of phage lysates

The extent to which phages for therapeutic applications have to be purified is a disputable issue. Requirements which should be met depend mainly on safety concerns, the route of phage delivery and local regulations. Additionally, purification costs and simplicity of the procedure as well as the possibility to scale it up are important factors that may decide on the applicability of a given phage.

Cell lysates, that have been traditionally used as phage providing therapeutics in the past and are still in common use in several countries, are not purified from contaminants before their application (Chanishvili, 2012; Górski *et al.*, 2012). What is more, in the past they were only deprived of solid cell remnants by centrifugation and treated with heat to kill the unlysed bacterial cells (D'Hérelle, 1938). Nowadays, heat treatment has been replaced by filtering through 0.22 μm low protein-affinity filters (see, for example, Letkiewicz *et al.*, 2009; Zimecki *et al.*, 2003). The lysates purified this way and produced under good manufacturing practice conditions have been used at the Phage Therapy Unit of the Ludwik Hirszfeld Institute of Immunology and Experimental Therapy in Wrocław, Poland, in experimental therapy of genital and urinary tract infections, soft tissue infections, skin infections, orthopaedic infections, respiratory tract infections as well as other infections caused by certain, most common bacterial pathogens (Międzybrodzki *et al.*, 2012). They contain bacterial components, endotoxins among them, which might be the reason behind potential adverse effects in the course of phage therapy. Infective particles of temperate phages could be additional lysate contaminants, if lysogens of these phages were used for therapeutic phage propagation (see 'Reasons and methods to avoid temperate phages'). Despite that, incidents of adverse effects of phage therapy have been surprisingly rare, in cases when such phage suspensions have been applied by topical, oral, rectal or vaginal route or by inhalation in aerosols (Międzybrodzki *et al.*, 2012).

Certain cell remnants present in lysates are claimed to have an immunostimulatory effect and thus are believed to contribute to phage therapy efficacy (Sulakvelidze and Kutter, 2005). However, results of recent studies indicate that their influence on the outcome of phage therapy is complex and can vary dependent on the phage, infecting bacterium, and other factors (Górski *et al.*, 2012). Thus, the content of such contaminants in therapeutic phage preparations should be entirely avoided or at least minimized.

The presence of certain toxins in cell lysates appears unavoidable. Typical examples of the latter are endotoxins – lipopolysaccharide components (LPS) of outer cell membrane in Gram-negative bacteria. When released from cells, they act as elicitors of the innate immunity system and can cause various biological effects, including fever, hypotension and endotoxic shock in humans and animals (reviewed by De Castro *et al.*, 2012; Mayeux, 1997). Lipid A – the part of LPS which is responsible for its toxic and immunostimulatory properties is an essential component of Gram-negative bacteria cells (reviewed by Raetz and Whitfield, 2002). Thus, viable, unconditional mutants that lack lipid A basic precursor (lipid IV_A) and could possibly serve for certain phage propagation do not exist (Galloway and Raetz, 1990; Klein *et al.*, 2009). Theoretically, bacterial mutants that synthesize specifically changed lipid A molecules of decreased toxicity, similar to the recently isolated *E. coli* mutants of such properties, could serve for certain therapeutic phage propagation (Ingram *et al.*, 2010). However, these mutants could not be used for the propagation of phages that use lipid A as a part of their receptor (Boman and Monnar, 1975; Ishiguro *et al.*, 1983). Additionally, the changes in the structure of lipid A may influence the overall structure of LPS and hence the adsorption of phages which use other LPS residues (reviewed by Rakhuba *et al.*, 2010). Thus, commercially available methods of endotoxin removal still remain a strategy of choice in lysates purification (Merabishvili *et al.*, 2009; Gill and Hyman, 2010).

In a recently described clinical trial of phage therapy in burn victims Merabishvili *et al.* (2009) used a mixture of phages against certain *P. aeruginosa* and *S. aureus* strains as a therapeutic preparation. It was only deprived of bacterial cells and solid cell debris by standard filtration and centrifugation procedures, and additionally purified from endotoxins by a passage through a commercial endotoxin affinity column. This simple purification procedure sufficed to deplete the mixture of any measurable effect on human keratinocytes and any pyrogenic effect after the injection into rabbits. Thus, it is recommended for the production of phage suspensions free of endotoxin-associated activities (Gill and Hyman, 2010).

Further purification of phage suspensions requires the use of more sophisticated methods. Phages in lysates can be concentrated by PEG/NaCl precipitation and recovered in a pellet. Cell and medium derived contaminants in a resuspended pellet can be separated from phage particles based on buoyant density differences, by caesium chloride (CsCl) gradient centrifugation. While single centrifugation on preformed CsCl step gradients suffices to get rid of most of the contaminants, an additional equilibrium centrifugation on self-forming gradient is recommended to obtain highly pure and concentrated phage preparations (Boulanger, 2009; Carlson, 2005; see also 'Protein analysis-based methods of phage grouping' above). However, the yield of phage recovery after CsCl gradient centrifugation is generally low. Additionally, certain phages cannot be recovered in an active form due to aggregation, virion damage or other reasons. In such cases, sucrose gradient centrifugation can be used as well (Serwer *et al.*, 1978; Pathria *et al.*, 2012).

An alternative to conventional ultracentrifugation-based methods of phage purification is ion exchange high-performance liquid chromatography (HPLC), using recently developed Convective Interactive Media[R] (CIM[R]) monolithic supports (Barut *et al.*, 2008; Smrekar *et al.*, 2008). Commercially available CIM[R] monoliths are rigid, porous, highly cross-linked glycidyl methacrylate-*co*-ethylene dimethacrylate polymers with homogeneous interconnected pore structures oriented in all directions and active groups bound to flow channel surfaces (BIA Separations, www.biaseparations.com/sp/209/cim-technology). They are cast as a single sponge-like block and inserted into a chromatography housing. Their large-diameter flow-through channels (1.2 mm or more, dependent on the monolith) in association with large surface areas for bound active groups and the lack of stagnant zones during flow, make them superior over conventional chromatographic supports for the separation of large biomolecules, including viruses. Numerous representatives of different phage families, mostly tailed phages, have been successfully concentrated and purified by using ion exchange HPLC on CIM^R monolithic columns or discs (Adriaenssens *et al.*, 2012b; Kramberger *et al.*, 2010; Liu *et al.*, 2012; Oksanen *et al.*, 2012; Smrekar *et al.*, 2008, 2011). The disadvantage of this method is the necessity to optimize the purification procedure for each bacteriophage individually, with respect to the anion exchanger used, the buffer set as well as loading, washing and elution conditions. However, the purity of final phage preparations has been comparable to that obtained after CsCl gradient centrifugation. The volume of loaded phage suspension is only limited by the column or disc capacity, and thus large volumes of low-titre lysates of poorly propagating phages can be used as a starting material, without a need for prior phage concentration. Active phage recovery typically varies between 40–70%, but in certain cases it can even reach 99.9% (Adriaenssens *et al.*, 2012b). Once optimized, the chromatographic phage purification is fast (about 1 hour) and can be easily scaled up.

Bacteriophage stability

The most potent therapeutic phages are expected to have the ability to survive and multiply under various growth conditions, and to be stable for a long time when stored. Thus, phages, whose lysates lose the titre quickly during storage, cannot be considered for therapeutic use. Certain storage methods can prolong the survival of certain phages, but they do not work equally well for all phages, due to variations in the sensitivity of phage virions to different environmental factors (reviewed by Ackermann *et al.*, 2004). Some phages can be stored as broth lysates at 4°C for years, with only a moderate loss of PFU/ml, while the titres of others drop dramatically even after overnight storage under similar conditions. In the case of many phages, freezing of phage suspensions with glycerol at −80°C or freeze drying slows down phage inactivation to values that can warrant at least a few months of shelf-life. Puapermpoonsiri *et al.* (2009; 2010) tested the influence of different lyophilization conditions as well as lyophilization buffer components on the activity of phages that were recovered from lyophilized samples. Although some of these components prolonged the lyophilized phages shelf-lives, none of them could prevent a phage titre drop with time. Alfadhel *et al.* (2011) showed that lyophilization of *S. aureus* bacteriophages in Eppendorf tubes in a viscous solution of hydroxypropyl methylcellulose yields pieces of highly porous, leaflet-like, dry matrix, which could be potentially used as nasal inserts, and which can release viable phages. However, the number of viable bacteriophages decreased at least an order of magnitude during lyophilization and then a further two to three orders of magnitude over a few months of storage. The only reliable method for phage storage, which does not depend on the phage virion structure and appears applicable perhaps for all tailed phages, whose hosts can be stored at −80°C, has been described recently (Golec *et al.*, 2011a). It employs bacterial cells as shelters for phage DNA and hence a reservoir of phages that can infect sensitive cells at the release. It has been used in the authors' laboratories e.g. for the storage of newly isolated phages, prior to a detailed examination of their stability in lysates. Plausibly, the freeze-dried preparations of therapeutic phage-infected cells of 'surrogate', non-pathogenic hosts could be used in the future as stable vectors for delivery of therapeutic phages into their target sites, e.g. via the oral, rectal or other routes. The effectiveness of such an approach has been already demonstrated in the case of an avirulent mycobacterium – *Mycobacterium smegmatis*, and its two virulent relatives *M. tuberculosis* and *M. avium*. *M. smegmatis* infected with a lytic mycobacteriophage TM4, could deliver TM4 to the inside of *M. tuberculosis*- and *M. avium*-infected macrophages, which resulted in a significant reduction of intracellular bacilli (Broxmeyer *et al.*, 2002).

Host range determination

Collections of phages include many representatives that differ in their host range. Some of phages in addition to being able to propagate in bacteria of one species can propagate in bacteria of related species or even related genera (e.g. Kelln and Warren, 1971; Kuhn *et al.*, 2002; Wommack and Colwell, 2000); while others can be limited to single strains of particular bacterial species (e.g. Bigby and Kropinski, 1989; Horiuchi *et al.*, 2012).

The ability of a given phage to propagate in the cells of a given bacterium depends on several factors (reviewed by Hyman and Abedon, 2010). Phages whose host specificity is wide with respect to strains representing a species of bacterial pathogen or a certain pathovar(s) of this pathogen are the most precious in therapeutic phage collections. Thus, determination of phage host range is of primary importance in the selection of phages for potential therapeutic use.

Determination of phage host range requires the access to collections of bacteria containing differentiated species representatives (Kutter, 2009). Ideally, strains in these collections should be well characterized by standard molecular typing methods for a given species, serotyping, multilocus sequence typing (MLST), pulsed-field gel electrophoresis of chromosomal DNA restriction fragments (PFGE), multiple loci variable number tandem repeat analysis (MLVA) among them, to ensure that they differ significantly, and do not represent, for example, different descendants of a single clone which predominates in a given hospital or area. However, standard molecular markers that are used for strain typing are rarely correlated with strain susceptibility to certain phages. An exception is PFGE, as some PFGE patterns, like the patterns of whole genome mapping (WGM, formerly optical mapping) may be indicative of certain integrated prophages (Borecka *et al.*, 1996; Cooke *et al.*, 2007; Shukla *et al.*, 2012), which in turn can cause strain resistance to similar or certain other phages. Additionally, it is recommended to differentiate strains in such collections by testing their susceptibility to phages of a standard phage typing set, if any exist for that bacterial species (Kutter, 2009). The latter helps to group the strains into subgroups whose members share similar features important for the infection with certain phages, and narrow the strain collection for initial phage testing by selecting single representatives of particular subgroups.

A typical spot test using the double agar overlay plaque assay (reviewed by Kropinski *et al.*, 2009b) with a given phage and each of the selected collection strains is a standard to begin the phage host range determination. In the case of a large number of strains to test it is convenient to prepare one Petri dish with a bacterial lawn for each strain, and place 5 µl drops of phage suspension in different dilutions, plus a control without phages, in different lawn regions radially from the Petri dish centre. Usually six drops of this volume can be safely placed on one Petri dish without the danger of them merging. Thus, when working with phage lysates, which typically do not exceed 10^{11} PFU/ml, dilutions 10^{0}, 10^{-2}, 10^{-4}, 10^{-6}, and 10^{-8} are sufficient for preliminary testing.

It is important to see single plaques to classify a strain as a sensitive one. In many cases, a zone of a complete cell lawn clearing can be visible in sites of high-titre lysate droplets, but with any lower-titre droplets there is no single plaque formation. Usually, this may be caused either by cell lysis with lysins or bacteriocins, if their active forms are present in the lysate, or by an abortive infection (reviewed by Abedon, 2011). The former may result from many reasons (Hyman and Abedon, 2010; Labrie at al., 2010). One of them is the so-called virion-mediated 'lysis from without', which is manifested by rapid lysis of cells infected with a phage at high multiplicity of infection (MOI >100) without any phage production (Delbrück, 1940; Young, 1992; Abedon, 2011). This phenomenon depends on the activity of phage tail lytic enzymes. They can cause cell lysis by puncturing the cell wall peptidoglycan at multiple sites of phage adsorption and weakening its structure so that it cannot counteract the osmotic pressure in a cytoplasm.

Bacterial strains for therapeutic phage propagation

The use of 'surrogate', non-pathogenic hosts is generally recommended for the propagation of therapeutic phages to limit the contamination of lysates with exotoxins or other virulence factors (Balogh *et al.*, 2010; Gill and Hyman, 2010). For instance, certain phages of such bacterial pathogens as *M. tuberculosis, Listeria monocytogenes* or *Erwinia amylovora* can be efficiently propagated in cells of related non-pathogenic bacterial species (*M. smegmatis, L. innocua* and *Pantoea agglomerans*, respectively) (Hatfull, 2012; Carlton *et al.*, 2005; Lehman *et al.*, 2009). However, this strategy can be only used for a limited number of phages, for which a 'surrogate' host can be found.

A more universal approach that has been used in the authors' laboratory (our unpublished results) is based on the construction of phage hosts that are depleted of certain mobile genetic elements: prophages, plasmids and pathogenicity islands among them. These elements are commonly present in numerous pathogenic bacterial strains and are carriers of genes that encode virulence-associated treats (reviewed by Boyd *et al.*, 2012; Jackson *et al.*, 2011). For instance, virtually all exotoxins of bacterial pathogens are encoded by mobile genetic elements or have been proposed to be of plasmid or prophage origin (Casas *et al.*, 2006; Novick, 2003; Peck, 2009). Thus, initially pathogenic bacterial strains that are attenuated by the depletion of these elements are an alternative choice for the propagation of therapeutic phages for which one cannot find a 'surrogate' host. Additionally, the use of such strains can minimize the cost of phage purification substantially.

Acquisition of prophage-free bacterial strains

The presence of prophages in bacterial hosts for therapeutic phage propagation poses a double safety risk. First, the spontaneous induction of prophage lytic development, even in sparse lysogens from a population, is a source of contamination of therapeutic phage preparations with infective particles of temperate phages. Although it has not been proven, in certain cases, contaminating temperate phages could theoretically worsen the outcome of infection when applied with therapeutic phage preparations, by lysogenizing an infecting bacterial strain and arming it with additional virulence factors. Second, prophages typically encode toxins or other virulence factors. Thus, protein products of certain prophage genes could either contribute to the adverse effects of phage therapy, when unpurified lysates are used as a source of therapeutic phages, or increase the cost of phage purification and purity control assays. Moreover, contamination of lysates with temperate phages that were released from a strain used to propagate new phage isolates may lead to the premature elimination of new phages from further studies due, for example, to the detection in the lysate of contaminating phage integrase gene. Thus, in the authors' opinion, the development of collections of non-lysogens for therapeutic phage propagation is a necessary step to increase the safety of phage therapy and to introduce it for a wider use.

The problem of prophages in strains for therapeutic phage propagation is difficult to overcome by a selection of non-lysogenic bacterial strains from environmental samples due to the ubiquity of lysogeny, especially among bacterial pathogens (see 'Reasons and methods to avoid temperate phages'). For the aforementioned reasons, in certain cases, the identification of prophages in the genomes of bacterial strains and their subsequent removal may be a necessary step in the acquisition of hosts that will be suitable for therapeutic phage propagation.

To retain its ability to develop lytically and to produce infective progeny phages, an integrated prophage has to contain all essential phage genes as well as genes that encode prophage integration/excision and regulatory functions. In the majority of cases they are lost from bacterial genomes during the course of evolution since they are an unnecessary extra genetic load. Exceptions are prophage genes that encode bacterial adaptive function and provide a selective advantage of lysogens over non-lysogens. As a result of the concomitant loss of integrated prophages or their fragments and integration of new prophages, bacterial chromosomes may contain complete as well as incomplete prophages, or even single genes of phage origin (Casjens, 2003; Brüssow *et al.*, 2004). The incomplete (cryptic, defective) prophages are permanently integrated into a bacterial chromosome and cannot by themselves be a source of mature particles of temperate phages that contaminate therapeutic phage preparations. However, defective prophages of bacterial pathogens usually retain at least some bacterial-virulence-associated genes, and thus their elimination from phage-propagating strains may be advisable.

The choice of prophage detection strategy depends on the availability of a prophage and/or prophage host's genomic sequence as well as the knowledge of a given prophage's biology.

Several bioinformatic methods have been developed to identify prophages in bacterial genomes by *in silico* DNA sequence analysis. The most commonly used programs that implement these methods can be downloaded from internet resources or used online through web servers (see, for example, PHAST: Zhou *et al.*, 2011; Prophinder: Lima-Mendez *et al.*, 2008; Phage_Finder: Fouts, 2006; and Prophage Finder: Bose and Barber, 2006). Their advantages and disadvantages have been discussed in detail by Zhou *et al.* (2011). From their users' point of view, these bioinformatic tools differ in terms of prediction accuracy as well as the input/output data format and speed of execution. PHAST appears to be superior as it can use both crude and annotated sequences as input data files, is supported by a customized phage and prophage database, allows for the prediction of the prophage sequence completeness and has the fastest execution time.

Prophage sequences are considered to be gene clusters of phage origin in bacterial genomes. As such, they are identified based on the comparison of amino acid sequences of their predicted products with amino acid sequences of database proteins, comparison of their GC content, dinucleotide frequencies, codon usage frequencies and other sequence parameters with those of other genes in the genome, and homology-based predictions of certain genes or sites in DNA (e.g. tRNA genes, phage attachment sites). In general, dinucleotide biases evaluated for successive genome segments of about an average prophage size are significantly more similar to each other than to those of sequences from more distant organisms. Thus, it is possible to identify prophages in bacterial DNA, often just on the basis of GC content differences (Karlin, 1998; Srividhya *et al.*, 2007). However, the GC content of certain phages is so similar to that of their host (Nelson *et al.*, 2002) that to identify their prophages several DNA sequence parameters should be taken into consideration. Comparative analysis of a phage propagation strain genome with those of other species representatives can be also used to identify prophages, based on the identification of genes that are interrupted by potential prophage sequences. In over 60% of cases, prophages use intragenic regions as their attachment sites in bacterial DNA (Fouts, 2006). Moreover, of almost 300 putative prophage attachment sites analysed, 33% were found in tRNA genes with a high preference to Arg, Leu, Ser, and Thr-tRNA. Results of the analysis of 302 bacterial genomes for the presence of prophages are available on the Internet (http://phage-finder.sourceforge.net/302_bact_gen.pdf). They can be helpful in the identification of prophages in the genomes of phage propagation strains as well.

Certain prophages can be identified prior to bacterial genome sequence determination, using similar PCR-based detection methods of highly conserved prophage sequences as those that are used for temperate phage identification and are described in 'Reasons and methods to avoid temperate phages' (see also Goerke *et al.*, 2009). Standard methods of prophage identification can be used as well, provided that one can induce a prophage. Induction of phage development in lysogens can occur spontaneously or in response to certain environmental factors, such as UV light, reactive oxygen species, homoserine lactone, antibiotics, mitomycin C or other mutagens, or a change of growth conditions, e.g. from low to high temperature (Affolter *et al.*, 1983; Bossi *et al.*, 2003; Cavalcanti and Siqueira Júnior, 1995; Clarke, 1998; Clark *et al.*, 1986; Duval-Iflah, 1972; Figueroa-Bossi and Bossi, 1999; Gasson and Davies, 1980; Ghosh *et al.*, 2009; Livny and Friedman, 2004; Madera *et al.*, 2009; Raya and H'bert, 2009; Selva *et al.*, 2009; Shimizu-Kadota and Sakurai, 1982). Typically, cells treated with these factors are grown in a liquid medium until the observable signs of lysis, if any occurs. They can also be streaked immediately on a soft agar medium containing suspended phage-sensitive cells (in a Petri plate). If the prophage is inducible and can initiate phage lytic development, incubation of plates at an appropriate temperature allows phage production in lysogens and lysis of the underlying sensitive cells. The lysis can be detected as a transparent or semi-transparent zone. Standard methods do not always allow detecting a functional prophage, and are useless for the detection of defective prophages.

Prophage induction can be used in curing bacterial strains of integrated prophages, provided that the prophage excision system is functional. Often, more than one inducing factor needs to be tested as not necessarily similar induction conditions may be suitable for all prophages. Typically, following prophage induction, cells are plated on solid media and tested for lysogeny. The presence of prophages in bacterial cells is sometimes correlated with a specific phenotype resulting from the disruption of certain host genes. This enables a simple phenotype-based distinction of lysogens from non-lysogens. For instance, certain *S. aureus* temperate phages integrate as prophages into the lipase (*geh*) or β-haemolysin (*hlb*) gene. Thus, prophage loss results in the restoration of the cells' ability to precipitate calcium salts of hydrolysed fatty acids in egg yolk containing media or in the restoration of haemolytic activity, respectively (Goerke *et al.*, 2009; Lee and Iandolo, 1985). Subsequently, the loss of prophage can be verified in each strain by testing its sensitivity to a parental strain phage (see the paragraph above) or by PCR with a prophage-specific primer pair.

Alternatively, one can search for colonies of spontaneously cured cells from among lysogens by plating tens to thousands of cells from a lysogen culture onto a solid medium, growing them, and testing by colony blot for the prophage loss (Loeffler and Fischetti, 2006). A PCR product of a gene, specific for a given prophage, can be used as a probe in these experiments. Therefore, this method is suitable for curing bacteria of prophages for which at least part if not all of the sequence is known.

To increase the frequency of prophage-cured cells formation one can clone the prophage excisionase gene in a plasmid and overexpress it in a lysogen, as it was shown in the case of lambda or KplE1 prophage-carrying *E. coli* cells (Leffers and Gottesmann, 1998, Panis *et al.*, 2007). This method is especially designed for the removal of those prophages which lost the ability to excise in response to inducing factors or for which such factors could not be identified. Its limitations may be difficulties in the introduction of plasmid vectors to phage hosts that are natural isolates and contain functional restriction systems, as well as the lack of plasmid expression vectors that can replicate in phage hosts of certain species.

Molecular biology tools that are typically used to delete single chromosomal genes can be used to remove integrated prophages as well. They are based on the homologous recombination-based methods of *in vivo* replacement of prophage-containing chromosomal DNA fragments with the corresponding fragments without the prophage. For instance, Bae *et al.* (2006) used a temperature-sensitive replicon-based suicidal plasmids containing the cloned DNA fragments that corresponded to prophage integration region, to cure a *S. aureus* strain of four prophages. Wang *et al.* (2010) utilized a curable plasmid expressing phage λ Red recombination system genes (reviewed by Murphy, 2012) to replace four prophages in the *E. coli* chromosome with a PCR-amplified antibiotic resistance cassette, which was then eliminated with the help of other curable plasmid, as described by Datsenko and Wanner (2000). Although the recombinational methods are elaborate they are suitable for the removal from bacterial chromosomes of fully functional as well as defective prophages.

In certain cases, the acquisition of prophage free hosts may appear to be a difficult task. Chromosomes of some pathogenic bacteria contain multiple prophages (Boyd and Brüssow, 2002). For example, *S. pyogenes* strain MGAS315 contains six prophages (13.6% of its genome), while *E. coli* O157:H7 strain Sakai contains 18 prophages (16% of its genome) (Fouts, 2006; Ohnishi *et al.*, 2001). In some cases, e.g. in *S. aureus* NCTC 8325 strain, the integrated form of the prophage can coexist with the excised one within the same cell (Tao *et al.*, 2010). Moreover, certain prophages can integrate into secondary attachment sites after their excision from the primary sites. Thus, phenotype changes of prophage containing strains do not always correspond to prophage loss (authors' unpublished results). Additionally, multiple prophages that are present in one cell can support each other's functions, which, in certain cases, may hinder the construction of prophage free bacteria. For example, the curing of double Gifsy-1 Gifsy-2 lysogen of *Salmonella typhimurium* from the Gifsy-2 prophage was found to cause a four orders of magnitude drop in the frequency of Gifsy-1 excision (Figueroa-Bossi *et al.*, 1997).

Another difficulty may be the impossibility of propagating a therapeutic phage in a prophage free host. This may occur in the case of defective therapeutic phages, whose lacking functions are complemented by prophage genes. Polyvalent staphylococcal phage 44AHJD appeared unable to form plaques on the lawn of prophage free host, unless the host cells were supplemented with a plasmid expressing the prophage endolysin gene (Nirmal Kumar *et al.*, 2012).

Acquisition of plasmid-free bacterial strains

Plasmids are autonomously replicating DNA molecules that are common in bacteria and do not encode essential functions. They can be maintained in cells in single or multiple copies, in a form of circular or linear DNA, that vary in size from about 1 kb to even as much as over 1 Mb, depending on a plasmid. In the majority of cases plasmids are stably inherited by daughter cells at cell division (reviewed by Zielenkiewicz and Cegłowski, 2001). There are at least two reasons to remove naturally occurring plasmids from bacteria that serve as phage propagation hosts. Firstly, plasmids usually carry genes adaptive for bacteria. In the case of plasmids from bacterial pathogens, toxins and other virulence factors are most often among products of these genes and will be lysate contaminants. Secondly, cases of plasmid transduction by those obligately virulent phages, which are commonly known as non-transducing have been described to occur (reviewed by Masters, 1996). Some are associated with the use of mutant phages (Morgan, 1979; Tianiashin *et al.*, 2003; see also 'Reasons and methods to avoid transducing phages'). The molecular basis of others may be associated with the integration of small circular plasmid DNA into phage DNA. Hence, this mechanism of phage-mediated plasmid transfer can overcome the requirement for sequence-specificity in the DNA packaging of certain phages. It depends on the recombination between a plasmid and a homologous fragment of phage DNA, if such exists (Masters, 1996; Piepersberg *et al.*, 1983; Raya and Klaenhammer, 1992). In a new phage-infected cell, the integrated plasmid can be resolved from the recombinant molecule by recombinational excision.

Plasmid identification in bacterial cells is most often based on the detection of plasmid DNA. Plasmids of less than ca. 25 kb in size can be detected in preparations of total cellular DNA that are separated electrophoretically in a standard agarose gel and stained with ethidium bromide. They form bands migrating in front of the major band, which represents chromosomal DNA. Larger plasmids can be detected in a similar way by PFGE, which serves to separate large DNA molecules. Additionally, plasmid DNA can be purified from bacteria with the use of conventional methods or commercially available kits, and analysed separately (Sambrook *et al.*, 1989; Carbone *et al.*, 2012, and references therein).

The association of a plasmid with any phenotypic trait it encodes provides a means to distinguish plasmid bearing cells from plasmid free cells (reviewed by Stanisich, 1988). The latter appear in a population spontaneously, but with frequencies that vary from quite high (within the range of % values) to astonishingly low (below 10^{-6} per cell), dependent on the plasmid. Different methods have been used to increase the frequency of plasmid loss. They are based mainly on the application of compounds that intercalate with higher efficiency into plasmid DNA than into chromosomal DNA and thus preferentially interfere with plasmid replication (Darini, 1996; Spengler *et al.*, 2003; reviewed by Stanisich, 1988; Spengler *et al.*, 2006). Successful plasmid elimination in the presence of ionic detergents (such as SDS) at sublethal concentrations, or under conditions of fast protoplast regeneration have also been described (Gruss and Novick, 1986; Kai *et al.*, 2002). As long as the frequency of plasmid loss is sufficiently high, plasmid-free cell clones can be identified based on the analysis of their total DNA, or by PCR with a plasmid-specific primer pair, provided that the latter can be designed.

The most efficient but laborious methods of plasmid elimination rely on the introduction of an incompatible plasmid that carries an antibiotic resistance marker to plasmid-bearing cells (Uraji *et al.*, 2002; Ni *et al.*, 2008). The selection of transformants for the incoming plasmid and their growth for several generations with the antibiotic selective for this plasmid will lead to resident plasmid loss. The disadvantage of this method is the necessity of prior classification of the resident plasmid to a recognized incompatibility group and the necessity to cure the resident plasmid-free cells from the incoming plasmid. The latter can be achieved, e.g. by using plasmids that have controllable replicons for curing.

Certain plasmids cannot be easily eliminated from cells, because they encode a post-segregational killing system, which ensures the death of cells that lost the plasmid. Their efficient curing strategy, which has been described recently, requires the construction of an unstable plasmid that contains their cloned replicon and post-segregational killing system, and the *sacB* gene, which confers the sensitivity to sucrose (Hale *et al.*, 2010). Such an unstable plasmid can displace the resident plasmid, and cells that subsequently lost it can be selected as able to form colonies on sucrose-containing media.

Bacterial phage resistance – is it a concern?

Even an obligately lytic and non-transducing phage, which propagates efficiently in its host strains, may appear unsuitable for therapeutic applications when used alone in the case when bacteria can gain resistance to this phage with a high frequency. The problem of emerging bacterial phage resistance in the course of phage therapy in humans does exist, as reported in the recent summary by Międzybrodzki *et al.* (2012).

Several types of phage resistance mechanisms are known. The readers who are interested in a detailed description of their molecular basis can find it in a recent review by Labrie *et al.* (2010) and in Chapter 4 of this volume. The resistance problem cannot be entirely overcome by the selection of proper phages, as it depends on both the phage and its host strain. While certain strains may remain susceptible to a given phage for tens of generations, cells resistant to the same phage may appear in populations of other strains of the same species after a few generations (authors' unpublished results). A recommended way to overcome the bacterial phage resistance problem is to use for therapeutic purposes cocktails that are composed of more than one phage (Gill and Hyman, 2010). The use of such cocktails was recommended by d'Hérelle, based on results of his early studies (d'Hérelle, 1938), and has been practiced with positive effects in Georgia and Russia (Chanishvili, 2012).

Concluding remarks

The emergence and spread of antibiotic resistant bacterial pathogens demands a search for infection treatment strategies alternative to antibiotics. One of them may be phage therapy. It was widely practiced in the past, but was abandoned in well developed countries with the introduction of antibiotics. Poor knowledge of phage structure and biology was one of the reasons behind numerous failures of phage therapy at that time. Nowadays, the antibiotic therapy crisis and the increasing number of well documented publications demonstrating general safety of phage therapy have rejuvenated the interest in this treatment option. Moreover, the access to sophisticated methods of molecular biology and electron microscopy, as well as high throughput 'omic technologies, allows one to study each phage in detail prior to deciding about its possible therapeutic application. In this chapter we try to convince the reader that despite the enormous variation among phages targeting different bacteria, one can formulate general recommendations which should enable choosing correct phages for phage therapy studies, based on the results of selected analysis methods. We summarize these recommendations in Table 2.1. Whereas some of them concern safety requirements, certain others may help predict, at least to some extent, a possible efficacy of new phage isolates in therapeutic applications prior to their testing in animals and in humans.

Table 2.1 General recommendations for the analysis and choice of candidate phages for phage therapy studies

Bacteriophage properties		Suitability for therapy	Possible verification methods [comments]
Lifestyle	Obligately lytic	+	1) Plaque clarity testing 2) Testing for the resistance or immunity to the same phage among infection surviving bacteria (if there are any) 3) Testing for the induction of phage lytic development in infection surviving bacteria (if there are any) 4) Searching for prophage lifestyle associated genes by PCR with specific primer pairs or, better, by the DNA sequence analysis 5) Identifying the phage's closest relatives based on morphological, genomic or/and proteomic analysis
	Temperate	–	
Ability to participate in horizontal gene transfer by transduction	Non-transducing	+	1) Testing for the ability to transfer antibiotic resistance or other selectable markers between bacterial strains 2) Testing for the presence of bacterial DNA in phage virions by PCR with a bacterium DNA-specific primer pair 3) Determining the type of virion DNA ends
	Transducing	–	
Homologs of bacterial pathogenicity-associated genes	Absent	+	1) Searching for bacterial pathogenicity-associated genes by PCR with specific primer pairs or, better, by the DNA sequence analysis [Certain obligately lytic phages contain seldom bacterial pathogenicity-associated gene homologs that apparently serve phage propagation functions. They do not have to disqualify these phages from therapeutic use.]
	Present	– (–/+)	
Spectrum of activity	Broad within a genus	+/–	1) Testing for the ability of a phage to plaque on cell layers of various strains representing well differentiated species or genus representatives, or various isolates of a given pathovar or serovar within a species, dependent on needs [Phages of too wide host-range within a genus may not be optimally adapted for propagation in any of their hosts. Phages of narrow host range within a species are recommended for therapy of infections with specific pathovars or serovars of a given species, such as e. g. *E. coli* O157:H7]
	Broad within a species	+	
	Narrow	– (+)	
Fitness under conditions resembling the conditions of therapy target site	High	+	1) Stability testing[a] 2) Adsorption kinetics testing with cells at different physiological states[a] 3) Assays of latent period and burst size (one step growth curve)[a]
	Low	–	
Host cells killing efficiency under conditions resembling the conditions of therapy target site	Fast	+	1) Killing-titer assays with cells of different strains that can be hosts for a given phage, and with cells at different physiological states[a] [High killing titers/ml as compared to the number of pfu/ml can limit the use of a phage to passive therapy only.]
	Slow	–	
Cultivation	Easy	+	1) Testing for the possibility to obtain lysates of high titers by phage propagation in bacteria grown under standard laboratory conditions (when optimized for a given phage) 2) Assays of reproducibility of phage propagation efficiency in bacteria grown under laboratory conditions optimized for the propagation of a given phage
	Difficult	–	
Purification	Simple	+	1) Assays of active phage recovery efficiency in preparations obtained from lysates that were purified by different methods
	Difficult	–	
Stability during storage	High	+	1) Measurements of phage decay during storage under different conditions (in refrigerated, frozen or liophilized lysates, in the presence of stabilizing factors; in refrigerated, frozen or liofilized, freshly infected, bacterial cells)
	Low	–	

Bacteriophage properties		Suitability for therapy	Possible verification methods [comments]
Emergence of resistance in target bacteria	Rare	–	1) Measurement of the time of appearence and the relative number of phage resistant cells in a mixed host-phage population
	Frequent	+/–	[The frequent emergence of phage resistant clones among host bacteria does not have to be a problem if a phage is used in coctails with other phages]

[a]To give informative results, all the assays indicated should be performed at conditions as close as possible to the conditions of a therapy target site.

Acknowledgements

The authors thank Andrew Kropinski for his helpful comments on this manuscript. This work was supported by funds from the Operational Program 'Innovative Economy, 2007–2013' (Priority axis 1. Research and Development of Modern Technologies, Measure 1.3 Support for R&D projects for entrepreneurs carried out by scientific entities, Submeasure 1.3.1, Development project no. POIG 01.03.01-02-003/08 entitled 'Optimization of characteristic and production of phage preparations for therapeutic purposes', from the targeted grant no. PBZ-MNiSW-04/I/2007 from the Ministry of Science and Higher Education and by statutory funds for the Faculty of Agriculture and Biology of the Warsaw University of Life Sciences.

References

Abedon, S.T. (2011). Lysis from without. Bacteriophage *1*, 46–49.

Abedon, S.T., and Thomas-Abedon, C. (2010). Phage therapy pharmacology. Curr. Pharm. Biotechnol. *11*, 28–47.

Abedon, S.T., Herschler, T.D., and Stopar, D. (2001). Bacteriophage latent-period evolution as a response to resource availability. Appl. Environ. Microbiol. *67*, 4233–4241.

Ackermann, H.W. (2003). Bacteriophage observations and evolution. Res. Microbiol. *154*, 245–251.

Ackermann, H.W. (2005). Bacteriophage classification. In Bacteriophages: Biology and Application, E.B. Kutter and A. Sulakvelidze, eds. (Boca Raton, FL, CRC Press), pp. 67–90.

Ackermann, H.W. (2007). 5500 Phages examined in the electron microscope. Arch. Virol. *152*, 227–243.

Ackermann, H.W. (2009a). Phage classification and characterization. Methods Mol. Biol. *501*, 127–140.

Ackermann, H.W. (2009b). Basic phage electron microscopy. Methods Mol. Biol. *501*, 113–126.

Ackermann, H.W. (2012). Bacteriophage electron microscopy. Adv. Virus Res. *82*, 1–32.

Ackermann, H.W., and Prangishvili, D. (2012). Prokaryote viruses studied by electron microscopy. Arch. Virol. *157*, 1843–1849.

Ackermann, H.W., and Tiekotter, K.L. (2012). Murphy's law-if anything can go wrong, it will: Problems in phage electron microscopy. Bacteriophage *2*, 122–129.

Ackermann, H.W., Tremblay, D., and Moineau, S. (2004). Long-term bacteriophage preservation. World Fed. Cult. Collection Newslett. *38*, 35–40.

Adams, M.H. (1959). Bacteriophages. Interscience Publishers, Inc. New York.

Adriaenssens, E.M., Van Vaerenbergh, J., Vandenheuvel, D., Dunon, V., Ceyssens, P.J., De Proft, M., Kropinski, A.M., Noben, J.P., Maes, M., and Lavigne, R. (2012a). T4-related bacteriophage LIMEstone isolates for the control of soft rot on potato caused by 'Dickeya solani'. PLoS One *7*, e33227.

Adriaenssens, E.M., Lehman, S.M., Vandersteegen, K., Vandenheuvel, D., Philippe, D.L., Cornelissen, A., Clokie, M.R., García, A.J., De Proft, M., Maes, M., and Lavigne, R. (2012b). CIM(®) monolithic anion-exchange chromatography as a useful alternative to CsCl gradient purification of bacteriophage particles. Virology *434*, 265–270.

Affolter, M., Parent-Vaugeois, C., and Anderson, A. (1983). Curing and induction of the Fels 1 and Fels 2 prophages in the Ames mutagen tester strains of *Salmonella typhimurium*. Mutat. Res. *110*, 243–262.

Alfadhel, M., Puapermpoonsiri, U., Ford, S.J., McInnes, F.J., and van der Walle, C.F. (2011). Lyophilized inserts for nasal administration harboring bacteriophage selective for *Staphylococcus aureus*: *in vitro* evaluation. Int. J. Pharm. *416*, 280–287.

Altschul, S.F., Gish, W., Miller, W., Myers, E.W., and Lipman, D.J. (1990). Basic local alignment search tool. J. Mol. Biol. *215*, 403–410.

Ambrozic, J., Ferme, D., Grabnar, M., Ravnikar, M., and Avgustin, G. (2001). The bacteriophages of ruminal prevotellas. Folia Microbiol. (Praha) *46*, 37–39.

Ammann, A., Neve, H., Geis, A., and Heller, K.J. (2008). Plasmid transfer via transduction from *Streptococcus thermophilus* to *Lactococcus lactis*. J. Bacteriol. *190*, 3083–3087.

Ashelford, K.E., Day, M.J., and Fry, J.C. (2003). Elevated abundance of bacteriophage infecting bacteria in soil. Appl. Environ. Microbiol. *69*, 285–289.

Askora, A., Kawasaki, T., Fujie, M., and Yamada, T. (2011). Resolvase-like serine recombinase mediates integration/excision in the bacteriophage φRSM. J. Biosci. Bioeng. *111*, 109–116.

Atterbury, R.J., Connerton, P.L., Dodd, Ch.E.R., Rees, C.E.D., and Connerton, I.F. (2003). Application of host-specific bacteriophages to the surface of chickens skin leads to a reduction in recovery of *Campylobacter jejuni*. Appl. Environ. Microbiol. *69*, 6302–6306.

Ayers, A.R., Ayers, S.B., and Goodman, R.N. (1979). Extracellular polysaccharide of *Erwinia amylovora*: a correlation with virulence. Appl. Environ. Microbiol. *38*, 659–666.

Bächi, B., and Arber, W. (1977). Physical mapping of BglII, BamHI, EcoRI, HindIII and PstI restriction fragments of bacteriophage P1 DNA. Mol. Gen. Genet. *153*, 311–324.

Bachrach, G., Leizerovici-Zigmond, M., Zlotkin, A., Naor, R., and Steinberg, D. (2003). Bacteriophage isolation from human saliva. Lett. Appl. Microbiol. *36*, 50–53.

Bae, T., Baba, T., Hiramatsu, K., and Schneewind, O. (2006). Prophages of *Staphylococcus aureus* Newman and their contribution to virulence. Mol. Microbiol. *62*, 1035–1047.

Balding, C., Bromley, S.A., Pickup, R.W., and Saunders, J.R. (2005). Diversity of phage integrases in Enterobacteriaceae: development of markers for environmental analysis of temperate phages. Environ. Microbiol. *7*, 1558–1567.

Balogh, B., Jones, J.B., Iriarte, F.B., and Momol, M.T. (2010). Phage therapy for plant disease control. Curr. Pharm. Biotechnol. *11*, 48–57.

Barrow, P., Lovell, M., and Berchieri, A. Jr. (1998). Use of lytic bacteriophage for control of experimental *Escherichia coli* septicemia and meningitis in chickens and calves. Clin. Diagn. Lab. Immunol. *5*, 294–298.

Bartell, P.F., Orr, T.E., Reese, J.F., and Imaeda, T. (1971). Interaction of *Pseudomonas* bacteriophage 2 with the slime polysaccharide and lipopolysaccharide of *Pseudomonas aeruginosa* strain B1. J. Virol. *8*, 311–317.

Barut, M., Podgornik, A., Urbas, L., Gabor, B., Brne, P., Vidic, J., Plevcak, S., and Strancar, A. (2008). Methacrylate-based short monolithic columns: enabling tools for rapid and efficient analyses of biomolecules and nanoparticles. J. Sep. Sci. *31*, 1867–1880.

Bazaka, K., Crawford, R.J., Nazarenko, E.L., and Ivanova, E.P. (2011). Bacterial extracellular polysaccharides. Adv. Exp. Med. Biol. *715*, 213–226.

Beard-Pegler, M.A., Gabelish, C.L., Stubbs, E., Harbour, C., Robson, J., Falk, M., Benn, R., and Vickery, A. (1989). Prevalence of peritonitis-associated coagulase-negative staphylococci on the skin of continuous ambulatory peritoneal dialysis patients. Epidemiol. Infect. *102*, 365–378.

Belcaid, M., Bergeron, A., and Poisson, G. (2010). Mosaic graphs and comparative genomics in phage communities. J. Comput. Biol. *17*, 1315–1326.

Bigby, D., and Kropinski, A.M. (1989). Isolation and characterization of a *Pseudomonas aeruginosa* bacteriophage with a very limited host range. Can. J. Microbiol. *35*, 630–635.

Biswas, B., Adhya. S., Washart, P., Paul, B., Trostel, A.N., Powell, B., Carlton, R., and Merril, C.R. (2002). Bacteriophage therapy rescues mice bacteremic from a clinical isolate of vancomycin-resistant *Enterococcus faecium*. Infect. Immun. *70*, 204–210.

Biswas, T., Aihara, H., Radman-Livaja, M., Filman, D., Landy, A., and Ellenberger, T. (2005). A structural basis for allosteric control of DNA recombination by lambda integrase. Nature *435*, 1059–1066.

Boman, H.G., and Monner, D.A. (1975). Blocking of bacteriophages phi W and phi 5 with lipopolysaccharides from *Escherichia coli* K-12 mutants. J. Bacteriol. *121*, 465–470.

Borecká, P., Rosypal, S., Pantůcek, R., and Doskar, J. (1996). Localization of prophages of serological group B and F on restriction fragments defined in the restriction map of *Staphylococcus aureus* NCTC 8325. FEMS Microbiol. Lett. *143*, 203–210.

Borodovsky, M., and Lomsadze, A. (2011). Gene identification in prokaryotic genomes, phages, metagenomes, and EST sequences with GeneMarkS suite. Curr. Protoc. Bioinformatics. Chapter *4*, Unit 4.5.1–17.

Borysowski, J., Łobocka, M., Międzybrodzki, R., Weber-Dąbrowska, B., and Górski, A. (2011). Potential of bacteriophages and their lysins in the treatment of MRSA: current status and future perspectives. BioDrugs. *25*, 347–355.

Bose, M., and Barber, R.D. (2006). Prophage Finder: a prophage loci prediction tool for prokaryotic genome sequences. In Silico Biol. *6*, 223–227.

Bossi, L., Fuentes, J.A., Mora, G., and Figueroa-Bossi, N. (2003). Prophage contribution to bacterial population dynamics. J. Bacteriol. *185*, 6467–6471.

Boulanger, P. (2009). Purification of bacteriophages and SDS-PAGE analysis of phage structural proteins from ghost particles. Methods Mol. Biol. *502*, 227–238.

Bourhy, P., Frangeul, L., Couvé, E., Glaser, P., Saint Girons, I., and Picardeau, M. (2005). Complete nucleotide sequence of the LE1 prophage from the spirochete *Leptospira biflexa* and characterization of its replication and partition functions. J. Bacteriol. *187*, 3931–3940.

Boussard, P., Pithsy, A., and Devleeschouwer, M.J. (1993). Relationship between slime production, antibiotic sensitivity and the phage type of coagulase-negative staphylococci. J. Clin. Pharm. Ther. *18*, 271–274.

Boyd, E.F. (2012). Bacteriophage-encoded bacterial virulence factors and phage-pathogenicity island interactions. Adv. Virus. Res. *82*, 91–118.

Boyd, E.F., and Brüssow, H. (2002). Common themes among bacteriophage-encoded virulence factors and diversity among the bacteriophages involved. Trends Microbiol. *10*, 521–529.

Boyd, E.F., Carpenter, M.R., and Chowdhury, N. (2012). Mobile effector proteins on phage genomes. Bacteriophage *2*, 139–148.

Brabban, A.D., Hite, E., and Callaway, T.R. (2005). Evolution of food-borne pathogens via temperate bacteriophage-mediated gene transfer. Foodborne Pathog. Dis. *2*, 287–303.

Breitbart, M. (2012). Marine viruses: truth or dare. Annu. Rev. Mar. Sci. *4*, 425–448.

Broxmeyer, L., Sosnowska, D., Miltner, E., Chacón, O., Wagner, D., McGarvey, J., Barletta, R.G., and Bermudez, L.E. (2002). Killing of *Mycobacterium avium* and *Mycobacterium tuberculosis* by a mycobacteriophage delivered by a nonvirulent mycobacterium: a model for phage therapy of intracellular bacterial pathogens. J. Infect. Dis. *186*, 1155–1160.

Brüssow, H. (2005). Phage therapy: the *Escherichia coli* experience. Microbiology *151*, 2133–2140.

Brüssow, H. (2007). Bacteria between protists and phages: from antipredation strategies to the evolution of pathogenicity. Mol. Microbiol. *65*, 583–589.

Brüssow, H., Canchaya, C., and Hardt, W.D. (2004). Phages and the evolution of bacterial pathogens: from genomic rearrangements to lysogenic conversion. Microbiol. Mol. Biol. Rev. *68*, 560–562.

Bryner, J.H., Ritchie, A.E., Booth, G.D., and Foley, J.W. (1973). Lytic activity of *Vibrio* phages on strains of Vibrio fetus isolated from man and animals. Appl. Microbiol. *26*, 404–409.

Bukhari, A.I., and Curtin, P. (1974). Temperature-sensitive mutants of bacteriophage Mu. J. Virol. *14*, 1615–1616.

Bull, J.J., and Regoes, R.R. (2006). Pharmacodynamics of non-replicating viruses, bacteriocins and lysins. Proc. Biol. Sci. *273*, 2703–2712.

Bull, J.J., Levin, B.R., DeRouin, T., Walker, N., and Bloch, C.A. (2002). Dynamics of success and failure in phage and antibiotic therapy in experimental infections. BMC Microbiol. *26*, 35.

Bull, J.J., Vimr, E.R., and Molineux, I.J. (2010). A tale of tails: Sialidase is key to success in a model of phage therapy against K1-capsulated *Escherichia coli*. Virology *398*, 79–86.

Bull, J.J., Otto, G., and Molineux, I.J. (2012). *In vivo* growth rates are poorly correlated with phage therapy success in a mouse infection model. Antimicrob. Agents Chemother. *56*, 949–954.

Callaway, T.R., Edrington, T.S., Brabban, A., Kutter, E., Karriker, L., Stahl, C., Wagstrom, E., Anderson, R.C., Genovese, K., McReynolds, J., *et al.* (2010). Occurrence of *Salmonella*-specific bacteriophages in swine feces collected from commercial farms. Foodborne Pathog. Dis. *7*, 851–856.

Canchaya, C., Fournous, G., Chibani-Chennoufi, S., Dilmann, M.L., and Brüssow, H. (2003). Phage as agents of lateral gene transfer. Curr. Opin. Microbiol. *6*, 417–424.

Cantalupo, P.G., Calgua, B., Zhao, G., Hundesa, A., Wier, A.D., Katz, J.P., Grabe, M., Hendrix, R.W., Girones, R., Wang, D., and Pipas, J.M. (2011). Raw sewage harbors diverse viral populations. mBio *2*, e00180–11.

Carbone, A., Fioretti, F.M., Fucci, L., Ausió, J., and Piscopo, M. (2012). High efficiency method to obtain supercoiled DNA with a commercial plasmid purification kit. Acta Biochim. Pol. *59*, 275–278.

Carlson, K. (2005). Working with bacteriophages: Common techniques and methodological approaches. In Bacteriophages: Biology and Application, E.B. Kutter and A. Sulakvelidze, eds. (Boca Raton, FL, CRC Press), pp. 437–494.

Carlson, N.G., and Little, J.W. (1993). A novel antivirulence element in the temperate bacteriophage HK022. J. Bacteriol. *175*, 7541–7549.

Carlton, R.M., Noordman, W.H., Biswas, B., de Meester, E.D., and Loessner, M.J. (2005). Bacteriophage P100 for control of *Listeria monocytogenes* in foods: genome sequence, bioinformatic analyses, oral toxicity study, and application. Regul. Toxicol. Pharm. *43*, 301–312.

Carrolo, M., Frias, M.J., Pinto, F.R., Melo-Cristino, J., and Ramirez, M. (2010). Prophage spontaneous activation promotes DNA release enhancing biofilm formation in *Streptococcus pneumoniae*. Plos One *5*, e15678.

Carvalho, C., Susano, M., Fernandes, E., Santos, S., Gannon, B., Nicolau, A., Gibbs, P., Teixeira, P., and Azeredo, J. (2010). Method for bacteriophage isolation against target *Campylobacter* strains. Lett. Appl. Microbiol. *50*, 192–197.

Casas, V., and Maloy, S. (2011). Role of bacteriophage-encoded exotoxins in the evolution of bacterial pathogens. Future Microbiol. *6*, 1461–1473.

Casas, V., Miyake, J., Balsley, H., Roark, J., Telles, S., Leeds, S., Zurita, I., Breitbart, M., Bartlett, D., Azam, F., and Rohwer, F. (2006). Widespread occurrence of phage-encoded exotoxin genes in terrestrial and aquatic environments in Southern California. FEMS Microbiol. Lett. *261*, 141–149.

Casjens, S. (2003). Prophages and bacterial genomics: what have we learned so far? Mol. Microbiol. *49*, 277–300.

Casjens, S.R., and Gilcrease, E.B. (2009). Determining DNA packaging strategy by analysis of the termini of the chromosomes in tailed-bacteriophage virions. Methods Mol. Biol. *502*, 91–111.

Casjens, S., Winn-Stapley, D., Gilcrease, E., Moreno, R., Kuhlewein, C., Chua, J.E., Manning, P.A., Inwood, W., and Clark, A.J. (2004a). The chromosome of *Shigella flexneri* bacteriophage Sf6: complete nucleotide sequence, genetic mosaicism, and DNA packaging. J. Mol. Biol. *339*, 379–394.

Casjens. S.R., Gilcrease, E.B., Huang, W.M., Bunny, K.L., Pedulla, M.L., Ford, M.E., Houtz, J.M., Hatfull, G.F., and Hendrix, R.W. (2004b). The pKO2 linear plasmid prophage of *Klebsiella oxytoca*. J. Bacteriol. *186*, 1818–1832.

Casjens, S.R., Gilcrease, E.B., Winn-Stapley, D.A., Schicklmaier, P., Schmieger, H., Pedulla, M.L., Ford, M.E., Houtz, J.M., Hatfull, G.F., and Hendrix, R.W. (2005). The generalized transducing Salmonella bacteriophage ES18: complete genome sequence and DNA packaging strategy. J. Bacteriol. *187*, 1091–1094.

Castillo, F.J., and Bartell, P.F. (1974). Studies on the bacteriophage 2 receptors of *Pseudomonas aeruginosa*. J. Virol. *14*, 904–909.

Cavalcanti, S.M., and Siqueira Júnior, J.P. (1995). Cure of prophage in *Staphylococcus aureus* by furocoumarin photoadditions. Microbios. *81*, 85–91.

Cerca, N., Oliveira, R., and Azeredo, J. (2007). Susceptibility of *Staphylococcus epidermidis* planktonic cells and biofilms to the lytic action of staphylococcus bacteriophage K. Lett. Appl. Microbiol. *45*, 313–317.

Ceyssens, P.J., Noben, J.P., Ackermann, H.W., Verhaegen, J., De Vos, D., Pirnay, J.P., Merabishvili, M., Vaneechoutte,

M., Chibeu, A., Volckaert, G., and Lavigne, R. (2009). Survey of *Pseudomonas aeruginosa* and its phages: de novo peptide sequencing as a novel tool to assess the diversity of worldwide collected viruses. Environ. Microbiol. *11*, 1303–1313.

Chanishvili, N. (2012). Phage therapy-history from Twort and d'Hérelle through Soviet experience to current approaches. Adv. Virus Res. *83*, 3–40.

Chibani-Chennoufi, S., Sidoti, J., Bruttin, A., Dillmann, M., Kutter, E., Qadri, F., Sarker, S.A., and Brüssow, H. (2004). Isolation of *Escherichia coli* bacteriophages from the stool of pediatric diarrhea patients in Bangladesh. J. Bacteriol. *186*, 8287–8294.

Christensen, G.D., Parisi, J.T., Bisno, A.L., Simpson, W.A., and Beachey, E.H. (1983). Characterization of clinically significant strains of coagulase-negative staphylococci. J. Clin. Microbiol. *18*, 258–269.

Chung, Y.B., and Hinkle, D.C. (1990a). Bacteriophage T7 DNA packaging. I. Plasmids containing a T7 replication origin and the T7 concatemer junction are packaged into transducing particles during phage infection. J. Mol. Biol. *216*, 911–926.

Chung, Y.B., and Hinkle, D.C. (1990b). Bacteriophage T7 DNA packaging. II. Analysis of the DNA sequences required for packaging using a plasmid transduction assay. J. Mol. Biol. *216*, 927–938.

Clark, D.W., Meyer, H.-P., Leist, C., and Fiechter, A. (1986). Effects of growth medium on phage production and induction in *Escherichia coli* K-12 lambda lysogens. J. Biotechnol. *3*, 271–280.

Clark, J.R., and March, J.B. (2006). Bacteriophages and biotechnology: vaccines, gene therapy and antibacterials. Trends Biotechnol. *24*, 212–218.

Clarke, K.J. (1998). Virus particle production in lysogenic bacteria exposed to protozoan grazing. FEMS Microbiol. Lett., *166*, 177–180.

Clockie, M. (2009). PCR and partial sequencing of bacteriophage genomes. Methods Mol. Biol. *502*, 47–55.

Cole, J.N., Pence, M.A., von Köckritz-Blickwede, M., Hollands, A., Gallo, R.L., Walker, M.J., and Nizet, V. (2010). M protein and hyaluronic acid capsule are essential for *in vivo* selection of *covRS* mutations characteristic of invasive serotype M1T1 group A *Streptococcus*. mBio *1*, e00191–10.

Comeau, A.M., Short, S., and Suttle, C.A. (2004). The use of degenerate-primed random amplification of polymorphic DNA (DP-RAPD) for strain-typing and inferring the genetic similarity among closely related viruses. J. Virol. Methods. *118*, 95–100.

Comeau, A.M., Chan, A.M., and Suttle, C.A. (2006). Genetic richness of vibriophages isolated in a coastal environment. Environ. Microbiol. *8*, 1164–1176.

Comeau, A.M., Tetart, F., Trojet, S.N., Prere, M.-F., and Krisch, H.M. (2007). Phage-antibiotic synergy (PAS): beta-lactam and quinolone antibiotics stimulate virulent phage growth. PLoS One *8*, e799.

Cooke, F.J., Wain, J., Fookes, M., Ivens, A., Thomson, N., Brown, D.J., Threlfall, E.J., Gunn, G., Foster, G., and Dougan, G. (2007). Prophage sequences defining hot spots of genome variation in *Salmonella enterica* serovar Typhimurium can be used to discriminate between field isolates. J. Clin. Microbiol. *45*, 2590–2598.

Coren, J.S., Pierce, J.C., and Sternberg, N. (1995). Headful packaging revisited: the packaging of more than one DNA molecule into a bacteriophage P1 head. J. Mol. Biol. *249*, 176–184.

Cornelissen, A., Ceyssens, P.J., T'Syen, J., Van Praet, H., Noben, J.P., Shaburova, O.V., Krylov, V.N., Volckaert, G., and Lavigne, R. (2011). The T7-related *Pseudomonas putida* phage φ15 displays virion-associated biofilm degradation properties. PLoS One *6*(4), e18597.

Cortez, D., Forterre, P., and Gribaldo, S.A. (2009). Hidden reservoir of integrative elements is the major source of recently acquired foreign genes and ORFans in archaeal and bacterial genomes. Genome Biol. *10*, R65.

Court, D.L., Oppenheim, A.B., and Adhya, S.L. (2007). A new look at bacteriophage lambda genetic networks. J. Bacteriol. *189*, 298–304.

Cresawn, S.G., Bogel, M., Day, N., Jacobs-Sera, D., Hendrix. R.W., and Hatfull, G.F. (2011). Phamerator: a bioinformatic tool for comparative bacteriophage genomics. BMC Bioinformatics *12*, 395.

Danz, F.J., and Schultz, E.W. (1949). Gelatinous variants of *Pseudomonas aeruginosa*. J. Bacteriol. *58*, 367–378.

D'Ari, R. (1977). Effects of mutations in the immunity system of bacteriophage P1. J. Virol. *23*, 467–475.

Darini, A.L. (1996). A genetic study of a *Staphylococcus aureus* plasmid involving cure and transference. Sao Paulo Med. J. *114*, 1068–1072.

Datsenko, K.A., and Wanner, B.L. (2000). One-step inactivation of chromosomal genes in *Escherichia coli* K-12 using PCR products. Proc. Natl Acad. Sci. USA *97*, 6640–6645.

De Castro, C., Holst, O., Lanzetta, R., Parrilli, M., and Molinaro, A. (2012). Bacterial lipopolysaccharides in plant and mammalian innate immunity. Protein Pept Lett. *19*, 1040–1044.

Delbrück, M. (1940). The growth of bacteriophage and lysis of the host. J. Gen. Physiol. *23*, 643–660.

De Mello Varani, A., Souza, R.C., Nakaya, H.I., de Lima, W.C., Paula de Almeida, L.G., Kitajima, E.W., Chen, J., Civerolo, E., Vasconcelos, A.T., and Van Sluys, M.A. (2008). Origins of the *Xylella fastidiosa* prophage-like regions and their impact in genome differentiation. PLoS One 3, e4059.

Deschavanne, P., DuBow, M.S., and Regeard, C. (2010). The use of genomic signature distance between bacteriophages and their hosts displays evolutionary relationships and phage growth cycle determination. Virol. J. *7*, 163.

Deutzmann, R. (2004). Structural characterization of proteins and peptides. Methods Mol. Med. *94*, 269–297.

D'Hérelle F. (1917). Sur un microbe invisible antagoniste des bacilles dysentériques. C. R. Acad. Sci. Ser D. *165*, 373–375.

D'Hérelle. (1938). Preparation of Therapeutic Bacteriophages, Appendix 1 from: Le Phénomène de la Guérison dans les maladies infectieuses. Masson et Cie, 1938, Paris-OCLC 5784382. (translation from Bacteriophages 2011; 1, 55–65).

Dhillon, T.S., and Dhillon, E.K. (1976). Temperate coliphage HK022. Clear plaque mutants and preliminary vegetative map. Jpn. J. Microbiol. *20*, 385–396.

Dinsdale, E.A., Edwards, R.A., Hall, D., Angly, F., Breitbart, M., Brulc, J.M., Furlan, M., Desnues, C., Haynes, M., Li, L., McDaniel, L., Moran, M.A., Nelson, K.E., Nilsson, C., Olson, R., Paul, J., Brito, B.R., Ruan, Y., Swan, B.K., Stevens, R., Valentine, D.L., Thurber, R.V., Wegley, L., White, B.A., and Rohwer, F. (2008). Functional metagenomic profiling of nine biomes. Nature *452*, 629–632.

Donlan, R.M. (2009). Preventing biofilms of clinically relevant organisms using bacteriophage. Trends Microbiol. *17*, 66–72.

Drahovska, H., Mikasova, E., Szemes, T., Ficek, A., Sasik, M., Majtan, V., and Turna, J. (2007). Variability in occurrence of multiple prophage genes in *Salmonella typhimurium* strains isolated in Slovak Republic. FEMS Microbiol. Lett. *270*, 237–244.

Dreher, T.W. (2010). Viral tRNAs and tRNA-like structures. Wiley Interdiscip. Rev. RNA. *1*, 402–414.

Drexler, H. (1970). Transduction by bacteriophage T1. Proc. Natl. Acad. Sci. U.S.A. *66*, 1083–1088.

Dubos, R.J., Straus, J.H., and Pierce, C. (1943). The multiplication of bacteriophage *in vivo* and its protective effect against an experimental infection with *Shigella dysenteriae*. J. Exp. Med. *20*, 161–168.

Duval-Iflah, Y. (1972). Lysogenic conversion of the lipase gene in *Staphylococcus pyogenes* group III strains. Can. J. Microbiol. *18*, 1491–1497.

Eggers, C.H., and Samuels, D.S. (1999). Molecular evidence for a new bacteriophage of *Borrelia burgdorferi*. J. Bacteriol. *181*, 7308–7313.

Eggers, C.H., Casjens, S., Hayes, S.F., Garon, C.F., Damman, C.J., Oliver, D.B., and Samuels, D.S. (2000). Bacteriophages of spirochetes. J. Mol. Microbiol. Biotechnol. *2*, 365–373.

Etienne, J., Brun, Y., el Solh, N., Delorme, V., Mouren, C., Bes, M., and Fleurette, J. (1988). Characterization of clinically significant isolates of *Staphylococcus epidermidis* from patients with endocarditis. J. Clin. Microbiol. *26*, 613–617.

Eyer, L., Pantůcek, R., Zdráhal, Z., Konecná, H., Kaspárek, P., Růzicková, V., Hernychová, L., Preisler, J., and Doskar, J. (2007). Structural protein analysis of the polyvalent staphylococcal bacteriophage 812. Proteomics *7*, 64–72.

Fang, C.T., Chuang, Y.P., Shun, C.T., Chang, S.C., and Wang, J.T. (2004). A novel virulence gene in *Klebsiella pneumoniae* strains causing primary liver abscess and septic metastatic complications. J. Exp. Med. *199*, 697–705.

Fang, N.X., Huang, B., Hiley, L., Bates, J., and Savill, J. (2012). A rapid multiplex DNA suspension array method for *Salmonella typhimurium* subtyping using prophage-related markers. J. Microbiol. Methods. *88*, 19–27.

Farrar, M.D., Howson, K.M., Bojar, R.A., West, D., Towler, J.C., Parry, J., Pelton, K., and Holland, K.T. (2007). Genome sequence and analysis of a *Propionibacterium acnes* bacteriophage. J. Bacteriol. *189*, 4161–4167.

Figueroa-Bossi, N., and Bossi, L. (1999). Inducible prophages contribute to *Salmonella* virulence in mice. Mol. Microbiol. *33*, 167–176.

Figueroa-Bossi, N., Coissac, E., Netter, P., and Bossi, L. (1997). Unsuspected prophage-like elements in *Salmonella typhimurium*. Mol. Microbiol. *25*, 161–173.

Flemming, H.C., and Wingender, J. (2010). The biofilm matrix. Nat. Rev. Microbiol. *8*, 623–633.

Fogg, P.C., Hynes, A.P., Digby, E., Lang, A.S., and Beatty, J.T. (2011). Characterization of a newly discovered Mu-like bacteriophage, RcapMu, in *Rhodobacter capsulatus* strain SB1003. Virology *421*, 211–221.

Fouts, D.E. (2006). Phage_Finder: automated identification and classification of prophage regions in complete bacterial genome sequences. Nucleic Acids Res. *34*, 5839–5851.

Gallet, R., Kannoly, S., Wang, I.N.(2011). Effects of bacteriophage traits on plaque formation. BMC Microbiol. *11*, 181.

Galloway, S.M., and Raetz, C.R. (1990). A mutant of *Escherichia coli* defective in the first step of endotoxin biosynthesis. J. Biol. Chem. *265*, 6394–6402.

Gantzer, C., Henny, J., and Schwartzbrod, L. (2002). *Bacteroides fragilis* and *Escherichia coli* bacteriophages in human faeces. Int. J. Hyg. Environ. Health. *205*, 325–328.

Gasson, M.J., and Davies, F.L. (1980). Prophage-cured derivatives of *Streptococcus lactis* and *Streptococcus cremoris*. Appl. Environ. Microbiol. *40*, 964–966.

Ghosh, D., Roy, K., Williamson, K.E., White, D.C., Wommack, K.E., Sublette, K.L., and Radosevich, M. (2008). Prevalence of lysogeny among soil bacteria and presence of 16S rRNA and trzN genes in viral-community DNA. Appl. Environ. Microbiol. *74*, 495–502.

Ghosh, D., Roy, K., Williamson, K.E., Srinivasiah, S., Wommack, K.E., and Radosevich, M. (2009). Acyl-homoserine lactones can induce virus production in lysogenic bacteria: an alternative paradigm for prophage induction. Appl. Environ. Microbiol. *75*, 7142–7152.

Gill, J.J., and Hyman, P. (2010). Phage choice, isolation, and preparation for phage therapy. Curr. Pharm. Biotechnol. *11*, 2–14.

Gill, J.J., Sabour, P.M., Leslie, K.E., and Griffiths, M.W. (2006). Bovine whey proteins inhibit the interaction of *Staphylococcus aureus* and bacteriophage K. J. Appl. Microbiol. *101*, 377–386.

Gödeke, J., Paul, K., Lassak, J., and Thormann, K.M. (2011). Phage-induced lysis enhances biofilm formation in *Shewanella oneidensis* MR-1. ISME J. *5*, 613–626.

Goerke, C., Pantucek, R., Holtfreter, S., Schulte, B., Zink, M., Grumann, D., Bröker, B.M., Doskar, J., and Wolz, C. (2009). Diversity of prophages in dominant *Staphylococcus aureus* clonal lineages. J. Bacteriol. *191*, 3462–3468.

Golec, P., Dąbrowski, K., Hejnowicz, M.S., Gozdek, A., Loś, J.M., Węgrzyn, G., Lobocka, M.B., and Loś, M. (2011a). A reliable method for storage of tailed phages. J. Microbiol. Methods. *84*, 486–489.

Golec, P., Wiczk, A., Łoś, J.M., Konopa, G., Węgrzyn, G., and Łoś, M. (2011b). Persistence of bacteriophage T4 in a starved *Escherichia coli* culture: evidence for the presence of phage subpopulations. J. Gen. Virol. *92*, 997–1003.

Górski, A., and Weber-Dąbrowska, B. (2005). The potential role of endogenous bacteriophages in controlling invading pathogens. Cell. Mol. Life Sci. *62*, 511–519.

Górski, A., Międzybrodzki, R., Borysowski, J., Dąbrowska, K., Wierzbicki, P., Ohams, M., Korczak-Kowalska, G., Olszowska-Zaremba, N., Łusiak-Szelachowska, M., *et al.* (2012). Phage as a modulator of immune responses: practical implications for phage therapy. Adv. Virus Res. *83*, 41–71.

Grkovic, S., O'Callaghan, M., and Mahanty, H.K. (1995). Characterization of *Serratia entomophila* bacteriophages and the phage-resistant mutant strain BC4B. Appl. Environ. Microbiol. *61*, 4160–4166.

Grodzicker, T., Arditti, R.R., and Eisen, H. (1972). Establishment of repression by lambdoid phage in catabolite activator protein and adenylate cyclase mutants of *Escherichia coli*. Proc. Natl. Acad. Sci. U.S.A. *69*, 366–370.

Groth, A.C., and Calos, M.P. (2004). Phage integrases: biology and applications. J. Mol. Biol. *335*, 667–678.

Gruss, A., and Novick, R. (1986). Plasmid instability in regenerating protoplasts of *Staphylococcus aureus* is caused by aberrant cell division. J. Bacteriol. *165*, 878–883.

Güssow, D., and Clackson, T. (1989). Direct clone characterization from plaques and colonies by the polymerase chain reaction. Nucleic Acids Res. *17*, 4000.

Gutiérrez, D., Martín-Platero, A.M., Rodríguez, A., Martínez-Bueno, M., García, P., and Martínez, B. (2011). Typing of bacteriophages by randomly amplified polymorphic DNA (RAPD)-PCR to assess genetic diversity. FEMS Microbiol. Lett. *322*, 90–97.

Guttman, B., Raya, P., and Kutter, E. (2005). Basic phage biology. In Bacteriophages: Biology and Application, E.B. Kutter and A. Sulakvelidze, eds. (Boca Raton, FL, CRC Press), pp. 29–66.

Hagens, S., Habel, A., von Ahsen, U., von Gabain, A., and Bläsi, U. (2004). Therapy of experimental pseudomonas infections with a nonreplicating genetically modified phage. Antimicrob. Agents Chemother. *48*, 3817–3822.

Hale, L., Lazos, O., Haines, A., and Thomas, C. (2010). An efficient stress-free strategy to displace stable bacterial plasmids. Biotechniques *48*, 223–228.

Hall, R.T., Hall, S.L., Barnes, W.G., Izuegbu, J., Rogolsky, M., and Zorbas, I. (1987). Characteristics of coagulase-negative staphylococci from infants with bacteremia. Pediatr. Infect. Dis. J. *6*, 377–383.

Hankin, E.H. (1896). 'L'action bactericide des eaux de la Jumna et du Gange sur le vibrion du cholera' (in French). *Annales de l'Institut Pasteur 10*, 511–523.

Hanks, M.C., Newman, B., Oliver, I.R., and Masters, M. (1988). Packaging of transducing DNA by bacteriophage P1. Mol. Gen. Genet. *214*, 523–532.

Hashimoto, C., and Fujisawa, H. (1988). Packaging and transduction of non-T3 DNA by bacteriophage T3. Virology *166*, 432–439.

Hatfull, G.F. (2008). Bacteriophage genomics. Curr. Opin. Microbiol. *11*, 447–453.

Hatfull, G.F. (2012). The secret lives of mycobacteriophages. Adv. Virus Res. *82*, 179–288.

Hatfull, G.F., and Hendrix, R.W. (2011). Bacteriophages and their genomes. Curr. Opin. Virol. *1*, 298–303.

Heintschel von Heinegg, E., Nalik, H.P., and Schmid, E.N. (1993). Characterisation of a *Helicobacter pylori* phage (HP1). J. Med. Microbiol. *38*, 245–249.

Hendrix, R.W. (2002). Bacteriophages: evolution of the majority. Theor. Popul. Biol. *61*, 471–480.

Hendry, G.S., and Fitz-James, P.C. (1974). Characteristics of phi T, the temperate bacteriophage carried by *Bacillus megaterium* 899a. J. Virol. *13*, 494–499.

Henn, M.R., Sullivan, M.B., Stange-Thomann, N., Osburne, M.S., Berlin, A.M., Kelly, L., Yandava, C., Kodira, C., Zeng, Q., Weiand, M., Sparrow, T., Saif, S., Giannoukos, G., Young, S.K., Nusbaum, C., Birren, B.W., and Chisholm, S.W. (2010). Analysis of high-throughput sequencing and annotation strategies for phage genomes. PLoS One 5, e9083.

Hens, D.K., Chatterjee, N.C., and Kumar, R. (2006). New temperate DNA phage BcP15 acts as a drug resistance vector. Arch. Virol. *151*, 1345–1353.

Hertwig, S., Popp, A., Freytag, B., Lurz, R., and Appel, B. (1999). Generalized transduction of small *Yersinia enterocolitica* plasmids. Appl. Environ. Microbiol. *65*, 3862–3866.

Higgins, M.L., and Lechevalier, M.P. (1969). Poorly lytic bacteriophage from *Dactylosporangium thailandensis* (Actinomycetales). J. Virol. *3*, 210–216.

Hitch, G., Pratten, J., and Taylor, P.W. (2004). Isolation of bacteriophages from the oral cavity. Lett. Appl. Microbiol. *3*, 215–219.

Horiuchi, T., Sakka, M., Hayashi, A., Shimada, T., Kimura, T., and Sakka, K. (2012). Complete genome sequence of bacteriophage BC-611 specifically infecting *Enterococcus faecalis* strain NP-10011. J. Virol. *86*, 9538–9539.

Howe, M.M. (1973). Transduction by bacteriophage MU-1. Virology 55, 103–117.

Hranueli, D., Pigac, J., and Vesligaj, M. (1979). Characterization and persistence of actinophage RP2 isolated from *Streptomyces rimosus* ATCC 10970. J. Gen. Microbiol. *114*, 295–303.

Huang, L.H., Farnet, C.M., Ehrlich, K.C., and Ehrlich, M. (1982). Digestion of highly modified bacteriophage DNA by restriction endonucleases. Nucleic Acids Res. *10*, 1579–1591.

Hurwitz, B.L., Deng, L., Poulos, B.T., and Sullivan, M.B. (2013). Evaluation of methods to concentrate and purify ocean virus communities through comparative, replicated metagenomics. Environ. Microbiol. *15*, 1428–1440.

Hwang, S., Yun, J., Kim, K., Heu, S., Lee, S., and Ryu, S. (2009). Isolation and characterization of bacteriophages specific for *Campylobacter jejuni*. Microbiol. Immunol. *53*, 559–566.

Hyman, P., and Abedon, S.T. (2009). Practical methods for determining phage growth parameters. Methods Mol. Biol. *501*, 175–202.

Hyman, P., and Abedon, S.T. (2010). Bacteriophage host range and bacterial resistance. Adv. Appl. Microbiol. *70*, 217–248.

Ingram, B.O., Masoudi, A., and Raetz, C.R. (2010). *Escherichia coli* mutants that synthesize dephosphorylated lipid A molecules. Biochemistry. *49*, 8325–8337.

Ishiguro, E.E., Ainsworth, T., Shaw, D.H., Kay, W.W., and Trust, T.J. (1983). A lipopolysaccharide-specific bacteriophage for *Aeromonas salmonicida*. Can. J. Microbiol. *29*, 1458–1461.

Jackson, R.W., Vinatzer, B., Arnold, D.L., Dorus, S., and Murillo, J. (2011). The influence of the accessory genome on bacterial pathogen evolution. Mob. Genet. Elements. *1*, 55–65.

Jannière, L., McGovern, S., Pujol, C., Petit, M.A., and Ehrlich, S.D. (1996). *In vivo* analysis of the plasmid pAM beta 1 resolution system. Nucleic Acids Res. *24*, 3431–3436.

Jensen, E.C., Schrader, H.S., Rieland, B., Thompson, T.L., Lee, K.W., Nickerson, K.W., and Kokjohn, T.A. (1998). Prevalence of broad-host-range lytic bacteriophages of *Sphaerotilus natans, Escherichia coli* and *Pseudomonas aeruginosa*. Appl. Environ. Microbiol. *64*, 575–580.

Jiang, S.C., and Paul, J.H. (1998). Gene transfer by transduction in the marine environment. Appl Environ Microbiol. *64*, 2780–2787.

Jin, T., Bokarewa, M., McIntyre, L., Tarkowski, A., Corey, G.R., Reller, L.B., and Fowler, V.G., Jr. (2003). Fatal outcome of bacteraemic patients caused by infection with staphylokinase-deficient *Staphylococcus aureus* strains. J. Med. Microbiol. *52*, 919–923.

John, S.G., Mendez, C.B., Deng, L., Poulos, B., Kauffman, A.K., Kern, S., Brum, J., Polz, M.F., Boyle, E.A., and Sullivan, M.B. (2011). A simple and efficient method for concentration of ocean viruses by chemical flocculation. Environ. Microbiol. Rep. *3*, 195–202.

Jones, J.B., Jackson, L.E., Balogh, B., Obradovic, A., Iriarte, F.B., and Momol, M.T. (2007). Bacteriophages for plant disease control. Annu. Rev. Phytopathol. *45*, 245–262.

Jordana-Lluch, E., Martró Català, E., and Ausina Ruiz, V. (2012). [Mass spectrometry in the clinical microbiology laboratory]. Enferm Infecc Microbiol Clin. *30*, 635–644. [article in Spanish]

Jothikumar, N., Reddy, C.G., Sundari, R.B., and Saigopal, D.V. (2000). Isolation of coliphages specific to enterotoxigenic *E. coli* (ETEC). J. Environ. Monit. *2*, 372–374.

Kahánková, J., Pantůček, R., Goerke, C., Růžičková, V., Holochová, P., and Doškař, J. (2010). Multilocus PCR typing strategy for differentiation of *Staphylococcus aureus* siphoviruses reflecting their modular genome structure. Environ. Microbiol. *12*, 2527–2538.

Kai, L., Rui, B., and Xue-ming Z. (2002). Trans. Tianjin Univ. *8*, 148–151.

Kaiser, D., and Dworkin, M. (1975). Gene transfer to myxobacterium by *Escherichia coli* phage P1. Science *187*, 653–654.

Karlin, S. (1998). Global dinucleotide signatures and analysis of genomic heterogeneity. Curr. Opin. Microbiol. *1*, 598–610.

Kelln, R.A., and Warren, R.A. (1971). Isolation and properties of a bacteriophage lytic for a wide variety of pseudomonads. Can. J. Microbiol. *17*, 677–682.

Kenzaka, T., Tani, K., Sakotani, A., Yamaguchi, N., and Nasu, M. (2007). High-frequency phage-mediated gene transfer among *Escherichia coli* cells, determined at the single-cell level. Appl. Environ. Microbiol. *73*, 3291–3299.

Klein, G., Lindner, B., Brabetz, W., Brade, H., and Raina, S. (2009). *Escherichia coli* K-12 suppressor-free mutants lacking early glycosyltransferases and late acyltransferases: minimal lipopolysaccharide structure and induction of envelope stress response. J. Biol. Chem. *284*, 15369–15389.

Klumpp, J., Dorscht, J., Lurz, R., Bielmann, R., Wieland, M., Zimmer, M., Calendar, R., and Loessner, M.J. (2008). The terminally redundant, nonpermuted genome of *Listeria* bacteriophage A511: A model for the SPO1-like myoviruses of Gram-positive bacteria. J. Bacteriol. *190*, 5753–5765.

Knezevic, P., Kostanjsek, R., Obreht, D., and Petrovic, O. (2009). Isolation of *Pseudomonas aeruginosa* specific phages with broad activity spectra. Curr. Microbiol. *59*, 173–180.

Knirel, Y.A. (1990). Polysaccharide antigens of *Pseudomonas aeruginosa*. Crit. Rev. Microbiol. *17*, 273–304.

Kokjohn, T.A. (1989). Transduction: Mechanism and potential for gene transfer in the environment. In Gene transfer in the environment. S.B. Levy and R.V. Miller, eds. (New York, NY, McGraw-Hill), pp. 73–93.

Koransky, I.H., and Anderson, D.A. (1973). Persistance to coliphage infection induced in *Escherichia coli* by growth in the presence of a surfactant. Appl. Microbiol. *25*, 509–510.

Kramberger, P., Honour, R.C., Herman, R.E., Smrekar, F., and Peterka, M. (2010). Purification of the *Staphylococcus aureus* bacteriophages VDX-10 on methacrylate monoliths. J. Virol. Methods. *166*, 60–64.

Kropinski, A.M. (2006). Phage therapy – everything old is new again. Can. J. Infect. Dis. Med. Microbiol. *17*, 297–306.

Kropinski, A.M. (2009a). Internet resources of interest to bacteriophage workers. Methods Mol. Biol. *502*, 365–370.

Kropinski, A.M. (2009b). Measurement of the rate of attachment of bacteriophage to cells. Methods Mol. Biol. *501*, 151–155.

Kropinski, A.M., Borodovsky, M., Carver, T.J., Cerdeño-Tárraga, A.M., Darling, A., Lomsadze, A., Mahadevan, P., Stothard, P., Seto, D., Van Domselaar, G., and Wishart, D.S. (2009a). *In silico* identification of genes in bacteriophage DNA. Methods Mol. Biol. *502*, 57–89.

Kropinski, A.M., Mazzocco, A., Waddell, T.E., Lingohr, E., and Johnson, R.P. (2009b). Enumeration of bacteriophages by double agar overlay plaque assay. Methods Mol. Biol. *501*, 69–76.

Krüger, D.H., and Schroeder, C. (1981). Bacteriophage T3 and bacteriophage T7 virus–host cell interactions. Microbiol. Rev. *45*, 9–51.

Krupovic, M., Prangishvili, D., Hendrix, R.W., and Bamford, D.H. (2011). Genomics of bacterial and archaeal viruses: dynamics within the prokaryotic virosphere. Microbiol. Mol. Biol. Rev. *75*, 610–635.

Krzywy, T., Kucharewicz-Krukowska, A., and Slopek, S. (1976). Morphologic homogeneity of bacteriophages and their biological activity. Arch. Immunol. Ther. Exp. *24*, 29–37.

Kuhn, J., Suissa, M., Chiswell, D., Azriel, A., Berman, B., Shahar, D., Reznick, S., Sharf, R., Wyse, J., Bar-On, T. *et al.* (2002). A bacteriophage reagent for *Salmonella*: molecular studies on Felix 01. Int. J. Food Microbiol. *74*, 217–227.

Kumari, S., Harjai, K., and Chhibber, S. (2010). Isolation and characterization of *Klebsiella pneumoniae* specific bacteriophages from sewage samples. Folia Microbiol (Praha). *55*, 221–227.

Kunisawa, T., Kanaya, S., and Kutter, E. (1998). Comparison of synonymous codon distribution patterns of bacteriophage and host genomes. DNA Res. *5*, 319–326.

Kutter, E. (2009). Phage host range and efficiency of plating. Methods Mol. Biol. *501*, 141–149.

Kutter, E., Kellenberger, E., Carlson, K., Eddy, S., Neitzel, J., Messinger, L., North, J., and Guttman, B. (1994). Effects of bacterial growth conditions and physiology on T4 infection. In Molecular Biology of Bacteriophage T4, J.D., Karam, ed. (ASM Press, Washington, DC), pp. 406–418.

Kuzio, J., and Kropinski, A.M. (1983). O-antigen conversion in *Pseudomonas aeruginosa* PAO1 by bacteriophage D3. J. Bacteriol. *155*, 203–212.

Kwan, T., Liu, J., DuBow, M., Gros, P., and Pelletier, J. (2005). The complete genomes and proteomes of 27 *Staphylococcus aureus* bacteriophages. Proc. Natl. Acad. Sci. U.S.A. *102*, 5174–5179.

Kwiatek, M., Parasion, S., Mizak, L., Gryko, R., Bartoszcze, M., and Kocik, J. (2012). Characterization of a bacteriophage, isolated from a cow with mastitis, that is lytic against *Staphylococcus aureus* strains. Arch. Virol. *157*, 225–234.

Kwoh, D.Y., and Kemper, J. (1978). Bacteriophage P22-mediated specialized transduction in *Salmonella typhimurium*: high frequency of aberrant prophage excision. J. Virol. *27*, 519–534.

Kylberg, K.J., Bendig, M.M., and Drexler, H. (1975). Characterization of transduction by bacteriophage T1: time of production and density of transducing particles. J. Virol. *16*, 854–858.

Labrie, S.J., Samson, J.E., and Moineau, S. (2010). Bacteriophage resistance mechanisms. Nat. Rev. Microbiol. *8*, 317–327.

Lamont, I., Kalionis, B., and Egan, J.B. (1988). Control of gene expression in the P2-related temperate coliphages. V. The use of sequence analysis of 186 *vir* mutants to indicate presumptive repressor binding sites. J. Mol. Biol. *199*, 379–382.

Lan, R., Stevenson, G., Donohoe, K., Ward, L., and Reeves, P.R. (2007). Molecular markers with potential to replace phage typing for *Salmonella enterica* serovar *typhimurium*. J. Microbiol. Methods *68*, 145–156.

Lavigne, R., Sun, W.D., and Volckaert, G. (2004). PHIRE, a deterministic approach to reveal regulatory elements in bacteriophage genomes. Bioinformatics *20*, 629–635.

Lavigne, R., Noben, J.P., Hertveldt, K., Ceyssens, P.J., Briers, Y., Dumont, D., Roucourt, B., Krylov, V.N., Mesyanzhinov, V.V., Robben, J., and Volckaert, G. (2006). The structural proteome of *Pseudomonas aeruginosa* bacteriophage phiKMV. Microbiology *152*, 529–534.

Lavigne, R., Seto, D., Mahadevan, P., Ackermann, H.W., and Kropinski, A.M. (2008). Unifying classical and molecular taxonomic classification: analysis of the *Podoviridae* using BLASTP-based tools. Res. Microbiol. *159*, 406–414.

Lavigne, R., Ceyssens, P.J., and Robben, J. (2009a). Phage proteomics: applications of mass spectrometry. Methods Mol. Biol. *502*, 239–251.

Lavigne, R., Darius, P., Summer, E.J., Seto, D., Mahadevan, P., Nilsson, A.S., Ackermann, H.W., and Kropinski, A.M. (2009b). Classification of *Myoviridae* bacteriophages using protein sequence similarity. BMC Microbiol. *9*, 224–241.

Lavigne, R., Villegas, A., and Kropinski, A.M. (2009c). In silico characterization of DNA motifs with particular reference to promoters and terminators. Methods Mol. Biol. *502*, 113–129.

Lazzaro, B.P., Sceurman, B.K., Carney, S.L., and Clark, A.G. (2002). fRFLP and fAFLP: medium-throughput genotyping by fluorescently post-labeling restriction digestion. Biotechniques. *33*, 539–540, 542, 545–546.

Lee, C.Y., and Iandolo, J.J. (1985). Mechanism of bacteriophage conversion of lipase activity in *Staphylococcus aureus*. J. Bacteriol. *164*, 288–293.

Leffers, G.G. Jr., and Gottesman, S. (1998). Lambda Xis degradation *in vivo* by Lon and FtsH. J. Bacteriol. *180*, 1573–1577.

Lehman, S.M., Kropinski, A.M., Castle, A.J., and Svircev, A.M. (2009). Complete genome of the broad-host-range *Erwinia amylovora* phage phiEa21-4 and its relationship to *Salmonella* phage Félix O1. Appl. Environ. Microbiol. *75*, 2139–2147.

Lehnherr, H. (2005). Bacteriophage P1. In The bacteriophages, 2nd edn, R. Calendar, ed. (Oxford University Press, New York, N.Y.), pp. 350–364.

Leplae, R., Lima-Mendez, G., and Toussaint, A. (2010) ACLAME: a CLAssification of Mobile genetic Elements, update 2010. Nucleic Acids Res. *38*, D57–61.

Letkiewicz, S., Międzybrodzki, R., Fortuna, W., Weber-Dąbrowska, B., and Górski, A. (2009). Eradication of *Enterococcus faecalis* by phage therapy in chronic prostatitis – case report. Folia Microbiol. *54*, 457–461.

Levin, B.R., and Bull, J.J. (2004). Population and evolutionary dynamics of phage therapy. Nat. Rev. Microbiol. *2*, 166–173.

Li, L., Yang, H., Lin, S., and Jia, S. (2010). Classification of 17 newly isolated virulent bacteriophages of *Pseudomonas aeruginosa*. Can. J. Microbiol. *56*, 925–933.

Liebeschuetz, J., and Ritchie, D.A. (1986). Phage T1-mediated transduction of a plasmid containing the T1 *pac* site. J. Mol. Biol. *192*, 681–692.

Lima-Mendez, G., Toussaint, A., and Leplae, R. (2011). A modular view of the bacteriophage genomic space: identification of host and lifestyle marker modules. Res. Microbiol. *162*, 737–746.

Lima-Mendez, G., Van Helden, J., Toussaint, A., and Leplae, R. (2008). Prophinder: a computational tool for prophage prediction in prokaryotic genomes. Bioinformatics *24*, 863–865.

Lingohr, E.J., Villegas, A., She, Y.M., Ceyssens, P.J., and Kropinski, A.M. (2008). The genome and proteome of the *Kluyvera bacteriophage* Kvp1-another member of the T7-like *Autographivirinae*. Virol. J. *5*, 122.

Liu, K., Wen, Z., Li, N., Yang, W., Hu, L., Wang, J., Yin, Z., Dong, X., and Li, J. (2012). Purification and concentration of mycobacteriophage D29 using monolithic chromatographic columns. J. Virol. Methods. *186*, 7–13.

Livny, J., and Friedman, D.I. (2004). Characterizing spontaneous induction of Stx encoding phages using a selectable reporter system. Mol. Microbiol. *51*, 1691–1704.

Ljungquist, E., Kockum, K., and Bertani, L.E. (1984). DNA sequences of the repressor gene and operator region of bacteriophage P2. Proc. Natl. Acad. Sci. U.S.A. *81*, 3988–3992.

Łobocka, M.B., Svarchevsky, A.N., Rybchin, V.N., and Yarmolinsky, M.B. (1996). Characterization of the primary immunity region of the *Escherichia coli* linear plasmid prophage N15. J. Bacteriol. *178*, 2902–2910.

Łobocka, M.B., Rose, D.J., Plunkett, G. 3rd, Rusin, M., Samojedny, A., Lehnherr, H., Yarmolinsky, M.B., and Blattner, F.R. (2004). Genome of bacteriophage P1. J. Bacteriol. *186*, 7032–7068.

Łobocka, M., Hejnowicz, M.S., Dąbrowski, K., Gozdek, A., Kosakowski, J., Witkowska, M., Ulatowska, M.I., Weber-Dąbrowska, B., Kwiatek, M., Parasion, S., Gawor, J., Kosowska, H., and Głowacka, A. (2012). Genomics of staphylococcal Twort-like phages – potential therapeutics of the post-antibiotic era. Adv Virus Res. *83*, 143–216.

Lockington, R.A., Attwood, G.T., and Brooker, J.D. (1988). Isolation and characterization of a temperate bacteriophage from the ruminal anaerobe *Selenomonas ruminantium*. Appl. Environ. Microbiol. *54*, 1575–1580.

Loeffler, J.M., and Fischetti, V.A. (2006). Lysogeny of *Streptococcus pneumoniae* with MM1 phage: improved adherence and other phenotypic changes. Infect. Immun. *74*, 4486–4495.

Łoś, J.M., Golec, P., Węgrzyn, G., Węgrzyn, A., and Łoś, M. (2008). Simple method for plating *Escherichia coli* bacteriophages forming very small plaques or no plaques under standard conditions. Appl. Environ. Microbiol. *74*, 5113–5120.

Luchansky, J.B., Kleeman, E.G., Raya, R.R., and Klaenhammer, T.R. (1989). Genetic transfer systems for delivery of plasmid deoxyribonucleic acid to *Lactobacillus acidophilus* ADH: conjugation, electroporation, and transduction. J. Dairy Sci. 72, 1408–1417.

Lucks, J.B., Nelson, D.R., Kudla, G.R., and Plotkin, J.B. (2008). Genome landscapes and bacteriophage codon usage. PLoS Comput. Biol. *4*(2):e1000001.

Luo, C.H., Chiou, P.Y., Yang, C.Y., and Lin, N.T. (2012). Genome, integration and transduction of a novel temperate phage of *Helicobacter pylori*. J. Virol. *86*, 8781–8792.

Łusiak-Szelachowska, M., Annabhani, A., Weber-Dąbrowska, B., Górski, A., Bębenek, M., Pudełko, M., Strutyńska-Karpińska, M., Muszyński, J., and Paradowski, L. (2008). *Escherichia coli* bacteriophages in human stool of patients with gastrointestinal tract diseases. Polish Gastroenterol. *15*, 87–90.

McLaughlin, M.R. (2007). Simple colorimetric microplate test of phage lysis in *Salmonella enterica*. J. Microbiol. Methods *69*, 394–398.

McLaughlin, M.R., and Balaa, M.F. (2006). Enhanced contrast of bacteriophage plaques in *Salmonella* with ferric ammonium citrate and sodium thiosulfate (FACST) and tetrazolium red (TZR). J. Microbiol. Methods. *65*, 318–323.

McNair, K., Bailey, B.A., and Edwards, R.A. (2012). PHACTS, a computational approach to classifying the lifestyle of phages. Bioinformatics *28*, 614–618.

McShan, W.M., and Feretti, J.J. (2007). Prophages and their contribution to host cell phenotype. In Bacteriophage Genetics and Molecular Biology, S. McGrath and D. van Sideren, eds. (Norfolk, UK: Caister Academic Press), pp. 229–250.

Madera, C., García, P., Rodríguez, A., Suárez, J.E., and Martínez, B. (2009). Prophage induction in *Lactococcus lactis* by the bacteriocin Lactococcin 972. Int. J. Food Microbiol. *129*, 99–102.

Mankiewicz, E., and Liivak, M. (1967). Mycobacteriophages isolated from human sources. Nature *216*, 485–486.

Mann, B.A., and Slauch, J.M. (1997). Transduction of low-copy number plasmids by bacteriophage P22. Genetics *146*, 447–456.

Mann, E.E., and Wozniak, D.J. (2012). *Pseudomonas* biofilm matrix composition and niche biology. FEMS Microbiol. Rev. *36*, 893–916.

Marrero, R., and Lovett, P.S. (1982). Interference of plasmid pCM194 with lysogeny of bacteriophage SP02 in *Bacillus subtilis*. J. Bacteriol. *152*, 284–290.

Masters, M. (1996). Generalized transduction. In *Escherichia coli* and *Salmonella typhimurium*: Cellular and Molecular Biology, 2nd edn., F.C. Neidhardt, R. Curtiss III, J.L. Ingraham, E.C.C. Lin, K.B. Low, B. Magasanik, W.S. Reznikopp, M. Riley, M. Schaechter, and H.E. Umbarger, eds. (Washington, DC: American Society for Microbiology), pp. 2421–2448.

Masters, M. (2004). Transduction: host DNA transfer by bacteriophages. In The Desk Encyclopedia of Microbiology, M. Schaechter, ed. (London, UK: Elsevier Academic Press), pp. 1000–1012.

Maxted, W.R. (1952). Enhancement of streptococcal bacteriophage lysis by hyaluronidase. Nature *170*, 1020–1021.

Maxwell, K.L., and Frappier, L. (2007). Viral proteomics. Microbiol. Mol. Biol. Rev. *71*, 398–411.

Mayeux, P.R. (1997). Pathobiology of lipopolysaccharide. J. Toxicol. Environ. Health. *51*, 415–435.

Mazzocco, A., Waddell, T.E., Lingohr, E., and Johnson, R.P. (2009a). Enumeration of bacteriophages using the small drop plaque assay system. Methods Mol. Biol. *501*, 81–85.

Mazzocco, A., Waddell, T.E., Lingohr, E., and Johnson, R.P. (2009b). Enumeration of bacteriophages by the direct plating plaque assay. Methods Mol. Biol. *501*, 77–80.

Merabishvili, M., Verhelst, R., Glonti, T., Chanishvili, N., Krylov, V., Cuvelier, C., Tediashvili, M., and

Vaneechoutte, M. (2007). Digitized fluorescent RFLP analysis (fRFLP) as a universal method for comparing genomes of culturable dsDNA viruses: application to bacteriophages. Res. Microbiol. *158*, 572–581.

Merabishvili, M., Pirnay, J.P., Verbeken, G., Chanishvili, N., Tediashvili, M., Lashkhi, N., Glonti, T., Krylov, V., Mast, J., Van Parys, L., Lavigne, R., Volckaert, G., Mattheus, W., Verween, G., De Corte, P., Rose, T., Jennes, S., Zizi, M., De Vos, D., and Vaneechoutte, M. (2009). Quality-controlled small-scale production of a well-defined bacteriophage cocktail for use in human clinical trials. PLoS One *4*, e4944.

Mercanti, D.J., Carminati, D., Reinheimer, J.A., and Quiberoni, A. (2011). Widely distributed lysogeny in probiotic lactobacilli represents a potentially high risk for the fermentative dairy industry. Int. J. Food Microbiol. *144*, 503–510.

Mercenier, A., Slos, P., Faelen, M., and Lecocq, J.P. (1988). Plasmid transduction in *Streptococcus thermophilus*. Mol. Gen. Genet. *212*, 386–389.

Międzybrodzki, R., Borysowski, J., Weber-Dąbrowska, B., Fortuna, W., Letkiewicz, S., Szufnarowski, K., Pawełczyk, Z., Rogóż, P., Kłak, M., Wojtasik, E., and Górski, A. (2012). Clinical aspects of phage therapy. Adv. Virus Res. *83*, 73–121.

Miller, E.S., Heidelberg, J.F., Eisen, J.A., Nelson, W.C., Durkin, A.S., Ciecko, A., Feldblyum, T.V., White, O., Paulsen, I.T., Nierman, W.C., Lee, J., Szczypinski, B., and Fraser, C.M. (2003a). Complete genome sequence of the broad-host-range vibriophage KVP40: comparative genomics of a T4-related bacteriophage. J. Bacteriol. *185*, 5220–5233.

Miller, E.S., Kutter, E., Mosig, G., Arisaka, F., Kunisawa, T., and Rüger, W. (2003b). Bacteriophage T4 genome. Microbiol. Mol. Biol. Rev. *67*, 86–156.

Miller, R.V. (2001). Environmental bacteriophage–host interactions: factors contribution to natural transduction. Antonie Van Leeuwenhoek. *79*, 141–147.

Miwa, T., and Matsubara, K. (1983). Formation of oligomeric structures from plasmid DNA carrying *cos* lambda that is packaged into bacteriophage lambda heads. J. Bacteriol. *153*, 100–108.

Moreno, F. (1977). Development of bacteriophage phi29 in sporulating and non-sporulating cells of *Bacillus subtilis* 168. Ann. Microbiol. (Paris). *128B*, 3–18.

Morgan, A.F. (1979). Transduction of *Pseudomonas aeruginosa* with a mutant of bacteriophage E79. J. Bacteriol. *139*, 137–140.

Morino, T., and Takahashi, H. (1996). Transduction of a cosmid with the R4 phage *cos* sequence by heterogeneous actinophages, SPA10 and SPA38. Biosci. Biotechnol. Biochem. *60*, 2076–2077.

Morino, T., Takahashi, H., and Saito, H. (1985). Construction and characterization of a cosmid of *Streptomyces lividans*. Mol. Gen. Genet. *198*, 228–233.

Mullan, M.W.A. (1979). Lactic streptococcal bacteriophage enumeration. A review of factors affecting plaque formation. Dairy Industries International *44*, 11–14.

Murphy, K.C. (2012). Phage recombinases and their applications. Adv. Virus Res. *83*, 367–414.

Murray, N.E., Brammar, W.J., and Murray, K. (1977). Lambdoid phages that simplify the recovery of *in vitro* recombinants. Mol. Gen. Genet. *150*, 53–61.

Nakaminami, H., Noguchi, N., Nishijima, S., Kurokawa, I., So, H., and Sasatsu, M. (2007). Transduction of the plasmid encoding antiseptic resistance gene *qacB* in *Staphylococcus aureus*. Biol. Pharm. Bull. *30*, 1412–1415.

Nakamura, M., Horiuchi, S., and Nakaya, R. (1975). Comparative studies on generalized transducing bacteriophages of *Proteus mirabilis*, phim and pi1. Jpn. J. Microbiol. *19*, 123–131.

Nelson, K.E., Weinel, C., Paulsen, I.T., Dodson, R.J., Hilbert, H., Martins dos Santos, V.A., *et al.* (2002). Complete genome sequence and comparative analysis of the metabolically versatile *Pseudomonas putida* KT2440. Environ. Microbiol. *4*, 799–808.

Ni, B., Du, Z., Guo, Z., Zhang, Y., and Yang, R. (2008). Curing of four different plasmids in *Yersinia pestis* using plasmid incompatibility. Lett. Appl. Microbiol. *47*, 235–240.

Nirmal Kumar, G.P., Sundarrajan, S., Paul, V.D., Nandini, S., Saravanan, R.S., Hariharan, S., Sriram, B., and Padmanabhan, S. (2012). Use of prophage free host for achieving homogenous population of bacteriophages: New findings. Virus Res. *169*, 182–187.

Nordström, K., Forsgren, A., and Cox, P. (1974). Prevention of bacteriophage adsorption to *Staphylococcus aureus* by immunoglobulin G. J. Virol. *14*, 203–206.

Novick, R.P. (2003). Mobile genetic elements and bacterial toxinoses: the superantigen-encoding pathogenicity islands of *Staphylococcus aureus*. Plasmid. *49*, 93–105.

Nunes-Düby, S.E., Kwon, H.J., Tirumalai, R.S., Ellenberger, T., and Landy, A. (1998). Similarities and differences among 105 members of the Int family of site-specific recombinases. Nucleic Acids Res. *26*, 391–406.

Ogunseitan, O.A., Sayler, G.S., and Miller, R.V. (1992). Application of DNA probes to analysis of bacteriophage distribution patterns in the environment. Appl. Environ. Microbiol. *58*, 2046–2052.

Ohnishi, M., Kurokawa, K., and Hayashi, T. (2001). Diversification of *Escherichia coli* genomes: are bacteriophages the major contributors? Trends Microbiol. *9*, 481–485.

Oksanen, H.M., Domanska, A., and Bamford, D.H. (2012). Monolithic ion exchange chromatographic methods for virus purification. Virology *434*, 271–277.

Oliveira, A., Sillankorva, S., Quinta, R., Henriques, A., Sereno, R., and Azeredo, J. (2009). Isolation and characterization of bacteriophages for avian pathogenic *E. coli* strains. J. Appl. Microbiol. *106*, 1919–1927.

Olszewski, J., Winona, L., and Oshima, K.H. (2005). Comparison of 2 ultrafiltration systems for the concentration of seeded viruses from environmental waters. Can. J. Microbiol. *51*, 295–303.

Oppenheim, A.B., Kobiler, O., Stavans, J., Court, D.L., and Adhya, S. (2005). Switches in bacteriophage lambda development. Annu. Rev. Genet. *39*, 409–429.

Panis, G., Mejean, V., and Ansaldi, M. (2007). Control and regulation of KplE1 prophage site-specific recombination: a new recombination module analyzed. J. Biol. Chem. *282*, 21798–21809.

Pantůcek, R., Doskar, J., Růzicková, V., Kaspárek, P., Orácová, E., Kvardová, V., and Rosypal, S. (2004). Identification of bacteriophage types and their carriage in *Staphylococcus aureus*. Arch. Virol. *149*, 1689–1703.

Pathria, S., Rolando, M., Lieman, K., Hayes, S., Hardies, S., and Serwer, P. (2012). Islands of non-essential genes, including a DNA translocation operon, in the genome of bacteriophage 0305ϕ8-36. Bacteriophage *2*, 25–35.

Paul, J.H. (2008). Prophages in marine bacteria: dangerous molecular time bombs or the key to survival in the seas? ISME J. *2*, 579–589.

Payne, R.J.H., and Jansen, V.A.A. (2000). Phage therapy: The peculiar kinetics of self-replicating pharmaceuticals. Clin. Pharmacol. Ther. *68*, 225–230.

Peck, M.W. (2009). Biology and genomic analysis of *Clostridium botulinum*. Adv. Microb. Physiol. *55*, 183–265, 320.

Pedulla, M.L., Ford, M.E., Houtz, J.M., Karthikeyan, T., Wadsworth, C., Lewis, J.A., Jacobs-Sera, D., Falbo, J., Gross, J., Pannunzio, N.R., Brucker, W., Kumar, V., Kandasamy, J., Keenan, L., Bardarov, S., Kriakov, J., Lawrence, J.G., Jacobs, W.R., Hendrix, R.W., and Hatfull, G.F. (2003). Origins of highly mosaic mycobacteriophage genomes. Cell *113*, 171–182.

Peña, C.E., Stoner, J., and Hatfull, G.F. (1998). Mycobacteriophage D29 integrase-mediated recombination: specificity of mycobacteriophage integration. Gene. *225*, 143–151.

Pham, T.T., Jacobs-Sera, D., Pedulla, M.L., Hendrix, R.W., and Hatfull, G.F. (2007). Comparative genomic analysis of mycobacteriophage Tweety: evolutionary insights and construction of compatible site-specific integration vectors for mycobacteria. Microbiology *153*, 2711–2723.

Pickard, D., Thomson, N.R., Baker, S., Wain, J., Pardo, M., Goulding, D., Hamlin, N., Choudhary, J., Threfall, J., and Dougan, G. (2008). Molecular characterization of the *Salmonella enterica* serovar Typhi Vi-typing bacteriophage E1. J. Bacteriol. *190*, 2580–2587.

Piepersberg, W., Bräu, B., and Davies, J. (1983). High frequency transduction of R-factor encoded gentamicin resistance by bacteriophage P1Cm. J. Antibiot. (Tokyo). *36*, 588–594.

Prell, H.H., Auer, J., Soska, J., and Hava, P. (1984). The immI region of *Salmonella* phage P22. 1. Characterisation of the mutants P22 cir4–1, P22 cir5–1 and P22 cir6–1. Mol. Gen. Genet. *198*, 105–109.

Proux, C., van Sinderen, D., Suarez, J., Garcia, P., Ladero, V., Fitzgerald, G.F., Desiere, F., and Brüssow, H. (2002). The dilemma of phage taxonomy illustrated by comparative genomics of Sfi21-like *Siphoviridae* in lactic acid bacteria. J. Bacteriol. *184*, 6026–6036.

Puapermpoonsiri, U., Spencer, J., and van der Walle, C.F. (2009). A freeze-dried formulation of bacteriophage encapsulated in biodegradable microspheres. Eur. J. Pharm. Biopharm. *72*, 26–33.

Puapermpoonsiri, U., Ford, S.J., and van der Walle, C.F. (2010). Stabilization of bacteriophage during freeze drying. Int. J. Pharm. *389*, 168–175.

Raetz, C.R., and Whitfield, C. (2002). Lipopolysaccharide endotoxins. Annu. Rev. Biochem. *71*, 635–700.

Rakhuba, D.V., Kolomiets, E.I., Dey, E.S., and Novik, G.I. (2010). Bacteriophage receptors, mechanisms of phage adsorption and penetration into host cell. Pol. J. Microbiol. *59*, 145–155.

Ravin, N.V. (2011). N15: the linear phage-plasmid. Plasmid. *65*, 102–109.

Raya, R.R., and H'bert, E.M. (2009). Isolation of phage via induction of lysogens. Methods Mol. Biol. *501*, 23–32.

Raya, R.R., and Klaenhammer, T.R. (1992). High-frequency plasmid transduction by *Lactobacillus gasseri* bacteriophage phiadh. Appl. Environ. Microbiol. *58*, 187–193.

Reese, J.F., Dimitracopoulos, G., and Bartell, P.F. (1974). Factors influencing the adsorption of bacteriophage 2 to cells of *Pseudomonas aeruginosa*. J. Virol. *13*, 22–27.

Rippon, J.E. (1956). The classification of bacteriophages lysing staphylococci. J. Hyg. (Lond.) *54*, 213–226.

Ristl, R., Steiner, K., Zarschler, K., Zayni, S., Messner, P., and Schäffer, C. (2011). The s-layer glycome-adding to the sugar coat of bacteria. Int. J. Microbiol. *2011*, pii: 127870.

Robb, S.M., Woods, D.R., and Robb, F.T. (1978). Phage growth characteristics on stationary phase *Achromobacter* cells. J. Gen. Virol. *41*, 265–272.

Roberts, I.S. (1995). Bacterial polysaccharides in sickness and in health. The 1995 Fleming Lecture. Microbiology *141*, 2023–2031.

Roberts, M.D., and Drexler, H. (1981a). Isolation and genetic characterization of T1-transducing mutants with increased transduction frequency. Virology *112*, 662–669.

Roberts, M.D., and Drexler, H. (1981b). T1 mutants with increased transduction frequency are defective in host chromosome degradation. Virology *112*, 670–677.

Rohwer, F. (2003). Global Phage Diversity. Cell, *113*, 141.

Rohwer, F., and Edwards, R. (2002). The Phage Proteomic Tree: a genome-based taxonomy for phage. J. Bacteriol. *184*, 4529–4535.

Roman, B.L., Pham, V.N., Bennett, P.E., and Weinstein, B.M. (1999). Non-radioisotopic AFLP method using PCR primers fluorescently labeled with Cy5. Biotechniques. *26*, 236–238.

Romero, P., Croucher, N.J., Hiller, N.L., Hu, F.Z., Ehrlich, G.D., Bentley, S.D., García, E., and Mitchell, T.J. (2009). Comparative genomic analysis of ten *Streptococcus pneumoniae* temperate bacteriophages. J. Bacteriol. *191*, 4854–4862.

Ronning, C.M., Losada, L., Brinkac, L., Inman, J., Ulrich, R.L., Schell, M., Nierman, W.C., and Deshazer, D. (2010). Genetic and phenotypic diversity in *Burkholderia*: contributions by prophage and phage-like elements. BMC Microbiol. *10*, 202.

Ross, I.L., and Heuzenroeder, M.W. (2005). Discrimination within phenotypically closely related definitive types of *Salmonella enterica* serovar Typhimurium by the multiple amplification of phage locus typing technique. J. Clin. Microbiol. *43*, 1604–1611.

Ruhfel, R.E., Robillard, N.J., and Thorne, C.B. (1984). Interspecies transduction of plasmids among *Bacillus anthracis*, *B. cereus*, and *B. thuringiensis*. J. Bacteriol. *157*, 708–711.

Rychlik, I., Hradecka, H., and Malcova, M. (2008). *Salmonella enterica* serovar Typhimurium typing by prophage-specific PCR. Microbiology *154*, 1384–1389.

Sakaguchi, Y., Hayashi, T., Kurokawa, K., Nakayama, K., Oshima, K., Fujinaga, Y., Ohnishi, M., Ohtsubo, E., Hattori, M., and Oguma, K. (2005). The genome sequence of *Clostridium botulinum* type C neurotoxin-converting phage and the molecular mechanisms of unstable lysogeny. Proc. Natl. Acad. Sci. U.S.A. *102*, 17472–17477.

Sambrook, J., Fritsch, E.F., and Maniatis, T. (1989). Molecular Cloning: a Laboratory Manual, 2nd edn. Cold Spring Harbor Laboratory, Cold Spring Harbor, New York.

Sander, M., and Schmieger, H. (2001). Method for host-independent detection of generalized transducing bacteriophages in natural habitats. Appl. Environ. Microbiol. *67*, 1490–1493.

Sau, K., Sau, S., Mandal, S.C., and Ghosh, T.C. (2005). Factors influencing the synonymous codon and amino acid usage bias in AT-rich *Pseudomonas aeruginosa* phage PhiKZ. Acta Biochim. Biophys. Sin. (Shanghai). *37*, 625–633.

Saussereau, E., and Debarbieux, L. (2012). Bacteriophages in the experimental treatment of *Pseudomonas aeruginosa* infections in mice. Adv. Virus Res. *83*, 123–141.

Säwström, C., Lisle, J., Anesio, A.M., Priscu, J.C., and Laybourn-Parry, J. (2008). Bacteriophage in polar inland waters. Extremophiles *12*, 167–175.

Saye, D.J., Ogunseitan, O., Sayler, G.S., and Miller, R.V. (1987). Potential for transduction of plasmids in a natural freshwater environment: effect of plasmid donor concentration and a natural microbial community on transduction in *Pseudomonas aeruginosa*. Appl. Environ. Microbiol. *53*, 987–995.

Schmid, E.N., von Recklinghausen, G., and Ansorg, R. (1990). Bacteriophages in *Helicobacter* (*Campylobacter*) *pylori*. J. Med. Microbiol. 32, 101–104.

Scholl, D., and Merril, C. (2005). Polysaccharide-degrading phages. In Phages. Their role in bacterial pathogenesis and biotechnology. M.K. Waldor, D.I. Friedman, and S.L. Adhya, eds. (Washington DC, ASM Press), pp. 400–414.

Scholl, D., Adhya, S., and Merril, C. (2005). *Escherichia coli* K1's capsule is a barrier to bacteriophage T7. Appl. Environ. Microbiol. *71*, 4872–4874.

Schrader, H.S., Schrader, J.O., Walker, J.J., Wolf, T.A., Nickerson, K.W., and Kokjohn, T.A. (1997). Bacteriophage infection and multiplication occur in *Pseudomonas aeruginosa* starved for 5 years. Can. J. Microbiol. *43*, 1157–1163.

Scott, J.R. (1970). Clear plaque mutants of phage P1. Virology *41*, 66–71.

Selva, L., Viana, D., Regev-Yochay, G., Trzcinski, K., Corpa, J.M., Lasa, I., Novick, R.P., and Penadés, J.R. (2009). Killing niche competitors by remote-control bacteriophage induction. Proc. Natl. Acad. Sci. U.S.A. *106*, 1234–1238.

Serwer, P., Graef, P.R., and Garrison, P.N. (1978). Use of ethidium bromide fluorescence enhancement to detect duplex DNA and DNA bacteriophages during zone sedimentation in sucrose gradients: molecular weight of DNA as a function of sedimentation rate. Biochemistry *17*, 1166–1170.

Serwer, P., Hayes, S.J., Zaman, S., Lieman, K., Rolando, M., and Hardies, S.C. (2004). Improved isolation of undersampled bacteriophages: finding of distant terminase genes. Virology *329*, 412–424.

Serwer, P., Hayes, S.J., Thomas. J.A., Demeler, B., and Hardies, S.C. (2009). Isolation of novel large and aggregating bacteriophages. Methods Mol. Biol. *501*, 55–66.

Shaburova, O.V., Krylov, S.V., Veĭko, V.P., Pleteneva, E.A., Burkal'tseva, M.V., Miroshnokov, K.A., Cornelissen, A., Lavigne, R., Sykilinda, N.N., Kadykov, V.A., Mesianzhinov, V.V., Volckaert, G., and Krylov, V.N. (2009). Search for destruction factors of bacterial biofilms: comparison of phage properties in a group of *Pseudomonas putida* bacteriophages and specificity of their halo-formation products. Genetika. *45*, 185–193. Russian.

Shao, Y., and Wang, I.N. (2008). Bacteriophage adsorption rate and optimal lysis time. Genetics *180*, 471–482.

Sharp, P.M. (1986). Molecular evolution of bacteriophages: evidence of selection against the recognition sites of host restriction enzymes. Mol. Biol. Evol. *3*, 75–83.

Shevchenko, A., Sunyaev, S., Loboda, A., Shevchenko, A., Bork, P., Ens, W., and Standing, K.G. (2001). Charting the proteomes of organisms with unsequenced genomes by MALDI-quadrupole time-of-flight mass spectrometry and BLAST homology searching. Anal. Chem. *73*, 1917–1926.

Shi, T., and Friedman, D.I. (2001). The operator-early promoter regions of Shiga-toxin bearing phage H-19B. Mol. Microbiol. *41*, 585–599.

Shimizu-Kadota, M., and Sakurai, T. (1982). Prophage curing in *Lactobacillus casei* by isolation of a thermoinducible mutant. Appl. Environ. Microbiol. *43*, 1284–1287.

Shivu, M.M., Rajeeva, B.C., Girisha, S.K., Karunasagar, I., Krohne, G., and Karunasagar, I. (2007). Molecular characterization of *Vibrio harveyi* bacteriophages isolated from aquaculture environments along the coast of India. Environ. Microbiol. *9*, 322–331.

Shope, R.E. (1927). Bacteriophages isolated from the common house fly (*Musca dometica*). J. Exp. Med. *45*, 1037–1044.

Shukla, S.K., Pantrangi, M., Stahl, B., Briska, A.M. Stemper, M.E., Wagner, T.K., Zentz, E.B., Callister S.M., Lovrich, S.D., Henkhaus, J.K., and Dykes C.W. (2012). Comparative whole genome mapping to determine *Staphylococcus aureus* genome size, virulence motifs and clonality. J. Clin. Microbiol. *50*, 3526–3533.

Smith, H.W., Huggins, M.B., and Shaw, K.M. (1987). Factors influencing the survival and multiplication of bacteriophages in calves and in their environment. J. Gen. Microbiol. *133*, 1127–1135.

Smith, M.C., and Thorpe, H.M. (2002). Diversity in the serine recombinases. Mol. Microbiol. *44*, 299–307.

Smrekar, F., Ciringer, M., Peterka, M., Podgornik, A., and Strancar, A. (2008). Purification and concentration of bacteriophage T4 using monolithic chromatographic supports. J. Chromatogr. B Analyt. Technol. Biomed. Life Sci. *861*, 177–180.

Smrekar, F., Ciringer, M., Jančar, J., Raspor, P., Strancar, A., and Podgornik, A. (2011). Optimization of lytic phage manufacturing in bioreactor using monolithic supports. J. Sep. Sci. *34*, 2152–2158.

Soothill, J.S. (1992). Treatment of experimental infections of mice with bacteriophages. J. Med. Microbiol. *37*, 258–261.

Spengler, G., Miczák, A., Hajdú, E., Kawase, M., Amaral, L., and Molnár, J. (2003). Enhancement of plasmid curing by 9-aminoacridine and two phenothiazines in the presence of proton pump inhibitor 1-(2-benzoxazolyl)-3,3,3-trifluoro-2-propanone. Int. J. Antimicrob. Agents. *22*, 223–227.

Spengler, G., Molnár, A., Schelz, Z., Amaral, L., Sharples, D., and Molnár, J. (2006). The mechanism of plasmid curing in bacteria. Curr. Drug Targets. *7*, 823–841.

Srinivasiah, S., Bhavsar, J., Thapar, K., Liles, M., Schoenfeld, T., and Wommack, K.E. (2008). Phages across the biosphere: contrasts of viruses in soil and aquatic environments. Res. Microbiol. *159*, 349–357.

Srividhya, K.V., Alaguraj, V., Poornima, G., Kumar, D., Singh, G.P., Raghavenderan, L., Katta, A.V., Mehta, P., and Krishnaswamy, S. (2007). Identification of prophages in bacterial genomes by dinucleotide relative abundance difference. PLoS One *2*, e1193.

Stanisich, V. (1988). Identification and analysis of plasmids at the genetic level. In Methods in Microbiology, vol. 21, J. Grinsted and P.M. Bennett, eds. (Academic Press Inc., San Diego, CA), pp. 11–48.

Stepanov, A.S., Puzanova, O.B., Gavrilov, S.V., Brandzishevskiĭ, Iu.V., Bragin, I.V., and Anisimov, P.I. (1989). Transduction and conjugation transfer of the pXO2 plasmid in *Bacillus anthracis*. Mol. Gen. Mikrobiol. Virusol. *12*, 39–43. [Russian]

Sternberg, N.L., and Maurer, R. (1991). Bacteriophage-mediated generalized transduction in *Escherichia coli* and *Salmonella typhimurium*. Methods Enzymol. *204*, 18–43.

Steward, G.F., and Culley, A.I. (2010). Extraction and purification of nucleic acids from viruses. In Manual of Aquatic Viral Ecology. S.W. Wilhelm, M.G. Weinbauer, and C.A. Suttle, eds. (Waco, TX: ASLO), pp. 154–165.

Stewart, C.R., Casjens, S.R., Cresawn, S.G., Houtz, J.M., Smith, A.L., Ford, M.E., Peebles, C.L., Hatfull, G.F., Hendrix, R.-W., Huang, W.M., and Pedulla, M.L. (2009). The genome of *Bacillus subtilis* bacteriophage SPO1. J. Mol. Biol. *388*, 48–70.

Sulakvelidze, A., and Barrow, P. (2005). Phage therapy in animals and agribusiness. In Bacteriophages: Biology and Applications. E.B. Kutter and A. Sulakvelidze, eds. (Boca Raton, FL, CRC Press), pp. 335–380.

Sulakvelidze, A., and Kutter, E. (2005). Phage therapy in humans. In Bacteriophages: Biology and Applications. E.B. Kutter and A. Sulakvelidze, eds. (Boca Raton, FL, CRC Press), pp. 381–436.

Szczepańska, A.K. (2009). Bacteriophage-encoded functions engaged in initiation of homologous recombination events. Crit. Rev. Microbiol. *35*, 197–220.

Takahashi, A., Yomoda, S., Kanda, T., Fukumura, Y., Ohkubo, T., Inoue, M., and Kobayashi, I. (1997). Slime formation as a marker of serious infection with methicillin-resistant *Staphylococcus aureus*. J. Med. *28*, 87–98.

Takahashi, H., and Saito, H. (1982). Mechanism of pBR322 transduction mediated by cytosine-substituting T4 bacteriophage. Mol. Gen. Genet. *186*, 497–500.

Tao, L., Wu, X., and Sun, B. (2010). Alternative sigma factor sigmaH modulates prophage integration and excision in *Staphylococcus aureus*. PLoS Pathog. *6*, 1000888.

Taylor, C.M., and Roberts, I.S. (2005). Capsular polysaccharides and their role in virulence. Contrib. Microbiol. *12*, 55–66.

Thierauf, A., Perez, G., and Maloy, A.S. (2009). Generalized transduction. Methods Mol. Biol. *501*, 267–286.

Thiyagarajan, S., Chrisolite, B., Alavandi, S.V., Poornima, M., Kalaimani, N., and Santiago, T.C. (2011). Characterization of four lytic transducing bacteriophages of luminescent *Vibrio harveyi* isolated from shrimp (*Penaeus monodon*) hatcheries. FEMS Microbiol. Lett. *325*, 85–91.

Thomson, J.G., Yau, Y.Y., Blanvillain, R., Nunes, W.M., Chiniquy, D., Thilmony, R., and Ow, D.W. (2009). ParA resolvase catalyzes site-specific excision of DNA from the *Arabidopsis* genome. Transgenic Res. *18*, 237–248.

Thomson, N., Baker, S., Pickard, D., Fookes, M., Anjum, M., Hamlin, N., Wain, J., House, D., Bhutta, Z., Chan, K., Falkow, S., Parkhill, J., Woodward, M., Ivens, A., and Dougan, G. (2004). The role of prophage-like elements in the diversity of *Salmonella enterica* serovars. J. Mol. Biol. *339*, 279–300.

Tianiashin, V.I., Zimin, V.I., and Boronin, A.M. (2003). The cotransduction of pET system plasmids by mutants of T4 and RB43 bacteriophages. Mikrobiologiia. *72*, 785–791. [Russian]

Ticknor, L.O., Kolstø, A.B., Hill, K.K., Keim, P., Laker, M.T., Tonks, M., and Jackson, P.J. (2001). Fluorescent amplified fragment length polymorphism analysis of Norwegian *Bacillus cereus* and *Bacillus thuringiensis* soil isolates. Appl. Environ. Microbiol. *67*, 4863–4873.

Tiekotter, K.L., and Ackermann, H.W. (2009). High-quality virus images obtained by transmission electron microscopy and charge coupled device digital camera technology. J. Virol. Methods. *159*, 87–92.

Tinsley, C.R., Bille, E., and Nassif, X. (2006). Bacteriophages and pathogenicity: more than just providing a toxin? Microbes Infect. *8*, 1365–1371.

Tokuno, S.I., and Gough, M. (1977). Regulation of bacteriophage P22 DNA synthesis and repressor levels in P22cly infections. J. Virol. *21*, 956–964.

Twort, F.W. (1915). An investigation on the nature of ultra-microscopic viruses. Lancet *186*, 1241–1243.

Tye, B.K., Huberman, J.A., and Botstein, D. (1974). Non-random circular permutation of phage P22 DNA. J. Mol. Biol. *85*, 501–528.

Uchiyama, J., Maeda, Y., Takemura, I., Gamoh, K., Matsuzaki, S., and Daibata, M. (2012). Analysis of deoxynucleosides in bacteriophages ϕEF24C and K and the frequency of a specific restriction site in the genomes of members of the bacteriophage subfamily *Spounavirinae*. Arch. Virol. *157*, 1587–1592.

Umene, K., Shimada, K., Tsuzuki, T., Mori, R., and Takagi, Y. (1979). Lambda bacteriophage-mediated transduction of ColE1 deoxyribonucleic acid having a lambda bacteriophage-cohesive end site: selection of packageable-length deoxyribonucleic acid. J. Bacteriol. *139*, 738–747.

Uraji, M., Suzuki, K., and Yoshida, K. (2002). A novel plasmid curing method using incompatibility of plant pathogenic Ti plasmids in *Agrobacterium tumefaciens*. Genes Genet. Syst. *77*, 1–9.

Van Twest, R., and Kropinski, A.M. (2009). Bacteriophage enrichment from water and soil. Methods Mol. Biol. *501*, 15–21.

Van Wamel, W.J.B., Rooijakkers, S.H.M., Ruyken, M., van Kessel, K.P.M., and van Strijp, J.A.G. (2006). The innate immune modulators staphylococcal complement inhibitor and chemotaxis inhibitory protein of *Staphylococcus aureus* are located on {beta}-hemolysin-converting bacteriophages. J. Bacteriol. *188*, 1310–1315.

Varga, M., Kuntová, L., Pantůček, R., Mašlaňová, I., Růžičková, V., and Doškař, J. (2012). Efficient transfer of antibiotic resistance plasmids by transduction within methicillin-resistant *Staphylococcus aureus* USA300 clone. FEMS Microbiol. Lett. *332*, 146–152.

Vetcher, A.A., Lushnikov, A.Y., Navarra-Madsen, J., Scharein, R.G., Lyubchenko, Y.L., Darcy, I.K., and Levene, S.D. (2006). DNA topology and geometry in Flp and Cre recombination. J. Mol. Biol. *357*, 1089–1104.

Vogel, J.L., Li, Z.J., Howe, M.M., Toussaint, A., and Higgins, N.P. (1991). Temperature-sensitive mutations in the bacteriophage Mu c repressor locate a 63-amino-acid DNA-binding domain. J. Bacteriol. *173*, 6568–6577.

Voroshilova, N.N., and Kazakova, T.B. (1983). Effect of the development phase and growth rate of a *Shigella sonnei* population on the reproduction of homologous bacteriophage [in Russian]. Zh. Mikrobiol. Epidemiol. Immunobiol. *4*, 42–45.

Vos, P., Hogers, R., Bleeker, M., Reijans, M., van de Lee, T., Hornes, M., Frijters, A., Pot, J., Peleman, J., Kuiper, M., *et al.* (1995). AFLP: a new technique for DNA fingerprinting. Nucleic Acids Res. *23*, 4407–4414.

Waddell, T.E., Franklin, K., Mazzocco, A., Kropinski, A.M., and Johnson, R.P. (2009). Generalized transduction by lytic bacteriophages. Methods Mol. Biol. *501*, 293–303.

Walker, J.R., Ussery, C.L., and Allen, J.S. (1973). Bacterial cell division regulation: lysogenization of conditional cell division *lon*⁻ mutants of *Escherichia coli* by bacteriophage. J. Bacteriol. *113*, 1326–1332.

Wan, X.Q., Tang, D.S., Liu, A.P., Tan, S.Y., Li, W.K., Kuang, J., and Li, H.M. (2011). Isolation of a wild-type virulent phage of *Helicobacter pylori* and its simulated treatments of gastrointestinal Hp *in vitro* [in Chinese]. Nan Fang Yi Ke Da Xue Xue Bao. *31*, 304–307.

Wang, I.N. (2006). Lysis timing and bacteriophage fitness. Genetics *172*, 17–26.

Wang, I.-N., Dykhuizen, D.E., and Slobodkin, L.B. (1996). The evolution of phage lysis timing. Evol. Ecol. *10*, 545–558.

Wang, H., and Cutler, A.J. (1992). A simple, efficient PCR technique for characterizing bacteriophage plaques. PCR Methods Appl. *2*, 93–95.

Wang, X., Kim, Y., Ma, Q., Hong, S.H., Pokusaeva, K., Sturino, J.M., and Wood, T.K. (2010). Cryptic prophages help bacteria cope with adverse environments. Nature Communications *1*, 147.

Weinbauer, M.G., and Rassoulzadegan, F. (2004). Are viruses driving microbial diversification and diversity? Environ. Microbiol. *6*, 1–11.

Weld, R.J., Butts, C., and Heinemann, J.A. (2004). Models of phage growth and their applicability to phage therapy. J. Theor. Biol. *227*, 1–11.

Welker, N.E., and Campbell, L.L. (1965). Induction and properties of a temperature bacteriophage from *Bacillus stearothermophilus*. J. Bacteriol. *89*, 175–184.

Wiggins, B.A., and Alexander, M. (1985). Minimum bacterial density for bacteriophage replication: implications for significance of bacteriophages in natural ecosystems. Appl. Environ. Microbiol. *49*, 19–23.

Williamson, K.E., Wommack, K.E., and Radosevich, M. (2003). Sampling natural viral communities from soil for culture-independent analyses. Appl. Environ. Microbiol. *69*, 6628–6633.

Williamson, K.E., Radosevich, M., Smith, D.W., and Wommack, K.E. (2007). Incidence of lysogeny within temperate and extreme soil environments. Environ. Microbiol. *9*, 2563–2574.

Wilson, G.G., Young, K.K.Y., and Edlin, G.J. (1979). High-frequency generalized transduction by bacteriophage T4. Nature *280*, 80–82.

Wilson, S.E., Ingham, C.J., Hunter, I.S., and Smith, M.C. (1995). Control of lytic development in the *Streptomyces* temperate phage phi C31. Mol. Microbiol. *16*, 131–143.

Wommack, K.E., and Colwell, R.R. (2000). Virioplankton: viruses in aquatic ecosystems. Microbiol. Mol Biol. Rev. *64*, 69–114.

Woody, M.A., and Cliver, D.O. (1997). Replication of coliphage Q beta as affected by host cell number, nutrition, competition from insusceptible cells and non-FRNA coliphages. J. Appl. Microbiol. *82*, 431–440.

Yarmolinsky, M.B. (2004). Bacteriophage P1 in retrospect and in prospect. J. Bacteriol. *186*, 7025–7028.

Yordpratum, U., Tattawasart, U., Wongratanacheewin, S., and Sermswan, R.W. (2011). Novel lytic bacteriophages from soil that lyse *Burkholderia pseudomallei*. FEMS Microbiol. Lett. *314*, 81–88.

Young, K.K., and Edlin, G. (1983). Physical and genetical analysis of bacteriophage T4 generalized transduction. Mol. Gen. Genet. *192*, 241–246.

Young, K.K., Edlin, G.J., and Wilson, G.G. (1982). Genetic analysis of bacteriophage T4 transducing bacteriophages. J. Virol. *41*, 345–347.

Young, R. (1992). Bacteriophage lysis: mechanism and regulation. Microbiol. Rev. *56*, 430–481.

Yuan, P., Gupta, K., and Van Duyne, G.D. (2008). Tetrameric structure of a serine integrase catalytic domain. Structure. *16*, 1275–1286.

Zabala, B., Hammerl, J.A., Espejo, R.T., and Hertwig, S. (2009). The linear plasmid prophage Vp58.5 of *Vibrio parahaemolyticus* is closely related to the integrating phage VHML and constitutes a new incompatibility group of telomere phages. J. Virol. *83*, 9313–9320.

Zhong, L., Cheng, Q., Tian, X., Zhao, L., and Qin, Z. (2010). Characterization of the replication, transfer, and plasmid/lytic phage cycle of the *Streptomyces* plasmid-phage pZL12. J. Bacteriol. *192*, 3747–3754.

Zhou, Y., Liang, Y., Lynch, K.H., Dennis, J.J., and Wishart, D.S. (2011). PHAST: a fast phage search tool. Nucleic Acids Res. *39*, W347–52.

Zielenkiewicz, U., and Cegłowski, P. (2001). Mechanisms of plasmid stable maintenance with special focus on plasmid addiction systems. Acta Biochim. Pol. *48*, 1003–1023.

Ziermann, R., and Calendar, R. (1990). Characterization of the *cos* sites of bacteriophages P2 and P4. Gene. *96*, 9–15.

Zimecki, M., Weber-Dąbrowska, B., Łusiak-Szelachowska, M., Mulczyk, M., Boratyński, J., Syper, D., and Górski, A. (2003). Bacteriophages provide regulatory signals in mitogen-induced murine splenocyte proliferation. Cell. Mol. Biol. Lett. *8*, 700–711.

Zuber, S., Boissin-Delaporte, C., Michot, L., Iversen, C., Diep, B., Brüssow, H., and Breeuwer, P. (2008). Decreasing *Enterobacter sakazakii* (*Cronobacter* spp.) food contamination level with bacteriophages: prospects and problems. Microb. Biotechnol. *1*, 532–543.

Bacteriophages as Drugs: The Pharmacology of Phage Therapy

3

Stephen T. Abedon

Abstract

Pharmacology can be differentiated into two key aspects, pharmacodynamics and pharmacokinetics. Pharmacodynamics describes a drug's impact on the body while pharmacokinetics describes, instead, the body's impact on drugs. Another way of considering these terms is that pharmacodynamics is a description of both the positive and negative consequences of drugs reaching certain densities in the body while pharmacokinetics is concerned with the ability of drugs to reach and sustain those densities. Bacteriophages, or phages, are the viruses of bacteria. Here I consider the pharmacology of phages as drugs or – as they are applied to treat bacterial infections (phage therapy) – at least as drug-like entities. I suggest that phages, contrasting other narrow-spectrum but chemical antibacterials, possess a unique combination of properties including ease of discovery and laboratory characterization, low toxicity and minimal side-effects, and the potential to amplify their numbers *in situ*. The result of these somewhat un-chemotherapeutic-like aspects can mean that detailed exploration especially of phage therapy pharmacokinetics in many instances may be somewhat superfluous. Nonetheless, an improved pharmacological understanding of phage therapy should allow for more informed development, as well as rational *post hoc* debugging, of phage therapy experiments and improved phage therapy design.

Introduction

The words 'drug' and 'chemotherapeutic' can be used interchangeably, and historically this is especially so when drugs were employed to combat roguely replicating entities such as cancer cells, pathogenic bacteria, and viruses. More generally, a drug can be viewed as a bioactive but non-food substance that provides some benefit to the body. Defining 'substance' broadly, what medical drugs and chemotherapeutics have in common is that both are therapeutic while, historically, only chemotherapeutics strictly are 'chemicals'. That is, drugs existed prior to the medical use of synthetic, organic chemicals, such as Ehrlich's anti-syphilis chemotherapeutic, Salvarsan. Consequently, the concept of 'drugs' as equivalent to chemotherapeutic agents at best does not predate the mid to late 19th century (Sneader, 2005). Indeed, the word 'chemotherapy' was not coined until the early 20th century.

Bacteriophages, or phages, are the viruses of bacteria (Kutter and Sulakvelidze, 2005; Calendar and Abedon, 2006; Mc Grath and van Sinderen, 2007; Hyman and Abedon, 2012a,b). Since phages can display bioactivity upon association with bodies, they can be considered to be either drugs or drug-like, just as the word 'drug' itself seems to have originated as a reference (in French) to similarly natural products, i.e. dried herbs (Harper, 2010). The bioactivity of phages predominantly, though not exclusively, involves phage infection and subsequent lysis of the bacteria that exist in association with bodies. Phage therapy thus is the application of phages to bodies as either drugs or drug-like entities to combat bacterial infections. Antibacterial *chemo*therapeutics similarly can be applied to bodies as antibiotics, antiseptics can be used on body surfaces, and harsher disinfectants are employed to treat non-body environments (Cerf *et al.*, 2010). For discussion of advantages of employing phages as antibacterial agents, particularly in comparison to antibacterial chemotherapeutics, see Loc-Carrillo and Abedon (2011), Sulakvelidze and Kutter (2005), and Sulakvelidze *et al.* (2001). More recently, Brüssow (2012) has noted (p. 138) that, 'The FDA sees the following advantages with phage: Phages are highly specific, they work where alternatives have failed, are self-dosing, can be highly purified, and are generally non-toxic.' On the other hand, in the same piece Brüssow also suggests, 'The pharmacokinetics of phages might not be ideal…', etc.

Phages – due to a combination of their complexity and ability to *replicate* at the expense of their therapeutic targets – are more than simply chemicals. Because of this ability to replicate, phages can increase in number on their own accord, *in situ*, as they indeed have been doing for billions of years. The result are interactions with bodies that are more complicated than one finds with less vibrant therapeutics, though among 'drugs' this ability to replicate following application to the body is also seen, for example, with live-attenuated vaccines (Chan *et al.*, 2013). Notwithstanding this ability to replicate within the body, phage antibacterial activity is limited by a combination of (1) phage ability to kill target bacteria, defining their spectrum of activity (d'Hérelle and Smith, 1930), and (2) phage potential to be delivered to the vicinity of those bacteria, as analogous to drug delivery. In considering phages in this role as the equivalent of antibacterial drugs, phage therapy can be framed in terms of the science of pharmacology, the study of drug–body interactions.

I have recently published a chapter considering the 'Impact of phage properties on bacterial survival' (Abedon, 2009a), a book along with a related open-access chapter that deals extensively with the phage therapy especially of bacterial biofilms (Abedon, 2010a; 2011b; for a PDF of the former, the chapter, go to www.novapublishers.com), and, with the help of collaborators, six additional reviews that deal extensively with various aspects of phage therapy pharmacology (Abedon and Thomas-Abedon, 2010; Abedon, 2011a, 2012a; Curtright and Abedon, 2011; Chan and Abedon, 2012b; Chan *et al.*, 2013). My goal in this chapter is to address the subject of phage therapy pharmacology in a manner that does not simply repeat these other efforts. In addition, I limit my consideration of topics that are covered by other authors in this volume such as phage host range (see instead Chapters 1 and 2), phage impact on biofilms (Azeredo and Sillankorva, this volume), or indeed phage therapy efficacy (as covered by multiple chapters).

See also Abedon (2010b) for a six-chapter edited volume reviewing various aspects of phage therapy, Hyman and Abedon (2012a) for a more recent volume that also extensively considers various aspects of phage therapy, two recently published reviews of phage therapy of humans that discuss phage therapy pharmacology (Kutter *et al.*, 2010; Abedon *et al.*, 2011), a phage therapy pharmacology-related chapter by Gill (2010) as applied to biocontrol of bacteria in food, an overview of phage therapy pharmacology recently published by Ryan *et al.* (2011), Table 3.1 for a list of terms and definitions relevant to phage therapy pharmacology, and phage-therapy.org for an ongoing effort to broadly document the field of phage therapy.

Pharmacology

Pharmacology is the study of drug functionality. Typically a drug will display greater functionality when *in vivo* concentrations are higher but one must contrast this with greater negative effects. The science of pharmacology thus combines issues of drug design or discovery with the development of drug delivery strategies that are effective, convenient, minimally invasive, and relatively inexpensive. The general goal is to sustain drugs such that concentrations *in vivo* can have substantial positive effects while negative effects are at most minor. Pharmacological considerations thus can be greatly simplified for drugs that display both low inherent toxicities along with a relative lack of other side-effects. In particular, a drug that has a potential toxic concentration that is substantially higher than its effective concentration can be taken at higher doses and/or with longer intervals between doses than a drug that does not have these properties. A drug for which this so-called therapeutic *window* or *index* is much smaller, by contrast, may require constant monitoring, particularly if the drug's side-effects are severe.

The science of pharmacology traditionally is differentiated into two major components, pharmacodynamics and pharmacokinetics. Primary pharmacodynamic effects are the positive impacts of a drug, such as the bactericidal activity of an antibacterial agent. A great deal of pharmacological research is devoted to identifying agents that can potentially deliver primary pharmacodynamic effects and then testing those agents for such efficacy. Secondary pharmacodynamic effects instead are negative consequences, a.k.a. side-effects or toxicities. These side-effects often can be viewed as 'emergent' properties associated with pharmaceuticals (Curtright and Abedon, 2011). That is, because of the often complex interactions between traditional drugs and bodies, it can be difficult to predict the occurrence of all possible side-effects associated with a given drug based solely on drug chemistry, thus typically necessitating extensive *in vivo* and clinical safety determination before such secondary pharmacodynamic effects can be identified.

Pharmacokinetics, contrasting pharmacodynamics, is the study of the reaching and sustaining of effective drug concentrations at target sites, within bodies, of drug activity (Levison and Levison, 2009). The latter traditionally is differentiated into four components known as absorption, distribution, metabolism, and excretion. Particularly given systemic drug delivery, as opposed to topical application, the 'life cycle' of a drug is one of uptake into the systemic circulation of the blood (absorption), movement out of the blood and into the tissues where it will have an effect (distribution), reductions in blood densities as a consequence of metabolism (breaking down of the drug, often by the liver), and excretion, typically as effected by the kidneys. Absorption and distribution together are responsible for spatially delivering drugs to the sites in the body where they will display their primary pharmacodynamic effects. A general overview of these concepts is provided in Fig. 3.1. A specifically phage-associated consideration of pharmacology is provided in Fig. 3.2 and there are a number of publications that provide general overviews of these and other pharmacological issues pertaining to phage therapy (Payne and Jansen, 2001, 2003; Payne *et al.*, 2000; Levin and Bull, 2004; Bull and Regoes, 2006; Górski *et al.*, 2006; Abedon, 2009b, 2010a, 2011a, 2012a; Abedon and Thomas-Abedon, 2010; Kutter *et al.*, 2010; Abedon *et al.*, 2011; Curtright and Abedon, 2011; Jonczyk *et al.*, 2011; Loc-Carrillo and Abedon, 2011; Ryan *et al.*, 2011; Chan and Abedon, 2012b; Olszowska-Zaremba *et al.*, 2012; Chan *et al.*, 2013).

Properties of an ideal drug

In some circumstances it is possible to devise strategies where drug delivery can occur automatically, on an as-needed basis. For example, insulin pumps in principle can be programmed to deliver insulin to the blood in direct response to glucose levels. Such delivery is not only convenient but also can be more precise, better approximating the homeostatic mechanisms that normally maintain the body's chemistry and function within well-defined limits. That is, drug densities, *in situ*, ideally can be made to *increase* in concentration automatically in response to need. Alternatively, when drugs are not delivered either continuously or otherwise immediately on demand, then body concentrations of a drug will tend to vary to a greater extent, that is, between relatively high levels immediately post dosing to relatively low levels immediately prior to dosing (Fig. 3.1). This places a limit on the drug doses that can be delivered at one time since too-high concentrations are more likely to produce side-effects. Similarly, a failure to adequately sustain concentrations within the body can result in reduced efficacy, which places limitations on the length of intervals between dosing. In principle, therefore, automatic delivery not only can be more convenient than standard, not-automatic dosing protocols, but also may more readily sustain drugs at effective concentrations.

The ideal drug is one that can be delivered conveniently and relatively infrequently, is highly efficacious, minimally dangerous, and with concentrations that can be automatically adjusted *in situ* in real time and even on a tissue-by-tissue basis. To these beneficial drug properties one can also add ease of discovery, straightforward *in vitro* as well as *in vivo* characterization, low cost of manufacture, and reasonable stability during storage. Phages, with relatively few caveats, can meet many or all of these standards.

Table 3.1 Concepts and terms relevant to phage therapy pharmacology

Term	Definition
Absorption	Movement of drug into the blood following dosing; not *ad*sorption
Active penetration	Layer-by-layer phage clearance of bacterial biofilms as mediated by active phage replication, or at least as associated with phage-mediated bacterial lysis
Active treatment	Clearance of bacterial populations *in situ* that is dependent upon auto-dosing
Adsorption	Strictly the attachment of a phage to a bacterium but also typically describes as well the local movement of phages to the vicinity of targeted bacteria
Auto dosing	Drug dosing that is *not* a consequence of extrinsic application; phage *in situ* increases in density that are a consequence of phage replication; a.k.a., self amplification
Bacteriophage translocation	Defined after the concept of bacterial translocation, describes the potential for phages to reach systemic circulation particularly starting from the gut
Body	Body tissues as well as associated microflora
Decimal reduction time (D-value; *D*)	The length of time required to reduce viability of target microorganisms by 90% as a result of antimicrobial application of a given type, density, and under a given set of conditions
Distribution	Drug movement out of systemic circulation and into other body tissues
Dosing	Mechanism initiating the elevation of drug concentrations *in situ*; here 'application' and 'delivery' as resulting in dosing are used synonymously
Efficacy (1)	Positive outcome of an antibacterial procedure; this is not, as I use the term here, a description of intrinsic properties of a drug that may be involved in effecting that outcome (see following definition)
Efficacy (2)	Inherent ability of a drug to produce an effect following binding to its specific target, which for phages would be bacterial killing following adsorption; this definition of efficacy is *not* employed in this chapter other than in this table
Excretion	Drug removal from the body and systemic circulation as typically mediated by kidneys
Inundation threshold	Phage density required to exactly balance bacterial replication with bacterial deaths, the latter including as mediated by phages
Killing titre	Measure of phage density based on bacterial-killing ability; assumes adsorption by approximately 100% of the adsorption-competent phage population
Lytic phage	A phage that, at the end of its infection period, lyses its bacterial host, resulting in the death of its host bacterium and loss of continued infection productivity; as most temperate phages are also lytic phages, the concept of lytic phage contrasts instead with that of chronically (continuously) released phages (e.g. such as the filamentous phage M13)
Metabolism	Chemical modification of drugs typically resulting in drug inactivation though in the case of phages can instead result in both activation and auto-dosing; inactivation for phages can also result from immune system action
Minimum effective concentration	That drug density, as measured at the site of drug action, that achieves desired levels and rates of bacterial killing
Minimum toxic concentration	That drug density as measured *in situ* at which toxicity is apparent
Multiplicity of infection (MOI)	Strictly the ratio of *adsorbed* or infecting phages to target bacteria; too often used incorrectly as synonymous with the ratio of *added* phages to target bacteria; by either measure MOI should never be employed as the sole dosing descriptor during phage therapy
MOI_{actual}	Ratio of adsorbed phages to target bacteria; the actual meaning of MOI
MOI_{input}	Ratio of added phages to target bacteria; under conditions of higher bacterial densities MOI_{input} can approximate MOI_{actual} but this is not the case at lower bacterial densities
Obligately lytic	Phage property describing both how phage progeny are released (lytic) and the likelihood of that release given successful infection (obligate); contrast temperate
Passive treatment	Clearance of bacterial populations *in situ* that occurs without dependence on auto-dosing

Term	Definition
Penetration	Phage movement to the vicinity of bacteria including as mediated by bacterial lysis
Pharmacodynamics	Drug impact on bodies
Pharmacokinetics	Body impact on drugs; the factors impacting the attainment of drug densities necessary to effect primary pharmacodynamic effects
P_{max}	Maximal (peak) phage density, often as observed particularly at the site of phage bioactivity; contrasts C_{max} which is maximal drug density as typically observed in blood plasma
P_{min}	Minimum phage density experienced over the course of treatment following the ramp up in density associated with initial dosing; like P_{max}, P_{min} is most productively appreciated as phage densities found within target tissues, particularly infections, rather than in terms of plasma densities
Potency	Amount of drug activity corresponding to drug concentration *in situ*; with phages this will be a function of the phage adsorption rate constant as well as the phage potential to kill bacteria following binding, with the latter equivalent to efficacy (2)
Primary pharmacodynamic effects	The positive impact of drugs on bodies, ideally resulting in efficacy (1)
Secondary pharmacodynamic effects	The negative impact of drugs on bodies, i.e. toxicity, side-effects, or destruction of normal flora bacteria
Temperate phage	Phage that is able to initiate non-defective, ongoing, and metabolically active infections that produce no virion progeny but instead generate bacterial lysogens; temperate phages can carry bacterial virulence factor genes and bacterial lysogens typically display an acquired resistance to the associated temperate phage (or phages); note that the term 'lysogenic phage' is a misnomer and should never be used
Titre	Phage density, usually as provided in per ml units; titres are the preferred phage dosing unit of measurement, with dosing ideally reported in terms of both volumes applied and titres of phages found within those volumes
Trough concentration	Minimum drug concentration experienced between dosing; see P_{min}
Virulence (1)	Description of a combination of a phage's anti-bacterial killing power (efficacy '2'), kinetics of bacterial acquisition, and rate at which new progeny are generated (auto-dosing) given bacterial acquisition, where greater killing power, faster acquisition of target bacteria, and greater auto-dosing proficiency combine to result in greater phage antibacterial virulence (though with the caveat that phage virulence can vary as a function of properties that often differ both with infection conditions and the bacterial strain that is being targeted); generally during phage therapy higher antibacterial virulence is preferable to lower virulence
Virulence (2)	Description of the infection properties of a non-temperate and non-chronically infecting phage, that is, lytic phages that are unable to display lysogeny are often described as being virulent; this definition is equivalent to that of obligately lytic (as used here), seems to have been derived historically from that of virulence (1), and is not otherwise employed in this chapter

Primary pharmacodynamics

A drug's most salient feature is its primary pharmacodynamics, i.e. its positive effects. These positive effects will not come about, however, until sufficient concentrations are present in target tissues, where targets in the case of antibacterials are tissues infected by bacteria. With phages these sufficient densities can be achieved by either of two means (Payne and Jansen, 2003; Cairns *et al.*, 2009; Abedon and Thomas-Abedon, 2010; Letarov *et al.*, 2010): via phage application alone (so-called passive treatment) or via a combination of phage application and *in situ* amplification (so-called active treatment). In this section I discuss the basics of phage dosing in terms of positive microbiological outcomes, that is, reductions in

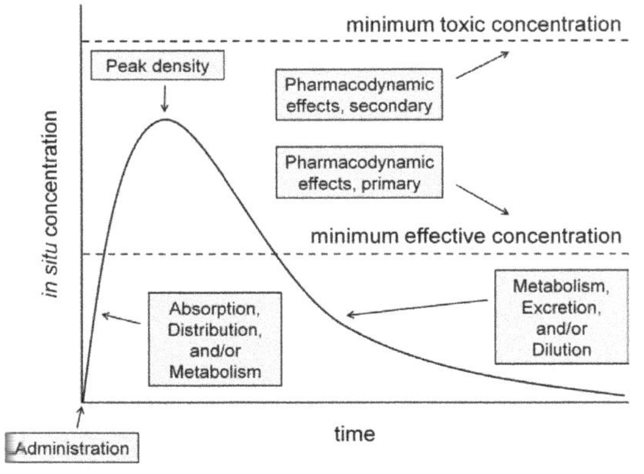

Figure 3.1 Representative drug time course highlighting pharmacodynamic issues (as explicitly indicated) and pharmacokinetic issues (absorption, distribution, metabolism, and excretion). Metabolism for phages, as indicated to the left, is replication and is relevant particularly given active treatment. Metabolism, as indicated to the right, is immune-system sequestration or inactivation of phage particles. Minimum effective concentrations and toxic concentrations refer to bacterial killing efficacy and side-effects, respectively.

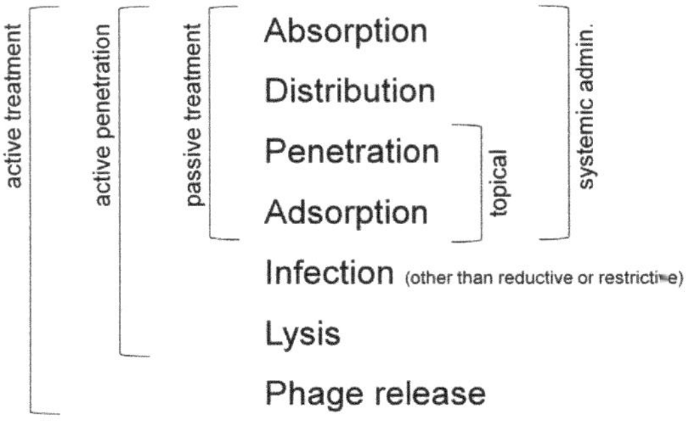

Figure 3.2 Steps of phage therapy treatment as viewed from a pharmacological perspective, not all of which necessarily take place during a given treatment. Absorption is phage movement into the blood. Distribution is phage movement into other tissues typically as from the blood. Penetration, here, can be viewed as into bacterial biofilms. Adsorption is phage attachment to bacteria and infection is the post-adsorption stage of phage–bacterial interactions. Lysis is the destruction of both infected bacteria and a bacterial infection. Phage release in turn follows lysis. Note that passive treatment requires only phage adsorption along with associated bacterial killing; the latter can be a consequence of lysis though also can result from other mechanisms. Active penetration of phages into biofilms probably requires, at least, phage-induced bacterial lysis or other means of bacterial removal (e.g. as mediated by extracellular polysaccharide depolymerase enzymes) but may require productive phage infection as well. Topical application at a minimum requires penetration to target bacteria and subsequent phage adsorption to attain efficacy. Systemic administration involves also the absorption and distribution steps, i.e. access to systemic circulation and is followed, in many though not all scenarios, by movement out of systemic circulation (distribution).

bacterial densities. For extensive mathematical treatment of this issue, see Abedon (2011a). See also Abedon (2012a).

Using titres to describe phage potential to kill bacteria

Most drugs are dosed in a manner that is intended to avoid exceeding toxic densities while at the same time reaching or exceeding effective densities (Fig. 3.1). For phages the development of dosing strategies can be greatly simplified due to the low toxicity that is a typical phage characteristic (Abedon and Thomas-Abedon, 2010; Curtright and Abedon, 2011; Olszowska-Zaremba *et al.*, 2012). To a substantial extent, therefore, emphasis during the design of phage therapy protocols can be placed on enhancing antibacterial activity rather than the distraction of needing to strive towards avoiding the induction of excessive side-effects. Specifically, to be effective antibacterial agents, phages must reach densities such that the rate of phage adsorption to bacteria is faster than the rate of bacterial replication and, ideally, quite a bit faster. Bacteria, as a consequence of exposure to sufficient phage densities, are killed and then cleared, and this clearance preferably occurs over relatively short intervals.

Assuming that there is a good bactericidal match between applied phages and target bacteria, then the extent of phage adsorption to individual bacteria, what can be described as an *actual* multiplicity of phage infection (MOI_{actual}) – or what more accurately can be described as a multiplicity of *adsorption* – is the primary determinant of bacterial eradication. MOI_{actual}, in turn, is a function of phage titres, traditionally measured as phages per ml, since the rate that individual bacteria become adsorbed by phages is a direct function of how many phages are found in a bacterium's immediate vicinity (Abedon, 2012b,c). Titre-based descriptions of phage-mediated bacterial killing include inundation thresholds, killing titres, and decimal reduction times (Payne and Jansen, 2001; Abedon and Thomas-Abedon, 2010; Abedon, 2011a,b). These measures should be used as guides towards appreciating what phage densities may be required to achieve the larger goal of bacterial eradication.

Inundation thresholds

The inundation threshold represents the minimum phage density that can keep a bacterial infection from getting worse. More precisely, it is the phage density that can keep a bacterial culture from becoming more concentrated (Payne and Jansen, 2001; Abedon and Thomas-Abedon, 2010; Letarov *et al.*, 2010). This, in turn, is equivalent to the phage density that achieves bacterial killing at a rate that equals that of bacterial replication. Though this density can be thought of as a minimum inhibitory phage concentration (Cairns and Payne, 2008), in terms of the design of phage therapy protocols what instead should be achieved are net *reductions* in bacterial densities. In addition, as noted more generally by Cerf *et al.* (2010), 'minimum inhibitory concentration (MIC) is never the concentration at which disinfectants should be applied' because '[t]ime is a fundamental dimension in the disinfectant context' (pp. 247 and 250). Nonetheless, the inundation threshold can provide an illustration of what phage concentrations at best only kill bacteria as fast as new bacteria are produced, that is, which are minimally effective but still inadequate to actually reduce bacterial loads.

By way of example, just to counter the replication of a bacterium that divides once every 30 min, a phage with an adsorption rate constant of 2.5×10^{-9} ml/min (k) (Stent, 1963) would have to be present at a concentration of approximately 10^7 phages/ml. This number is found via the equation, $P_I = \ln(2)/(30 \times 2.5 \times 10^{-9})$, where the inundation threshold, P_I, is equal to 10^7 phages/ml given these parameter values. Note that ln(2)/(doubling time) is the Malthusian parameter, here in minute time units. That is, ln(2)/(doubling time) = ln(2)/(30 min) = r, where r is the growth-rate term as found in the exponential growth equation. Furthermore, $N_t = N_0 e^{rt}$, where N_0 is the initial bacterial density, N_t the bacterial density at time, t, and t is in min.

Another way of making the same point is to note that the expression $dN/dt = N(r - Pk) = 0$ is true – meaning no net bacterial population growth nor decline is occurring – when P is the phage inundation threshold (P_I), that is, when phage adsorption exactly balances bacterial growth. Rearranging, this is where P_I equals r/k (Payne and Jansen, 2001; Levin and Bull, 2004). For a vigorously growing infection one might view 10^7 phages/ml thus as an *inadequate* phage dose for passive treatment even if the number of phages applied should greatly exceed the number of bacteria present, and even if that phage density should hold steady rather than decline over time. An additional consideration is that calculation of inundation thresholds for phages in reality is not nearly as straightforward as the above example might suggest since neither bacterial replication rates nor phage adsorption rates are easily determined *in situ*. For additional discussion of these issues, see Abedon (2011a).

Killing titres

A killing titre measures the bactericidal density of phages found in a given stock or formulation. Killing titres are a more practical as well as more direct measure of what phage densities can be required to clear bacteria, though with the caveat that killing titres do not take into account the issue of ongoing bacterial replication, that is, as considered in the previous section. Thus, phage adsorption must be relatively fast for killing titre calculations to reasonably hold.

To determine a killing titre, one mixes a phage population with a bacterial population in such a manner that all or nearly all phages adsorb, and ideally under conditions in which bacterial replication is minimal. Measurement is made of bacterial viability both before and after phage adsorption. The ratio of bacterial viability after phage adsorption (N_t) to that before phage exposure (N_0) is then assumed to be equal to e^{-M}, where M is the ratio of adsorbed phages to bacteria present, i.e. MOI_{actual} (that is, a Poisson distribution of phage adsorption to bacteria is assumed). Rearranging, $M = -\ln(N_t/N_0)$. From this calculation it should be clear that the greater the fraction of bacteria that one wants to kill, i.e. the smaller that N_t is relative to N_0, then the greater the number of phages that must adsorb, on average, to each bacterium that is present in a culture (greater M). By way of example, for $N_t/N_0 = 10^{-1}$, 10^{-2}, 10^{-3}, 10^{-4}, 10^{-5} or 10^{-6}, then M must be equal to 2.3, 4.6, 6.9, 9.2, 11.5 or 13.8 adsorbed phages/bacterium, respectively. See Abedon (2011a; 2011b) for additional discussion of this concept.

Note that killing titre calculations assume equivalent phage penetration to all bacteria, which is unlikely given heterogeneity in bacterial availability to phages. This assumption of equivalent penetration therefore can lead to an underestimation of what ratios of adsorbed phages to bacteria are necessary to achieve sufficient bacterial killing *in situ*. To some degree this issue may be mitigated, however, if replication can increase phage numbers in the immediate vicinity of less phage-available bacteria, such as during phage penetration into biofilms (Abedon and Thomas-Abedon, 2010; Abedon, 2010a, 2011b, 2012b). Alternatively, extremely high multiplicities of phage adsorption can result in at least some cases in non-productive phage infections (Abedon, 2011d). A yet additional issue is that killing titres do not explicitly take into account *rates* of phage impact on bacteria. This is because killing titres instead are determined by how many phages *adsorb* to bacteria, and phage adsorption generally takes time.

Decimal reduction time

The rate at which bacteria are acquired by phages is a function of three variables: adsorption properties of individual phages, phage densities, and the availability of bacteria to phages. Here I focus on phage properties and densities, and for convenience assume that all bacteria are equally available to phages. If that assumption is false, then it is a pharmacokinetic rather than pharmacodynamic concern (i.e. phage penetration ability to bacteria) or it is an issue of phage host range (see 'Spectrum of activity', below). The measure I consider is decimal reduction time or D-value. D-value, or D, is the time it takes to reduce bacterial viability by one log, e.g. such as from 10^6 bacteria/ml to 10^5. Assuming a constancy in phage densities, then $D = -\ln(0.1)/Pk$ where k is the phage adsorption rate constant, P is phage density, and 0.1 represents the decimal reduction.

Employing the same phage adsorption rate constant used to illustrate inundation thresholds, i.e. 2.5×10^{-9} ml/min, note that the titre of 10^7 phages/ml would result in a D-value of approximately 92 min. Thus, it would take over 350 min or about six hours under ideal conditions, and assuming that bacteria are not replicating, to reduce the viability of a bacterial culture approximately 10,000-fold. This reduction is found to be $e^{-9.2}$-fold when calculating killing titres, that is, where $9.2 = M = \text{MOI}_{\text{actual}}$, or approximately $M = 10$, a multiplicity threshold that I will take up subsequently. At 10^5 phages/ml, by contrast, that same degree of reduction in bacterial viability would take instead approximately 25 days. If faster reductions in bacterial densities are desired, if not all bacteria are equally available to phages (Barrow and Soothill, 1997), or if phage densities should decline over the course of treatment, then supplying higher phage numbers may be necessary to achieve desired rates of killing. See Abedon (2011a,b) for additional discussion.

From these calculations we can see that bacterial killing may not be sufficiently fast if phage densities are not sufficiently high. By 'sufficiently high' I mean phage densities that are near to or in excess of about 10^8 phages/ml (by contrast, note the calculations for 10^7 phages/ml, above). Derivation along with experimental application of this idea – that it is phage density that determines rates of bacterial adsorption by phages along with a utility of phage application at titres in the vicinity of 10^8/ml – can be found, in the modern phage literature, in the following references: Abedon (1990, 1999), Payne and Jansen (2001), Kasman *et al.* (2002), Goodridge (2008), Bigwood *et al.* (2009), Hyman and Abedon (2009) Abedon and Thomas-Abedon (2010), Hagens and Loessner (2010) and Abedon (2011a–c); see also Spouge (1994) and Payne *et al.* (2000). For $k = 2.5 \times 10^{-9}$ ml/min, as above, then 10^8 phages/ml corresponds to a D-value of approximately 10 min. Note that in cases where excessively rapid bacterial lysis might be a concern, e.g. such as phage treatment of Gram-negative infections of the blood, then lower phage densities per dose might be more appropriate, though with the caveat that it may not always be possible to control just what densities phages attain nor how rapidly cultures are lysed.

Peculiarities of single-hit killing kinetics

Phages display single-hit killing kinetics, meaning that generally only a single adsorption by a properly functioning, obligately lytic phage will be necessary to result in bacterial death. On the one hand, this is a positive aspect of phages as antibacterial drugs (Loc-Carrillo and Abedon, 2011). That is, it means that the number of units needed to reach bacteria to effect killing is lower than that seen with most other antibacterial drugs. On the other hand, single-hit killing kinetics can lead to overly simplistic perceptions of the phage potential to eliminate targeted bacteria. Specifically, it is an incorrect inference that supplying one phage for every bacterium alone should inevitably achieve complete bacterial killing and this is so even if all targeted bacteria are highly phage susceptible (or, indeed, even if all supplied phages adsorb).

There are two reasons why single-hit killing kinetics can lead to fallacious understanding of phage therapy primary pharmacology. The first is that multiple phages can adsorb to a single bacterium and the second is that phages must traverse what for them are vast distances before they can encounter bacteria (Hagens and Loessner, 2010; Abedon, 2011b; 2011c). Both of these considerations result in a need for more phages to effect sufficient bacterial killing than one might assume based solely on the (false) idea that every phage adsorption results in bacterial killing. The first issue, that of multiple phage adsorption of individual bacteria, I take up in the following paragraph. The second consideration is covered above under the heading of D-values. In particular, even if enough phages are supplied to kill off all of the bacteria currently found within a culture, that killing will not necessarily be achieved over short or even reasonable intervals because inevitably there will be delays between phage administration and bacterial adsorption.

Because more than one phage can adsorb to a single bacterium, this means that already-adsorbed bacteria can serve as anti-phage antagonists. In other words, since phages adsorb irreversibly and only a single phage is required to effect bacterial killing, in effect all subsequently adsorbing phages to a given bacterium are killed without contributing to bacterial-killing efficacy. To control for this loss of activity, many more phages may be required to achieve a desired level of bacterial killing than there are bacteria, e.g. approximately ten-fold more as calculated under the heading of killing titres. This phage excess is required even if one ignores delays in phage adsorption due either to mass-action issues or problems in phage penetration to bacteria. See Gill (2010) for discussion of these latter issues and Bull and Regoes (2006) as well as Abedon (2011a) for additional discussion of single-hit killing kinetics.

Phage dosing rules of thumb

Given $N_t/N_0 = e^{-M}$, then a phage density of $P_0 = -N_0 \ln(N_t/N_0) = N_0 M$ should be seen as a reasonable goal of phage dosing, that is, the product of the initial bacterial density, N_0, and the ratio of adsorbed phages to bacteria, M, where N_t/N_0 represents the extent of bacterial survival that is sought. Thus, 10^6 bacteria/ml reduced to a density of 10^{-6} bacteria/ml – a drop in density to 10^{-12} of the original – will require a minimum of 3×10^7 phages/ml ($M \approx 30$). Alternatively, such substantial bacterial overkill may not be necessary to the extent that phage-mediated disruption of bacterial infections is followed by immune system clearance (Levin and Bull, 2004). Nonetheless, this phage density corresponds to an initial D-value, as seen prior to phage losses due to bacterial adsorption, of approximately 30 min, assuming a phage adsorption rate constant of 2.5×10^{-9} ml/min. This means that, if conditions for bacterial acquisition are poorer than anticipated, then this phage density, sufficient as it may be for bacterial overkill, may in fact not result in timely bacterial killing.

Perhaps consistent with that concern, note that some efforts to model phage therapy mathematically, as reviewed by Gill (2008), have employed adsorption constants of much lower magnitude than that used in the above calculations, such as $\sim 10^{-11}$ ml/min or lower (often expressed, though, in per hour rather than per min units, i.e. a constant of 10^{-9} ml/h $\approx 2 \times 10^{-11}$ ml/min). With a phage density of 3×10^7/ml and an adsorption rate constant of 10^{-11} ml/min, then the calculated D-value is 5 days! Consistently, during the simulations reviewed by Gill, phage densities often reach approximately 10^{10}/ml before one sees substantial declines in bacterial densities, which can serve as an example of how poor phage penetration to individual bacterial cells can negatively impact phage therapy efficacy (Barrow and Soothill, 1997). Also consistently, Levin and Bull (1996) note in their model that, with lower phage adsorption rates, bacteria are more likely to exceed their 'lethal threshold'.

D'Hérelle and Smith (1930, p. 170) provide a similar observation: 'In the case of acute diseases, powerful bacteriophage must be brought into contact with the pathogenic bacteria to destroy them before they have had time to occasion lesions sufficient to cause death. Bacteriophage therapy should be instituted as early in the infection as possible.' Further down on the same page, d'Hérelle (original emphasis) also seems to allude to the concept of active treatment for these 'acute' infections: 'In acute diseases, it is sufficient to apply, *as soon as the first symptoms are noticed*, or early during the course of the disease, a small quantity of a potent 'stock' bacteriophage in order to occasion the destruction of the bacteria and thus bring about recovery.' Accordingly, even if substantial overkill of bacteria is not necessary to achieve desired levels of treatment efficacy, positive results still may require either sufficient phage numbers or sufficiently early phage application if only for the sake of achieving timely reductions in bacterial density such that mortality is averted. See similarly a brief discussion by Carmody *et al.* (2010) on this subject.

Payne *et al.* (2000) posit an alternative scenario in which too early treatment, followed by substantial phage decay, could give rise to a 'failure threshold time', which is defined as a time during infection development before which active treatment is not expected to provide adequate antibacterial efficacy. One can also view this perspective, however, as pointing to a problem of supplying too few phages under conditions where in fact therapy is initiated too early to support phage population growth. It is also an open question as to when, under clinical conditions, one in fact might catch an infection too early in its development to support active treatment. A possible answer is during prophylactic treatment, though presumably under such circumstances an important consideration would be to supply as well as retain sufficient phage densities that infections are brought under control early in their development, i.e. before active treatment is possible (Abedon, 2010a, 2011b).

The utility of active phage replication, in part, is that when bacterial densities are sufficiently high, then adequate D-values can be sustained through replicative replacement of those phages lost due to adsorption or other mechanisms. When bacterial densities are low, however, then substantial increases in phage densities will not proceed as a consequence of phage replication, though neither will substantial decreases in D-values unless free phage losses occur for reasons other than bacterial adsorption. With this latter case, treatments are predominantly passive with outcomes almost entirely dependent on what phage densities are supplied along with how often these phages are applied.

For these various reasons, striving for phage densities in the vicinity of 10^8 phages/ml, or higher, and then sustaining those phage densities over the course of treatment – whether through passive or active means – should be the default dosing goal in phage therapy protocols. Lower densities may be employed or achieved only to the extent that they are demonstrably and consistently effective in reducing bacterial densities to desirable extents. Substantially higher densities may be employed should 10^8 phages/ml be found to be inadequate in achieving desirable levels of bacterial killing, with the caveat that an alternative explanation for poor efficacy, to that of too few phages, can be poor phage choice. Keep in mind, however, that it is always important to monitor for toxicity during animal testing and particularly so given increased levels of dosing since the standard pharmacodynamic assumption is that greater drug densities will be associated with greater drug toxicity.

Striving for substantial bacterial killing

A reasonable dividing line, following phage application, between substantial and insubstantial reductions in bacterial densities, whether *in vitro* or *in vivo*, is approximately 10,000-fold. Experimentation that fails to achieve this substantial bacterial killing should be deemed to be inadequate unless clinically satisfactory endpoints nonetheless are achieved. Experimentation in which neither clinically positive outcomes nor substantial reductions in bacterial densities are observed, however, should not be used as evidence either for or against the potential for phage therapy to control a specific bacterial infection and particularly not unless higher phage density 'positive controls' are attempted. Examples of such bacterial-killing positive controls would be 10^9, 10^{10}, or even higher phage titres (phages/ml) achieved within the vicinity of target bacteria *in situ*.

If microbiologically or clinically positive results still are not achieved then either phage choice or dosing conditions should be questioned or, instead, effort should be made to supply phages in doses that are as high as can be reasonably accomplished. This includes as may be attained through multiple or continuous dosing (below) and so long as no more than mild side-effects occur. In other words, it is not reasonable to declare that any drug is inefficacious so long as higher or otherwise more extensive dosing may yet be *safely* employed. On the other hand, for reasons of economy, safety, perhaps improved efficacy (next section), and also to better appreciate the primary pharmacodynamics of specific phage therapy protocols, in some cases it may be relevant also to explore dosing at the other end of the spectrum, that is, whether achievement of less than 10^8 phages/ml at the site of phage activity too can result in reasonable efficacy. For additional considerations of 'best practices' during phage therapy experimentation, see Abedon (2012a).

MOI_{input}

There has been a tendency in numerous phage therapy studies to employ not phage titres as a dosing measure but instead what can be described as an MOI_{input}, that is, the ratio of *added* phages to target bacteria (Abedon, 2010a, 2011a,b, 2012a). This is unfortunate because MOI_{input}, in and of itself, is an almost meaningless dosing measure. An exception to this claim of meaninglessness, however, occurs when this ratio is lower than, say, ten (Kasman *et al.*, 2002; Abedon and Thomas-Abedon, 2010) since then it can indicate that phage replication (active treatment) may be required to achieve substantial levels of bacterial killing. The converse, $MOI_{input} > 10$, however, is *not* in itself an indication that dosing is sufficient to achieve reasonable bacterial killing. For example, if bacterial densities are only 10^2/ml then a ratio of 1000:1 added phages to bacteria translates to a still insufficient density of 10^5 phages/ml (see 'Decimal reduction time', above). That is, 10^5 phages/ml, with its impressive MOI_{input} of 1000, still is 1000-fold smaller than the 10^8 phages/ml target density suggested above. Additional discussion of, in particular, the merits of MOI_{actual} (which is preferred) versus MOI_{input} (which is difficult to justify as a sole dosing measure), can be found in Kasman *et al.* (2002), Abedon and Thomas-Abedon (2010), and Gill (2010), as well as Abedon (Abedon, 2010a, 2011a,b, 2012a).

Secondary pharmacodynamics

Side-effects define the upper limits of what drug concentrations can be tolerated by bodies. With antibacterials these secondary pharmacodynamic effects consist of a combination of unintended drug interactions with body tissues, including associated microbiomes, and the consequence of intended drug interaction with target bacteria. The latter can include toxin release during bactericidal activity but also modifications that can occur when drug–target interactions are not sterilizing, such as lysogenic conversion in the case of temperate phages (Krylov, 2001; Abedon and LeJeune, 2005; Hyman and Abedon, 2008; Los *et al.*, 2010; Christie *et al.*, 2012; Kuhl *et al.*, 2012). When properly chosen, however, phages can display minimal impacts on body tissues (see below). Even so, most phage therapeutics still will display, along with lytic antibiotics such as penicillin, the potential to solubilize substantial amounts of bacterial toxins in the course of bacterial killing.

Notwithstanding this release of bacterial toxins in the course of phage-induced bacterial lysis, phage use as antibiotic equivalents typically has not been associated with side-effects that are either significant (Sulakvelidze and Kutter, 2005; Kutter, 2008; Housby and Mann, 2009; Letkiewicz *et al.*, 2010; Olszowska-Zaremba *et al.*, 2012) or long-lasting (Chanishvili *et al.*, 2009b); for further discussion of phage safety, see Sulakvelidze and Pasternack (2010). The mildness of phages in terms of their impact on bodies, though not entirely understood, nonetheless greatly simplifies phage therapy pharmacological development because it allows researchers to focus especially on improving efficacy rather than concentrating instead on reducing toxicity. The result is a potential for dosing with greater phage numbers than may be minimally necessary to achieve antibacterial efficacy, or simply to allow dosed phages to increase in number to those higher levels on their own (Payne and Jansen, 2003). To a large extent, in other words, one can attain pharmacodynamically effective phage densities without substantial concern that those doses are going to translate into toxicity (though toxicity testing nonetheless is still an important component of the development of phage therapeutics). By contrast, Cerf *et al.* (2010, p. 252) note that (emphasis mine), 'because of their toxicity, the concentration of *antibiotics* can seldom be increased to improve the outcome of the treatment.' Though not all phages nor all circumstances of phage therapy are lacking in risk, nonetheless it is important to emphasize that for phage therapy as actually practiced, secondary pharmacodynamic concerns seem to not readily translate into any more than minor side-effects.

In Curtright and Abedon (2011) we provide a conceptual discussion of just why it is that phages in many cases can be viewed as qualitatively different from more typical chemotherapeutics in terms of their secondary pharmacodynamics. The specific issue addressed is the question of why or indeed how a more chemically complex entity (phages) could also display fewer 'emergent properties' of side-effects and toxicity in the course of application to, for example, animal tissues. We argue in particular that natural selection has tended to enhance in obligately lytic phages those properties that result in effective exploitation of *bacteria*, especially over shorter time scales. The resulting adaptations tend to not inherently result in toxicity to organisms other than bacteria, nor to organisms other than those specific bacteria that are targeted by specific phages. In other words, obligately lytic phages as a product of evolution are highly specialized though otherwise benign delivery devices of antibacterial agents precisely to bacterial cytoplasms, where they can then be effective even at extremely low overall concentrations (i.e. as measured per kilogram of body weight of phage-treated individuals).

As a consequence of these considerations, a general expectation when using phages as antibacterial agents is that adverse impacts of phage particles along with phage infections on animal tissues will be relatively slight and thus will *not* tend to emerge upon animal or clinical testing. This suggestion does not imply that such testing may be superfluous, and also is highly dependent on proper phage choice (Gill and Hyman, 2010, see also Chapter 2), particularly such that phages carrying bacterial virulence factors genes are not employed for phage therapy. Nonetheless, it provides what we believe to be a reasonable argument for why the complexity of phage-based antibacterials, as a product of evolution, may actually provide a utility in terms of reducing secondary pharmacodynamic issues rather than a hindrance.

Accumulation rates

The rate at which phages, as drugs, can increase in density, *in situ*, can be described as P' (Chan and Abedon, 2012a), which is the first derivative of the variable, P, for phage density. $P' > 0$ thus describes an increase in phage numbers, which can occur as a consequence of phage replication. Though replication is unusual for a drug, nonetheless a similar amplification in densities can occur simply through dosing by conventional means, i.e. by administering more drug, including automatic dosing in response to *in situ* conditions (Abedon and Thomas-Abedon, 2010). The resulting increases in drug concentrations following conventional dosing versus this phage 'auto dosing' (Loc-Carrillo and Abedon, 2011), however, can result in changes in concentrations in distinctly different locations, with phage self amplification tending to occur precisely in the vicinity of phage bioactivity. Pharmacologically, the mechanisms giving rise to these increases in phage densities can be categorized under the heading of 'metabolism' (Abedon and Thomas-Abedon, 2010), that is, the pharmacokinetic modification of drugs by body tissues, with microflora considered to be aspects of bodies. In this case, though, metabolism results in drug 'activation', such as is also the case for prodrugs, rather than the more typical inactivation (Villalobos *et al.*, 1998; Levison and Levison, 2009).

Phage population growth with and without complications

Drug accumulation involves diffusion and other movement processes that follow drug dosing. Alternatively, mechanisms exist by which certain drugs, such as the anti-tuberculosis drug isoniazid (Then and Sahl, 2010), become activated in the course of metabolism. Similarly, phages can increase in numbers when in the presence of susceptible bacteria, resulting in phage population growth. All three of these processes can be viewed as pharmacokinetic issues, i.e. absorption, distribution, and metabolism; even excretion can increase phage numbers to the extent that the site of drug action is reached via passage through the kidneys (Keller and Engley, 1958; Zobnina, 1963; Weber-Dabrowska *et al.*, 1987; Dabrowska *et al.*, 2005; Górski *et al.*, 2007; Nishikawa *et al.*, 2008; Letarov *et al.*, 2010). Notwithstanding how we classify *in situ* increases in phage titres, under most circumstances the magnitude of P' is relevant only to the extent that it is too small, that is, if there are substantial delays in necessary phage accumulation.

In this section I focus on phage population growth and the related consideration of phage therapy by active treatment. Obligately lytic phage populations in particular grow through repeated rounds of bacterial adsorption, infection, and lysis. Anything that speeds up adsorption, decreases the time spent infecting, or increases the yield from infections should to a first approximation increase P', that is, greater adsorption constants, shorter latent periods, or larger burst sizes (Levin and Bull, 2004). Phage survival is also important though alternatively can be viewed as a component of an 'effective' burst size, that is, the number of phages not only produced by an infected bacterium but which also manage to survive to infect new bacteria (Abedon, 2009c; Chan and Abedon, 2012a). Rates of adsorption, rates of lysis, and effective burst sizes thus determine the magnitude of P' that results from phage population growth.

Addressed in part by Abedon (2009a), a number of considerations exist that serve to complicate the above ideas that increasing burst size, adsorption rates, or rates of phage-induced bacterial lysis will tend to result in faster phage population growth. First, faster adsorption may not contribute to faster phage population growth within spatially structured environments, such as one sees during phage plaque formation or perhaps during phage exploitation of biofilms (Gallet *et al.*, 2009; Abedon, 2011b; Gallet *et al.*, 2011). Second, trade-offs can exist where shortening of latent periods, all else held constant, results in reductions in phage burst sizes simply because less time is available during shorter latent periods to make more phages; see, for example, Abedon (2009c). Third, the utility of latent-period shortening is seen predominantly at higher rather than lower bacterial densities (Abedon, 2009b; Cairns *et al.*, 2009). Fourth, and similarly, the utility of increasing a phage's adsorption affinity for bacteria, thereby increasing rates of phage adsorption, is seen particularly at lower rather than higher bacterial densities (Abedon, 2009b; Cairns *et al.*, 2009). Fifth, P' can be dependent on *in situ* conditions that can be difficult to fully appreciate (Bull *et al.*, 2010), including in terms of the impact of bacterial physiology on phage growth parameters (Carlton, 1999; Merril *et al.*, 2003, 2006), that is, on latent period, burst size, or adsorption constant (Hadas *et al.*, 1997). Sixth, target bacteria, as they exist *in situ*, will not necessarily be identical to bacterial hosts employed to characterize phages *in vitro*, again potentially affecting phage growth parameters, perhaps for the worse.

The result of these various issues is that *in vitro* antibacterial outcomes as well as theoretical calculations of P' may not precisely reflect actual, *in situ* phage population growth rates. Similarly, *in vitro* selection schemes for improved phage growth performance will not necessarily improve P' *in situ*. Even if care is taken to properly match *in vitro* to *in situ* conditions, variance in these conditions between infections could serve to reduce these levels of similarity. Indeed, one concern with many phage therapy animal experiments is that variance typically is suppressed and particularly so in terms of the genotype of target bacteria. Therefore, it often is hard to say whether adequate phage population growth may occur to effect phage therapy efficacy under real world conditions and this is so even if sufficient bacterial densities appear to be present to support such growth.

Active treatment (with some consideration of passive treatment)

Active treatment both involves and requires active phage replication, resulting minimally as well as necessarily in $P' > 0$. Active treatment thus requires a phage ability to replicate upon bacterial infection. In addition, and perhaps less intuitively, active treatment is absolutely dependent on bacterial density: If there are not enough bacteria present, less than the so-called proliferation threshold (Payne *et al.*, 2000; Payne and Jansen, 2001, 2003; Thomas-Abedon and Abedon, 2012; Abedon, 2011a) or what others refer to simply as a threshold (Brüssow, 2007; Gill, 2008; Letarov *et al.*, 2010), then phage population growth and therefore active treatment simply will not work (Levin and Bull, 2004; Bigwood *et al.*, 2009; Capparelli *et al.*, 2010). With careful reading of the phage therapy literature (see, for example, Abedon, 2010a, 2011b), one notices a number of studies that achieve less efficacy than would seem to be desired seemingly because the phages employed did not replicate to sufficient densities, either because the phages themselves were not able to infect with sufficient vigour or because insufficient bacterial densities were present throughout treatments to support sufficient, ongoing phage production.

Summers (2001) provides a wonderfully illustrative example of this problem. In the late 1960s microbiologically positive results were obtained in the phage treatment of cholera, that is, 'very high dose phage therapy was comparable to tetracycline in reducing the excretion of vibrios in the stool' (p. 6). This smaller study was followed up with a larger one but where the dose used was 1000-fold lower. In the larger study, however, there were 'no significant effects of phage treatment' (p. 7), with an obvious inference that the lower phage doses could have contributed to the associated failure to observe a positive outcome. Sulakvelidze and Barrow (2005) similarly note, in their review of treatment of *Salmonella* infections in animals, a requirement for 'high-titre phage preparations . . for a positive therapeutic effect' (p. 339). It does not appear to be standard practice in phage therapy experimentation, however, to follow up active treatments that display poor antibacterial efficacy with, for example, intentionally passive treatments, that is, higher titre bacterial killing positive controls as discussed above. Just because active or otherwise low-titre treatments may be ineffective below certain bacterial densities, in other words, does not mean that phage therapy, defined more broadly to include also passive and/or higher-titre treatments, inherently will not work.

Note additionally that in many cases what authors describe as high-titre treatments do not necessarily supply sufficient phage densities to achieve passive treatment. That is, as a rule of thumb I would not consider phage densities to likely result in passive rather than active treatment unless supplied phage titres, in the vicinity of bacteria, are at least in the range of 10^8/ml. Even then, at such titres I will accept only that *plausibly* passive treatment may have been achieved. In fact, I tend to be particularly sceptical that anything other than active treatment has been attempted unless overwhelming phage densities are supplied, i.e. $>> 10^8$/ml achieved in the vicinity of target bacteria. This is particularly so if bacteria are biofilm associated or otherwise difficult for phages to penetrate to, and also if multiple or continuous dosing has not been attempted.

An additional concern is that active treatment can fail if rates of phage population growth are too slow. This is true even if that slowness is a result of initial phage application being too small in number, resulting in excessive lags between phage addition and the generation of sufficient phages to bring bacterial infections under phage control (Levin and Bull, 1996). The concern, however, presumably is somewhat limited in its applicability since it demands that bacterial populations be substantially growing in density even after treatment has commenced and, of course, is somewhat dependent on the application of too few phages. Acute infections caught very early in their development thus could be more difficult to treat actively, as Payne, Jansen, and colleague also discuss (Payne *et al.*, 2000; Payne and Jansen, 2001; 2003). It is notable therefore, and contrasting the typical clinical application of phages, that many phage therapy laboratory *experiments* may provide just such a situation, that is, of phage application while bacteria are replicating to higher densities, and this occurs particularly when there is little or no delay between bacterial challenge and phage application (Levin and Bull, 2004). This point presumably still holds even if bacteria are supplied initially at levels that approximate or exceed lethal doses since resulting lethality in the absence of phages likely is not completely independent of bacterial growth to higher densities (Levin and Bull, 1996). Thus, when phage therapy is studied using experimental infections of animals, the number of phages applied could be highly relevant to outcomes, even if active treatment is anticipated.

Active treatment also can fail for reasons that are independent of P'. This is because killing efficacy is primarily a function of peak phage densities (next section) rather than necessarily a consequence of how long phage populations take to reach their peak densities. It is important to realize, therefore, that for active treatment to achieve efficacy then sufficient phage numbers ultimately must be present in the vicinity of target bacteria, just as also is true with passive treatment. Furthermore, this is true even if those phage numbers are supplied in part via *in situ* phage amplification. Nonetheless, what active treatment can supply to phage therapy protocols, as discussed in greater detail below, is the presentation of large numbers of phages to bacteria that otherwise are difficult to 'passively' deliver phages directly to. 'Auto dosing' also can provide 'a "margin of safety" towards attaining phage therapy efficacy' (Abedon and Thomas-Abedon, 2010). Whether actively or passively supplied, ultimately phage therapy efficacy is dependent on the generation in the vicinity of target bacteria of peak phage concentrations that are sufficient to result in substantial bacterial eradication (Abedon, 2011a, 2012a).

Concentrations

What phage concentrations, or densities, are sufficient for or perhaps even excessive towards achieving phage therapy efficacy is a somewhat unresolved question. As noted, though, perhaps 10^8 phages/ml could constitute a reasonable target phage density under many circumstances, achieved either through conventional dosing or instead via *in situ* phage replication. This titre, as found within the vicinity of target bacteria, might then be described as approximating a minimal effective phage density. By 'effective' I mean ability of phages at that density to impact bacteria to some 'reasonable' extent and at some reasonable rate, that is, rather than in terms of the phage potential to self-amplify to yet higher densities. Effective densities need not be global to the system being treated, but instead can be immediately local to target bacteria. Similarly, densities that are effective are not those at which phages are dosed but rather are the actual densities achieved within target tissues.

Effective densities may not change appreciably over the course of treatments since the rate at which bacteria are adsorbed by phages is a function of phage density and is otherwise independent of bacterial densities. Alternatively, if different bacterial populations are targeted – ones where lower phage densities are more efficacious (see 'Eagle effect', below) versus ones where higher phage densities are more important – then minimum effective phage densities could possibly change over the course of treatment. For example, it might be permissible to initiate treatments with lower densities and then to follow up with higher densities later in treatment to 'mop up' remaining bacteria that are less available to phages. It also may be permissible to rely on active treatment earlier on during treatment and then follow this up with passive treatment. This approach similarly could involve lower phage doses earlier versus later in treatment, or alternatively less frequent dosing also earlier versus later. My impression in going through the phage-biofilm literature is that just such a staged approach might be useful particularly towards complete biofilm elimination as mediated by phage action alone (Abedon, 2010a, 2011b). Minimum effective densities additionally will change depending on the rate at which bacterial elimination is desired, with killing over minutes or hours requiring higher phage densities than killing that instead is envisaged to involve weeks or months of treatment (Abedon, 2011a).

Phage densities ideally will remain at or higher than their minimum effective density through much of or the entire course of treatment. This need for ongoing phage presence at their minimum effective density is particularly important to the extent that bacteria (1) are able to increase in density, (2) enter into phage-refractory physiological state such as forming into biofilms or entering stationary phase, or (3) otherwise increase the degree to which they negatively impact tissues if phage densities are below this minimum effective density. In other words, if phage densities drop sufficiently during treatment then that could result in a prolongation of patient suffering or indeed a worsening of disease. Furthermore, failure to achieve minimally effective densities with therapeutic phages could result in treatment failure altogether.

Peak concentrations (P_{max})

For a standard drug, its peak concentration or density, that is, C_{max}, is a consequence of conventional dosing alone. However, peak densities for phages, here described instead as P_{max}, also can be a consequence of *in situ* replication, the result of which can be described as a yield. In addition, while C_{max} typically refers to drug plasma concentrations, by P_{max}, or P_{min}, here I am referring to drug concentrations within target tissues. The reason for this distinction is that conventional drugs will typically achieve higher concentrations in blood than they will achieve following distribution outside of the blood, that is, unless drugs are preferentially concentrated in specific tissues. Highest *in situ* concentrations for phages, however, may be achieved instead at the site of infection, rather than within plasma, and this is due to the phage ability to replicate when associated with target bacteria.

Consistent with this view, one can conceivably consider phage replication *in situ* as equivalent to the above-noted preferential concentration of a drug within specific tissues, though the concentration of phages comes about via a combination of affinity for specific tissues, particularly target bacteria, and subsequent amplification (except when non-productive phages are employed, as discussed by Manoharadas and Bläsi, this volume). By contrast, enrichment of conventionally targeted drugs within specific tissues would be a consequence of affinity only. Nonetheless, a standard and reasonable default assumption should be that greater drug densities, as achieved *in situ* and concentrated in the presence of target tissues, not only may result in a greater positive drug impact but might also reduce a drug's negative impact, that is, to the extent that those higher phage densities are limited to target tissues. The issue of why secondary pharmacodynamic issues often are not expected to emerge in the course of animal or clinical testing of phages – despite the high complexity of both phage virions and their infection process – similarly stems, at least in part, from higher phage concentration occurring particularly in the immediate vicinity of their bacterial targets, with typical phage dosing otherwise often quite low in terms of total mass of bioactive material (Harper *et al.*, 2011).

Sufficient phage densities to result in substantial bacterial killing are not always trivial to achieve. Difficulties involved, however, can be different when employing active-treatment versus passive-treatment strategies. For active treatment, the problem is a combination of phage failure to display adequate P' in combination with failure to achieve adequate P_{max}, where optimizing phage population growth rates *in situ*, resulting in increases in P', will not always also maximize yield (P_{max}; increasing burst sizes of course is an obvious exception to this claim, though only if burst sizes are increased without substantially lengthening phage latent periods). For passive treatment, by contrast, the issue solely is one of achieving an adequate P_{max}. As also is the case with drug accumulation rates, pharmacokinetic concerns resulting in phage losses or dilution can be crucial determinants of whether sufficient P_{max} can be achieved. On its face it seems logical that should phage access to bacteria be limited, but still greater than zero, then the application of more phages should result in greater bacterial killing (Abedon, 2006). To my knowledge, however, this concept has not been systematically explored, particularly using phage–bacterial combinations that have been well-defined *in vitro*. As a result, there is uncertainty as to just what P_{max} may be necessary, in different systems, to provide optimal levels of bacterial killing.

Passive treatment (with some consideration of active treatment)

With passive treatment, phage adsorption need only result in the killing of target bacteria rather than also the production of additional phages. From Barrow and Soothill (1997), p. 271, '... sufficiently large dose of phage could be given that might be able to overwhelm the bacterial pathogens without any significant phage multiplication taking place.' That is, with passive treatment, by definition (Payne *et al.*, 2000, 2001; Payne and Jansen, 2003), doses supplied must be sufficiently large that a good approximation of all target bacteria are adsorbed without relying on *in situ* phage replication to bolster phage numbers. Phages must also adsorb bacteria reasonably quickly, i.e. see discussion of D-values above, with the more phages supplied then the faster bacteria will be adsorbed. Passive treatment is thus dependent on four criteria: (i) Phages adequately reaching target bacteria, (ii) phages adsorbing the bacteria they reach, (iii) phages killing the bacteria they have adsorbed, and (iv) all of this occurring over reasonable time frames. Though these criteria seemingly are at least modestly stringent, in fact they can be less stringent in terms of phage properties than the requirements for active treatment since the latter by definition requires, in addition, phage population growth. Thus, passive treatment can demand only relatively low levels of performance on the part of the phages employed, i.e. just adsorption and killing. Furthermore, passive treatment can be effected by phages that have been engineered to kill bacteria without lysing them (Goodridge, 2010a; Manoharadas and Bläsi, this volume).

On the other hand, the first criterion, can phages reach target bacteria, particularly in adequate numbers, can be *more* stringent with passive treatment than active treatment. This is because every phage necessary to effect bacterial inundation must be supplied via external dosing rather than generated *in situ*. Fortunately, for the utility of phage therapy, this can be an only academic concern: To a large degree naturally occurring phages typically will have the potential to display both passive and active treatments, killing bacteria based solely on those phage numbers supplied while at the same time either replicating to higher numbers *in situ* or at least sustaining those numbers over time through ongoing replication (Cairns *et al.*, 2009). This idea is distinct from what Payne and Jansen (2003) refer to as a 'Mixed passive and active therapy', where failures in passive treatment are followed, after some time, with a phage resurgence that is due to active replication. What I am referring to instead is the possibility of near-immediate augmentation of passive dosing that is a consequence of active phage replication. These issues point to the importance of performing adequate *in vitro* growth and host range determinations with all potential therapeutic phages even if only passive treatment is envisaged, though phage growth characterization will be more relevant the more that active treatment is relied upon rather than mostly or purely passive treatment; see, however, Bull *et al.* (2012) for additional caveats. Together these are points that were also made by d'Hérelle and Smith (1930, pp. 175–176):

> With regard to ... the quantity of bacteriophage introduced, it must be stated that this is of less importance in the case of bacteriophage therapy than in most other therapeutic procedures. The important consideration is the virulence of the bacteriophage, rather than the amount. This is true, because... bacteriophage in the presence of susceptible organisms perpetuates itself and the amount administered does not determine the amount of bacteriophage ultimately to develop.

For additional consideration of phage choice for phage therapy, see Chapter 2, as well as Gill and Hyman (2010).

Multiple or continuous dosing

Various factors can conspire to make it necessary, particularly given passive treatment, to supply more than just the minimum number of phages thought to be adequate to achieve sufficient bacterial killing. One way of augmenting phage numbers, a means that should be assumed to be required unless it can be demonstrated experimentally to not be necessary (Abedon, 2012a), is to supply phages either in multiple doses (Summers, 2001; Capparelli *et al.*, 2007) or continuously (Kutter *et al.*, 2010). An exception to this suggestion that multiple or continuous dosing might always be assumed to be necessary is in the treatment of non-human subjects where the limitations of convenience or economic feasibility can be much more severe. This results in greater impetus to design treatments in such a way that a single dose is sufficient (Carvalho *et al.*, 2010b). Under those circumstances, dosing with higher phage titres may be more feasible, at least from a regulatory or safety perspective. See, however, discussion of phage use in agribusiness by Sulakvelidze and Pasternack (2010). In Georgia and Poland, where phage therapy of humans has been routinely employed (Kutter *et al.*, 2010; Abedon *et al.*, 2011), multiple or continuous dosing is typical. Alternatively, it is important to keep in mind that single doses at sufficient phage titres certainly can be effective and potentially even more effective than multiple dosing with phages at lower titres – just not necessarily *always* more effective nor necessarily *always* more feasible.

A strategy of multiple dosing may be viewed as an extension of Payne and Jansen's (2001) argument that circumstances can exist in which phage densities are not adequately sustained *in situ*. This could be, for example, given phage application too early during an infection to achieve active treatment ('it is possible for treatment to occur so early that no therapeutic effect at all is achieved', p. 44). Indeed, a need for multiple or continuous dosing potentially can exist both if or if not active treatment is anticipated (Payne and Jansen, 2003) or, as Lampert *et al.* put it over 75 years ago (1935, p. 443), 'We consider our successful results to be due in large measure to the technique of application which we employ routinely, and which is based upon direct contact of the bacteriophage with the infected tissues by means of generous daily applications.' An additional issue is the duration of treatment, which in some instances, particularly against chronic infections, can take weeks (Slopek *et al.*, 1983; 1987; Duckworth and Gulig, 2002). Similarly, treatment of intestinal ailments using the formulation Intestiphage involves ingestion of many tens of millilitres multiple times per day (Kutter *et al.*, 2010). Quoting d'Hérelle and Smith (1930) yet again (p. 176):

> The number of administrations essential to induce a therapeutic effect varies with the type of condition under treatment; in acute infections or in infectious conditions that have not become fully chronic, a single administration or, perhaps, two administrations should overcome the organisms. In chronic conditions, on the other hand, it may be necessary to continue bacteriophage therapy over a relatively long period.

More generally, quoting Adembri and Novelli (2009), and presumably obvious though nonetheless worth stressing (p. 523): 'Microbiological and clinical success are less likely when suboptimal exposure and/or incorrect duration of antimicrobial therapy occur...'

Trough concentrations (P_{min})

An additional dosing consideration and one that is related to multiple dosing is what is known as trough concentrations. From Czock *et al.* (2009, p. 484):

> Clinically, the maximum concentration should be the pharmacodynamic target for concentration-dependent antimicrobials (e.g. aminoglycosides, fluoroquinolones) .. For time-dependent antimicrobials (e.g. penicillins, cephalosporins, carbapenems, and possibly also anti-viral drugs), the trough concentration (C_{min}) should be the target because a threshold concentration must be assumed and the trough concentration should not fall below this threshold at the end of the administration interval.

In other words, the goal of dosing should be to sustain drug concentrations at or above those required to achieve efficacy, and, for at least some antibacterials, as high as possible (Cerf *et al.*, 2010). Whether these considerations apply to phages, however, is an open question, particularly given their single-hit killing kinetics along with the potential for phages to amplify their numbers when in the presence of target bacteria. The resulting active treatment in other words could under at least some circumstances have the effect of reducing the occurrence of troughs in phage concentrations. Nevertheless, if there exists a minimum *effective* density of phages that is necessary to achieve a degree of bacterial killing, killing that also occurs at a reasonable rate, then the success or at least duration of phage therapy treatments likely will be a function of the extent of time over which peak phage concentrations meet or exceed this density, whether those densities are achieved by active or instead by passive means.

Payne and Jansen (2003) provide a formula for how often one should repeat dosing, for a given dose applied, to sustain an effective phage density *in situ*. This formula is $t_i = \ln(P_0/P_e)/d$, where t_i is the interval between dosing, P_0 is the phage density applied (or, more realistically, the phage density that actually reaches target bacteria), P_e is the effective phage density (that is, the target phage density where 'e' stands for effective, which is the minimum trough density that should be allowed), and d is a description of how fast phages decay *in situ*. Note that the formula loses meaning if phage *in situ* replication sufficiently counters phage decay such that d is a negative number. That is, with negative decay, i.e. sufficient phage population growth, repeated phage dosing may not be necessary. Indeed, if phage replication exactly balances phage decay then $d = 0$ and therefore t_i equals an expected infinity, meaning that no redosing is required. Redosing also may not necessarily be crucial for the treatment of all patients and infections. Nonetheless multiple dosing still represents a critical component of at least some treatment success.

Payne and Jansen additionally suggest that their formula implies a utility for more frequent dosing rather than less frequent but larger doses. This is a perspective that, if taken to its logical conclusion, would point to a utility to employing continuous dosing and otherwise not using doses containing extremely high phage numbers. Instead, if such continuous dosing is possible, then P_e would simply be sustained at the site of target bacteria through ongoing influx of phages found at that density. In other words, if $P_0 = P_e$ then $t_i = 0$, which by definition is continuous dosing. See Abedon (2011a) for further consideration of frequency of phage dosing.

An additional consideration, addressed in the above quotation, is the question of what factors might contribute to an *upper* limit on phage dosing, particularly a maximum concentration (P_{max} or C_{max}). That is, is P_{max} limited by safety concerns, regulatory considerations, economic issues, phage replication performance *in situ*, other pharmacokinetic considerations, or instead due to a potential for antibacterial efficacy to decline as a consequence of greater levels of dosing? Generally, for phages, the answer to this question is not known. Note that the latter point, i.e. reduced bacterial killing or subject survival upon the application of higher phage numbers, such as observed by Callaway *et al.* (2008) as well as Hung *et al.* (2011), is known with chemical antibacterials as the Eagle effect or phenomenon (Levison and Levison, 2009). See Abedon (1994, 2011d) and Thomas-Abedon and Abedon (2012) for discussion of possible causes of phage therapy Eagle effects. Additional possible causes include the systemic application of insufficiently purified phage formulations, that is, as due to the toxic impact of impurities particularly of bacterial origin (Merril *et al.*, 2003), and the problem of excessive reduction to lysogeny should temperate phages be allowed to adsorb with high multiplicities (Weitz *et al.*, 2008).

Absorption and distribution

A drug has to reach its target or targets to be effective. Consequently, the ability of drugs to move between as well as within the different 'compartments' making up bodies is crucial to their *in situ* activity. For many drugs this movement will consist of uptake into the bloodstream (absorption) along with movement into other tissues (distribution), although for phages the terms translocation and penetration to a degree have been used synonymously. The phage ability to pass from compartment to compartment within bodies, such as from the gastrointestinal lumen into the blood, for example, has been described by Górski *et al.* (2006) in their review of the subject as 'bacteriophage translocation'; see also Olszowska-Zaremba *et al.* (2012). Penetration is used to describe phage movement into localized infections, such as infiltration into biofilms (Azeredo and Sillankorva, this volume), or into animal tissues in general (Dabrowska *et al.*, 2005). See also Letarov *et al.* (2010) for review of this subject. Distribution, or penetration, also can refer to movement from extracellular to intracellular compartments, though the phage potential to impact intracellular bacterial infections has not been extensively explored (Broxmeyer *et al.*, 2002; Danelishvili *et al.*, 2006; Capparelli *et al.*, 2007).

Issues associated with gastrointestinal (enteric) delivery of phages along with the issue of bacterial translocation will not be extensively addressed here. I consider instead a subset of topics pertaining to phage absorption and distribution, including examination of the extent to which phage therapy researchers should be concerned about these issues.

Importance, or lack thereof, of phage therapy pharmacokinetics

While the pharmacokinetics of chemotherapeutics can differ between species being treated (e.g. mouse vs. human) and also in terms of the health of the recipient (healthy volunteers versus patients), phage therapy so far has not been studied sufficiently robustly to understand its pharmacokinetics in any one system, much less comparatively. Indeed, such determinations are not fully worked out even for standard chemotherapeutic antibacterials (Adembri and Novelli, 2009; Czock *et al.*, 2009). Furthermore, because the phages in phage therapy tend to be diverse, differing depending on their narrow-spectrum targets, and even among the phages sharing a particular target pathogen, it may be that data collected for any one phage–bacterium combination will not be greatly applicable to other phage–bacterial combinations. For example, *per os* delivery, that is, oral delivery that results in phage absorption into the blood, appears to be possible for some phages, as reviewed by Górski *et al.* (2006) and Olszowska-Zaremba *et al.* (2012), but perhaps not for certain T4-like and other large phages (Krylov, 2001; Bruttin and Brüssow, 2005; Denou *et al.*, 2009). Alternatively, the phage potential for translocation from the gut might differ depending on the health state of the subject (Letarov *et al.*, 2010).

These problems also exist for antibiotics, but the number of antibiotics for which pharmacokinetic information is desired seemingly is far smaller than is the case for phages (Harper and Kutter, 2009), where the phage therapy literature is full of research groups isolating novel phage strains that are then characterized primarily in terms of their primary pharmacodynamics (i.e. efficacy). As a consequence of this diversity of phages that potentially could be used therapeutically, one could readily argue that performing detailed pharmacokinetic analyses of all phages of interest would either represent a daunting task or instead is something that should represent a lower priority than efficacy determination. The latter point may be particularly valid (1) to the extent that concerns of phage absorption and distribution are less relevant given phage application directly to bacterial infections (such as topical application), (2) if phages can be applied in great numbers due to relatively low costs in combination with their seemingly low toxicity, or indeed (3) so long as phages are capable of amplifying their numbers to effective densities when in contact with target bacteria. Thus, while at the same time pharmacokinetic issues may be more difficult to address for phages as a whole, under many treatment scenarios they also may be less important to address.

Phage routes to different body compartments

Phage distribution to different areas in the body, that is, different compartments, is reviewed by Górski *et al.* (2006); see also Letkiewicz *et al.* (2010). The more compartments a drug must pass through to reach its site of activity, then the more steps during which drug concentration can be reduced (Levison and Levison, 2009), such as through dilution, 'translocation' inefficiencies, and phage loss or inactivation (Donlan, 2009).

The latter can be a consequence of adaptive immunity mounted against phages to which an animal is not naive (Huff *et al.*, 2010) or via action of the animal reticulo-endothelial system (Merril *et al.*, 2006; Merril, 2008) Note that general reviews considering mechanisms of phage inactivation are available (Carlton, 1999; Merril *et al.*, 2003; Dabrowska *et al.*, 2005; Sulakvelidze and Kutter, 2005) and Kutter (2008) provides an overview especially of the early literature. Letarov *et al.* (2010), however, point out that the amount of protein found in phages against which antibodies might form, particularly whose binding can result in phage inactivation, is relatively low. Nonetheless, when drug inactivation is an issue then higher amounts of drug may need to be supplied to reach minimum effective concentrations. As Carlton (1999) consistently though more narrowly suggests (p. 270): '... it may be possible to administer a higher dose of phage, to compensate for those that are rendered non-viable by interaction with neutralizing antibodies.'

Such concerns often can be mitigated by the phage potential to replicate once it reaches target bacteria. Nonetheless, inefficiencies in transfer between compartments can still result in fewer phages reaching their target than is optimal for treatment, which in a worst-case scenario could mean no phages at all reaching that point. For example, the effective densities for chemical antibiotics can be difficult to achieve via systemic application given poor blood circulation to burns (Ahmad, 2002), osteomyelitis, or diabetic ulcers (Stone, 2002). In addition, if strictly passive treatment is to be undertaken, then phage losses may be of even greater concern since subsequent phage replication, by definition, is not counted upon given passive treatment for the achievement of efficacy. Indeed, if saturation of translocation mechanisms between compartments can occur (Górski *et al.*, 2006; Abedon and Thomas-Abedon, 2010), then passive treatment through systemic application may not even be possible via certain dosing routes. See Duerr *et al.* (2004) for consideration of how phages might be 'designed' to more effectively translocate from the digestive tract into the blood and Letarov *et al.* (2010) for further review of this subject.

The greater the number of phages that must be delivered to target bacteria to achieve desired levels of killing or control, then either the more directly phages must be applied to infections – such as directly to burns (Kumari *et al.*, 2011), ear infections (Hawkins *et al.*, 2010), wounds (Kutter *et al.*, 2010), bacteraemias (Gupta and Prasad, 2010), or nasal cavities (Mann, 2008) – or simply the more phages that must be supplied to the patient, or both. As a result, failures in phage therapy efficacy can be due to failures in adequate dosing or instead a consequence of a need for phage application more directly to infections. These failures would be rather than because of poor intrinsic phage potential to clear bacterial infections but instead due to insufficient numbers of phages encountering target bacteria. As Cerf *et al.* (2010) put it for antibacterial therapy in general, 'bacteriological failure is usually the result of suboptimal therapy.' It is crucial for researchers, therefore, to employ pharmacologically rigorous follow up should they observe disappointing phage performance *in situ* (Abedon, 2012a).

'Guesstimating' dosing

Non-topical phage application includes those to the gastrointestinal lumen (enteral) or instead directly to internal body tissues (parenteral); note that application to the lungs via inhalation is considered to be a form of topical drug application, i.e. such as via phage nebulization (Golshahi *et al.*, 2010). For gastrointestinal targeting, for example, the key parameters determining phage density at the site of target bacteria are phage dose, phage survival particularly during passage through the stomach (Smith *et al.*, 1987; Ma *et al.*, 2008; Carvalho *et al.*, 2010b; Wang and Sabour, 2010), and phage dilution at various steps in the process. If active treatment is possible, that is, if sufficient bacterial densities exist to support substantial (and sufficient) phage population growth, then these dosing concerns may not be highly relevant to treatment success. If however active phage replication is insufficient to reach minimally effective phage densities, that is, those necessary to achieve substantial bacterial eradication, then phages must be supplied up front in sufficiently high numbers and in such a manner that their numbers or densities are not considerably reduced by a combination of inactivation, dilution, and removal.

Let's consider an illustrative example of the consequences of not being able to rely upon *in situ* phage replication to bolster phage densities: One can easily envisage situations such that a minimum of 10^8 phages/ml may be required to effect reasonably rapid bacterial eradication, with a 10-fold increase employed to make up for presumed *in situ* inefficiencies in phage acquisition of bacteria (e.g. lower phage adsorption rate constants or bacterial location within difficult to penetrate sites), another ten-fold increase to take into account phage dilution such as within the gastrointestinal lumen, and perhaps another ten-fold increase to make up for losses during passage such as through the stomach. The result is a not unrealistic expectation that 10^{11} phages per dose might be required to actually achieve some reasonable approximation of bacterial eradication via passive treatment. See Johnson *et al.* (2008) for a nice overview of the use of such high-dosage oral application of phages to food animals (which, as reviewed, results in varying success).

Hung *et al.* (2011) were successful – 44% survival vs. 10% – in treating *Klebsiella* liver abscesses after a 30-min delay with as few as 2×10^5 plaque-forming units (PFU) per mouse given via intra-gastric delivery that followed 0.2 ml of 0.2 M $NaHCO_3$ to neutralize gastric acid. A dose of 2×10^6 PFU/mouse in turn was highly effective (100% survival). On the other hand, given a 6-hour delay an intragastric dose of 2×10^8 PFU/mouse resulted in only 75% survival, potentially demonstrating the tendency for higher bacterial loads or greater establishment of infections to require higher phage numbers to bring infections under control (though note that no $NaHCO_3$-positive, mock phage-treatment negative control seems to have been performed in this study, meaning that it is conceivable that protective effects were due to some other aspect of treatments besides administration of infective phage virions). After 24-hours post bacterial challenge there was 56% survival at that higher phage dose and after 48 h 25% survival, which statistically the authors described as failing to significantly protect. It would be of interest to see whether higher phage densities and/or multiple phage dosing could reduce mortality for any of these multi-hour delays in treatment.

Fairly high phage densities, that is, might be required – whether achieved via active or passive means – given greater bacterial densities, such as the 10^8 to 10^{10} per gram of infected tissue suggested by Levison and Levison (2009). The bottom line, therefore, is that it is easy to imagine that an application of 10^9 phages to a human subject in order to eradicate a particular strain of, for example, gastrointestinal bacteria, could be grossly inadequate if active treatment does not occur. Such concerns are particularly relevant if efforts are not made to assure and ideally to also characterize intact phage passage through the stomach as well as phage potential to replicate to sufficiently high densities once target bacteria are contacted.

Systemic phage application – into the blood

The pharmacokinetic issues associated with systemic phage application are more diverse than those of gastrointestinal application, if only because of the greater variety of compartments in addition to the gastrointestinal tract within which infections can flourish. Secondary pharmacodynamic issues, i.e. side-effects, also can be more varied given systemic application, though secondary pharmacodynamics by definition are not pharmacokinetic issues. As with gastrointestinal application, pharmacokinetic issues are greatly simplified if active treatment is sufficient to achieve bacterial eradication since the number of phages that must enter bacteria-containing compartments is much lower than if passive treatment instead is relied upon. In either case, phage delivery can be simplified conceptually by employing the standard pharmacokinetic concepts of absorption and distribution. That is, separating phage movement into two basic steps, into the blood (absorption) and out of the blood to the rest of the body (distribution). These pharmacokinetic concepts are in addition to those resulting in phage loss or inactivation, 'activation', or the issues associated with phage penetration into bacterial biofilms.

Consideration of the movement of phages into blood is an important issue for a variety of reasons. First, it is a primary concern in the pharmacokinetic analysis of drugs in general, that is, essentially the default pathway that drugs take through the body can be broken down into application that is followed by absorption. Second, blood has the utility of being easily sampled, though plasma drug levels are not always the most meaningful measure of drug pharmacokinetics (Levison and Levison, 2009). Nonetheless, time courses of drug movement especially into and out of the blood can be easily obtained, particularly from larger animals and without animal sacrifice. The result not only is greatly reduced expense but also ready investigation of this aspect of drug pharmacokinetics in humans. Three, for targets that are limited to systemic circulation, drug access to the blood is of course of primary concern. Lastly, to study drug distribution, that is, out of the blood and into other tissues, it is essential to first deliver a drug to the blood.

Systemic phage application – out of the blood

There exist a variety of pathways by which phage absorption can be initiated, the most straightforward of which is intravenous (i.v.) delivery. Such parenteral delivery not only results in direct application to the systemic circulation but also in principle can be applied at such a rate so as to achieve steady state levels in plasma (Levison and Levison, 2009). Intravenous application should be employed for the study of phage distribution – movement out of the blood into different locations – that is unless there is suspicion that alternative absorption protocols could provide qualitatively different phage distribution patterns. This point is also relevant for issues of decline in phage concentrations in the blood where if phages are not first delivered directly to the blood then it may be difficult to distinguish slow phage movement into the blood from fast phage movement out of the blood (Uchiyama *et al.*, 2009). Following intraperitoneal (i.p.) injection, for example, Chhibber *et al.* (2008) did not obverse peak phage densities in the blood until six hours post phage delivery. Cerveny *et al.* (2002) by contrast employed i.v. inoculation but also followed distribution primarily in terms of blood concentrations, noting that the phage (p. 6259) 'was unable to escape the vasculature into the tissues of these animals not infected with *V. vulnificus*', suggesting that (p. 6260) 'as the vibrios reach sufficient numbers in the tissues, they induce vasodilation and vascular permeability … enabling the phages access to the bacteria as oedema floods the infected tissues.'

While i.v. delivery provides an obvious simplification particularly in terms of the study of subsequent distribution, it is not necessarily ideal for phage delivery. Concerns include its overly invasive nature, the rapidity with which absorption is achieved (though absorption rates given intravenous delivery can be as slow as desired), concerns associated with carryover of contaminating bacterial toxins such as endotoxin directly into the blood (Carlton, 1999; Boratynski *et al.*, 2004; Merril *et al.*, 2006; Gill and Hyman, 2010), and a lack of convenience, particularly in terms of its limiting phage delivery to clinical settings. As a consequence, alternative approaches to phage absorption have been developed by various groups, though it is important to reiterate that for the study particularly of phage distribution, simplified delivery of phages intravenously can be an asset.

Intravenous delivery of bacteriophages has been used for experimental treatment of, for example, skin lesions in mice (Cerveny *et al.*, 2002) or i.v. bacterial challenge also in mice (Capparelli *et al.*, 2010; Zimecki *et al.*, 2010). Alternatives to i.v. application include *per os* (below); i.p. delivery to mice used to treat lung infection (Carmody *et al.*, 2010; Chhibber *et al.*, 2008), burn infection (Kumari *et al.*, 2009, 2010), urinary tract infection (Nishikawa *et al.*, 2008), bacteraemia (Sunagar *et al.*, 2010), and liver abscesses (Hung *et al.*, 2011); subcutaneous (s.c.) injection used to treat mouse bacteraemia (Gupta and Prasad, 2010); intramuscular (i.m.) administration for thoracic air sac infection in chickens (Huff *et al.*, 2010), etc.

Delivery routes employed can matter, as Heo *et al.* (2009) observed. In five of six measures, better absorption as well as distribution to liver and lungs was obtained given i.m. versus i.p. administration of *Pseudomonas* phages. Mouse survival, however, was greater with i.p. versus i.m. application against an i.p. bacterial challenge. Kumari *et al.* (2009) similarly suggest a trend in the literature for greater efficacy following i.p. phage administration versus i.m., or s.c. Kutateladze and Adamia (2010) list phage application to humans via rectal, intrapleural, and i.v. means that are in addition to oral and various topical routes; see also Chanishvili *et al.* (2001). Letkiewicz *et al.* (2010) describe rectal as well as *per os* administration to treat prostatitis, also in humans; see also Górski *et al.* (2006) as well as Dabrowska *et al.* (2005). Note that here as well as in various other places in this review I have limited referencing to relatively recent studies; see Sulakvelidze and Kutter (2005) as well as Abedon *et al.* (2011) for review of older studies.

Breaching the blood–brain barrier?

Phages have at least some potential to penetrate to brain tissue. In an oft-cited phage therapy study, Dubos *et al.* (1943) found that i.p.-delivered phages were able to reach *Shigella* injected into the brains of mice. Phages also were found in not-infected brains. What is not clear, however, is whether phages in either case were crossing *intact* blood–brain barrier (Merril *et al.*, 2006). As noted by Dubos *et al.* (p. 165):

> It is clear from the results presented in Table II that, in uninfected mice, the bacteriophage in the blood rapidly reaches a titre compatible with the numbers that could be expected if the total number of particles injected were simply diluted in the total fluid volume of the mouse. The brain level remains at all times below that of the blood, perhaps due to the small blood content of the brain.

Consistently, a number of mechanisms can allow large, hydrophilic substances, such as antibiotics or animal viruses, to pass from the blood through the blood–brain barrier including inflammation, leucocytes, or exposure to various substances (Brightman and Kaya, 2000). See Pouillot *et al.* (2012) for a more recent as well as careful analysis of issues of phage distribution including in terms of crossing the blood–brain barrier.

Smith and Huggins (1982) observed similar results, treating mice that had been challenged intracerebrally with *E. coli* using i.m.-delivered phages, as also have Barrow *et al.* (1998) with experimentally infected poultry. Tellingly, with regard to the potential for phages to cross intact blood–brain barrier, Smith and Huggins observed that phage densities in uninoculated muscle were similar to those in uninoculated brain, again suggesting that phages were found within the vascularization of these tissues rather than within the interstitial fluid of either (see too the discussion by Barrow *et al.*). Phage densities also were much higher in blood, liver, and especially spleen (see also, Nishikawa *et al.*, 2008), where the latter, contrasting the continuous endothelium associated with brain tissue, contain what are known as discontinuous or sinusoidal capillaries in which endothelial pores, or fenestrae, are particularly large. The spleen, in fact, can be a major site of phage loss from blood as phages can be sequestered there by what is known as the reticulo-endothelial system (Merril, 2008), a.k.a., the mononuclear phagocyte system, but so too can the liver sequester phages (Merril *et al.*, 2006; Letarov *et al.*, 2010). Merril (2008) also reviews the surprising phage potential to access the central nervous system following nasal application.

In the Dubos *et al.* (1943) and other studies just reviewed a common feature is the issue of phage potential to penetrate into the brains of uninfected animals versus animals that were infected with phage-susceptible bacteria. For the sake of gaining pharmacokinetic information, however, it is not essential to employ phage–host combinations that are phage productive. Indeed, it is a not unreasonable strategy towards phage pharmacokinetic characterization to produce experimental infections by bacteria that do not support adsorption by the phage of interest (that is, adsorption-resistant strains or mutants). It is thus possible, at least in principle, to vary studies considering phage access to central nervous system tissues in terms of animal health – bacterially infected versus not infected – as well as in terms of infection susceptibility to the delivered phage.

Spectrum of activity

The spectrum of a phage's antibacterial activity can vary depending on the genotype of target bacteria, the physiology of those same bacteria, and as a consequence of environmental factors that can affect not only bacterial physiology but also phage adsorption characteristics. The latter includes in ways that are independent of the bacterial state

(Hyman and Abedon, 2009). It is an important consideration therefore not just whether a given phage can kill a given bacterium *in vitro* but also whether that killing as well as phage population growth can satisfactorily occur *in situ* (Cerveny *et al.*, 2002; Gill *et al.*, 2006; Bull *et al.*, 2010, 2012), e.g. such as against bacteria found in various phases of growth or arrayed into biofilms. This concern also seems to have been noted by d'Hérelle and Smith (1930, p. 170): 'In the case of chronic diseases, the patient must receive a bacteriophage having a destructive power so great that the already resisting bacteria are destroyed. In the latter case, the uncertainty of the treatment is evidently greater than is the case in acute infections.'

For phages, rather than characterizing individual isolates *in situ* as a late step in development such as in the course of animal testing, it can be more effective, as Bull *et al.* (2010) point out, to employ enrichment schemes such that phages with desirable *in situ* characteristics are isolated from the start. This in some cases might have the added benefit of permitting more facile phage isolation, perhaps particularly when working with fastidious bacteria (Carvalho *et al.*, 2010a). In addition, such enrichment schemes potentially could attain end points in terms of phage suitability to *in situ* environments that are similar to what can be achieved by isolating phages directly from those or associated environments, e.g. (Kutter, 2009; Carvalho *et al.*, 2010b; Miller *et al.*, 2010).

Levin and Bull (2004) likewise suggest treating infections with a cocktail of phages and then isolating for subsequent treatment those phages whose numbers increase. This is an approach that, if combined with molecular techniques for identification of enriched phages, could allow for rapid choice for subsequent treatment from a pre-existing bank of treatment-ready phage formulations. Alternatively there exist a number of potential strategies for rapid identification of bacterial susceptibility to specific phages, e.g. Merril *et al.* (2003) or, more broadly in terms of bacterial identification, Cox (2012). Combining these ideas, I envisage, for example, multi-phage rapid-identification testing, with the actual phage detection performed either on-site or off, perhaps combined with overnight delivery of specific phage formulations from a central phage repository. With the latter, clinics then would not need to stock up on a great diversity of either narrow-spectrum formulations or cocktails, particularly when targeting chronic infections since under such circumstances time between presentation and subsequent treatment is less dear. See Pirnay *et al.* (2011) for discussion of similar issues as well as Chan and Abedon (2012) and Chan *et al.* (2013) for additional discussion of phage banks versus phage cocktails.

Of course, the above-noted enrichment and subsequent phage identification need not be dependent on this idea of *in vivo* phage enrichment. Indeed, utilization of enrichment schemes that display high fidelity to *in situ* conditions is not universally necessary for subsequent phage therapy success, nor even for ease of discovery of effective phages, as numerous authors have demonstrated, e.g. (Kumari *et al.*, 2009; Capparelli *et al.*, 2010; Gupta and Prasad, 2010; Sunagar *et al.*, 2010). For related consideration, particularly of phage host range, see reviews by myself and colleagues (Hyman and Abedon, 2010; Chan and Abedon, 2012a). A recent article by Then and Sahl (2010) also provides perspective on the utility of narrow-spectrum antimicrobials in general. See Chapter 2 for further discussion of phage choice for phage therapy.

Development issues

Development costs are a consequence of all those things that are required to bring a drug to the point where it is employed in clinical practice, typically described as bringing a drug to market. These costs are associated with discovery, characterization, testing for efficacy as well as toxicity, development of methods of manufacture and delivery, and various additional aspects associated with gaining regulatory approval. Phages are products of microorganisms and therefore the discovery process conceptually is similar to that of traditional antibiotics, while the characterization process, as it occurs *in vitro*, tends to be specific for how phages are normally characterized in the laboratory, e.g. see Clokie and Kropinski (2009). In particular, comparative genomics has become an important component of phage characterization, particularly to avoid phages that are either temperate or which encode bacterial virulence factors. See Brüssow (2012) and Chan *et al.* (2013) for these and additional considerations. Phenotypic phage characterization is also necessary, e.g. such as in terms of antibacterial virulence characteristics.

Phages too can be differentiated in terms of their ability to display generalized transduction, though operationally this has been of lower priority in terms of phage screening than issues of ability to display lysogenic cycles or expression of virulence-factor encoding genes (Thiel, 2004). Phages in addition can be modified in various ways, genetically as well as chemically (Krylov, 2001; Goodridge, 2010b). A further consideration is the potential for phages to display synergistic, additive, or antagonistic interactions with other phages or with other substances that phages may be delivered in association with. These other substances can include the use of phages in conjunction with treatment with antibiotics (Chanishvili *et al.*, 2001; Weber-Dabrowska *et al.*, 2003; Kutter, 2008; Chanishvili *et al.*, 2009a).

The process of phage characterization should again converge with that of antibiotics in terms of *in situ* efficacy and toxicity testing. The generally low toxicity of phages in combination with their typical ease of discovery, however, can lead to a need to modify these latter steps in order to fully take advantage of the phage utility as antagonists of bacterial infections. In this way especially, phages may differ in terms of their development from other narrow spectrum anti-bacterial strategies (Then and Sahl, 2010) and particularly so to the extent that phage secondary pharmacodynamic (side) effects are slight, generalizable, and/or easily avoided through straightforward *in vitro* characterization. Additional issues include what form phages should assume for their distribution and delivery (liquid, dried, or associated with solid or semi-solid matrices), how they will be preserved, and determination of expected shelf life. Studies have looked at freeze drying phages, for example, e.g. Alfadhel *et al.* (2011), though with substantial phage loss. See Chapter 2 for consideration of issues surrounding phage choice for therapy purposes and Chapter 12 for those stemming from the regulation of phages as drugs; for the latter, see also Sulakvelidze and Pasternack (2010). For discussions of phage therapy-related patents, see Gill *et al.* (2007), Courchesne *et al.* (2009), and Sulakvelidze and Pasternack (2010). See also Krylov (2001).

Conclusion

Pharmacology is the study of drug impact on bodies as well as body impact on drugs, where body is defined broadly to include normal flora and bacterial pathogens as well as normal body tissues. With phage therapy, pharmacology is complicated due to the potential for phages to replicate as well as the large size and complexity of their virions. The latter, for example, makes the phage ability to translocate across certain tissues somewhat surprising. Though the phage potential to replicate *in situ* is a conceptual complication on their pharmacokinetics, alternatively this ability can serve to reduce the importance of greatly understanding certain other aspects of phage pharmacology, such as more quantitative measures of phage competence to move between body compartments. Indeed, though not covered here in great detail, it would appear to be far more important to develop realistic animal models of disease for testing phage therapeutic potential (Abedon, 2012a) rather than to ponder exhaustively on the relative efficiency with which phages can cross the various barriers found within the body. The latter point, however, is less valid given circumstances where phage *in situ* self-amplification is not possible.

Hugely important to the practice of phage therapy and somewhat independent of their ability to replicate is that the use of phages appears to be fairly safe, generating seemingly at worst only minimal side-effects. As a consequence, the pharmacology of phage therapy is greatly simplified, often permitting achievement of efficacy despite pharmacological ignorance and with high *in situ* phage concentrations attained via a blend of conventional and auto dosing. Notwithstanding this phage potential to not only amplify their numbers *in situ* and seeming ability to do so rather safely, for the sake of improvement of efficacy, regulatory approval, and simply rigorous experimentation, it is desirable that pharmacological considerations not be ignored during phage therapy development. In short, it is important to know, *in situ*, where in bodies phages are, in what densities, and, especially, to what if any pharmacodynamic effect. This is just as knowledge of these things is desirable for any drug and also just as 'strong understanding of phage biology has the potential to facilitate more rational thinking about the therapeutic process and the selection of therapeutic phages' (Kutter, 2008). Such understanding is particularly important when phage therapy efforts result in disappointing primary pharmacodynamics.

Ultimately the properties that distinguish obligately lytic phages, as drugs, from other especially narrow spectrum antimicrobials – thereby contributing greatly to the phage potential as antibacterial agents – principally are a combination of their ease of discovery, generally low toxicity, and, of course, potential to self amplify *in situ*. Commercially as well as practically, it is also highly relevant that the evolution of bacterial resistance to chemotherapeutic antibacterials, such as antibiotics, is not known to result generally in cross-resistance to phages, e.g. Sulakvelidze (2005) or Międzybrodzki *et al.* (2007).

Acknowledgements

Thank you to Andrew Curtright and Bob Blasdel for their excellent comments.

References

Abedon, S.T. (1990). Selection for lysis inhibition in bacteriophage. J. Theor. Biol. *146*, 501–511.

Abedon, S.T. (1994). Lysis and the interaction between free phages and infected cells. In The Molecular Biology of Bacteriophage T4, J.D. Karam, E. Kutter, K. Carlson, and B. Guttman, eds. (Washington, DC, ASM Press), pp. 397–405.

Abedon, S.T. (1999). Bacteriophage T4 resistance to lysis-inhibition collapse. Genet. Res. *74*, 1–11.

Abedon, S.T. (2006). Phage ecology. In The Bacteriophages, R. Calendar and S.T. Abedon, eds. (Oxford: Oxford University Press), pp. 37–46.

Abedon, S.T. (2009a). Impact of phage properties on bacterial survival. In Contemporary Trends in Bacteriophage Research, H.T. Adams, ed. (Hauppauge, New York, Nova Science Publishers), pp. 217–235.

Abedon, S.T. (2009b). Kinetics of phage-mediated biocontrol of bacteria. Foodborne Pathog. Dis. *6*, 807–815.

Abedon, S.T. (2009c). Phage evolution and ecology. Adv. Appl. Microbiol. *67*, 1–45.

Abedon, S.T. (2010a). Bacteriophages and biofilms. In Biofilms: Formation, Development and Properties, W.C. Bailey, ed. (Hauppauge, NY, Nova Science Publishers), pp. 1–58.

Abedon, S.T. (2010b). The 'nuts and bolts' of phage therapy. Curr. Pharm. Biotechnol. *11*, 1.

Abedon, S. (2011a). Phage therapy pharmacology: calculating phage dosing. Adv. Appl. Microbiol. *77*, 1–40.

Abedon, S.T. (2011b). Bacteriophages and Biofilms: Ecology, Phage Therapy, Plaques (Hauppauge, NY, Nova Science Publishers).

Abedon, S.T. (2011c). Envisaging bacteria as phage targets. Bacteriophage *1*, 228–230.

Abedon, S.T. (2011d). Lysis from without. Bacteriophage *1*, 46–49.

Abedon, S.T. (2012a). Phage therapy best practices. In Bacteriophages in Health and Disease, P. Hyman and S.T. Abedon, eds. (Wallingford, UK: CABI Press), pp. 256–272.

Abedon, S.T. (2012b). Spatial vulnerability: bacterial arrangements, microbiolonies, and biofilms as responses to low rather than high phage densities. Viruses *4*, 663–687.

Abedon, S.T. (2012c). Thinking about microcolonies as phage targets. Bacteriophage *2*, 200–204.

Abedon, S.T., and LeJeune, J.T. (2005). Why bacteriophage encode exotoxins and other virulence factors. Evol. Bioinform. Online *1*, 97–110.

Abedon, S.T., and Thomas-Abedon, C. (2010). Phage therapy pharmacology. Curr. Pharm. Biotechnol. *11*, 28–47.

Abedon, S.T., Kuhl, S.J., Blasdel, B.G., and Kutter, E.M. (2011). Phage treatment of human infections. Bacteriophage *1*, 66–85.

Adembri, C., and Novelli, A. (2009). Pharmacokinetic and pharmacodynamic parameters of antimicrobials: potential for providing dosing regimens that are less vulnerable to resistance. Clin. Pharmacokinet. *48*, 517–528.

Ahmad, S.I. (2002). Treatment of post-burns bacterial infections by bacteriophages, specifically ubiquitous Pseudomonas spp. notoriously resistant to antibiotics. Med. Hypoth. *58*, 327–331.

Alfadhel, M., Puapermpoonsiri, U., Ford, S.J., McInnes, F.J., and van der Walle, C.F. (2011). Lyophilized inserts for nasal administration harboring bacteriophage selective for *Staphylococcus aureus*: *in vitro* evaluation. Int. J. Pharm. *416*, 280–287.

Barrow, P.A., and Soothill, J.S. (1997). Bacteriophage therapy and prophylaxis: rediscovery and renewed assessment of potential. Trends Microbiol. *5*, 268–271.

Barrow, P., Lovell, M., and Berchieri, A. jr. (1998). Use of lytic bacteriophage for control of experimental *Escherichia coli* septicemia and meningitis in chickens and calves. Clin. Diag. Lab. Immunol. *5*, 294–298.

Bigwood, T., Hudson, J.A., and Billington, C. (2009). Influence of host and bacteriophage concentrations on the inactivation of food-borne pathogenic bacteria by two phages. FEMS Microbiol. Lett. *291*, 59–64.

Boratynski, J., Syper, D., Weber-Dabrowska, B., Lusiak-Szelachowska, M., Pozniak, G., and Górski, A. (2004). Preparation of endotoxin-free bacteriophages. Cell. Mol. Biol. Lett. *9*, 253–259.

Brightman, M.W., and Kaya, M. (2000). Permeable endothelium and the interstitial space of brain. Cell Mol Neurobiol. *20*, 111–130.

Broxmeyer, L., Sosnowska, D., Miltner, E., Chacón, O., Wagner, D., McGarvey, J., Barletta, R.G., and Bermudez, L.E. (2002). Killing of *Mycobacterium avium* and *Mycobacterium tuberculosis* by a mycobacteriophage delivered by a nonvirulent mycobacterium: A model for phage therapy of intracellular bacterial pathogens. J. Infect. Dis. *186*, 1155–1160.

Brüssow, H. (2007). Phage therapy: the Western perspective. In Bacteriophage: Genetics and Microbiology, S. McGrath and D. van Sinderen, eds. (Norfolk, UK: Caister Academic Press), pp. 159–192.

Brüssow, H. (2012). What is needed for phage therapy to become a reality in Western medicine? Virology *434*, 138–142.

Bruttin, A., and Brüssow, H. (2005). Human volunteers receiving *Escherichia coli* phage T4 orally: a safety test of phage therapy. Antimicrob. Agents Chemother. *49*, 2874–2878.

Bull, J.J., and Regoes, R.R. (2006). Pharmacodynamics of non-replicating viruses, bacteriocins and lysins. Proc. R. Soc. Lond. B Biol. Sci. *273*, 2703–2712.

Bull, J.J., Vimr, E.R., and Molineux, I.J. (2010). A tale of tails: Sialidase is key to success in a model of phage therapy against K1-capsulated *Escherichia coli*. Virology *398*, 79–86.

Bull, J.J., Otto, G., and Molineux, I.J. (2012). *In vivo* growth rates are poorly correlated with phage therapy success in a mouse infection model. Antimicrob. Agents Chemother. *56*, 949–954.

Cairns, B.J., and Payne, R.J.H. (2008). Bacteriophage therapy and the mutant selection window. Antimicrob. Agents Chemother. *52*, 4344–4350.

Cairns, B.J., Rimms, A.R., Jansen, V.A.A., Connerton, I.F., and Payne, R.J.H. (2009). Quantitative models of *in vitro* bacteriophage-host dynamics and their application to phage therapy. PLoS Path. *5*, e1000253.

Calendar, R., and Abedon, S.T. (2006). The Bacteriophages (Oxford: Oxford University Press).

Callaway, T.R., Edrington, T.S., Brabban, A.D., Anderson, R.C., Rossman, M.L., Engler, M.J., Carr, M.A., Genovese, K.J., Keen, J.E., Looper, M.L., Kutter, E.M., and Nisbet, D.J. (2008). Bacteriophage isolated from feedlot cattle can reduce *Escherichia coli* O157:H7 populations in ruminant gastrointestinal tracts. Foodborne Pathog. Dis. *5*, 183–191.

Capparelli, R., Parlato, M., Borriello, G., Salvatore, P., and Iannelli, D. (2007). Experimental phage therapy against *Staphylococcus aureus* in mice. Antimicrob. Agents Chemother. *51*, 2765–2773.

Capparelli, R., Nocerino, N., Iannaccone, M., Ercolini, D., Parlato, M., Chiara, M., and Iannelli, D. (2010). Bacteriophage therapy of *Salmonella enterica*: a fresh appraisal of bacteriophage therapy. J. Infect. Dis. *201*, 52–61.

Carlton, R.M. (1999). Phage therapy: past history and future prospects. Arch. Immunol. Ther. Exp. *47*, 267–274.

Carmody, L.A., Gill, J.J., Summer, E.J., Sajjan, U.S., Gonzalez, C.F., Young, R.F., and LiPuma, J.J. (2010). Efficacy of bacteriophage therapy in a model of *Burkholderia cenocepacia* pulmonary infection. J. Infect. Dis. *201*, 264–271.

Carvalho, C., Susano, M., Fernandes, E., Santos, S., Gannon, B., Nicolau, A., Gibbs, P., Teixeira, P., and Azeredo, J. (2010a). Method for bacteriophage isolation against target *Campylobacter* strains. Lett Appl. Microbiol. *50*, 192–197.

Carvalho, C.M., Gannon, B.W., Halfhide, D.E., Santos, S.B., Hayes, C.M., Roe, J.M., and Azeredo, J. (2010b). The *in vivo* efficacy of two administration routes of a phage cocktail to reduce numbers of *Campylobacter coli* and *Campylobacter jejuni* in chickens. BMC. Microbiol. *10*, 232.

Cerf, O., Carpentier, B., and Sanders, P. (2010). Tests for determining in-use concentrations of antibiotics and disinfectants are based on entirely different concepts: 'resistance' has different meanings. Int. J. Food Microbiol. *136*, 247–254.

Cerveny, K.E., Depaola, A., Duckworth, D.H., and Gulig, P.A. (2002). Phage therapy of local and systemic disease caused by *Vibrio vulnificus* in iron-dextran-treated mice. Infect. Immun. *70*, 6251–6262.

Chan, B.K., and Abedon, S.T. (2012a). Bacteriophage adaptation, with particular attention to issues of phage host range. In Bacteriophages in Dairy Processing, A. Quiberoni and J. Reinheimer, eds. (Hauppauge, NY, Nova Science Publishers), pp. 25–52.

Chan, B.K., and Abedon, S.T. (2012b). Phage therapy pharmacology: phage cocktails. Adv. Appl. Microbiol. *78*, 1–23.

Chan, B.K., Abedon, S.T., and Loc-Carrillo, C. (2013). Phage cocktails and the future of phage therapy. Future Microbiol. *8*, 769–783.

Chanishvili, N., Chanishvili, T., Tediashvili, M., and Barrow, P.A. (2001). Phages and their application against drug-resistant bacteria. J. Chem. Technol. Biotechnol. *76*, 689–699.

Chanishvili, N., Khurtsia, N., and Malkhazova, I. (2009a). Combined use of phages and antibiotics. In A Literature Review of the Practical Application of Bacteriophage Research, N. Chanishvili and R. Sharp, eds. (Tbilisi, Georgia: Eliava Institute), pp. 74–81.

Chanishvili, N., Malkhazova, I., and Khurtsia, N. (2009b). Phage therapy against intestinal infections. In A Literature Review of the Practical Application of Bacteriophage Research, N. Chanishvili and R. Sharp, eds. (Tbilisi, Georgia: Eliava Institute), pp. 33–58.

Chhibber, S., Kaur, S., and Kumari, S. (2008). Therapeutic potential of bacteriophage in treating *Klebsiella pneumoniae* B5055-mediated lobar pneumonia in mice. J. Med. Microbiol. *57*, 1508–1513.

Christie, G.E., Allison, H.A., Kuzio, J., McShan, M., Waldor, M.K., and Kropinski, A.M. (2012). Prophage-induced changes in cellular cytochemistry and virulence. In Bacteriophages in Health and Disease, P. Hyman and S.T. Abedon, eds. (Wallingford, UK: CABI Press), pp. 33–60.

Clokie, M., and Kropinski, A. (2009). Bacteriophages: Methods and Protocols (U.S.A.: Humana Press).

Courchesne, N.M., Parisien, A., and Lan, C.Q. (2009). Production and application of bacteriophage and bacteriophage-encoded lysins. Recent Pat. Biotechnol. 3, 37–45.

Cox, C.R. (2012). Bacteriophage-based methods of baterial detection and identification. In Bacteriophages in Health and Disease, P. Hyman and S.T. Abedon, eds. (Wallingford, UK: CABI Press).

Curtright, A.J., and Abedon, S.T. (2011). Phage therapy: emergent property pharmacology. J. Bioanal. Biomed. S6, 002.

Czock, D., Markert, C., Hartman, B., and Keller, F. (2009). Pharmacokinetics and pharmacodynamics of antimicrobial drugs. Expert Opin. Drug Metab. Toxicol. *5*, 475–487.

Dabrowska, K., Switala-Jelen, K., Opolski, A., Weber-Dabrowska, B., and Górski, A. (2005). Bacteriophage penetration in vertebrates. J. Appl. Microbiol. *98*, 7–13.

Danelishvili, L., Young, L.S., and Bermudez, L.E. (2006). *In vivo* efficacy of phage therapy for *Mycobacterium avium* infection as delivered by a nonvirulent mycobacterium. Microb. Drug Res. *12*, 1–6.

Denou, E., Bruttin, A., Barretto, C., Ngom-Bru, C., Brüssow, H., and Zuber, S. (2009). T4 phages against *Escherichia coli* diarrhea: potential and problems. Virology *388*, 21–30.

Donlan, R.M. (2009). Preventing biofilms of clinically relevant organisms using bacteriophage. Trends Microbiol. *17*, 66–72.

Dubos, R.J., Straus, J.H., and Pierce, C. (1943). The multiplication of bacteriophage *in vivo* and its protective effects against an experimental infection with Shigella dysenteriae. J. Exp. Med. *20*, 161–168.

Duckworth, D.H., and Gulig, P.A. (2002). Bacteriophages: potential treatment for bacterial infections. BioDrugs *16*, 57–62.

Duerr, D.M., White, S.J., and Schluesener, H.J. (2004). Identification of peptide sequences that induce the transport of phage across the gastrointestinal mucosal barrier. J. Virol. Meth. *116*, 177–180.

Gallet, R., Kannoly, S., and Wang, I.N. (2011). Effects of bacteriophage traits on plaque formation. BMC Microbiol. *11*, 181.

Gallet, R., Shao, Y., and Wang, I.N. (2009). High adsorption rate is detrimental to bacteriophage fitness in a biofilm-like environment. BMC Evol. Biol. 9, 241.

Gill, J.J. (2008). Modeling of bacteriophage therapy. In Bacteriophage Ecology, S.T. Abedon, ed. (Cambridge, UK: Cambridge University Press), pp. 439–464.

Gill, J.J. (2010). Practical and theoretical considerations for the use of bacteriophages in food systems. In Bacteriophages in the Control of Food- and Waterborne Pathogens, P.M. Sabour and M.W. Griffiths, eds. (Washington, DC: ASM Press), pp. 217–235.

Gill, J.J., Hollyer, T., and Sabour, P.M. (2007). Bacteriophages and phage-derived products as antibacterial therapeutics. Expert Opin. Ther. Pat. *17*, 1341–1350.

Gill, J.J., and Hyman, P. (2010). Phage choice, isolation and preparation for phage therapy. Curr. Pharm. Biotechnol. *11*, 2–14.

Gill, J.J., Sabour, P.M., Leslie, K.E., and Griffiths, M.W. (2006). Bovine whey proteins inhibit the interaction of *Staphylococcus aureus* and bacteriophage K. J. Appl. Microbiol. *101*, 377–386.

Golshahi, L., Seed, K.D., Dennis, J.J., and Finlay, W.H. (2010). Toward modern inhalational bacteriophage therapy: nebulization of bacteriophages of *Burkholderia cepacia* complex. J. Aerosol Med. Pulm. Drug Deliv. *21*, 351–360.

Goodridge, L.D. (2008). Phages, bacteria, and food. In Bacteriophage Ecology, S.T. Abedon, ed. (Cambridge, UK: Cambridge University Press), pp. 302–331.

Goodridge, L.D. (2010a). Designing phage therapeutics. Curr. Pharm. Biotechnol. *11*, 15–27.

Goodridge, L.D. (2010b). Designing phage therapeutics. Curr. Pharm. Biotechnol. *11*, 15–27.

Górski, A., Borysowski, J., Miedzybrodzki, R., and Weber-Dabrowska, B. (2007). Bacteriophages in medicine. In Bacteriophage: Genetics and Microbiology, S. McGrath and D. van Sinderen, eds. (Norfolk, UK: Caister Academic Press), pp. 125–158.

Górski, A., Wazna, E., Dabrowska, B.-W., Switala-Jelen, K., and Miedzybrodzki, R. (2006). Bacteriophage translocation. FEMS Immunol. Med. Microbiol. *46*, 313–319.

Gupta, R., and Prasad, Y. (2010). Efficacy of polyvalent bacteriophage P-27/HP to control multidrug resistant *Staphylococcus aureus* associated with human infections. Curr. Microbiol. *62*, 255–260.

Hadas, H., Einav, M., Fishov, I., and Zaritsky, A. (1997). Bacteriophage T4 development depends on the physiology of its host *Escherichia coli*. Microbiology *143*, 179–185.

Hagens, S., and Loessner, M.J. (2010). Bacteriophage for biocontrol of food-borne pathogens: calculations and considerations. Curr. Pharm. Biotechnol. *11*, 58–68.

Harper, D. Online Etymology Dictionary. 2010. www.etymonline.com.

Harper, D.R., Anderson, J., and Enright, M.C. (2011). Phage therapy: delivering on the promise. Ther. Deliv. *2*, 935–947.

Harper, D.R., and Kutter, E. (2009). Bacteriophage: therapeutic use. In Encyclopedia of Life Sciences, John Wiley & Sons, Ltd.), pp. 1–7.

Hawkins, C., Harper, D., Burch, D., Anggard, E., and Soothill, J. (2010). Topical treatment of *Pseudomonas aeruginosa* otitis of dogs with a bacteriophage mixture: A before/after clinical trial. Vet. Microbiol. *145*, 309–313.

Heo, Y.J., Lee, Y.R., Jung, H.H., Lee, J., Ko, G., and Cho, Y.H. (2009). Antibacterial efficacy of phages against *Pseudomonas aeruginosa* infections in mice and Drosophila melanogaster. Antimicrob. Agents Chemother. *53*, 2469–2474.

d'Hérelle, F., and Smith, G.H. (1930). The Bacteriophage and its Clinical Application (Springfield, IL: Charles C. Thomas, Publisher).

Housby, J.N., and Mann, N.H. (2009). Phage therapy. Drug Discov. Today *14*, 536–540.

Huff, W.E., Huff, G.R., Rath, N.C., and Donoghue, A.M. (2010). Immune interference of bacteriophage efficacy when treating colibacillosis in poultry. Poult. Sci. *89*, 895–900.

Hung, C.H., Kuo, C.F., Wang, C.H., Wu, C.M., and Tsao, N. (2011). Experimental phage therapy in treating *Klebsiella pneumoniae*-mediated liver abscesses and bacteremia in mice. Antimicrob. Agents Chemother. *55*, 1358–1365.

Hyman, P., and Abedon, S.T. (2008). Phage ecology of bacterial pathogenesis. In Bacteriophage Ecology, S.T. Abedon, ed. (Cambridge, UK: Cambridge University Press), pp. 353–385.

Hyman, P., and Abedon, S.T. (2009). Practical methods for determining phage growth parameters. Meth. Mol. Biol. *501*, 175–202.

Hyman, P., and Abedon, S.T. (2010). Bacteriophage host range and bacterial resistance. Adv. Appl. Microbiol. *70*, 217–248.

Hyman, P., and Abedon, S.T. (2012a). Bacteriophages in Health and Disease (Wallingford, UK: CABI Press).

Hyman, P., and Abedon, S.T. (2012b). Smaller fleas: viruses of microorganisms. Scientifica *2012*, 734023.

Johnson, R.P., Gyles, C.L., Huff, W.E., Ojha, S., Huff, G.R., Rath, N.C., and Donoghue, A.M. (2008). Bacteriophages for prophylaxis and therapy in cattle, poultry and pigs. Anim. Health Res. Rev. *9*, 201–215.

Jonczyk, E., Klak, M., Miedzybrodzki, R., and Gorski, A. (2011). The influence of external factors on bacteriophages-review. Folia Microbiol. (Praha) *56*, 191–200.

Kasman, L.M., Kasman, A., Westwater, C., Dolan, J., Schmidt, M.G., and Norris, J.S. (2002). Overcoming the phage replication threshold: a mathematical model with implications for phage therapy. J. Virol. *76*, 5557–5564.

Keller, R., and Engley, F.B. (1958). Fate of bacteriophage particles introduced into mice by various routes. Proc. Soc. Exp. Biol. Med. *98*, 577–580.

Krylov, V.N. (2001). Phage therapy in terms of bacteriophage genetics: Hopes, prospects, safety, limitations. Rus. J. Genet. *37*, 715–730.

Kuhl, S., Hyman, P., and Abedon, S.T. (2012). Diseases caused by phages. In Bacteriophages in Health and Disease, P. Hyman and S.T. Abedon, eds. (Wallingford, UK: CABI Press), pp. 21–32.

Kumari, S., Harjai, K., and Chhibber, S. (2009). Efficacy of bacteriophage treatment in murine burn wound infection induced by *Klebsiella pneumoniae*. J. Microbiol. Biotechnol. *19*, 622–628.

Kumari, S., Harjai, K., and Chhibber, S. (2010). Evidence to support the therapeutic potential of bacteriophage Kpn5 in burn wound infection caused by *Klebsiella pneumoniae* in BALB/c mice. J. Microbiol. Biotechnol. *20*, 935–941.

Kumari, S., Harjai, K., and Chhibber, S. (2011). Bacteriophage versus antimicrobial agents for the treatment of murine burn wound infection caused by *Klebsiella pneumoniae* B5055. J. Med. Microbiol. *60*, 205–210.

Kutateladze, M., and Adamia, R. (2010). Bacteriophages as potential new therapeutics to replace or supplement antibiotics. Trends Biotechnol. *28*, 581–585.

Kutter, E. (2008). Phage therapy: bacteriophages as naturally occurring antimicrobials. In Practical Handbook of Microbiology, E. Goldman and L.H. Green, eds. (Boca Raton, FL, CRC Press), pp. 713–730.

Kutter, E.M. (2009). Bacteriophage therapy: past and present. In Encyclopedia of Microbiology, M. Schaecter, ed. (Oxford: Elsevier), pp. 258–266.

Kutter, E., and Sulakvelidze, A. (2005). Bacteriophages: Biology and Application (Boca Raton, FL: CRC Press).

Kutter, E., De Vos, D., Gvasalia, G., Alavidze, Z., Gogokhia, L., Kuhl, S., and Abedon, S.T. (2010). Phage therapy in clinical practice: treatment of human infections. Curr. Pharm. Biotechnol. *11*, 69–86.

Lampert, R., Boyce, F.F., and McFetridge, E.M. (1935). Bacteriophage therapy: A clinical study, with special reference to the technique of application. Am. J. Surgery *29*, 436–443.

Letarov, A.V., Golomidova, A.K., and Tarasyan, K.K. (2010). Ecological basis of rational phage therapy. Acta Naturae *2*, 60–71.

Letkiewicz, S., Miedzybrodzki, R., Klak, M., Jonczyk, E., Weber-Dabrowska, B., and Górski, A. (2010). The perspectives of the application of phage therapy in chronic bacterial prostatitis. FEMS Immunol. Med. Microbiol. *60*, 99–112.

Levin, B.R., and Bull, J.J. (1996). Phage therapy revisited: the population biology of a bacterial infection and its treatment with bacteriophage and antibiotics. Am. Nat. *147*, 881–898.

Levin, B.R., and Bull, J.J. (2004). Population and evolutionary dynamics of phage therapy. Nat. Rev. Microbiol. *2*, 166–173.

Levison, M.E., and Levison, J.H. (2009). Pharmacokinetics and pharmacodynamics of antibacterial agents. Infect. Dis. Clin. North Am. *23*, 791–815, vii.

Loc-Carrillo, C., and Abedon, S.T. (2011). Pros and cons of phage therapy. Bacteriophage *1*, 111–114.

Los, M., Kuzio, J., McConnell, M.R., Kropinski, A.M., Wegrzyn, G., and Christie, G.E. (2010). Lysogenic conversion in bacteria of importance to the food industry. In Bacteriophages in the Control of Food- and Waterborne Pathogens, P.M. Sabour and M.W. Griffiths, eds. (Washington, DC: ASM Press), pp. 157–198.

Ma, Y., Pacan, J.C., Wang, Q., Xu, Y., Huang, X., Korenevsky, A., and Sabour, P.M. (2008). Microencapsulation of bacteriophage felix O1 into chitosan-alginate microspheres for oral delivery. Appl. Environ. Microbiol. *74*, 4799–4805.

Mc Grath, S., and van Sinderen, D. (2007). Bacteriophage: Genetics and Molecular Biology (Norfolk, UK: Caister Academic Press).

Mann, N.H. (2008). The potential of phages to prevent MRSA infections. Res. Microbiol. *159*, 400–405.

Merril, C.R. (2008). Interaction of bacteriophages with animals. In Bacteriophage Ecology, S.T. Abedon, ed. (Cambridge, UK: Cambridge University Press), pp. 332–352.

Merril, C.R., Scholl, D., and Adhya, S.L. (2003). The prospect for bacteriophage therapy in Western Medicine. Nat. Rev. Drug Discov. *2*, 489–497.

Merril, C.R., Scholl, D., and Adhya, S. (2006). Phage therapy. In The Bacteriophages, R. Calendar and S.T. Abedon, eds. (Oxford: Oxford University Press), pp. 725–741.

Miedzybrodzki, R., Fortuna, W., Weber-Dabrowska, B., and Gorski, A. (2007). Phage therapy of staphylococcal infections (including MRSA) may be less expensive than antibiotic treatment. Postepy Hig. Med. Dosw. (Online) *61*, 461–465.

Miller, R.W., Skinner, E.J., Sulakvelidze, A., Mathis, G.F., and Hofacre, C.L. (2010). Bacteriophage therapy for control of necrotic enteritis of broiler chickens experimentally infected with *Clostridium perfringens*. Avian Dis. *54*, 33–40.

Nishikawa, H., Yasuda, M., Uchiyama, J., Rashel, M., Maeda, Y., Takemura, I., Sugihara, S., Ujihara, T., Shimizu, Y., Shuin, T., and Matsuzaki, S. (2008). T-even-related bacteriophages as candidates for treatment of *Escherichia coli* urinary tract infections. Arch. Virol. *153*, 507–515.

Olszowska-Zaremba, N., Borysowski, J., Dabrowska, K.., and Górski, A. (2012). Phage translocation, safety, and immunomodulation. In Bacteriophages in Health and Disease, P. Hyman and S.T. Abedon, eds. (Wallingford, UK: CABI Press).

Payne, R.J.H., and Jansen, V.A.A. (2001). Understanding bacteriophage therapy as a density-dependent kinetic process. J. Theor. Biol. *208*, 37–48.

Payne, R.J.H., and Jansen, V.A.A. (2003). Pharmacokinetic principles of bacteriophage therapy. Clin. Pharmacokinet. *42*, 315–325.

Payne, R.J.H., Phil, D., and Jansen, V.A.A. (2000). Phage therapy: The peculiar kinetics of self-replicating pharmaceuticals. Clin. Pharmacol. Ther. *68*, 225–230.

Pirnay, J.P., De Vos, D., Verbeken, G., Merabishvili, M., Chanishvili, N., Vaneechoutte, M., Zizi, M., Laire, G., Lavigne, R., Huys, I., Van den Mooter, G., Buckling, A., Debarbieux, L., Pouillot, F., Azeredo, J., Kutter, E., Dublanchet, A., Gorski, A., and Adamia, R. (2011). The phage therapy paradigm: prêt-à-porter or sur-mesure? Pharm. Res *28*, 934–937.

Pouillot, F., Chomton, M., Blois, H., Courroux, C., Noelig, J., Bidet, P., Bingen, E., and Bonacorsi, S. (2012). Efficacy of bacteriophage therapy in experimental sepsis and meningitis caused by a clone O25b:H4-ST131 *Escherichia coli* strain producing CTX-M-15. Antimicrob. Agents Chemother. *56*, 3568–3575.

Ryan, E.M., Gorman, S.P., Donnelly, R.F., and Gilmore, B.F. (2011). Recent advances in bacteriophage therapy: how delivery routes, formulation, concentration and timing influence the success of phage therapy. J. Pharm. Pharmacol. *63*, 1253–1264.

Slopek, S., Durlakova, I., Weber-Dabrowska, B., Kucharewicz-Krukowska, A., Dabrowski, M., and Bisikiewicz, R. (1983). Results of bacteriophage treatment of suppurative bacterial infections. I. General evaluation of the results. Arch. Immunol. Ther. Exp. *31*, 267–291.

Slopek, S., Weber-Dabrowska, B., Dabrowski, M., and Kucharewicz-Krukowska, A. (1987). Results of bacteriophage treatment of suppurative bacterial infections in the years 1981–1986. Arch. Immunol. Ther. Exp. *35*, 569–583.

Smith, H.W., and Huggins, M.B. (1982). Successful treatment of experimental *Escherichia coli* infections in mice using phage: Its general superiority over antibiotics. J. Gen. Microbiol. *128*, 307–318.

Smith, H.W., Huggins, M.B., and Shaw, K.M. (1987). Factors influencing the survival and multiplication of bacteriophages in calves and in their environment. J. Gen. Microbiol. *133*, 1127–1135.

Sneader, W. (2005). Drug Discovery: A History (West Sussex, England, UK: John Wiley & Sons).

Spouge, J.L. (1994). Viral multiplicity of attachment and its implications for human immunodeficiency virus therapies. J. Virol. *68*, 1782–1789.

Stent, G.S. (1963). Molecular Biology of Bacterial Viruses (San Francisco, CA: WH Freeman and Co.).

Stone, R. (2002). Bacteriophage therapy. Stalin's forgotten cure. Science *298*, 728–731.

Sulakvelidze, A. (2005). Phage therapy: an attractive option for dealing with antibiotic-resistant bacterial infections. Drug Discov. Today *10*, 807–809.

Sulakvelidze, A., and Barrow, P. (2005). Phage therapy in animals and agribusiness. In Bacteriophages: Biology and Applications, E. Kutter and A. Sulakvelidze, eds. (Boca Raton, FL, CRC Press), pp. 335–380.

Sulakvelidze, A., and Kutter, E. (2005). Bacteriophage therapy in humans. In Bacteriophages: Biology and Applications, E. Kutter and A. Sulakvelidze, eds. (Boca Raton, FL, CRC Press), pp. 381–436.

Sulakvelidze, A., and Pasternack, G.R. (2010). Industrial and regulatory issues in bacteriophage applications in food production and processing. In Bacteriophages in the Control of Food- and Waterborne Pathogens, P.M. Sabour and M.W. Griffiths, eds. (Washington, DC: ASM Press), pp. 297–326.

Sulakvelidze, A., Alavidze, Z., and Morris, J.G., Jr. (2001). Bacteriophage therapy. Antimicrob. Agents Chemother. *45*, 649–659.

Summers, W.C. (2001). Bacteriophage therapy. Annu. Rev. Microbiol. *55*, 437–451.

Sunagar, R., Patil, S.A., and Chandrakanth, R.K. (2010). Bacteriophage therapy for *Staphylococcus aureus* bacteremia in streptozotocin-induced diabetic mice. Res. Microbiol. *161*, 854–860.

Then, R.L., and Sahl, H.G. (2010). Anti-infective strategies of the future: is there room for species-specific antibacterial agents? Curr. Pharm. Des. *16*, 555–566.

Thiel, K. (2004). Old dogma, new tricks—21st Century phage therapy. Nat. Biotech. *22*, 31–36.

Thomas-Abedon, C., and Abedon, S.T. (2013). Wild-type phage T4 plaques are (almost) as large as phage-T4 r-mutant plaques. Manuscript in preparation.

Uchiyama, J., Maeda, Y., Takemura, I., Chess-Williams, R., Wakiguchi, H., and Matsuzaki, S. (2009). Blood kinetics of four intraperitoneally administered therapeutic candidate bacteriophages in healthy and neutropenic mice. Microbiol. Immunol. *53*, 301–304.

Villalobos, T.I., Renfro, B., and Rathore, M.H. (1998). Pharamacokinetics and pharmacodynamics of antibacterial agents in pediatrics: a practical approach. Jacksonville Medicine.

Wang, Q., and Sabour, P.M. (2010). Encapsulation and controlled release of bacteriophages for food animal production. In Bacteriophages in the Control of Food- and Waterborne Pathogens, P.M. Sabour and M.W. Griffiths, eds. (Washington, DC: ASM Press), pp. 237–255.

Weber-Dabrowska, B., Dabrowski, M., and Slopek, S. (1987). Studies on bacteriophage penetration in patients subjected to phage therapy. Arch. Immunol. Ther. Exp. *35*, 563–568.

Weber-Dabrowska, B., Mulczyk, M., and Górski, A. (2003). Bacteriophages as an efficient therapy for antibiotic-resistant septicemia in man. Transplant. Proc. *35*, 1385–1386.

Weitz, J.S., Mileyko, Y., Joh, R.I., and Voit, E.O. (2008). Collective decision making in bacterial viruses. Biophys. J. *95*, 2673–2680.

Zimecki, M., Artym, J., Kocieba, M., Weber-Dabrowska, B., Borysowski, J., and Górski, A. (2010). Prophylactic effect of bacteriophages on mice subjected to chemotherapy-induced immunosuppression and bone marrow transplant upon infection with *Staphylococcus aureus*. Med. Microbiol. Immunol. *199*, 71–79.

Zobnina, K.S. (1963). Excretion of dysentery bacteriophage by the kidneys of mice during experimental dysentery infection. Bull. Exp. Biol. Med. *56*, 1008–1011.

Fighting Bacteriophage Infection: Mechanisms of Bacterial Resistance

4

Anneleen Cornelissen, Rob Lavigne and Sylvain Moineau

Abstract

In man-made and natural environments, there is a continuous ongoing interaction between phages and their bacterial hosts, a co-evolutionary arms race between two competing organisms which contributes enormously to their diversity. During continuous cycles of co-evolution, phage-resistant bacterial hosts emerge aiming at preservation of their bacterial lineages. For every step in the phage infection cycle, bacteria have evolved various defence mechanisms, passive or active, to evade phage propagation and subsequent spreading of phage progeny in the surrounding environment. However, when facing this selective pressure imposed by the host, phages have developed different strategies to subvert these defence systems in order to thrive in these new bacterial populations. Knowledge of these phage-host dynamics represents a vital tool for phage therapeutic purposes in which the emergence of phage-resistant bacterial pathogens forms a notable disadvantage. In contrast, in the fermentation industries, bacteriophages themselves pose a contamination problem, which can be relieved by selection of phage-insensitive bacterial strains.

Introduction

Survival is one of nature's primary driving forces for all living organisms, either as individuals or as a species. Bacteriophage infections represent one of the main challenges for a microbial community, since infection of one bacterium and subsequent propagation can rapidly destroy the bacterial population. Bacteriophages are now recognized as the most abundant biological entities on Earth, as the almost inconceivably large size of their global population is estimated at $>10^{30}$ phages outnumbering the global bacterial population by about 10-fold (Brüssow and Hendrix, 2002). Their remarkable diversity, illustrated by the high number of novel genes found in newly analysed phage genomes, is a reflection of the diversity of their bacterial hosts and the evolutionary arms race between phages and bacteria (Okafor, 2007). This evolutionary arms race is often referred to as the Red Queen hypothesis (van Valen, 1973). This hypothesis is based on what the Red Queen said to Alice in Lewis Carroll's *Through the Looking Glass*: 'In this place it takes all the running you can do, to keep in the same place. If you want to get somewhere else, you must run at least twice as fast as that!' This hypothesis serves as metaphor for his proposed 'Law of Extinction': The extinction probability of a species does not improve over time despite evolutionary adjustments, for the simple reason that other species are co-evolving. Due to co-evolutionary prey–predator interactions, progress ('running faster') of one species may confer this species' selective advantages compared to the other species and may hereby possibly lead to extinction of this second species. The only way by which the other species can maintain its fitness ('keeping in the same place') relative to the species it is co-evolving with, is through counter-adaptation ('running twice as fast'). In sum, in a competitive world, continuing development is necessary in order to maintain its fitness relative to the systems it is co-evolving with. This evolutionary trend is nowhere this distinct as in phage–host interactions. Due to the extremely rapid evolution of phage particles, acute pressure is posed on the bacterial population to evade phage infection and subsequent killing (Fuhrman, 1999). This arms race between phages and bacteria is thought to have a key regulatory function of bacterial populations, and thereby influences the global ecosystem (Suttle, 2007) and climate (Furhman, 1999; Suttle, 2007), the evolution of the biosphere (Comeau and Krisch, 2005) and virulence factors in human pathogens (Brüssow *et al.*, 2004).

This chapter focuses on the bacterial resistance mechanisms against phages and the phage counter-adaptations to escape the bacterial defence. Each step of the phage replication cycle can be targeted by bacterial resistance mechanisms, either passive or active (Table 4.1) (Labrie *et al.*, 2010).

Preventing phage adsorption

The first step in the phage replication cycle is the adsorption of the bacteriophage particle to the host cell surface receptor. Within the astonishing diversity of bacterial cell surface components, phages must recognize their one particular receptor (Rakhuba *et al.*, 2010). Phages are known to bind pili, flagella, capsular exopolysaccharides, surface-exposed proteins, lipopolysaccharide structures of Gram-negative bacteria and peptidoglycan and teichoic acids of Gram-positive bacteria (Rakhuba *et al.*, 2010). These structures are often recognized by phage tail fibres, tail spikes or baseplates attached to the distal end of the tail, which serve as 'sensory peptides' directing the phage to a specific bacterial target. The nature of the host cell receptor determines the adsorption specificity of the phage and concomitantly its ability to infect a single strain or a broader bacterial spectrum of multiple strains or related species (Blair and Williams, 1961; Jacobs-Sera *et al.*, 2012; O'Flaherty *et al.*, 2005).

In response, bacteria have evolved an array of passive defence mechanisms to prevent phage adsorption. These adsorption-blocking

Table 4.1 Overview of the bacterial resistance mechanisms against phages

Mechanism	Example
Prevention of phage adsorption	
Preventing adsorption to protein receptors	
Masking or (lipo)protein-mediated blockage	T5-encoded Llp inactivates the FhuA receptor (Pedruzzi *et al.*, 1998)
Production of competitive inhibitors	FhuA serves as receptor for Microcin J25 and phage T5 (Destoumieux-Garzón *et al.*, 2005)
Phase variation	BvgAS phosphorelay signal transduction system in Bordetella subspecies (Cotter and Miller, 2001)
Formation of a physical barrier by lipopolysaccharide molecules or by the extracellular matrix	Polysialic acid capsule of *E. coli* K1 strains (e.g. Whitfield and Roberts, 1999)
Preventing phage DNA entry	
Physical barrier formed by the Gram-negative peptidoglycan layer	TM4 tape measure protein with peptidoglycan hydrolase activity (Piuri and Hatfull, 2006)
Phage-mediated superinfection exclusion (Sie)	T4-encoded Imm and Sp block superinfection by other T-even phages (e.g. Lu *et al.*, 1993)
Plasmid-encoded phage defense system	ColIB inhibits infection of T-odd phages (Duckworth and Pinkerton, 1988)
Degradation of phage genomic DNA	
Restriction-modification systems (Type I–IV)	Type IA: EcoKI (Webb *et al.*, 1996)
CRISPR/Cas system	E. coli K12: "core" Cas proteins (Cas1-Cas3 and Cas5e) and the subtype-specific Cse1- Cse4 proteins (e.g. Sinkunas *et al.*, 2011)
Abortive infection systems	F exclusion of *E. coli* phage T7 (e.g. Garcia and Molineux, 1995)

mechanisms are discussed here for each of the different host cell components which can be recognized by a specific bacteriophage. In turn, phage will often evolve specific strategies to subvert these adsorption-blocking mechanisms.

Protein receptors

Masking or (lipo) protein-mediated blockage of proteic receptors

By changing the structure or three-dimensional conformation of the host cell receptors, reducing the number of surface-exposed receptors or a specific receptor-masking strategy, bacteria can limit phage infection efficiently.

Staphylococcus aureus possesses a cell-wall anchored immunoglobulin G-binding protein A, a virulence factor, with either four or five domains that each bind to the Fc region of immunoglobulin IgG (Uhlen *et al.*, 1984). Nordström and Forsgren (1974) demonstrated a negative correlation between the protein A content of *S. aureus* strains and the quantity of adsorbed phages. Protein A as a surface component probably renders bacteria resistant to phage adsorption by masking the phage receptor. The M-protein of *Streptococcus pyogenes* likely serves a similar receptor-masking function (Maxted, 1955).

In the early stages of coliphage T5 infection, a T5-encoded lipoprotein Llp is produced which inactivates its own receptor, the ferrichrome-iron protein FhuA (Pedruzzi *et al.*, 1998). Probably, several Llp proteins cluster around the FhuA protein in the lipid bilayer, hereby shielding (directly or indirectly) this cell receptor from phage recognition. The Llp protein not only prevents superinfection but also prevents inactivation of phage progeny by binding receptors present in the cell wall of lysed host bacteria (Decker *et al.*, 1994; McCorquodale and Warner, 1988).

The *Escherichia coli* host cells also use lipoproteins for phage surface exclusion, as demonstrated for *E. coli* F^+ strains (Riede and Eschbach, 1986). The F-factor-encoded outer membrane protein TraT likely interacts directly with the major *E. coli* outer membrane protein OmpA, which serves as a receptor for different T-even-like phages (Perumal and Minkley, 1984; Riede and Eschbach, 1986). TraT can hereby partially inhibit OmpA-specific phage binding, but not completely block this binding.

Production of competitive inhibitors

Molecules which are naturally present in the microbial environment and which bind specifically to the phage receptors render these receptors unavailable for phage adhesion.

For example, the *E. coli* outer-membrane protein FhuA serves as a high-affinity transporter for iron chelated to the siderophore ferrichrome (Braun and Braun, 2002). Besides this physiological function, FhuA serves as a receptor for microcin J25 and for the unrelated phages T1, T5 and Φ80 (Coulton *et al.*, 1983; Destoumieux-Garzón *et al.*, 2005). Microcin J25 is naturally secreted under conditions of nutrient depletion by the faecal *E. coli* AY25 strain and is active against the phylogenetically related microbial strains (Salomón and Farías, 1992). Microcin J25 is also able to inhibit *in vivo* adsorption of phage T5 to the FhuA receptor in a dose-dependent manner with full inhibition observed at 10 μM Microcin J25 (Destoumieux-Garzón *et al.*, 2005). Wayne and Neilands (1975) have shown that ferrichrome at micromolar concentrations strongly inhibits plaque formation of phage Φ80 on *E. coli* K12 due to competition for the same FhuA receptor site.

E. coli phage BF23 and vitamin B_{12} also compete for occupancy of the same receptor site (Bradbeer *et al.*, 1976). Vitamin B_{12}, which is transported for adequate cell growth in the absence of methionine, can inhibit 50% of phage BF23 adsorption at concentrations in the range of 0.5 to 2.0 nM, thereby protecting susceptible *E. coli* cells from phage infection.

Phase variation and tropism switching

The infectious cycle of *Bordetella* subspecies is controlled by the BvgAS phosphorelay signal transduction system. This system mediates transition between the Bvg^+ phase (adapted to colonization of the respiratory tract through expression of virulence and colonization factors) and the Bvg^- phase (adapted to *ex vivo* survival) (Uhl and Miller, 1996; Cotter and Miller, 2001). Temperate bacteriophage BPP-1 ('Bvg plus tropic phage-1') initiates infection of *Bordetella* species

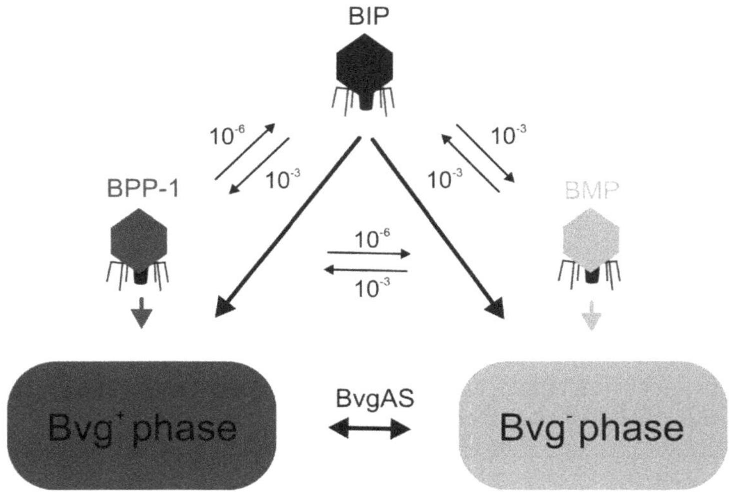

Figure 4.1 Overview of tropism switching frequencies by *Bordetella* phages.

by binding the surface receptor, the autotransporter pertactin (Prn) which is only expressed in the Bvg$^+$ phase (Uhl and Miller, 1996). However, its infection specificity or tropism is not absolute. At a frequency of 10^{-6}, BPP-1 variants, called BMP ('Bvg minus-tropic phage'), form plaques with a normal morphology on Bvg$^-$ phase cells (Fig. 4.1). Additionally, some variants, called BIP ('Bvg indiscriminant phage'), were identified which form plaques with nearly equal efficiency on Bvg$^+$ and Bvg$^-$ phase cells (Liu *et al.*, 2002). BPP-1 can switch its tropism by altering the *mtd* ('major tropism determinant') gene, which encodes the distal tail fibre protein necessary for binding receptor molecules on host *Bordetella* species (Dia *et al.*, 2010). Mtd variants are created at high frequency by a diversity-generating retroelement (DGR) (Uhl and Miller, 1996; Liu *et al.*, 2002; 2004; McMahon *et al.*, 2005). This way, DGR generates a mutant copy of an invariant template repeat (TR) in a process of transcription and reverse transcription. This mutant TR-derived copy replaces the variability region (VR1), which encodes the C-lectin-type fold at the C terminus of the Mtd. DGR hereby diversifies the receptor-binding protein and provides alongside a vast repertoire of ligand–receptor interactions (Liu *et al.* 2002). Using comparative genome analysis, DGR systems may be also present in the *Vibrio harveyi* VHML phage and the *Magnetospirillum magneticum* phage, as in phages infecting *Bifidobacterium* spp. (Ventura *et al.*, 2005; Medhekar and Miller, 2007).

Lipopolysaccharide structures

Lipopolysaccharide structures or the presence of an extracellular matrix can also block adsorption at proteic receptors by physically blocking the receptor site. However these surface polysaccharide structures can become a target of phages which have gained the capacity to specifically recognize, bind to and in some cases enzymatically cleave these polysaccharide receptors (Takeda and Uetake, 1973; Lindberg *et al.*, 1978; Eriksson *et al.*, 1979; Geyer *et al.*, 1983).

Lipopolysaccharides (LPS) represent the outermost layer of the outer membrane of Gram-negative bacteria. LPS are strongly negatively charged molecules which incorporate three parts: a hydrophobic lipid A region is via an anionic oligosaccharide core covalently linked to a O-polysaccharide chain, also called the O-antigen distal to the outer membrane. Only smooth type LPS contains this typical tripartite structure, while rough type LPS lacks the O-antigen but contains a complete lipid A region and an oligosaccharide core (Wilkinson, 1996). Bacteriophages recognizing smooth type LPS show a very narrow host range specificity due to the large diversity of O-antigen structures between different bacterial species and/or strains. However, phages specific for rough type LPS display a broader host range since the oligosaccharide core is generally well conserved in various bacterial strains/species (Rakhuba *et al.*, 2010)

Characteristic for bacteriophages recognizing the O-antigen of smooth type LPS, is that phage adsorption leads to specific enzymatic cleavage of this polysaccharide. Phages ε15 and P22 possess this endorhamnosidase activity within their associated tail spike and hereby cleave the bond Rha-1→3-Gal in the O-antigen of *Salmonella anatum* and *Salmonella typhimurium* respectively (Takeda and Uetake, 1973; Kanegasaki and Wright, 1973; Iwashita and Kanegasaki, 1973; Eriksson *et al.*, 1979). The endorhamnosidase activity of phage Sf6, isolated from *Salmonella flexneri* serotype 3a, is associated with hydrolysis of the Rha–1→3-Rha bond (Lindberg *et al.*, 1978). Phage H-F6S infects *S. flexneri* strains containing smooth type LPS, while mutants lacking the O-antigen are resistant to phage H-F6S. Absence of the O-antigen however, renders these *S. flexneri* mutants sensitive to phages like T3, T4 and T7, which recognize receptors of the LPS oligosaccharide core normally covered by this O-antigen (Rakhuba *et al.*, 2010). This indicates that acquisition of an O-antigen may form a first line of defence rendering strains resistant to phages recognizing underlying LPS core structures. Phages with endorhamnosidase activity in tail-associated enzymes have gained the capacity to subvert this physical O-antigen blockage. On the other hand, mutant strains with an incomplete LPS structures resulting from disruption of one of the steps in the biosynthesis steps, could also block bacteriophage infection. For example, absence of the terminal N-acetylglucosamine of the LPS oligosaccharide core of wild-type *Salmonella* strains generates resistance to phage Felix O-1 infection, which necessitates a complete LPS core for adsorption (Lindberg, 1967; Lindberg and Hellerovist, 1971).

Phase variation represents another bacterial defence system against phages that recognize LPS structures. In *Haemophilus influenzae*, the surface-exposed epitopes of the LPS oligosaccharide core are subjected to high-frequency phase variation (Hood *et al.*, 1996a; Bayliss *et al.*, 2002; De Bolle *et al.*, 2000). This process of switching surface epitopes is mediated by highly mutable loci, usually by changing the number of tetranucleotide repeats within the coding sequence of a specific gene (Hood *et al.*, 1996b, 2004; Bayliss *et al.*, 2001, 2002, 2004). Infection of *H. influenzae* Rd by phage HP1c1 is dependent on the presence of the first two sugars (Glc-Gal) extending from the third heptose of the LPS oligosaccharide core, which serves as bacterial receptor. Phage-resistant mutants lack one of the tetranucleotide repeats (5′-CAAT-3′) within a LPS biosynthetic gene, namely gene *lic2A*, which leads to the disruption of the phage receptor. Reversion to the original number of repeats restores phage sensitivity. The absence of Dam methylation destabilizes the tetranucleotide repeat tract, hereby increasing the switch from phage sensitive to resistant strains (Zaleski *et al.*, 2005).

In *Vibrio cholerae* a major target for bacteriophages is LPS O1 antigen. Initial steps of O1 antigen biosynthesis depend on functions of two phase variable genes (Seed *et al.*, 2012). The diversification of O1 antigens within *V. cholerae* subpopulations provides a means for escaping predation by certain O1-specific phages.

Variation in O-antigen structure may also be the result of lysogenization by temperate LPS-specific phages which provoke surface alteration of their bacterial host to prevent superinfection by the same temperate phage. For example, lysogenization of *S. anatum* by temperate phage ε15, which carries endorhamnosidase activity within its tail spikes, not only changes the glycosidic linkages between O-antigen units from alpha to beta, but also prevents O-acetylation (Robbins *et al.*, 1965; Losick, 1969). This surface alteration blocks the endorhamnosidase activity of the superinfecting ε15 phage. The lambdoid phage D3, which lysogenizes *P. aeruginosa* strain PAO1 is able to convert the O-antigen linkages from α1–4 to β1–4 and causes O-acetylation of the N-acetylfucosamine moiety (Kuzio and Kropinski, 1983). This serotype conversion, which blocks superinfection by other D3 phages is mediated by a three-component system encoded by bacteriophage D3 itself: the α-polymerase inhibitor (Iap) prevents the assembly of α-linked B-band LPS and allows the phage-encoded β-polymerase (Wzyβ) to form new β-linked B-band LPS. Furthermore, D3 adds O-acetyl groups to the N-acetylfucosamine sugar residues of the O-antigen by the action of the D3-encoded *O*-acetylase (Oac) (Newton *et al.*, 2001). Similarly, temperate phage φV10, closely related to phage ε15, encodes an *O*-acetyltransferase, which modifies the O157 antigen of *E. coli* O157:H7. This modification is sufficient to block φV10 superinfection by preventing adsorption to O157 antigen receptor (Perry *et al.*, 2009).

Extracellular matrix

By nature, bacteria preferentially form biofilms, bacterial communities encased in a self-produced exopolysaccharide (EPS) matrix, which is commonly strengthened by proteins and extracellular DNA (Mann and Wozniak, 2012). This biofilm-environment protects the enclosed bacteria against harsh environmental conditions like desiccation, antimicrobial agents (Mah, 2012), as well as phage infection by protecting the bacterial surface receptor underneath this 'physical' EPS barrier (Sutherland, 2001). Moreover, the low metabolic activity of the biofilm-enclosed cells (Costerton, 1995; Beveridge *et al.*, 1997) slows down the infection process and thus postpones bacterial cell lysis (Sillankorva *et al.*, 2004; Cerca *et al.*, 2007).

EPS are essentially very long, thin molecular chains with molecular masses ranging from 0.5 to 2.0×10^6 Da, which vary greatly in composition and mutual association depending on the bacterial species or strain. Some EPS molecules are neutral or polycationic, but the majority are polyanionic with uronic acid or ketal-linked pyruvate building blocks (Mack *et al.*, 1996; Sutherland, 2001). Depending on the linkage type, a secondary configuration organized in aggregated helices or more flexible structures is formed. The final tertiary biofilm network depends on further intermolecular interactions with other EPS-structures as well as other molecules, including extracellular DNA and proteins (Mayer *et al.*, 1999; Sutherland, 2001).

Wilkinson and Holmes (1979) noted a reduced phage adsorption of phages 84 and 52A for capsulated *S. aureus* strains M and Smith. Similarly, three *Streptococcus pneumoniae* phages (ω2, ω3 and ω8) were able to lyse nine non-capsulated *S. pneumoniae* strains, but blockage of their receptor by EPS rendered 41 capsulated strains insensitive to all three phages (Bernheimer and Tiraby, 1976). More recently, Scholl *et al.* (2005) demonstrated that the polysialic acid capsule of *E. coli* K1 strains physically blocks infection by phage T7, which recognizes a LPS structure beneath it.

Within mesophilic lactic acid bacteria, this EPS production is plasmid-linked. The non-covalently bound extracellular material encoded by plasmid DNA has been shown to inhibit phage adsorption to lactococcal cells. For example, *Lactococcus lactis* ssp. *cremoris* MG1363 harbouring plasmid pCI658 produces a hydrophilic EPS, which shields the cell surface receptors and prevents phage adsorption (Forde and Fitzgerald, 1999a). The pCI658-encoded EPS molecules were depicted as strands of material extending out of the cell surface with galactose and glucuronic acid as putative components (Forde *et al.*, 1999). A similar adsorption blocking mechanism was identified for *L. lactis* spp. *cremoris* UC503 that acquired plasmid pCI528 which directs the synthesis of a non-covalently bound hydrophilic EPS rich in galactose and rhamnose (Lucey *et al.*, 1992). However, Deveau *et al.* (2002) observed that eight distinct phages (Q61 to Q68) within the 936 species specifically infect EPS containing *Lactococcus lactis* strains. EPS of these strains does not confer phage resistance, and is moreover not necessary for phage infection as observed for the cured derivatives of the susceptible strains.

Some phages have evolved specifically to recognize and cleave EPS structures. Recognition of a specific EPS structure generally occurs by a tail spike or tail fibre which will bind the EPS receptor and subsequently cleave it by its associated enzymatic activity (Stirm and Freund-Mölbert, 1971). Besides the tail-associated enzymes which are covalently bound to the phage particle or attached to it via an adapter molecule, free soluble enzymes, possibly caused by an excess of enzymes not incorporated in viral particles during phage replication, are generally found in the environment after phage-mediated cell lysis (Eklund and Wyss, 1962; Bessler *et al.*, 1975; Castillo and Bartell, 1976). Those free enzymes can therefore cleave the EPS molecules and render the phage receptor accessible. Another common characteristic of EPS-specific phages is that their interaction with an EPS structure is reversible. EPS serves as a receptor for initial phage attachment, whereas bacterial cell wall components are obligatory for irreversible adsorption (Taylor, 1966; Stirm *et al.*, 1971). However, loss of the EPS receptor will generally lead to an inhibition of phage adsorption and concomitant bacterial resistance against this EPS-specific phage (Gross *et al.*, 1977; Albert *et al.*, 1996; Cornelissen *et al.*, 2011). As described further below, EPS-specific enzymes can be classified into hydrolases (also called polysaccharases) and lyases (Geyer *et al.*, 1983). Hydrolases are either glycoside hydrolases cleaving the glycosidic linkages of EPS or they are deacetylases cleaving acetyl substituents of EPS units. Lyases will introduce a double bond between the C4 and C5 of this uronic acid after cleavage of the glycosidic linkage between a monosaccharide and the C4 of uronic acid (Sutherland, 1995).

Deacetylases

The Vi-antigen in the capsule of *Salmonella*, *Citrobacter* and *E. coli* consists of a polymer of *N*-acetyl-D-galactosaminuronic acid linked by α1,4-bonds and is partially O-acetylated (Luderitz *et al.*, 1968). Adsorption of phage II to the Vi-antigen by a tail-associated enzyme is accompanied by enzymatic cleavage of side acetyl groups without complete depolymerization of the EPS backbone (Taylor, 1965; 1966). Deacetylation of the Vi-antigen inhibits adsorption of the phage to the Vi-polysaccharide and thus phage infection.

EPS-degrading enzymes: glycoside hydrolases and lyases

Phages harbouring these EPS-degrading enzymes will depolymerize the EPS backbone and are able to tunnel their way through the thick protective EPS layer to reach the bacterial cell surface (Bayer *et al.*, 1979). Identification of phages harbouring these enzymes is based on their specific plaque morphology: a clear lytic zone is surrounded by an opaque halo zone, which increases in diameter over time (Fig. 4.2) (Geyer *et al.*, 1983; Hughes *et al.*, 1998a,b; Cornelissen *et al.*, 2011). Within this halo zone, bacteria are devoid of their EPS layer due to the action of phage and free enzymatically active tail spikes or fibres, which diffused out of the primary zone of phage infection into the halo zone. Since the bacteria within the halo zone have entered the stationary phase of growth, infection will be prevented or slowed down. Due to the high structural diversity of the EPS of different bacterial species/strains and the restricted specificity of EPS-degrading enzymes for closely related polysaccharide structures, these EPS-degrading phages generally harbour a narrow host range (Reiger-Hug and Stirm, 1981; Sutherland, 1995, 1999).

Hughes *et al.* (1998a; 1998b) demonstrated that bacteriophages with EPS-degrading enzymes infecting *Pantoea agglomerans* strain 53b and Ent were common in the environments tested. Moreover, the phage-associated EPS-degrading enzyme of phage SF153b plays a major role in the degradation of existing biofilms, since the largest part of biofilm-associated bacteria are removed by the action of the

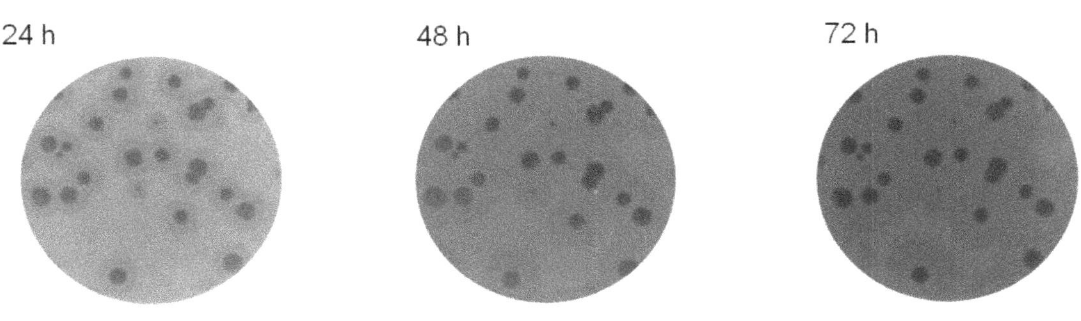

Figure 4.2 Plaques of *P. putida* phage ϕ15 on the host PpG1. With increasing period of time, the diameter of the halo zone, surrounding the plaque with constant diameter, increases. This observation is indicative of the presense of EPS-degrading enzymes associated with virus particles.

EPS-degrading enzyme long before lysis of the infected cells.

At least three chemically distinct EPS structures – alginate, PEL and PSL – have been identified for the species *P. aeruginosa*, the production of which is dependent on the strain and environmental conditions. The *pel* and *psl* loci are involved in the production of a glucose- and mannose-rich EPS of the non-mucoid strains PA14 and PAO1, respectively (Friedman and Kolter, 2004a,b; Jackson *et al.*, 2004; Matsukawa and Greenberg, 2004). Furthermore, the reference strain PAO1 probably contains two additional loci (PA1381-PA1398 and PA3552-PA3558) involved in EPS biosynthesis.

Alginate is probably the best characterized *P. aeruginosa* EPS, but is only produced by mucoid *P. aeruginosa* strains which colonize the lungs of cystic fibrosis patients during the chronic phase of infection (Wong *et al.*, 2000). This alginate, also excreted by *Azotobacter* spp. and some other *Pseudomonas* spp., is a copolymer composed of 1,4-linked *α*-L-guluronate (G) and its C5 epimer β-D-mannuronate (M), arranged in block structures with homopolymeric G- or M-blocks or heteropolymeric blocks (i.e. containing random blocks of M/G). This bacterial alginate is characterized by O-acetylation, often on the 2 and/or 3 positions of β-D-mannuronate (Skjåk-Bræk *et al.*, 1985; Gacesa, 1988). Phages infecting these alginate producing bacteria have an alginate lyase which catalyses the cleavage of the copolymer (Bartell *et al.*, 1966; Glonti *et al.*, 2009). The specificity of the alginate lyases is determined by M/G-block structure and by the degree of O-acetylation. Despite elaborate studies on alginate lyases of bacteria, algae and marine molluscs, only a few articles describe alginate lyases encoded by phage genomes. The alginate lyase of phage F116 reduces the viscosity and facilitates its dispersion through alginate-containing matrices by enzymatic degradation of the alginate polymer. Hanlon *et al.* (2001) hypothesized that this alginate lyase activity can assist in phage migration through the mucoid *P. aeruginosa* biofilms. More recently, Glonti *et al.* (2009) identified four *P. aeruginosa* phages (PT-1, -6, -7 and -12), representing the different families of the *Caudovirales* order, which are able to degrade three structurally different pseudomonal alginic acid preparations. Older studies of phage-associated EPS-degrading enzymes specific for *P. aeruginosa* EPS do not specify the exact chemical composition of the EPS structure. For example, phage 2 depolymerase decreases the viscosity of the host EPS with 20 to 25% and releases monosaccharides from the substrate (Bartell *et al.*, 1966; Castillo and Bartell, 1976). Lysogenization of *P. aeruginosa* strain BI with phage 2 changes its EPS composition, hereby protecting its host against superinfection. In addition, absence of a susceptible EPS structure in enzymatically pretreated host cells or in phage-resistant mutants protect the cells against phage 2 infection (Castillo and Bartell, 1974; 1976). Its enzymatic activity is located in a 180 kDa component, which probably corresponds with the adsorption apparatus of six droplike tail spikes (Bartell *et al.*, 1968; Castillo and Bartell, 1976).

A well-studied example of EPS-degrading phages is formed by a panel of tailed phages specific for capsulated *E. coli* K1, a neuroinvasive pathogen with a α-2,8-linked polysialic acid EPS with up to two hundred residues per polymer chain. This K1 antigen is just one of the at least 80 different *E. coli* EPS structures, type K antigens, known to date (Whitfield and Roberts, 1999). The polysialic acid capsule is the major virulence factor of *E. coli* K1 because it is weakly immunogenic due to structural homology with the host polysialic acid of the neural cell adhesion molecule and it is essential for viable passage across the blood–brain barrier (Mühlenhoff *et al.*, 1998; Kim *et al.*, 2003). Double-stranded DNA phages, infecting *E. coli* K1 possess a tail-associated endosialidase, a glycoside hydrolase which selectively degrades α 2,8-linked polysialic acid (Finne and Mäkelä, 1985; Hallenbeck *et al.*, 1987; Pelkonen *et al.*, 1989). K1-specific phages of different families (*Podoviridae* and *Myoviridae*) as well as virulent and temperate phages have been identified (Deszo *et al.*, 2005; Stummeyer *et al.*, 2006). All endosialidases share a conserved central catalytic part followed by a closely related C-terminal domain essential for proper folding and trimerization, but differ remarkably in their N-terminal domain (Mühlenhoff *et al.*, 2003).

Three representatives of the K1-specific phages, the temperate phage CUS-3 and virulent phages K1E and K1F, are related to a different LPS-specific progenitor type phage, P22, SP6 and T7, respectively. Horizontal gene transfer of a bacterial sialidase gene probably changed the host specificity of these phages: the new tail spikes emerged due to association of this endosialidase with the capsid binding module of the respective progenitor phage (Stummeyer *et al.*, 2005; 2006). For phages CUS-3 and K1F, the endosialidase enzyme is neatly fused to the conserved N-terminal capsid-binding domain used respectively by the 'P22-like phages' (P22, Sf6 and HK620) and 'T7-like phages' (T7, T3 and φYeO3-12) for attaching their endorhamnosidase activity and LPS oligosaccharide core-binding unit to the phage tail (Steven *et al.*, 1988). For 'SP6-like phages', including *E. coli* phage K1E, the degenerated tail spike protein (Gp37) serves as versatile adapter protein interconnecting the tail and the conserved undecapeptide in the N-terminus of the catalytic module (Gp47) (Stummeyer *et al.*, 2006). The SP6 catalytic module is homologous to the other progenitor phage P22 and has endorhamnosidase activity associated with its tail spike, while the catalytic module of *E. coli* phage K1E carries the endosialidase activity (Gp47) (Fig. 4.3). Another SP6-like *E. coli* phage K1-5 encodes besides an endosialidase (Gp47) also a lyase (Gp46) specific for the chemically different capsular polysaccharide of *E. coli* K5. This gene product, which is also present in phage K1E but in a truncated inactive form, is linked to the phage tail via the same adapter protein and enables the phage to infect both *E. coli* K1 and K5 strains (Scholl *et al.*, 2001).

Co-evolutionary arms race

It can be hypothesized that the emergence of the tremendous chemical diversity of bacterial polysaccharide structures, both LPS and EPS, and the diversity of polysaccharide-specific phages are the consequence of the co-evolutionary arms race between phages and bacteria. Pressure imposed by polysaccharide-degrading phages can provide the necessary selection for bacteria to evolve different polysaccharide structures. These new polysaccharide structures are no longer susceptible to the polysaccharide-degrading phages to which they are exposed to, but still protect them from phages which recognize structures underneath the EPS structure for which the protective EPS layer was originally developed. In turn, this would create a selective advantage for phages that gain new catalytic modules which specifically cleave the newly acquired polysaccharide structure.

However the acquisition of new bacterial polysaccharides, often encoded by large gene clusters or new phage-encoded catalytic modules is more

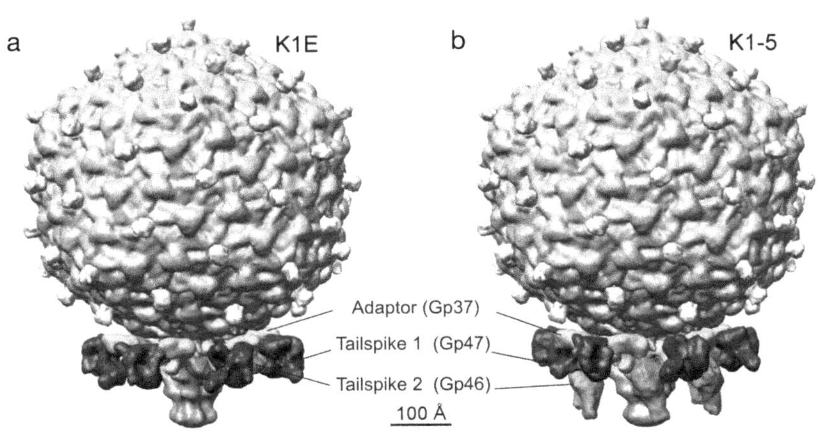

Figure 4.3 Combined cryoEM map of the structure of *E. coli* phages K1E (a) and K1–5 (b). The tailspikes with endosialidase (Gp47) and lyase activity (Gp46) are linked to the phage tail *via* the same adapter protein (Gp37) (adapted from Leiman *et al.*, 2007).

complex than generating simple mutations, but requires horizontal gene transfer between other bacteria and phages.

The presence of a similar tripartite build-up of bacteriophage tail-associated enzymes with a N-terminal domain for particle binding, a central catalytic domain and C-terminal folding domain, enables phages to easily acquire new host specificities. Horizontal gene transfer of the gene encoding the catalytic domains confers new host specificity, while a conserved N-terminal capsid binding domain or conserved N-terminal peptide interacting with the phage adapter ensures proper connection to the phage particle.

Preventing phage DNA entry

DNA entry-blocking represents an active resistance mechanism which inhibits the entry of phage genomic DNA into the host cell after its successful adsorption to the bacterial surface receptors.

Physical barrier formed by the Gram-negative peptidoglycan layer

At the transition from exponential growth to stationary phase, the peptidoglycan layer thickens and the level of cross-linking of peptidoglycan molecules increases (Pisabarro *et al.*, 1985; Templin *et al.*, 1999). This enforces the bacterial cell wall which phages need to cross in order to inject their genome in the bacterial cell. Mycobacteriophage TM4 possesses a tape measure protein with a peptidoglycan hydrolase activity in motif 3 (Pedulla *et al.*, 2003). Although this motif is not necessary for phage viability, it promotes efficient DNA injection into stationary phase cells. As such, wild-type TM4 have a similar efficiency of plating on stationary phase and exponentially growing cells, while the efficiency of plating of the motif 3-deletion mutant is reduced by about 50% on stationary phase cells compared to the exponentially growing cells (Piuri and Hatfull, 2006). Similarly, mutants of *E. coli* phage T7 defective in murein hydrolase activity show defects in latent period or efficiency of plating. These defects are more pronounced in the stationary growth phase which has an increased level of peptidoglycan cross-linking (Pisabarro *et al.*, 1985; Moak and Molineux, 2000).

Phage-mediated superinfection exclusion

Superinfection exclusion (Sie) function requires proteins anchored to the membrane or associated with membrane components. Although this phenomenon is also observed in virulent phages, Sie systems are often encoded by temperate phages. Their (pro)phage origin suggests that Sie systems may be more important for phage–phage interactions than for phage–host interactions.

Sie systems in Gram-negative bacteria

The lytic *E. coli* phage T4 contains two Sie systems encoded by the genes *imm* and *sp*, which are necessary for prevention of superinfections by other T-even phages. These two Sie systems function independently of each other and use a different mechanism of action. Imm, a 83-residues protein, is suggested to be located at the cytoplasmic membrane and with the help of another membrane protein, indirectly blocks DNA transfer by inducing a conformational change of a component at the phage DNA injection site (Lu *et al.*, 1993). The other Sie system, the host membrane protein Sp, inhibits peptidoglycan hydrolysis activity of the T4 lysozyme (Gp5). This T4 lysozyme is located at the extreme tip of the phage tail and creates a hole in the bacterial cell wall through which phage DNA is injected (Lu and Henning, 1994; Moak and Molineux, 2000).

Another prophage P22, infecting *Salmonella enterica* subsp. *enterica* serovar *typhimurium* codes for a Sie system (SieA) directed at the exclusion of the heteroimmune *Salmonella* phages L, MG178, and MG40 and superinfecting P22 (Susskind *et al.*, 1971). SieA is an inner membrane protein which blocks the transfer of the phage DNA in the cytoplasm with a –to date- unknown molecular mechanism of action (Susskind *et al.*, 1974; Hofer *et al.*, 1995). Initially, SieB was also believed to be a superinfection exclusion system independently of SieA, but is now recognized to be involved in abortive infection (Hofer *et al.*, 1995; Susskind and Botstein, 1978).

The temperate phage P1 encodes a superimmunity function (Sim) located in the periplasm of the lysogenic cell (Kliem and Dreiseikelmann, 1989). Mature Sim (24 kDa) is processed from a precursor protein by removal of a hydrophobic leader peptide of a 20-residue in a SecA-dependent manner (Maillou and Dreiseikelmann, 1990). Sim confers immunity against wild-type phage P1 and also c1, c4 and vir mutants. As for SieA, the molecular mechanism of Sim's DNA blocking activity is uncharacterized. Evidence for this activity is deduced from the fact that phage adsorption is unaffected by Sim and the bacterium can be successfully transformed with phage genomic DNA (Kliem and Dreiseikelmann, 1989).

Sie systems in Gram-positive bacteria

Sie systems have mainly been identified in prophages of Gram-positive strains (e.g. *L. lactis*, *Streptococcus thermophilus*) used for milk fermentation processes in the dairy industry. Sie_{2009} of the temperate lactococcal phage Tuc2009 mediates a phage-resistance phenotype in *L. lactis* UC509 through a DNA injection blocking mechanism. High constitutive expression of Sie_{2009} from two different promoters confers complete resistance against a certain subset of the 936-type lactococcal phage group, which is predominantly present in the dairy industry (McGrath *et al.*, 2002). Other Sie systems in *L. lactis* strains have been identified in P335-type prophages (McGrath *et al.*, 2002; Mahony *et al.*, 2008). They are all represented by small proteins with a hydrophobic amino terminus and one or two transmembrane domains. Furthermore, they possess a relatively low G+C content (27–34%) compared with the lactococcal genome (~36%) and a remarkably high isoelectric point (from 9.5 to 9.8) (Wegmann *et al.*, 2007). These lactococcal Sie proteins are localized to the membrane and inhibit phage DNA entry into the host cell without affecting phage adsorption (McGrath *et al.*, 2002; Mahony *et al.*, 2008).

The *ltp* (lipoprotein of temperate phage) gene of *S. thermophilus* prophage TP-J34 encoding a signal-peptide containing lipoprotein, represents a different Sie system which blocks the entry of phage DNA (Neve *et al.*, 1998). The Ltp protein protects *S. thermophilus* J34 against TP-J34 infection. Interestingly, when expressed in *L. lactis* Bu2–60, it also provides resistance against the 936-type lactococcal phage P008, which does not belong to the 936-type subset susceptible for Sie_{2009} (Sun *et al.*, 2006).

Plasmid-encoded phage defence system

Plasmid pN40 encodes besides two abortive infection systems (AbiE and AbiF) an early-acting phage resistance mechanism in the *L. lactis* subsp. *lactis* MG1614 background. This mechanism mediates insensitivity to φc2 and specifically blocks DNA penetration of the phage, probably at the membrane level (Garvey *et al.*, 1996).

Colicins are encoded by colicinogenic plasmids (pCol) which are widely distributed in natural *E. coli* populations (Riley and Gordon, 1992). As bactericidal protein toxins, they are lethal to sensitive *Enterobacteriaceae* (Lazdunski *et al.*, 1998). Bacteriophage defence might be one of colicins' additional functions as evidenced in the inhibitory effect conferred by ColIB against T-odd phages and the limited protection provided by ColE7-K317 against phage M13K07 and λ (Reakes *et al.*, 1987; Duckworth and Pinkerton, 1988; Lin *et al.*, 2004). Despite the fact that no complete resistance is observed, colicin-producing cells display a slower multiplication rate of phage M13 progeny than non-producing cells. Chen *et al.* (2010) observed that interaction of the receptor-binding and toxicity domains of colicin E7 with the M13 major coat protein (gp8) interferes with the gp8 depolymerization into the cytoplasmic membrane prior to phage infection. Although the details of bacteriophage DNA translocation are still unclear, a normal gp8 depolymerization is a prerequisite for phage DNA entry to the host cell (Click and Webster, 1998).

Degradation of phage nucleic acids

After successful adsorption and genome entry, the phage replication cycle can be actively interrupted by the presence of restriction-modification systems or by the CRISPR/Cas system.

Restriction–modification systems

Probably the best-studied anti-phage barrier is the restriction–modification (R–M) system, present in more than 90% of sequenced bacterial and archaeal genomes (Roberts *et al.*, 2010). Their main function is to defend the host against foreign DNA, including phage genomic DNA.

This is achieved through cleavage of incoming foreign DNA by host-encoded restriction endonucleases, which is recognized by the absence of a characteristic modification, a methylation. This methylation takes place at defined sites within a recognition sequence and protects the host DNA from cleavage. When unmethylated phage genomic DNA enters a bacterial cell harbouring a specific R–M system, it will either be recognized by the restriction endonuclease and rapidly degraded or methylated by the bacterial methylase hereby escaping restriction and going on to the next step in the phage replication cycle. The severity of restriction depends on the R–M system and the phage. Wilson and Murray (1991) observed that the efficiency of plating decreases logarithmically with increasing number of recognition sites. When the phage genomic DNA is methylated, it will be insensitive for the cognate restriction endonuclease and the phage will therefore be able to infect neighbouring hosts carrying the same R–M system. However, when infecting new hosts with different R–M system, the phages will be restricted unless they become methylated by the methylase of the new host as well. Based on subunit structure and cofactor requirements, R–M systems have been classified into three main groups (Types I–III).

- Type I R–M systems are hetero-oligomeric enzymes of three subunits: HsdR ('host specificity for DNA recognition'), HsdM ('modification/methylation') and HsdS ('specificity'). *In vivo*, two different oligomeric forms are observed: M_2S and R_2M_2S. The multimer M_2S containing two HsdM subunits and one HsdS subunit is responsible for DNA methylation in the presence of the cofactors *S*-adenosyl-methionine (AdoMet) and Mg^{2+}, while R_2M_2S (M_2S with two HsdR subunits) requires three co-factors (AdoMet, Mg^{2+} and ATP) for restriction of unmethylated DNA. After ATP-dependent DNA translocation, R_2M_2S cuts DNA approximately halfway between the two bipartite asymmetrical recognition sites (Murray, 2000; Dryden *et al.*, 2001; Bourniquel and Bickle, 2002). Based on genetic complementation and molecular evidence, five families (IA-IE) of Type I R–M systems have been identified (Murray, 2001; Chin *et al.*, 2004). An example is EcoKI (Type IA), which is able to limit phage propagation by factors of 10^3–10^8 (Webb *et al.*, 1996).
- Type II R–M system comprises two separate and independently working restriction endonucleases and methylases with an identical, single, 4–8 bp palindromic recognition sequence. Independent of ATP or GTP, the restriction enzymes cleave DNA within or close to the recognition site generating defined restriction products with 5′-PO_4 and 3′-OH termini using Mg^{2+} as a sole co-factor (Pingoud and Jeltsch, 1997; Pingoud *et al.*, 2005). Type II methylases, as monomers, methylate a specific base of their recognition sequence on both DNA strands. Methylation activity requires the co-factor AdoMet and occurs at cytosine (N^4 of C^5 position) or at the N^6 position of adenine (Sistla and Rao, 2004). Despite high conservation of amino acid motifs of Type II methylases, Type II restriction endonucleases are very dissimilar and have been classified into eleven overlapping subclasses (Roberts *et al.*, 2010). These restriction endonucleases, such as EcoRI (subtype P), SfiI (subtype F) and FokI (subtype S), are used in every day recombinant DNA work.
- The Type III multisubunit complex is composed of the Res (DNA cleavage) and Mod (DNA recognition and modification) subunits, encoded by the *mod* and *res* genes (Janscak *et al.*, 2001). A stable dimer formed by the Mod subunit acts as an independent methylase in the presence of AdoMet and methylates the N^6 position of adenine on one strand, creating hemimethylated DNA. As only completely unmethylated DNA is cut by Type III restriction endonucleases, hemimethylation offers rapid protection. The Res_2Mod_2 complex mediates DNA cleavage in the presence of ATP and Mg^{2+} and is stimulated by AdoMet. Cleavage by the Res_2Mod_2 nuclease complex is preceded by DNA-translocation, and occurs 25–27 bases from one of the two head-to-head arranged asymmetrical recognition sites (Dryden *et al.*, 2001; Boruniquel and Bickle, 2002). The contiguous genes encoding the Type III R–M enzymes display a high degree

of sequence similarity. The best-studied Type III examples are EcoP1I and EcoP15I (Bickle and Krüger, 1993).

- These three R–M systems are assisted by modification-dependent restriction (MDR) systems, which specifically recognize and cut modified DNA. These MDR endonucleases, referred to as Type IV restriction endonucleases have however no associated methylase. The best characterized representative is the two-subunit complex McrBC from *E. coli* K12 consisting of McrB and McrC, which are responsible for DNA recognition and cleavage, respectively. McrBC binds two recognition sites which are (hydroxy)methylated at the cytosine residue (position N^4 or C^5) and separated by 40 to 3000 nucleotides. Cleavage occurs preferentially 30 bp from one of these sites (Sutherland *et al.*, 1992). The presence of the co-factors, Mg^{2+} and GTP, is required for DNA translocation and cleavage (Dryden *et al.*, 2001; Bourniquel and Bickle, 2002). McrA and Mrr represent two other MDR endonucleases from *E. coli* (Raleigh and Wilson, 1986), while DpnI was identified in *Streptococcus pneumoniae* (Vovis and Lacks, 1977).

Phages have responded to the threat of the R–M systems through the evolution of many different strategies:

- Counter-selection against recognition sites. Selective pressure imposed by the bacterial host may lead to the total loss of or reduction in the number of recognition sites specific for the restriction endonucleases of the bacterial host. This can be due to accumulation of point mutations in the phage genome. For example, the BsuRI recognition site (5′-GGCC-3′) is statistically predicted to occur four hundred times in the *Bacillus subtilis* phage φ1 genome, however this recognition site is completely absent (Kawamura *et al.*, 1981). Reduction in the number of recognition sites leads to EcoRII resistance for phages T3 and T7. EcoRII must bind two target sequences before restriction is able to occur. Because the distance between *Eco*RII sites in the genomes of phages T3 and T7 is too large, their genomic DNA is insensitive to restriction (Krüger *et al.*, 1988; Bickle and Krüger, 1993). Additionally, T7 possesses a second feature to subvert Type III restriction enzymes. Instead of two head-to-head arranged recognition sites, all the EcoP1I sites in the T7 genome are in the same orientation (Meisel *et al.*, 1992).
- Stimulation of host modification functions. *E. coli* phage λ alleviates restriction of Type I R–M systems due to the presence of the gene Ral that stimulates the modification reaction (Zabeau *et al.*, 1980), leading to an increased proportion of methylated genome in phage λ progeny. The effect of the Ral function is not observed upon primary infection, but upon reinfection of the same host strain with the methylated λ progeny from the first infection. Moreover, unmodified phage DNA will avoid the restriction activity in a larger proportion when a helper phage expressing the Ral protein is provided *in trans*.
- Inhibition of restriction endonucleases. The genome of *E. coli* phage T7 contains gene *0.3*, often called *ocr* ('overcome classical restriction'), which encodes a protein dimer of two 13.5 kDa subunits. This dimer allocates a negative charge distribution complementary to the charge within the DNA binding site of Type I R–M systems (Mark and Studier, 1981; Dunn *et al.*, 1981; Atanasiu *et al.*, 2001). Since it also mimics the size and shape of phage double-stranded B-fold DNA, Ocr blocks the active site of Type I R–M systems and prevents phage genome restriction (Mark and Studier, 1981; Bandypadhay *et al.*, 1985; Atanasiu *et al.*, 2001). Ocr operates in a non-sequence-specific manner, targeting R–M systems with different recognition sequences (Moffatt and Studier, 1988; Bandyopadhay *et al.*, 1985; Kruger *et al.*, 1977; 1983). Following adsorption, phage T7 injects the left end of its linear genome, which comprises the *ocr* gene (Krüger and Schroeder, 1981; Moffat and Studier, 1988). Transcription and translation of this region is a prerequisite for injection of the remaining phage DNA, which is protected by the Ocr protein from restriction. Coliphage T3 possesses a protein similar to phage T7 Ocr, blocking Type I and Type III R–M systems. It can also hydrolyse

AdoMet, thereby subverting restriction endonucleases that require AdoMet for activity.

- Virus-encoded DNA modification. *B. subtilis* temperate phages SPR and φ3T carry their own methylase with the same sequence specificity as the host methylase (Cregg *et al.*, 1980; Warren, 1980; Noyer-Weidner *et al.*, 1981). The prophage-encoded methylase gene is only expressed after prophage induction and during lytic infection, so that the phage genomic DNA produced by both replication cycles is protected against the host restriction endonucleases. Resistance through acquisition of the cognate methylase gene was also observed for a virulent phage (Hill *et al.*, 1991) as well as for a temperate lactococcal bacteriophage Tuc2009 (McGrath *et al.*, 1999).
- Incorporation of unusual nucleotides. Many *B. subtilis* bacteriophages are able to subvert R–M systems recognizing sequences containing thymidine by the complete replacement of this base in their DNA. In this way, phages SPOI and SP8 have replaced thymidine by 5-hydroxymethyluracil, and phages PBS1 and PBS2 by uracil (Krüger and Bickle, 1983). The latter phages also possess an inhibitor (Ugi) of the host uracil-DNA glycosylase, a DNA base-excision repair enzyme that removes uracil produced by spontaneous deamination of cytosine (Lindahl, 1979), using the concept of protein mimicry of DNA (Putman *et al.*, 1999) (see also Ocr of bacteriophage T3, T7). Coliphage T4 incorporates the unusual base hydroxymethylcytosine (HMC) instead of cytosine into its genomic DNA, hereby subverting R–M systems with cytosine-based recognition sequences (Carlson *et al.*, 1994). Bacteria, which have acquired the MDS system (see above: R–M systems) can attack this modified phage DNA. In turn, phage T4 is resistant to the MDS systems due to glucosylation of the HMC residues (Carlson *et al.*, 1994). A second counter-attack of bacterial origin in this co-evolutionary arms race with phage T4 is observed in *E. coli* CT596. A prophage-encoded two-component system constituting the glucose-modified restriction proteins, GmrS and GmrD, specifically recognizes and cleaves DNA with glucosylated HMC (Bair and Black, 2007). In another twist, T4 blocks this GmrS-GmrD system with internal protein IPI* (see 'Coinjection of phage internal proteins and genomic DNA', below). The *mom* gene product protects phage Mu against Type I and Type II R–M systems by acetimidation of the N^6 position of adenine residues within the recognition sites (Hattman, 1979, 1980).
- Coinjection of phage internal proteins and genomic DNA. Coliphage T4 encodes an internal protein I (IPI), which in its mature form IPI* is coinjected with the phage genome into the *E. coli* host, in several hundreds of copies. IPI* directly interacts with the *E. coli* CT596 GmrS-GmrD restriction system to inhibit its enzymatic activity (Bair *et al.*, 2007; Rifat *et al.*, 2008). In a new counter-attack, the GmrSD enzyme of *E. coli* UT189 evolved to a single fused polypeptide which is immune to IPI* (Rifat *et al.*, 2008). Temperate phage P1 mediates efficient generalized transduction between different *E. coli* hosts carrying Type I restriction enzymes through injection of two phage capsid internal proteins, DarA and DarB. It is hypothesized that the DarA and DarB ('defence against restriction') proteins bind the phage genomic DNA and remain associated with it after injection (Iida *et al.*, 1987). The presence in the DarB protein sequence of motifs characteristic of DNA methyltransferases suggests that DarB methylates phage DNA at infection (Łobocka *et al.*, 2004).
- Destruction of restriction endonuclease co-factors. Since Type I and Type III R–M systems require AdoMet as an essential or stimulating co-factor for stable binding phage DNA, a reduction of the intracellular AdoMet concentration should prevent restriction (Sistla and Rao, 2004). The *ocr* gene product of coliphage T3 also has AdoMet hydrolase activity. This activity can prevent newly synthesized Type I and Type III restriction enzymes from acquiring AdoMet, but cannot inhibit restriction enzymes that are already bound to the co-factor (Hausmann, 1967; Spoerel *et al.*, 1979; Krüger and Schroeder, 1981). AdoMet hydrolase also improves the chances of phage survival in cells with the Type III EcoP1 system, by alleviating

the repressor-like action of EcoP1 on phage gene expression (Krüger *et al.*, 1982).

- Phage-mediated homologous recombination. Lambdoid phages like λ and Rac, repair the double-stranded restriction break by a phage-encoded homologous recombination mechanism in which they copy intact homologous DNA with or without crossing-over of the flanking sequences (Takahashi and Kobayashi, 1990; Kobayashi, 1998). Only one progeny DNA is generated out of two recombining DNAs (Takahashi *et al.*, 1997). This mechanism is mediated by the RecE and RecT proteins of the Rac prophage, and by Redα and Redβ of phage λ (Takahashi and Kobayashi, 1990; Kusano *et al.*, 1994). For example, the RecE pathway alleviates Type III restriction of EcoP1 and EcoP15. The Type I restriction of EcoKI is only slightly alleviated by the RecE pathway, while the Type II restriction enzyme *Eco*RI is not (Handa and Kobayashi, 2005).

CRISPR/Cas system

After the first description of a CRISPR-locus, consisting of 14 repeats of 29 base pairs interspersed by 32–33 base pairs of non-repeating sequence in *E. coli*, comparative analysis revealed that approximately 40% and 90% of all sequenced bacterial and archaeal genomes contain at least one CRISPR locus (Ishino *et al.*, 1987; Grissa *et al.*, 2007). These clustered regularly interspaced short palindromic repeats (CRISPR) typically consist of 23–50 bp direct repeats interspaced by non-repetitive spacers (17–100 bp) of similar length, and are usually flanked by a varying number of *cas* ('CRISPR-associated') genes. Several hypotheses for the biological role of the CRISPR-Cas system have been proposed. It was not until 2007 that experimental evidence demonstrated that the acquisition of spacers in *Streptococcus thermophilus* was associated with the acquisition of resistance against virulent phage(s) carrying the identical sequence (Barrangou *et al.*, 2007). The CRISPR/Cas system is now defined as an adaptive immune system which provides acquired immunity against foreign DNA (Richter *et al.*, 2012; Westra *et al.*, 2012).

Structural features of the CRISPR/Cas system

The number of repeat/spacer units in a single locus can be as high as 587 and up to 23 loci can be found in a single organism (Grissa *et al.*, 2007). Commonly, these loci consist of fewer than 50 units as observed for the lactic acid bacteria (Horvath *et al.*, 2009).

- Repeats. Within a single CRISPR locus repeat sequences are highly conserved in sequence and size, except for the last repeat, which usually slightly diverges from the other repeats. The repeats are also highly divergent between bacterial species (Jansen *et al.*, 2002; Kunin *et al.*, 2007). Most repeat sequences contain a short (5–7 bp) palindrome, hereby contributing to an RNA stem–loop secondary structure of the repeat (Jansen *et al.*, 2002). This structural feature, together with the conserved motif (GAAA(C/G)) at the 3′ repeat terminus, are suggested to act as binding sites for the Cas proteins and are involved in the processing of the CRISPR transcript (see below) (Kunin *et al.*, 2007).
- Spacers. In contrast to the repeats of a given CRISPR locus, spacers are generally unique and many spacers match with high sequence identity to foreign mobile genetic elements, like phage and plasmid-derived sequences. For example in lactic acid bacteria approximately 40% of all spacers have a homologue, called the protospacer, matching either phage (75%) or plasmid sequences (20%) (Bolotin *et al.*, 2005). Within phages, spacers are found on both strands and are evenly distributed across the genome (Mojica *et al.*, 2005; Pourcel *et al.*, 2005; Barrangou *et al.*, 2007; Lillestøl *et al.*, 2006).
- Leader. The leader sequence, a usually AT-rich sequence of up to 550 bp is located directly upstream from the first repeat-spacer unit at the CRISPR 5′ terminus (Jansen *et al.*, 2002; Lillestøl *et al.*, 2006). As observed for repeat sequences, leaders are highly divergent across species boundaries and contain no coding sequences. Leaders serve at least two possible functions: a recognition sequence for addition of new spacers (Pourcel *et al.*, 2005; Barrangou

et al., 2007) and a promoter for the transcribed CRISPR array (Tang *et al.*, 2002; 2005).

- Cas proteins. The genes coding for the CRISPR-associated proteins are located immediately upstream or downstream from the leader-CRISPR locus. Cas gene products represent a large, heterogeneous protein family carrying functional domains typical for nucleases, helicases, polymerases, polynucleotide-binding proteins and proteins involved in transcriptional regulation (Haft *et al.*, 2005). The variability in *cas* genes associated with each CRISPR locus is extremely high. Haft *et al.* (2005) identified 45 Cas protein families associated with CRISPR loci, which can be classified into eight subtypes each associated with a reference organism (Ecoli, Ypest, Nmeni, Dvulg, Tneap, Hmari, Apern, Mtube) (van der Oost *et al.*, 2009). In addition, *Cas* genes can be classified in one of three categories, based on the type of proteins they encode: the core (or signature) proteins, the subtype specific proteins or the Repeat-Associated Mysterious Proteins (RAMP). The 'core' *cas* genes, *cas1* and *cas2* (sometimes encoded as a fused domain of *cas3*), form the universal markers of the CRISPR-Cas system. Subtype-specific *cas* genes are diverse and form clusters across multiple genomes and signature *cas* genes for each subtype are in the process of being identified (Haft *et al.*, 2005). The Cas module RAMP ('repeat associated mysterious protein') containing six genes (*cmr1-cmr6*), is only present in genomes with one of the eight CRISPR-Cas subtypes, but is not necessarily located near the CRISPR locus (Makarova *et al.*, 2002; Haft *et al.*, 2005).

Bacterial species are not limited to one subtype of CRISPR/Cas, making CRISPR loci very heterogeneous. The variability in both repeat/spacer content and in *cas* gene organization is responsible for the extreme CRISPR/Cas diversity and versatility. Given the overall complexity of the CRISPR/Cas systems, a 'polythetic' (based on several criteria) classification of the CRISPR/Cas systems is under discussion (Makarova *et al.*, 2011). The highly dynamic evolution of CRISPR/Cas systems is also complicating the development of a uniform classification and nomenclature (see below).

CRISPR-Cas mechanism of action

The exact mechanism by which the CRISPR/Cas system provides resistance against foreign extra-chromosomal DNA is not yet fully understood. However already two phases in the CRISPR/cas function have been recognized. A primary immunization step against specific foreign DNA is followed by an immunity process, which confers resistance against a next infection.

Immunization (adaptation stage)

Following phage infection, a bacterium with a specific CRISPR/Cas system will integrate a new spacer which is generally 100% identical to a phage genomic sequence, called the protospacer, at the 5′ end of the repeat-spacer region of the CRISPR locus. Short conserved regions, termed 'CRISPR motifs' or 'protospacer-adjacent motifs' (PAM motifs), located on either side of the protospacer may serve as a recognition signal for protospacer selection and indicate that these sequences are not randomly selected out of the phage genome (Deveau *et al.*, 2008; Mojica *et al.*, 2009; Semenova *et al.*, 2009). These PAM motifs can vary between CRISPR/Cas systems. For example, the CRISPR1 and CRISPR3 loci of *S. thermophilus* contain a AGAA and GGNG motif, respectively. Generally, one single repeat-spacer unit is added, but up to four new units have been observed (Deveau *et al.*, 2008). Moreover, different phage-resistant hosts can acquire distinct spacers from the same infecting phage. CRISPR loci will however not grow indefinitely, as internal spacer deletions, likely via homologous recombination between CRISPR repeats, have been observed (Tyson *et al.*, 2007; Deveau *et al.*, 2008, Horvath *et al.*, 2008). A metagenome sequencing project of two distinct *Leptospirillum* sp. populations over a period of months indicated that spacer diversity is highly polarized with the cluster being more conserved at the distal half and highly divergent at the proximal half from the Leader region. The appearance of new spacers was accompanied by the loss of more conserved ones, suggesting that both populations, which are essentially identical, had diverged in their CRISPR content to adapt to a new environment with a distinct phage population (Tyson *et al.*, 2007). This probably permits the host to limit the locus expansion so that the relative locus does not increase to a detrimental size without influencing the host fitness. The mechanisms behind the different steps of recognition of invasive foreign DNA, spacer selection and integration into the CRISPR locus and novel repeat manufacturing remains so far uncharacterized. Nonetheless the functions of some Cas gene products have been inferred. At least three *cas* genes have been identified to be involved in this immunization process. Inactivation of the *cas2* gene or the CRISPR1-associated *cas7* of *S. thermophilus* gene prevents the integration of new spacers after phage exposure, implying that these gene products are necessary for recognizing foreign DNA and/or generation of novel spacer-repeat units (Barrangou *et al.*, 2007). *P. aeruginosa* Cas1 is a sequence-non-specific dsDNase generating ~80-nucleotide fragments which possibly function as initial sources of new spacers (Wiedenheft *et al.*, 2009).

Immunity (interference stage)

For the spacer-encoded defence to become available to the Cas proteins involved in CRISPR-encoded phage resistance, the CRISPR locus is transcribed into a single CRISPR RNA, the pre-crRNA. In this process, the leader sequence acts as promoter for a constitutive and unidirectional transcription (Cui *et al.*, 2008; Hale *et al.*, 2008; Lillestøl *et al.*, 2009). The full-length pre-crRNA is subsequently processed by the Cas proteins into specific small crRNA units corresponding to one spacer flanked by two partial repeats (Makarova *et al.*, 2006; Brouns *et al.*, 2008). In a next step, the crRNAs use their base-pairing potential to guide the Cas interference machinery towards foreign DNA (Marraffini and Sontheimer, 2008; Garneau *et al.*, 2010) or RNA (Hale *et al.*, 2009) that matches its sequence, ultimately leading to the degradation of that foreign genetic element (Barrangou *et al.*, 2007).

One example is the CRISPR/Cas system of *E. coli* K12, composed of the 'core' Cas gene products (Cas1-Cas3 and Cas5e) and the subtype-specific Cse1-Cse4 proteins. The CRISPR–associated complex for antiviral defence (Cascade), composed of Cas5e and Cse1-Cse4, specifically cleave the pre-crRNA within each repeat (8 nucleotides upstream of the spacer) creating mature crRNAs (a 61-nucleotide CRISPR RNA), which remain associated with Cascade. The crRNAs are then trimmed at the 3′ terminus, giving rise to crRNAs with a well-defined 5′ end which begins ~ 8 nucleotides in the upstream repeat sequence and with a more heterogeneous 3′ end (Brouns *et al.*, 2008). These Cascade–crRNA complexes together with Cas3 are required in the interference step (Brouns *et al.*, 2008) (Fig. 4.4). Cas3 combines ssDNA nuclease and ATPase/helicase activity within a single protein, thereby probably contributing to protospacer sequence recognition by the Cascade–crRNA complex and subsequent DNA cleavage (Sinkunas *et al.*, 2011). It is suggested that interactions of the ternary Cascade–crRNA-Cas3 complex with the foreign DNA promotes strand separation, followed by crRNA hybridization to the complementary protospacer leading to the

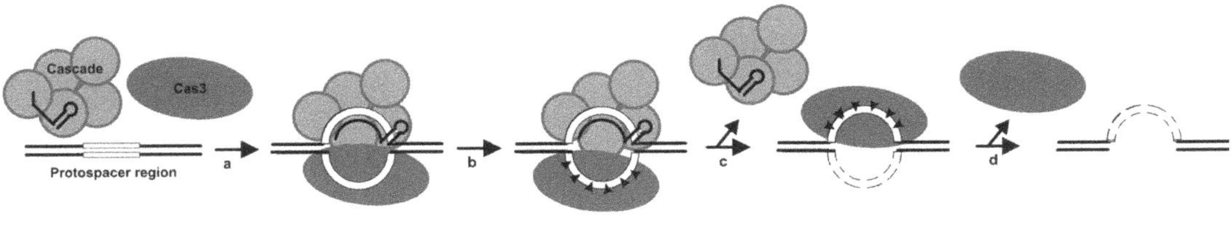

Figure 4.4 CRISPR/Cas system of *E. coli* K12. After cleavage of the pre-crRNA by Cascade, the crRNas remain associated. (a) Interaction of the ternary Cascade–crRNA-Cas3 complex with protospacer promotes strand separation and crRNA hybridization, hereby generating the R-loop structure. (b) The Cas3 nuclease cuts ssDNA in the R-loop structure at multiple sites creating a single-stranded break. (c) Cas3 helicase further remodels the Cascade–crRNA complex (e.g. by displacing it) in order to cleave the second DNA strand and (d) create a double-strand break.

generation of R-loop structures (Camps and Loeb, 2005; Jore *et al.*, 2011). Formation of these loops may be promoted by the helicase activity of Cas3, which subsequently creates a single-stranded break in the protospacer region with its nuclease activity. One may speculate that the Cas3 helicase further remodels the Cascade–crRNA complex by displacing the complex, hereby creating a platform for the second DNA strand cleavage to generate a double-stranded break (Garneau *et al.*, 2010).

Since the CRISPR spacer and the protospacer are very similar, the CRISPR/Cas system must discriminate between self and non-self in order to cope with invasions without triggering an 'autoimmune' response and killing itself. Compensatory mutagenic analysis in *Staphylococcus epidermidis* revealed that CRISPRs are responsible for differential recognition of self and non-self DNA: the ~8-nucleotide repeat sequence of the crRNA 5′ terminus is only able to pair with the repeat sequence of the CRISPR locus of the host genome. The absence of this complementarity allows CRISPR/Cas-mediated degradation of the foreign DNA (Marraffini and Sontheimer, 2010). In other CRISPR/Cas systems, the presence of the PAM in the invading DNA (but absent in CRISPR loci) is important in target recognition. As with mutations in the targeted proto-spacer, mutations in the PAM can critically affect interference (Deveau *et al.*, 2008). Finally, we now know that spacers can also match host genes such as 16S rRNA, DNA polymerase or tRNA synthetases (Stern *et al.*, 2010). Mutations in either the PAM (when identified), the adjacent repeat, the spacer itself, the *cas* genes or the partial or total deletion of the CRISPR/Cas locus are ways to evade autoimmunity.

Several lines of evidence suggest that most CRISPR/Cas system recognizes dsDNA. First of all, phage genome analysis indicated that targets are distributed on both the sense and antisense strands (Cui *et al.*, 2008; Deveau *et al.*, 2008; Horvath *et al.*, 2008; van der Ploeg *et al.*, 2009). Since transcription of the repeat-spacer array occurs in a unidirectional mode, recognition of antisense sequences is only possible with a dsDNA target. Moreover, no spacers matching RNA viruses have been identified to date (Mojica *et al.*, 2009; Wiedenheft *et al.*, 2009). Thirdly, many protospacers occur in phage genes which are expressed late in the infection cycle (Deveau *et al.*, 2008; Semenova *et al.*, 2009). If mRNA were the target, host viability at this point would already be too compromising for the bacterial cell to survive the phage infection, which suggests that dsDNA is a more reliable target. Recently, it was shown that, in *S. thermophilus*, the CRISPR/Cas machinery cleaves phage and plasmid dsDNA in the protospacer *in vivo*, also at a specific distance from the 3′-end of the heteroduplex (Garneau *et al.*, 2010). The cleavage occurs quickly and cleaved phage/plasmid DNA can be isolated from the host cytoplasm.

On the other hand, the possibility of a translational shutdown using mRNA targets is supported by two studies of archaeal CRISPR/Cas systems. A possible solution for mRNA targets described from the foreign antisense strand is the conversion of the unidirectionally transcribed host spacers into dsRNA (Lillestøl *et al.*, 2006). A second observation by Hale *et al.* (2009) was the crRNA-guided RNA cleavage of the crRNAs, complexed with the RAMP module Cas proteins (Cmr1-Cmr6), while no DNase activity was observed. RNA cleavage at 14 nucleotides from the target nucleotide opposite the 3′-terminal base of the crRNA generated 3′-P and 5′-OH ends. This activity was observed with purified and synthetic crRNAs as well as with purified recombinant Cmr proteins.

Given the large diversity of CRISPR-Cas systems in bacteria and Archaea (Haft *et al.*, 2005; Makarova *et al.*, 2006), it is possible that both DNA and RNA are likely targets, depending on the subtype of CRISPR/Cas system present.

Circumventing CRISPR-based immunity and evolution

Although CRISPR/Cas systems can provide high levels of phage resistance, a relatively small proportion of phages retain the ability to infect the 'immunized' host. Within *S. thermophilus*, a 100% sequence match between the CRISPR spacer and the protospacer is obligatory for CRISPR-based immunity. A single point mutation or deletion within the protospacer permits the phage to subvert the CRISPR/Cas mechanism (Barrangou *et* al., 2007; Deveau *et al.*, 2008; Heidelberg *et al.*, 2009). Furthermore, mutation of the CRISPR-motif results in phage sensitivity, despite the presence of an identical protospacer (Semenova *et al.*, 2009). These mutations can, however, alter the amino acid sequence by the introduction of non-silent mutations or premature stop codons that truncate the phage protein (Deveau *et al.*, 2008). In addition to mutations, phages may have lost the target protospacer to subvert the CRISPR/Cas mechanism (Deveau *et al.*, 2008; Andersson and Banfield, 2008). Within environmental phage populations, recombination of previously established spacer polymorphisms constitute a better defence against CRISPR/Cas systems than mutation or deletion, as the risk for altering amino acid sequence and corresponding protein function is reduced (Andersson and Banfield, 2008). Alternative anti-CRISPR strategies also likely exist. For example, the T7 phage kinase (Gp0.7) phosphorylates serine and threonine residues of the CasB protein (Qimron *et al.*, 2010). Phosphorylation of the host RNA polymerase and RNaseE by Gp0.7 inhibits their function, but its influence on the CRISPR/Cas mechanism is not yet proven. Some *Pseudomonas* bacteriophages also carry genes that inactivate the CRISPR/Cas immune system by an unknown mechanism (Bondy-Denomy *et al.*, 2013).

As a counter-response, bacteria can iteratively add new spacers to the CRISPR locus targeting the phage mutants. In addition, horizontal gene transfer of CRISPR/Cas loci may be desirable in a hostile environment as evidenced by the remarkable differences in codon bias and G+C content between the host genome and CRISPR/Cas loci, their presence on mobile elements and their variable presence and location in closely related genomes (Godde and Bickerton, 2006; Horvath *et al.*, 2009). This is supported by the lack of congruence between the phylogenetic relation of CRISPR/Cas subtypes and the established bacterial or archaeal taxonomy. The same CRISPR/Cas subtypes are observed within phylogenetically distant species, while different subtypes can be present within one species (Haft *et al.*, 2005; Godde and Bickerton, 2006; Makarova *et al.*, 2006). This horizontal gene transfer is likely mediated by (mega)plasmids and phages (Godde and Bickerton, 2006; Sebaihia *et al.*, 2006; Sorek *et al.*, 2008). Furthermore, high levels of spacer polymorphism are generated due to CRISPR locus expansion via spacer acquisition and contraction via spacer loss in response to a continuously varying phage population. This suggests a rapid and dynamic turnover of CRISPR loci. As such, these CRISPR loci represent the most hypervariable genomic regions in bacteria (Andersson and Banfield, 2008; Horvath *et al.*, 2008).

Regulation of the CRISPR/Cas system

Genome-wide transcription profiling of *Thermus thermophilus* infected with phage ΦYS40 illustrated that the cyclic AMP receptor protein upregulates *cas* operons. The cAMP signal transduction pathway is activated during carbon source limitation when bacteria may be more susceptible for phage infection. However, expression of several *cas* genes and CRISPR transcription were upregulated independently of the cyclic AMP receptor protein, indicating other unidentified regulatory factors being involved (Shinkai *et al.*, 2007; Agari *et al.*, 2010). In *Streptococcus mutans*, transcriptome analysis of a ClpP protease mutant strain revealed an upregulation of the cas genes associated with a CRISPR2 locus, while the expression levels of the second CRISPR locus, CRISPR1, cas genes were unaltered (Chattoraj *et al.*, 2010).

The CRISPR/Cas system of *E. coli* K12 is regulated by the antagonists, LeuO and the heat-stable nucleoid-structuring (H-NS) protein (Westra *et al.*, 2010). Transcription of the Cascade genes (*casABCDE*) and the CRISPR array is repressed by the H-NS protein, a global transcriptional repressor in many Gram-negative bacteria (Hommais *et al.*, 2001; Oshima *et al.*, 2006; Pul *et al.*, 2010). Since the H-NS protein binds incoming DNA (phage or plasmid) directly (Navarre *et al.*, 2006; 2007), a decreased concentration of the H-NS repressor will be created inside the cell hereby permitting expression of the Cascade genes and *leuO*, the latter being negatively regulated by H-NS (Hommais *et al.*, 2001; Chen *et al.*, 2005). LeuO, belonging to the LysR family of transcription factors, relieves the H-NS mediated repression upstream of *casA* through direct DNA binding. Since *leuO* is furthermore positively regulated by LeuO itself (Hommais *et al.*, 2001; Chen *et al.*, 2005), an amplification of the upregulation of *cas* gene transcription is achieved.

Abortive infection systems

When a phage has successfully subverted the early-acting resistance mechanisms, the last crucial steps in the phage lytic cycle involve its intracelllular development and final release of its progeny. Abortive infection (Abi) is a collective term used for a broad range of host resistance mechanisms that interrupt these late stages of phage development (genome transcription, translation, replication, and phage packaging). It is assumed that these Abi systems ultimately lead to the death of the infected cell since phage-mediated corruption of host functions has already been initiated before kick-off of the defence mechanism. The bacterial population benefits from the suicidal sacrifice of the infected cell as it restricts the infection to the sacrificed cell and inhibits further virus spreading to nearby susceptible cells. Abi systems are encoded by a heterologous set of proteins with the majority of them are plasmid- or prophage-encoded. However, many details of their complex molecular mechanism are not yet fully understood.

Rex system of *E. coli* prophage λ interrupts phage replication

The rexA ('rII exclusion') and rexB genes of prophage λ encode a two-component system and represent one of the best characterized Abi systems (Benzer, 1955; Toothman and Herskowith, 1980). When infection with *rII* mutants of *E. coli* bacteriophage T4 occurs, infection proceeds normally until T4 DNA replication initiates. At that time, a phage protein–DNA complex is formed as a replication or recombination intermediate, which activates RexA, a 31 kDa intracellular sensor-activator (Parma *et al.*, 1992). At least two RexA proteins are needed to activate one response regulator RexB protein, suggesting that the RexA/RexB ratio is a crucial factor in Rex-mediated Abi activity (Snyder and McWilliams, 1989). RexB forms an ion channel which is opened upon interaction with RexA proteins and permits the passage of monovalent cations, thereby destroying the membrane potential and causing a cellular drop in ATP-levels, termination of macromolecular synthesis and ultimately cell death.

Wild-type bacteriophage T4 encodes two proteins, RIIA and RIIB, which help to subvert the Rex-mediated Abi system. However, no complete protection is offered as overproduction of both Rex proteins causes wild-type T4 phage to be excluded (Shinedling *et al.*, 1987). Also, T4 phages with mutations in the *rII* and *motA* genes may become insensitive for the Rex system (Molineux, 1991). The transcription factor MotA activates the middle gene promoters during the transcriptional switch from early to middle gene expression of phage T4 (Hinton *et al.*, 2005). Following the induction of prophage λ a mechanism is required to prevent self-exclusion by the Rex system during the lytic growth phase. Parma *et al.* (1992) suggest that overexpression of RexB accomplished through transcriptional regulation, forms the basis for this mechanism as it alters the crucial RexA/RexB ratio for Abi activity (Parma *et al.*, 1992).

Inhibition of the translational machinery by two *E. coli* Abi systems: Lit and PrrC

The defective prophage e14 located in the isocitrate dehydrogenase (*icd*) gene of *E. coli* K12 strain interferes with T4 (and other T-even phages) infection by expression of the Lit ('late inhibitor of T4') protein (Hill *et al.*, 1989; Kao and Snyder, 1988; Linder *et al.*, 1994). Late during T4 infection, the Gol peptide (named after the ability of *gp23* mutants to 'Grow On Lit producing bacteria'), located in the major capsid protein (Gp23) of phage T4 activates Lit (Champness and Snyder, 1982; Bingham *et al.*, 2000). The Gol peptide binds specifically to domains II and III of elongation factor Tu (EF-Tu), creating the uniquely activated substrate for the Lit metalloprotease, which subsequently cleaves the RGITI motif of the effector I region, the guanine nucleotide binding domain, of EF-Tu (Vallee and Auld, 1990; Yu and Snyder, 1994). This stops protein synthesis resulting in bacterial cell death and abortion of T4 infection (Yu and Snyder, 1994).

The PrrC protein represents another Abi system causing abortion of phage infection in *E.*

coli strain CT196 (Depew and Cozzarelli, 1974; Levitz *et al.*, 1990). Gene *prrC* encodes a ribonuclease which specifically cleaves the anticodon loop of the host lysine tRNA (Amitsur *et al.*, 1987; Levitz *et al.*, 1990). As the combined action of polynucleotide kinase and RNA ligase restores $tRNA^{Lys}$, only T4 mutants defective in the corresponding genes are translationally blocked in their development, while wild-type T4 is not (Amitsur *et al.*, 1987). The *prrC* gene is located in a cryptic genetic element sandwiched between the *hsdMS* (*prrAB*) and *hsdR* (*prrD*) genes of the EcoprrI Type I R–M system (Abdal Jabbar and Snyder, 1984; Levitz *et al.*, 1990; Tyndall *et al.*, 1994). The PrrC gene product also physically interacts with these Hsd proteins of the EcoprrI system, hereby masking its ribonuclease activity. The meaning of this genetic linkage and physical coupling of a DNA and tRNA restriction Prr system is explained by the dual function of the phage T4 protein Stp ('Suppressor of the Three-prime Phosphatase'). The primary function of the Stp polypeptide is probably the formation of an anti-restriction system that dissociates the EcoprrI restriction modification complex thereby inactivating its DNA restricting activity. However, the T4 Stp protein is turned against the phage as dissociation of the EcoprrI complex also activates the PrrC ribonuclease (Amitsur *et al.*, 1992). As such, the PrrC ribonuclease seems to form a second line of phage defence that kills the cells and prevents the spread of T4 to other Prr-encoding cells, if the restriction-modification system is inactivated by T4 Stp.

F exclusion of *E. coli* phage T7

Replication of bacteriophage T7 in *E. coli* strains harbouring the F plasmid is interrupted midway through the phage life cycle. The membrane-associated protein PifA, whose gene is located on the F plasmid, mediates this phage exclusion. After a normal early phage gene transcription, transcription of the T7 late region is notably reduced (Britton and Haselkorn, 1975; Condit and Steitz, 1975; Young and Menard, 1975; Beck and Molineux; 1991; Molineux, 1991). This reduction is caused by premature inhibition of macromolecular synthesis (Garcia and Molineux, 1995). Furthermore, leakage of small molecules like intracellular nucleotides is observed, while little or no phage DNA is replicated and the host chromosome is not extensively degraded (Schmitt *et al.*, 1991). Although the cells do not lyse, this complex combination of physiological changes leads to death of the infected cell. Two T7 phage proteins, Gp1.2 and Gp10, trigger exclusion (Molineux *et al.*, 1989; Schmitt and Molineux, 1991). While Gp10 codes for the major capsid protein, Gp1.2 ensures phage propagation by inhibition of cellular dGTPase in mutants overexpressing this protein (Huber *et al.*, 1988; Nakai and Richardson, 1990). Subversion of the PifA resistance system is accomplished by a null or missense mutation in gene 1.2 or a double missense mutation in gene 10. Gene 1.2 mutation and one gene 10 mutation reduce interaction with PifA, while the other mutation in gene 10 reduces Gp10 production (Schmitt *et al.*, 1991; Molineux *et al.*, 1989). In addition, inhibition of the T7 growth by the F plasmid is alleviated in *E. coli* mutants, which shows overexpression of FxsA, a protein of unknown function. FxsA probably sequesters PifA hereby reducing F exclusion.

Lactococcal Abi systems

Lactococcus lactis is a Gram-positive bacterium widely used by the dairy industry to manufacture an array of fermented dairy products including various cheeses. In the non-sterile environment of pasteurized milk, the added lactococcal cells will come into contact with virulent phages found in milk. Thus, these *L. lactis* strains must be phage-resistant to drive the milk fermentation process. Several of these phage-resistant lactococcal strains carry Abi systems. Almost all lactococcal Abi systems are plasmid-encoded and the Abi phenotype is most frequently provided by a single gene, although in a few cases the systems consist of two to four distinct gene products (e.g. AbiE, AbiG, AbiL and AbiT). Lactococcal *abi* genes have a lower G+C content than their host genome and generally share a limited amino acid similarity (Chopin *et al.*, 2005). So far, at least 23 different *L. lactis* Abi systems interfering with distinct steps of the phage multiplication cycle have been identified (Chopin *et al.*, 2005). The absence of homology with proteins of known function hampers investigation of their molecular mechanism.

For this reason, knowledge of most lactococcal Abi systems is limited to their general effect on the phage life cycle. For example, AbiA, AbiF, AbiK, AbiP and AbiR interfere with DNA replication, while RNA transcription is interrupted by AbiB, AbiG, and AbiU (Garvey *et al.*, 1995; Parreira *et al.*, 1996; Emond *et al.*, 1997; O'Connor *et al.*, 1996; Twomey *et al.*, 2000; Dai *et al.*, 2001; Domingues *et al.*, 2004). AbiC was observed to reduce major capsid protein synthesis, AbiE, AbiI and AbiQ affect the phage DNA packaging (Durmaz *et al.*, 1992; Emond *et al.*, 1998) while AbiZ and AbiT cause premature cell death.

Combined genetic and biochemical studies provided more detailed information regarding the mode of action of some lactococcal Abi systems.

- Transcription of AbiB initiated in an upstream promoter of a IS*S1* element, is constitutive and does not increase after phage infection (Cluzel *et al.*, 1991). AbiB is only active against 936-type phages (*Siphoviridae* family). Its activity is mediated by an early phage protein that either induces the synthesis or stimulates the activity of an RNase, hereby causing a remarkable decay of mRNA ten minutes post phage infection (Parreira *et al.*, 1996).
- AbiD1, active against the two main groups of virulent lactococcal phages (936 and c2), interferes with a RuvC-like endonuclease, which removes branched-DNA structures during phage multiplication. This interference blocks resolution of phage DNA structures and packaging. An increase in AbiD1 mRNA translation is mediated in phage bIL66 by Orf1 gene product, expressed in mid-infection (Bidnenko *et al.*, 1995, 2002).
- AbiK (599 amino acids) encoded by plasmid pSRQ800 (Boucher *et al.*, 2000), is able to abort infection of the three most prevalent lactococcal phage groups found in the dairy industry (P335-, 936- and c2-type phages) (Emond *et al.*, 1997). AbiK transcription is initiated by two promoters, P*abiK* and P*orf3*, with the latter being repressed by Orf4 (Fortier *et al.*, 2005). AbiK targets a phage protein named Sak (for Sensitivity to AbiK) leading to the abortion of the phage infection (Bouchard and Moineau, 2004). Characterization of Sak proteins indicated that they are single-strand annealing proteins (Ploquin *et al.*, 2008; Scaltriti *et al.*, 2010, 2011). It was recently found that AbiK polymerizes long DNAs of 'random' sequence, analogous to a terminal transferase (Wang *et al.*, 2011). The DNAs are covalently attached to the AbiK protein. Mutagenesis of AbiK confirms that the polymerase activity resides in a reverse transcriptase motif and is essential for phage resistance. These results establish an unprecedented property for a polymerase. However, the exact mode of action remains unknown.
- AbiP, encoded by a single *orf*, is effective against some 936-type phages. The AbiP mechanism inhibits the temporal transcription switch off of phage early genes and arrests phage bIL66M1 DNA replication around ten minutes after infection (Domingues *et al.*, 2004). AbiP is a membrane-anchored protein with a N-terminal membrane-spanning domain and a putative nucleic acid binding domain facing the cytoplasm. It binds double- and single-stranded nucleic acids in a sequence-independent manner with a preference for RNA (Domingues *et al.*, 2008).
- The chromosome-encoded AbiV system is silent in the wild-type phage-sensitive strain *L. lactis* subsp. *cremoris* MG1363, but can be activated by spontaneous reorganization in the *abiV* upstream region (Haaber *et al.*, 2008; 2009a). AbiV and the phage-encoded protein SaV ('sensitive to AbiV') form homodimers, which strongly and specifically interact with each other to form a stable $AbiV_2$-SaV_2 protein complex (Haaber *et al.*, 2009b; 2010). The small amount of SaV produced early in the phage infection cycle, rapidly interacts with the host AbiV to form an active complex, which prevents further protein synthesis by the cell. As also the expression of phage-encoded proteins including activators of middle and late gene transcription are hindered, this blocks the transcriptional processes of most middle but in particular the late phage genes (Haaber *et al.* 2010).
- The conjugative plasmid pTR2030 confer phage resistance to commercial *Lactococcus* starter cultures (Durmaz and Klaenhammer

2006). This plasmid contains genes coding for the R–M system *Lla*I and two abortive infections systems, AbiA and AbiZ. While AbiA, which shares amino acid similarities with AbiK, interferes with the early stages of phage DNA replication, AbiZ interacts cooperatively with the phage holin during the late infection phase. This causes premature lysis of the infected cell, hereby reducing the burst size of the phage (Durmaz and Klaenhammer, 2007).

- The AbiT system, as AbiZ, may also be membrane-anchored, causes premature cell death in 936- and P335-type phages. AbiT system is encoded by two constitutively cotranscribed genes, *abiTi* and *abiTii*, of plasmid pED1 (Bouchard *et al.*, 2002).

Lactococcal P335-type phages can subvert the Abi systems by acquisition of point mutations or by DNA exchange with a resident prophage (Labrie and Moineau, 2007), while virulent c2- and 936-type phages, which have no homologous prophage in the host genome, evolve resistance only by point mutation or by recombination during co-infection (Forde and Fitzgerald, 1999b; Dinsmore and Klaenhammer, 1994, 1997; Bidnenko *et al.*, 1995; Bouchard *et al.*, 2002; Bouchard and Moineau, 2000, 2004). For example, the presence of a point mutation in the 5′ region of *orf1* renders phage bIL66 resistant to AbiD1. Expression of orf1 *in trans* increases AbiD1 efficiency and will hereby abort infection of the AbiD1 resistant phage mutants (Bidnenko *et al.*, 2002). Mutants of phage ul36 insensitive to AbiK were shown to have one point mutation in gene *sak*. Altogether only three different amino acid substitutions were observed in the Sak gene product of ul36 mutants (Bouchard and Moineau, 2004).

Toxin–antitoxin (TA) systems

TA systems are widespread throughout prokaryotes on large, low-copy plasmids, conjugative transposons, superintegrons, temperate phages, and on bacterial chromosomes (Pandey and Gerdes, 2005). They play a role in a diverse array of biological processes, including plasmid maintenance, stress responses, persistence, and as an antiphage strategy (Magnuson, 2007). A TA system most often consists of a toxin gene preceded by an antitoxin gene which usually overlap or are separated by a small intergenic region. Both genes are co-transcribed from a single promoter and in most cases also co-translated. Transcription is typically autoregulated by the antitoxin, while the toxin functions as a co-repressor. This homeostatic negative-feedback loop ensures the maintenance of steady state levels of toxin and antitoxin (Magnuson and Yarmolinsky, 1998). Intracellular interaction of the stable toxin and the labile antitoxin suppresses toxicity. However, reduced expression or proteolytic degradation of the antitoxin in response to certain stimuli unveils the stable toxin, leading to an increased free toxin concentration, which induces bacteriostasis and cell death (Aizenman *et al.*, 1996; Christensen *et al.*, 2004).

Three different types of TA systems have been identified and each type of TA system can mediate abortive phage infection (Fig. 4.5).

- Type I systems are based on sequence complementarity of the toxin mRNA and the *cis*-encoded antisense antitoxin mRNA which are able to form a double-stranded RNA duplex that is targeted for degradation. When the antitoxin mRNA concentration decreases, translation of the toxin mRNA occurs, whose protein activity mediates cell death (Gerdes *et al.*, 1986; Gerdes and Wagner, 2007). Besides maintenance of the low-copy plasmid R1, the presence of the *hok* ('host killing') and *sok* ('suppressor of killing') genes in *E. coli* cells also mediates disruption of the later stages of phage T4 development (assembly, packaging and lysis) (Gerdes *et al.*, 1990; Pecota and Wood, 1996). Before the toxin Hok is translated, the Sok mRNA pool (the Hok mRNA antitoxin) must be depleted. Inhibition of the host RNA polymerase (RNAP) and consequently cessation of *sok* transcription starts after coinjection of 25 to 50 copies of the Alt protein with the T4 genomic DNA. The Alt protein ADP-ribosylates the host RNAP, causing a twofold reduction in RNAP activity (Kutter *et al.*, 1994; Wilkens *et al.*, 1994). Host transcription is further halted by a set of phage proteins (ModA, Alc and AsiA) and by endonucleolytic

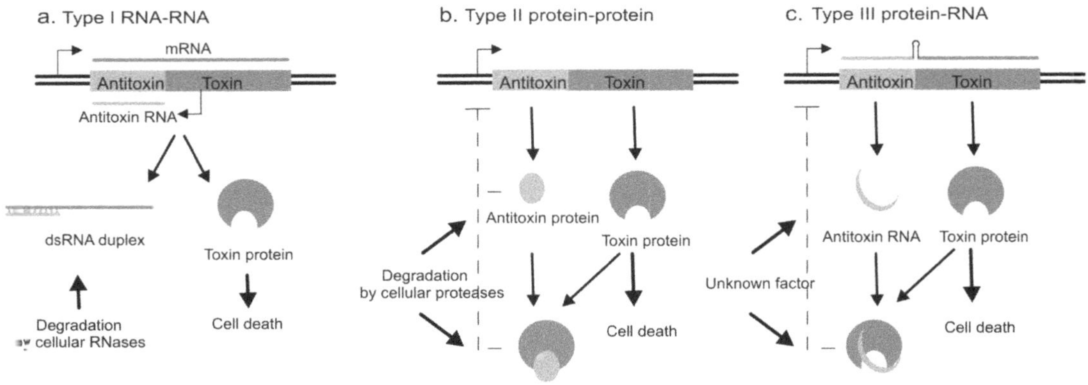

Figure 4.5 Schematic overview of toxin–antitoxin systems. (a) In type I TA systems, the antisense antitoxin RNA forms a dsRNA duplex with a short region of the full-length mRNA. This prevents translation of the toxin gene and targets its mRNA for degradation by a cellular RNase. (b) In type II TA systems, the antitoxin protein interacts with the toxin protein. Cellular proteases can degrade the antitoxin protein, hereby releasing the toxin. The antitoxin protein or the antitoxin–toxin complex often negatively regulates transcription of the locus. (c) In type III TA systems, the antitoxin RNA forms a complex with the toxin protein. Transcriptional levels of ToxI and ToxN are regulated by a transcriptional terminator between the toxin and antitoxin genes, while the complex negatively regulates transcription. Unknown cellular factors can trigger ToxN activity by dissociating the ToxN-ToxI inactive complex, degrading the antitoxin RNA or decreasing the transcriptional level.

degradation of host DNA (Rabussay, 1983; Jaffé *et al.*, 1985; Kutter *et al.*, 1994; Wilkens *et al.*, 1994; Roucourt and Lavigne, 2009). Finally, translation of Hok mRNA, in addition to elimination of Sok antisense RNA, requires truncation by host RNase III (Gerdes *et al.*, 1990). Within fifteen minutes after T4 infection, the Hok protein destructs the cell.

- Type II TA systems are proteins with a high diversity of antitoxins binding to similar core toxin folds. Commonly, Type II toxins are endoRNases, as free enzymes or associated with ribosomes, or DNA gyrase inhibitors (Jiang *et al.*, 2002; Hargraeves *et al.*, 2002; Liu *et al.*, 2008; Miallau *et al.*, 2009; Neubauer *et al.*, 2009). In addition, protease and phosphotransferase activity of type II toxins as well as global inhibition of peptidoglycan synthesis through phosphorylation have been observed (Meinhart *et al.*, 2003; Yamamoto *et al.*, 2009; Mutschler *et al.*, 2011). The gene pair *mazEF* located in the *relA* operon of the *E. coli* chromosome represents such a proteic TA module (Aizenman *et al.*, 1996). The co-expression of MazEF is negatively autoregulated at the transcriptional level by the combined action of both MazE and MazF proteins on the *mazEF* promoter P_2 (Marianovsky *et al.*, 2001). MazF is a stable toxin with endoRNase activity which cleaves cellular mRNAs between A and C at the ACA sequence (Christensen *et al.*, 2003; Zhang *et al.*, 2003). MazE is a labile antitoxin that is degraded *in vivo* by the ClpPA serine protease (Aizenman *et al.*, 1996). MazE and MazF form a heterohexamer inactive complex constituting of a MazE dimer sandwiched between two MazF dimers (Kamada *et al.*, 2003; Lah *et al.*, 2003; Loris *et al.*, 2003). Zhang *et al.* (2003) hypothesized that the highly negatively charged C-terminal region of MazE (antitoxin) mimics single-stranded RNA, hereby permitting binding of the MazF (toxin) RNA-binding site and blocking of its endoRNase activity. As MazE is continuously being degraded by ClpPA, prevention of MazF-mediated cell death requires a continuous production of MazE. A wide range of stressful conditions (amino acid starvation, antibiotics, UV irradiation, oxidative stress, elevated temperature) leading to the inhibition of MazE expression and consequently to cell death, have been identified (Aizenman *et al.*, 1996; Cashel *et al.*, 1996; Hazan *et al.*,

2001, 2004; Sat *et al.*, 2001). MazEF acts as a defence mechanism to protect the bacterial population against phage P1. The presence of MazEF reduces the number of phage progeny both upon prophage induction and infection with virulent P1 phage (Hazan and Engelberg-Kulka, 2004), the latter possibly disrupting the MazE and MazF ratio.

- ToxIN encoded by a cryptic plasmid of the phytopathogen *Pectobacterium atrosepticum* (formerly *Erwinia carotovora* subsp. *atroseptica*) is the first example of the Type III TA system (Blower *et al.*, 2009; Fineran *et al.*, 2009). Here, a protein toxin (ToxN denoting 'toxin') interacts with a non-coding RNA pseudoknot, the antitoxin (ToxI for 'ToxN inhibitor'). The antitoxin is encoded by a repetitive DNA sequence upstream from a transcriptional terminator, which is followed by the *toxN* gene. A single constitutive promoter upstream of the antitoxin drives transcription of the TA locus with the intervening transcriptional terminator regulating the relative antitoxin RNA and toxin mRNA levels. The full-length antitoxin transcript is cleaved by the endoRNase ToxN to generate 36-nucleotide repeat units, which fold as an interdigitated hairpin-type pseudoknot, ToxI. Three ToxI pseudoknot monomers interact with three ToxN monomers to generate a trimeric ToxN–ToxI inactive complex. During phage infection, an unknown mechanism destabilizes the ToxN–ToxI complex, releasing active ToxN endoRNase, which cleaves cellular mRNAs in a sequence-specific manner hereby inducing bacteriostasis and leading to bacterial cell death (Blower *et al.*, 2011). This ToxIN system provides resistance against a range of phages within multiple enteric bacteria (Fineran *et al.*, 2009).

Conclusion

The diverse array of bacterial resistance mechanisms at the different steps of the phage replication cycle, as enumerated within this chapter, can provide the bacteria with a whole line of defence when facing phages. As such, bacteria carrying one or multiple resistance mechanisms are more likely to survive in these hostile environments and can guarantee the survival of the bacterial population. As logic precludes, the more resistance mechanisms present within one bacterium the higher the chance of its survival. Moreover, these mechanisms can also confer the host with some additional benefits besides resistance. The presence of these resistance mechanisms is however also accompanied with a cost for the host and the risk of autoimmunity. Finally, many questions are raised whether these resistance mechanisms are rather selfish mobile elements with the associated bacterial resistance as a side-effect.

Costs and other benefits of a bacterial defense mechanism

One of the primary costs of a resistance mechanism to a bacterial cell is the energy cost linked to this additional genetic cargo, particularly when it is associated with a high-copy plasmid. Evidence for this is provided by the fact that only a small subset of a bacterial population contains a plasmid or prophage encoding a bacterial defence mechanism. This is furthermore strengthened by the fact that the repeat-spacer array for CRISPR-mediated resistance does not grow indefinitely. 'Older' spacers are lost over time, while newly added spacers, which confer resistance to phages in the present environment, are added to the sequence. Besides, these 'older' spacers are often no longer needed as the original phage populations has changed.

A second important cost is the risk of autoimmunity. For example, in R–M systems, the restriction enzyme is often more stable than the accompanying methylase. When a bacterium loses its methylase gene it will not be able any more to discern self and non-self and eventually restrict its own genomic DNA (Naito *et al.*, 1995). For CRISPR mechanisms, the accidental incorporation of self-genetic material within the CRISPR array, results in spacers targeting their own bacterial genome and thus in autoimmunity (Marraffini and Sontheimer, 2010; Stern *et al.*, 2010). Also, any errant function of the Abi system, particularly in the case of TA systems, will lead to autoimmunity and death of the bacterial cell.

Bacterial resistance mechanisms, which evolutionary changed their function from a defence system to a regulatory host mechanism, can benefit the bacterial host. Such evolutionary events are called 'exaptations' which describe usage of a biological structure or function for another than its initial purpose (Brosius and Gould, 1992).

Some R–M systems have lost their restriction enzyme, leaving an 'orphan' methylase now devoted to epigenetic modifications. The Dam ('DNA adenine methyltransferase') methylase of *E. coli* is such an example. Methylation by the Dam methylase is important for binding of the replication–initiation complex to the methylated *oriC*, mismatch repair by the MutHLS complex and (post)transcriptional regulation of gene expression, especially genes involved in bacterial pathogenicity (Marinus and Casadesus, 2009). The concentration of another orphan methylase called the CcrM ('cell cycle-regulated methyltransferase') methylase of *Caulobacter crescentus* increases towards the end of the bacterial cell cycle and prepares the origin of replication for replication initiation through methylation (Marinus and Casadesus, 2009).

Methylase of type III R–M systems in pathogenic bacteria are subject to reversible high-frequency on/off switching of expression ('phase variation') mediated by mutations in DNA repeats. This phase variable expression of the methylase gene may upregulate or downregulate expression of multiple genes depending on the site of methylation. This methylase activity thus mediates phase variation of multiple genes in bacterial pathogens (Srikhanta *et al.*, 2005). Generally, this phase variation forms a strategy for rapid adaptation to changes within the host environment through creation of a genetically and phenotypically diverse population (Moxon and Thaler, 1997).

Also components of TA-systems can participate in host cell regulatory processes. The bacterium *Myxococcus xanthus* shows multicellular fruiting body formation during which ~80% of bacterial cells undergo obligatory cell lysis. This programmed cell death is mediated by a solitary mazF gene which lacks the cognate mazE-like antitoxin gene and whose expression is developmentally regulated (Nariy and Inouye, 2008).

Finally, CRISPR/Cas system may also be involved in other functions. In *Myxococcus xanthus*, *cas* genes are co-transcribed with the *dev* operon. This operon is responsible for triggering spore differentiation and, as for the *cas* genes, is subject to strong negative autoregulation by DevS (Viswanathan *et al.*, 2007). In *E. coli*, *cas1* may also have a function in DNA repair (Babu *et al.*, 2011). In the biofilm-producing strain *Pseudomonas aeruginosa* PA14, CRISPR activity was shown to decrease swarming mobility and biofilm production of the lysogens (Zegans *et al.*, 2009).

Selfish mobile elements?

Like viruses, transposons and homing endonucleases, R–M and Abi systems can operate on a primary 'selfish' level to promote their own survival (Nakayama and Kobayashi, 1998; Kobayashi, 2001; Makarova *et al.*, 2009). According to this hypothesis, the bacterial defence conferred by these systems is only a by-product of the fact that these systems defend themselves. Specifically for R–M systems, three main strategies aimed at increasing their relative frequency within a bacterial cell population have been identified. First, they defend themselves from invaders carrying possibly other R–M systems by attacking 'non-self' DNA. Hereby, they are also able to protect host cells from invading foreign DNA, like phage genomic DNA, and represent one of the first lines of bacterial defence. Secondly, any attempt to eliminate the R–M system gene is complex. After a cell has lost its R–M system, its descendants contain fewer and fewer restriction enzymes and methylases as this R–M system is no longer coded for in the descendants. At a certain moment, the residual methylase pool will not be able to modify all recognition sites on the newly replicated chromosomes. Restriction of only one unmodified recognition site will kill the progeny cell (Naito *et al.*, 1995). Thirdly, R–M systems undergo extensive horizontal transfer, as they are often associated with plasmids, transposons and integrons. Additionally, R–M gene complexes behave as mobile elements themselves and occasionally cause evolutionary changes within genomes (Kobayashi, 2001).

Future trends

Despite our growing knowledge of phages, bacteria and their accompanying interactions, much still remains obscure. The recently uncovered CRISPR/Cas resistance mechanism, accentuates the fact that our understanding of the complex array of bacterial resistance mechanisms is far from complete. Many more resistance mechanisms are likely to be revealed. This could be expected based on the large phage and bacterial diversity, the large number of hypothetical genes in sequenced phage and microbial genomes and by the essential role of resistance in the phage-host balance in the different environments explored. For this reason, one needs to go beyond classical phage host range analyses and search for the bacterial systems underlying resistance of phage unsusceptible strains. These systems need in a next step to be studied at the molecular level. Furthermore, most phage resistance mechanisms have only been identified in double-stranded DNA phages. Bacterial defence systems against single-stranded DNA and RNA phages as well as double-stranded RNA phages need yet to be discovered.

Knowledge at the molecular level of this arms race between phages and bacteria is of significant importance with regard to problems one is faced with in industrial and hospital environments. Within the milk fermentation industry, phages pose a contamination problem and one is interested in selection and/or construction of *L. lactis* strains which are able to subvert this phage attack in order to maintain a stable fermentation process. However, within other industrial settings (e.g. agriculture, food industry, livestock) problems are caused by bacteria and often difficult to resolve as use of antibiotics in most cases is prohibited. Moreover, within the hospital environment bacterial infections can become life threatening as a consequence of the developed multidrug resistance of the infecting bacteria. Despite the first approvals for use of phage cocktails in agriculture (as biopesticides), in the food industry and for the treatment of live stock, phage cocktails are still not allowed in the hospital environment. Not only is the concept of using 'viruses' against bacteria one of the many hurdles to tackle, but also the rapid development of bacterial resistance to phages poses many questions. The more we know about bacterial resistance mechanisms, the better we can create specific phage therapies able to conquer this developing resistance.

Acknowledgements

The authors gratefully acknowledge J.E. Samson (Université Laval, Québec, Canada) and E. Adriaenssens and K. Vandersteegen (Laboratory of Gene Technology, K.U.Leuven, Belgium) for carefully reading this book chapter. Anneleen Cornelissen holds a predoctoral fellowship of the Instituut voor de aanmoediging van Innovatie door Wetenschap en Technologie in Vlaanderen (IWT, Belgium).

References

Abdul Jabbar, M., and Snyder, L. (1984). Genetic and physiological studies of an *Escherichia coli* locus that restricts polynucleotide Chinese and RNA ligase-deficient mutants of bacteriophage T4. J. Virol. *51*, 522–529.

Agari, Y., Sakamoto, K., Tamakoshi, M., Oshima, T., Kuramitsu, S., and Shinkai, A. (2010). Transcription profile of *Thermus thermophilus* CRISPR systems after phage infection. J. Mol. Biol. 395, 270–281.

Aizenman, E., Engelberg-Kulka, H., and Glaser, G. (1996). An *Escherichia coli* chromosomal 'addiction module' regulated by guanosine 3′–5′-bispyrophosphate: a model for programmed bacterial cell death. Proc. Natl. Acad. Sci. U.S.A. *93*, 6059–6063.

Albert, M.J., Bhuiyan, N.A., Rahman, A., Ghosh, A.N., Hultenby, K., Weintraub, A., Nahar, S., Kibriya, A.K., Ansaruzzaman, M., and Shimada, T. (1996). Phage specific for *Vibrio cholerae* O139 Bengal. J. Clin. Microbiol. *34*, 1843–1845.

Amitsur, M., Morad, I., and Kaufmann, G. (1987). *In vitro* reconstitution of anticodon nuclease from components encoded by phage T4 and *Escherichia coli* CTr5X. EMBO J. *8*, 2411–2415.

Amitsur, M., Morad, I., Chapman-Shimshoni, D., and Kaufmann, G. (1992). HSD restriction-modification proteins partake in latent anticodon nuclease. EMBO J. *11*, 3129–3134.

Andersson, A.F., and Banfield, J.F. (2008). Virus population dynamics and acquired virus resistance in natural microbial communities. Science *320*, 1047–1050.

Atanasiu, C., Byron, O., Mcmiken, H., Sturrock, S.S., and Dryden, D.T.F. (2001). Characterization of the structure of Ocr, the gene 0.3 of bacteriophage T7. Nucleic Acids Res. *14*, 3059–3068.

Babu, M., Beloglazova, N., Flick, R., Graham, C., Skarina, T., Nocek, B., Gagarinova, A., Pogoutse, O., Brown, G., Binkowski, A., Phanse, S., Joachimiak, A., Koonin, E.V., Savchenko, A., Emili, A., Greenblatt, J., Edwards, A.M., and Yakunin, A.F. (2011). A dual function of the CRISPR–Cas system in bacterial antivirus immunity and DNA repair. Mol. Microbiol. 79, 484–502.

Bair, C.L., and Black, L.W. (2007). A type IV modification dependent restriction nuclease that targets glucosylated hydroxymethyl cytosine modified DNAs. J. Mol. Biol. *366*, 768–778.

Bair, C.L., Rifat, D., and Black, L.W. (2007). Exclusion of glucosyl-hydroxymethylcytosine DNA containing bacteriophages is overcome by the injected protein inhibitor IPI*. J. Mol. Biol. *366*, 779–789.

Bandyopadhay, P.K., Studier, F.W., Hamilton, D.L., and Yuan, R. (1985). Inhibition of the type I restriction-modification enzyme *Eco*B and *Eco*K by the gene 0.3 protein of bacteriophage T7. J. Mol. Biol. *182*, 567–578.

Barrangou, R., Fremaux, C., Deveau, H., Richards, M., Boyaval, P., Moineau, S., Romero, D.A., and Horvath, P. (2007). CRISPR provides acquired resistance against viruses in prokaryotes. Science *315*, 1709–1712.

Bartell, P.F., Orr, T.E., and Lam, G.K.H. (1966). Polysaccharide depolymerase associated with bacteriophage infection. J. Bacteriol. *92*, 56–62.

Bartell, P.F., Orr, T.E., and Lam, G.K.H. (1968). Purification and properties of polysaccharide depolymerase associatd with phage-infected *Pseudomonas aeruginosa*. J. Biol. Chem. *243*, 2077–2080.

Bayer, M.E., Thurow, H., and Bayer, M.H. (1979). Penetration of the polysaccharide capsule of *Escherichia coli* (Bil 62/42) by bacteriophage K29. Virology *94*, 95–118.

Bayliss, C.D., Field, D., and Moxon, E.R. (2001). The simple sequence contingency loci of *Haemophilus influenzae* and *Neisseria meningitidis*. J. Clin. Invest. *107*, 657–662.

Bayliss, C.D., van de Ven, T., and Moxon, E.R. (2002). Mutations in *polI* but not *mut*SLH destabilize *Haemophilus influenzae* tetranucleotide repeats. EMBO J. *21*, 1465–1476.

Bayliss, C.D., Sweetman, W.A., and Moxon, E.R. (2004). Mutations in *Haemophilus influenzae* mismatch repair genes increase mutation rates of dinucleotide repeat tracts but not dinucleotide repeat-driven pili phase variation rates. J. Bacteriol. *186*, 2928–2935.

Beck, P.J., and Molineux, I.J. (1991). Defective transcription of the right end of bacteriophage T7 DNA during an abortive infection of F plasmid-containing *Escherichia coli*. J. Bacteriol. *173*, 947–954.

Bernheimer, H.P., and Tiraby, J.-G. (1976). Inhibition of phage infection by *Pneumococcus* capsule. Virology *73*, 308–309.

Bessler, W., Fehmel, F., Freund-Mölbert, Knüfermann, H., and Stirm, S. (1975). *Escherichia coli* capsule bacteriophages. J. Virol. *15*, 976–984.

Beveridge, T., Makin, S., Kadurugamuwa, J., and Li, Z. (1997). Interactions between biofilms and the environment. FEMS Microbiol. Rev. *20*, 291–303.

Bickle, T.A., and Krüger, D.H. (1993). Biology of DNA restriction. Microbiol. Rev. *57*, 434–450.

Bidnenko, E., Ehrlich, S.D., and Chopin, M.-C. (1995). Phage operon involved in sensitivity to the *Lactococcus lactis* abortive infection mechanism AbiD1. J. Bacteriol. *177*, 3824–3829.

Bidnenko, E., Chopin, M.-C., Ehrlich, S.D., and Anba, J. (2002). *Lactococcus lactis* AbiD1 abortive infection efficiency is drastically increased by a phage protein. FEMS Microbiol. Lett. *214*, 283–287.

Bingham, R., Ekunwe, S.I.N., Falk, S., Snyder, L., and Kleanthous, C. (2000). The major head protein of bacteriophage T4 binds specifically to elongation factor Tu. J. Biol. Chem. *275*, 23219–23226.

Blair, J.E., and Williams, E.O. (1961). Phage typing of staphylococci. Bull. World Health Organ. *24*, 771–784.

Blower, T.R., Fineran, P.C., Johnson, M.J., Toth, I.K., Humphreys, D.P., and Salmond, G.P.C. (2009). Mutagenesis and functional characterization of the RNA and protein components of the *toxIN* abortive infection and toxin–antitoxin locus of *Erwinia*. J. Bacteriol. *191*, 6029–6039.

Blower, T.R., Pei, X.Y., Short, F.L., Fineran, P.C., Humphreys, D.P., Luisi, B.F., and Salmond, G.P.C. (2011). A processed noncoding RNA regulates an altruistic bacterial antiviral system. Nat. Struct. Mol. Biol. *18*, 185–191.

Bolotin, A., Quinquis, B., Sorokin, A., and Ehrlich, S.D. (2005). Clustered regularly interspaced short palindrome repeats (CRISPRs) have spacers of extrachromosomal origin. Microbiology *151*, 2551–2561.

Bondy-Denomy, J., Pawluk, A., Maxwell, K.L., and Davidson, A.R. (2013). Bacteriophage genes that inactivate the CRISPR/Cas bacterial immune system. Nature *493*, 429–432.

Bouchard, J.D., and Moineau, S. (2000). Homologous recombination between a lactococcal bacteriophage and the chromosome of its host strain. Virology *270*, 65–75.

Bouchard, J.D., and Moineau, S. (2004). Lactococcal phage genes involved in sensitivity to AbiK and their relation to single-strand annealing proteins. J. Bacteriol. *186*, 3649–3652.

Bouchard, J.D., Dion, E., Bissonnette, F., and Moineau, S (2002). Characterization of the two-component abortive phage infection mechanism AbiT from *Lactococcus lactis*. J. Bacteriol. *184*, 6325–6332.

Boucher, I., Emond, E., Dion, E., Montpetit, D., and Moineau, S. (2000). Microbiological and molecular impacts of AbiK on the lytic cycle of *Lactococcus lactis* phages of the 936 and P335 species. Microbiology *146*, 445–453.

Bourniquel, A.A., and Bickle, T.A. (2002). Complex restriction enzymes: NTP-driven molecular motors. Biochimie *84*, 1047–1059.

Bradbeer, C., Woodrow, M.L., and Khalifah, L.I. (1976). Transport of vitamin B_{12} in *Escherichia coli*: common receptor system for vitamin B_{12} and bacteriophage BF23 on the outer membrane of the cell envelope. J. Bacteriol. *125*, 1032–1039.

Braun, V., and Braun, M. (2002). Active transport of iron and siderophore antibiotics. Curr. Opin. Microbiol. *5*, 194–201.

Britton, J.R., and Haselkorn, R. (1975). Macromolecular synthesis in T7 infected F' cells. Virology *67*, 264–275.

Brosius, J., and Gould, S.J. (1992). On 'genomeclature': a comprehensive (and respectful) taxonomy for pseudogenes and other 'junk DNA'. Proc. Natl. Acad. Sci. U.S.A. *89*, 10706–10710.

Brouns, S.J., Jore, M.M., Lundgren, M., Westra, E.R., Slijkhuis, R.J.H., Snijders, A.P.L., Dickman, M.J., Makarova, K.S., Koonin, E.V., and van der Oost, J. (2008). Small CRISPR RNAs guide antiviral defense in prokaryotes. Science *321*, 960–964.

Brüssow, H., and Hendrix, R.W. (2002). Phage genomics: small is beautiful. Cell. *108*, 13–16.

Brüssow, H., Canchaya, C., and Hardt, W.D. (2004). Phages and the evolution of bacterial pathogens: from genomic rearrangements to lysogenic conversion. Microbiol. Mol. Biol. Rev. *68*, 560–602.

Cashel, M., Gentry, D.R., Hernandez, V.Z., and Vinella, D. (1996). The stringent response. In *Escherichia coli* and *Salmonella*: Cellular and Molecular Biology, F.C. Neidhardt, R. Curtiss III, J.L. Ingraham, E.C.C. Lin, K.B. Low, B. Magasanik, W.R. Reznikoff, M. Riley, M Schaechter, and H.E. Umbarger, eds. (Washington, DC, American Society for Microbiology), pp. 1458–1496.

Camps, M., and Loeb, L.A. (2005). Critical role of R-loops in processing replication blocks. Front. Biosci. *10*, 689–698.

Carlson, K., Raleigh, E.A., and Hattman, S. (1994). Restriction and Modification. In The Molecular Biology of Bacteriophage T4, K. Carlson, ed. (Washington, DC, American Society for Molecular Biology, Cold Spring Harbor Library Press), pp. 369–381.

Castillo, F.J., and Bartell, P.F. (1974). Studies on the bacteriophage 2 receptors of *Pseudomonas aeruginosa*. J. Virol. *14*, 904–909.

Castillo, F.J., and Bartell, P.F. (1976). Localization and functional role of the *Pseudomonas* bacteriophage 2 depolymerase. J. Virol. *18*, 701–708.

Cerca, N., Oliveira, R., and Azeredo, J. (2007). Susceptibility of *Staphylococcus epidermidis* planktonic cells and biofilms to the lytic action of staphylococcus bacteriophage K. Lett. Appl. Microbiol. *45*, 313–317.

Champness, W.C., and Snyder, L. (1982). The gol site: a cis-acting bacteriophage T4 regulatory region that can affect expression of all the T4 late genes. J. Mol. Biol. *155*, 395–407.

Chattoraj, P., Banerjee, A., Biswas, S., and Biswas, I. (2010). ClpP of *Streptococcus mutans* differentially regulates expression of genomic islands, mutacin production, and antibiotic tolerance. J. Bacteriol. *192*, 1312–1323.

Chen, C.C., and Wu, H.Y. (2005). LeuO protein delimits the transcriptionally active and repressive domains on the bacterial chromosome. J. Biol. Chem. *280*, 15111–15121.

Chen, Y.-R., Yang, T.-Y., Lei, G.-S., Liao, C.-C., and Chak, K.-F. (2010). Interaction of colicin E7 with the major coat protein (g8p) may confer limited protection on colicinogenic *Escherichia coli* against M13 bacteriophage infection. Microbiology *156*, 3379–3385.

Chin, V., Valinluck, V., Magaki, S., and Ruye, J. (2004). KpnBI is the prototype of a new family (IE) of bacterial type I restriction-modification system. Nucleic Acids Res. *32*, e138.

Chopin, M.-C., Chopin, A., and Bidnenko, E. (2005). Phage abortive infection in lactococci: variations on a theme. Curr. Opin. Microbiol. *8*, 473–479.

Christensen, S.K., Pedersen, K., Hensen, F.G., and Gerdes, K. (2003). Toxin–antitoxin loci as stress-response elements: ChpAK/MazF and ChpBK cleave translated mRNAs and are counteracted by tmRNA. J. Mol. Biol. *332*, 809–819.

Christensen, S.K., Maenhaut-Michel, G., Mine, N., Gottesman, S., Gerdes, K., and Van Melderen, L. (2004). Overproduction of the Lon protease triggers inhibition of translation in *Escherichia coli*: involvement of the *yefM-yeoB* toxin–antitoxin system. Mol. Microbiol. *51*, 1705–1717.

Condit, R.C., and Steitz, J.A. (1975). F factor-mediated inhibition of bacteriophage T7 growth: analysis of T7 RNA and protein synthesis *in vivo* and *in vitro* using male and female *Escherichia coli*. J. Mol. Biol, *98*: 31–43.

Cornelissen, A., Ceyssens, P.-J., T'Syen, J., Van Praet, H., Noben, J.-P., Shaburova, O.V., Krylov, V.N., Volckaert, G., and Lavigne, R. The T7-Related *Pseudomonas putida* phage φ15 displays virion-associated biofilm degradation properties. PLoS One *6*, e18597.

Cotter, P.A., and Miller, J.F. (2001). *Bordetella*. In Principles of Bacterial Pathogenesis, E.A. Groisman, ed. (San Diego, Academic Press), pp. 620–674.

Coulton, J.W., Mason, P., and Dubow, M.S. (1983). Molecular cloning of the ferrichrome-iron receptor of *Escherichia coli* K-12. J. Bacteriol. *156*, 1315–1321.

Cregg, J.M., Nguyen, A.H., and Ito, J. (1980). DNA modification induced during infection of *Bacillus subtilis* by phage φ3T. Gene. *12*, 17–24.

Click, E.M., and Webster, R.E. (1998). The TolQRA proteins are required for membrane insertion of the major capsid protein of the filamentous phage f1 during infection. J. Bacteriol. *180*, 1723–1728.

Cluzel, P.J., Chopin, A., Ehrlich, S.D., and Chopin, M.C. (1991). Phage abortive infection mechanism from *Lactococcus lactis* subsp. *lactis*, expression of which is mediated by an iso-ISS1 element. Appl. Environ. Microbiol. *57*, 3547–3551.

Comeau, A.M., and Krisch, H.M. (2005). War is peace – dispatches from the bacterial and phage killing fields. Curr. Opin. Microbiol. *8*, 488–494.

Costerton, J.W. (1995). Overview of microbial biofilms. J. Ind. Microbiol. *15*, 137–140.

Cui, Y., Li, Y., Gorgé, O., Platonov, M.E., Yan, Y., Guo, Z., Pourcel, C., Dentovskaya, S.V., Balakhonov, S.V., Wang, X., Song, Y., Anisimov, A.P., Vergnaud, G., and Yang, R. (2008). Insight into microevolution of *Yersinia pestis* by clustered regularly interspaced short palindromic repeats. PLoS One 3, e2652.

Dai, G., Su, P., Allison, G.E., Geller, B.L., Zhu, P., Kim, W.S., and Dunn, N.W. (2001). Molecular characterization of a new abortive infection system (AbiU) from *Lactococcus lactis* LL51-1. Appl. Environ. Microbiol. *67*, 5225–5232.

Dai, W., Hodes, A., Hui, W.H., Gingery, M., Miller, J.F., and Zhou, Z.H. (2010). Three-dimensional structure of tropism-switching *Bordetella* bacteriophage. Proc. Natl. Acad. Sci. U.S.A. *107*, 4347–4352.

De Bolle, X., Bayliss, C.D., Field, D., van de Ven, T., Saunders, N.J., Hood, D.W., and Moxon, E.R. (2000). The length of a tetranucleotide repeat tract in *Haemophilus influenzae* determines the phase variation rate of a gene with homology to type III DNA methyltransferases. Mol. Microbiol. 35, 211–222.

Decker, K., Krauel, V., Meesmann, A., and Heller, K. (1994). Lytic conversion of *Escherichia coli* by bacteriophage T5: blocking of the FhuA receptor protein by a lipoprotein expressed early during infection. Mol. Microbiol. *12*, 321–332.

Depew, R.E., and Cozzarelli, N.R. (1974). Genetics and physiology of bacteriophage T4 3′-phosphatase: evidence for involvement of the enzyme in T4 DNA metabolism. J. Virol. *13*, 888–897.

Deszo, E.L., Steenbergen, S.M., Freedberg, D.I., and Virmr, E.R. (2005). *Escherichia coli* K1 polysialic acid O-acteyltransferase gene, neuO, and the mechanism of capsule form variation involving a mobile contingency locus. Proc. Natl. Acad. Sci. U.S.A. *102*, 5564–5569.

Deveau, H., Van Calsteren, M.-R., and Moineau, S. (2002). Effect of exopolysaccharides on phage–host interactions in *Lactococcus lactis*. Appl. Environ. Microbiol. *68*, 4364–4369.

Deveau, H., Barrangou, R., Garneau, J., Labonté, J., Fremaux, C., Boyaval, P., Romero, D.A., Horvath, P., and Moineau, S. (2008). Phage response to CRISPR-encoded resistance in *Streptococcus thermophilus*. J. Bacteriol. *190*, 1390–1400.

Destoumieux-Garzón, D., Duquesne, S., Peduzzi, J., Goulard, C., Desmadril, M., Letellier, L., Rebuffat, S., and Boulanger, P. (2005). The iron-siderophore transporter FhuA is the receptor for the antimicrobial peptide microcin J25: role of the microcin Val[11]-Pro[16] β-hairpin region in the recognition mechanism. *Biochem.* J. *389*, 869–876.

Dinsmore, P.K., and Klaenhammer, T.R. (1994). Phenotypic consequences of altering the copy number of AbiA, a gene responsible for aborting bacteriophage infection in *Lactococcus lactis*. Appl. Environ. Microbiol. *60*, 1129–1136.

Dinsmore, P.K., and Klaenhammer, T.R. (1997). Molecular characterization of a genomic region in a *Lactococcus* bacteriophage that is involved in its sensitivity to the phage defense mechanism AbiA. J. Bacteriol. *179*, 2949–2957.

Domingues, S., Chopin, A., Ehrlich, S.D., and Chopin, M.-C. (2004). The lactococcal abortive phage infection system AbiP prevents both phage DNA replication and temporal transcription switch. J. Bacteriol. *186*, 713–721.

Domingues, S., McGovern, S., Plochocka, D., Santos, M.A., Ehrlich, S.D., Polard, P., and Chopin, M.-C. (2008). The lactococcal abortive infection protein AbiP is membrane-anchored and binds nucleic acids. Virology *373*, 14–24.

Dryden, D.T., Murray, N.E., and Rao, D.N. (2001). Nucleoside triphosphate-dependent restriction enzymes. Nucleic Acids Res. *29*, 3728–3741.

Duckworth, D.H., and Pinkerton, T.C. (1988). CoIIb plasmid genes that inhibit the replication of T5 and T7 bacteriophage. Plasmid. *20*, 182–193.

Dunn, J.J., Elzinga, M., Mark, K.K., and Studier, F.W. (1981). Amino acid sequence of the gene 0.3 protein of bacteriophage T7 and nucleotide sequence of its messenger RNA. J. Biol. Chem. *256*, 2579–2585.

Durmaz, E., and Klaenhammer, T.R. (2007). Abortive phage resistance mechanism AbiZ speeds the lysis clock to cause premature lysis of phage-infected *Lactococcus lactis*. J. Bacteriol. *189*, 1417–1425.

Durmaz, E., Higgins, D.L., and Klaenhammer, T.R. (1992). Molecular characterization of a second abortive phage resistance gene present in *Lactococcus lactis* subsp. *lactis* ME2. J. Bacteriol. *174*, 7463–7469.

Emond, E., Holler, B.J., Boucher, I., Vandenbergh, P.A., Vedamuthu, E.R., Kondo, J.K., and Moineau, S. (1997). Phenotypic and genetic characterization of the bacteriophage abortive infection mechanism AbiK from *Lactococcus lactis*. Appl. Environ. Microbiol. *63*, 1274–1283.

Emond, E., Dion, E., Walker, S.A., Vedamuthu, E.R., and Kondo, J.K., and Moineau, S. (1998). AbiQ, an abortive infection mechanism from *Lactococcus lactis*. Appl. Environ. Microbiol. *64*, 4748–4756.

Eriksson, U., Svenson, S.B., Lönngren, J., and Lindberg, A.A. (1979). *Salmonella* phage glycanases: substrate specificity of the phage P22 endorhamnosidase. J. Gen. Virol. *43*, 503–511.

Fineran, P.C., Blower, T.R., Foulds, I.J., Humphreys, D.P., Lilley, K.S., and Salmond, G.P.C. (2009). The phage abortive infection system, ToxIN, functions as a protein-RNA toxin–antitoxin pair. Proc. Natl. Acad. Sci. U.S.A. *106*, 894–899.

Finne, J., and Mäkelä, P.H. (1985). Cleavage of the polysialosyl units of brain glycoproteins by a bacteriophage endosialidase. Involvement of a long oligosaccharide segment in molecular interactions of polysialic acid. J. Biol. Chem. *260*, 1265–1270.

Forde, A., and Fitzgerald, G.F. (1999). Analysis of exopolysaccharide (EPS) production mediated by the bacteriophage adsorption blocking plasmid, pCI658, isolated from *Lactococcus lactis* ssp. *cremoris* HO2. Int. Dairy J. *9*, 465–472.

Forde, A., Daly, C., and Fitzgerald, G.F. (1999). Identification of four resistance plasmids form *Lactococcus lactis* ssp. *cremoris* HO2. Appl. Environ. Microbiol. *65* 1540–1547.

Forde, A., and Fitzgerald, G.F. (1999b). Bacteriophage defence systems in lactic acid bacteria. Antonie Van Leeuwenhoek. *76*, 89–113.

Fortier, L.-C., Bouchard, J.D., and Moineau, S. (2005) Expression and site-directed mutagenesis of the lactococcal abortive phage infection protein AbiK. J. Bacteriol. *187*, 3721–3730.

Friedman, L., and Kolter, R. (2004a). Two genetic loci produce distinct carbohydrate-rich structural components of the *Pseudomonas aeruginosa* biofilm matrix. J. Bacteriol. *186*, 4457–4465.

Friedman, L., and Kolter, R. (2004b). Genes involved in matrix formation in *Pseudomonas aeruginosa* PA14 biofilms. Mol. Microbiol. *51*, 675–690.

Fuhrman, J.A. (1999). Marine viruses and their biogeochemical and ecological effect. Nature *399*, 541–548.

Glonti, T., Chanishvili, N., and Taylor, P.W. (2010). Bacteriophage-derived enzyme that depolymerizes the alginic acid capsule associated with cystic fibrosis isolates of *Pseudomonas aeruginosa*. J. Appl. Microbiol. *108*, 695–702.

Gacesa, P. (1988). Alginates. Carbohydr. Polym. *8*, 161–182.

Garneau, J.E., Dupuis, M., Villion, M., Romero, D.A., Barrangou, R., Boyaval, P., Fremaux, C., Horvath,

P., Magadán, A.H., and Moineau, S. (2010). The CRISPR/Cas bacterial immune system cleaves bacteriophage and plasmid DNA. Nature *468*, 67–71.

Garcia, L.R., and Molineux, I.J. (1995). Incomplete entry of bacteriophage T7 DNA into F plasmid-containing *Escherichia coli*. J. Bacteriol. *177*, 4077–4083.

Garvey, P., Fitzgerald, G.F., and Hill, C. (1995). Cloning and DNA sequence analysis of two abortive infection phage resistance determinants from the lactococcal plasmid pNP40. Appl. Environ. Microbiol. *61*, 4321–4328.

Gerdes, K., and Wagner, E.G. (2007). RNA antitoxins. Curr. Opin. Microbiol. *10*, 117–124.

Gerdes, K., Rasmussen, P.B., and Molin, S. (1986). Unique type of plasmid maintenance function: postsegregational killing of plasmid-free cells. Proc. Natl. Acad. Sci. U.S.A. *83*, 3116–3120.

Gerdes, K., Thisted, T., and Martinussen, J. (1990). Mechanism of post-segregational killing by the *hok/sok* system of plasmid R1: *sok* antisense RNA regulates formation of *hok* mRNA species correlated with killing of plasmid-free cells. Mol. Microbiol. *4*, 1807–1818.

Geyer, H., Himmelspach, K., Kwiatkowski, B., Schlecht, S., and Stirm, S. (1983). Degradation of bacterial surface carbohydrates by virus-associated enzymes. Pure Appl. Chem. *55*, 637–653.

Godde, J.S., and Bickerton, A. (2006). The repetitive DNA elements called CRISPRs and their associated genes: evidence of horizontal transfer among prokaryotes. J. Mol. Evol. *62*, 718–729.

Grissa, I., Vergnaud, G., and Pourcel, C. (2007). The CRISPRdb database and tools to display CRISPR and to generate dictionaries of spacers and repeats. BMC Bioinformatics *8*, 172.

Gross, R.J., Cheasty, T., and Rowe, B. (1977). Isolation of bacteriophages specific for the K1 polysaccharide antigen of *Escherichia coli*. J. Clin. Microbiol. *6*, 548–550.

Haaber, J., Fortier, L.C., Moineau, S., and Hammer, K. (2008). AbiV, a novel abortive infection mechanism on the chromosome of *Lactococcus lactis* subsp. *cremoris* MG1363. Appl. Environ. Microbiol. *74*, 6528–6537.

Haaber, J., Moineau, S., and Hammer, K. (2009a). Activation and transfer of the chromosomal phage resistance mechanism AbiV in *Lactococcus lactis*. Appl. Environ. Microbiol. *75*, 3358–3361.

Haaber, J., Rousseau, G.M., Hammer, K., and Moineau, S. (2009b). Identification and characterization of the phage gene *sav*, involved in sensitivity to the lactococcal abortive infection mechanism AbiV. Appl. Environ. Microbiol. *75*, 2484–2494.

Haaber, J., Samson, J.E., Labrie, S.J., Campanacci, V., Cambillau, C., Moineau, S., and Hammer, K. (2010). Lactococcal abortive infection protein AbiV interacts directly with the phage protein SaV and prevents translation of phage proteins. Appl. Environ. Microbiol. *76*, 7085–7092.

Haft, D.H., Selengut, J., Mongodin, E.F., and Nelson, K.E. (2005). A guild of 45 CRISPR-associated (Cas) protein families and multiple CRISPR/Cas subtypes exist in prokaryotic genomes. Plos Comput. Biol. *1*, e60.

Hale, C., Kleppe, K., Terns, R.M., and Terns, M.P. (2008). Prokaryotic silencing (psi)RNAs in *Pyrococcus furiosus*. RNA. *14*, 2572–2579.

Hallenbeck, P.C., Vimr, E.R., Yu, F., Bassler, B., and Troy, F.A. (1987). Purification and properties of a bacteriophage-induced endo-N-acteylneuraminidase specific for poly-alpha-2,8-sialosyl carbohydrate units. J. Biol. Chem. *262*, 3553–3561.

Handa, N., and Kobayashi, I. (2005). Type III restriction is alleviated by bacteriophage (RecE) homologous recombination function but enhanced by bacterial (RecBCD) function. J. Bacteriol. *187*, 7362–7373.

Hanlon, G.W., Denyer, S.P., Olliff, C.J., and Ibrahim, L.J. (2001). Reduction in exopolysaccharide viscosity as an aid to bacteriophage penetration through *Pseudomonas aeruginosa* biofilms. Appl. Environ. Microbiol. *67*, 2746–2753.

Hargreaves, D., Santos-Sierra, S., Giraldo, R., Sabariegos-Jareño, R., de la Cueva-Méndez, G., Boelens, R., Díaz-Orejas, R., and Rafferty, J.B. (2002). Structural and functional analysis of the Kid toxin protein from *E. coli* plasmid R1. Structure. *10*, 1425–1433.

Hattman, S. (1979). Unusual modification of bacteriophage Mu DNA. J. Virol. *32*, 468–475.

Hattman, S. (1980). Specificity of the bacteriophage Mu *mom*$^{+}$-controlled DNA modification. J. Virol. *34*, 277–279.

Hausmann, R. (1967). Synthesis of an S-adenosylmethionine-cleaving enzyme in T3-infected *Escherichia coli* and its disturbance by coinfection with enzymatically incompetent bacteriophage. J. Virol. *1*, 57–63.

Hazan, R., and Engelberg-Kulka, H. (2004). *Escherichia coli mazEF*-mediated cell death as a defense mechanism that inhibits the spread of phage P1. Mol. Gen. Genomics. *272*, 227–234.

Hazan, R., Sat, B., Reches, M., and Engelberg-Kulka, H. (2001). Postsegregational killing mediated by the P1 phage 'addiction module' *phd-doc* requires the *Escherichia coli* programmed cell death system mazEF. J. Bacteriol. *183*, 2046–2050.

Hazan, R., Sat, B., and Engelberg-Kulka, H. (2004). *Escherichia coli mazEF* mediated cell death is triggered by various stressful conditions. J. Bacteriol. *186*, 3663–3669.

Heidelberg, J.F., Nelson, W.C., Schoenfeld, T., Bhaya, D., and Ahmed, N. (2009). Germ warfare in a microbial mat community: CRISPRs provide insights into the co-evolution of host and viral genomes. PLoS One *4*, e4169.

Hill, C.W., Gray, J.A., and Brody, H. (1989). Use of the isocitrate dehydrogenase structural gene for attachment of e14 in *Escherichia coli* K-12. J. Bacteriol. *171*, 4083–4084.

Hill, C., Miller, L.A., and Klaenhammer, T.R. (1991). *In vivo* genetic exchange of a functional domain from a type II A methylase between lactococcal plasmid pTR2030 and a virulent bacteriophage. J. Bacteriol. *173*, 4363–4370.

Hinton, D.M., Pande, S., Wais, N., Johnson, X.B., Vuthoori, M., Makela, A., and Hook-Barnard, I. (2005). Transcriptional takeover by σ appropriation: remodeling of the σ^{70} subunit of *Escherichia coli* RNA polymerase by the bacteriophage T4 activator MotA and co-activator AsiA. Microbiology *151*, 1729–1740.

Hofer, B., Ruge, M., and Dreiseikelmann, B. (1995). The superinfection exclusion gene (SieA) of bacteriophage P22: identification and overexpression of the gene and localization of the gene product. J. Bacteriol. *177*, 3080–3086.

Hommais, F., Krin, E., Laurent-Winter, C., Soutourina, O., Malpertuy, A., Le Caer, J.P., Danchin, A., and Bertin, P. (2001). Large-scale monitoring of pleiotropic regulation of gene expression by the prokaryotic nucleoid-associated protein, H-NS. Mol. Microbiol. *40*, 20–36.

Hood, D.W., Deadman, M.E., Allen, T., Masoud, H., Martin, A., Brisson, J.R., Fleischmann, R., Venter, J.C., Richards, J.C., and Moxon, E.R. (1996a). Use of the complete genome sequence information of *Haemophilus influenzae* strain Rd to investigate lipopolysaccharide biosynthesis. Mol. Microbiol. *22*, 951–965.

Hood, D.W., Deadman, M.E., Jennings, M.P., Bisceric, M., Fleischmann, R.D., Venter, J.C., and Moxon, E.R. (1996b). DNA repeats identify novel virulence genes in *Haemophilus influenzae*. Proc. Natl. Acad. Sci. U.S.A. *93*, 11121–11125.

Hood, D.W., Deadman, M.E., Cox, A.D., Makepeace, K., Martin, A., Richards, J.C., and Moxon, E.R. (2004). Three genes, *lgtF*, *lic2C* and *lpsA*, have a primary role in determining the pattern of oligosaccharide extension from the inner core of *Haemophilus influenzae* LPS. Microbiology *150*, 2089–2097.

Horvath, P., Romera, D.A., Coûte-Monvoisin, A.-C., Richards, M., Deveau, H., Moineau, S., Boyaval, P., Fremaux, C., and Barrangou, R. (2008). Diversity, activity, and evolution of CRISPR loci in *Streptococcus thermophilus*. J. Bacteriol. *190*, 1401–1412.

Horvath, P., Coûte-Monvoisin, A.C., Romero, D.A., Boyaval, P., Fremaux, C., and Barrangou, R. (2009). Comparative analysis of CRISPR loci in lactic acid bacteria genomes. Int. J. Food Microbiol. *131*, 62–70.

Huber, H.E., Beauchamp, B.B., and Richardson, C.C. (1988). *Escherichia coli* dGTP triphosphohydrolase is inhibited by gene *1.2* protein of bacteriophage T7. J. Biol. Chem. *263*, 13549–13556.

Hughes, K.A., Sutherland, I.W., and Jones, M.V. (1998a). Biofilm susceptibility to bacteriophage attack: the role of phage-borne polysaccharide depolymerase. Microbiology *144*, 3039–3047.

Hughes, K.A., Sutherland, I.W., and Clark, J., and Jones, M.V. (1998b). Bacteriophage and associated polysaccharide depolymerases – novel tools for study of bacterial biofilms. J. Appl. Microbiol. *85*, 583–590.

Iida, S., Streiff, M.B., Bickle, T.A., and Arber, W. (1987). Two DNA antirestriction systems of bacteriophage P1, *dar*A, and *dar*B: characterization of *dar*A- phages. Virology *157*, 156–166.

Ishino, Y., Shinagawa, H., Makino, K., Amemura, M., and Nakata, A. (1987). Nucleotide sequence of the iap gene, responsible for alkaline phosphatase isozyme conversion in *Escherichia coli*, and identification of the gene product. J. Bacteriol. *169*, 5429–5433.

Iwashita, S., and Kanegasaki, S. (1973). Smooth specific phage adsorption: endorhamnosidase activity of tail part of P22. Biochem. Biophys. Res. Commun. *5*, 103–109.

Jackson, K.D., Starkey, M., Kremer, S., Parsek, M.R., and Wozniak, D.J. (2004). Identification of psl, a locus encoding a potential exopolysaccharide that is essential for *Pseudomonas aeruginosa* PAO1 biofilm formation. J. Bacteriol. *186*, 4466–4475.

Jacobs-Sera, D., Marinelli, L.J., Bowman, C., Broussard, G.W., Guerrero Bustamante, C., Boyle, M.M., Petrova, Z.O., Dedrick, R.M., Pope, W.H.; Science Education Alliance Phage Hunters Advancing Genomics And Evolutionary Science Sea-Phages Program, Modlin, R.L., Hendrix, R.W., and Hatfull, G.F. (2012). On the nature of mycobacteriophage diversity and host preference. Virology *434*, 187–201.

Jaffé, A., Ogura, T., and Hirage, S. (1985). Effects of the ccd function of the F plasmid on bacterial growth. J. Bacteriol. *163*, 841–849.

Janscak, P., Sandmeier, U., Szczelkun, M.D., and Bickle, T.A. (2001). Subunit assembly and mode of DNA cleavage of the type III restriction endonucleases *Eco*P1I and *Eco*P15I. J. Mol. Biol. *306*, 417–431.

Jansen, R., Embden, J.D., Gaastra, W., and Schouls, L.M. (2002). Identification of genes that are associated with DNA repeats in prokaryotes. Mol. Microbiol. *43*, 1565–1575.

Jiang, Y., Pogliano, J., Helinski, D.R., and Konieczny, I. (2002). ParE toxin encoded by the broad-host-range plasmid RK2 is an inhibitor of *Escherichia coli* gyrase. Mol. Microbiol. *44*, 971–979.

Jore, M.M., Lundgren, M., van Duijn, E., Bultema, J.B., Westra, E.R., Waghmare, S.P., Wiedenheft, B., Pul, U., Wurm, R., Wagner, R., Beijer, M.R., Barendregt, A., Zhou, K., Snijders, A.P., Dickman, M.J., Doudna, J.A., Boekema, E.J., Heck. A.J., van der Oost, J., and Brouns, S.J. (2011). Structural basis for CRISPR RNA-guided DNA recogniton by cascade. Nat. Struct. Mol. Biol. *18*, 529–536.

Kamada, K., Hanaoka, F., and Burley, S.K. (2003). Crystal structure of the MazE/MazF complex: molecular bases of antidote-toxin recognition. Mol. Cell. *11*, 875–884.

Kao, C., and Snyder, L. (1988). The lit gene product which blocks bacteriophage T4 late gene expression is a membrane protein encoded by a cryptic DNA element, e14. J. Bacteriol. *170*, 2056–2062.

Kawamura, F., Mizukami, T., Shimotsu, H., Anzai, H., Takahashi, H., and Saito, H. (1981). Unusually infrequent cleavage with several endonucleases and physical map construction of *Bacillus subtilis* bacteriophage φ1 DNA. J. Virol. *37*, 1099–1102.

Kim, K.S. (2003). Pathogenesis of bacterial meningitis: from bacteraemia to neuronal injury. Nat. Rev. Neurosci. *4*, 376–385.

Kleim, M., and Dreiseikelmann, B. (1989). The superimmunity gene *sim* of bacteriophage P1 causes superinfection exclusion. Virology *171*, 350–355.

Knegasaki, S., and Wright, A. (1973). Studies on the mechanism of phage adsorbtion: interaction between phage ε15 and its cellular receptor. Virology *52*, 160–173.

Kobayashi, I. (1998). Selfishness and death: raison d'etre of restriction, recombination and mitochondria. Trends Genet. *14*, 368–374.

Kobayashi, I. (2001). Behavior of restriction-modification systems as selfish mobile elements and their impact on genome evolution. Nucleid Acids Res. *29*, 3742–3756.

Krüger, D.H., and Bickle, T.A. (1983). Bacteriophage survival: multiple mechanisms for avoiding the deoxyribonucleic acid restriction systems of their hosts. Microbiol. Rev. *47*, 345–360.

Krüger, D.H., and Schroeder, C. (1981). Bacteriophage T3 and bacteriophage T7 virus–host cell interactions. Microbiol. Rev. *45*, 9–51.

Krüger, D.H., Schroeder, C., Hansen, S., and Rosenthal, H.A. (1977). Active protection by bacteriophages T3 and T7 against *E.coli* B- and K-specific restriction of their DNA. Mol. Gen. Genet. *153*, 99–106.

Krüger, D.H., Reuter, M., Hansen, S., and Schroeder, C. (1982). Influence of phage T3 and T7 gene functions on a type III (EcoP1) DNA restriction-modification system *in vivo*. Mol. Gen. Genet. *185*, 457–461.

Krüger, D.H., Hansen, S., and Reuter, M. (1983). The *ocr* gene function of bacteriophages T3 and T7 counteracts the *Salmonella typhimurium* DNA restriction systems SA and SB. J. Virol. *45*, 1147–1149.

Krüger, D.H., Barcak, G.J., Reuter, M., and Smith, H.O. (1988). *Eco*RII can be activated to cleave refractory DNA recognition sites. Nucleic Acids Res. *16*, 3997–4008.

Kunin, V., Sorek, R., and Hugenholtz, P. (2007). Evolutionary conservation of sequence and secondary structures in CRISPR repeats. Genome Biol. *8*, R61.

Kusano, K., Sunohara, Y., Takahashi, N., Yoshikura, H., and Kobayashi, I. (1994). DNA double-strand break repair: genetic determinants of flanking crossing-over. Proc. Natl. Acad. Sci. U.S.*A*. *91*, 1173–1177.

Kutter, E., White, T., Kashlev, M., Uzan, M., McKinney, J., and Guttman, B. (1994). Effects on host genome structure and expression. In Molecular Biology of Bacteriophage T4, J.D. Karam, ed. (Washington, DC, American Society for Microbiology), pp. 357–368.

Kuzio, J., and Kropinski, A.M. (1983). O-antigen conversion in *Pseudomonas aeruginosa* PAO1 by bacteriophage D3. J. Bacteriol. *155*, 203–212.

Labrie, S.J., and Moineau, S. (2007). Abortive infection mechanisms and prophage sequences significantly influence the genetic makeup of emerging lytic lactococcal phages. J. Bacteriol. *189*, 1482–1487.

Labrie, S.J., Samson, J.E., and Moineau, S. (2010). Bacteriophage resistance mechanisms. Nat. Rev. Microbiol. *8*, 317–327.

Lah, J., Marianovsky, I., Engelberg-Kulka, H., Glaser, H., and Loris, R. (2003). Interaction of the addiction antidote MazE with dromedary single domain antibody fragment: structure, thermodynamics of binding, stability and influence of DNA recognition. J. Biol. Chem. *278*, 14101–14111.

Lazdunski, C.J., Bouveret, E., Rigal, A., Journet, L., Lloubes, R., and Benedetti, H. (1998). Colicin import into *Escherichia coli* cells. J. Bacteriol. *180*, 4993–5002.

Leiman, P.G., Battisti, A.J., Bowman, V.D., Stummeyer, K., Mühlenhoff, M., Gerardy-Schahn, R., Scholl, D., and Molineux, I.J. (2007). The structures of bacteriophages K1E and K1-5 explain processive degradation of polysaccharide capsules and evolution of new host specificities. J. Mol. Biol. *371*, 836–849.

Levitz, R., Chapman, D., Amitsur, M., Green, R., Snyder, L., and Kaufmann, G. (1990). The optional *E. coli prr* locus encodes a latent form of phage T4-induced anticodon nuclease. EMBO J. *9*, 1383–1389.

Lillestøl, R.K., Redder, P., Garrett, R.A., and Brügger, K. (2006). A putative viral defence mechanism in archaeal cells. Archaea. *2*, 59–72.

Lillestøl, R.K., Shah, S.A., Brügger, K., Redder, P., Phan, H., Christiansen, J., and Garret, R.A. (2009). CRISPR families of the crenarchaeal genus *Sulfolobus*: bidirectional transcription and dynamic properties. Mol. Microbiol. *72*, 259–272.

Lin, Y.H., Liao, C.C., Liang, P.H., Yuan, H.S., and Chak, K.F. (2004). Involvement of colicin in the limited protection of the colicin producing cells against bacteriophage. Biochem. Biophys. Res. Commun. *318*, 81–87.

Lindahl, T. (1979). DNA glycosylases, endonucleases for apurinic/apyrimidinic sites and base excision repair. Prog. Nucleic Acids Res. Mol. Biol. *22*, 135–192.

Lindberg, A.A. (1967). Studies of a receptor for Felix O-1 phage in *Salmonella minnesota*. J. Gen. Microbiol. *48*, 225–233.

Lindberg, A.A., and Hellerovist, C.G. (1971). Bacteriophage attachment sites, serological specificity and chemical composition of the lipopolysaccharides of semi-rough and rough mutants of *Salmonella typhimurium*. J. Bacteriol. *105*, 57–64.

Lindberg, A.A., Wollin, R., Gemski, P., and Wohlhieter, J.A. (1978). Interaction between bacteriophage Sf6 and *Shigella flexneri*. J. Virol. *27*, 38–44.

Linder, C.H., Carlson, K., Albertioni, F., Söderström, J., and Påhlson, C. (1994). A late exclusion of bacteriophage T4 can be suppressed by *Escherichia coli* GroEL or Rho. Genetics. *137*, 613–625.

Liu, M., Deora, R., Doulatov, S.R., Gingery, M., Eiserling, F.A., Preston, A., Maskell, D.J., Simons, R.W., Cotter, P.A., Parkhill, J., and Miller, J.F. (2002). Reverse transcriptase-mediated tropism switching in *Bordetella* bacteriophage. Science *295*, 2091–2094.

Liu, M., Gingery, M., Doulatov, S.R., Liu, Y., Hodes, A., Baker, S., Davis, P., Simmonds, M., Churcher, C., Mungall, K., Quail, M.A., Preston, A., Harvill, E.T., Maskell, D.J., Eiserling, F.A., Parkhill, J., and Miller, J.F. (2004). Genomic and genetic analysis of *Bordetella* bacteriophages encoding reverse transcriptase-mediated tropism-switching cassettes. J. Bacteriol. *186*, 1503–1517.

Liu, M., Zhang, Y., Inouye, M., and Woychik, N.A. (2008). Bacterial addiction module toxin Doc inhibits translation elongation through its association with the 30S ribosomal subunit. Proc. Natl. Acad. Sci. U.S.A. *105*, 5885–5890.

Łobocka, M.B., Rose, D.J., Plunkett, G. 3rd, Rusin, M., Samojedny, A., Lehnherr, H., Yarmolinsky, M.B., and Blattner, F.R. (2004). Genome of bacteriophage P1. J. Bacteriol. *186*, 7032–7068.

Loris, R., Marianovsky, I., Lah, J., Laermans, T., Engelberg-Kulka, H., Glaser, G., Muylderman, S., and Wyne, L. (2003). Crystal structure of the intrinsically flexible addiction antidote MazE. J. Biol. Chem. *278*, 28252–28257.

Losick, R. (1969). Isolation of a trypsin-sensitive inhibitor of O-antigen synthesis involved in lysogenic conversion by bacteriophage ε^{15}. J. Mol. Biol. *42*, 237–246.

Lu, M.-J., and Henning, U. (1994). Superinfection exclusion by T-even-type coliphages. Trends Microbiol. *2*, 137–139.

Lu, M.-J., Stierhof, Y.-D., and Henning, U. (1993). Location and unusual membrane topology of the immunity protein of the *Escherichia coli* phage T4. J. Virol. *67*, 4905–4913.

Lucey, M., Daly, C., and Fitzgerald, G.F. (1992). Cell surface characteristics of *Lactococcus lactis* harbouring pCI828, a 46 kb plasmid encoding inhibition of bacteriophage adsorption. J. Gen. Microbiol. *138*, 2137–2143.

Lüderitz, O., Jann, K., and Wheat, R. (1968). Somatic and capsular antigens of Gram-negative bacteria. Comp. Biochem. *26A*, 105–228.

McCorquodale, D.J., and Warner, H.R. (1988). Bacteriophage T5 and related phages. In The Bacteriophages, R. Calendar, ed. (New York, Plenum Press), pp. 439–475.

McGrath, S., Seegers, J.F.M.L., Fitzgerald, G.F., and van Sinderen, D. (1999). Molecular characterization of a phage-encoded resistance system in *Lactococcus lactis*. Appl. Environ. Microbiol. *65*, 1891–1899.

McGrath, S., Firzgerald, G., and van Sinderen, D. (2002). Identification and characterization of phage-resistance gene in temperate lactococcal bacteriophages. Mol. Microbiol. *43*, 509–520.

Mack, D., Fischer, W., Krokotsch, A., Leopold, K., Hartmann, R., Egge, H., and Laufs, R. (1996). The intercellular adhesin involved in biofilm accumulation of *Staphylococcus epidermidis* is a linear β-1,6-linked glucosaminoglycan: purification and structural analysis. J. Bacteriol. *178*, 175–183.

McMahon, S.A., Miller, J.L., Lawton, J.A., Kerkow, D.E., Hodes, A., Marti-Renom, M.A., Doulatov, S., Narayanan, E., Sali, A., Miller, J.F., and Ghosh, P. (2005). The C-type lectin fold as an evolutionary solution for massive sequence variation. Nat. Struct. Mol. Biol. *12*, 886–892.

Magnuson, R.D. (2007). Hypothetical functions of toxin–antitoxin systems. J. Bacteriol. *189*, 6089–6092.

Magnuson, R., and Yarmolinsky, M.B. (1998). Corepression of the P1 addiction operon by Phd and Doc. J. Bacteriol. *180*, 6342–6351.

Mah, T.F. (2012). Biofilm-specific antibiotic resistance. Future Microbiol. *7*, 1061–1072.

Mahony, J., McGrath, S., Fitzgerald, G.F., and van Sinderen, D. (2008). Identification and characterization of lactococcal-prophage-carried superinfection exclusion genes. Appl. Environ. Microbiol. *74*, 6206–6215.

Maillou, J., and Dreiseikelmann, B. (1990). The *sim* gene of *Escherichia coli* phage P1: nucleotide sequence and purification of the processed protein. Virology *175*, 500–507.

Maillou, L., Faller, M., Chiang, J., Arbing, M., Guo, F., Cascio, D., and Eisenberg, D. (2009). Structure and proposed activity of a member of the VapBC family of toxin–antitoxin systems. VapBC-5 from *Mycobacterium tuberculosis*. J. Biol. Chem. *284*, 276–283.

Makarova, K.S., Aravind, L., Grishin, N.V., Rogozin, I.B., and Koonin, E.V. (2002). A DNA repair system specific for thermophilic Archaea and bacteria predicted by genomic context analysis. Nucleic Acids Res. *30*, 482–496.

Makarova, K.S., Grishin, N.V., Shabalina, S.A., Wolf, Y.I., and Koonin, E.V. (2006). A putative RNA-interference–based immune system in prokaryotes: computational analysis of the predicted enzymatic machinery, functional analogies with eukaryotic RNAi, and hypothetical mechanisms of action. Biol. Direct. *1*, 7.

Makarova, K.S., Wolf, Y.I., and Koonin, E.V. (2009). Comprehensive comparative-genomic analysis of Type 2 toxin–antitoxin systems and related mobile stress response systems in prokaryotes. Biol. Direct. *4*, 19.

Makarova, K.S., Haft, D.H., Barrangou, R., Brouns, S.J., Charpentier, E., Horvath, P., Moineau, S., Mojica, F.J., Wolf, Y.I., Yakunin, A.F., van der Oost, J., and Koonin, E.V. (2011). Evolution and classification of the CRISPR-Cas systems. Nat. Rev. Microbiol. *9*, 467–477.

Mann, E.E., and Wozniak, D.J. (2012). Pseudomonas biofilm matrix composition and niche biology. FEMS Microbiol. Rev. *36*, 893–916.

Marianovsky, I., Aizenman, E., Engelberg-Kulka, H., and Glaser, G. (2001). The regulation of the *Escherichia coli mazEF* promoter involves an unusual alternating palindrome. J. Biol. Chem. *276*, 5975–5984.

Marinus, M.G., and Casadesus, J. (2009). Roles of DNA adenine methylation in host–pathogen interactions: mismatch repair, transcriptional regulation, and more. FEMS Microbiol. Rev. *33*, 488–503.

Mark, K..-K., and Studier, F.W. (1981). Purification of the gene 0.3 protein of bacteriophage T7, an inhibitor of the DNA restriction system of *Escherichia coli*. J. Biol. Chem. *256*, 2573–2578.

Marraffini, L.A., and Sontheimer, E.J. (2008). CRISPR interference limits horizontal gene transfer in staphylococci by targeting DNA. Science *322*, 1843–1845.

Marraffini, L.A., and Sontheimer, E.J. (2010). Self versus non-self discrimination during CRISPR-RNA directed immunity. Nature *463*, 568–572.

Matsukawa, M., and Greenberg, E.P. (2004). Putative exopolysaccharide synthesis genes influence *Pseudomonas aeruginosa* biofilm development. J. Bacteriol. *186*, 4449–4456.

Mayer, C., Moritz, R., Kirschner, C., Borchard, W., Maibaum, R., Wingender, J., and Flemming, H.C. (1999) The role of intermolecular interactions: studies on model systems for bacterial biofilms. Int. J. Biol. Macromol. *26*, 3–16.

Maxted., W.R. (1955). The influence of bacteriophage on *Streptococcus pyogenes*. J. Gen. Microbiol. *12*, 484–495.

Medhekar, B., and Miller, J.F. (2009). Diversity-generating retroelements. Curr. Opin. Microbiol. *10*, 388–395.

Meinhart, A., Alonso, J.C., Strater, N., and Saenger, W. (2003). Crystal structure of the plasmid maintenance sytem ε/ζ: functional mechanism of toxin ζ and inactivation by $\varepsilon_2\zeta_2$ complex formation. Proc. Natl. Acad. Sci. U.S.A. *100*, 1661–1666.

Meisel, A., Bickle, T.A., Krüger, D.H., and Schroeder, C. (1992). Type III restriction enzymes need tow inversely oriented recognition sites for DNA cleavage. Nature *355*, 467–469.

Moak, M., and Molineux, I.J. (2000). Role of the Gp16 lytic transglycosylase motif in bacteriophage T7 virions at the initiation of infection. Mol. Microbiol. *37*, 345–355.

Moffatt, B.A., and Studier, F.W. (1988). Entry of bacteriophage T7 DNA into the cell and escape from restriction. J. Bacteriol. *170*, 2095–2105.

Mojica, F.J., Diez-Villasenor, C., Garcia-Martinez, J., and Soria, E. (2005). Intervening sequences of regularly spaced prokaryotic repeats derive from foreign genetic elements. J. Mol. Evol. *60*, 174–182.

Mojica, F.J., Diez-Villasenor, C., Garcia-Martinez, J., and Almendros, C. (2009). Short motif sequences determine the targets of the prokaryotic CRISPR defence system. Microbiology *155*, 733–740.

Molineux, I.J. (1991). Host–parasite interactions: recent developments in the genetics of abortive phage infections. New. Biol. *3*, 230–236.

Molineux, I.J., Schmitt, C.K., and Condreay, J.P. (1989). Mutants of bacteriophage T7 that escape F restriction. J. Mol. Biol. *207*, 563–574.

Moxon, E.R., and Thaler, D.S. (1997). The tinkerer's evolving toolbox. Nature *387*, 659–662.

Mühlenhoff, M., Eckhardt, M., and Gerardy-Schahn, R. (1998). Polysialic acid: three-dimensional structure, biosynthesis and function. Curr. Opin. Struct. Biol. *8*, 558–564.

Mühlenhoff, M., Stummeyer, K., Grove, M., Sauerborn, M., and Gerardy-Schahn, R. (2003). Proteolytic processing and oligomerization of bacteriophage-derived endosialidases. J. Biol. Chem. *278*, 12634–12644.

Murray, N.E. (2000). Type I restriction systems: sophisticated molecular machines (a legacy of Bertani and Weigle). Microbiol. Mol. Biol. Rev. *64*, 412–434.

Murray, N.E. (2001). Immigration control of DNA in bacteria: self versus non-self. Microbiology *148*, 3–20.

Mutschler, H., Gebhardt, M., Shoeman, R.L., and Meinhart, A. (2011). A novel mechanism of programmed cell death in bacteria by toxin–antitoxin systems corrupts peptidoglycan synthesis. PLoS Biol. *9*: e1001033.

Naito, T., Kusano, K., and Kobayashi, I. (1995). Selfish behavior of restriction-modification systems. Science *267*, 897–899.

Nakai, H., and Richardson, C.C. (1990). The gene *1.2* protein of bacteriophage T7 interacts with the *Escherichia coli* dGTP triphosphohydrolase to form a GTP-binding protein. J. Biol. Chem. *265*, 4411–4419.

Nakayama, Y., and Kobayashi, I. (1998). Restriction-modification gene complexes as selfish gene entities: roles of a regulatory system in their establishment, maintenance, and apoptotic mutual exclusion. Proc. Natl. Acad. Sci. U.S.A. *95*, 6442–6447.

Nariya, H., and Inouye, M. (2008). MazF, an mRNA interferase, mediates programmed cell death during multicellular *Myxococcus* development. Cell. *132*, 55–66.

Navarre, W.W., Porwollik, S., Wang, Y., McClelland, M., Rosen, H., Libby, S.J., and Fang, F.C. (2006). Selective silencing of foreign DNA with low GC content by the H-NS protein in *Salmonella*. Science *313*, 236–238.

Navarre, W.W., McClelland, M., Libby, S.J., and Fang, F.C. (2007). Silencing of xenogeneic DNA by H-NS-facilitation of lateral gene transfer in bacteria by a defense system that recognizes foreign DNA. Genes. Dev. *21*, 1456–1471.

Neubauer, C., Gao, Y.-G., Andersen, K.R., Dunham, C.M., Kelley, A.C., Hentschel, J., Gerdes, K., Ramakrishnan, V., and Brodersen, D.E. (2009). The structural basis for mRNA recognition and cleavage by the ribosome-dependent endonuclease RelE. Cell. *139*, 1084–1095.

Neve, H., Zenz, K.I., Desiere, F., Koch, A., Heller, K.J., and Brüssow, H. (1998). Comparison of the lysogeny modules from the temperate *Streptococcus thermophilus* bacteriophages TP-J34 and Sfi21: implication for the modular theory of phage evolution. Virology *241*, 61–72.

Newton, G.J., Daniels, C., Burrows, L.L., Kropinski, A.M., Clarke, A.J., and Lam, J.S. (2001). Three-component-mediated serotype conversion in *Pseudomonas aeruginosa* by bacteriophage D3. Mol. Microbiol. *39*, 1237–1247.

Nordström, K., and Forsgren, A. (1974). Effect of protein A on adsorption of bacteriophages to *Staphylococcus aureus*. J. Virol. *14*, 198–202.

Noyer-Weidner, M., Pawlek, B., Jentsch, S., Günthert, U., and Trautner, T.A. (1981). Restriction and modification in *B. subtilis*: gene coding for a BsuR-specific modification methyltransferase in the temperate bacteriophage φ3T. J. Virol. *38*, 1077–1080.

O'Connor, L., Coffey, A., Daly, C., and Fitzgerald, G.F. (1996). AbiG a genotypically novel abortive infection mechanism encoded by plasmid pCI750 of *Lactococcus lactis* ssp. *cremoris* UC653. Environ. Microbiol. *62*, 3075–3082.

O'Flaherty, S., Ross, R.P., Meaney, W., Fitzgerald, G.F., Elbreki, M.F., and Coffey, A. (2005). Potential of the polyvalent anti-*Staphylococcus* bacteriophage K for control of antibiotic-resistant staphylococci from hospitals. Appl. Environ. Microbiol. *71*, 1836–1842.

Okafor, N. (2007). Modern industrial microbiology and biotechnology (Enfield, NH, Science publishers).

van der Oost, J., Jore, M.M., Westra, E.R., Lundgren, M., and Brouns, S.J. (2009). CRISPR-based adaptive and heritable immunity in prokaryotes. Trends Biochem. Sci. *34*, 401–407.

Oshima, T., Ishikawa, S., Kurokawa, K., Aiba, H., and Ogasawara, N. (2006). *Escherichia coli* histone-like protein H-NS preferentially binds to horizontally acquired DNA in association with RNA polymerase. DNA Res. *13*, 141–153.

Pajunen, M.I., Kiljunen, S.J., Söderholm, M.E.-L., and Skurnik, M. (2001). Complete genomic sequence of the lytic bacteriophage φYe-O12 of *Yersinia enterocolitica* serotype O:3. J. Bacteriol. *183*, 1928–1937.

Pajunen, M.I., Elizondo, M.R., Skurnik, M., Kieleczawa, J., and Molineux, I.J. (2002). Complete nucleotice sequence and likely recombinatorial origin of bacteriophage T3. J. Mol. Biol. *319*, 1115–1132.

Pandey, D.P., and Gerdes, K. (2005). Toxin–antitoxin loci are highly abundant in free-living but lost from host-associated prokaryotes. Nucleic Acids Res. *33*, 966–976.

Parma, D.H., Snyder, M., Sobolevski, S., Brody, E., and Gold, L. (1992). The Rex system of bacteriophage λ: tolerance and altruistic cell death. Genes Dev. *6*, 497–510.

Parreira, R., Ehrlich, S.D., and Chopin, M.C. (1996). Dramatic decay of phage transcripts in lactococcal cells carrying the abortive infection determinant AbiB. Mol. Microbiol. *19*, 221–230.

Pecota, D.C., and Wood, T.K. (1996). Exclusion of T4 phage by the *hok/sok* killer locus from plasmid R1. J. Bacteriol. *178*, 2044–2050.

Pedruzzi, I., Rosenbusch, J.P., and Locher, K.P. (1998). Inactivation *in vitro* of the *Escherichia coli* outer membrane protein FhuA by a phage T5-encoded lipoprotein. FEMS Microbiol. Lett. *168*, 119–125.

Pedulla, M.L., Ford, M.E., Houtz, J.M., Karthikeyan, T., Wadsworth, C., Lewis, J.A., Jacobs-Sera, D., Falbo, J., Gross, J., Pannunzio, N.R., Brucker, W., Kumar, V., Kandasamy, J., Keenan, L., Bardarov, S., Kriakov, J., Lawrence, J.G., Jacobs, W.R. Jr, Hendrix, R.W., and Hatfull, G.F. (2003). Origins of highly mosaic mycobacteriophage genomes. Cell. *113*, 171–182.

Pelkonen, S., Pelkonen, J., and Finne, J. (1989). Common cleavage pattern of polysialic acid by bacteriophage endosialidase of different properties and origin. J. Virol. *63*, 4409–4416.

Perry, L.L., SanMiguel, P., Minocha, U., Terekhov, A.I., Shroyer, M.L., Farris, L.A., Bright, N., Reuhs, B.L., and Applegate, B.M. (2009). Sequence analysis of *Escherichia coli* O157:H7 bacteriophage φV10 and identification of a phage-encoded immunity protein that modifies the O157 antigen. FEMS Microbiol. Lett. *292*, 182–186.

Perumal, N., and Minkley, E.G. (1984). The product of the F sex factor traT surface exclusion gene is a lipoprotein. J. Biol. Chem. *259*, 5357–5360.

Pingoud, A., and Jeltsch, A. (1997). Recognition and cleavage of DNA by type-II restriction endonucleases. Eur. J. Biochem. *246*, 1–22.

Pingoud, A., Fuxreiter, M., Pingoud, V., and Wende, W. (2005). Type II restriction endonucleases: structure and mechanism. Cell. Mol. Life Sci. *62*, 685–707.

Pisabarro, A.G., de Pedro, M.A., and Vazquez, D. (1985). Structural modifications in the peptidoglycan of *Escherichia coli* associated with changes in the state of growth of the culture. J. Bacteriol. *161*, 238–242.

Piuri, M., and Hatfull, G.F. (2006). A peptidoglycan hydrolase motif within the mycobacteriophage TM4 tape measure protein promotes efficient infection of stationary phase cells. Mol. Microbiol. *62*, 1569–1585.

van der Ploeg, J.R. (2009). Analysis of CRISPR in *Streptococcus mutans* suggests frequent occurrence of acquired immunity against infection by M102-like bacteriophages. Microbiology *155*, 1966–1976.

Ploquin, M., Bransi, A., Paquet, E.R., Stasiak, A.Z., Stasiak, A., Yu, X., Cieslinska, A.M., Egelman, E.H., Moineau, S., and Masson, J.Y. (2008). Functional and structural basis for a bacteriophage homolog of human RAD52. Curr. Biol. *18*, 1142–1146.

Poteete, A.R., and Fenton, A.C. (1993). Efficient double-strand break-stimulated recombination promoted by the general recombination systems of phages lambda and P22. Genetics. *134*, 1013–1021.

Pourcel, C., Salvignol, G., and Vergnaud, G. (2005). CRISPR elements in *Yersinia pestis* acquire new repeats by preferential uptake of bacteriophage DNA, and provide additional tools for evolutionary studies. Microbiology *151*, 653–663.

Pul, U., Wurm, R., Arslan, Z., Geissen, R., Hofmann, N., and Wagner, R. (2010). Identification and characterization of *E. coli* CRISPR-cas promoters and their silencing by H-NS. Mol. Microbiol. *75*, 1495–1512.

Putman, C.D., Shroyer, M.J.N., Lundquist, A.J., Mol, C.D., Arvai, A.S., Mosbaugh, D.W., and Tainer, J.A. (1999). Protein mimicry of DNA from crystal structures of the uracil-DNA glycosylase inhibitor protein and its complex with *Escherichia coli* uracil-DNA glycosylase. J. Mol. Biol. *287*, 331–346.

Qimron, U., Tabor, S., and Richardson, C.C. (2010). New details about bacteriophage T7-interactions. Microbe *5*, 117–120.

Rabussay, D. (1983). Regulation of gene expression transcription: phage-evoked changes in RNA polymerase. In Bacteriophage T4, C. Mathews, E.M. Kutter, G. Mosig, and P.B. Berget, eds. (Washington, DC, American Society for Microbiology), pp. 167–173.

Rakhuba, D.V., Kolomiets, E.I., Szwajcer, Dey, E., and Novik, G.I. (2010). Bacteriophage receptors, mechanisms of phage adsorption and penetration into host cell. Pol. J. Microbiol. *59*, 145–155.

Raleigh, E.A., and Wilson, G. (1986). *Escherichia coli* K-12 restricts DNA containing 5-methylcytosine. Proc. Natl. Acad. Sci. U.S.A. *83*, 9070–9074.

Reakes, C.F., Gann, A.A., Rossouw, F.T., and Rowbury, R.J. (1987). Abortive infection by bacteriophage Me1 of *Escherichia coli* K12 strains bearing the plasmid ColV, I-K94. J. Gen. Virol. *68*, 263–272.

Richter, C., Chang, J.T., and Fineran, P.C. (2012). Function and regulation of clustered regularly interspaced short palindromic repeats (CRISPR)/CRISPR associated (Cas) systems. Viruses. *4*, 2291–2311.

Riede, I., and Eschbach, M.-L. (1986). Evidence that TraT interacts with OmpA of *Escherichia coli*. FEBS Lett. *205*, 241–245.

Rieger-Hug, D., and Stirm, S. (1981). Comparative study of host capsule depolymerases associated with *Klebsiella* bacteriophages. Virology *113*, 363–378.

Rifat, D., Wright, N.T., Varney, K.M., Weber, D.J., and Black, L.W. (2008). Restriction endonuclease inhibitor IPI* of bacteriophage T4: a novel structure for a dedicated target. J. Mol. Biol. *375*, 720–734.

Riley, M.A., and Gordon, D.M. (1992). A survey of Col plasmids in natural isolates of *Escherichia coli* and an investigation into the stability of Col-plasmid lineages. J. Gen. Microbiol. *138*, 1345–1352.

Robbins, P.W., Keller, J.M., Wright, A., and Bernstein, R.L. (1965). Enzymatic and kinetic studies on the mechanism of O-antigen conversion by bacteriophage ε^{15}. J. Biol. Chem. *240*, 384–390.

Roberts, R.J., Vincze, T., Posfai, J., and Macelis, D. (2010). REBASE – a database for DNA restriction and modification: enzymes, genes and genomes. Nucleic Acids Res. *38*, D234–D236.

Roucourt, B., and Lavigne, R. (2009). The role of interactions between phage and bacterial proteins within the infected cell: a diverse and puzzling interactome. Environ. Microbiol. *11*, 2789–2805.

Salomón, R.A., and Farías, R.N. (1992). Microcin 25, a novel antimicrobial peptide produced by *Escherichia coli*. J. Bacteriol. *174*, 7428–7435.

Sat, B., Hazan, R., Fisher, T., Khaner, H., Glaser, G., and Engelberg-Kulka, H. (2001). Programmed cell death in *Escherichia coli*: some antibiotics can trigger *mazEF* lethality. J. Bacteriol. *183*, 2041–2045.

Scaltriti, E., Moineau, S., Launay, H., Masson, J.Y., Rivetti, C., Ramoni, R., Campanacci, V., Tegoni, M., and Cambillau, C. (2010). Deciphering the function of lactococcal phage ul36 Sak domains. J. Struct. Biol. *170*, 462–469.

Scaltriti, E., Launay, H., Genois, M.M., Bron, P., Rivetti, C., Grolli, S., Ploquin, M., Campanacci, V., Tegoni, M., Cambillau, C., Moineau, S., and Masson, J.Y. (2011). Lactococcal phage p2 ORF35-Sak3 is an ATPase involved in DNA recombination and AbiK mechanism. Mol. Microbiol. *80*, 102–116.

Schmitt, C.K., and Molineux, I.J. (1991). Expression of gene *1.2* and gene *10* of bacteriophage T7 is lethal in F plasmid-containing *Escherichia coli*. J. Bacteriol. *173*, 1536–1543.

Schmitt, C.K., Kemp, P., and Molineux, I.J. (1991). Genes *1.2* and *10* of bacteriophages T3 and T7 determine the permeability lesions observed in infected cells of *Escherichia coli* expressing the F plasmid gene *pifA*. J. Bacteriol. *173*, 6507–6514.

Scholl, D., Rogers, S., Adhya, S., and Merril, C.R. (2001). Bacteriophage K1-5 encodes two different tail fiber proteins, allowing it to infect and replicate on both K1 and K5 strains of *Escherichia coli*. J. Virol. *75*, 2509–2515.

Scholl, D., Adhya, S., and Merril, C. (2005). *Escherichia coli* K1's capsule is a barrier to bacteriophage T7. Appl. Environ. Microbiol. *71*, 4872–4874.

Sebaihia, M., Wren, B.W., Mullany, P., Fairweather, N.F., Minton, N., Stabler, R., Thomson, N.R., Roberts, A.P., Cerdeño-Tárraga, A.M., Wang, H., *et al.* (2006). The multidrug-resistant human pathogen *Clostridum difficile* has a highly mobile mosaic genome. Nature Genet. *38*, 779–786.

Seed, K.D., Faruque, S.M., Mekalanos, J.J., Calderwood, S.B., Qadri, F., and Camilli, A. (2012). Phase variable O antigen biosynthetic genes control expression of the major protective antigen and bacteriophage receptor in Vibrio cholerae O1. PLoS Pathog. *8*, e1002917.

Semenova, E., Nagornykh, M., Pyatnitskiy, M., Artamonova, I.I., and Severinov, K. (2009). Analysis of CRISPR system function in plant pathogen *Xanthomonas oryzae*. FEMS Microbiol. Lett. *296*, 110–116.

Shinedling, S., Parma, D., and Gold, L. (1987). Wild type bacteriophage T4 is restricted by the lambda *rex* genes. J. Virol. *61*, 3790–3794.

Shinkai, A., Kira, S., Nakagawa, N., Kashihara, A., Kuramitsu, S., and Yokoyama, S. (2007). Transcription activation mediated by a cyclic AMP receptor protein from *Thermus thermophilus* HB8. J. Bacteriol. *189*, 3891–3901.

Sillankorva, S., Oliveira, R., Vieira, M., Sutherland, I.W., and Azeredo, J. (2004). Bacteriophage f-S1 infection of *Pseudomonas fluorescens* planktonic cells versus biofilms. Biofouling. *20*, 133–138.

Sinkunas, T., Gasiunas, G., Fremaux, C., Barrangou, R., Horvath, P., and Siksnys, V. (2011). Cas3 is a single-stranded DNA nuclease and ATP-dependent helicase in the CRISPR/Cas immune system. EMBO J. *30*, 1335 – 1342.

Sistla, S., and Rao, D.N. (2004). *S*-adenosyl-L-methionine-dependent restriction enzymes. Crit. Rev. Biochem. Mol. Biol. *39*, 1–19.

Skjåk-Bræk, G., Larsen, B., and Grasdalen, H. (1985). The role of *O*-acetyl groups in the biosynthesis of alginate by *Azotobacter vinelandii*. Carbohydr. Res. *145*, 169–174.

Snyder, L., and McWilliams, K. (1989). The *rex* genes of bacteriophage lambda can inhibit cell function without phage superinfection. Gene. *81*, 17–24.

Spoerel, N., Herrlich, P., and Bickle, T.A. (1979). A novel bacteriophage defence mechanism: the anti-restriction protein. Nature *278*, 30–34.

Sorek, R., Kunin, V., and Hugenholtz, P. (2008). CRISPR – a widespread system that provides acquired resistance against phages in bacteria and Archaea. Nat. Rev. Microbiol. *6*, 181.

Srikhanta, Y.N., Maguire, T.L., Stacey, K.J., Grimmond, S.M., and Jennings, M.P. (2005). The phasevarion: a genetic system controlling coordinated, random switching of expression of multiple genes. Proc. Natl. Acad. Sci. U.S.A. 102: 5547–5551.

Stern, A., Keren, L., Wurtzel, O., Amitai, G., and Sorek, R. (2010). Self-targeting by CRISPR: gene regulation or autoimmunity? Trends Genet. *26*: 335–340.

Steven, A.C., Trus, B.L., Maizel, J.V., Unser, M., Parry, D.A., and Wall, J.S. (1988). Molecular substructure of a viral receptor-recognition protein. The gp17 tail-fiber of bacteriophage T7. J. Mol. Biol. *200*, 351–365.

Stirm, S., and Freund-Mölbert, E. (1971). *Escherichia coli* capsule bacteriophages. II. Morphology. J. Virol. *8*, 330–342.

Stirm, S., Bessler, W., Fehmel, F., Freund-Mölbert, E., and Thurow, H. (1971). Isolation of spike-formed particles from bacteriophage lysates. Virology *45*, 303–308.

Stummeyer, K., Dickmanns, A., Mühlenhoff, M., Gerardy-Schahn, R., and Ficner, R. (2005). Crystal structure of the polysialic acid-degrading endosialidase of bacteriophage K1F. Nat. Struct. Mol. Biol. *12*, 90–96.

Stummeyer, K., Schwarzer, D., Claus, H., Vogel, U., Gerardy-Schahn, R., and Mühlenhof, M. (2006). Evolution of bacteriophages infecting encapsulated bacteria: lessons from *Escherichia coli* K1-specific phages. Mol. Microbiol. *60*, 1123–1135.

Sun, X., Göghler, A., Heller, K.J., and Neve, H. (2006). The *ltp* gene of temperate *Streptococcus thermophilus* phage TP-J34 confers superinfection exclusion to *Streptococcus thermophilus* and *Lactococcus lactis*. Virology *350*, 146–157.

Susskind, M.M., and Botstein, D. (1978). Molecular genetics of bacteriophage P22. Microbiol. Rev. *42*, 385–413.

Susskind, M.M., Wright, A., and Botstein, D. (1971). Superinfection exclusion by P22 prophage in lysogens of *Salmonella typhimurium*. II. Genetic evidence for two exclusion systems. Virology *45*, 638–652.

Susskind, M.M., Botstein, D., and Wright, A. (1974). Superinfection exclusion by P22 bacteriophage in lysogens of *Salmonella typhimurium*. III. Failure of superinfecting phage DNA to enter sieA+ lysogens. Virology *62*, 350–366.

Sutherland, E., Coe, L., and Raleigh, E.A. (1992). McrBC: a multi-subunit GTP-dependent restriction endonuclease. J. Mol. Biol. *225*, 327–348.

Sutherland, I.W. (1995). Polysaccharide lyases. FEMS Microbiol. Rev. *16*, 323–347.

Sutherland, I.W. (1999). Polysaccharases for microbial polysaccharides. Carbohydr. Polym. *38*, 319–328.

Sutherland, I.W. (2001). Biofilm exopolysaccharides: a strong and sticky framework. Microbiology *147*, 3–9.

Suttle, C.A. (2007). Marine viruses – major players in the global ecosystem. Nat. Rev. Microbiol. *5*, 801–812.

Takahashi, N., and Kobayashi, I. (1990). Evidence for the double-strand break repair model of bacteriophage lambda recombination. Proc. Natl. Acad. Sci. U.S.A. *87*, 2790–2794.

Takahashi, N.K., Sakagami, K., Kusano, K., Yamamoto, K., Yoshikura, H., and Kobayashi, I. (1997). Genetic recombination through double-strand break repair: shift from two-progeny mode to one-progeny mode by heterologous inserts. Genetics *146*, 9–26.

Takeda, K., and Uetake, H. (1973). In vitro interaction between phage and lipopolysaccharide: a novel glycosidase associated with *Salmonella* phage ε15. Virology *52*, 148–159.

Tang, T.-H. (2005). Identification of novel non-coding RNAs as potential antisense regulators in the archaeon *Sulfolobus solfataricus*. Mol. Microbiol. *55*, 469–481.

Tang, T.-H., Bachellerie, J.-P., Rozhdestvensky, T., Bortolin, M.-L., Huber, H., Drungowski, M., Elge, T., Brosius, J., and Hüttenhofer, A. (2002). Identification of 86 candidates for small non-messenger RNAs from the archaeon *Archaeoglobus fulgidus*. Proc. Natl. Acad. Sci. U.S.A. *99*, 7536–7541.

Taylor, K. (1965). Enzymatic deacetylation of Vi-polysaccharide by Vi-phage. II. Biochem. Biophys. Res. Commun. *20*, 752–756.

Taylor, K. (1966). Physical and chemical changes of Vi-polysaccharide due to Vi-phage II action. Acta. Biochim. Polon. *13*, 79–106.

Templin, M.F., Ursinus, A., and Holtje, J.V. (1999). A defect in cell wall recycling triggers autolysis during the stationary growth phase of *Escherichia coli*. EMBO J. *18*, 4108–4117.

Tyndall, C., Meister, J., and Bickle, T.A. (1994). The *Escherichia coli prr* region encodes a functional type IC DNA restriction system closely integrated with an anticodon nuclease gene. J. Mol. Biol. *237*, 266–274.

Tyson, G.W., and Banfield, J.F. (2007). Rapidly evolving CRISPRs implicated in acquired resistance of microorganisms to viruses. Environ. Microbiol. *10*, 200–207.

Twomey, D.P., De Urraza, P.J., McKay, L.L., and O'Sullivan, D.J. (2000). Characterization of AbiR, a novel multicomponent abortive infection mechanism encoded by plasmid pKR223 of *Lactococcus lactic* subsp. *lactis* KR2. Appl. Environ. Microbiol. *66*, 2647–2651.

Uhl, M.A., and Miller, J.F. (1996). Integration of multiple domains in a two-component sensor protein: the *Bordetella pertussis* BvgAS phosphorelay. EMBO J. *15*, 1028–1036.

Uhlén, M., Guss, B., Nilsson, B., Gatenbeck, S., Philipson, L., and Lindberg, M. (1984). Complete sequence of the staphylococcal gene encoding protein A. A gene evolved through multiple duplications. J. Biol. Chem. *259*, 1695–1702.

Vallee, B.L., and Auld, D.S. (1990). Zinc coordination, function, and structure of zinc enzymes and other proteins. Biochem. *29*, 5647–5659.

Van Valen, L. (1973). A new evolutionary law. Evol Theor. *1*, 1–30.

Ventura, M., Lee, J.-H., Canchaya, C., Zink, R., Leahy, S., Moreno-Munoz, J.A., O'Connell-Motherway, M., Higgins, D., Fitzgerald, G.F., O'Sullivan, D.J., and van Sinderen, D. (2005). Prophage-like elements in bifidobacteria: insights from genomics, transcription, integration, distribution, and phylogenetic analysis. Appl. Environ. Microbiol. *71*, 8692–8705.

Viswanathan, P., Murphy, K., Julien, B., Garza, A.G., and Kroos, L. (2007). Regulation of *dev*, an operon that includes genes essential for *Myxococcus xanthus* development and CRISPR-associated genes and repeats. J. Bacteriol. *189*, 3738–3750.

Vovis, G.F., and Lacks, S. (1977). Complementary action of restriction enzymes endo R-*Dpn*I and endo R-*Dpn*II on bacteriophage-f1 DNA. J. Mol. Biol. *115*, 525–538.

Wang, C., Villion, M., Semper, C., Coros, C., Moineau, S., and Zimmerly, S. (2011). A reverse transcriptase-related protein mediates phage resistance and polymerizes untemplated DNA *in vitro*. Nucleic Acid Res. *39*, 7620–7629.

Warren, R.A.J. (1980). Modified bases in bacteriophage DNAs. Annu. Rev. Microbiol. *34*, 137–158.

Wayne, R., and Neilands, J.B. (1975). Evidence for common binding sites for ferrichrome compounds and bacteriophage φ80 in the cell envelope of *Escherichia coli*. J. Bacteriol. *121*, 497–503.

Webb, J.L., King, G., Ternent, D., Titheradge, A.J., and Murray, N.E. (1996). Restriction by *Eco*KI is enhanced by co-operative interactions between target sequences and is dependent on DEAD box motifs. EMBO J. *15*, 2003–2009.

Wegmann, U., O'Connell-Motherway, M., Zomer, A., Buist, G., Shearman, C., Canchaya, C., Ventura, M., Goesmann, A., Gasson, M.J., Kuipers, O.P., *et al.* (2007). Complete genome sequence of the prototype lactic acid bacterium *Lactococcus lactis* subps. *cremoris* MG1363. J. Bacteriol. *189*, 3256–3270.

Westra, E.R., Pul, U., Heidrich, N., Jore, M.M., Lundgren, M., Stratmann, T., Wurm, R., Raine, A., Mescher, M., van Heereveld, L., *et al.* (2010). H-NS-mediated repression of CRISPR-based immunity in *Escherichia coli* K12 can be relieved by the transcription activator LeuO. Mol. Microbiol. *77*, 1380–1393.

Westra, E.R., Swarts, D.C., Staals, R.H., Jore, M.M., Brouns, S.J., and van der Oost, J. (2012). The CRISPRs, they are a-changin': how prokaryotes generate adaptive immunity. Annu. Rev. Genet. *46*, 311–339.

Whitfield, C., and Roberts, I.S. (1999). Structure, assembly, and regulation of expression of capsules in *E. coli*. Mol. Microbiol. *31*, 1307–1319.

Wiedenheft, B., Zhou, K., Jinek, M., Coyle, S.M., Ma, W., and Doudna, J.A. (2009). Structural basis for DNase activity of a conserved protein implicated in CRISPR-mediated genome defense. Structure. *17*, 904–912.

Wilkens, K., and Rüger, W. (1994). Transcription from early promoters. In Molecular Biology of Bacteriophage T4, J.D. Karam, ed. (Washington, DC, American Society for Microbiology), pp. 132–141.

Wilkinson, B.J., and Holmes, K.M. (1979). *Staphylococcus aureus* cell surface: capsule as a barrier to bacteriophage adsorption. Infect. Immun. *23*, 549–552.

Wilkinson, S.G. (1996). Bacterial lipopolysaccharides – themes and variations. Prog. Lipid. Res. *35*, 283–343.

Wilson, G.G., and Murray, N.M. (1991). Restriction and modification systems. Annu. Rev. Genet. *25*, 585–627.

Wong, T.Y., Preston, L.A., and Schiller, N.L. (2000). Alginate lyase: review of major sources and enzyme characteristics, structure–function analysis, biological roles, and applications. Annu. Rev. Microbiol. *54*, 289–340.

Yamamoto, S., Kiyokawa, K., Tanaka, K., Moriguchi, K., and Suzuki, K. (2009). Novel toxin–antitoxin system composed of serine protease and AAA-ATPase homologues determines the high level of stability and incompatibility of the tumor-inducing plasmid pTiC58. J. Bacteriol. *191*, 4656–4666.

Young, E.T., and Menard, R.C. (1975). Analysis of the template activity of bacteriophage T7 messenger RNAs during infection of male and female strains of *Escherichia coli*. J. Mol. Biol. *99*, 167–184.

Yu, Y.T., and Snyder, L. (1994). Translation elongation factor Tu cleaved by a phage-exclusion system. Proc. Natl. Acad. Sci. U.S.A. *91*, 802–806.

Zabeau, M., Friedman, S., Van montagu, M., and Schell, J. (1980). The *ral* gene of phage λ. I. Identification of a nonessential gene that modulates restriction and modification in *E. coli*. Mol. Gen. Genet. *179*, 63–73.

Zaleski, P., Wojciechowski, M., and Piekarowics, A. (2005). The role of Dam methylation in phase variation of *Haemophilus influenzae* genes involved in defence against phage infection. Microbiology *151*, 3361–3369.

Zegans, M.E., Wagner, J.C., Cady, K.C., Murphy, D.M., Hammond, J.H., and O'Toole, G.A. (2009). Interaction between bacteriophage DMS3 and host CRISPR region inhibits group behaviors of *Pseudomonas aeruginosa*. J. Bacteriol. *191*, 210–219.

Zhang, Y., Zhang, J., Hoeflich, K.P., Ikura, M., Quing, G., and Inouye, M. (2003). MazF cleaves cellular mRNAs specifically at ACA to block protein synthesis in *Escherichia coli*. Mol. Cell. *12*, 913–923.

Non-bactericidal Effects of Phages in Mammals

5

Krystyna Dąbrowska, Ryszard Międzybrodzki, Paulina Miernikiewicz, Grzegorz Figura, Marlena Klak and Andrzej Górski

Abstract

Bacteriophages, although unable to propagate in eukaryotic cells, may induce physiological effects in mammalian organisms. Phage impact on the immune system as well as phage interactions with its elements may decide on the final outcome of phage therapy, thus being of a great practical importance for medical applications of phages. The most spectacular but also expected effect of phages in living organisms is the induction of anti-phage antibodies. These have been showed to play a role in phage clearance. However, phages have been shown to be able to modify ROS and cytokine production in mammalian immunological cells. Phage interaction with mammalian cells can be, at least partially, mediated by direct adhesion of bacterial viruses to the cells. Phages are abundant parasites of symbiotic or pathogenic bacteria in animals and humans. Consequently, mammals have become an 'environment' for phages. This environment is a multi-factor system able to induce strong pressure. Phages are complex structures, potentially able to evolve new means of interacting with their environment. Apart from natural phages, engineered phage particles are gaining an important position in biotechnology and medicine. They can be applied as carriers for vaccines or other biologically active agents. These applications, together with phage therapy of bacterial infections, induce constantly growing interest in interactions of phages with mammalian systems.

Introduction

Interactions of phages with mammalian systems depend on effective contact of phage virions with mammalian organisms. Phage therapy as well as natural phages' abundant existence expose humans and animals to interaction with phages. Since bacterial viruses are estimated to be the most numerous entities in the world, this contact seems to be constant and intensive. However, are phages able to penetrate mammalian organisms and affect internal organs and systems directly?

Data on phage penetration in mammals are not uniform. While phages can be artificially introduced into a mammalian system via traditional application routes, preferably injections, data on phage penetration from the gastrointestinal tract are not clear. Some authors have reported that no phages were detected in animal blood after oral phage administration (Brüssow, 2011). However, others have reported the presence of orally applied bacteriophages in the blood and urinary tract of humans (Weber-Dabrowska *et al.*, 1987) and in mice (Keller and Engley, 1958; Hoffmann, 1965). These conflicting results suggest that phage ability to penetrate from the gastrointestinal tract into the blood and, as a result, into internal organs depends on phage particle physical properties, i.e. virion diameter, shape, surface charge value, etc. Additionally, it might be influenced by the host (mammal) physiology, including the species, which is related to diet and intestinal conditions, but also by an animal's health, age, feeding, etc. It is probable that a phage's fate in a mammalian intestine depends, regardless of other factors, on whether the phage is a newly introduced strain or part of the normal microbial flora. It seems that newly introduced phages tend to propagate and show a substantial titre in an intestine in contrast to phages being established components of microbial flora whose propagation is balanced and their presence in the blood less probable (Chibani-Chennoufi *et al.*, 2004, Letarov and Kulikov, 2009, Letarova *et al.*, 2011). Thus, it is difficult to give a general schema of 'natural' phage penetration into higher organisms, even though phages are known to be abundant and to constitute an important part of mammalian microbial flora (an extensive review on bacteriophages in human and animal body associated microbial communities has been presented recently by Letarov and Kulikov [2009]). A hypothesis on the physiological state of natural 'phagaemia' which results from constant phage presence and translocation from the gastrointestinal tract has recently been presented by Gorski *et al.* (2006).

Once introduced into the blood, phages seem to migrate immediately to internal organs, even crossing the blood–brain barrier and penetrating back to the intestine (as a result being present in faeces) or secreted by kidneys to urine (Chibani-Chennoufi *et al.*, 2004; Dabrowska *et al.*, 2005). Thus, penetration of natural phages as well as the very easy and immediate penetration of artificially introduced therapeutic phages results in direct contact of bacterial viruses with mammalian cells and tissues. It used to be generally accepted that phages have little effect on mammalian organisms because they are unable to infect eukaryotic cells. However, most bacteriophages are complex structures with a substantial number of capsid proteins and large genomes. These make phages multi-antigenic objects with considerable potential of interacting with the mammalian system. Even if they are not able to propagate in eukaryotic cells, they may induce physiological effects in mammalian organisms. These effects are of a great importance for proper design and development of phage therapy of bacterial infections, since not only antibacterial activity of phages may contribute to a final effect of the treatment. Phage impact on the immune system as well as phage interactions with its elements may decide on the final outcome of the therapy.

Phages and cell adhesion

The very first observation of phage adhesiveness towards some mammalian cells was described as early as 1940s by Bloch who showed that phages can accumulate in cancer tissue (Bloch, 1940). Not much later, Kantoch and Mordarski (1958) demonstrated that cancer cells bind phages both *in vitro* and *in vivo*. Wenger *et al.* (1978) suggested that phages may attach to the plasma membrane of lymphocytes. The main question that arises on the adhesive effects of phages is whether this phenomenon is specific or does it result merely from non-specific interactions of the capsid surface with cell membranes. However, in the reports on the very first studies of phages in cancer models there is no specification (name, taxonomy) of the phages used. This means that it is not possible to formulate further hypotheses on phage proteins involved in those interactions.

Recently, binding of phages to the mammalian cell membrane was shown by Dabrowska *et al.* (2004). This research was inspired by the early observations of Bloch and Kańtoch. In this work weak adhesion of T4 phage to malignant cells was revealed. Furthermore, a T4 phage mutant with increased adhesiveness to the cells was developed (high-affinity phage: HAP1). It was eight times more effective in binding to cells than the parental strain.

T4 phage and HAP1 phage were then investigated and compared in a murine model of intravenously injected melanoma. HAP1 preparation was significantly more active in suppressing melanoma colonies in murine lungs, the most probably because of its higher adhesion potential in comparison to the wild T4 strain (Dabrowska *et al.*, 2004; Dabrowska *et al.*, 2007). This was in line with early observations of Bloch (1944), who correlated phage accumulation in tumour with antitumour effects. Since adhesion processes as well as metastasis are related to cell migration, T4 and HAP1 phages were also investigated regarding their ability to interfere with cancer cells' migration in vitro. Migration assays were carried out on Matrigel, an extracellular matrix (ECM) substitute, or fibronectin. In both cases, phages were able to diminish cell migration, but no significant differences between T4 and HAP1 were observed (Dabrowska *et al.*, 2009).

Studies of HAP1 phage revealed that one of its genes coding for a head protein was damaged with a nonsense codon that prevented synthesis of a functional product. Interestingly, this damaged non-essential protein which was lacked by HAP1 was gpHoc, the most exposed decorative element of the phage head, containing immunoglobulin-like folds of eukaryotic origin in its structure (Bateman *et al.*, 1997; Dabrowska *et al.*, 2007; Fokine *et al.*, 2011). Elevated adhesion properties of Hoc-lacking phage are consistent with observations of Sathaliyawala *et al.* (2010), who reported that hoc$^-$soc$^-$ phages frequently aggregate, as visualized in cryo-electron microscopic images, apparently through interactions between capsids. This author proposed that Hoc might diminish the aggregation of phage particles in infected cells where the concentration of newly assembled phage can be quite high, or, oppositely, the Ig-like domains of gpHoc might weakly and non-specifically interact with surface carbohydrates and other molecules of bacteria. Adhesion to bacteria or biofilms might provide phage with a survival advantage (Sathaliyawala *et al.*, 2010). In these cases, phage adhesion to mammalian cells as well as its control by gpHoc would be a 'side-effect' resulting from phage control of some infection stages. Recent bioinformatic studies revealed that Ig-like domains are quite frequently present in phage genomes (Fraser *et al.*, 2006), which may suggest that a similar adhesion-modulating effect of some capsid proteins can be expected in other phage strains.

Increased activity of the phage lacking gpHoc supports the hypothesis that other capsid elements can adhere to and affect mammalian cells, possibly by specific means of interaction. Hoc protein is located in the phage head; thus other head proteins are probably involved. The question on the specificity of the observed phenomenon is still open, since non-specific adhesion ability can also be differentiated between capsid proteins, i.e. some of them can be more 'sticky' than others. However, differences in adhesion capability between T4 phage and HAP1 phage supported with *in silico* analysis of T4 capsid proteins inspired a hypothesis on potential specific motifs able to mediate the effect (Gorski *et al.*, 2003). Potential activity of the KGD motif (a homologue of RGD) that is present in gp24 (head vertex protein) of T4 phage was proposed. The motif is well exposed at the phage head as revealed by the gp24 structure data (Protein Data Bank ID: 1YUE). KGD is known to interact with beta-3 integrins. The hypothesis was partially supported by observations that phage adhesion was decreased by typical beta-3 integrin inhibitors such as eptifibatide or antibodies (Dabrowska, 2004).

Potential interaction of phage KGD sequence with cell beta-3 integrins may explain both adhesive properties of phages and antimetastatic effects. Beta-3 integrins are a class of adhesion receptors which mediate cell–cell and cell-extracellular matrix (ECM) interactions. Thus they play a role in a number of processes, including maintaining tissue integrity, cellular migration, cell survival, adhesion and differentiation. Disintegrins able to bind this class of molecules were reported to interfere with tumour growth, progression, and metastasis (Adair and Yeager, 2002; Giancotti and Ruoslahti, 1999). Thus, gp24 was proposed as the protein active in phage adhesion to mammalian cells and in cancer processes (Dabrowska *et al.*, 2004). However, final verification of gp24 interactions with integrins is still to be done.

Overview of main aspects of phage antigenicity

The most spectacular but also expected effect of phages in living organisms is the induction of anti-phage antibodies. Bacteriophages are complex structures characterized by antigenic diversity of the whole phage particles. Therefore they can be specifically recognized by more than one type of antibody (Plancon *et al.*, 2002). Anti-phage antibodies were employed in medicine as a test for immune competence of immunodeficient patients. For example, HIV-infected patients can be examined through phiX174 phage immunization for the ability of antibody production, amplification, and isotype switching (Fogelman, 2000). Similar assays have been successfully used in the diagnostics of a number of other primary and secondary immunodeficiencies (Górski *et al.*, 2007).

Phage immunogenicity was one of the very first means for phage classification about five decades ago. Phage inactivation with anti-phage sera enabled searching for cross-reaction of sera with related phages (heterologous inactivation).

Intensiveness of cross-reaction depends on the relation between phages, generally decreasing with growing evolutionary distance between species (Stent, 1963). For example, among T-even phages, four serological groups were found. Within these groups cross-reactions were observed while serum obtained by immunization with one group of bacteriophages was inactive on another group (Kantoch, 1956). Cross-reactions of anti-phage sera can be explained by similarities between phage capsid proteins and thus between capsid antigens. These similarities result from homologies that now can be observed in phage genomes. The arrangement of antigens on capsids is not even, which is in line with reported differentiated arrangement of the regions of homology in bacteriophage genomes (Krisch and Comeau, 2008). Some capsid components are antigenically specific for a phage, while others can be shared by different phage species (Tikhonenko, 1976). Antibodies against selected phage gene products were also one of the main tools for recognition of gene function in phages. Structural proteins can be localized in capsids by means of specific antibodies giving accurate and exact information on a virus particle arrangement (Ishii and Yanagida, 1975). Neutralization of bacteriophages with antisera was demonstrated to be time and dose dependent (Stent, 1963). Interestingly, anti-phage antibodies can be used as diagnostic tools of bacterial infections since their titre increases during the time of infection. As expected, elevated number of bacteria is correlated with appearance of specific phages.

Research on engineered phages can be helpful in studies of phage–antibody interactions. Although the results of these studies cannot always be simply transferred to wild phages, they give information about molecular details that possibly determine phage effects on immunity. T7 bacteriophage displaying short peptides were studied by Sokoloff *et al.* (2000), who reported a relationship between phage surface peptide properties and natural (i.e. preexisting) antibody response. Carboxy-terminal lysine or arginine residues introduced on a phage capsid protect the phage against inactivation in rats (Sokoloff *et al.*, 2000).

Neutralization of bacteriophages by anti-phage antibodies has been postulated to be a limiting factor for phage therapy effectiveness (Carlton, 1999; Górski *et al.*, 2012). Studies on antibodies neutralizing bacteriophages were also performed by Smith *et al.* (1987), who showed that pre-immunization of mice with a phage may suppress its antibacterial action and reduce the therapeutic effect. Specific antibodies were detected in human sera both following phage treatment as well as in non-treated patients, i.e. they had to result from natural contact of a patient with a phage or with closely related ones. Probably the first observation of natural anti-phage antibodies in human sera was reported by Kamme for staphylococcal phages (Kamme, 1973). Later, Kucharewicz-Krukowska and Ślopek (1987) observed natural antibodies in 13 out of 57 patients (22.8%) before initiation of phage therapy. As a result of phage therapy, antibody titre on the 10th day of therapy increased in 56% of examined sera, regardless of the presence of antibodies against a definite bacteriophage in patient's serum before the treatment; the results were obtained in a differentiated phage set (Kucharewicz-Krukowska and Ślopek, 1987). Anti-phage antibodies seem to be the most expected and well-known effect of phages in mammals.

A spectacular sign of interaction between phages and the immune system is phage removal from an organism. In pre-immunized individuals, this is probably mainly due to the action of antibodies. However, even in naïve animals clearance is usually quite rapid, and it is known to be dependent on phage capsid molecular properties. The general mechanism of phage clearance is based on reticulo-endothelial system (RES) filtration (Geier *et al.*, 1973). Probably the very first report on molecular motifs regulating phage clearance from mammals was presented by Merril *et al.* (1996), who investigated the clearance of phage lambda derivatives in naïve germ-free mice. A point mutation in the major capsid protein gene (G→A transition) caused a substitution of the amino acid lysine for glutaminic acid, resulting in a long-circulating phenotype. The long-circulating mutants were able to persist in mice circulation much longer and they also had higher anti-bacterial activity in bacteraemic mice than the corresponding parental strain (Merril *et al.*, 1996). A 'short circulating' phage was also described. It was a T4 phage mutant which lost its non-essential head protein Hoc, which resulted in rapid removal of the phage from murine circulation (Dabrowska *et al.*, 2007). Chemically modified phage M13 was also investigated in terms of its time of clearance. Molenaar *et al.* (2002) showed that either conjugation of galactose or succinic acid groups to phage coat proteins resulted in substantially reduced plasma half-life of the phage. These reports show that also phage interactions with innate immunity in mammals depend on molecular properties of a phage and that there is a potential of their modification and design. On the other hand, phage display of random peptides (C-X7-C library) on the T7 phage coat was shown not to influence the clearance of phages from murine blood (Srivastava *et al.*, 2004).

The influence of phages on cytokine production

Cytokines are members of a large and diverse family of proteins secreted by specific cells of the immune system. These small, pleiotropic molecules mediate and regulate the immune responses affecting many cell functions, including proliferation, apoptosis, and differentiation. An analysis of the cytokine profile in serum may describe the immunological status of an organism. Cytokines adjust the host response to infection, inflammation and trauma. Some of them are pro-inflammatory, e.g. interleukin 1 (IL-1), IL-6 and tumour necrosis factor alpha (TNF-α), whereas others (anti–inflammatory cytokines) suppress inflammation (Dinarello, 2000).

Data on phage effects on cytokine production are quite scanty. However, recent findings suggest that bacteriophages may inhibit production of IL-6 and TNF-α *in vitro* in peripheral blood mononuclear cells (PBMCs). Korczak-Kowalska *et al.* (2010) showed that purified preparation of T4 phage did not cause increased secretion of IL-6 and TNF-α from resting PBMCs. Pseudomonas phage PAK-P1 did not affect the production of these proinflammatory cytokines in uninfected BALB/c mice, either. Moreover, it reduced IL-6 and TNF-α level in the lungs of infected mice (Debarbieux *et al.*, 2010). T4 phage may also inhibit interleukin (IL)–2 and to some extent interferon gamma (IFN-γ) production by human leucocytes (Przerwa *et al.*, 2005). IFN-γ exhibits pleiotropic features: it has antiviral activity (like INF-α and INF-β) and induces the production of TNF-α and nitric oxide (NO). Both IL-2 and IFN-γ increase macrophage activation. These data indicate that administration of T4 bacteriophage preparations does not lead to inflammatory reactions, at least not in a Il-2- or TNF-α-dependent way. However, it was observed that purified M13 phages (without LPS contamination) may induce secretion of IFN-γ and IL-12 in mouse splenocyte cultures. These Th1 cytokines can affect the antitumour immune responses (Eriksson *et al.*, 2006). Moreover, Eriksson *et al.* (2009) reported that M13 phages are also considered as potential antiviral agents since they demonstrate the ability to increase IFN production. Studies of phages' influence on cytokine production seem to be crucial for the ongoing discussion on the safety of phage therapy and can serve as an important indicator of phage-mammalian immune system interaction.

Highly purified T4 phage capsid proteins were also investigated in terms of their influence on cytokine production. Interestingly, gpHoc, one of the head unessential proteins, reduced the level of lymphotactin *in vivo* in murine blood (Paulina Miernikiewicz, unpublished observations). This cytokine is a key regulator of lymphocyte trafficking during acute graft rejection (Wang *et al.*, 1998), thus suggesting the potential ability of the phage to moderate the immunological reaction by modification of this cytokine's secretion. Importantly, this effect may be mediated by a particular protein type (gpHoc), which is present in many copies on the phage head, especially well exposed to external interactions and non-essential for the phage infection cycle. Hoc protein is also distinguished by its characteristic structure, especially the presence of immunoglobulin-like domains, which supports the hypothesis of Hoc's capability to interact with mammalian cells (Bateman *et al.*, 1997; Dabrowska *et al.*, 2006; Fokine *et al.*, 2011).

Phage influence on phagocytosis and oxidative burst

Due to the fundamental role of phagocytosis in antibacterial defence, studies of potential phage–phagocyte interactions and the influence of phages on the elimination of bacteria by phagocytes are relevant to phage therapy. Félix d'Hérelle (1922) thought that phages act as specific opsonins markedly facilitating bacterial phagocytosis. He observed an increase in phagocytic index when *Shigella* phage were added to the suspension of phagocytes incubated with *Shigella*. However, the reports of Kańtoch and coworkers showed no phage effect on phagocytosis in the case of T5 phage (Kańtoch and Szalaty, 1960), or even an inhibitory effect in the case of T2 phage (Kańtoch *et al.*, 1958). Similarly, in a study published by Przerwa *et al.* (2006), both T4 and *Pseudomonas* F8 phages preincubated *in vitro* with neutrophils or monocytes significantly inhibited the phagocytosis of live *E. coli* irrespective of homologous (T4) or heterologous (F8) relation between the phage and bacteria in a dose-dependent manner (interestingly, the preincubation of bacteria with a homologous phage at a lower dose slightly increased the phagocytic index in neutrophils). However, *in vivo* experiments did not confirm that T4 could significantly influence phagocytic activity of neutrophils or monocytes in healthy or *E. coli* infected mice. Whether these interactions may have clinical implications still needs evaluation (Weber-Dąbrowska *et al.*, 2002).

It is well known that pathogenic mammalian viruses stimulate phagocytes to excessive production of reactive oxygen species (ROS) which may exert harmful effects on mammalian cells (Vlahos *et al.*, 2011; Gonzalez-Dosal *et al.*, 2011; Akaike, 2001). However, *in vitro* studies of the direct interactions of phagocytes with bacterial viruses reveal that phages are much weaker inducers of the oxidative burst than pathogenic viruses at comparable titres (Przerwa *et al.*, 2006). In another study, the effects of four different phage preparations on the oxidative burst of monocytes and neutrophils were evaluated. This study showed that only purified preparation of T4 phage induces a relatively weak oxidative burst in phagocytic cells, while neither purified preparation of the staphylococcal phage A3/R nor lysates of T4 and A3/R phages significantly stimulate the production of ROS (Borysowski *et al.*, 2010). A study by Borysowski *et al.* (2008) revealed that purified T4 phage does not significantly induce *in vivo* the oxidative burst in peripheral blood neutrophils of mice when administered intraperitoneally.

Interestingly, Przerwa *et al.* (2006) observed that production of ROS by phagocytes (both monocytes and neutrophils) stimulated with viable bacteria (*E. coli B*) preincubated with their homologous phage T4 was much weaker than ROS production by phagocytes stimulated only with bacteria, and it was inversely dependent on the dose of T4. When heterologous *Pseudomonas* phage F-8 was used instead of T4 this effect was not observed. Międzybrodzki *et al.* (2008) showed that T4 phage could inhibit ROS production by neutrophils stimulated with both homologous and heterologous bacteria (*E. coli* B and *E. coli* R4, respectively) and with endotoxins derived from these bacteria as well. This study showed that phage inhibition of ROS production resulted not only from the phage-mediated bacterial lysis and reduction of the number of bacteria in the suspension used for stimulation, but also from interactions between phages and phagocytic cells. Moreover, the experiments in which lipopolysaccharides (LPSs) isolated from homologous and heterologous *E. coli* strains were used for neutrophil stimulation also revealed a significant inhibition of the oxidative burst by T4 phage in a dose-dependent way. Further experiments, in which phorbol 12-myristate 13-acetate (PMA) or different doses of LPSs were used for stimulation, suggested that this phenomenon results not only from simple LPS binding by phage but also from the interactions between phage and neutrophils and may depend on the kind of activator (the phage did not influence the oxidative burst induced by PMA). Taking into consideration that F-8 phage did not influence the oxidative burst in phagocytes (Przerwa *et al.*, 2006), one cannot exclude that inhibition of ROS production by neutrophils stimulated with heterologous bacteria or LPS may be specific for the T4 phage.

These observations are of great importance for phage therapy because some authors were concerned about potential toxic effects of ROS release by phagocytes (for example endotoxic shock) in response to excessive bacterial lysis during therapeutic phage application (Dixon, 2004). One may hypothesize that mammalian evolution would result in a mechanism which during bacterial infection would diminish production of ROS (which are essential components of the innate immune response against bacteria but may also be toxic for mammalian cells) when a sufficiently high titre of the phage is detected (as a result of effective phage propagation in pathogenic bacteria). This could protect mammalian cells from cytotoxic effects of ROS when other effective mechanisms of antibacterial defence are triggered in the body. Taken together, these data seem to support the hypothesis of a possible protective role of endogenous phages (Górski and Weber-Dabrowska, 2005).

Engineered phages as vaccines

Besides effects that can be induced by wild-type phages, often rather weak and not well recognized, phages can be employed as effective tools in many biotechnology disciplines. These are for example platforms presenting proteins/peptides or DNA carriers (phage vectors). Their effects are specific, according to a designed direction and therefore often very intensive. They often result from changes in bacteriophage capsid arrangement and components, but not only from these. As a result of certain modifications, bacteriophages can gain new possibilities to interact with mammalian organisms.

A very promising application of engineered phages is their use in the new vaccine technologies. Modified bacteriophages can be used as vehicles for both antigens and antigen-encoding DNA (Gao *et al.*, 2010; Clark and March, 2004a).

Immunization/vaccination with a specific antigen employs phage display technology. It relies on a phage presenting antigen proteins or peptides on its surface. This is achieved by modification of the phage genome; the sequence coding for a protein to be presented is fused with the sequence of a selected gene coding for a phage capsid protein. During infection and phage replication in bacteria, fusion proteins are expressed (phage protein in combination with the introduced protein/peptide which is an antigen), then incorporated into the capsid and presented on its surface (Mutuberria *et al.*, 2001; Manoutcharian, 2005). Once engineered, phages are easy and cheap to produce. They are much more stable than antigens alone. They also seem to be effective adjuvants, since some antigens presented on phage capsids were much more immunogenic than antigens applied as purified preparations alone (Synnott *et al.*, 2009). After entering the blood stream they undergo phagocytosis by antigen-presenting cells, and are removed by the spleen and Kupffer liver cells. A significant advantage of phages is that, as antigen particles, they can be presented by both MHC class I and MHC class II molecules; thus they can stimulate the cellular and humoral immune response (Manoutcharian, 2005; Clark and March, 2006).

There are a few kinds of phage display system. The classification is based on which bacteriophage is used for protein presentation. The four main systems rely on M13, lambda, T4 and T7 bacteriophages (Gao *et al.*, 2010). M13 is a filamentous phage, in its system the most often used protein is p3, which is present in 3–5 copies per phage particle. Proteins as big as 100 kDa can be linked to this element of the phage capsid. P8, the phage M13 main capsid protein, can also be used as the fusion basis. Because of its high number of copies (2670 in one virion) and tight assembly, only peptides not longer than six amino acids can be presented with this protein. However, presentation of some small peptides as high-copy capsid elements can be an advantage, so this protein may also be a convenient tool of peptide presentation (Benhar, 2001; Smith, 1985; Gao *et al.*, 2010).

In the lambda phage system, head protein D or tail protein pV is used for presentation. However, in the case of presentation on pV protein not only fewer copies of fusion protein can be created, but phage yield is also decreased. It is assumed that this results from disruption of tail formation during phage assembly. In comparison with M13 phage the advantage of the system based on lambda phage is that translocation of proteins through *Escherichia coli* membrane is not required. Moreover, lambda phage capsid is bigger, so larger and more complex proteins can be displayed

(Maruyama *et al.*, 1994; Gao *et al.*, 2010; Cicchini *et al.*, 2002).

In the system based on T4 phage, gpSoc or gpHoc proteins are used for phage-display technology. The main advantage of the system is a possibility of presenting large proteins in a high copy number. Furthermore, presence of two decorating non-essential proteins on the T4 head allowed a bipartite system. This system enables display of proteins simultaneously on gpSoc and gpHoc. It may be used for multiple presentation of a single antigen, but a more interesting possibility is simultaneous presentation of two different antigens on the same capsid (Wu *et al.*, 2007). This may be crucial for effective vaccination and for cooperative immunization with more than one antigen at the same time.

In the T7 system, the presentation technology employs a protein that exists in two different forms: 10A and 10B. Form 10B is produced from translation frame shift of protein 10A and makes up 10% of T7 capsid (Condron *et al.*, 1991). Small proteins (less than 50 amino acids) are displayed in this system in a high copy number, while larger proteins are displayed in a medium or small amount, which allows regulation, to some extent, of the final yield of the targeted antigen. T7 virions are also very stable, which is an important advantage of the system.

Each system has its own advantages and drawbacks, but the drawback of all of them is lack of post-translation modifications occurring in eukaryotic cells (Gao *et al.*, 2010). Therefore effective immunization with a eukaryotic antigen (e.g. a parasite antigen) may be difficult. Nevertheless, phage display systems seem to be potent tools for antigen presentation and successful vaccination. They have been shown to be effective, e.g. as platforms for anthrax and HIV vaccines (Shivachandra *et al.*, 2007; Sathaliyawala *et al.*, 2006).

Immunization with DNA relies on introduction of an expression cassette which controls the expression of an antigen-coding gene in antigen-presenting cells. The 'vaccinating gene' must be cloned into a phage genome under control of a eukaryotic promoter. After entering the blood stream phage particles are phagocytosed by antigen-presenting cells thus delivering 'vaccinating genes' to them. Next the genes undergo expression. Antigens produced this way are subsequently presented on cell surfaces and they were shown to be able to induce both humoral and cellular immune responses, a capacity which makes phages an effective vaccine carrier (Clark and March, 2004a,b).

The advantages of using bacteriophages as vaccine carriers in case of immunization by both DNA and a recombined antigen include lower costs, ease of production on a large scale and their high stability. Although phages are immunogenic, anti-phage antibodies do not interfere with the immune response against a targeted antigen of the phage vaccine (Manoutcharian, 2005; Gao *et al.*, 2010). Application of bacteriophages as carriers of an expression cassette with antigen-coding genes (DNA vaccines) is easier than application of other carriers/techniques, which would allow DNA to be introduced into eukaryotic cells. It is also assumed that bacteriophage carriers are relatively safe, since phages are known to be unable to infect or proliferate in eukaryotic cells (Clark and March, 2004a; Stone, 2002).

Recently a number of studies on vaccines based on modified bacteriophages were carried out. A good example of these studies is the work on a vaccine for infectious bursal disease virus (IBDV). In the late 1980s, a very virulent form of this virus (vvIBDV) emerged, which made control of infectious bursal disease even more difficult than it was before. IBDV is a virus infecting chickens, which causes infectious bursal disease and other diseases as a result of lack of immune defence. In the phage-vaccine approach, bacteriophages were used because other vaccines (for example, those produced in bacterial, insect or yeast system, or a vaccine based on attenuated virus) were expensive, hard to obtain, or potentially hazardous. The major immunogenic protein of vvIBDV is VP2, so it was chosen for presentation on bacteriophage T4 gpSoc protein. This system was applied because it allows for presentation of larger proteins in a high copy number. Effectiveness of the obtained vaccine was tested *in vivo*. Young chickens were immunized for eight weeks. Four weeks after, birds were challenged with vvIBDV and observed for seven days. All animals which received bacteriophage vaccine survived and did not show any symptoms of the disease, in contrast to control groups with visible disease signs and some deaths. This study showed that bacteriophages displaying VP2 protein as vaccine can be safe and able to provide effective protection against vvIBDV. Further advantages of this vaccine are its cheap production, convenient transport and storage, and ease of application (Cao *et al.*, 2005).

Other important studies were performed on cancer vaccine. Immunization of mice was carried out by delivery of an expression cassette, in which E7 protein from HPV-16 was placed under control of the cytomegalovirus (CMV) promoter. HPV-16 is a virus which causes most of the cervical cancer cases. Its proteins E6 and E7 are oncogenic and have an important function in inducing and maintaining cell transformation. They are expressed on most cancer cells containing HPV. Bacteriophage lambda was used as a carrier for expression cassette delivery. The phage was chosen because of its genetic flexibility, high stability, and low costs of production and purification. As the cancer model, the tumour cell line TC-1was used. The cells were transformed so they expressed E7 protein on their surfaces. Mice were administered with cancer cells and after 1 week they were immunized with the modified lambda phage. Mice administered with modified bacteriophage showed significantly reduced tumour volume in comparison with the control (Ghaemi *et al.*, 2010).

Application of bacteriophages as vaccines undoubtedly has many advantages over commonly used methods of immunization. However, more studies are needed before widespread use of such vaccines is possible.

Phages and pathogenic viruses of mammals

Shortly after the discovery of phages they were widely used not only for treatment of bacterial infections but also other conditions which later were recognized as unjustified (Barrow and Soothill, 1997). The use of phages for treatment of viral infections was one of them; however, there are some experimental data which support this application (Międzybrodzki *et al.*, 2005).

Centifanto (1965) observed that lambda phage lysate of *Escherichia coli* K-12 was active against herpes simplex virus (HSV) and vaccinia virus in a plaque inhibition assay on chick embryo monolayer cultures (*in vitro*), and in a herpetic corneal ulcer model in rabbit (*in vivo*). The phage lysate did not induce interferon (IFN) in the cells and extract from uninfected bacteria was not active. This activity was not attributed to the infective lambda particles; an antiviral agent (named by the author 'phagicin') was rather associated with phage replication and it could be obtained by disruption of phage particles as well (Centifanto, 1968). Phagicin could not inhibit the activity of some other viruses, being specific against HSV and vaccinia virus. It was sensitive to trypsin and pepsin but its activity was not inhibited after treatment with deoxyribonuclease, ribonuclease, and ultraviolet irradiation; therefore, Centifanto hypothesized that phagicin might be a phage internal protein which interferes with the intracellular replication of viral DNA. These results were supported by Meek and Takahashi (1968), who also showed that phagicin inhibited the synthesis of viral DNA but not the host DNA. However, Brown and Cohen (1973) failed to isolate phagicin from lambda phage lysates of *E. coli*.

The antiviral action of purified lambda phage was reported by Merril (1977). The phage showed a protective anti-HSV and anti-vesicular stomatitis virus (VSV) action *in vitro* on a culture of chick embryo fibroblasts infected with viruses. The same effect was reported by Kleinschmidt *et al.* (1970) for *E. coli* T4 phage, which inhibited VSV in monolayers of primary rabbit kidney cells. Keyhani (1969) performed a protective experiment *in vivo* in which phage lysate of *Brucella abortus* applied orally or orally and intramuscularly successfully rescued chickens after challenge with a virulent Newcastle disease virus (over fivefold increase in survival in comparison with the control group).

It is known that nucleic acids of animal viruses are IFN inducers. Similarly, some antiviral activity of phages has been attributed in part to interferon (IFN) induction by phage-derived nucleic acids. Kleinschmidt and colleagues showed that whole T4 phage particles administered intravenously to mice were capable of inducing IFN in blood, while T4 ghosts and extracted T4 DNA failed (Kleinschmidt *et al.*, 1970). The authors came to the conclusion that DNA was the T4 molecule responsible for the induction of IFN and that the encapsulation of the DNA in the intact phage particle enabled the delivery of the DNA to the inducer recognition site. An argument for viral character of T4 phage preparation-induced IFN production was the stronger and more prolonged IFN induction by T4 when compared to the effect exerted by endotoxin used in a dose comparable to the endotoxin concentration in the phage preparation. Iizuka *et al.* (1994) studied the activity of single-stranded DNA (ssDNA) extracted from M13 phage in a model of duck hepatitis B virus (DHBV) infection. He observed that daily intravenous injections of nucleic acid significantly decreased the cumulative level of serum DHBV DNA, and reduced the viral DNA level in the livers of infected animals. Although M13 ssDNA was superior to acyclovir, the treatment had to be continued for prolongation of the antiviral effect. Antiviral activity of M13 ssDNA was confirmed by Mori *et al.* (1996), who showed that its intravenous administration one day before vaccinia virus infection could significantly reduce the formation of tail lesions caused by virus in mice. They also observed raised IFN (mainly IFN-β) levels in mouse serum after injection of the phage DNA. Antiviral activity of double-stranded RNA (dsRNA) was also observed (Vales *et al.*, 1991). Specifically, it was shown that dsRNA from φ6 bacteriophage was able to inhibit HSV type 1 (HSV-1) and HSV type 2 (HSV-2) infection in fetal rhesus monkey kidney cells (MA104) *in vitro* but this activity was cell specific because it could not protect African green monkey kidney (Vero) cells against cytopathic effect of the virus. The φ6 phage dsRNA was also used *in vivo* for treatment of guinea pigs with genital HSV-2 infection. Its single intravaginal dose given five days after virus challenge completely cured animals while in the control group 39% of animals died. The systemic (intraperitoneal) administration of nucleic acid was less effective. In the study of Feldmane *et al.* (1977), dsRNA extracted from the amB_{11} mutant of the R17 *E. coli* phage was shown to induce IFN in mice after intravenous injection. It could also rescue mice from death due to infection with VSV, encephalomyocarditis, and Moloney mice sarcoma viruses.

The efficacy of the clinical application of the native phage nucleic acid in the treatment of viral infections such as dermatoses and viral eye diseases (herpes simplex recidivans, herpes zoster, male genital keratoconjunctivitis herpetica, verruca vulgaris and conjunctivitis lignose) was investigated by Borecky *et al.* (1978). Over 200 patients were treated (no longer than 20 days) in the open study with ointment containing dsRNA derived from a suppressor-sensitive (amber) mutant of f2 coliphage. Although the treatment was harmless to the patients, caused a rapid feeling of relief in the majority of cases, and shortened the duration of the illness, the study could not provide a clear answer as to its therapeutic value due to great variations in the severity of the dermatoses, difficult control of the therapeutic regimen, and the insufficient number of patients for statistical evaluation. Interestingly, the authors also presented results of a double blind study involving HSV-infected patients treated topically with f2-RNA for 10 days (12 subjects in the study group, and 25 subjects receiving placebo). The treatment proved to be significantly effective in 94.1% of cases compared with 25% in the control group ($P < 0.001$).

A hypothesis of bacteriophage–virus interactions through specific cellular receptors (e.g integrins) was presented by Gorski *et al.* (2003). Hartley *et al.* (2003), using a library of phage-displayed RANTES mutants, isolated a phage clone that could inhibit infection of peripheral blood mononuclear cells by the R5-tropic strains of HIV type 1 (HIV-1) *in vitro*. Krichevsky *et al.* (2005) demonstrated that filamentous fd bacteriophages that are used to construct phage-display peptide libraries recognize the nuclear localization signal (NLS) sequence of the HIV-1 Tat protein. The Tat-NLS-mediated nuclear import was specifically inhibited by fd phage, and a peptide bearing the first 20 amino acids of the p8 protein (NTP8) inhibited HIV-1 propagation in cultured cells.

Remarks on bacterial remains in phage preparations

Immunological studies of bacteriophage effects on mammalian immunity require removal of bacterial 'debris' released during bacterial cell lysis to the phage lysate. Immunogenic and stimulating components of bacterial cells include LPS, bacterial proteins, bacterial DNA (non-methylated CpG), and peptidoglycan. They are known to contain conserved pathogen-associated molecular patterns (PAMPs) which interact with specific classes of pattern recognition receptors. An important class of these receptors are Toll-like receptors (TLRs) which mediate the innate immune responses to invading pathogens by recognizing PAMPs. Innate immune cells in turn activate adaptive immune cells, thus inducing a general immunological response (Declue *et al.*, 2011; Hayashi *et al.*, 2011). Since phage lysates contain bacterial remains some studies of phages' effects in mammals should be interpreted with caution. Therefore purification of bacteriophages has become one of the most important aspects of studies on phage–mammal interactions.

The traditional approach in phage purification, based on caesium chloride gradient centrifugation, has been recently supported by new techniques. Most of them involve chromatography, including size-exclusion chromatography, as well as ion-exchange or affinity chromatography. Novel methodologies apply hydrophobic sulfur resins, monolith columns, modifications of phage capsid by phage-display technique, or bacteriophage-derived proteins of high affinity to LPS (Boratyński *et al.*, 2003; Hyglos GmbH, 2011; Kramberger *et al.*, 2011; Oślizło *et al.*, 2011). The variety of phage strains and extreme diversity of phage molecular characteristics necessitates diversification of phage purification methods, since a method effective for one phage type may be ineffective for another virus. Although new purification procedures have been recently developed, it is possible that phages are capable of interacting with and binding some bacterial components. These interactions may include both specific binding and unspecific reactions, resulting in substantial difficulties in the removal of impurities. Additionally, aggressive methods of purification cause difficulties with some phage strains' stability. Therefore in many cases there is still only the possibility of substantial decrease of the impurities content rather than complete removal (Olchawa *et al.*, 2010).

Conclusions

The number of novel applications of phages is constantly growing. All biomedical applications – phage therapy, engineered vaccine technologies, as well as food processing applications – require that the highest safety standards be met. This means that uncertainty about the impact of phage on the human body must be dispelled. At first, it was generally accepted that phages have little effect on mammalian organisms because they are unable to infect eukaryotic cells. Antigenic properties of phage proteins have been acknowledged and studied, but data on other possible effects of phage on human or animal bodies are very scanty and remain to be thoroughly investigated. Recent studies of phage effects on some physiological processes showed that phages are not neutral for mammalian organisms. Phages are abundant parasites of symbiotic or pathogenic bacteria in animals and humans. Consequently, mammals have become an 'environment' for phages. This environment is a multi-factor system able to induce strong pressure. Phages are complex structures, potentially able to evolve new means of interacting with their environment. If so, these effects should be extensively studied, taking into account separation of phage and bacterial effects, and differences in the effects induced by individual phage groups, since phages are probably not only the most widespread entities in the world, but also the most differentiated ones.

References

Adair, B.D., and Yeager, M. (2002). Three-dimensional model of the human platelet integrin $\alpha_{IIb}\beta_3$ based on electron cryomicroscopy and x-ray crystallography. Proc. Natl. Acad. Sci. U.S.A. *99*, 14059–14064.

Akaike, T. (2001). Role of free radicals in viral pathogenesis and mutation. Rev. Med. Virol. *11*, 87–101.

Barrow, P.A., and Soothill, J.S. (1997). Bacteriophage therapy and prophylaxis: rediscovery and renewed assessment of potential. Trends Microbiol. *5*, 268–271.

Bateman, A., Eddy, S.R., and Mesyanzhinov, V.V. (1997). A member of the immunoglobulin superfamily in bacteriophage T4. Virus Genes *14*, 163–165.

Benhar, I. (2001). Biotechnological applications of phage and cell display. Biotechnol. Adv. *19*, 1–33.

Bloch, H. (1940). Experimental investigation on the relationships between bacteriophages and malignant tumors. Arch. Virol. (in German) *1*, 481–496.

Boratyński, J., Syper, D., Weber-Dabrowska, B., Łusiak-Szelachowska, M., Poźniak, G., and Górski, A. (2004).

Preparation of endotoxin-free bacteriophages. Cell. Mol. Biol. Lett. *9*, 253–259.

Borecky, L., Buchvald, J., Adlerova, E., Stodola, I., Obrucnikova, E., Gruntova, Z., Lackovic, V., and Doskocil, J. (1978). Results of a five-year study of the curative effect of double stranded ribonucleic acid in viral dermatoses and eye diseases. Adv. Exp. Med Biol. *110*, 175–191.

Borysowski, J., Wierzbicki, P., Kłosowska, D., Przerwa, A., and Gorski, A. (2008). T4 phage does not induce a strong oxidative burst in peripheral blood neutrophils in mice [abstract]. Edinburgh International Phage Conference, 26–29 July, 2008, Edinburgh, UK, Abstract book p. 75.

Borysowski J., Wierzbicki, P., Kłosowska, D., Korczak-Kowalska G., Weber-Dąbrowska B., and Górski, A. (2010). The effects of T4 and A3/R phage preparations on whole-blood monocyte and neutrophil respiratory burst. Viral Immunol. *23*, 541–544.

Brown, A., and Cohen, S.N. (1973). Lack of specific inhibition of Vaccinia plaque formation by bacteriophage lambda products. Infect. Immun. *7*, 862–864.

Brussow, H. (2011). Safety evaluations of a T4 *Escherichia coli* phage cocktail. 1st Oxford International Bacteriophage Conference Phage2011, 19–21 September, 2011, St Hilda's College, Oxford, UK, Abstract book p. 24.

Bruttin, A., and Brüssow, H. (2005). Human volunteers receiving *Escherichia coli* phage T4 orally: a safety test of phage therapy. Antimicrob. Agents Chemother. *49*, 2874–2878.

Cao, Y.C., Shi, Q., C., Ma, J.Y., Xie, Q.M., and Bi, Y.Z. (2005). Vaccination against very virulent infectious bursal disease virus using recombinant T4 bacteriophage displaying viral protein VP2. Acta Biochim. Biophys. Sin. *37*, 657–664.

Carlton, R.M. (1999). Phage therapy: past history and future prospects. Arch. Immunol. Ther. Exp, *47*, 267–274.

Centifanto, Y.M. (1965). A therapeutic antiviral from infected E coli (phagicin). Proc. Soc. Exp. Biol. Med. *120*, 607–611.

Centifanto, Y.M. (1968). Antiviral agent from lambda-infected *Escherichia coli* K-12. I. Isolation. Appl. Microbiol. *16*, 827–834.

Chibani-Chennoufi, S., Sidoti, J., Bruttin, A., Dillmann, M.L., Kutter, E., Qadri, F., Sarker, S.A., and Brüssow, H. (2004). Isolation of *Escherichia coli* bacteriophages from the stool of pediatric diarrhea patients in Bangladesh. J. Bacteriol. *186, 24*, 8287–8294.

Cicchini, C., Ansuini, H., Amicone, L., Alonzi, T., Nicosia, A., Cortese, R., Tripodi, M., and Luzzago, A. (2002). Searching for DNA–protein interactions by lambda phage display. J. Mol. Biol. *322*, 697–706.

Clark, J.R., and March, J.B. (2004a). Bacteriophage-mediated nucleic acid immunization. FEMS Immunol. Med. Microbiol. *40*, 21–26.

Clark, J.R., and March, J.B. (2004b). Bacterial viruses as human vaccines? Expert Rev. Vaccines 3, 463–476.

Clark, J.R., and March, J.B. (2006). Bacteriophages and biotechnology: vaccines, gene therapy and antibacterials. Trends Biotechnol. *24*, 212–218.

Condron, B.G., Atkins, J.F., and Gesteland, R.F. (1991). Frameshifting in gene 10 of bacteriophage T7. J. Bacteriol. *173*, 6998–7003.

Dabrowska, K., Opolski, A., Wietrzyk, J., Switala-Jelen, K., Boratynski, J., Nasulewicz, A., Lipinska, L., Chybicka, A., Kujawa, M., Zabel, M., *et al.* (2004). Antitumor activity of bacteriophages in murine experimental cancer models caused possibly by inhibition of beta3 integrin signaling pathway. Acta Virol. *48*, 241–248.

Dabrowska, K., Switała-Jelen, K., Opolski, A., Weber-Dabrowska, B., and Gorski, A. (2005). Bacteriophage penetration in vertebrates. J. Appl. Microbiol. *98*, 7–13.

Dabrowska, K., Switała-Jeleń, K., Opolski, A., and Górski, A. (2006). Possible association between phages, Hoc protein, and the immune system. Arch. Virol. *151*, 209–215.

Dabrowska, K., Zembala, M., Boratynski, J., Switala-Jelen, K., Wietrzyk, J., Opolski, A., Szczaurska, K., Kujawa, M., Godlewska, J., and Gorski, A. (2007). Hoc protein regulates the biological effects of T4 phage in mammals. Arch. Microbiol. *187*, 489–498.

Dąbrowska, K., Skaradziński, G., Jończyk, P., Kurzępa, A., Wietrzyk, J., Owczarek, B., Żaczek, M., Świtała-Jeleń, K., Boratyński, J., Poźniak, G., *et al.* (2009). The effect of bacteriophages T4 and HAP1 on *in vitro* melanoma migration. BMC Microbiol. *9*, 13.

Debarbieux, L., Leduc, D., Maura, D., Morello, E., Criscuolo, A., Grossi, O., Balloy, V., and Touqui, L. (2010). Bacteriophages can treat and prevent *Pseudomonas aeruginosa* lung infections. J. Infect. Dis. *201*, 1096–1104.

Declue, A.E., Johnson, P.J., Day, J.L., Amorim, J.R., and Honaker, A.R. (2011). Pathogen associated molecular pattern motifs from Gram-positive and Gram-negative bacteria induce different inflammatory mediator profiles in equine blood. Vet. J. *192*, 455–460.

Dinarello, C.A. (2000). Proinflammatory cytokines. Chest *118*, 503–508.

Dixon, B. (2004). New dawn for phage therapy. Lancet Infect. Dis. *4*, 186.

Eriksson, F., Culp, W.D., Massey, R., Egevad, L., Garland, D., Persson, M.A.A., and Pisa, P. (2006). Tumor specific phage particles promote tumor regression in a mouse melanoma model. Cancer Immunol. Immunother. *56*, 677–687.

Eriksson, F., Tsagozis, P., Lundberg, K., Parsa, R., Mangsbo, S.M., Persson, M.A.A., Harris, R.A., and Pisa, P. (2009). Tumor-specific bacteriophages induce tumor destruction through activation of tumor-associated macrophages. J. Immunol. *182*, 3105–3111.

Feldmane, G., Loza, V., Duks, A., Yesyacova, I., Vinkele, R., Buike, A., Meldrajs, J., and Erele, I. (1977). Nucleic acid as interferon inducer. Its antiviral and antitumor activity. Arch. Immunol. Ther. Exp. *25*, 693–700.

Fogelman, I., Davey, V., Ochs, H.D., Elashoff, M., Feinberg, M.B., Mican, J., Siegel, J.P., Sneller, M., and Lane. H.C. (2000). Evaluation of CD4+ T cell function *in vivo* in HIV-infected patients as measured by bacteriophage phiX174 immunization. J. Infect. Dis. *182*, 435–441.

Fokine, A., Islam, M.Z., Zhang, Z., Bowman, V.D., Rao, V.B., and Rossmann, M.G. (2011). Structure of the three N-terminal immunoglobulin domains of the highly immunogenic outer capsid protein from a T4-like bacteriophage. J. Virol. *85*, 8141–8148.

Fraser, J.S., Yu, Z., Maxwell, K.L., and Davidson, A.R. (2006). Ig-like domains on bacteriophages: a tale of promiscuity and deceit. J. Mol. Biol. *359*, 496–507.

Gao, J., Wang, Y., Liu, Z., and Wang, Z. (2010). Phage display and its application in vaccine design. Ann. Microbiol. *60*, 13–19.

Geier, M.R., Trigg, M.E., and Merril, C.R. (1973). Fate of bacteriophage lambda in non-immune germ-free mice. Nature *246*, 221–223.

Ghaemi, A., Soleimanjahi, H., Gill, P., Hassan, Z., Razeghi, S., Jahromi, M., and Roohvand, F. (2010). Recombinant λ-phage nanobioparticles for tumor therapy in mice models. Genet. Vaccines Ther. *8*, 3.

Giancotti, F.G., and Ruoslahti, E. (1999). Integrin signaling. Science *285*, 1028–1032.

Gonzalez-Dosal R., Horan, K.A., Rahbek, S.H., Ichijo, H., Chen, Z.J., Mieyal, J.J., Hartmann, R., and Paludan, S.R. (2011). HSV infection induces production of ROS, which potentiate signaling from pattern recognition receptors: role for S-glutathionylation of TRAF3 and 6. PLoS Pathog. *7*, e1002250.

Górski, A., and Weber-Dabrowska, B. (2005). The potential role of endogenous bacteriophages in controlling invading pathogens. Cell. Mol. Life Sci. *62*, 511–519.

Gorski, A., Dabrowska, K., Switala-Jelen, K., Nowaczyk, M., Weber-Dabrowska, B., Boratynski, J., Wietrzyk, J., and Opolski, A. (2003). New insights into the possible role of bacteriophages in host defense and disease. Med. Immunol. *2*, 2–6.

Górski, A., Wazna, E., Dabrowska, B.W., Dabrowska, K., Switała-Jeleń, K., and Miedzybrodzki, R. (2006). Bacteriophage translocation. FEMS Immunol. Med. Microbiol. *46*, 313–319.

Górski A., Borysowski, J., Międzybrodzki, R., and Weber-Dąbrowska B. (2007). Bacteriophages in medicine. In Bacteriophage: Genetics and Molecular Biology, S. McGrath, D. van Sinderen, eds. (Horizon Scientific Press), pp. 125–158.

Górski, A., Międzybrodzki, R., Borysowski, J., Dąbrowska, K., Wierzbicki, P., Ohams, M., Korczak-Kowalska, G., Olszowska-Zaremba, N., Łusiak-Szelachowska, M., Kłak, M., *et al.* (2012). Phage as a modulator of immune responses: practical implications for phage therapy. Adv. Virus Res. *83*, 39–68.

Hartley, O., Dorgham, K., Perez-Bercoff, D., Cerini, F., Heimann, A., Gaertner, H., Offord, R.E., Pancino, G., Debre, P., and Gorochov, G. (2003). Human immunodeficiency virus type 1 entry inhibitors selected on living cells from a library of phage chemokines. J. Virol. *77*, 6637–6644.

Hayashi, T., Nakamura, T., and Takaoka, A. (2011). Pattern recognition receptors. Nihon Rinsho Meneki Gakkai Kaishi *34*, 329–345.

d'Hérelle, F. Opsonic power of the lysins. In Bacteriophage. Its role in Immunity. Williams & Wilkins, Baltimore, USA, 1922, p.125.

Hoffmann, M. (1965). Animal experiments on the mucosal passage and absorption viremia of T3 phages after oral, tracheal and rectal administration. Zentralblatt fur Bakteriologie, Parasitenkunde, Infektionskrankheiten und Hygiene. 1. Abt. Medizinisch-hygienische Bakteriologie, Virusforschung und Parasitologie. Originale *198*, 371–390.

Hyglos GmbH, EndoTrap technology (2011). on-line: http://www.hyglos.de/?id=20

Iizuka, A., Watanabe, T., Kubo, T., Yamamoto, M., Ogawa, K., Ohkuma, T., and Kaji, A. (1994). M13 bacteriophage DNA inhibits duck hepatitis B virus during acute infection. Hepatology *19*, 1079–1087.

Ishii, T., and Yanagida, M. (1975). Molecular organization of the shell of the T-even bacteriophage head. J. Mol. Biol. *97*, 655–660.

Kamme, C. (1973). Antibodies against Staphylococcal bacteriophages in human sera. Acta Pathol. Microbiol. Scand. Sect. B. *81*, 741–748.

Kańtoch, M. (1956). Wybrane zagadnienia z nauki o bakteriofagach; Post. Hig. i Med. Dosw *10*, 47–86.

Kańtoch, M., and Mordarski, M. (1958). Binding of bacterial viruses by cancer cells in vitro. Post. Hig. Med. Dosw. *12*, 191–192.

Kańtoch, M., and Szalaty, H. (1960). Lytic activity of bacteriophages toward phagocytized bacteria. Arch. Immunol. Ther. Exp. *8*, 407–416.

Kańtoch, M., Skurski, A., and Wieczorek, Z. (1958). In vitro blockade of bacterial phagocytosis of leukocytes by means of bacterial viruses. Schweiz. Z. Pathol. Mikrobiol. *21*, 1106–1119.

Keller, R., and Engley, F.B. (1958). Fate of bacteriophage particles introduced into mice by various routes. Proc. Soc. Exp. Bio. Med. *98*, 577–579.

Keyhani, M. (1969). Protection against experimental Newcastle disease in chickens given *Brucella* bacteriophage. Vet. Rec. *84*, 657–658.

Kleinschmidt, W.J., Douthart, R.J., and Murphy, E.B. (1970). Interferon production by T4 coliphage. Nature *228*, 27–30.

Korczak-Kowalska, G., Wierzbicki. P., Borysowski, J., Kłosowska, D., and Górski A.(2010). T4 phage preparations do not induce the production of IL-6 and TNF-α in human monocytes in vitro. Viruses of Microbes Conference, 21–25 June 2010, Paris, France, Abstract Book p. 221.

Kramberger, P., Honour, R.C., Herman, R.E., Strancar, A., Smrekar, F., and Peterka, M. (2011). Single-step purification of *Staphylococcus aureus* bacteriophage VDX-10 on methacrylate monoliths. Phages Meeting, 19–21 September 2011, St Hilda's College, Oxford, UK, Abstract Book p. 21.

Krichevsky, A., Rusnati, M., Bugatti, A., Waigmann, E., Shohat, S., and Loyter, A. (2005). The fd phage and a peptide derived from its p8 coat protein interact with the HIV-1 Tat-NLS and inhibit its biological functions. Antiviral Res. *66*, 67–78.

Krisch, H.M., and Comeau, A.M. (2008). The immense journey of bacteriophage T4 – from d'Hérelle to Delbrück and then to Darwin and beyond. Res. Microbiol. *59*, 314–324.

Kucharewicz-Krukowska, A., and Slopek, S. (1987). Immunogenic effect of bacteriophage in patients subjected to phage therapy. Arch. Immunol. Ther. Exp. *35*, 553–561.

Letarov, A., and Kulikov, E. (2009). The bacteriophages in human- and animal body-associated microbial communities. J. Appl. Microbiol. *107*, 1–13.

Letarova, M., Strelkova, D., and Letarov, A. (2011). The test for 'physiological phagemia' hypothesis in horses. 19th International Phage Biology Meeting, 7–12 August 2011, Evergreen State College, Olympia, Washington, U.S.A., Abstract Book p. 56.

Manoutcharian, K. (2005). Bacteriophages as tools for vaccine and drug development. Expert Rev. Vaccines *4*, 5–7.

Maruyama, I.N., Maruyama, H.I., and Brenner, S. (1994). Lambda foo: a lambda phage vector for the expression of foreign proteins. Proc. Natl. Acad. Sci. U.S.A. *91*, 8273–8277.

Meek, E.S., and Takahashi, M. (1968). Differential inhibition by phagicin of DNA synthesis in cells infected with vaccinia. Nature *220*, 822.

Merril, C.R. (1977). Interactions of bacterial viruses and bacterial genes with animal systems. In Molecular Genetic Modification of Eukaryotes, pp. 83–100. Edited by Irwin Rubenstein, New York, Academic Press.

Merril, C.R., Biswas, B., Carlton, R., Jensen, N.C., Creed, G.J., Zullo, S., and Adhya, S. (1996). Long-circulating bacteriophage as antibacterial agents. Proc. Natl. Acad. Sci. U.S.A. *93*, 3188–3192.

Międzybrodzki, R., Fortuna, W., Weber-Dąbrowska, B., and Górski, A. (2005). Bacterial viruses against viruses pathogenic for man? Virus Res. *110*, 1–8.

Międzybrodzki R., Świtała-Jeleń K., Fortuna, W., Weber-Dąbrowska B., Przerwa, A., Kurzępa, A., Łusiak-Szelachowska, M., Dąbrowska, K., Boratyński, J., *et al.* (2008). Bacteriophage inhibition of reactive oxygen species generation by endotoxin-stimulated polymorphonuclear leukocytes. Virus Res. *131*, 233–242.

Molenaar, T.J., Michon, I., de Haas, S.A., van Berkel, T.J., Kuiper, J., and Biessen, E.A. (2002). Uptake and processing of modified bacteriophage M13 in mice: implications for phage display. Virology *293*, 182–191.

Mori, K., Kubo, T., Kibayashi, Y., Ohkuma, T., and Kaji, A. (1996). Anti-vaccinia virus effect of M13 bacteriophage DNA. Antiviral Res. *31*, 79–86.

Mutuberria, R., Arends, J.W., Griffioen, A.W., and Hoogenboom, H.R. (2001). Phage display technology for target discovery in drug delivery research. In Drug Targeting Organ-Specific Strategies, G. Molema and D.K.F. Meijer, eds. (Wiley-VCH Verlag GmbH), pp. 255–273.

Olchawa, E., Międzybrodzki, R., Weber-Dąbrowska, B., and Górski, A. (2010). Endotoxin removal from *Staphylococcus aureus* phage lysate. Viruses of Microbes Conference, Jun 21–25.2010, Paris, France, Abstract Book p. 284.

Oslizlo, A., Miernikiewicz, P., Piotrowicz, A., Owczarek, B., Kopciuch, A., Figura, G., and Dabrowska, K. (2011). Purification of phage display-modified bacteriophage T4 by affinity chromatography. BMC Biotechnol. *11*, 59.

Plancon, L., Janmot, C., le Maire, M., Desmadril, M., Bonhivers, M., Letellier, L., and Boulanger, P. (2002). Characterization of a high-affinity complex between the bacterial outer membrane protein FhuA and the phage T5 protein pb5; J. Mol. Biol. *318*, 557–569.

Przerwa, A., Kniotek, M., Nowaczyk, M., Weber-Dabrowska, B., Switała-Jelen, K., Dabrowska, K., and Gorski, A. (2005). Bacteriophages inhibit IL-2 production by human T lymphocytes. Abstracts of the 12th Congress of European Society of Organ Transplantation, October 16–20.2005, Geneva.

Przerwa, A., Zimecki, M., Świtała-Jeleń K., Dąbrowska, K., Krawczyk, E., Łuczak, M., Weber-Dąbrowska E., Międzybrodzki, R., and Górski, A. (2006). Effects of bacteriophages on free radical production and phagocytic functions. Med. Microbiol. Immunol. *195*, 143–150.

Sathaliyawala, T., Rao, M., Maclean, D.M., Birx, D.L., Alving, C.R., and Rao, V.B. (2006). Assembly of HIV antigens on bacteriophage T4: a novel *in vitro* approach to construct multicomponent HIV vaccines. J. Virol. *80*, 7688–7698.

Sathaliyawala, T., Islam, M.Z., Li, Q., Fokine, A., Rossmann, M.G., and Rao, V.B. (2010). Functional analysis of the highly antigenic outer capsid protein, Hoc, a virus decoration protein from T4-like bacteriophages. Mol. Microbiol. 77, 444–455.

Shivachandra, S.B., Li, Q., Peachman, K.K., Matyas, G.R., Leppla, S.H., Alving, C.R., Rao, M., and Rao, V.B. (2007). Multicomponent anthrax toxin display and delivery using bacteriophage T4. Vaccine *25*, 1225–1235.

Smith, G.P. (1985). Filamentous fusion phage: novel expression vectors that display cloned antigens on the virion surface. Science *228*, 1315–1317.

Smith, H.W., Huggins, M.B., and Shaw, K.M. (1987). Factors influencing the survival and multiplication of bacteriophages in calves and in their environment. J. Gen. Microbiol. *133*, 1127–1135.

Sokoloff, A.V., Bock, I., Zhang, G., Sebestyén, M.G., and Wolff, J.A. (2000). The interactions of peptides with the innate immune system studied with use of T7 phage peptide display. Mol. Ther. 2, 131–139.

Srivastava, A.S., Kaido, T., and Carrier, E. (2004). Immunological factors that affect the *in vivo* fate of T7 phage in the mouse. J. Virol. Methods. *115*, 99–104.

Stent, G.S. (1963). Molecular Biology of Bacterial Viruses (San Francisco, W.H. Freeman and Co.).

Stone, R. (2002). Bacteriophage therapy. Stalin's forgotten cure. Science *298*, 728–731.

Synnott, A., Ohshima, K., Nakai, Y., and Tanji, Y. (2009). IgA response of BALB/c mice to orally administered *Salmonella typhimurium* flagellin-displaying T2 bacteriophages. Biotechnol. Prog. *25*, 552–558.

Tikhonenko, A.S., Gachechiladze, K.K., Bespalova, I.A., Kretova, A.F., and Chanishvili, T.G. (1976). Electron-microscopic study of the serological affinity between the antigenic components of phages T4 and DDVI. Mol. Biol. Mosk. *10*, 667–673.

Vales, N., Barron, B.L., Padilla, J.A., Sandoval, E., and Sotelo, J. (1991). Effect of dsRNA from phi 6 bacteriophage on herpetic infection in cell culture and an animal model. J. Interferon Res. *11*, 271–274.

Vlahos, R., Stambas, J., Bozinovski, S., Broughton, B.R., Drummond, G.R., and Selemidis, S. (2011). Inhibition of Nox2 oxidase activity ameliorates influenza A virus-induced lung inflammation. PLoS Pathog. *7*, e1001271.

Wang, J.D., Nonomura, N., Takahara, S., Li, B.S., Azuma, H., Ichimaru, N., Kokado, Y., Matsumiya, K., Miki, T., Suzuki, S., and Okuyama, A. (1998). Lymphotactin: a key regulator of lymphocyte trafficking during acute graft rejection. Immunology *95*, 56–61.

Weber-Dabrowska, B., Dabrowski, M., and Slopek, S. (1987). Studies on bacteriophage penetration in patients subjected to phage therapy. Arch. Immunol. Ther. Exp. *35*, 563–568.

Weber-Dąbrowska, B., Zimecki, M., Mulczyk, M., and Górski, A. (2002). Effect of phage therapy on the turnover and function of peripheral neutrophils. FEMS Immunol. Med. Microbiol. *34*, 135–138.

Wenger, S.L., Turner, J.H., and Petricciani, J.C. (1978). The cytogenic, proliferative and viability effects of four bacteriophages on human lymphocytes. In Vitro *14*, 543–549.

Wu, J., Tu, C., Yu, X., Zhang, M., Zhang, N., Zhao, M., Nie, W., and Ren, Z. (2007). Bacteriophage T4 nanoparticle capsid surface SOC and HOC bipartite display with enhanced classical swine fever virus immunogenicity: a powerful immunological approach. J. Virol. Methods *139*, 50–60.

Part II

Applications of Phages and Phage-derived Enzymes as Antibacterials

Overview of Therapeutic Applications of Bacteriophages

6

David Kelly, Olivia McAuliffe, R. Paul Ross, Jim O'Mahony and Aidan Coffey

Abstract

This chapter will give an overview of the different applications of bacteriophages (phages). These include the use of whole virulent phages as antibacterials, recent advances in the development and applications of genetically modified phages, the use of phages as delivery vehicles and vaccines, and recent developments in the study and use of purified endolysins. Generally, it is clear that phages have a considerable array of potential applications as therapeutics in the modern medical and veterinary fields. In particular, the continued antibiotic resistance problem in clinically relevant micro-organisms calls for the exploitation of cheap, natural, readily available, safe and efficient therapeutic agents. Phages exhibit characteristics which satisfy all of these criteria and their reintroduction as medical treatment options is worthy of strong consideration.

The use of lytic phages

The discovery of phages came about around 20 years before penicillin, the first antibiotic, was used for practical applications. This discovery is attributed to two scientists, Frederick Twort (1915) and Félix d'Hérelle (1917), with the latter credited as the one to have thoroughly investigated, recognized and named them. However antibacterial activity, thought now to be associated with bacteriophages, had also been recognized by Hankin and Gamaleya (Fig. 6.1) as far back as 1896 and 1898 respectively (Sulakvelidze *et al.*, 2001).

D'Hérelle worked and published extensively on phage after their discovery and helped in the establishment of the International Bacteriophage Institute, later named Eliava Institute of Bacteriophages, Microbiology, and Virology in Tbilisi, Georgia in 1923 (Summers, 1999; Sulakvelidze *et al.*, 2001). Several reviews describe the history of phage therapy, including early uses for human applications (Alisky *et al.*, 1998; Summers, 1999, 2001; Sulakvelidze *et al.*, 2001; Sulakvelidze and Kutter, 2005; Dublanchet and Bourne, 2007; O'Flaherty *et al.*, 2009; Górski *et al.*, 2009; Kutter *et al.*, 2010; Abedon *et al.*, 2011; Chanishvili, 2012). In addition, a review of former USSR research on the use of phages in medical trials in infants and adults has been translated into English and published in a monograph by the Eliava Institute (Chanishvili, 2009). Results of many of the earlier phage therapy experiments were inconsistent, with reports of both success and failure. Failures were generally attributed to several factors including (1) a poor understanding of general phage biology, (2) poor quality experimental techniques, (3) low-quality phage preparations, and (4) a lack of understanding of the illnesses being treated. The emergence of the antibiotic era in the 1940s led to a decline in the use of phages in the West. However in Poland and the former Soviet Union, especially in Georgia and Russia, significant use of phages either as stand alone treatments, or in combination with antibiotics continued (Sulakvelidze *et al.*, 2001). The history of phage therapy in the West has been viewed as composing four periods (1) early enthusiasm (2) crucial scepticism (3) abandonment and (4) recent interest and reappraisal (Summers, 2001). This recent reappraisal of phages as viable options in treating bacterial infections has no doubt arisen due to the widespread problem of antibiotic resistance among several bacterial pathogens. A recent

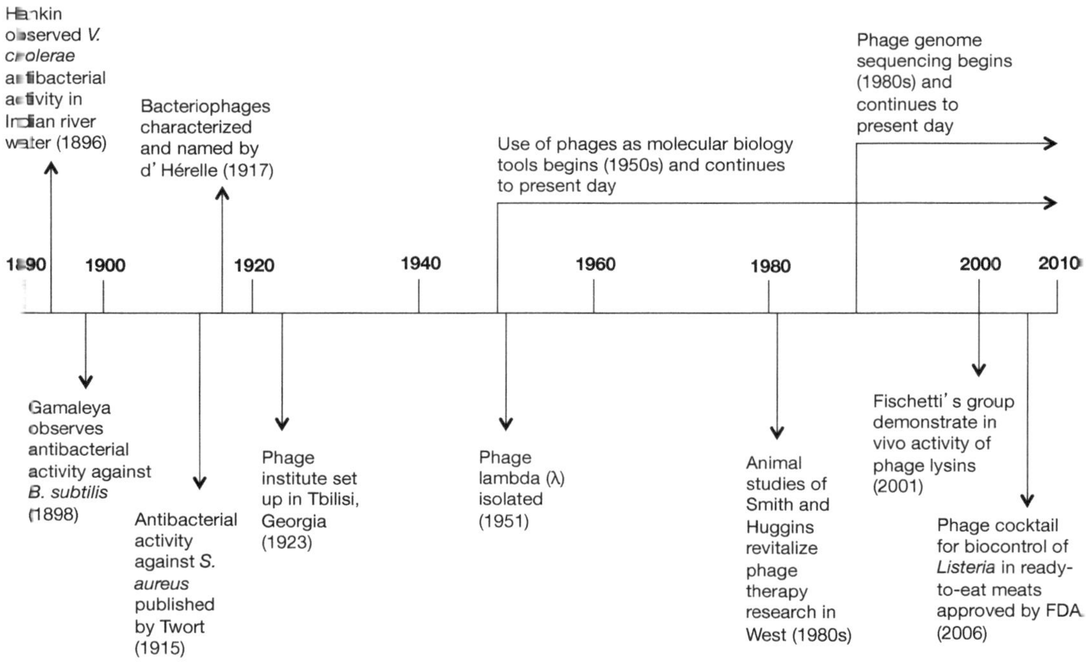

Figure 6.1 Major timelines in the history of phage therapy.

review lists a plethora of bacterial pathogens amongst which antibiotic resistance has become common (O'Flaherty *et al.*, 2009). The emergence of pathogenic bacteria which are resistant to the majority of currently available antimicrobial therapies, along with the increase in numbers of immunosuppressed hospital patients has become a major problem in the modern medical world (Abhlilash *et al.*, 2009). In recent years, pharmaceutical companies have been investing less in the discovery and development of novel antibiotics, due to poor investment returns (Projan and Shales, 2004; Norrby *et al.*, 2005). This trend of increasing antibiotic resistance, combined with the financial burden on health systems has led to a renewed interest in the use of alternative antibacterial agents, including phages (Shasha *et al.*, 2004; Gorski *et al.*, 2009).

General characteristics of lytic phages as antibacterial agents

The aforementioned ability of phages to infect a specific bacterial culture, replicate exponentially within the culture, and ultimately kill them, is the cornerstone of their use as antibacterial agents for phage therapy. Phage therapy itself can be described as the application of phages to control or eliminate bacterial infections and it is a very promising means of complementing the use of chemical antibiotics to treat infections, particularly where antibiotic resistance is prevalent (Abedon, 2010). In this regard, one of the most important features of phages as antibacterial agents is their capacity to kill antibiotic-resistant bacteria (Górski *et al.*, 2009; Borysowski *et al.*, 2011). A critical step in phage therapy is the choice of phage and this requires knowledge of the life cycles of phages whether virulent (lytic) or temperate (Kropinski, 2006; Hanlon, 2007). The latter persist as prophages, quiescent in the host and only replicate along with the host genome. These are poor candidates for phage therapy and are usually avoided for a number of reasons:

1. They do not kill 100% of the bacteria which they infect.
2. They may harbour genes that render the bacterium more virulent, a process known as lysogenic conversion.

3 The establishment of lysogeny leads to super-infection immunity, where the lysogen becomes impervious to infection by the same or related phages (though the super-infection immunity can be overcome at high MOI values).
4 Temperate phages may also carry out both specialized and generalized transduction, and this may increase the disease-causing ability of host bacteria (Kropinski, 2006; Deresinski, 2009; Gill and Hyman, 2010).

However, in spite of apparent disadvantages of temperate phages as potential antibacterial agents, some authors reported their successful application to eliminate bacteria in experimental models of infection (Paul *et al.*, 2011; Chung *et al.*, 2012).

Virulent phages on the other hand directly exploit the bacterial cell's replication machinery, and having replicated and assembled into complete virions, cause lysis and cell death of the bacterium, with concomitant release of approximately 10–100 progeny phage, which subsequently go on to infect additional bacteria (however, virulent phages do not always kill 100% of the infected bacterial cells). The main characteristics of lytic phages, especially their target specificity (antibacterial range often limited to some strains of a single bacterial species), their rapid lysis and killing of bacterial hosts independent of their antibiotic resistance profile, and their ability to amplify at the site of the bacterial infection make them good candidates as therapeutic agents for treatment of bacterial infections (Sulakvelidze *et al.*, 2001; Matsuzaki *et al.*, 2005; Hanlon, 2007; Górski *et al.*, 2009). It is known, however, that some obligately lytic phages can also mediate generalized transduction (Waddell *et al.*, 2009). Importantly, it was shown that phages can exert additive, or even synergistic antibacterial effects with antibiotics both *in vitro* and *in vivo* (Comeau *et al.*, 2007; Kirby *et al.*, 2012; Chhibber *et al.*, 2013; Yilmaz *et al.*, 2013; Knezevic *et al.*, 2013). Use of a combination of a phage and an antibiotic can result not only in stronger antibacterial effects, but also may limit the evolution of bacterial antibiotic-resistance (Zhang and Buckling, 2012). Another important feature of at least some bacteriophages is their capacity to eliminate bacterial biofilms (Yilmaz *et al.*, 2013; see 'The use of phages in the treatment of biofilms'). On the other hand, the main factors, which can limit therapeutic efficacy of lytic phages include the development of phage resistance (Carlton *et al.*, 1999; see also Chapter 4), immunogenicity (Górski *et al.*, 2012), and relatively short serum half-life (Merril *et al.*, 1996). The development of an effective phage therapy protocol is a multistep process, namely (1) phage isolation against the target bacterial strain; (2) activity testing against this strain and also against other related strains; (3) screening for negative traits such as the ability to enhance bacterial virulence; (4) propagation and growth to high titres; (5) removal of any bacterial toxins present in phage lysates by purification, (6) stabilization of phages for storage, (7) efficient delivery of viable phages to target bacteria in the site of infection. Once phages have been delivered it is then desirable that they will survive, adsorb to and kill bacteria, while not significantly stimulating an immune response from the organism to which it was delivered (Abedon, 2010). It is generally agreed that all newly isolated phages which are to be used for therapy should be characterized in detail prior to use and this should include: genome sequencing, structural studies, and bioinformatic analysis. The phages should also be put through a proof of application process for their intended use (Carlton *et al.*, 2005).

In addition to developing the above protocol and defining desirable phage characteristics, the choice of phage strain and the methods of phage preparation and purification are also an important consideration for the success of phage therapy trials. The requirements for isolation, preparation and purification of phages to maximize their therapeutic potential have been reviewed recently by Gill and Hyman (2010); these issues are also discussed in detail in Chapter 2.

After taking the above into consideration, it is still necessary to outline the prerequisites which must be met before phage therapy itself is initiated: (1) the biology of the therapeutic phage(s) must be well understood as phage–host systems tend to be extremely complicated; (2) therapeutic phage preparations should comply to all safety requirements, and be free of bacteria and their toxic components; (3) phage preparations should contain infective particles at sufficiently high titre; (4) the efficacy of phage therapy should be tested in an animal model of infection as each phage may behave differently *in vivo* compared to *in vitro* (Levin and Bull, 2004).

Another critical step is the choice of route of administration of therapeutic phage to the patient. There are many potential routes including oral, rectal, parenteral and topical (such as the intra-pleural route, bladder irrigation, and direct administration to wounds). In general, by topical route, which is the most commonly used, phages can be administered by sprays, aerosols, mouthwash, suppositories, bandages, eye drops and tampons (Deresinski, 2009). One potential way of parenteral administration is the intravenous route, but it is associated with considerable concerns over the safety of phage preparations especially regarding their purity (e.g. a level of endotoxin contamination), and immune reactions to phages and to different components of the bacterial cell wall which may be rapidly released from bacteria lysed by phages. Since phages are considered self replicating particles, administration of phages to the systemic circulation therefore not only involves a choice of many modes of administration, but also necessitates knowledge of the pharmacokinetics and pharmacodynamics of each phage and these are considered to be complex due to the fact that both phage and bacterium are replicating at the site of infection, while the latter are also undergoing lysis (Cairns and Payne, 2008). Pharmacokinetics can be described as the impact of a body on a specific drug, while pharmacodynamics is a description of a drug's impact on a body (Abedon and Thomas-Abedon, 2010). It is clear therefore that while phage therapy is a two step process, involving phage penetration to reach their target bacteria, followed by bacterial cell killing, that analysis of both steps also needs to take into account phage environmental interactions. This indicates that the interactions between phages and body tissues, as well as between phages and their target bacteria need to be considered during phage therapy treatments (Abedon and Thomas-Abedon, 2010).

It is noteworthy that endogenous phages already present in the human body may play a crucial role in the pathology and physiology of the body system (Górski and Weber-Dabrowska, 2005; Górski *et al.*, 2006; Letarov and Kulikov, 2009; Kurzepa *et al.*, 2009). These studies suggest that endogenous phages may undergo translocation through the gut wall and play an important role in eliminating bacteria, regulating the body ecosystem, and stimulating the immune system, in the process helping to control local inflammatory and autoimmune reactions (Górski and Weber-Dabrowska, 2005; Górski *et al.*, 2006; Letarov and Kulikov, 2009; Kurzepa *et al.*, 2009). The significance of these interactions may be crucial for the further development of phage therapy, as increased knowledge of human-phage associated interactions may lead to fewer restrictions and obstacles in initiating phage therapy treatments. Indeed, the finite amount of data published on these interactions to date, coupled with the increasing significance of phages in modern contemporary medicine is a problem which needs to be addressed. Despite a small number of such studies it is thought that the process of phage therapy may be improved through a more detailed understanding of its pharmacodynamics and pharmacokinetics. Basically a drug's pharmacodynamics lead to two important concerns, its therapeutic effects versus its toxic effects along with the associated dosing considerations also. For phages, which can be considered self-amplifying drugs, proper dosing by clinicians and doctors may be less crucial than for other administered drugs since the phage itself is responsible for multiplying and attaining therapeutic levels at the site of infection. Phage therapy may indeed be successful due to two key phage properties, their self-amplification and their safety, which suggest that there may be less reason for concern about the pharmacodynamics. Phages undergo degradation by the immune system and this does not result in the release of toxic byproducts, unlike the metabolic decay of many other drugs. Phages do not bind to or manipulate body tissues; the human body is routinely exposed to phages and hence therapeutic application of phages is not an entirely foreign experience for the body. The ability of phages to multiply *in situ* in addition to the possibility of applying them at high titre is an important aspect of phage therapy as achieving bacterial lysis and death in the scale required often requires high titres of phages (> 10^8 PFU/ml), which would not be feasible with many other antimicrobial agents. It may be suggested therefore that some phage therapy failures, both past and present may be considered pharmacokinetic failures, resulting from the supply of insufficient number of phages to the site of infection to cause effective bacterial killing (Waldor *et al.*, 2005; Merril, 2008; Abedon and Thomas-Abedon, 2010; Gill and Hyman, 2010). Furthermore it has been noted that the *in vitro* growth data for any given phage cannot be applied to the *in vivo* situation, while likewise the *in vivo* data for any one phage cannot be considered as standard for another phage. Hence the therapeutic use of phages as drugs tends to differ significantly from pharmaceuticals due to the pharmacokinetic differences (Payne and Jansen, 2003). These authors reported that the critical parameters affecting phage therapy in this context are: initial phage dose, phage adsorption rate, burst size, latent period, density-dependent thresholds and associated critical times, and phage clearance from body fluids by the reticulo-endothelial system (Payne and Jansen, 2001). Phages thus can be considered a unique medicine, which challenge current pharmacokinetic concepts. In the absence of target bacteria, phages are quickly diluted, minimizing adverse effects, while in the presence of the target pathogen, they represent a biologically amplifiable drug. When the host bacterial population is then exhausted, the phages are quickly eliminated from the body. Overall a better understanding of phage therapy pharmacodynamics and pharmacokinetics will likely lead to improvements and innovative phage therapy protocols, and also it may be possible, using relatively basic mathematical tools, to estimate phage densities which may be appropriate for both *in vitro* and *in vivo* bacterial killing. This may see a decrease in phage therapy failures due to protocols attempted without prior knowledge of phage killing potential (Abedon and Thomas-Abedon, 2010). The pharmacokinetics and pharmacodynamics of phage preparations are presented in Chapter 3.

In general, like any medical treatment option, lytic phages have their perceived advantages and disadvantages (Loc-Carrillo and Abedon, 2010); these are summarized in Table 6.1. These also need to be taken into consideration before the initiation of any phage therapy treatments.

The use of phages as antibacterial agents in animal models of human infection

Before any new potential drug or antimicrobial agent is considered for use as a human therapeutic, its efficacy and safety are generally evaluated in animal models of infection and this has certainly been the case with phages as antimicrobials, where numerous studies have focused on animal models of infection and pharmacokinetics of phage preparations. Important lines of research are studies on tissue/organ distribution of phages. In general, it was shown that phage particles can penetrate in fairly high numbers to different organs, including the spleen, liver, kidney, thymus, brain and muscles following systemic administration to mice (Geier *et al.*, 1973; Smith and Huggins, 1982). Moreover, several studies were carried out in animals which indicated that phage can appear in the urine after administration by different routes (Letkiewicz *et al.*, 2010a, and references therein). One study reported that phages could accumulate in dog kidneys after intravenous administration (Keller and Zatzman, 1959). In a more detailed study, published six years later, it was reported that phages were detectable in the urine of mice and rats when serum levels exceeded 10^5 PFU/ml. This work also found that urine analysis did not show any changes, while no detrimental histological changes were seen in the urinary tract. Also phage activity was not compromised in urine (Schultz and Neva, 1965). A similar study in sharks demonstrated that phages were detected in the kidneys for as long as 1 month after administration (Russell *et al.*, 1976). These data are important because they may explain the efficacy of phages administered intraperitoneally in experimental models of urinary tract infections in mice (Nishikawa *et al.*, 2008; Tóthová *et al.*, 2011). Early phage studies in animals also gave an indication of the effectiveness of different routes of administration. Rectal administration in rabbits and mice resulted in penetration through the wall of the intestine to the circulatory system within a few minutes (Hoffmann, 1965; Sechter *et al.*, 1989). Hoffmann's study also found that the

Table 6.1 Advantages and disadvantages of lytic phages as antibacterial agents

Advantages	Disadvantages
Bactericidal agents	Not all phages make good therapeutics
'Auto dosing' agents	Narrow bactericidal spectrum
Capacity to kill antibiotic-resistant bacteria	Viral/virulent terminology could give them an unfavourable public image
Minimal disruption to normal bacterial flora	Possible safety concerns
Relatively low potential for inducing resistance[1]	Physiological barriers can prevent phages from reaching target bacteria
Ease of discovery and isolation	Phage resistance
Versatility in modes of formulation and application	Possible risk of phage conversion of bacterial targets where phage genome has not been characterized
Active against biofilms	Immunogenicity
Single dose potential[2]	Short serum half-life
Potential cross-transmission to untreated subjects, either from the environment or treated subjects	
Capacity for low dosage use[3]	
Low environmental impact	
Natural products	
Relatively low cost	

[1]According to some authors, the rate of developing resistance to phages is lower than the rate of developing resistance to antibiotics and can be further reduced by using phage cocktails (Carlton, 1999; Sulakvelidze *et al.*, 2001). [2]Especially in experimental models of acute infections (Capparelli *et al.*, 2007; Capparelli *et al.*, 2010) and a recent clinical trial (Wright *et al.*, 2009). However, in clinical phage therapy, usually multiple doses of phages have been used (Weber-Dąbrowska *et al.*, 2000; Międzybrodzki *et al.*, 2012). [3]Especially in cases in which phage preparations are administered topically (Wright *et al.*, 2009).

blood phage level was as much as two orders of magnitude higher by rectal administration than by oral feeding possibly due to the lack of inactivation of phage particles by gastric juice (Hoffmann, 1965). More recently, it has been reported that *E. coli* phage T4 and phages of enterococci are capable of penetrating rat prostate tissue after intravenous administration (Miedzybrodzki *et al.*, 2008). Data also indicate that therapeutic phages could be isolated from rodent prostate tissue and urine after rectal application (Letkiewicz *et al.*, 2010a). Although one study in 1928 showed the presence of phages in the urine of human patients with urinary tract infections (Caldwell, 1928), other early data on humans are scarce, and hence the previously mentioned animal studies were likely to have been important in encouraging the persistence of researchers with phage *in vivo* studies and the development of phage therapy itself.

Studies have also been conducted to address the problem of unfavourable pharmacokinetics of phage preparations, especially short serum half-life. For example, Merril *et al.* (1996) demonstrated that long circulating phage mutants of *E. coli* phage lambda and *S. typhimurium* phage P22 could be isolated by a serial passage technique. In addition, they showed that long circulating lambda mutants are more effective than the corresponding parental strain in a murine model of bacteraemia. Subsequent experiments confirmed that a substitution of the amino acid lysine for glutamic acid in the major phage capsid protein, was sufficient to confer a long circulating phenotype on phage lambda mutants (Vitiello *et al.*, 2005). Phage modification with the use of polyethylene glycol (PEG) has also been applied successfully to reduce immunogenicity and to improve the pharmacokinetics of therapeutic phages. Examples of this were demonstrated by Kim *et al.* (2008), who showed improved survival of *Salmonella* phage Felix-O1 and *Listeria* phage A511 in mice after PEGylation. However, the authors did not investigate whether PEGylation of phages results in their higher therapeutic efficacy *in vivo*.

Although the number of reports on pharmacokinetics of bacteriophages is rather limited, many studies have been conducted to evaluate the efficacy of experimental phage therapy. Recently, O'Flaherty *et al.* (2009) reviewed the *in vivo* studies carried out in animal models of human infection. Animal models used included mice, rabbits, guinea pigs and hamsters, and phage efficacy against many significant human pathogens including *E. coli, Salmonella, Enterococcus, Pseudomonas, S. aureus, C. difficile, Klebsiella* and *Vibrio* was discussed (O'Flaherty *et al.*, 2009). Several additional animal studies have been reported since then. Antibiotic resistant *S. aureus* in particular is now considered one of the most common causes of both community-acquired and nosocomial infections, which are difficult to eliminate (O' Flaherty *et al.*, 2005; Mann, 2008). Capparelli *et al.* (2007) showed the efficacy of the phage M^{sa} as a therapeutic agent against both local and systemic infections due to *S. aureus*, including methicillin resistant strains (MRSA). More recently Gupta and Prasad (2011) showed that the broad host range phage P-27/HP protected mice from *S. aureus* bacteraemia and subsequent death when a single subcutaneous injection was given. Hsieh *et al.* (2011) also isolated a broad host range phage, Stau2, whose host range against *S. aureus* isolates was wider than the well recognized polyvalent phage K (Kelly *et al.*, 2011). This phage provided 100% protection against lethal doses of *S. aureus* strain S23 in mice (Hsieh *et al.*, 2011). The efficacy of phage therapy was also shown in a murine model of diabetic foot infection due to MRSA (Chhibber *et al.*, 2013).

A number of studies were also conducted to evaluate the efficacy of phage therapy in experimental models of infections due to *P. aeruginosa* (for a review, see Saussereau and Debarbieux, 2012). For example, Watanabe *et al.* (2007) examined the efficacy of phage therapy against gut derived sepsis caused by *P. aeruginosa* in a murine model. Oral administration of the lytic phage KPP10 resulted in survival of almost 67% of mice, in comparison to a survival rate of 0% for a saline treated control group. The study provided a sound basis for the possible oral treatment of gut derived sepsis (Watanabe *et al.*, 2007). Interestingly, phage therapy has also been used in *Drosophila melanogaster* to effectively eliminate infections caused by *P. aeruginosa*. Phages MPK1 and MPK6 were fed to the fruit flies, and both significantly delayed the killing of *Drosophila* by the bacterium (Heo *et al.*, 2009). The phages used in this study also protected mice from peritonitis-sepsis induced by *Pseudomonas*, and phage-treated mice had lower bacterial numbers in the lungs, liver and spleen. The insect model of infection may perhaps prove attractive to researchers not in favour of animal trials. Hagens *et al.* (2006) investigated the use of filamentous phage treatment to augment the efficacy of antibiotics. In this study the *Pseudomonas* phage Pf1, when used in combination with low concentrations of gentamicin rescued mice from lethal *P. aeruginosa* infections. Interestingly, neither treatment when administered alone was capable of curing the infection. In addition, *P. aeruginosa* K strains which harboured a plasmid-borne gentamicin resistance gene could be re-sensitized to this antibiotic upon infection with phage Pf1. High efficacy of phages in the treatment of experimental infections due to *P. aeruginosa* was also confirmed by Marza *et al.* (2006) who used phages to treat chronic otitis externa in a dog (this was not an experimental infection, but rather spontaneous antibiotic-refractory infection). This infection had failed to be treated effectively by topical and systemic administration of antibiotics, however inoculation of 4×10^2 PFU of phage into the auditory canal resulted in a significant clinical improvement only 27 h after the treatment was initiated.

Several studies report the efficacy of phages in murine models of infection with *Klebsiella pneumoniae*, a major causative agent of nosocomial infections, particularly in patients with burns (Malik and Chhibber, 2007; Kumari *et al.*, 2011; Hung *et al.*, 2011; Gu *et al.*, 2012). Another recent mouse study reported successful phage therapy against *B. cenocepacia* lung infection (Carmody *et al.*, 2010). The authors demonstrated that systemic administration of phages was more effective than administration by inhalation, indicating that circulating phages can gain better access to bacteria in the lungs than phages administered topically (Carmody *et al.*, 2010). In a fresh appraisal of phage therapy, Capparelli *et al.* (2010) carried out an important murine infection study which explored different aspects of phage therapy including the generation of anti-phage antibodies as well as the effects of phages on the virulence and the persistence of phage-resistant *S. enterica* in the circulatory system. These authors found that the phage-induced antibodies were non-neutralizing, and that phages could cure infection even when administered 2 weeks after bacterial inoculation into mice. Moreover, this study showed that the development of phage resistance results in a decrease in virulence and a faster clearance of bacteria from the circulation of mice (Caparelli *et al.*, 2010). The efficacy of phage therapy was also shown in experimental models of infections due to other important bacterial pathogens including *Enterococcus faecium* (Biswas *et al.*, 2002), *Escherichia coli* (Pouillot *et al.*, 2012), *Acinetobacter baumanii* (Soothill *et al.*, 1992), and *Listeria monocytogenes* (Mai *et al.*, 2010).

Although the majority of studies on experimental phage therapy were performed on immunocompetent animals, there are also reports on the successful use of phages in immunocompromised animals (Zimecki *et al.*, 2009, 2010). This thus supports the application of phage therapy for bacterial infections in immunocompromised patients.

Animal models of infection are an essential prelude to phage clinical trials in human patients, as they give an indication of the optimum conditions necessary to effect the maximum beneficial outcome. Experimental phage therapy performed on animal models of infections is discussed in detail in Chapter 10.

Use of lytic phages as antimicrobial agents in humans

Several reviews and book chapters have documented work on the application of lytic phage therapy in humans especially in Poland and the former Soviet Union (Alisky *et al.*, 1998; Carlton, 1999; Sulakvelidze and Morris, 2001; Summers, 2001; Sulakvelidze and Kutter, 2005; Chanishvili, 2012). In addition to the above reviews, the most prominent and comprehensive research in recent years has been carried out and also reviewed by the Polish group from the Institute of Immunology and Experimental Therapy (IIET) in Wrocław, Poland, headed by Ślopek and then by Górski (Slopek *et al.*, 1987; Weber-Dabrowska *et al.*, 2000; Górski *et al.*, 2007, 2009; Fortuna *et al.*, 2008; Miedzybrodzki *et al.*, 2012) and also by the phage group at the Eliava Institute in Tbilisi, Georgia (Chanishvili *et al.*, 2001; Chanishvili *et al.*, 2009). Today phage therapy is considered to be an experimental treatment in Poland, and it is administered to patients in whom antibiotic therapy has generally failed (Górski *et al.*, 2009; Miedzybrodzki *et al.*, 2012). The average success rate of phage therapy has been reported to be 85–92% in earlier trials (Slopek *et al.*, 1987; Weber-Dabrowska *et al.*, 2000) and even though these trials were not set up to be randomized clinical trials, they still clearly indicate high safety of phage therapy and a high capability of phages to eliminate bacterial infections in cases where antibiotic treatment was not effective. These articles detail the use of phages against several pathogens including *Staphylococcus, Klebsiella, E. coli, Proteus,* and *Pseudomonas,* and in the treatment of illnesses such as osteomyelitis, prosthetic joint infections, lower respiratory tract infections, skin and soft tissue infections (Slopek *et al.*, 1987; Weber-Dabrowska *et al.*, 2000). Significantly, the Polish Institute was the first institute to use phage lysates for the treatment of chronic bacterial prostatitis where other treatments had failed (Letkiewicz *et al.*, 2009; Letkiewicz *et al.*, 2010b). Recently the group have also published the results of a retrospective analysis of phage treatment of 153 patients conducted between 2008 and 2010. This analysis revealed that the percentage of good responses to the treatment ranged between 29.2% and 48.3% depending on the kind of infection, the aetiological factor, and the route of administration of phage preparations. Practically no serious side-effects were observed in the patients (Miedzybrodzki *et al.*, 2012). The efficacy rate revealed by this analysis is substantially lower than the efficacy rates reported earlier (Slopek *et al.*, 1987; Weber-Dabrowska *et al.*, 2000) likely owing to different ways of monitoring of patients as well as different types of treated infections (acute vs. chronic infections). In addition, the Polish group have summarized four small clinical trials which were carried out to evaluate the therapeutic efficacy of phage therapy in immunocompromised patients suffering from cancer, leukaemia and chronic urinary tract infections (Borysowski and Górski, 2008).

Other notable work was carried out in the former Soviet Union at the Eliava Institute in Tbilisi, Georgia (Chanishvili *et al.*, 2001; Chanishvili *et al.*, 2009; Sulakvelidze and Kutter, 2005; Chanishvili, 2012). Phages were mass produced at this institute and used for therapy throughout the entire former Soviet Union. Phage therapy experience itself at the Eliava Institute was recently published as a short communication (Kutateladze and Adamia, 2008), and describes a brief history of the institute, and examples of preparations used there, of which the most prominent are Pyophage (a cocktail of phages against purulent septic infections), Intestiphage (a phage preparation used against pathogens which cause intestinal infections), staphylococcal bacteriophage used against various staphylococcal infections, and PhageBioDerm – a biodegradable, non-toxic polymer impregnated with Pyophage and different phage components for treatment of *P. aeruginosa*, *S. aureus*, *Proteus*, *E. coli* and *Streptococcus* wound infections. A review by Kutter *et al.* (2010) focuses on lessons learned from the experience of practitioners which was obtained from both personal communications and publications. Discussion of therapeutic phage use in Georgia focused on two main areas of application: the use of phages for the treatment of wounds and surgical infections, and their use in the treatment and prophylaxis of enteric infections. The authors also discussed specific phage therapy approaches which are applied at the Hirszfeld Institute in Poland, and concluded that the results reviewed are highly suggestive that the Institute's approach to phage therapy has the potential for substantial efficacy, although the protocols employed would require rigorous double blind clinical testing (Kutter *et al.*, 2010).

It is also noteworthy that the phage lysate Stafal has been registered for the treatment of staphylococcal infections by topical application in the Czech Republic and Slovakia. This preparation could be used for the elimination of skin infections, as well as in potential reservoirs of infection such as the nasal passage and the intestinal and urinary tracts. It may also be used to prevent pyogenic infections in operation wounds. These applications have been reviewed by Górski *et al.* (2009).

However, despite these strongholds of phage therapy practice in eastern Europe, relatively little work on phage therapy has been undertaken in the western world, and commercial phage preparations had not been available since the early phage products of companies like Eli Lilly and L'Oreal went into decline after the advent of antibiotics. Nevertheless, as discussed in a recent review, phage therapy has been used in many patients also in France; in fact, several articles about phage therapy have been published in France (Abedon *et al.*, 2011, and references therein). It is noteworthy that phage therapy was also used in the US especially in the 1920s and 1930s, where several papers on therapeutic use of phages in humans were published (Abedon *et al.*, 2011, and references therein). In addition, some studies showed that the administration to patients by a variety of routes of a phage preparation called Staphage Lysate (SPL) (Delmont Laboratories, USA) led to only minor side-effects (Sulakvelidze and Barrow, 2005). More recently, a number of trials have been carried out in the West by commercial phage companies as discussed later. Indeed several companies are now active in the commercialization of phage products as the interest in phage as an alternative therapy gathers pace in other countries including the UK, Canada, The Netherlands, India, USA, Portugal, Israel, and Australia (Housby and Mann, 2009; Monk *et al.*, 2010). These companies utilize phage technologies ranging from whole phage to phage DNA and peptides, and have identified the large niche markets for antibacterial products against the common antibiotic resistant infectious bacteria. Three companies, Novolytics limited (UK), Phico Therapeutics (UK) and BioPhage Pharma Inc. (Canada), are close to the stage of clinical trials. Phage nasal decolonization of *Staphylococcus aureus* is an area which is currently under investigation and the fact that a product for this type of medical problem would not require intravenous use will potentially get products to market far more quickly due to there being fewer regulatory hurdles to overcome (Housby and Mann, 2009).

One of the greatest challenges of phage therapy is to verify by controlled clinical trials the safety and efficacy of phages found in numerous earlier less controlled studies performed in Poland, Georgia, and Russia (Parracho *et al.*, 2012). Recently three controlled safety trials of phage preparations have been carried out. The first two trials showed that the oral administration of coliphage T4 and a T4-like phage cocktail to healthy volunteers does not result in any serious side-effects (Bruttin and Brussow, 2005; Sarker *et al.*, 2012). The third study was an FDA-approved phase I physician-led trial carried out at a wound centre in Lubbock, Texas using a mixture of phages targeting *P. aeruginosa, S. aureus,* and *E. coli.* No ill effects were noted in this trial (Rhoads *et al.*, 2009). In addition, two safety trials of phage preparations were reportedly performed in healthy children, but their results have not been published yet (Brüssow, 2012). Moreover, a laboratory based quality control experiment with a phage preparation prepared at the Eliava Institute has been conducted at the Queen Astrid Military Hospital in Brussels, Belgium. This study confirmed the absence of temperate phages and also determined the stability, pyrogenicity, cytotoxicity and sterility of the preparation (Merabishvili *et al.*, 2009). The phage mixture was evaluated by all stringent safety measures and it was also evaluated for therapeutics in that it was applied topically to burn wounds of eight patients. No adverse reactions were reported and the experiment was seen as an important step in the promotion of phage treatment in Western Europe, at the same time creating a discussion platform for a regulatory framework for the approval of phage therapy. In the UK, Wright *et al.* (2009) described a fully regulated, placebo controlled, double blind, randomized phase I/II clinical trial of the efficacy and the safety of a bacteriophage therapeutic against chronic otitis involving the private company Biocontrol Limited. It is also worth noting that a large trial of 450 patients evaluating the safety and the efficacy of a phage cocktail against diarrhoea caused by *E. coli* is currently being carried out in Bangladesh by Nestle (http://clinicaltrials.gov/ct2/show/NCT00937274).

It is clear from the above that phage products against human infections are now being taken through fully regulated controlled human trials. Rigorous assessments of safety and efficacy are being included. The majority of the activity involves small companies established over the last 20 years and the area is set to grow in the future (Housby and Mann, 2009; Loc Carrillo and Abedon, 2010; Monk *et al.*, 2010; Pirnay *et al.*, 2010).

While a considerable body of evidence has been published on the safety and efficacy of phage therapy in immunocompetent patients, data remains scarce with regard to its use in immunocompromised patients. The group of Górski has been active in addressing this knowledge deficit and recently published an article summarizing the findings of a range of experimental and clinical studies, which could be of relevance in the utilization of phage therapy in immunocompromised human patients. In this article they discussed different issues relevant to the use of phages in immunocompromised patients. They concluded that the direct killing of target bacteria by the phages was the major mechanism by which the *in vivo* therapeutic effect was mediated and not by any immunostimulatory activity. In addition it was suggested that in immunosuppressed subjects, phage clearance from the circulatory system may be slowed down, enhancing their survival, and availability to host bacteria in the bloodstream for longer periods (Borysowski and Górski, 2008).

Phage therapy in humans, especially clinical experiences of the two major centres (the IIET and the Eliava Institute), as well as the results of recent controlled clinical trials of phage preparations are discussed in detail in Chapter 11. Regulatory issues relevant to therapeutic use of phages are presented in Chapter 12.

The use of phages in the treatment of biofilms

Biofilms are agglomerations of bacterial cells and their excreted products attached to both living and inert surfaces. Secreted products include exopolysaccharide (EPS), enzymes, bacteriocins, low molecular mass solutes and nucleic acid. Several bacteria have been implicated in biofilm formation including staphylococci, enterococci, *E. coli, Proteus,* klebsiellae, streptococci, pseudomonads, and acinetobacters (Donlan, 2009). Many of these are associated with a range of pathologies in humans including burn infections, chronic otitis media, chronic bacterial prostatitis, periodontitis, respiratory infections, urinary tract infections and endocarditis. Biofilms can also be formed on indwelling and subcutaneous medical implants such as pacemakers, heart valves, catheters, grafts and stents, and joint prostheses which may lead to persistent infections (Costerton *et al.*, 1999; Davey and O'Toole, 2000). The multi-species nature of natural biofilms makes their elimination difficult. Interestingly, some phages have been shown to produce enzymes which can degrade the EPS matrix of a biofilm (Hughes *et al.*, 1998; Hanlon *et al.*, 2001). Sutherland *et al.* (2004), Azeredo and Sutherland (2008), and Donlan (2009) reviewed a range of studies, each of which illustrated the successful removal of biofilms with phages. The applications of phages vary among studies; for example, Curtin and Donlan (2006) used pre-treatment with phages to eliminate *S. epidermidis* from a hydrogel-coated catheter whereas Carson *et al.* (2010) successfully used phages to remove established biofilms of *P. mirabilis* and *E. coli* from indwelling urological devices and urinary tract catheters. Goldman *et al.* (2009) suggested that a combination of several phage types may be used to prevent adhesion and biofilm formation and successfully utilized phages to control biofilm fouling of ultrafiltration membranes resulting in significantly improved membrane permeability of effluent treatments. Other authors have also used phage mixtures for biofilm treatment. A cocktail composed of five *Pseudomonas* phages was used to pretreat catheters prior to inoculation with biofilm forming *P. aeruginosa* strains. In this case biofilm density was significantly reduced. Phage treatment post inoculation was also successful in reducing biofilm counts, although supplemental inoculation was required to address biofilm regrowth. This study also demonstrated that fewer resistant cells were isolated when treated with this phage cocktail than when treated with an individual *Pseudomonas* phage (M4) (Fu *et al.*, 2010). Lu and Collins (2007) designed an enzymatic phage T7 derivative, which expressed dispersin B, a polysaccharide degrading enzyme. This engineered phage was subsequently found to have a biofilm dispersing ability two orders of magnitude higher than that of the non-engineered phage and it reduced *E. coli* biofilm counts by over 99%. In recent studies bacteriophages were successfully used to eliminate bacterial biofilms in a combination with antibiotics (Yilmaz *et al.*, 2013) and sharp debridement (Seth *et al.*, 2013). The above selection of studies provide clear evidence for the potential of phages in the removal of problematic biofilms, and also emphasize the need for continued research into the improvement of this application.

The use of phages to eliminate bacterial biofilms is presented in detail in Chapter 13.

Phages for veterinary applications

In addition to the use of phages in animal models of human infections, phages have been widely researched for veterinary applications. The important work of Smith and Huggins at the Institute for Animal Disease Research in Cambridgeshire in the 1980s may have been a catalyst for the reinvestigation of phage therapy for medical treatments in general, including those for veterinary applications (Smith and Huggins, 1982, 1983; Smith *et al.*, 1987a,b). O'Flaherty *et al.* (2009) comprehensively reviewed the literature on the veterinary applications of phages against a range of pathogens and in a range of animals. In addition to those discussed in that review, a number of additional studies have been published in recent years involving *E. coli*, *Salmonella*, *Clostridium*, and *Campylobacter*. Oliveira *et al.* (2009, 2010) investigated the effect of the mode of administration of three phages (phi F78E, phi F258E, and phi F61E) on their dissemination in chickens, and also the efficacy of a cocktail of these three phages in the control of colibacillosis in poultry whether delivered orally or by aerosol. Subsequently, the authors found that the success of phage treatment was dosage dependent, and indicated that 10^9 PFU/ml of phage phi F78E was more effective than 10^7 PFU/ml for elimination of infection (Oliveira *et al.*, 2009). They also indicated that a three phage cocktail was more effective than individual phages (Oliveira *et al.*, 2010). A recent study by Huff *et al.* (2013) showed that only the route of administration which enables high numbers of phages to be delivered to the site of infection is effective in eliminating bacteria. Huff and colleagues (2010) investigated also the effect of prior exposure to phages on the ability of that same phage to treat chicken *E. coli* infection. The study found that mortality in untreated chickens was 55% and this was reduced to 8% following phage treatment, but interestingly pretreatment by the same phage reduced the efficacy of phage therapy in that mortality was 33%. The authors reported that serum IgG levels against the phages were high when phage pretreatment was employed. Thus, prior exposure to phage can limit the efficacy of phage therapy (Huff *et al.*, 2010). Control of *E. coli* O157 carriage in beef cattle using phages has been exploited by a number of authors (Rivas *et al.*, 2010; Coffey *et al.*, 2011). Rozema *et al.* (2009) compared the efficacy of oral and rectal administration of anti-O157 phages and reported that the latter gave rise to improved reduction of O157 levels. Phages have also been employed to help eliminate *Salmonella* in chickens (Andreatti *et al.*, 2007; Borie *et al.*, 2008, 2009; Bardina *et al.*, 2012). In all cases phages were shown to be effective. An interesting study by Waseh *et al.* (2010) demonstrated that truncated and functionally equivalent versions of phage P22 tailspike proteins (Tsp) significantly reduced *S. enterica* colonization in the chicken gut. Significantly, they indicated that the proteins were not degraded by chicken gastro-intestinal fluids. *Campylobacter* carriage in chickens has also been an important focus for phage therapy research as reviewed by O'Flaherty *et al.* (2009). More recently, Carvalho *et al.* (2010) investigated phage administration in feed and oral gavage and found that the former gave a more sustainable reduction in *Campylobacter* numbers. Miller *et al.* (2010) utilized a phage cocktail for treatment of necrotic enteritis (NE) caused by *C. perfringens* in broiler chickens and found that administration via feed or drinking water was effective in the control of this pathogen. Phage therapy was also investigated for use as a low environmental impact option to treat and inactivate bacteria in fish farming plants (Almeida *et al.*, 2009). These authors reported success with phages against *Lactococcus garvieae*, *Pseudomonas plecoglossicida*, *Vibrio harveyi*, and *Aeromonas salmonicida*. In addition, phages specific for other fish pathogens were reported, including *Flavobacterium psychrophilum*, *Flavobacterium columnare*, *Edwardsiella ictaluri*, and *Vibrio parahaemolyticus*; these also could find some therapeutic applications (Pereira *et al.*, 2011; Prasad *et al.*, 2011; Carrias *et al.*, 2011; Castillo *et al.*, 2012). Johnson *et al.* (2008) discussed the literature on the use of phages for both prophylaxis and therapy in cattle, pigs and poultry. These authors indicated that the most promising animal health applications of phage therapy are for acute intestinal, systemic and respiratory infections of the above animals. They also suggested that phage therapy for animal diseases needs to be developed further, although its progression may be held back by improved management practices, vaccines, and prophylactic use of various antimicrobial drugs. However they also stressed that the limitations of phage therapy itself must be considered, and these included knowledge of the epidemiology of the disease agent, the efficacy of phages in fluid as opposed to other matrices, and the timing and mode of administration. In general these authors postulate that if phage therapy is to realize its potential, it may depend on its complementarity with other treatment approaches (Johnson *et al.*, 2008).

The use of phages in treatment of infections in production animals is discussed in detail in Chapter 8.

Additional applications of phages

In addition to the above, phages have also been used for the control of disease-causing phytobacteria, as reviewed by Jones *et al.* (2007), Balogh *et al.* (2010), Frampton *et al.* (2012) and Jones *et al.* (2012); this topic is discussed in detail in Chapter 7. Withey *et al.* (2005) discussed the use of phages for improvement of the quality of effluent in waste-water treatment processes and Cherwonogrodzky (2005) has reviewed their possible role in defence against biothreats. Hagens and Loessner (2007; 2010) have described the application of phages for both the detection and control of pathogens in food systems, a topic which was also reviewed by the European Food Safety Authority (Anonymous, 2009), Coffey *et al.* (2010), Mahony *et al.* (2011), Sillankorva *et al.* (2012), and Brovko *et al.* (2012). The use of phages against food-borne pathogens is presented in detail in Chapter 9.

Phage genome analysis has also been exploited with the aim of discovering novel antimicrobial agents (Liu *et al.*, 2004). In this study 26 *S. aureus* phages were sequenced and phage proteins that inhibited cellular growth upon expression in *S. aureus* host cells were identified. Cellular targets for many of these proteins were subsequently identified in these bacteria, and were shown to be essential components of host DNA replication and transcription machineries. Interaction between a phage protein and a putative helicase loader was then used to screen for inhibitory molecules, and several were identified which inhibited bacterial growth and DNA synthesis. This suggested that mimicking the inhibitory effect of phage polypeptides on bacterial growth by chemical compounds may yield novel antimicrobials to combat disease causing pathogens.

Interestingly, there are some data to show that phages can exert anticancer activity. This activity can result both from direct interactions of phages with cancer cells and from inducing anticancer immune responses (for a review, see Budynek *et al.*, 2010). In the case of phage lysates, anticancer activity can be mediated not only by phage virions, but also by some components of bacterial cells (Dąbrowska *et al.*, 2010). It appears that these effects of phages may have implications for phage therapy in cancer patients. Phages can also interact with some populations of immune cells including T and B cells, neutrophils, and macrophages; these interactions, as well as their implications for phage therapy, were discussed in a recent review (Górski *et al.*, 2012). Miedzybrodzki *et al.* (2005) explored three mechanisms whereby phages could be exploited to inhibit human virus infections. These were (1) interferon induction by phage nucleic acid causing prevention of pathogenic virus infection (2) direct competition of phages and viruses for cellular receptors; and (3) phage-mediated display or induction of antibodies, which may have subsequent antiviral action. These authors postulated that these antiviral applications may be a cost-effective alternative to conventional antiviral drugs. Interactions of bacteriophages with eukaryotic cells which can be of importance to phage therapy, as well as anti-viral effects of phages are discussed in detail in Chapter 5.

The therapeutic use of phage lysins

The work cited above concentrates on the use of whole phages as antibacterial agents. Recent research on phage genomes has increased our knowledge of the mechanisms by which phages infect host cells and carry out their own replication, and has created opportunities to scan the genomes of therapeutic phages for possible novel antibacterials. The best example of such work is the utilization of the lytic enzymes of phages, which mediate bacterial cell death. These enzymes, also known as virolysins, endolysins, or simply lysins, have been extensively evaluated for therapeutic applications since the first *in vivo* trial in 2001 (Nelson *et al.*, 2001). The progress in the area has been evaluated recently by Fischetti (2010), Nelson *et al.* (2012), and Schmelcher *et al.* (2012). Lysins are peptidoglycan hydrolases encoded by double stranded DNA phages and are expressed at the end of the phage lytic cycle, facilitating host cell lysis and phage progeny release (Young *et al.*, 2000; Catalão *et al.*, 2013). The name endolysin was first used by Jacob and Fuerst (1958) to designate a lytic substance of protein origin, which was produced in bacterial cells during phage multiplication, and which acted on the cell wall internally (Jacob and Fuerst, 1958). The first lysins however were only reported to be active against dead cells (Ralston *et al.*, 1955), and it was not until the 1970s that a lysin (PAL) active against both live and dead *S. aureus* cells was identified (Sonstein *et al.*, 1971). This early work stimulated the identification of lysins from phages of *Staphylococcus, Bacillus, Listeria* and *Clostridium* in the 1990s, and in turn encouraged the recent surge of interest in these naturally occurring phage enzymes as therapeutic agents (Borysowski *et al.*, 2006; Courchesne *et al.*, 2009, Fenton *et al.*, 2010a, Fischetti, 2010; Nelson *et al.*, 2012; Schmelcher *et al.*, 2012).

Today we know that lysins generally display a dual domain modular structure consisting of an N-terminal catalytic domain and a C-terminal cell wall binding domain (Perez-Dorado *et al.*, 2007; Fischetti, 2010). While recombinant lysins can kill Gram-positive bacteria when acting on bacterial cells from outside ('lysis from without'), this approach is not as straightforward in Gram-negative bacteria, given that they possess an outer membrane which prevents access of exogenous lysin to the peptidoglycan. Nevertheless, for Gram-positives, lysins represent a viable genus-specific antibacterial agent. Indeed the investigations and studies conducted to date suggest that the advantages of the use of phage lysins for therapeutic purposes far outweigh the disadvantages (Table 6.2).

Table 6.2 Advantages and disadvantages of endolysins as antibacterials

Advantages	Disadvantages
Can be used both as a prophylactic and for treatment	Susceptible to inactivation due to their protein nature
Can be considered protein therapeutics, as compared to viral nature of phages	No action against Gram-negative bacteria[1]
Capacity to kill antibiotic-resistant bacteria	Costs related to production
Lysin resistance has not been reported to date	Immunogenicity
Genus-specific antibacterial, which will not eliminate commensal flora	Short serum half-life
Can be engineered and improved	
Non-toxic	
Can potentially be used in a range of environments (Human, animal, food, biofilm)	

[1] Although generally lysins are active against Gram-positive bacteria only, some lysins were reported that are also capable of killing Gram-negative bacteria (Borysowski *et al.*, 2006). Besides, recently some modifications of lysin molecules were described that enable these enzymes to kill Gram-negative bacteria (Lukacik *et al.*, 2012).

Lysins possess several attributes as novel antibacterials. These include especially a novel mode of action, the capacity to kill antibiotic-resistant bacteria, and a narrow antibacterial range (Borysowski *et al.*, 2006; Fischetti *et al.*, 2010; Schmelcher *et al.*, 2012). In addition, some lysins were shown to be capable of eliminating bacterial biofilms (Son *et al.*, 2010; Shen *et al.*, 2013; Fenton *et al.*, 2013). Nevertheless, several factors must be taken into account if these enzymes are to realize their full potential. First, lysin preparations should not pose a toxicity risk when used in animals or humans. To date the research reports indicate that no toxicity or side-effects have been observed in treatment of topical or systemic infections in murine models (Loessner, 2005; Fenton *et al.*, 2010a). One of the problems associated with lysin treatment could be production of pro-inflammatory cytokines following lysis of bacterial cells (Entenza *et al.*, 2005). However, increased cytokine production in lysin treatment of systemic infections may be reduced by administering regulated smaller doses of enzyme (Fischetti, 2008). Second, immunogenicity must be considered, as lysins are capable of eliciting an immune response when administered mucosally or systematically (Fischetti, 2008). This could potentially be detrimental to their own activity. Several *in vitro* and *in vivo* trials have been undertaken to investigate this and these generally found that while lytic activity was diminished by antibodies, it was not completely inhibited (Loeffler *et al.*, 2003; Fischetti, 2005). Attempts were also made to reduce immunogenicity of the Cpl-1 lysin by conjugating to polyethylene glycol (PEG); however, conjugation of PEG molecules resulted in a significant decrease in lysin activity (Resch *et al.*, 2011). Possible resistance to lysins must also be taken into account when considering their potential therapeutic use. Since phages have evolved over millions of years to utilize their lysins to kill their host bacteria, bacterial resistance to these enzymes is improbable. Indeed research to date has indicated that the lysins have evolved to target specific molecules in the peptidoglycan, which are essential for bacterial viability (Loessner, 2005; Fischetti, 2008). An example of this is choline, which is essential for pneumococcal viability and which is the cell wall receptor for the pneumococcal lysin (Garcia *et al.*, 1983; Lopez *et al.*, 2004). Other studies have shown that repeated exposure of *B. cereus* and *S. pneumoniae* to low concentrations of lysin on both agar plates and in liquid culture did not give rise to resistant mutants even after numerous challenges (Loeffler *et al.*, 2001; Schuch *et al.*, 2002). In addition, polysaccharide capsules of *S. pneumoniae* and *B. anthracis*, which could potentially mask the peptidoglycan were reported not to inhibit lysin activity (Loeffler *et al.*, 2001; Schuch *et al.*, 2002). In theory, the use of lysins with two catalytic domains, each targeting different peptidoglycan bonds could potentially act as a safeguard against resistance to one enzymatic activity, if this arose (Fischetti, 2008; Becker *et al.*, 2008). Importantly, lysins may be applied in synergy with other lysins or conventional antibiotics; this has been demonstrated in both *in vitro* and *in vivo* studies by several authors (Jado *et al.*, 2003; Loeffler and Fischetti, 2003 Becker *et al.*, 2008).

Several animal trials have revealed the ability of lysins to successfully eradicate, or at least reduce, pathogenic bacterial colonization of mucosa which can provide a starting point for infection (Nelson *et al.*, 2001; Loeffler *et al.*, 2001, 2003; Cheng *et al.*, 2005; McCullers *et al.*, 2007; Fenton *et al.*, 2010b; Pastagia *et al.*, 2011). A recent review from our laboratory summarizes the effective use of lysins against a range of bacterial pathogens including streptococci, *S. aureus*, enterococci, *B. anthracis*, and *Clostridium* (Fenton *et al.*, 2010a). It needs to be stressed that lysins can be used to eliminate antibiotic-resistant strains of important pathogens, including MRSA (Borysowski *et al.*, 2011). In the context of dental caries, the use of bacterial autolysins has been explored by Yoshimura *et al.* (2006) and Thanyasrisung *et al.* (2009) who demonstrated that N-acetylmuraminidases specific for *S. sobrinus* and *S. mutans* might be used to eliminate the causative agents of dental caries and periodontal disease. This could be achieved by incorporation of the lysin into chewing gum, mouthwash or dentifrice. Thus, it appears that also bacteriophage lysins could find similar applications. One such enzyme could be a lysin encoded by *Actinomyces naeslundii* phage Av-1 which is capable of eliminating the causative agent of gingivitis and root surface caries (Courchesne *et al.*, 2009).

Fenton *et al.* (2010a) summarized the different lysins targeting pathogenic bacteria. In addition to those, a number of additional lysin studies have appeared in the recent literature. Daniel *et al.* (2010) constructed a chimeric lysin (ClyS) by fusing the N-terminal domain of the *S. aureus* phage Twort lysin with the C-terminal domain from a different *S. aureus* phage lysin (phiNM3). This chimeric enzyme lysed all *S. aureus* regardless of antibiotic resistance profile in vitro. In an *in vivo* murine study, it protected against MRSA induced septic death when used in combination with oxacillin. Interestingly, the doses of ClyS and oxacillin, which protected mice synergistically did not protect these mice when used alone (Daniel *et al.*, 2010). Subsequently, this chimeric lysin was formulated into an ointment and tested as a topical skin decolonization treatment in a mouse model of infection. ClyS effected a greater log reduction in *S. aureus* than the standard topical antibacterial agent mupirocin and in *in vitro* tests demonstrated a lower potential for the development of resistant bacteria than mupirocin (Pastagia *et al.*, 2011). The lysin LysGH15 was also recently investigated by Gu *et al.* (2011) and a single intraperitoneal injection of 50 μg of lysin was sufficient to protect mice who had been administered MRSA injections at double the lethal dose, highlighting that lysins can be effective therapeutics even when administered at relatively low doses (Gu *et al.*, 2011). Other lysins with potential therapeutic applications which were reported recently include enzymes specific for *Bacillus cereus* (Park *et al.*, 2012), *Listeria monocytogenes* (Zhang *et al.*, 2012), *C. perfringens* (Seal, 2013), *S. aureus* (Mishra *et al.*, 2013), as well as a broad-range lysin capable of lysing several different Gram-positive pathogens such as *Streptococcus pyogenes* and *S. aureus* including MRSA (Gilmer *et al.*, 2013). In recent times genetic screening strategies have also been increasingly used for the identification of novel lysins. Fischetti and colleagues (Schuch *et al.*, 2009) used this approach to successfully identify the lysin PlyG from a *B. anthracis* phage. This method is technically applicable to the cloning, expression, and analysis of lysins from any other phage. The group of Fischetti has also developed a two-step screening technique for the cloning of phage lysins from uncultured viral DNA (Schmitz *et al.*, 2010). When they applied this approach to a viral metagenomic library constructed from animal faecal samples, 26 lysins were expressed and cloned. Interestingly these included Gram-positive-like enzymes, Gram negative-like enzymes and several additional enzymes whose predicted structures were less similar to those from known phages. The genomes of strains of *C. perfringens* were also computationally screened for lysins and lysin-like open reading frames (ORFs) in resident prophages and several enzymes were identified. This study clearly demonstrated that the field of genomics/metagenomics can serve as an abundant source of novel lysins (Schmitz *et al.*, 2011). It is also possible to engineer lysins with different combinations of catalytic and binding domains, thus increasing their potential for application. O'Flaherty *et al.* (2009) and Schmelcher *et al.* (2012) recently indicated the possibilities for engineering lysins and these can include (1) an exchange of the binding and catalytic domains (2) construction of a chimeric lysin with two distinct lytic activities, (3) construction of a chimeric lysin with two distinct genus specificities, and (4) removal of the C-terminal domain. By engineering lysins, one can obtain enzymes with higher antibacterial activity. Indeed, in a recent study from our laboratory, it was shown that a truncated derivative of the staphylococcal phage lysin (LysK) showed increased lytic activity against live *S. aureus* cells (Horgan *et al.*, 2009). This truncated derivative CHAP$_K$ was subsequently used in murine decolonization trials (Fenton *et al.*, 2010b).

In addition to their use for therapeutic purposes in animals and humans, lysins, like their parent phages, also have many other potential uses including food safety, control of phytopathogens, veterinary applications and also as a diagnostic tool in pathogen detection. These applications have been described in detail in Fenton *et al.* (2010a), Schmelcher *et al.* (2012) and in Nelson *et al.* (2012). Recent developments in the area include exploring the possibility of using lysins as agents for the elimination of bacteria, which have an outer membrane including mycobacteria and Gram-negative genera. Briers *et al.* (2011) tested the combination of *P. aeruginosa* endolysin EL188 and several OM permeabilizing compounds on *P. aeruginosa* strains and showed a 4-log reduction in the target bacterium. Lukacik *et al.* (2012) reported some modifications of a lysin's molecule which also enable killing of Gram-negative bacteria. Thus the application of phage lysins need not be limited to Gram-positive bacteria. The Gram-positive pathogen *Mycobacterium* presents a similar problem for exogenous endolysins in that they must circumvent a mycolic acid-rich outer membrane in order to gain access to the peptidoglycan layer. Recent studies have shown that mycobacteriophages encode a lipolytic enzyme (termed LysB) in addition to the cell-wall degrading lysin (Gill *et al.*, 2008, 2010). A similar enzyme was also identified in other mycobacteriophage genomes including Ardmore phage (Henry *et al.*, 2010). Mycobacteriophage endolysins were discussed in a recent review (Payne and Hatfull, 2012).

The use of lysins to eliminate bacteria is presented in detail in Chapter 14.

Genetically engineered phages

In general, phage therapy exploits natural lytic phages for the elimination of the problematic bacteria. However another approach, which in recent times is becoming more common, is the genetic manipulation of phages towards improved therapeutic characteristics (Goodridge, 2010; Moradpour and Ghasemian, 2011). There are a number of approaches that can be used, and these are largely based on events which can occur naturally in phage evolution. These include:

1. Broadening the host range of therapeutic phages, which can be achieved by manipulations leading to novel or multiple receptor recognition by phages (Pouillot *et al.*, 2010); an alternative approach is the isolation of natural host range phage mutants (Morona and Henning, 1984).
2. Enabling phage to kill without bringing about cell lysis and release of endotoxins. This involves using phages to deliver genes encoding lethal agents to bacterial cells (Westwater *et al.*, 2003) or employing engineered lysis-deficient virulent or temperate phages (Matsuda *et al.*, 2005; Paul *et al.*, 2011).
3. Temperate phages being genetically modified to render them exclusively lytic. This can be achieved by inactivating genes involved in lysogeny (Rapson *et al.*, 2003).
4. Phages being modified to enable them to survive longer in the mammalian circulatory system, a process which can be achieved by altering amino acids in capsid proteins (Merril *et al.*, 1996).

In the context of broadening the host range of individual phages, Morona and Henning (1984) demonstrated that the T-even phage Ox2 can give rise to mutants with a broadened host range. These mutants were capable of using two bacterial receptors (OmpA and OmpC) rather than a single receptor (OmpA) used by the wild-type phage. Similarly, Moreno and Wandersman (1980) observed that mutants of another T4-type phage TuIa recognized the receptors OmpC and LamB, while the wild-type phage employed the OmpF receptor for recognition. An interesting mechanism of host range alteration has also been reported by Liu *et al.* (2002), who reported that the *Bordetella* phage BPP-1 could alter its tail fibre proteins using a reverse transcriptase enzyme. This resulted in the display of different surface proteins depending on whether the phages were actively infecting their host bacteria or surviving outside of their hosts. Tetart *et al.* (1996) isolated a host range mutant of T4 phage with a duplicated segment of the tail fibre adhesin gene, responsible for recognition of the bacterial surface receptor. This resulted in the phage being able to adsorb to *Yersinia* in addition to *E. coli* and closely related *Shigella* species.

A sophisticated method of generating phage variants with new host ranges was recently reported by Pouillot *et al.* (2010). These authors genetically engineered phage T4 with a view to detection and possible eradication of threats from uncharacterized emerging bacterial pathogens, which may be implicated in bioterrorism threats. They developed very large, genetically engineered lytic phage banks containing, in an *E. coli* host, a very wide spectrum of variants for any chosen phage-associated function, including phage host-range. Screening of such a bank should allow the rapid isolation of recombinant T4 particles capable of detecting, infecting, and destroying hosts belonging to Gram-negative bacterial species far removed from the original *E. coli.*

Exploitation of phages for the elimination of bacterial cells without potentially harmful phage-mediated lysis is also possible. In this approach, bacteria are killed as a result of multiplication of engineered non-lytic phage within bacterial cells (both virulent and temperate phage can be engineered this way; Matsuda *et al.*, 2005; Paul *et al.*, 2011) or alternatively the phage virion is exploited for the delivery of genes encoding inhibitory agents to bacterial cells. In the latter case, bacteriophages themselves are not supposed to kill bacteria, but rather to deliver the above-mentioned genes to bacterial cells. This is generally achieved by employing filamentous phages whose progeny are chronically released from the cell. Hagens *et al.* (2004) has described the successful utilization of this method for the elimination of *Pseudomonas.* They replaced an export protein gene in the filamentous *Pseudomonas* phage Pf3 with a gene for the restriction endonuclease *Bgl*II. Treatment of *Pseudomonas* infections in mice with phage Pf3 and with a lytic phage resulted in higher survival rates and reduced inflammatory responses in the mice treated with phage Pf3. The reduced inflammatory response was likely due to reduced endotoxin release from the bacteria where the filamentous phage was employed. A similar approach was used to eliminate *E. coli* (Hagens and Blasi, 2003). Phages have also been used to deliver to bacterial cells the gene encoding the *doc* toxic protein, which was derived from phage P1. This was shown by Norris *et al.* (2000) to exhibit bactericidal activity towards *E. coli* and bactericidal or bacteriostatic activity towards *P. aeruginosa*, *S. aureus*, and *E. faecalis*. In another study, Westwater *et al.* (2003) developed recombinant non-lytic phages based on the M13 phagemid system and the post-segregational killing toxins Gef and ChpBk encoded by the *gef* and *chpBk* genes which originated on the *E. coli* K12 chromosome. The recombinant phage was able to reduce target bacterial numbers by several orders of magnitude *in vitro* and in an *in vivo* mouse model (Westwater *et al.*, 2003). In another study, Moradpour *et al.* (2009) engineered a non-replicating M13 phagemid to express a lethal catabolite gene activator protein for the elimination of *E. coli* O157:H7. While in all above studies filamentous phages were used as vehicles for genes encoding antibacterial proteins, Fairhead (2004) removed the lysis genes from non-filamentous phages and replaced them with genes encoding small acid-soluble spore proteins (SASP) from *Bacillus*. These were then used to eliminate pathogenic bacteria after phage infection. The SASP proteins concerned were DNA-binding proteins which when expressed halted all cellular activity. This system provided few opportunities for bacterial resistance and is being investigated for the control of MRSA and *C. difficile* (Hanlon, 2007).

Lu and Collins (2007) engineered the *E. coli* phage T7 to express the enzyme dispersin B intracellularly during infection, and upon cell lysis to release it into the extracellular environment, where it was effective in degrading bacterial biofilms.

Subsequently they engineered phage M13mp18 to overexpress a repressor (*lex*A3) of the SOS response in bacteria which can reduce the therapeutic efficacy of some antibiotics. Suppression of the SOS network in *E. coli* with this engineered phage significantly enhanced the killing of these cells by quinolone antibiotics (Lu and Collins, 2009). Thus, such phage could be used to augment antibacterial effects of antibiotics. In the context of temperate phage being genetically modified to render them exclusively lytic, it is noteworthy that temperate phages, since they reside in bacteria, are theoretically more ubiquitous than virulent phages, and thus their exploitation and manipulation may prove useful in some circumstances. Examples of this include the work of Rapson *et al.* (2003), who successfully mutated the *vir* gene of a temperate phage which gave rise to a stable lytic phage. Atsumi and Little (2004) replaced the Cro repressor of coliphage lambda with a genetic module containing the Lac repressor and several Lac operators, giving rise to mutant phages with different regulatory behaviours, some of which behaved lytically under controlled conditions.

It is also possible to engineer temperate phages to express genes toxic to bacteria only after integration into the bacterial chromosome as shown by Schaak (2004), who used a recombinant phage employing a gene for a toxin protein, which was bactericidal intracellularly but not extracellularly.

Modification of phages enabling them to survive longer in the mammalian circulatory system can be very beneficial where phages are being exploited to eliminate infections. Merril and colleagues identified long circulating phages with enhanced survival in blood, which resulted from amino acid shifts in the major phage capsid (E) protein (Merril *et al.*, 1996). These phages were more effective than their parental strains in a murine model of bacteraemia (Merril *et al.*, 1996). This group subsequently showed that the change of glutamic acid to a lysine at residue 158 of this major capsid protein was sufficient to confer the 'long-circulating' phenotype (Vitiello *et al.*, 2005).

Non-recombinant modification of phage genomes can also be useful for improving the efficacy of phages in a therapeutic application. In a recent study, a serial passage technique was used to generate r/m (restriction modification) modified derivatives of the *S. aureus* phage K. These modified phage circumvented the r/m systems of several *S. aureus* strains which were previously impervious to lysis by the parental phage K. The combination of six modified phages in a cocktail with phage K resulted in the lysis of 24 of 29 strains, which had been previously resistant to phage K (Kelly *et al.*, 2011). A bioinformatic study by Skiena (2001) suggested that a large proportion of restriction sites could actually be removed without changing the phage phenotype. This would be achieved by exploiting the triplet codes redundancy and searching for the coding sequence which minimized the number of restriction sites while at the same time coding for the desired phage proteins. This approach would render phages capable of avoiding elimination by indigenous bacterial restriction modification systems.

As indicated by the examples above, genetic modification (as well as isolation of natural phage mutants) affords many opportunities to improve phages and these modifications have the potential to address many of the natural shortcomings of lytic phages themselves. The use of genetically modified phages to eliminate bacteria is discussed in detail in Chapter 15.

Phages as delivery vectors

The characteristics required of an ideal biological delivery vehicle include: (1) sufficient packaging size for the intended molecule(s), (2) specificity for its target cell, (3) efficient delivery of the cargo molecule(s), and (4) the ability to evade recognition by the host immune system. Currently, viral and liposomal vectors show the most promise as delivery vehicles for drugs to treat a wide range of diseases. However they have several limitations including limited packaging capacity, low integration probability, reduced efficacy upon repeat administration, risk of strong immune responses, inflammatory and toxicity risks, and insertional mutagenesis risks (Seow and Wood, 2009). To this end, alternative delivery vehicles are increasingly being investigated including phages and phage virus-like particles (VLP), which have some significant advantages as delivery vectors. These include (1) the ability to package RNA,

DNA, nanoparticles, chemotherapeutics, proteins, and antisense oligonucleotides, (2) relative resistance to non-enzymatic degradation including high temperatures and pH extremes, (3) the ability to engineer targeting molecules into phage capsid proteins, (4) cheap large scale purification processes compared with cell-based mammalian systems (Seow and Wood, 2009; Ashley *et al.*, 2011). In the context of the use of bacteriophages as delivery vehicles it is important that some phages have high stability and can withstand different unfavourable physicochemical conditions (Jończyk *et al.*, 2011). For example, Jepson and March (2004) reported that phage lambda, one of the phages used as a delivery vehicle could retain infectivity and structural integrity for as long as 6 months at 4°C in water, which is longer than any other biological delivery vehicle, including mammalian viruses. It is also stable over a pH range of 3–11, making it a suitable candidate for delivery of DNA vaccine using the oral route (Jepson and March, 2004).

Several studies showed that phages can be successfully used as vehicles to deliver to bacterial cells different compounds exerting antibacterial activity. For example, a study by Embleton *et al.* (2005) evaluated the use of photosensitizers delivered by phages as bactericidal agents. To that end *S. aureus* phage 75 was covalently cross-linked to the photosensitizer tin (IV) chlorin e6 (SnCe6). When exposed to red light, considerable reductions in MRSA and VISA titres were achieved using as little as 1.5 μg/ml of the conjugated drug. Yacoby *et al.* (2006) exploited the labile-linkage of toxins which are lethal to bacteria. In this case, filamentous phages were genetically modified to display a targeting molecule on their surface, and deliver a high concentration of chloramphenicol to the *S. aureus* target bacteria. They showed that drugs could be linked to the phage by chemical conjugation. Therefore controlled release was possible by mixing a constant ratio of chloramphenicol molecules per phage. The drug employed is devoid of cytotoxic activity and is not activated until it dissociates from the phage at its intended target site. This approach allows the drug to be brought specifically to the target bacterium, thus avoiding the elimination of non-target bacteria. This technique may allow for the reintroduction of non-specific drugs that may have been excluded from antibacterial use because of their toxicity or low selectivity (Yacoby *et al.*, 2006). This research group also addressed the limitation regarding the low number of drug molecules, which could be attached to each phage molecule (Yacoby *et al.*, 2007). This line of research is discussed in detail in Chapter 16.

Phages can also be used as delivery vehicles for vaccines, and this can be accomplished in two main ways. Direct vaccination involves the use of phages, which exploit phage display to carry vaccine antigens on their surface. In contrast, phages can also be used to facilitate DNA vaccination (Clark and March, 2004a,b; Thomas *et al.*, 2012), an indirect method involving the incorporation of a vaccine gene, which is regulated by a eukaryotic expression cassette into the phage DNA. The vaccine proteins are expressed following uptake of the whole phage and subsequent transcription and translation of the vaccine DNA by immune cells. In the first method, phages can be constructed to display specific antigens (Cui *et al.*, 2013), or alternatively phages which display libraries of peptides can be screened with antiserum sample or specific monoclonal antibodies to isolate novel peptides which mimic the structure of certain epitopes of an antigen; these peptides, termed mimotopes, can be used as surrogate antigens even in cases where the primary antigen has not been identified (Knittelfelder *et al.*, 2009; Gazarian *et al.*, 2012; Li *et al.*, 2012). Phages displaying proteins as antigens have been used as vaccines in different animal models of infection (Irving *et al.*, 2001; Wang and Yu, 2004; Cui *et al.*, 2013). For example, a recombinant phage based-vaccine was developed for the vaccination of pigs against *Taenia solium* induced cysticercosis, which is also the causative agent of neurocysticercosis in humans (Manoutcharian *et al.*, 2004). This same author also reports that a filamentous phage vaccine trial was carried out on a small group of multiple myeloma patients and demonstrated that phage vaccination induced tumour-specific immune responses exerting beneficial effects on human patients (Manoutcharian, 2005). The efficacy of the second approach, namely DNA vaccination using phage was demonstrated by March *et al.* (2004) in mice and rabbits. These authors used phage lambda containing the DNA vaccine cassette encoding either a hepatitis B surface antigen (HBsAg) or a green fluorescent protein antigen (EGFP). Upon delivery of the DNA using the lambda-EGFP phage in two separate vaccinations, the uptake and expression of the cassette brought about specific anti-EGFP responses up to almost one month post-vaccination in mice. Similarly, two vaccinations of rabbits using the lambda-HBsAg phage brought about high level anti-HBsAg responses in one out of four animals, while a third vaccination showed high level responses in all four animals which persisted for more than 6 months. Interestingly high anti-phage antibody levels were observed in all rabbits after the first immunization with lambda-HBsAg, however these did not prevent subsequent immune responses against the DNA vaccine components. Clark *et al.* (2011) compared the efficacy of a lambda phage delivered DNA vaccine and a commercial recombinant protein vaccine against hepatitis B and found that in general antibody responses in phage vaccinated rabbits were stronger than in those vaccinated with the recombinant protein vaccine. They suggested that due to the stability of phage vaccines at a range of temperatures, and their low cost of production that further studies into phage vaccination against hepatitis B were indeed warranted (Clark *et al.*, 2011). The stability and suitability of phage lambda as a workhorse for DNA vaccine delivery has also been reported by Jepson and March (2004). It is clear that phages possess undoubted potential as vaccine delivery vehicles. The fact that phages themselves are natural stimulants of the immune system means that any antigen molecule presented on the phage protein coat has natural adjuvant or synergistic activity and does not require conjugation to any carrier molecules before immunization (Clark and March, 2006; Sartorius *et al.*, 2011; Pouyanfard *et al.*, 2012). Some strategies of engineering phage were also developed to enhance immunogenicity of presented antigens (Henry *et al.*, 2011). Clark and March (2004a) have also reviewed the possibility of producing a hybrid phage vaccine for targeting of both the cellular and humoral arms of the immune system. This type of vaccine would be composed of a promoter driven DNA vaccine internally contained within a phage particle and a phage display variant of the same antigen on the phage surface. Successful use of such a vaccine in mice was recently reported by Tao *et al.* (2013).

Phages have also been used as vehicles for the delivery of labels for the detection of specific bacteria, as reviewed by Petty *et al.* (2006), Singh *et al.* (2012), and Smartt *et al.* (2012). An example of the successful application of this approach is the use of bioluminescent phage for the diagnostic detection of *Yersinia pestis* (Schofield *et al.*, 2009). Another potential application of phages is their use as imaging probes for cancer cells and as delivery vehicles for anticancer compounds. For example, Li *et al.* (2010) chemically modified reactive groups on the surface of phage M13 allowing attachment of molecules for targeting (folic acid) and also for imaging (a fluorescent probe) of cancer cells. One such modified M13 phage had applications for image analysis of human KB cancer cells. Phages were also employed as gene therapy vectors (Larocca *et al.*, 2002; Kia *et al.*, 2012). Dickerson *et al.* (2005) used filamentous phage as a vehicle for cocaine-sequestering antibodies and showed that the phage could penetrate the central nervous system and block some of the cocaine effects in a rodent model; such phage could be applied in the treatment of cocaine addiction. Overall there are a number of benefits of using phages as delivery vehicles. These include their potential to package novel functionalities into the particle, their significant packaging capacity, their safety in humans and animals, and their ability to package non-DNA loads. Phages are easy to purify and also stable, and, in some cases, have the capacity to cross the gastrointestinal tract and enter the bloodstream. This suggests that oral phage vaccines may potentially be viable options as vaccines for human and veterinary uses and also in the developing world (Clark and March, 2004a). The low cost is also an important advantage as companies are investing less in expensive conventional drug and vaccine developments, and thus phage-based technologies may indeed offer novel strategies for vaccine and drug delivery.

Overall, it is clear that the use of phages, whether whole phage themselves, their lytic enzymes, genetically modified phages, or phage delivery vehicles and derived vaccines have a considerable array of applications as therapeutics in the modern medical and veterinary fields. The continued antibiotic resistance problem coupled with the economic strain on global health systems necessitates the exploitation of cheap, natural, readily available, safe and efficient therapeutic agents. Phages satisfy all of these criteria and their reintroduction as medical treatment options in the western world is worthy of strong consideration.

References

Abedon, S.T. (2010). The 'nuts and bolts' of phage therapy. Curr. Pharm. Biotechnol. *11*, 1.

Abedon, S.T., and Thomas-Abedon, C. (2010). Phage therapy pharmacology. Curr. Pharm. Biotechnol. *11*, 28–47.

Abedon, S.T., Kuhl, S.J., Blasdel, B.G., and Kutter, E.M. (2011). Phage treatment of human infections. Bacteriophage. *1*, 66–85.

Abhilash, M., Vidya, A., and Jagadevi, T. (2009). Bacteriophage therapy: A war against antibiotic resistant bacteria. Internet J. Alt. Med. *2009*, 7.

Alisky, J., Iczkowski, K., Rapoport, A., and Troitsky, N. (1998). Bacteriophages show promise as antimicrobial agents. J. Infect. *36*, 5–15.

Almeida, A., Cunha, A., Gomes, N.C., Alves, E., Costa, L., and Faustino, M.A. (2009). Phage therapy and photodynamic therapy: low environmental impact approaches to inactivate microorganisms in fish farming plants. Mar. Drugs. *7*, 268–313.

Andreatti Filho, R.L., Higgins, J.P., Higgins, S.E., Gaona, G., Wolfenden, A.D., Tellez, G., and Hargis, B.M. (2007). Ability of bacteriophages isolated from different sources to reduce *Salmonella enterica* serovar enteritidis *in vitro* and *in vivo*. Poult. Sci. *86*, 1904–1909.

Anonymous. (2009). The use and mode of action of bacteriophages in food production. EFSA J. *1076*, 1–26.

Ashley, C.E., Carnes, E.C., Phillips, G.K., Durfee, P.N., Buley, M.D., Lino, C.A., Padilla, D.P., Phillips, B., Carter, M.B., Willman, C.L., *et al.* (2011). Cell-specific delivery of diverse cargos by bacteriophage MS2 virus-like particles. ACS Nano. *5*, 5729–5745.

Atsumi, S., and Little, J. (2004). Regulatory circuit design and evolution using phage λ. Genes Dev. *18*, 2086–2094.

Azeredo, J., and Sutherland, I.W. (2008). The use of phages for the removal of infectious biofilms. Curr. Pharm. Biotechnol. 9, 261–266.

Balogh, B., Jones, J.B., Iriarte, F.B., and Momol, M.T. (2010). Phage therapy for plant disease control. Curr. Pharm. Biotechnol. *11*, 48–57.

Bardina, C., Spricigo, D.A., Cortés, P., and Llagostera, M. (2012). Significance of the bacteriophage treatment schedule in reducing Salmonella colonization of poultry. Appl. Environ. Microbiol. *78*, 6600–6607.

Becker, S.C., Foster-Frey, J., and Donovan, D.M. (2008). The phage K lytic enzyme LysK and lysostaphin act synergistically to kill MRSA. FEMS Microbiol. Lett. *287*, 185–191.

Biswas, B., Adhya, S., Washart, P., Paul, B., Trostel, A.N., Powell, B., Carlton, R., and Merril, C.R. (2002). Bacteriophage therapy rescues mice bacteremic from a clinical isolate of vancomycin-resistant *Enterococcus faecium*. Infect. Immun. *70*, 204–210.

Borie, C., Albala, I., Sánchez, P., Sánchez, M.L., Ramírez, S., Navarro, C., Morales, M.A., Retamales, A.J., and Robeson, J. (2008). Bacteriophage treatment reduces Salmonella colonization of infected chickens. Avian Dis. *52*, 64–67.

Borie, C., Sánchez, M.L., Navarro, C., Ramírez, S., Morales, M.A., Retamales, J., and Robeson, J. (2009). Aerosol spray treatment with bacteriophages and competitive exclusion reduces *Salmonella enteritidis* infection in chickens. Avian Dis. *53*, 250–254.

Borysowski, J., and Górski, A. (2008). Is phage therapy acceptable in the immunocompromised host? Int. J. Infect. Dis. *12*, 466–471.

Borysowski, J., Weber-Dabrowska, B., and Górski, A. (2006). Bacteriophage endolysins as a novel class of antibacterial agents. Exp. Biol. Med. (Maywood). *231*, 366–377.

Borysowski, J., Lobocka, M., Miedzybrodzki, R., Weber-Dabrowska, B., and Górski, A. (2011). Potential of bacteriophages and their lysins in the treatment of MRSA: current status and future perspectives. Biodrugs. *25*, 347–350.

Branston, S., Stanley, E., Ward, J., and Keshavarz-Moore, E. (2011). Study of robustness of filamentous bacteriophages for industrial applications. Biotechnol Bioeng. *108*, 1468–1472.

Briers, Y., Walmagh, M., and Lavigne, R. (2011). Use of bacteriophage endolysin EL188 and outer membrane permeabilizers against *Pseudomonas aeruginosa*. J. Appl. Microbiol. *110*, 778–785.

Brovko, L.Y., Anany, H., and Griffiths, M.W. (2012). Bacteriophages for detection and control of bacterial pathogens in food and food-processing environment. Adv. Food Nutr. Res. *67*, 241–288.

Brüssow, H. (2012). What is needed for phage therapy to become a reality in Western medicine? Virology *434*, 138–142.

Brüssow, H. (2013). Bacteriophage–host interaction: from splendid isolation into a messy reality. Curr. Opin. Microbiol. *16*, 500–506.

Bruttin, A., and Brüssow, H. (2005). Human volunteers receiving *Escherichia coli* phage T4 orally: a safety test of phage therapy. Antimicrob. Agents Chemother. *49*, 2874–2878.

Budynek, P., Dabrowska, K., Skaradziński, G., and Górski, A. (2010). Bacteriophages and cancer. Arch. Microbiol. *192*, 315–320.

Cairns, B.J., and Payne, R.J. (2008). Bacteriophage therapy and the mutant selection window. Antimicrob. Agents Chemother. *52*, 4344–4350.

Caldwell, J. (1928). Bacteriologic and bacteriophagic study of infected urines. J. Infect. Dis. *39*, 343–362.

Capparelli, R., Parlato, M., Borriello, G., Salvatore, P., and Iannelli, D. (2007). Experimental phage therapy against *Staphylococcus aureus* in mice. Antimicrob. Agents Chemother. *51*, 2765–2773.

Capparelli, R., Nocerino, N., Iannaccone, M., Ercolini, D., Parlato, M., Chiara, M., and Iannelli, D. (2010). Bacteriophage therapy of *Salmonella enterica*: a fresh appraisal of bacteriophage therapy. J. Infect. Dis. *201*, 52–61.

Carlton, R.M. (1999). Phage therapy: past history and future prospects. Arch. Immunol. Ther. Exp. (Warsz). *47*, 267–274.

Carlton, R.M., Noordman, W.H., Biswas, B., de Meester, E.D., and Loessner, M.J. (2005). Bacteriophage P100 for control of *Listeria monocytogenes* in foods: genome sequence, bioinformatic analyses, oral toxicity study, and application. Regul. Toxicol. Pharmacol. *43*, 301–331.

Carmody, L.A., Gill, J.J., Summer, E.J., Sajjan, U.S., Gonzalez, C.F., Young, R.F., and LiPuma J.J. (2010). Efficacy of bacteriophage therapy in a model of *Burkholderia cenocepacia* pulmonary infection. J. Infect. Dis. *201*, 264–271.

Carrias, A., Welch, T.J., Waldbieser, G.C., Mead, D.A., Terhune, J.S., and Liles, M.R. (2011). Comparative genomic analysis of bacteriophages specific to the channel catfish pathogen *Edwardsiella ictaluri*. Virol. J. *8*, 6.

Carson, L., Gorman, S.P., and Gilmore, B.F. (2010). The use of lytic bacteriophages in the prevention and eradication of biofilms of *Proteus mirabilis* and *Escherichia coli*. FEMS Immunol. Med. Microbiol. *59*, 447–455.

Carvalho, C.M., Gannon, B.W., Halfhide, D.E., Santos, S.B., Hayes, C.M., Roe, J.M., and Azeredo, J. (2010). The *in vivo* efficacy of two administration routes of a phage cocktail to reduce numbers of *Campylobacter coli* and *Campylobacter jejuni* in chickens. BMC Microbiol. *10*, 232.

Castillo, D., Higuera, G., Villa, M., Middelboe, M., Dalsgaard, I., Madsen, L., and Espejo, R.T. (2012). Diversity of *Flavobacterium psychrophilum* and the potential use of its phages for protection against bacterial cold water disease in salmonids. J. Fish Dis. *35*, 193–201.

Catalão, M.J., Gil, F., Moniz-Pereira, J., São-José, C., and Pimentel, M. (2013). Diversity in bacterial lysis systems: bacteriophages show the way. FEMS Microbiol. Rev.. *37*, 554–571.

Chanishvili, N. (2009). A literature review of the practical application of bacteriophage research. Eliava Institute of Bacteriophages, Microbiology and Virology.

Chanishvili, N. (2012). Phage therapy – history from Twort and d'Hérelle through Soviet experience to current approaches. Adv. Virus Res. *83*, 3–40.

Chanishvili, N., Chanishvili, T., Tediashvili, M., and Barrow, P.A. (2001). Phages and their application against drug-resistant bacteria. J. Chem. Technol. Biot. *76*, 689–699.

Chatel, J.M., Pothelune, L., Ah-Leung, S., Corthier, G., Wal, J.M., and Langella, P. (2008). In vivo transfer of plasmid from food-grade transiting lactococci to murine epithelial cells. Gene Ther. *15*, 1184–1190.

Cheng, Q., Nelson, D., Zhu, S., and Fischetti, V.A. (2005). Removal of group B streptococci colonizing the vagina and oropharynx of mice with a bacteriophage lytic enzyme. Antimicrob. Agents Chemother. *49*, 111–117.

Cherwonogrodzky, J.W. (2005). Research strategies for the treatment of biothreats. Curr. Opin. Pharmacol. *5*, 465–472.

Chhibber, S., Kaur, T., and Sandeep Kaur. (2013). Co-therapy using lytic bacteriophage and linezolid: effective treatment in eliminating methicillin resistant *Staphylococcus aureus* (MRSA) from diabetic foot infections. PLoS One *8*, e56022.

Chung, I.Y., Sim, N., and Cho, Y.H. (2012). Antibacterial efficacy of temperate phage-mediated inhibition of bacterial group motilities. Antimicrob. Agents Chemother. *56*, 5612–5617.

Clark, J.R., and March, J.B. (2004a). Bacterial viruses as human vaccines? Expert Rev. Vaccines. *3*, 463–476.

Clark, J.R., and March, J.B. (2004b). Bacteriophage-mediated nucleic acid immunisation. FEMS Immunol. Med. Microbiol. *40*, 21–26.

Clark, J.R., Bartley, K., Jepson, C.D., Craik, V., and March, J.B. (2011). Comparison of a bacteriophage-delivered DNA vaccine and a commercially available recombinant protein vaccine against hepatitis B. FEMS Immunol. Med. Microbiol. *61*, 197–204.

Coates, T., Bax, R., and Coates, A. (2009). Nasal decolonization of *Staphylococcus aureus* with mupirocin: strengths, weaknesses and future prospects. J. Antimicrob. Chemother. *64*, 9–15.

Coffey, B., Mills, S., Coffey, A., McAuliffe, O., and Ross R.P. (2010). Phage and their lysins as biocontrol agents for food safety applications. Annu. Rev. Food Sci. Technol. *1*, 449–468.

Comeau, A.M., Tétart, F., Trojet, S.N., Prère, M.F., and Krisch, H.M. (2007). Phage–antibiotic synergy (PAS): beta-lactam and quinolone antibiotics stimulate virulent phage growth. PLoS One *2*, e799.

Costerton, J.W., Stewart, P.S., and Greenberg, E.P. (1999) Bacterial biofilms: a common cause of persistent infections. Science *284*, 1318–1322.

Courchesne, N.M., Parisien, A., and Lan, C.Q. (2009). Production and application of bacteriophage and bacteriophage-encoded lysins. Recent Pat. Biotechnol. *3*, 37–45.

Cui, J., Ren, H.J., Liu, R.D., Wang, L., Zhang, Z.F., and Wang, Z.Q. (2013). Phage-displayed specific polypeptide antigens induce significant protective immunity against Trichinella spiralis infection in BALB/c mice. Vaccine. *31*, 1171–1177.

Curtin, J.J., and Donlan, R.M. (2006). The interaction of phage and biofilms. FEMS Microbiol. Lett. *232*, 1–6.

Dabrowska, K., Skaradziński, G., Kurzepa, A., Owczarek, B., Zaczek, M., Weber-Dabrowska, B., Wietrzyk, J., Maciejewska, M., Budynek, P., and Górski, A. (2010). The effects of staphylococcal bacteriophage lysates on cancer cells in vitro. Clin. Exp. Med. *10*, 81–85.

Daniel, A., Euler, C., Collin, M., Chahales, P., Gorelick, K.J., and Fischetti, V.A. (2010). Synergism between a novel chimeric lysin and oxacillin protects against infection by methicillin-resistant *Staphylococcus aureus*. Antimicrob. Agents Chemother. *54*, 1603–1612.

Davey, M.E., and O'Toole, G.A. (2000). Microbial biofilms: from ecology to molecular genetics. Microbiol. Mol. Biol. Rev. *64*, 847–867.

Deresinski, S. (2009). Bacteriophage therapy: exploiting smaller fleas. Clin. Infect. Dis. *48*, 1096–1101.

D'Hérelle, F. (1917). Sur un microbe invisible antagoniste des bacilles dysenteriques. C. R. Acad. Sci. (Paris). *165*, 373–375.

Dickerson, T.J., Kaufmann, G.F., and Janda, K.D. (2005). Bacteriophage-mediated protein delivery into the central nervous system and its application in immunopharmacotherapy. Expert Opin. Biol. Ther. *5*, 773–781.

Donlan, R.M. (2009). Preventing biofilms of clinically relevant organisms using bacteriophage. Trends Microbiol. *17*, 66–72.

Dublanchet, A., and Bourne, S. (2007). The epic of phage therapy. Can. J. Infect. Dis. Med. Microbiol. *18*, 15–18.

Embleton, M.L., Nair, S.P., Heywood, W., Menon, D.C., Cookson, B.D., and Wilson, M. (2005). Development of a novel targeting system for lethal photosensitization of antibiotic-resistant strains of *Staphylococcus aureus*. Antimicrob. Agents Chemother. *49*, 3690–3696.

Entenza, J.M., Loeffler, J.M., Grandgirard, D., Fischetti, V.A., and Moreillon, P. (2005). Therapeutic effects of bacteriophage Cpl-1 lysin against *Streptococcus pneumoniae* endocarditis in rats. Antimicrob. Agents Chemother. *49*, 4789–4792.

Fairhead, H. (2004). Small acid-soluble proteins and uses thereof. U.S. patent 20040097705.

Fenton, M., Ross, R.P., McAuliffe, O., O'Mahony, J., and Coffey, A. (2010a). Recombinant bacteriophage lysins as antibacterials. Bioeng. Bugs. *1*, 9–16.

Fenton, M., Casey, P.G., Hill, C., Gahan, C.G., Ross, R.P., McAuliffe, O., O'Mahony, J., Maher, F., and Coffey, A. (2010b). The truncated phage lysin CHAP(k) eliminates *Staphylococcus aureus* in the nares of mice. Bioeng. Bugs. *1*, 404–407.

Fenton, M., Keary, R., McAuliffe, O., Ross, R.P., O'Mahony, J., and Coffey, A. (2013). Bacteriophage-derived peptidase CHAP(K) eliminates and prevents staphylococcal biofilms. Int. J. Microbiol. *2013*, 625341.

Fischetti, V.A. (2005). Bacteriophage lytic enzymes: novel anti-infectives. Trends Microbiol. *13*, 491–496.

Fischetti, V.A. (2008). Bacteriophage lysins as effective antibacterials. Curr. Opin. Microbiol. *11*, 393–400.

Fischetti, V.A. (2010). Bacteriophage endolysins: a novel anti-infective to control Gram-positive pathogens. Int. J. Med. Microbiol. *300*, 357–362.

Fortuna, W., Miedzybrodzki, R., Weber-Dabrowska, B., and Gorski, A. (2008). Bacteriophage therapy in children: facts and prospects. Med. Sci. Monitor *14*, 126–132.

Frampton, R.A., Pitman, A.R., and Fineran, P.C. (2012). Advances in bacteriophage-mediated control of plant pathogens. Int. J. Microbiol. *2012*, 326452.

Fu, W., Forster, T., Mayer, O., Curtin, J.J., Lehman, S.M., and Donlan, R.M. (2010). Bacteriophage cocktail for the prevention of biofilm formation by *Pseudomonas aeruginosa* on catheters in an *in vitro* model system. Antimicrob. Agents Chemother. *54*, 397–404.

Garcia, P., Lopez, R., Ronda, C., Garcia, E., and Tomasz, A. (1983). Mechanism of phage-induced lysis in pneumococci. J. Gen. Microbiol. *129*, 479–487.

Gazarian, K., Rowlay, M., Gazarian, T., Vazquez Buchelli, J.E., and Hernández Gonzáles, M. (2012). Mimotope peptides selected from phage display combinatorial library by serum antibodies of pigs experimentally infected with Taenia solium as leads to developing diagnostic antigens for human neurocysticercosis. Peptides. *38*, 381–388.

Geier, M.R., Trigg, M.E., and Merril, C.R. (1973). Fate of bacteriophage lambda in non-immune germ-free mice. Nature *246*, 221–223.

Gil, F., Catalao, M.J., Moniz-Pereira, J., Leandro, P., McNeil, M., and Pimentel, M. (2008). The lytic cassette of mycobacteriophage Ms6 encodes an enzyme with lipolytic activity. Microbiology *154*, 1364–1371.

Gil, F., Grzegorzewicz, A.E., Catalão, M.J., Vital, J., McNeil, M.R., and Pimentel, M. (2010). Mycobacteriophage Ms6 LysB specifically targets the outer membrane of *Mycobacterium smegmatis*. Microbiology *156* (Pt 5), 1497–1504.

Gill, J.J., and Hyman, P. (2010). Phage choice, isolation, and preparation for phage therapy. Curr. Pharm. Biotechnol. *11*, 2–14.

Goldman, G., Starosvetsky, J., and Armon, R. (2009). Inhibition of biofilm formation on UF membrane by use of specific bacteriophages. J. Membr. Sci. *342*, 145–152.

Goodridge, L.D. (2010). Designing phage therapeutics. Curr. Pharm. Biotechnol. *11*, 15–27.

Górski, A., and Weber-Dabrowska, B. (2005). The potential role of endogenous bacteriophages in controlling invading pathogens. Cell. Mol. Life Sci. *62*, 511–519.

Górski, A., Wazna, E., Dabrowska, B.W., Dabrowska, K., Switała-Jeleń, K., and Miedzybrodzki, R. (2006). Bacteriophage translocation. FEMS Immunol. Med. Microbiol. *46*, 313–319.

Górski, A., Borysowski, J., Międzybrodzki, R., and Weber-Dąbrowska, B. (2007). Bacteriophages in medicine. In Bacteriophage: genetics and molecular biology S. McGrath and D. van Sinderen, eds. (Caister Academic Press, Norfolk, UK) pp. 125–158.

Górski, A., Miedzybrodzki, R., Borysowski, J., Weber-Dabrowska, B., Lobocka, M., Fortuna, W., Letkiewicz, S., Zimecki, M., and Filby, G. (2009). Bacteriophage therapy for the treatment of infections. Curr. Opin. Investig. Drugs. *10*, 766–774.

Górski, A., Międzybrodzki, R., Borysowski, J., Dąbrowska, K., Wierzbicki, P., Ohams, M., Korczak-Kowalska, G., Olszowska-Zaremba, N., Łusiak-Szelachowska, M., Kłak, M., *et al.* (2012). Phage as a modulator of immune responses: practical implications for phage therapy. Adv. Virus Res. *83*, 41–71.

Greenstein, D., and Brent, R. (2001). Introduction to vectors derived from filamentous phages. Curr. Protoc. Mol. Biol. Chapter 1:Unit1.14.

Gu, J., Xu, W., Lei, L., Huang, J., Feng, X., Sun, C., Du, C., Zuo, J., Li, Y., Du, T., *et al.* (2011). LysGH15, a novel bacteriophage lysin, protects a murine bacteremia model efficiently against lethal methicillin-resistant *Staphylococcus aureus* infection. J. Clin. Microbiol. *49*, 111–117.

Gu, J., Liu, X., Li, Y., Han, W., Lei, L., Yang, Y., Zhao, H., Gao, Y., Song, J., Lu, R., *et al.* (2012). A method for generation phage cocktail with great therapeutic potential. PLoS One 7, e31698.

Gupta, R., and Prasad, Y. (2011). Efficacy of polyvalent bacteriophage P-27/HP to control multidrug resistant *Staphylococcus aureus* associated with human infections. Curr. Microbiol. 62, 255–260.

Hagens, S., and Bläsi, U. (2003). Genetically modified filamentous phage as bactericidal agents: a pilot study. Lett. Appl. Microbiol. 37, 318–323.

Hagens, S., and Loessner, M.J. (2007). Application of bacteriophages for detection and control of food-borne pathogens. Appl. Microbiol. Biotechnol. 76, 513–519.

Hagens, S., and Loessner, M.J. (2010). Bacteriophage for biocontrol of foodborne pathogens: calculations and considerations. Curr. Pharm. Biotechnol. 11, 58–68.

Hagens, S., Habel, A., von Ahsen, U., von Gabain, A., and Bläsi, U. (2004). Therapy of experimental pseudomonas infections with a non-replicating genetically modified phage. Antimicrob. Agents Chemother. 48, 3817–3822.

Hagens, S., Habel, A., and Blasi, U. (2006). Augmentation of the antimicrobial efficacy of antibiotics by filamentous phage. Microb. Drug Resist. *Fall. 12*, 164–168.

Hanlon, G.W. (2007). Bacteriophages: an appraisal of their role in the treatment of bacterial infections. Int. J. Antimicrob. Agents. 30, 118–128.

Hanlon, G.W., Denyer, S.P., Olliff, C.J., and Ibrahim, L.J. (2001). Reduction in exopolysaccharide viscosity as an aid to bacteriophage penetration through *Pseudomonas aeruginosa* biofilms. Appl. Environ. Microbiol. 67, 2746–2753.

Henry, K.A., Murira, A., van Houten, N.E., and Scott, J.K. (2011). Developing strategies to enhance and focus humoral immune responses using filamentous phage as a model antigen. Bioeng. Bugs. 2, 275–283.

Henry, M., O'Sullivan, O., Sleator, R.D., Coffey, A., Ross, R.P., McAuliffe, O., and O'Mahony, J.M. (2010). In silico analysis of Ardmore, a novel mycobacteriophage isolated from soil. Gene. 453, 9–23.

Heo, Y.J., Lee, Y.R., Jung, H.H., Lee, J., Ko, G., and Cho, Y.H. (2009). Antibacterial efficacy of phages against *Pseudomonas aeruginosa* infections in mice and Drosophila melanogaster. Antimicrob. Agents Chemother. 53, 2469–2474.

Hoffmann, M. (1965). [Animal experiments on the mucosal passage and absorption viremia of T3 phages after oral, tracheal and rectal administration]. Zentralbl. Bakteriol. Orig. 198, 371–390. [article in German]

Horgan, M., O'Flynn, G., Garry, J., Cooney, J., Coffey, A., Fitzgerald, G.F., Ross, R.P., and McAuliffe, O. (2009). Phage lysin LysK can be truncated to its CHAP domain and retain lytic activity against live antibiotic-resistant staphylococci. Appl. Environ. Microbiol. 75, 872–874.

Housby, J.N., and Mann, N.H. (2009). Phage therapy. Drug Discov. Today. 14, 536–540.

Hsieh, S.E., Lo, H.H., Chen, S.T., Lee, M.C., and Tseng, Y.H. (2011). Wide host range and strong lytic activity of *Staphylococcus aureus* lytic phage Stau2. Appl. Environ. Microbiol. 77, 756–761.

Huff, W.E., Huff, G.R., Rath, N.C., and Donoghue, A.M. (2010). Immune interference of bacteriophage efficacy when treating colibacillosis in poultry. Poult. Sci. 89, 895–900.

Huff, W.E., Huff, G.R., Rath, N.C., and Donoghue, A.M. (2013). Method of administration affects the ability of bacteriophage to prevent colibacillosis in 1-day-old broiler chickens. Poult. Sci. 92, 930–934.

Hughes, K.A., Sutherland, I.W., and Jones, M.V. (1998). Biofilm susceptibility to bacteriophage attack: the role of phage-borne polysaccharide depolymerase. Microbiology 144, 3039–3047.

Hung, C.H., Kuo, C.F., Wang, C.H., Wu, C.M., and Tsao, N. (2011). Experimental phage therapy in treating *Klebsiella pneumoniae*-mediated liver abscesses and bacteremia in mice. Antimicrob. Agents Chemother. 55, 1358–1356.

Irving, M.B., Pan, O., and Scott, J.K. (2001). Random-peptide libraries and antigen-fragment libraries for epitope mapping and the development of vaccines and diagnostics. Curr. Opin. Chem. Biol. 5, 314–324.

Jacob, F., and Fuerst, C.R. (1958). The mechanism of lysis by phage studied with defective lysogenic bacteria. J. Gen. Microbiol. 18, 518–526.

Jado, I., López, R., García, E., Fenoll, A., Casal, J., and García, P. (2003). Phage lytic enzymes as therapy for antibiotic-resistant *Streptococcus pneumoniae* infection in a murine sepsis model. J. Antimicrob. Chemother. 52, 967–973.

Jepson, C.D., and March, J.B. (2004). Bacteriophage lambda is a highly stable DNA vaccine delivery vehicle. Vaccine. 22, 2413–2419.

Johnson, R.P., Gyles, C.L., Huff, W.E., Ojha, S., Huff, G.R., Rath, N.C., and Donoghue, A.M. (2008). Bacteriophages for prophylaxis and therapy in cattle, poultry and pigs. Anim. Health Res. Rev. 9, 201–215.

Jones, J.B., Jackson, L.E., Balogh, B., Obradovic, A., Iriarte, F.B., and Momol, M.T. (2007). Bacteriophages for plant disease control. Annu. Rev. Phytopathol. 45, 245–262.

Jones, J.B., Vallad, G.E., Iriarte, F.B., Obradović, A., Wernsing, M.H., Jackson, L.E., Balogh, B., Hong, J.C., and Momol, M.T. (2012). Considerations for using bacteriophages for plant disease control. Bacteriophage. 2, 208–214.

Jończyk, E., Kłak, M., Międzybrodzki, R., and Górski, A. (2011). The influence of external factors on bacteriophages – review. Folia Microbiol. (Praha). 56, 191–200.

Keller, R., and Zatzman, M.L. (1959). Studies on the factors concerned in the disappearance of bacteriophage particles from the animal body. J. Immunol. 83, 167–172.

Kelly, D., McAuliffe, O., Ross, R.P., O'Mahony, J., and Coffey, A. (2011). Development of a broad-host-range phage cocktail for biocontrol. Bioeng. Bugs 2, 1–7.

Kia, A., Przystal, J.M., Nianiaris, N., Mazarakis, N.D., Mintz, P.J., and Hajitou, A. (2012). Dual systemic tumor targeting with ligand-directed phage and Grp78 promoter induces tumor regression. Mol. Cancer Ther. 11, 2566–2577.

Kim, K.P., Cha, J.D., Jang, E.H., Klumpp, J., Hagens, S., Hardt, W.D., Lee, K.Y., and Loessner, M.J. (2008). PEGylation of bacteriophages increases blood circulation time and reduces T-helper type 1 immune response. Microb. Biotechnol. *1*, 247–257.

Kirby, A.E. (2012). Synergistic action of gentamicin and bacteriophage in a continuous culture population of *Staphylococcus aureus*. PLoS One 7, e51017.

Knezevic, P., Curcin, S., Aleksic, V., Petrusic, M., and Vlaski, L. (2012). Phage-antibiotic synergism: a possible approach to combatting *Pseudomonas aeruginosa*. Res. Microbiol. *164*, 55–60.

Knittelfelder, R., Riemer, A.B., and Jensen-Jarolim, E. (2009). Mimotope vaccination – from allergy to cancer. Expert Opin. Biol. Ther. *9*, 493–506.

Kropinski, A.M. (2006). Phage therapy – everything old is new again. Can. J. Infect. Dis. Med. Microbiol. *17*, 297–306.

Kumari, S., Harjai, K., and Chhibber, S. (2011). Bacteriophage versus antimicrobial agents for the treatment of murine burn wound infection caused by *Klebsiella pneumoniae* B5055. J. Med. Microbiol. *60(Pt 2)*, 205–210.

Kurzepa, A., Dabrowska, K., Skaradziński, G., and Górski, A. (2009). Bacteriophage interactions with phagocytes and their potential significance in experimental therapy. Clin. Exp. Med. *9*, 93–100.

Kutateladze, M., and Adamia, R. (2008). Phage therapy experience at the Eliava Institute. Med. Mal. Infect. *38*, 426–430.

Kutter, E., De Vos, D., Gvasalia, G., Alavidze, Z., Gogokhia, L., Kuhl, S., and Abedon, S.T. (2010). Phage therapy in clinical practice: treatment of human infections. Curr. Pharm. Biotechnol. *11*, 69–86.

Larocca, D., Jensen-Pergakes, K., Burg, M.A., and Baird, A. (2001). Receptor-targeted gene delivery using multivalent phagemid particles. Mol. Ther. *3*, 476–484.

Larocca, D., Burg, M.A., Jensen-Pergakes, K., Ravey, E.P., Gonzalez, A.M., and Baird, A. (2002). Evolving phage vectors for cell targeted gene delivery. Curr. Pharm. Biotechnol. *3*, 45–57.

Li, Y., Liu, Y.H., Li, Z.J., Li, Y.G., Liu, M.Y., Liu, L.B., Wang, X.G., Wang, X.L., Suo, J., and Han, W.Y. (2013). Identification and characterization of antigenic epitope of *Staphylococcus aureus* ClfA adhesin. Res. Vet. Sci. *94*, 490–495.

Letarov, A., and Kulikov, E. (2009). The bacteriophages in human- and animal body-associated microbial communities. J. Appl. Microbiol. *107*, 1–13.

Letkiewicz, S., Miedzybrodzki, R., Fortuna, W., Weber-Dabrowska, B., and Górski, A. (2009). Eradication of *Enterococcus faecalis* by phage therapy in chronic bacterial prostatitis – case report. Folia Microbiol. (Praha). *54*, 457–461.

Letkiewicz, S., Międzybrodzki, R., Kłak, M., Jończyk, E., Weber-Dąbrowska, B., and Górski, A. (2010a). The perspectives of the application of phage therapy in chronic bacterial prostatitis. FEMS Immunol. Med. Microbiol. *60*, 99–112.

Letkiewicz, S., Miedzybrodzki, R., Klak, M., Weber-Dabrowska, B., and Górski, A. (2010b). Pathogen eradication by phage therapy in patients with chronic bacterial prostatitis (poster no. 374). 25th anniversary EAU congress in Barcelona, 2010. Available at http://www.postersessiononline.com/173580348_eu/congresos/25eau/aula/-P_374_25eau.pdf

Levin, B.R., and Bull, J.J. (2004). Population and evolutionary dynamics of phage therapy. Nat. Rev. Microbiol. *2*, 166–173.

Li, K., Chen, Y., Li, S., Nguyen, H.G., Niu, Z., You, S., Mello, C.M., Lu, X., and Wang, Q. (2010). Chemical modification of M13 bacteriophage and its application in cancer cell imaging. Bioconjug. Chem. *21*, 1369–1377.

Liu, M., Deora, R., Doulatov, S.R., Gingery, M., Eiserling, F.A., Preston, A., Maskell, D.J., Simons, R.W., Cotter, P.A., Parkhill, J., *et al.* (2002). Reverse transcriptase-mediated tropism switching in *Bordetella* bacteriophage. Science *295*, 2091–2094.

Liu, J., Dehbi, M., Moeck, G., Arhin, F., Bauda, P., Bergeron, D., Callejo, M., Ferretti, V., Ha, N., Kwan, T., *et al.* (2004). Antimicrobial drug discovery through bacteriophage genomics. Nat. Biotechnol. *22*, 185–191.

Loc-Carrillo, C., and Abedon, S.T. (2010). Pros and cons of phage therapy. Bacteriophage. *1*, 1–4.

Loeffler, J.M., and Fischetti, V.A. (2003). Synergistic lethal effect of a combination of phage lytic enzymes with different activities on penicillin-sensitive and -resistant *Streptococcus pneumoniae* strains. Antimicrob. Agents Chemother. *47*, 375–377.

Loeffler, J.M., Nelson, D., and Fischetti, V.A. (2001). Rapid killing of *Streptococcus pneumoniae* with a bacteriophage cell wall hydrolase. Science *294*, 2170–2172.

Loeffler, J.M., Djurkovic, S., and Fischetti, V.A. (2003). Phage lytic enzyme Cpl-1 as a novel antimicrobial for pneumococcal bacteremia. Infect. Immun. *71*, 6199–6204.

Loessner, M.J. (2005). Bacteriophage endolysins – current state of research and applications. Curr. Opin. Microbiol. *8*, 480–487.

López, R., and García, E. (2004). Recent trends on the molecular biology of pneumococcal capsules, lytic enzymes, and bacteriophage. FEMS Microbiol. Rev. *28*, 553–580.

Lu, T.K., and Collins, J.J. (2007). Dispersing biofilms with engineered enzymatic bacteriophage. Proc. Natl. Acad. Sci. U.S.A. *104*, 11197–11202.

Lu, T.K., and Collins, J.J. (2009). Engineered bacteriophage targeting gene networks as adjuvants for antibiotic therapy. Proc. Natl. Acad. Sci. U.S.A. *106*, 4629–4634.

Lukacik, P., Barnard, T.J., and Buchanan, S.K. (2012). Using a bacteriocin structure to engineer a phage lysin that targets Yersinia pestis. Biochem. Soc. Trans. *40*, 1503–1506.

McCullers, J.A., Karlström, A., Iverson, A.R., Loeffler, J.M., and Fischetti, V.A. (2007). Novel strategy to prevent otitis media caused by colonizing *Streptococcus pneumoniae*. PLoS Pathog. 3, 28.

Mahony, J., McAuliffe, O., Ross, R.P., and van Sinderen, D. (2011). Bacteriophages as biocontrol agents of food pathogens. Curr. Opin. Biotechnol. 22, 1–7.

Mai, V., Ukhanova, M., Visone, L., Abuladze, T., and Sulakvelidze, A. (2010). Bacteriophage administration reduces the concentration of *Listeria monocytogenes* in the gastrointestinal tract and its translocation to spleen and liver in experimentally infected mice. Int. J. Microbiol. *2010*, 624234.

Malik, R., and Chhibber, S. (2007). Protection with bacteriophage KØ1 against fatal *Klebsiella pneumoniae*-induced burn wound infection in mice. J. Microbiol. Immunol. Infect. *42*, 134–140.

Mann, N.H. (2008). The potential of phages to prevent MRSA infections. Res. Microbiol. *159*, 400–405.

Manoutcharian, K. (2005). Bacteriophages as tools for vaccine and drug development. Expert Rev. Vaccines. *4*, 5–7.

Manoutcharian, K., Díaz-Orea, A., Gevorkian, G., Fragoso, G., Acero, G., González, E., De Aluja, A., Villalobos, N., Gómez-Conde, E., and Sciutto, E. (2004). Recombinant bacteriophage-based multiepitope vaccine against *Taenia solium* pig cysticercosis. Vet. Immunol. Immunopathol. *99*, 11–24.

March, J.B., Clark, J.R., and Jepson, C.D. (2004). Genetic immunisation against hepatitis B using whole bacteriophage lambda particles. Vaccine. *22*, 1666–1671.

Marza, J.A., Soothill, J.S., Boydell, P., and Collyns, T.A. (2006). Multiplication of therapeutically administered bacteriophages in *Pseudomonas aeruginosa* infected patients. Burns. *32*, 644–646.

Matsuda, T., Freeman, T.A., Hilbert, D.W., Duff, M., Fuortes, M., Stapleton, P.P., and Daly, J.M. (2005). Lysis-deficient bacteriophage therapy decreases endotoxin and inflammatory mediator release and improves survival in a murine peritonitis model. Surgery *137*, 639–646.

Matsuzaki, S., Rashel, M., Uchiyama, J., Sakurai, S., Ujihara, T., Kuroda, M., Ikeuchi, M., Tani, T., Fujieda, M., Wakiguchi, H., *et al.* (2005). Bacteriophage therapy: a revitalized therapy against bacterial infectious diseases. J. Infect. Chemother. *11*, 211–219.

Merabishvili, M. et. al. (2009). Quality-controlled small-scale production of a well-defined bacteriophage cocktail for use in human clinical trials. PLoS One *4*, e4944.

Merril, C.R. (2008). Interaction of bacteriophages with animals. In Bacteriophage Ecology, S.T. Abedon, ed. (Cambridge, UK, Cambridge University Press), pp. 332–352.

Merril, C.R., Biswas, B., Carlton, R., Jensen, N.C., Creed, G.J., Zullo, S., and Adhya, S. (1996). Long-circulating bacteriophage as antibacterial agents. Proc. Natl. Acad. Sci. U.S.A. *93*, 3188–3192.

Miedzybrodzki, R., Fortuna, W., Weber-Dabrowska, B., and Gorski, A. (2005). Bacterial viruses against viruses pathogenic for man?. Virus Res. *110*, 1–8.

Miedzybrodzki, R., Letkiewicz, S., and Gorski, A. (2008). Pathogen eradication by phage therapy in patients with chronic bacterial prostatitis. Presented at the Edinburgh International Phage Conference, Edinburgh.

Miedzybrodzki, R., Borysowski, J., Weber-Dąbrowska, B., Fortuna, W., Letkiewicz, S., Szufnarowski, K., Pawełczyk, Z., Rogóż, P., Klak, M., Wojtasik, E., *et al.* (2012). Clinical aspects of phage therapy. Adv. Virus Res. *83*, 73–121.

Miller, R.W., Skinner, E.J., Sulakvelidze, A., Mathis, G.F., and Hofacre, C.L. (2010). Bacteriophage therapy for control of necrotic enteritis of broiler chickens experimentally infected with *Clostridium perfringens*. Avian Dis. *54*, 33–40.

Mishra, A.K., Rawat, M., Viswas, K.N., Abhisek, Kumar, S., and Reddy, M. (2013). Expression and lytic efficacy assessment of *Staphylococcus aureus* phage SA4 lysin gene. J. Vet. Sci. *14*, 37–43.

Molenaar, T.J., Michon, I., de Haas, S.A., van Berkel, T.J., Kuiper, J., and Biessen, E.A. (2002). Uptake and processing of modified bacteriophage M13 in mice: implications for phage display. Virology *293*, 182–191.

Monk, A.B., Rees, C.D., Barrow, P., Hagens, S., and Harper, D.R. (2010). Bacteriophage applications: where are we now? Lett. Appl. Microbiol. *51*, 363–369.

Moradpour, Z., and Ghasemian, A. (2011). Modified phages: novel antimicrobial agents to combat infectious diseases. Biotechnol. Adv. *29*, 732–738.

Moradpour, Z., Sepehrizadeh, Z., Rahbarizadeh, F., Ghasemian, A., Yazdi, M.T., and Shahverdi, A.R. (2009). Genetically engineered phage harbouring the lethal catabolite gene activator protein gene with an inducer-independent promoter for biocontrol of *Escherichia coli*. FEMS Microbiol. Lett. *296*, 67–71.

Moreno, F., and Wandersman, C. (1980). OmpC and LamB proteins can serve as substitute receptors for host range mutants of coliphage TuIa. J. Bacteriol. *144*, 1182–1185.

Morona, R., and Henning, U. (1984). Host range mutants of bacteriophage Ox2 can use two different outer membrane proteins of *Escherichia coli* K-12 as receptors. J. Bacteriol. *159*, 579–582.

Nelson, D., Loomis, L., and Fischetti, V.A. (2001). Prevention and elimination of upper respiratory colonization of mice by group A streptococci by using a bacteriophage lytic enzyme. Proc. Natl. Acad. Sci. U.S.A. *98*, 4107–4112.

Nelson, D.C., Schmelcher, M., Rodriguez-Rubio, L., Klumpp, J., Pritchard, D.G., Dong, S., and Donovan, D.M. (2012). Endolysins as antimicrobials. Adv. Virus Res. *83*, 299–365.

Nishikawa, H., Yasuda, M., Uchiyama, J., Rashel, M., Maeda, Y., Takemura, I., Sugihara, S., Ujihara, T., Shimizu, Y., Shuin, T., *et al.* (2008). T-even-related bacteriophages as candidates for treatment of *Escherichia coli* urinary tract infections. Arch. Virol. *153*, 507–515.

Norrby, S.R., Nord, C.E., and Finch, R. (2005). Lack of development of new antimicrobial drugs: a potential serious threat to public health. Lancet Infect. Dis. *5*, 115–119.

Norris, J.S., Westwater, C., and Schofield, D. (2000). Prokaryotic gene therapy to combat multidrug resistant bacterial infection. Gene Ther. *7*, 723–735.

O'Flaherty, S., Ross, R.P., Meaney, W., Fitzgerald, G.F., Elbreki, M.F., and Coffey, A. (2005). Potential of the polyvalent anti-*Staphylococcus* bacteriophage K for control of antibiotic-resistant staphylococci from hospitals. Appl. Environ. Microbiol. *71*, 1836–1842.

O'Flaherty, S., Ross, R.P., and Coffey, A. (2009). Bacteriophage and their lysins for elimination of infectious bacteria. FEMS Microbiol. Rev. *33*, 801–819.

Oliveira, A., Sereno, R., Nicolau, A., and Azeredo, J. (2009). The influence of the mode of administration on the dissemination of three coliphages in chickens. Poult. Sci. *88*, 728–733.

Oliveira, A., Sereno, R., and Azeredo, J. (2010). In vivo efficiency evaluation of a phage cocktail in controlling severe colibacillosis in confined conditions and experimental poultry houses. Vet. Microbiol. *146*, 303–308.

Park, J., Yun, J., Lim, J.A., Kang, D.H., and Ryu, S. (2012). Characterization of an endolysin, LysBPS13, from a Bacillus cereus bacteriophage. FEMS Microbiol. Lett. *332*, 76–83.

Parracho, H.M., Burrowes, B.H., Enright, M.C., McConville, M.L., and Harper, D.R. (2012). The role of regulated clinical trials in the development of bacteriophage therapeutics. J. Mol. Genet. Med. *6*, 279–286.

Pastagia, M., Euler, C., Chahales, P., Fuentes-Duculan, J., Krueger, J.G., and Fischetti, V.A. (2011). A novel chimeric lysin shows superiority to mupirocin for skin decolonization of methicillin-resistant and -sensitive *Staphylococcus aureus* strains. Antimicrob. Agents Chemother. *55*, 738–744.

Paul, V.D., Sundarrajan, S., Rajagopalan, S.S., Hariharan, S., Kempashanaiah, N., Padmanabhan, S., Sriram, B., and Ramachandran, J. (2011). Lysis-deficient phages as novel therapeutic agents for controlling bacterial infection. BMC Microbiol. *11*, 195.

Payne, K.M., and Hatfull, G.F. (2012). Mycobacteriophage endolysins: diverse and modular enzymes with multiple catalytic activities. PLoS One *7*, e34052.

Payne, R.J., and Jansen, V.A. (2001). Understanding bacteriophage therapy as a density-dependent kinetic process. J. Theor. Biol. *208*, 37–48.

Payne, R.J., and Jansen, V.A. (2003). Pharmacokinetic principles of bacteriophage therapy. Clin. Pharmacokinet. *42*, 315–325.

Pereira, C., Silva, Y.J., Santos, A.L., Cunha, A., Gomes, N.C., and Almeida, A. (2011). Bacteriophages with potential for inactivation of fish pathogenic bacteria: survival, host specificity and effect on bacterial community structure. Mar. Drugs. *9*, 2236–2255.

Perez-Dorado, I., Campillo, N.E., Monterroso, B., Hesek, D., Lee, M., Paez, J.A., Garcia, P., Martinez-Ripoll, M., Garcia, J.L., Mobashery, S., *et al.* (2007). Elucidation of the molecular recognition of bacterial cell wall by modular pneumococcal phage endolysin Cpl-1. J. Biol. Chem. *282*, 24990–24999.

Petty, N.K., Evans, T.J., Fineran, P.C., and Salmond, G.P. (2007). Biotechnological exploitation of bacteriophage research. Trends Biotechnol. *25*, 7–15.

Pirnay, J.P., De Vos, D., Verbeken, G., Merabishvili, M., Chanishvili, N., Vaneechoutte, M., Zizi, M., Laire, G., Lavigne, R., Huys, I., *et al.* (2011). The Phage Therapy Paradigm: Prêt-à-Porter or Sur-mesure? Pharm. Res. *28*, 934–937.

Pouillot, F., Blois, H., and Iris, F. (2010). Genetically engineered virulent phage banks in the detection and control of emergent pathogenic bacteria. Biosecur. Bioterror. *8*, 155–169.

Pouillot, F., Chomton, M., Blois, H., Courroux, C., Noelig, J., Bidet, P., Bingen, E., and Bonacorsi, S. (2012). Efficacy of bacteriophage therapy in experimental sepsis and meningitis caused by a clone O25b:H4-ST131 *Escherichia coli* strain producing CTX-M-15. Antimicrob. Agents Chemother. *56*, 3568–3575.

Pouyanfard, S., Bamdad, T., Hashemi, H., Bandehpour, M., and Kazemi, B. (2012). Induction of protective anti-CTL epitope responses against HER-2-positive breast cancer based on multivalent T7 phage nanoparticles. PLoS One *7*, e49539.

Prasad, Y., Arpana, Kumar, D., and Sharma, A.K. (2011). Lytic bacteriophages specific to *Flavobacterium columnare* rescue catfish, Clarias batrachus (Linn.) from columnaris disease. J. Environ. Biol. *32*, 161–168.

Projan, S.J., and Shlaes, D.M. (2004). Antibacterial drug discovery: is it all downhill from here? Clin. Microbiol. Infect. *10*, 18–22.

Ralston, D.J., Baer, B.S., Lieberman, M., and Krueger, A.P. (1955). Virolysin: a virus-induced lysin from staphylococcal phage lysates. Proc. Soc. Exp. Biol. Med. *89*, 502–507.

Rapson, M.E., Burdent, F.A., Glancy, L.P., Hodgson, D.A., and Mann, N.H. (2003). Bacteriophages useful for therapy and prophylaxis of bacterial infections. WO03080823.

Resch, G., Moreillon, P., and Fischetti, V.A. (2011). PEGylating a bacteriophage endolysin inhibits its bactericidal activity. AMB Express. *1*, 29.

Rhoads, D.D., Wolcott, R.D., Kuskowski, M.A., Wolcott, B.M., Ward, L.S., and Sulakvelidze, A. (2009). Bacteriophage therapy of venous leg ulcers in humans: results of a phase I safety trial. J. Wound Care. *18*, 237–243.

Rivas, L., Coffey, B., McAuliffe, O., McDonnell, M.J., Burgess, C.M., Coffey, A., Ross, R.P., and Duffy, G. (2010). In vivo and ex vivo evaluations of bacteriophages e11/2 and e4/1c for use in the control of *Escherichia coli* O157:H7. Appl. Environ. Microbiol. *76*, 7210–7216.

Rozema, E.A., Stephens, T.P., Bach, S.J., Okine, E.K., Johnson, R.P., Stanford, K., and McAllister, T.A. (2009). Oral and rectal administration of bacteriophages for control of *Escherichia coli* O157:H7 in feedlot cattle. J. Food Prot. 72, 241–250.

Russell, W.J., Taylor, S.A., and Sigel, M.M. (1976). Clearance of bacteriophage in poikilothermic vertebrates and the effect of temperature. J. Reticuloend. Soc. *19*, 91–96.

Sarker, S.A., McCallin, S., Barretto, C., Berger, B., Pittet, A.C., Sultana, S., Krause, L., Huq, S., Bibiloni, R., Bruttin, A., *et al.* (2012). Oral T4-like phage cocktail application to healthy adult volunteers from Bangladesh. Virology *434*, 222–232.

Sartorius, R., Bettua, C., D'Apice, L., Caivano, A., Trovato, M., Russo, D., Zanoni, I., Granucci, F., Mascolo, D., Barba, P., *et al.* (2011). Vaccination with filamentous bacteriophages targeting DEC-205 induces DC maturation and potent anti-tumor T-cell responses in the absence of adjuvants. Eur. J. Immunol. *41*, 2573–2584.

Saussereau, E., and Debarbieux, L. (2012). Bacteriophages in the experimental treatment of *Pseudomonas aeruginosa* infections in mice. Adv. Virus Res. *83*, 123–141.

Schaak, D.L. (2004). Toxin-phage bacteriocide antibiotic and uses thereof. Cambridge, MA/USA 20046759229.

Schmelcher, M., Donovan, D.M., and Loessner, M.J. (2012). Bacteriophage endolysins as novel antimicrobials. Future Microbiol. *7*, 1147–1171.

Schmitz, J.E., Schuch, R., and Fischetti, V.A. (2010). Identifying active phage lysins through functional viral metagenomics. Appl. Environ. Microbiol. *76*, 7181–7187.

Schmitz, J.E., Ossiprandi, M.C., Rumah, K.R., and Fischetti, V.A. (2011). Lytic enzyme discovery through multigenomic sequence analysis in *Clostridium perfringens*. Appl. Microbiol. Biotechnol. *89*, 1783–1795.

Schofield, D.A., Molineux, I.J., and Westwater, C. (2009). Diagnostic bioluminescent phage for detection of *Yersinia pestis*. J. Clin. Microbiol. *47*, 3887–3894.

Schuch, R., Nelson, D., and Fischetti, V.A. (2002). A bacteriolytic agent that detects and kills *Bacillus anthracis*. Nature *418*, 884–889.

Schuch, R., Fischetti, V.A., and Nelson, D.C. (2009). A genetic screen to identify bacteriophage lysins. Methods Mol. Biol. *502*, 307–319.

Schultz, I., and Neva, F.A. (1965). Relationship between blood clearance and viruria after intravenous injection of mice and rats with bacteriophage and polioviruses. J. Immunol. *94*, 833–841.

Seal, B.S. (2013). Characterization of bacteriophages virulent for *Clostridium perfringens* and identification of phage lytic enzymes as alternatives to antibiotics for potential control of the bacterium. Poult. Sci. *92*, 526–533.

Sechter, I., Touitou, E., and Donbrow, M. (1989). The influence of a non-ionic surfactant on rectal absorption of virus particles. Arch. Virol. *106*, 141–143.

Seow, Y., and Wood, M.J. (2009). Biological gene delivery vehicles: beyond viral vectors. Mol. Ther. *17*, 767–777.

Seth, A.K., Geringer, M.R., Nguyen, K.T., Agnew, S.P., Dumanian, Z., Galiano, R.D., Leung, K.P., Mustoe, T.A., and Hong, S.J. (2013). Bacteriophage therapy for *Staphylococcus aureus* biofilm-infected wounds: a new approach to chronic wound care. Plast. Reconstr. Surg. *131*, 225–234.

Shasha, S.M., Sharon, N., and Inbar, M. (2004). Bacteriophages as antibacterial agents. Harefuah. *143*, 121–125.

Shen, Y., Köller, T., Kreikemeyer, B., and Nelson, D.C. (2013). Rapid degradation of Streptococcus pyogenes biofilms by PlyC, a bacteriophage-encoded endolysin. J. Antimicrob. Chemother. *68*, 1818–1824.

Sillankorva, S.M., Oliveira, H., and Azeredo, J. (2012). Bacteriophages and their role in food safety. Int. J. Microbiol. *2012*, 863945.

Singh, A., Arutyunov, D., Szymanski, C.M., and Evoy, S. (2012). Bacteriophage based probes for pathogen detection. Analyst. *137*, 3405–3421.

Skiena, S.S. (2001). Designing better phages. Bioinformatics *17*, 253–261.

Slopek, S., Weber-Dabrowska, B., Dabrowski, M., and Kucharewicz-Krukowska, A. (1987). Results of bacteriophage treatment of suppurative bacterial infections in the years 1981–1986. Arch. Immunol. Ther. Exp. (Warsz). *35*, 569–583.

Smartt, A.E., Xu, T., Jegier, P., Carswell, J.J., Blount, S.A., Sayler, G.S., and Ripp, S. (2012). Pathogen detection using engineered bacteriophages. Anal. Bioanal. Chem. *402*, 3127–3146.

Smith, H.W., and Huggins, M.B. (1982). Successful treatment of experimental *Escherichia coli* infections in mice using phage: its general superiority over antibiotics. J. Gen. Microbiol. *128*, 307–318.

Smith, H.W., and Huggins, M.B. (1983). Effectiveness of phages in treating experimental *Escherichia coli* diarrhoea in calves, piglets and lambs. J. Gen. Microbiol. *129*, 2659–2675.

Smith, H.W., Huggins, M.B., and Shaw, K.M. (1987a). Factors influencing the survival and multiplication of bacteriophages in calves and in their environment. J. Gen. Microbiol. *133*, 1127–1135.

Smith, H.W., Huggins, M.B., and Shaw, K.M. (1987b). The control of experimental *Escherichia coli* diarrhoea in calves by means of bacteriophages. J. Gen. Microbiol. *133*, 1111–1126.

Son, J.S., Lee, S.J., Jun, S.Y., Yoon, S.J., Kang, S.H., Paik, H.R., Kang, J.O., and Choi, Y.J. (2010). Antibacterial and biofilm removal activity of a podoviridae *Staphylococcus aureus* bacteriophage SAP-2 and a derived recombinant cell-wall-degrading enzyme. Appl. Microbiol. Biotechnol. *86*, 1439–1449.

Sonstein, S.A., Hammel, J.M., and Bondi, A. (1971). Staphylococcal bacteriophage-associated lysin: a lytic agent active against *Staphylococcus aureus*. J. Bacteriol. *107*, 499–504.

Soothill, J.S. (1992). Treatment of experimental infections of mice with bacteriophages. J. Med. Microbiol. *37*, 258–261.

Sulakvelidze, A., and Barrow, P. (2005). Phage therapy in animals and agribusiness. In Bacteriophages: Biology and Applications, E. Kutter and A. Sulakvelidze, eds. (Boca Raton, FL, CRC Press), pp. 335–380.

Sulakvelidze, A., and Kutter, E. (2005). Bacteriophage therapy in humans. In Bacteriophages: Biology and Applications, E. Kutter and A. Sulakvelidze, eds. (Boca Raton, FL, CRC Press), pp 381–436.

Sulakvelidze, A., and Morris, J.G. Jr. (2001). Bacteriophages as therapeutic agents. Ann. Med. *33*, 507–509.

Sulakvelidze, A., Alavidze, Z., and Morris, J.G. Jr. (2001). Bacteriophage therapy. Antimicrob. Agents Chemother. *45*, 649–659.

Summers, W.C. (1999). Félix d'Hérelle and the origins of Molecular Biology. Yale University Press New Haven, CT.

Summers, W.C. (2001). Bacteriophage therapy. Annu. Rev. Microbiol. *55*, 437–451.

Sutherland, I.W., Hughes, K.A., Skillman, L.C., and Tait, K. (2004). The interaction of phage and biofilms. FEMS Microbiol. Lett. *232*, 1–6.

Tao, P., Mahalingam, M., Marasa, B.S., Zhang, Z., Chopra, A.K., and Rao, V.B. (2013). *In vitro* and *in vivo* delivery of genes and proteins using the bacteriophage T4 DNA packaging machine. Proc. Natl. Acad. Sci. U.S.A. *110*, 5846–5851.

Tetart, F., Repoila, F., Monod, C., and Krisch, H.M. (1996). Bacteriophage T4 host range is expanded by duplications of a small domain of the tail fiber adhesion. J. Mol. Biol. *258*, 726–731.

Thanyasrisung, P., Komatsuzawa, H., Yoshimura, G., Fujiwara, T., Yamada, S., Kozai, K., Eto, K., Izumi, Y., and Sugai, M. (2009). Automutanolysin disrupts clinical isolates of cariogenic streptococci in biofilms and planktonic cells. Oral. Microbiol. Immunol. *24*, 451–455.

Thomas, B.S., Nishikawa, S., Ito, K., Chopra, P., Sharma, N., Evans, D.H., Tyrrell, D.L., Bathe, O.F., and Rancourt, D.E. (2012). Peptide vaccination is superior to genetic vaccination using a recombineered bacteriophage λ subunit vaccine. Vaccine. *30*, 998–1008.

Tóthová, L., Celec, P., Bábíčková, J., Gajdošová, J., Al-Alami, H., Kamodyova, N., Drahovská, H., Liptáková, A., Turňa, J., and Hodosy, J. (2011). Phage therapy of Cronobacter-induced urinary tract infection in mice. Med. Sci. Monit. *17*, BR173–178.

Twort, F.W. (1915). An investigation on the nature of the ultramicroscopic viruses. Lancet *189*, 1241–1243.

Vitiello, C.L., Merril, C.R., and Adhya, S. (2005). An amino acid substitution in a capsid protein enhances phage survival in mouse circulatory system more than a 1000-fold. Virus Res. *114*, 101–103.

Waddell, T.E., Franklin, K., Mazzocco, A., Kropinski, A.M., and Johnson, R.P. (2009). Generalized transduction by lytic bacteriophages. Methods Mol. Biol. *501*, 293–303.

Wang, L.F., and Yu, M. (2004). Epitope identification and discovery using phage display libraries: applications in vaccine development and diagnostics. Curr. Drug Targets. *5*, 1–15.

Waseh, S., Hanifi-Moghaddam, P., Coleman, R., Masotti, M., Ryan, S., Foss, M., MacKenzie, R., Henry, M., Szymanski, C.M., and Tanha, J. (2010). Orally administered P22 phage tailspike protein reduces *salmonella* colonization in chickens: prospects of a novel therapy against bacterial infections. PLoS One 5, e13904.

Watanabe, R., Matsumoto, T., Sano, G., Ishii, Y., Tateda, K., Sumiyama, Y., Uchiyama, J., Sakurai, S., Matsuzaki, S., Imai, S., and Yamaguchi, K. (2007). Efficacy of bacteriophage therapy against gut-derived sepsis caused by *Pseudomonas aeruginosa* in mice. Antimicrob. Agents Chemother. *51*, 446–452.

Weber-Dabrowska, B., Mulczyk, M., and Górski, A. (2000). Bacteriophage therapy of bacterial infections: an update of our institute's experience. Arch. Immunol. Ther. Exp. (Warsz) *48*, 547–551.

Westwater, C., Kasman, L.M., Schofield, D.A., Werner, P.A., Dolan, J.W., Schmidt, M.G., and Norris, J.S. (2003). Use of genetically engineered phage to deliver antimicrobial agents to bacteria: an alternative therapy for treatment of bacterial infections. Antimicrob. Agents Chemother. *47*, 1301–1307.

Withey, S., Cartmell, E., Avery, L.M., and Stephenson, T. (2005). Bacteriophages – potential for application in wastewater treatment processes. Sci. Total Environ. *339*, 1–18.

Wright, A., Hawkins, C.H., Anggård, E.E., and Harper, D.R. (2009). A controlled clinical trial of a therapeutic bacteriophage preparation in chronic otitis due to antibiotic resistant *Pseudomonas aeruginosa;* a preliminary report of efficacy. Clin. Otolaryngol. *34*, 349–357.

Yacoby, I., Shamis, M., Bar, H., Shabat, D., and Benhar, I. (2006). Targeting antibacterial agents by using drug-carrying filamentous bacteriophages. Antimicrob. Agents Chemother. *50*, 2087–2097.

Yacoby, I., Bar, H., and Benhar, I. (2007). Targeted drug-carrying bacteriophages as antibacterial nanomedicines. Antimicrob. Agents Chemother. *51*, 2156–2163.

Yilmaz, C., Colak, M., Yilmaz, B.C., Ersoz, G., Kutateladze, M., and Gozlugol, M. (2013). Bacteriophage therapy in implant-related infections: an experimental study. J. Bone Joint Surg. Am. *95*, 117–125.

Yoshimura, G., Komatsuzawa, H., Hayashi, I., Fujiwara, T., Yamada, S., Nakano, Y., Tomita, Y., Kozai, K., and Sugai, M. (2006). Identification and molecular characterization of an N-Acetylmuraminidase, Aml, involved in *Streptococcus mutans* cell separation. Microbiol. Immunol. *50*, 729–742.

Young, I., Wang, I., and Roof, W.D. (2000). Phages will out: strategies of host cell lysis. Trends Microbiol. *8*, 120–128.

Zhang, Q.G., and Buckling, A. (2012). Phages limit the evolution of bacterial antibiotic resistance in experimental microcosms. Evol. Appl. *5*, 575–582.

Zhang, H., Bao, H., Billington, C., Hudson, J.A., and Wang, R. (2012). Isolation and lytic activity of the Listeria bacteriophage endolysin LysZ5 against *Listeria monocytogenes* in soya milk. Food Microbiol. *31*, 133–136.

Zimecki, M., Artym, J., Kocieba, M., Weber-Dabrowska, B., Borysowski, J., and Górski, A. (2009). Effects of prophylactic administration of bacteriophages to immunosuppressed mice infected with *Staphylococcus aureus*. BMC Microbiol. 9, 169.

Zimecki, M., Artym, J., Kocieba, M., Weber-Dabrowska, B., Borysowski, J., and Górski, A. (2010). Prophylactic effect of bacteriophages on mice subjected to chemotherapy-induced immunosuppression and bone marrow transplant upon infection with *Staphylococcus aureus*. Med. Microbiol. Immunol. *199*, 71–79.

Considerations for Using Bacteriophages in Plant Pathosystems

7

Jeffrey B. Jones, Aleksa Obradović and Botond Balogh

Abstract

Phages have the potential for controlling plant pathogens in the rhizosphere or phyllosphere. Success of phage in disease control requires that high populations of both phage and bacterium exist in order to initiate a chain reaction of bacterial lysis. Various factors exist that can hinder success of disease control. Physical factors in natural environments such as the presence of biofilms that trap bacteriophages, low soil pH which inactivates phages, low rates of diffusion of phages in soil that prevent contact with target bacteria, and inactivation of phages upon exposure to UV, all impact successful use of phages. Other considerations relate to the bacterial strains which exist in nature. The bacterial species may have a low or high degree of variation in sensitivity to bacteriophages. Therefore, phage selection for field use requires careful monitoring of strains in the field be done due to the potential for strain variation in the field and the likelihood for development of bacterial strains with resistance to the deployed bacteriophages. Application timing has also been shown to be an important factor in improving efficacy of bacteriophages. For instance, ultraviolet light is deleterious to bacteriophages and upon exposure phage populations plummet; therefore, evening applications of bacteriophages result in persistence of phages on leaf surfaces for longer periods of time and may result in improved disease control. Extending the period of time phages persist in the phyllosphere has been a major hurdle. Formulations have been identified which improve the persistence of bacteriophages on leaf surfaces; however, there is a need to identify superior formulations that extend the life on leaf surfaces from hours to days. Another strategy for maintaining high populations of phages has been to use non-pathogenic bacterial strains that are sensitive to the phage(s) or a closely related organism that does not cause disease on the plant host. Finally, bacteriophages may have value as part of an integrated management strategy.

Introduction

Bacterial plant pathogens account for significant losses to agricultural production systems. Disease control for many bacterial incited ailments is a major challenge affected by various factors including: pathogen variation; ability to overcome plant genetic resistance as a result of mutation or loss of plasmid; the lack of effective bactericides as a result of strains developing tolerance; and the pathogen's ability to reach high populations in a relatively short period of time when conditions are favourable for disease development. Since control of bacterial incited diseases is such a major challenge, integrated pest management strategies have been critical to reducing the impact from bacterial diseases (Obradovic *et al.*, 2004; Obradovic *et al.*, 2005). A major component of integrated pest management strategies has been the use of bactericides; however, control with bactericides has been extremely difficult because few effective bactericides are available. Antibiotics and copper-based bactericides have been the principal chemicals used for disease control. Copper has been the most extensively used bactericide; however, copper resistance is present in many plant pathogenic bacteria and is associated with plasmids and in rare situations the chromosome

(Bender and Cooksey, 1986, 1987; Stall *et al.*, 1986; Bender *et al.*, 1990; Lee *et al.*, 1994; Basim *et al.*, 1999; Canteros, 1999; Gutiérrez-Barranquero *et al.*, 2013). Transfer of this resistance is readily achieved (Stall *et al.*, 1986). Antibiotics have also been used as part of a management strategy for various bacterial diseases, since the 1950s. The aminoglycoside antibiotic streptomycin was used extensively for control of bacterial diseases and as a result of extensive use, streptomycin-resistant strains became prevalent, resulting in reduced disease control efficacy of bacterial spot of tomato and pepper (Thayer and Stall, 1961), fire blight of apple and pear (Manulis *et al.*, 1998), and many other bacterial plant pathogens (Cooksey, 1990). Systemic acquired resistance (SAR) compounds have provided a level of control against tomato and pepper bacterial diseases (Louws *et al.*, 2001, Romero *et al.*, 2001; Obradovic *et al.*, 2004; Huang *et al.*, 2012), *Xanthomonas* leaf blight on onion (Gent and Schwartz, 2005) and fire blight on apple (Maxson-Stein *et al.*, 2002); however these inducers may have negative physiological effects on plant growth and yield (Romero *et al.*, 2001; Gent and Schwartz, 2005).

Phage therapy has gained significant interest recently (Jones *et al.*, 2007, 2012; Balogh *et al.*, 2003, 2008, 2010; Obradovic *et al.*, 2004, 2005; Iriarte *et al.*, 2012; Frampton *et al.*, 2012; Lang *et al.*, 2007; Gent and Schwartz, 2005; Fujiwara *et al.*, 2011; Bae *et al.*, 2012; Lim *et al.*, 2013) as an alternative to the use of conventional bactericides, although its use goes back to the early days following the discovery of bacteriophages in the early 1900s (Twort, 1915; d'Hérelle, 1917). In 1924 bacteriophage was isolated from black rot infected cabbage and was shown to inhibit *in vitro* growth of the pathogen, *Xanthomonas campestris* pv. *campestris* (Mallmann and Hemstreet, 1924). Shortly thereafter bacteriophages were isolated from soil with activity against *Erwinia carotovora* subsp. *atroseptica*, causal agent of blackleg disease of potato (Kotila and Coons, 1925); in growth chamber experiments co-inoculation of *E. carotovora* subsp. *atroseptica* with phage prevented rotting of potato tubers (Kotila and Coons, 1925). Thomas (1935) treated maize seeds with a crude bacteriophage suspension isolated from Stewart's Wilt infected plant material and upon planting of the treated seed disease incidence was reduced from 18% to 1.4%.

More recently, Civerolo and Keil (1969) applied bacteriophage to peach foliage prior to inoculating with *Xanthomonas campestris* pv. *pruni* (Xcp) and reduced disease incidence from 58% on untreated plants to 22% on phage-treated plants. In another study Civerolo (1973) reported that treatment of peach seedlings with crude lysates of a phage mixture resulted in fewer infected leaves and a significant reduction of disease compared to controls. Civerolo speculated that phages may regulate populations of Xcp and in so doing reduce disease. In other studies with this bacterial pathogen, Saccardi *et al.* (1993) and Zaccardelli *et al.* (1992) reduced fruit spot incidence on peaches with biweekly spray applications of phage suspensions active against the target bacterium, Xcp. In their work, they isolated eight phages active against the pathogen, screened them for host range and lytic ability, and selected a lytic phage strain with the broadest host range for disease control.

Although the above studies and others provided positive results, phage therapy was not considered a good strategy for controlling plant pathogenic bacteria because of its unreliability. This is clearly the opinion in a review article in 1963 (Okabe and Goto, 1963). In a more recent review article it was stated that practical use of phages for control of bacterial plant disease in the field was unsuccessful (Goto, 1992). Furthermore, because of phages' narrow spectrum of activity, they were considered less likely to succeed compared to antibiotics (Summers, 2005). Therefore, phages were not considered a practical control strategy for bacterial plant diseases, with antibiotics and copper compounds becoming the principal control agents. Nonetheless in recent years several papers were published on the use of phages against different plant pathogens including *E. carotovora* subsp. *carotovora* (Lim *et al.*, 2013) and *Ralstonia solanacearum* (Fujiwara *et al.*, 2011; Iriarte *et al.*, 2012; Bae *et al.*, 2012). In addition, phages infecting *Xanthomonas arboricola* pv. *juglandis* (Romero-Suarez *et al.*, 2012) and *Erwinia amylovora* (Born *et al.*, 2011; Svircev *et al.*, 2006) were reported; these also hold promise for use for plant disease control.

Impediments in using phage therapy in plant pathology

Development of resistance in bacterial strains

The high probability of developing bacterial strains resistant to the phage is a real concern (Fujiwara *et al.*, 2011). It was calculated based on the classic experiments of Luria and Delbrück (1943) that the mutation rate of *E. coli* to resistance to phage T1 is 3.4×10^{-9} (Birge, 2000). Early on in studies utilizing bacteriophage the concern of host resistance was addressed by Katznelson (1937) and later as a concern in a review article (Okabe and Goto, 1963) and then more demonstratively as a significant impediment in the implementation of this strategy for control (Vidaver, 1976).

Phage selection

An important and oftentimes neglected aspect of phage-based biological control is pre-screening phages for their biocontrol value before application; that is, identifying specific bacteriophages with particular characteristics that may prove effective in control rather than arbitrarily selecting them based strictly on lytic activity for disease control studies. In two studies (Zaccardelli *et al.*, 1992, Saccardi *et al.*, 1993) the phage selected from a collection of eight phages for biological control experiments was a lytic phage with the broadest host range. Balogh (2006) observed that phages vary in their ability to interact with and multiply on their host bacterium on the grapefruit leafs, an ability that correlated with their disease control efficacy against citrus canker. However, he found no in-vitro characteristics that predicted disease control ability against bacterial spot on tomato (Balogh, unpublished results).

Application timing

Timing of bacteriophage applications relative to the arrival of the pathogen is a critical concern. Given the leaf surface being an extremely harsh environment for survival of bacteriophages (Gill and Abedon, 2003; Iriarte *et al.*, 2007) and the drastic reduction in bacteriophage populations on leaf surfaces as a result of ultraviolet light (Iriarte *et al.*, 2007), timing of bacteriophage applications is essential to optimize that high bacteriophage populations be present in close proximity to the target bacterium in order to maximize phage therapy efficacy. This has been shown to influence efficacy of disease control in several instances. Civerolo and Keil (1969) achieved a marked reduction of peach bacterial spot only if phage treatment was applied one hour or one day before inoculation with the pathogen. There was a slight disease reduction when phage was applied 1 hour after inoculation and no effect if applied 1 day later. Civerolo (1972) suggested that bacteria were inaccessible to phage in the intercellular spaces, or there were not enough phages reaching the pathogen. Schnabel *et al.* (1999) achieved a significant reduction of fire blight on apple blossoms when the phage mixture was applied at the same time as the pathogen, *Erwinia amylovora*. In contrast, disease reduction was not significant when phages were applied a day before inoculation. Likewise, Bae *et al.* (2011) and Iriarte *et al.* (2012) showed that phages were most effective when applied at the time of plant inoculation; use of phages up to a few days before or after inoculation resulted in a lower, or even no biocontrol efficacy. Bergamin Filho and Kimati (1981) investigated the effect of timing on the efficacy of phage treatment in greenhouse trials with two pathosystems: black rot of cabbage, caused by *Xanthomonas campestris* pv. *campestris* and bacterial spot of pepper, caused by *Xanthomonas campestris* pv. *vesicatoria*. Phage treatment was applied once varying from 7 days before to 4 days after pathogen inoculation. On cabbage significant disease reduction was achieved if the phage treatment was applied 3 days before to 1 day after inoculation, whereas on pepper from 3 days before to the day of inoculation. The greatest disease reduction occurred with application of phages the same day as inoculation in both pathosytems.

In terms of disease control in the field, the time of day when phages are applied also affects efficacy. Sunlight irradiation was identified as the single most detrimental factor reducing phage persistence in the tomato canopy (Iriarte *et al.*, 2007). When phage treatments were applied before dawn or after sunset, when they were not exposed immediately to direct sunlight, their residual activity was prolonged (Iriarte *et al.*, 2007) and resulted in more effective control of tomato bacterial spot, in comparison with daytime treatment applications (Balogh *et al.*, 2003).

Strain variation

Better understanding of bacterial strains associated with particular plant species has shown that significantly greater genetic diversity may exist than considered in earlier studies (Guy *et al.*, 2013). Much of our knowledge is based on experience with *Xanthomonas* spp. associated with tomato and pepper. Bouzar *et al.* (1991) used 26 bacteriophages to type approximately 130 *X. euvesicatoria* strains isolated from various countries in the Caribbean including Central America. There was a very high level of diversity in which at least 26 different reaction patterns were observed. Beyond the variation within a particular bacterial species, plant species may be infected by more than one bacterial species within the same genus and as a result in all likelihood will require different phages. For instance four different *Xanthomonas* species infect tomato (Jones *et al.*, 2004), and as a result require the use of different phages for disease control. In general, the phages that infect bacterial strains within one bacterial species do not infect strains within the other bacterial species.

Phage utilization in the rhizosphere

Phages have the potential for suppressing plant pathogens in the rhizosphere or phyllosphere. However, Gill and Abedon (2003) identified factors that can hinder success of disease control in the rhizosphere. The relatively low diffusion rate of phages through heterogeneous soil matrices changes as a function of available free water. Biofilms can trap phages (Storey and Ashbolt, 2001), soil clay particles can reversibly adsorb phages (Williams *et al.*, 1987), and low soil pH can inactivate phages (Sykes *et al.*, 1981; Bae *et al.*, 2012). In natural environments, as a result of low rates of phage diffusion and high rates of phage inactivation, low numbers of viable phages are available to lyse target bacteria (Gill and Abedon, 2003). One additional factor needed for a high degree of success is that high populations of both phage and bacterium exist in order to initiate a chain reaction of bacterial lysis (Gill and Abedon, 2003). However, in spite of the above factors that may limit phage stability in the rhizosphere, phages were found to be fairly stable in soil (Fujiwara *et al.*, 2011).

Persistence of phages on leaf surfaces

Biocontrol efficacy is influenced by the biocontrol agent and target pathogen densities (Johnson, 1994). Maintaining high populations of biocontrol agents in the target environment in close proximity to the target bacterium is of critical importance. Phage therapy requires high populations of phage and target bacterium (Gill and Abedon, 2003). Clearly, for good control, phages must exist above a threshold concentration; below that threshold phages will have a minor effect on bacterial populations. In support of this threshold effect hypothesis, Balogh (2002) observed that phage mixtures applied at 10^6 or 10^8 PFU/ml concentration provided similar levels of control of tomato bacterial spot, but at 10^4 PFU/ml were ineffective.

In the phyllosphere a harsh environment exists and phages in this location degrade extremely rapidly (Civerolo and Keil, 1969, McNeil *et al.* 2001; Balogh, 2002, Balogh *et al.*, 2003, Iriarte *et al.*, 2007; Jones *et al.*, 2012). This short-lived persistence on plant leaf surfaces is a limiting factor of phage treatment. In field and laboratory studies, insect viruses were shown to be inactivated by high temperatures, high and low pH and sunlight, and were readily dislodged by rain (Ignoffo *et al.*, 1989; Ignoffo and Garcia, 1992). Other factors that may contribute to phage inactivation include desiccation and exposure to certain chemical pesticides, such as copper bactericides commonly used in bacterial disease management strategies (Iriarte *et al.*, 2007). Under field conditions sunlight irradiation is the single major deleterious component affecting phage survival (Iriarte *et al.*, 2007). In particular UV-A and UV-B spectra of sunlight were determined to be the environmental factor most destructive to viruses (Ignoffo and Garcia, 1994). Initial phage studies on tomato showed that phage applied in the mid-morning was not effective in controlling bacterial spot of tomato (Jones, unpublished results).

Iriarte *et al.* (2007) demonstrated that phage populations plummeted precipitously during the early afternoon hours when UV radiation was high. However, if the bacteriophages were applied in the late afternoon or early evening close to sunset, phage populations did not plummet overnight and thus high phage concentrations persisted overnight and potentially could interact with bacterial strains on the leaf surface. Phage survival was highly correlated with the intensity of sunlight UV irradiation.

Future considerations for improving phage efficacy

As stated earlier biological control using bacteriophages is dependent on the ability of the biological control agent to persist at high levels in close proximity to the target bacterium (Johnson, 1994). Furthermore, the phages must reach and attach to their hosts before themselves being destroyed by adverse factors (Goodridge, 2004). Several considerations exist with regard to improving phage–bacterium interactions: concentration and accessibility of target bacteria; timing of phage application to optimize efficacy; phage infection and replication ability in the target environment; phage concentration at the site of interaction (i.e. phyllosphere or rhizosphere); rates of virion degradation (phage vary in degradative properties); and the presence of adequate moisture as a medium for phage diffusion (Gill and Abedon, 2003). Below, we will address several of the key issues which must be considered for improving phage efficacy.

Development of resistance in bacterial strains

Several strategies exist that address the issue phage resistance. One strategy, developed by Jackson (1989), minimizes the occurrence of phage-resistant mutants. This involved utilizing mixtures of wild-type and host-range mutant (h-mutant) phages. H-mutants are phages that lyse bacterial strains that are resistant to the parent phage (Adams, 1959), while maintaining the ability to lyse the wild-type bacterium. Using this strategy, a mixture of four phages including wild-type and host range-mutant (h-mutant) phages were applied twice weekly and provided significantly better disease control and produced greater yield of extra large fruits than the standard copper-mancozeb (Flaherty *et al.*, 2000).

A second strategy relies on constant monitoring of phage efficacy. Phage mixtures are applied on the target field, and bacteria are regularly isolated from phage-treated plants during the treatment period. The isolates are evaluated for their sensitivity to the utilized phages, and if resistance is observed, the phage in question is removed from the mix and replaced with a new one. This strategy relies on the existence of phage-banks, large collection of phages specific to the target bacterium.

Phage selection

The proper assay for phage selection is a critical and often overlooked factor in ensuring success of phage therapy in agriculture. Although *in vitro* assays are frequently used in the selection process for phages, these may not be good predictors of biological control ability. These assays provide optimal conditions for phage infections – such as exponentially growing susceptible bacterial culture, controlled constant temperature, constant level of free moisture, more or less constant pH, the availability of a wide range of nutrients, and protection from sunlight exposure – and as such, does not represent the 'reality of life' in the plant environment, where nutrients and water are scarce and the environmental conditions are in a constant limbo. Balogh (2006) found that two of three phages, which were active on *Xanthomonas citri* pv. *citri* in plate assays, were unable to lyse the bacterium on the grapefruit foliage. These two phages, not surprisingly, were not able to suppress citrus canker in greenhouse trials. Balogh evaluated eight bacteriophages that were active against *Xanthomonas perforans* for a number of *in vitro* characteristics, such as plaque size, antibacterial activity or phage multiplication rate, and found no correlations between these attributes and actual disease control efficacy (unpublished results). Based on these results, actual bioassays are unavoidable in order to gauge biocontrol activity. Such bioassays should evaluate the phages capacity to persist and replicate on the plant surface, and their effect on disease development.

Application timing

Given the difficulty in maintaining high phage populations in the phyllosphere or other aboveground plant parts, optimizing an application strategy that results in improved disease control from phage applications is necessary to essential. Using the bacterial spot of tomato system as a model, Balogh et al. (2003) and Iriarte et al. (2007) noted that in the phyllosphere, bacteriophage populations are drastically reduced when phages are applied in the morning or early afternoon. As a result very low phage populations are present on the leaf surface to interact with the target bacterium. Given that phage persistence on leaf surfaces is very poor over a 24 h period, frequent applications in the evening hours to maximize exposure of the target bacterium to high concentrations of phages are necessary to maximize disease control. Another factor one might consider when choosing when to apply (and what application volume to use), is the availability of free moisture on the leaves. Phages will only interact with their target bacterium if the leaf surface is wet, and consequently it seems advisable to apply them when free moisture is expected to stay on the leaves for an extended time period in order to ensure longer exposure time of the target bacterium.

Persistence of phages on leaf surfaces

Persistence on leaf surfaces is a major limiting factor in using phage therapy for disease control in the phyllosphere. Several strategies have been evaluated for increasing phage persistence, including the use of protective formulations, application scheduling for sunlight avoidance and co-application of bacterial hosts for *in vivo* phage propagation.

In several studies solar protectants were identified that increased biocontrol efficacy not only for bacteriophages, but also for entomopathogenic viruses and proteinaceous biopesticides (Behle *et al.*, 1996; Ignoffo *et al.*, 1997; Balogh *et al.*, 2003; Arthurs *et al.*, 2006; Reddy *et al.*, 2008; Behle and Popham, 2012). Balogh (2002) identified compounds that, when mixed with phage, extended the persistence of phage on the phyllosphere. Balogh *et al.* (2003) enhanced the efficacy of phage treatment with protective formulations that increased phage persistence on tomato foliage. Balogh *et al.* (2003) determined that phage populations persisted at significantly higher concentrations on leaf surfaces when the phage suspensions were applied in combination with skim milk alone or in combination with sucrose. Iriarte *et al.* (2007) corroborated that the combination of phages with skim milk was important for improved persistence on leaf surfaces, even under intense UV irradiation, although phage populations did drop to very low levels even with the formulated phage. Balogh *et al.* (2003) determined that addition of several formulations to the phage suspension led to enhanced disease control efficacy (Balogh *et al.*, 2003). In other field studies treatment of plants with formulated phages resulted in reduced disease and increased yield (Obradovic *et al.*, 2004, 2005). Although these studies improved phage persistence utilizing skim milk formulation, there is considerable need for identifying formulations that are superior to skim milk.

A second approach for improving persistence is based on the unique advantage of bacteriophages over chemical pesticides in the ability to increase their populations by multiplying on the target bacterial host. This ability could potentially be used if phages are applied into an environment where a phage-sensitive bacterium is present, or where the phages and bacterial host are delivered together (Bae *et al.*, 2012; Iriarte *et al.*, 2012). On leaf surfaces, where high host populations persist, phages persist at significantly higher levels than on surfaces without the host (Balogh, 2006). This approach was investigated for the soilborne bacterium, *Ralstonia solanacearum* in which Tanaka *et al.* (1990) used an avirulent strain of *Ralstonia solanacearum* and its phage that was active against both the virulent and avirulent strains to reduce tobacco bacterial wilt incidence. Although application of the avirulent strain alone caused significant disease control, co-application of phage with the avirulent strain significantly improved disease control beyond the avirulent strain alone. Svircev *et al.* (2006) employed a similar strategy for controlling fire blight of pear, by selecting phages based on the ability to lyse both the target organism, the pathogen *Erwinia amylovora* and also an antagonistic phyllosphere bacterium, *Pantoea agglomerans*. *P. agglomerans*, a biological control agent for *E. amylovora*, served as well as a phage carrier, and as a vehicle propagator of phage on the inoculation sites. While *P. agglomerans* significantly reduced disease, combining it with phage resulted in significantly better disease control; the combination resulted in disease control comparable to streptomycin treatment.

Different types of bacteria could be used for propagating phage on plants. The main requirements are that they persist in the plant environment relatively well and are hosts to the phages used. This would mean in practice that they have to be closely related to the target pathogen. They can be epiphytes, such as the *Pantoea agglomerans* used by Svircev *et al.* (2006), avirulent strains of the pathogen, such as the *Ralstonia solanacearum* strain utilized by Tanaka *et al.* (1990), or even attenuated strains of the pathogen (Iriarte *et al.*, 2012). Such pathogenically attenuated stains are under consideration for *Xanthomonas* phages (Iriarte *et al.*, 2012), as avirulent *Xanthomonas* strains are very weak epiphytes (Jones, unpublished results). The use of attenuated strains carries an inherent risk, since under favourable conditions, these strains will cause considerable level of disease.

Phages as part of integrated management strategy

In contrast to the long list of antifungal products, the search for bactericides suitable for crop protection turned out to be exceedingly difficult. In order to reduce losses in yield and quality of agricultural crops, continuous efforts have been made to identify chemicals and other strategies that could be effectively used for controlling plant pathogenic bacteria. Chemical control of plant bacterial diseases mostly relied on the application of the antibiotic streptomycin and copper-based compounds. However, efficacy of these compounds in plant protection was compromised by occurrence of resistant strains of pathogenic bacteria. Recently, plant activators, triggering natural plant defence mechanisms, have been introduced in disease management programs (Louws *et al.*, 2001; Qui *et al.*, 1997; Du *et al.*, 2012). One of them, acibenzolar-S-methyl (ASM), showed excellent potential as a part of integrated strategy for control of tomato bacterial spot (Louws *et al.*, 2001; Obradovic *et al.*, 2004, 2005). However, some results showed adverse effects of this compound on tomato and pepper growth and yield (Louws *et al.*, 2001). Due to lack of an efficient synthetic bactericide the most frequently implemented approaches in control of plant pathogenic bacteria were use of pathogen-free seed or planting material, genetic host resistance, cultural and sanitation practices, and biological control agents. In practice, none of the above non-chemical treatments have been able to stand alone and provide efficient control.

Multifaceted, integrated strategies carry the promise for effective, reliable and sustainable management of bacterial plant diseases. Several approaches have been explored for using phage treatment within an integrated management strategy. Tanaka *et al.* (1990) reduced tobacco bacterial wilt by co-application of an avirulent strain of the pathogen, *R. solanacearum*, with a phage that was active against both virulent and avirulent strains. Using a similar approach, Svircev *et al.* (2006) reduced fire blight of pear with co-application of an antagonistic epiphyte, *Pantoea agglomerans* and a phage that lysed both the antagonist and the pathogen, *Erwinia amylovora*.

Lang *et al.* (2007) evaluated phage treatment in combination with ASM or with copper-mancozeb for the control of *Xanthomonas* leaf blight of onion and found that both combinations resulted in enhanced disease control. To the contrary, Balogh *et al.* (2008) observed no improvement in control of citrus canker or citrus bacterial spot with the combination of bacteriophages with copper-mancozeb.

However, most of our knowledge is based on experience from several experiments of tomato bacterial spot control. In an effort to develop more sustainable strategies for reducing bacterial spot severity on tomato, Obradovic *et al.* (Obradovic *et al.*, 2004, 2005) compared efficacy of various combinations of unformulated phages, biocontrol agents, including strains of PGPR, bacterial antagonists, SAR inducers (harpin, ASM) and copper hydroxide in greenhouse experiments. The intention was to integrate some of these practices, optimizing their benefits in control of tomato bacterial spot in the greenhouse, aiming at developing a comprehensive phage-based integrated management strategy for disease control in commercial tomato fields. Several combinations of treatments that effectively controlled tomato bacterial spot in the greenhouse were tested in field trials carried out in north and central Florida during three consecutive seasons. Although copper sensitive strains were used, providing more effective control of bacterial spot with copper bactericides, application of formulated phages twice a week was either more effective or equally effective compared with the standard copper-mancozeb treatment. However, integration of phage treatments and ASM provided an additional reduction in disease pressure and resulted in more efficient foliar disease control than ASM, phage, or copper-mancozeb alone (Obradovic *et al.*, 2004).

Conclusions

Bacteriophage therapy is under intense evaluation for managing a number of bacterial plant diseases. The potential benefits are substantial: phage therapy would provide an environmentally friendly disease control strategy against a group of pathogens, bacteria, which are chronically hard to control. However, a number of hurdles exist: the short residual activity and narrow host range of these agents, the considerable effect of the environment and resistance development in the target bacteria. By now most of these issues have been addressed and it appears that bacteriophages do have a place in disease management as part of an integrated approach.

In the United States, bacteriophages are now commercially available for control of bacterial spot and speck on tomato and pepper (Agriphage from OmniLytics Inc., Salt Lake City, UT, EPA Registration # 67986-1).

References

Adams M.H. (1959). Bacteriophages. New York, Interscience Publishers.

Arthurs S.P., Lacey, L.A., and Behle, R.W. (2006). Evaluation of spray-dried lignin-based formulations and adjuvants as solar protectants for the granulovirus of the codling moth, *Cydia pomonella* (L). J. Inverteb. Pathol. 93, 88–95.

Bae, J.Y., Wu, J., Lee, H.J., Jo, E.J., Murugaiyan, S., Chung, E., and Lee, S.W. (2012). Biocontrol potential of a lytic bacteriophage PE204 against bacterial wilt of tomato. J. Microbiol. Biotechnol. 22, 1613–1620.

Balogh, B. (2002). Strategies of improving the efficacy of bacteriophages for controlling bacterial spot of tomato. MS Thesis, Gainesville, FL, University of Florida.

Balogh, B. (2006). Characterization and use of bacteriophages associated with citrus bacterial pathogens for disease control. PhD Dissertation, Gainesville, FL, University of Florida.

Balogh, B., Jones, J.B., Momol, M.T., Olson, S.M., Obradovic, A., King, P., and Jackson, L.E. (2003). Improved efficacy of newly formulated bacteriophages for management of bacterial spot on tomato. Plant Dis. 87, 949–954.

Balogh, B., Canteros, B.I., Stall, R.E., and Jones, J.B. (2008). Control of citrus canker and citrus bacterial spot with bacteriophages. Plant Dis. 92, 1048–1052.

Behle, R.W., and Popham, H.J. (2012). Laboratory and field evaluations of the efficacy of a fast-killing baculovirus isolate from *Spodoptera frugiperda*. J. Inverteb. Pathol. *109*, 194–200.

Behle, R.W., McGuire, M.R., and Shasha, B.S. (1996). Extending the residual toxicity of *Bacillus thuringiensis* with casein-based formulations. J. Econom. Entomol. *89*, 1399–1405.

Bender, C.L., and Cooksey, D.A. (1986). Indigenous plasmids in *Pseudomonas syringae* pv. *tomato*: Conjugative transfer and role in copper resistance. J. Bacteriol. *165*, 534–541.

Bender, C.L., and Cooksey, D.A. (1987). Molecular cloning of copper resistance genes from *Pseudomonas syringae* pv. *tomato*. J. Bacteriol. *169*, 470–474.

Bender, C.L., Malvick, D.K., Conway, K.E., George, S., and Cooksey, D.A. (1990). Characterization of pXv10A, a copper resistance plasmid in *Xanthomonas campestris* pv. *vesicatoria*. Appl. Environ. Microbiol. *56*, 170–175.

Bergamin Filho, A., and, Kimati, H. (1981). Estudos sobre um bacteriofago isolado de *Xanthomonas campestris*. II. Seu emprego no controle de *X. campestris* e *X. vesicatoria*. Summa Phytopath. 7, 35–43.

Birge, E.A. (2000). Bacterial and Bacteriophage Genetics, 4th edn (New York, Springler-Verlag), p. 74.

Born, Y., Fieseler, L., Marazzi, J., Lurz, R., Duffy, B., and Loessner, M.J. (2011). Novel virulent and broad-host-range *Erwinia amylovora* bacteriophages reveal a high degree of mosaicism and a relationship to Enterobacteriaceae phages. Appl. Environ. Microbiol. 77, 5945–5954.

Bouzar, H., Jones, J.B., Stall, R.E., Louws, F.J., Schneider, M., Rademaker, J.L.W., de Bruijn, F.J., and Jackson, L.E. (1999). Multiphasic analysis of xanthomonads causing bacterial spot disease on tomato and pepper in the Caribbean and Central America: Evidence for common lineages within and between countries. Phytopathology *89*, 328–335.

Canteros, B.I. (1999). Copper resistance in *Xanthomonas campestris* pv. *citri*. In Mahadevan A, ed. Plant Pathogenic Bacteria – Proceedings of the 9th International Conference, 26–29 August, 1996. Chennai, India, Centre for Advanced Study in Botany, University of Madras, 455–459.

Civerolo, E.L. (1972). Interaction between bacteria and bacteriophages on plant surfaces and in plant tissues. In Maas Geesteranus HP, ed. Proceedings of the Third International Conference of Plant Pathogenic Bacteria, April 14–21, 1971. Wageningen, Centre for Agricultural Publishing and Documentation, 25–37.

Civerolo, E.L. (1982). Disease management by cultural practices and environmental control. In Phytopathogenic Prokaryote, M.S. Mount and G.H. Lacy, eds. (New York, Academic Press), pp. 343–360.

Civerolo, E.L., and Keil, H.L. (1969). Inhibition of bacterial spot of peach foliage by *Xanthomonas pruni* bacteriophage. Phytopathology *59*, 1966–1967.

Cooksey, D.A. (1990). Genetics of bactericide resistance in plant pathogenic bacteria. Annu. Rev. Phytopathol. *28*, 201–219.

Coons, G.H., and Kotila, J.E. (1925). The transmissible lytic principle (bacteriophage) in relation to plant pathogens. Phytopathology *15*, 357–370.

Du, Q., Zhu, W., Zhao, Z., Qian, X., and Xu, Y. (2012). Novel benzo-1,2,3-thiadiazole-7-carboxylate derivatives as plant activators and the development of their agricultural applications. J. Agric. Food Chem. *60*, 346–353.

Flaherty, J.E., Jones, J.B., Harbaugh, B.K., Somodi, G.C., and Jackson, L.E. (2000). Control of bacterial spot on tomato in the greenhouse and field with H-mutant bacteriophages. Hort. Sci. *35*, 882–884.

Frampton, R.A., Pitman, A.R., and Fineran, P.C. (2012). Advances in bacteriophage-mediated control of plant pathogens. Int. J. Microbiol. *2012*, 326452.

Fujiwara, A., Fujisawa, M., Hamasaki, R., Kawasaki, T., Fujie, M., and Yamada, T. (2011). Biocontrol of Ralstonia solanacearum by treatment with lytic bacteriophages. Appl. Environ. Microbiol. *77*, 4155–4162.

Frey, P., Prior, P., Marie, C., Kotoujansky, A., Trigalet-Demery, D., and Trigalet, A. (1994). Hrp- mutants of *Pseudomonas solanacearum* as potential biocontrol agents of tomato bacterial wilt. Appl. Environ. Microbiol. *60*, 3175–3181.

Gent, D.H., and Schwartz, H.F. (2005). Management of xanthomonas leaf blight of onion with a plant activator, biological control agents, and copper bactericides. Plant Dis. *89*, 631–639.

Gill, J.J., and Abedon, S.T. (2003). Bacteriophage ecology and plants. http://www.apsnet.org/online/feature/pistachio/print.asp.

Goodridge, L.D. (2004). Bacteriophage biocontrol of plant pathogens: Fact or fiction? Trends Biotechnol. *22*, 384–385.

Goto, M. (1992). Fundamentals of Bacterial Plant Pathology (San Diego, CA, Academic Press).

Graham, J.H., and Leite, J.R.P. (2004). Lack of control of citrus canker by induced systemic resistance compounds. Plant Dis. *88*, 745–750.

Gutiérrez-Barranquero, J.A., de Vicente, A., Carrión, V.J., Sundin, G.W., and Cazorla, F.M. (2013). Recruitment and rearrangement of three different genetic determinants into a conjugative plasmid increase copper resistance in Pseudomonas syringae. Appl. Environ. Microbiol. *79*, 1028–1033.

Guy, E., Genissel, A., Hajri, A., Chabannes, M., David, P., Carrere, S., Lautier, M., Roux, B., Boureau, T., Arlat, M., *et al.* (2013). Natural genetic variation of Xanthomonas campestris pv. campestris pathogenicity on arabidopsis revealed by association and reverse genetics. MBio. *4*, e00538–12.

Hert, A.P. (2007). Evaluation of bacteriocins in *Xanthomonas perforans* for use in biological control of *Xanthomonas euvesicatoria*. PhD Dissertation, Gainesville, FL, University of Florida.

Huang, Ch.-H., Vallad, G.E., Zhang, S., Wen, A., Balogh, B., Figueiredo, J.F.L., Behlau, F., Jones, J.B., Momol, M.T., and Olson, S.M. (2012). Effect of application frequency and reduced rates of acibenzolar-s-methyl on the field efficacy of induced resistance against bacterial spot on tomato. Plant Dis. *96*, 221–227.

Ignoffo, C.M., and Garcia, C. (1992). Combinations of environmental factors and simulated sunlight affecting activity of inclusion bodies of the *Heliothis* (*Lepidoptera: Noctuidae*) nucleopolyhedrosis virus. Environ. Entomol. *21*, 210–213.

Ignoffo, C.M., and Garcia, C. (1994). Antioxidant and oxidative enzyme effects on the inactivation of inclusion bodies of the *Heliothis* baculovirus by simulated sunlight- UV. Environ. Entomol. *23*, 1025–1029.

Ignoffo, C.M., Rice, W.C., and McIntosh, A.H. (1989). Inactivation of nonoccluded and occluded baculoviruses and baculovirus-DNA exposed to simulated sunlight. Environ. Entomol. *18*, 177–183.

Ignoffo, C.M., Garcia, C., and Saathoff, S.G. (1997). Sunlight stability and rain-fastness of formulations of *Baculovirus heliothis*. Environ. Entomol. *26*, 1470–1474.

Iriarte, F.B., Balogh, B., Momol, M.T., Smith, L.M., Wilson, M., and Jones, J.B. (2007). Factors affecting survival of bacteriophage on tomato leaf surfaces. Appl. Environ. Microbiol. 73, 1704–1711.

Iriarte, F.B., Obradović, A., Wernsing, M.H., Jackson, L.E., Balogh, B., Hong, J.A., Momol, M.T., Jones, J.B., and Vallad, G.E. (2012). Soil-based systemic delivery and phyllosphere *in vivo* propagation of bacteriophages: Two possible strategies for improving bacteriophage persistence for plant disease control. Bacteriophage. *2*, 215–224.

Jackson, L.E. (1989). Bacteriophage prevention and control of harmful plant bacteria. US Patent No.4828999.

Ji, P., Campbell, H.L., Kloepper, J.W., Jones, J.B., Suslow, T.V., and Wilson, M. (2006). Integrated biological control of bacterial speck and spot of tomato under field conditions using foliar biological control agents and plant growth-promoting rhizobacteria. Biol. Control *36*, 358–367.

Johnson, K.B. (1994). Dose–response relationships and inundative biological control. Phytopathology *84*, 780–784.

Jones, J.B., Lacy, G.H., Bouzar, H., Stall, R.E., and Schaad, N.W. (2004). Reclassification of the xanthomonads associated with bacterial spot disease of tomato and pepper. Syst. Appl. Microbiol. *27*, 755–762.

Jones, J.B., Jackson, L.E., Balogh, B., Obradovic, A., Iriarte, F.B., and Momol, M.T. (2007). Bacteriophages for plant disease control. Annu. Rev. Phytopathol., *45*, 245–262.

Jones, J.B., Vallad, G.E., Iriarte, F.B., Obradović, A., Wernsing, M.H., Jackson, L.E., Balogh, B., Hong, J.C., and Momol, M.T. (2012). Considerations for using bacteriophages for plant disease control. Bacteriophage. *2*, 208–214.

Katznelson, H. (1937). Bacteriophage in relation to plant diseases. Botan. Rev. *3*, 499–521.

Kotila, J.E., and Coons, G.H. (1925). Investigations on the blackleg disease of potato. Mich. Agricult. Experiment. Stat. Technic. Bull. *67*, 3–29.

Lang, J.M., Gent, D.H., and Schwartz, H.F. (2007). Management of *Xanthomonas* leaf blight of onion with bacteriophages and a plant activator. Plant Dis. *91*, 871–878.

Lee, Y., Hendson, M., Ponopoulos, N.J., and Schroth, M.N. (1994). Molecular cloning, chromosomal mapping, and sequence analysis of copper resistance genes from *Xanthomonas campestris* pv. *juglandis*: Homology with small blue copper proteins and multicopper oxidase. J. Bacteriol. *176*, 173–188.

Lim, J.A., Jee, S., Lee, D.H., Roh, E., Jung, K., Oh, C., and Heu, S. (2013). Biocontrol of *Pectobacterium carotovorum* subsp. *carotovorum* using bacteriophage PP1. J. Microbiol. Biotechnol. *23*, 1147–1153.

Liu, T. (1998). Biological control with tomato bacterial spot with hrp- mutants of *Xanthomonas campestris* pv. *vesicatoria*. MS Thesis, Gainesville, FL, University of Florida.

Louws, F.J., Wilson, M., Campbell, H.L., Cuppels, D.A., Jones, J.B., Shoemaker, P.B., Sahin, F., and Miller, S.A. (2001). Field control of bacterial spot and bacterial speck of tomato using a plant activator. Plant Dis. *85*, 481–488.

Luria, S.E., and Delbrück, M. (1943). Mutations of bacteria from virus sensitivity to virus resistance. Genetics *28*, 491–511.

McNeil, D.L., Romero, S., Kandula, J., Stark, C., Stewart, A., and Larsen, S. (2001). Bacteriophages: a potential biocontrol agent against walnut blight (*Xanthomonas campestris* pv. *juglandis*). New Zeal. Plant Prot. *54*, 220–224.

Mallmann, W.L., and Hemstreet, C.J. (1924). Isolation of an inhibitory substance from plants. Agricult. Res., *28*, 599–602.

Manulis, S., Zutra, D., Kleitman, F., Dror, O., David, I., Zilberstaine, M., and Shabi, E. (1998). Distribution of streptomycin-resistant strains of *Erwinia amylovora* in Israel and occurrence of blossom blight in the autumn. Phytoparasitica *26*, 223–230.

Maxson-Stein, K., He, S.Y., Hammerschmidt, R., and Jones, A.L. (2002). Effect of treating apple trees with acibenzolar-*S*-methyl on fire blight and expression of pathogenesis-related protein genes. Plant Dis. *86*, 785–790.

Momol, M.T., Jones, J.B., Olson, S.M., Obradovic, A., Balogh, B., and King, P. (2002). Integrated management of bacterial spot on tomato in Florida. Online: EDIS, Institute of Food and Agricultural Sciences, University of Florida. http://edis.ifas.ufl.edu/PP110.

Moore, E.S. (1926). D'Hérelle's bacteriophage in relation to plant parasites. South African J. Sci. *23*, 306.

Obradović, A., Jones, J.B., Momol, M.T., Balogh, B., and Olson, S.M. (2004). Management of tomato bacterial spot in the field by foliar applications of bacteriophages and SAR inducers. Plant Dis. *88*, 736–740.

Obradović, A., Jones, J.B., Momol, M.T., Olson, S.M., Jackson, L.E., Balogh, B., Guven, K., and Iriarte, F.B. (2005). Integration of biological control agents and systemic acquired resistance inducers against bacterial spot on tomato. Plant Dis. *89*, 712–716.

Okabe, N., and Goto, M. (1963). Bacteriophages of plant pathogens. Ann. Rev. Phytopathol. *1*, 397–418.

Qui, D., Wei, Z.M., Bauer, D.W., and Beer, S.V. (1997). Treatment of tomato seed with harpin enhances germination and growth and induces resistance to *Ralstonia solanacearum*. Phytopathology, *87*, S80.

Reddy, N.P., A Khan, P.A., Devi, K.U., Victor, J.S., and Sharma, H.C. (2008). Assessment of the suitability of Tinopal as an enhancing adjuvant in formulations of the insect pathogenic fungus Beauveria bassiana (Bals.) Vuillemin. Pest Manag. Sci. *64*, 909–915.

Romero, A.M., Kousik, C.S., and Ritchie, D.F. (2001). Resistance to bacterial spot in bell pepper induced by acibenzolar-*S*-methyl. Plant Dis. *85*, 189–194.

Romero-Suarez, S., Jordan, B., and Heinemann, J.A. (2012). Isolation and characterization of bacteriophages infecting Xanthomonas arboricola pv. juglandis, the causal agent of walnut blight disease. World J. Microbiol. Biotechnol. *28*, 1917–1927.

Saccardi, A., Gambin, E., Zaccardelli, M., Barone, G., and Mazzucchi, U. (1993). *Xanthomonas campestris* pv. *pruni* control trials with phage treatments on peaches in the orchard. Phytopathol. Medit. *32*, 206–210.

Schnabel, E.L., Fernando, W.G.D., Meyer, M.P., Jones, A.L., and Jackson, L.E. (1999). Bacteriophage of *Erwinia amylovora* and their potential for biocontrol. Acta Hort. *489*, 649–654.

Stall, R.E., Loschke, D.C., and Jones, J.B. (1986). Linkage of copper resistance and avirulence loci on a self-transmissible plasmid in *Xanthomonas campestris* pv. *vesicatoria*. Phytopathology *76*, 240–243.

Storey, M.V., and Ashbolt, N.J. (2001). Persistence of two model enteric viruses (B40-8 and MS-2 bacteriophages) in water distribution pipe biofilms. Water Sci. Technol. *43*, 133–138.

Summers, W.C. (2005). Bacteriophage research: Early history. In Bacteriophages: Biology and Applications, E. Kutter and A. Sulakvelidze, eds. (Boca Raton, FL, CRC Press), 5–27.

Svircev, A.M., Lehman, S.M., Kim, W.S., Barszcz, E., Schneider, K.E., and Castle, A.J. (2006). Control of the fire blight pathogen with bacteriophages. In Proceedings of the 1st International Symposium on Biological Control of Bacterial Plant Diseases, October 23–26 2005, W. Zeller and C. Ullrich, eds. (Berlin, Germany, Die Deutsche Bibliothek – CIP-Einheitsaufnahme), 259–261.

Sykes, I.K., Lanning, S., and Williams, S.T. (1981). The effect of pH on soil actinophage. J. Gen. Microbiol. *122*, 271–280.

Tanaka, H., Negishi, H., and Maeda, H. (1990). Control of tobacco bacterial wilt by an avirulent strain of *Pseudomonas solanacearum* M4S and its bacteriophage. Annals Phytopathol. Soc. Japan *56*, 243–246.

Thayer P.L., and Stall, R.E. (1961). A survey of *Xanthomonas vesicatoria* resistance to streptomycin. Proc. Flor. State Horticul. Soc. *75*, 163–165.

Thomas, R.C. (1935). A bacteriophage in relation to Stewart's disease of corn. Phytopathology, *25*, 371–372.

Twort, F.W. (1915). An investigation on the nature of the ultra-microscopic viruses. Lancet *189*, 241–243.

Vidaver, A.K. (1976). Prospects for control of phytopathogenic bacteria by bacteriophages and bacteriocins. Annu. Rev. Phytopathol. *14*, 451–465.

Williams, S.T., Mortimer, A.M., and Manchester, L. (1987). Ecology of soil bacteriophages. In Phage Ecology, S.M. Goyal, C.P. Gerba, and G. Bitton, eds. (New York, John Wiley and Sons), pp. 157–179.

Zaccardelli, M., Saccardi, A., Gambin, E., and Mazzucchi, U. (1992). *Xanthomonas campestris* pv. *pruni* bacteriophages on peach trees and their potential use for biological control. Phytopathol. Medit. *31*, 133–140.

Bacteriophage Therapy in Animal Production

8

William E. Huff and Geraldine R. Huff

Abstract

Concerns over the consequences of bacterial resistance to antibiotics with the use of antibiotics in animal production have led to an increase in research on alternatives to antibiotics. Bacteriophages kill bacteria, are natural, safe, plentiful, self replicating, self limiting, can be used to specifically target pathogens without disruption of commensal bacteria, and have diverse biological properties. These properties make bacteriophages an attractive alternative to antibiotics, especially applicable for the control of antibiotic resistant bacteria. The efficacy of bacteriophages to prevent and treat animal diseases has been shown in almost all production animals in both laboratory and commercial field studies, without any adverse effects in the animals. Although the potential of bacteriophage to control significant diseases in animal production has been demonstrated, bacteriophage therapeutics do not represent a replacement of antibiotics. There are some applications in animal production systems where bacteriophage therapeutics have an advantage over the use of antibiotics and some applications where bacteriophage therapeutics are at a disadvantage over the use of antibiotics. In addition, the effectiveness of antibiotics and bacteriophage therapy can be enhanced when combined to treat animal diseases. The objectives of this chapter are to review the literature documenting the efficacy of bacteriophages to control diseases in animal production, to discuss the advantages and disadvantages of bacteriophage therapy, and to describe possible applications for the use of bacteriophages to control bacterial diseases in commercial poultry, swine, cattle, and aquaculture systems.

Introduction

Bacteriophages are viruses that kill bacteria, this fact makes them an attractive biological agent to prevent and treat bacterial diseases in animals. Bacteriophages have a long history having been co-discovered by Twort (1915) and d'Hérelle (1917). The potential use of bacteriophages as prophylactic and therapeutic agents was immediately realized with their discovery, and efforts to develop efficacious commercial products to treat bacterial diseases were pursued. In the 1930s d'Hérelle established a commercial laboratory in Paris and produced at least five phage preparations, and the pharmaceutical company Eli Lilly produced several bacteriophage products in 1940s (Sulakvelidze *et al.*, 2001). However, with the discovery and production of antibiotics, and the inconsistent results found with bacteriophage products, interest in bacteriophage as prophylactic and therapeutic agents waned.

There will always be a need to find novel approaches to prevent and treat bacterial diseases. However, growing concern in the scientific community and the general public over antibiotic resistance has heightened the awareness for the need to find alternatives to antibiotics. The debate on the contribution of how and which antibiotics used in the animal industry may have on the emergence of antibiotic resistant bacteria that result in human diseases has been intense. There is little or no debate that when any pressure is put on a population of bacteria, emergence of resistance is to be expected. The more difficult question to answer is, 'What is the relative significance to the emergence of antibiotic resistant bacteria that pose a threat to human health between the use of antibiotics in animal production versus the use of antibiotics in human medicine?' The consequences of these very real concerns with the use of antibiotics in animal production has been the ban of the use of subtherapeutic antibiotics as growth promoters by the European Union in 2006 (EU Regulation 1831/2003/EC), the prohibition by the US Food and Drug Administration of the use of fluoroquinolone antibiotics in animal production, and proposed legislation in the US Congress (bill HR 1549 and S 619) to restrict the use of all antibiotics in animal production. These concerns over the use of antibiotics in animal production have led to renewed interest in bacteriophages as an alternative to antibiotics.

Our laboratory, USDA/ARS, Fayetteville, AR, began a research programme in 2000 to critically evaluate the efficacy of bacteriophage as an alternative to antibiotics in poultry production. It is important to understand that bacteriophage should be viewed as an alternative to antibiotics, not a replacement for antibiotics. In general, our most important finding is that if high enough titres of a lytic bacteriophage with activity to the target bacteria can be delivered to the site of infection, the bacteriophage are very effective at resolving the infection. The purpose of this review is to present past and ongoing research on the use of bacteriophage to prevent and treat animal production diseases, and to address some important considerations on bacteriophage efficacy based on our own work as well as others.

Bacteriophage therapy in poultry production

One of the earliest attempts to use bacteriophage therapy was conducted by d'Hérelle for the treatment of fowl typhus soon after his discovery of bacteriophage (d'Hérelle, 1926). D'Hérelle summarized all of his work to this point in a fascinating book entitled 'Bacteriophage and its Behaviour', which can be downloaded from the website http://www.archive.org/details/bacteriophageits00dher. He isolated bacteriophages from the excreta of birds and scored the activity of the bacteriophages present on a scale of 0 to 4, and was able to show a correlation between birds that survived outbreaks of fowl typhus with the presence of bacteriophages. In controlled experiments he exposed birds for 3 days to a bird administered a bacteriophage with high activity and documented that the birds so exposed acquired this bacteriophage. These birds were then challenged for 21 consecutive days with *Salmonella gallinarum*, and remained healthy, while birds not exposed to bacteriophage and challenged with a single dose per os contracted fowl typhus and died within three days.

In more recent work (Berchieri *et al.*, 1991), bacteriophages were isolated to *S. typhimurium* strain F98 and tested for their efficacy to treat salmonellosis in poultry. In this work the birds were orally challenged with 10^6 CFU of *S. typhimurium* (F98) at one day of age and followed by an oral administration of 0.1 ml of 10^{12} PFU/ml of one of nine different bacteriophages. These birds were monitored over a 21-day period and three of the bacteriophage treatments significantly decreased mortality from 53% in the untreated birds to 28%, 22% and 16% with bacteriophages designated Φ1.1, ΦAB1, and Φ2.2, respectively. These data demonstrate that individual bacteriophage isolates can differ in their efficacy to treat salmonellosis. These researchers chose Φ2.2 for further work on the effect of bacteriophage titre on the efficacy to treat salmonellosis. Mortality in the untreated birds was 56%, which was significantly reduced to 20% only when undiluted bacteriophage (Φ2.2) was used, while a fourfold and twofold dilution, reduced mortality from 56% to 44% and 36%, respectively. These data demonstrate the importance of bacteriophage titre to treatment efficacy. Timing of bacteriophage treatment is also important when evaluating bacteriophage efficacy. Bardina and co-workers (2012) found that administering a bacteriophage cocktail orally was more effective at reducing colonization of the caecum in White Leghorn chickens when given prior to and immediately after an oral challenge with *Salmonella*.

Environmental augmentation with bacteriophage may be a practical way to reduce the infectious dose of pathogens and prevent disease outbreaks and colonization of animals with food borne pathogens. Lim *et al.* (2012) demonstrated that colonization of chicks with *Salmonella enteritidis* that were not challenged with *Salmonella* but were held in pens with chicks challenged with *Salmonella* could be significantly reduced with a bacteriophage fed at 10^7 or 10^9 PFU/g of feed. These same authors (Lim *et al.*, 2011) reported that bacteriophage fed to chickens 7 days prior to and 21 days after challenge with *Salmonella gallinarum* significantly reduced horizontal transmission of *Salmonella* based on mortality.

Necrotic enteritis in poultry caused by *Clostridium perfringens* type A, is a significant poultry disease causing high morbidity and mortality. *C. perfringens* is a ubiquitous bacteria in poultry facilities and in birds symptomatic and asymptomatic of the disease (Opengart, 2008). Although the use of coccidiostats and antibiotic growth promoters has limited the severity of necrotic enteritis, there is growing concern that restricting the use of antibiotic growth promoters will increase the incidence and severity of this disease in poultry production (Van Immerseel *et al.*, 2004). These concerns appear to be justified with a documented 25–40% increase in necrotic enteritis in poultry production facilities in Europe when some countries discontinued the use of antimicrobial feed additives (Kaldhusdal and Lovland, 2000; Van Immerseel *et al.*, 2004). Consumers and retailers are pressuring the animal industry to remove antibiotics from animal production, and there is a real concern that legislation in North America will restrict or eliminate the use of antibiotic growth promoters in animal production, creating a real and pressing need to find alternatives to antibiotic growth promoters for the control of necrotic enteritis.

Miller *et al.* (2010) compared the use of a toxoid vaccine to a bacteriophage product (INT-401) produced by Intralytix Corporation (Baltimore, MD, USA) to treat necrotic enteritis in poultry. The bacteriophage product contained five bacteriophages with activity to a broad range of *C. perfringens* strains. In two studies the birds were orally challenged with 5×10^3 oocysts of the coccidia *Eimeria maxima* at 14 days of age, which was followed at 19, 20, and 21 days of age (study 1) or 18, 19, and 21 day of age (study 2) with an oral challenge with 10^8 CFU of *C. perfringens*.

In general, the bacteriophage product controlled necrotic enteritis better than the toxoid vaccine, and the efficacy of this bacteriophage product to control necrotic enteritis was clearly demonstrated. In the first study there were two bacteriophage treatments that consisted of *in ovo* injection at 18 days of incubation followed by spray administration at hatching, and the second bacteriophage treatment consisted of oral gavage of bacteriophage administered on days 19–22, and the study was concluded when the birds were 28 days of age. Overall, the body weights of the birds in the two bacteriophage treatments were significantly improved compared to the *C. perfringens* challenged untreated birds, but were significantly lower then the body weights of the unchallenged birds, yet feed conversion in the two bacteriophage treatments were not significantly different from the controls and were significantly better than the *C. perfringens* untreated treatment. Total mortality over the course of the study was significantly decreased from 56% in the untreated *C. perfringens* challenge group to 18% and 16% in the two bacteriophage treatment groups, which was not significantly different from the negative controls, and mortality attributed only to necrotic enteritis was significantly decreased from 24% in the *C. perfringens* challenged untreated birds to 6 and 2% in the bacteriophage-treated birds, which again was not significantly different from the negative controls.

The second study was similar to the first but was concluded when the birds were 42 days of age with a bacitracin treatment and three bacteriophage treatments that differed in how bacteriophage were administered as either gavage, in the drinking water, or in the feed. Overall results found that body weights of birds given bacteriophage treatments administered by gavage or in the feed were significantly higher than the body weights of the *C. perfringens* control birds but significantly less than the body weights of the negative controls. When the bacteriophage treatment was administered in the drinking water the body weights were significantly higher than the *C. perfringens* control, but the body weights were not significantly different from the negative controls. Feed conversion in all of the bacteriophage treatments was significantly better compared to the *C. perfringens* control and significantly poorer compared to the negative control. Mortality from all causes was 67% in the *C. perfringens* challenged birds that was significantly decreased to 18%, 3% and 5%, when bacteriophage were administered by either gavage, drinking water, or feed, respectively. Mortality attributed to necrotic enteritis, was significantly decreased from 64% in the *C. perfringens* challenged birds to 14%, 0% and 0.7% in the bacteriophage treatments of gavage, drinking water, or feed, respectively.

Colibacillosis is considered to be the most significant bacterial disease throughout all poultry production (Barnes *et al.*, 2008). Colibacillosis starts as an *Escherichia coli* respiratory infection known as airsacculitis, primarily in the first week of life when the immunity of the chick is not fully developed. The infection can rapidly become systemic causing perihepatitis and pericarditis, resulting in high morbidity and mortality. Colibacillosis in poultry is primarily caused by a class of *E. coli* known as avian pathogenic *E. coli* (APEC), with five *E. coli* serotypes O1, O2, O35, O36, and O78 thought to be the principle cause of colibacillosis in poultry (Sojka, 1965; Heller and Drabkin, 1977).

Our laboratory, USDA/ARS, Fayetteville, AR, started a research programme several years ago to critically evaluate the efficacy of bacteriophages to prevent and treat colibacillosis. We isolated bacteriophages from municipal sewer treatment plants and poultry processing facilities with activity to an APEC strain of *E. coli* serotype O2 (Huff *et al.*, 2002a). Bacteriophage selection for *in vivo* efficacy testing was subjective and was based on the size and clarity of the plaques observed. Colibacillosis throughout this work was induced via a thoracic airsac injection of the *E. coli* serotype O2 strain.

Our initial studies represented an artificial experimental design where bacteriophage designated SPRO2 was mixed with our *E. coli* (10^4 CFU per ml) challenge strain immediately prior to challenging the birds at 1 week of age at different multiplicity of infection (MOI) ratios. Mortality in the *E. coli* challenged birds was 85% 2 weeks post challenge when the study was concluded. Mortality was significantly reduced to 35% when the challenge culture of *E. coli* was mixed with 10^4 PFU (MOI of 1) of bacteriophage, and there was no mortality in the birds challenged with *E. coli* mixed with 10^8 PFU of bacteriophage (MOI of 10^4). This experimental design, though artificial, does provide a relatively easy and quick *in vivo* test of efficacy when selecting bacteriophage for further study, and did demonstrate the importance of bacteriophage titre to treatment efficacy.

Our work on the prevention of colibacillosis was conducted by the administration of bacteriophage as an aerosol spray when the birds were 7 days of age followed by challenging the birds with *E. coli* (10^4 CFU) either immediately (7 days of age), 1 day (8 days of age), or 3 days (10 days of age) post bacteriophage administration. Three trials were conducted and mortality was significantly reduced in all three trials when the birds were challenged with *E. coli* immediately after bacteriophage pre-treatment. However, in only one of the trials was there a significant decrease in mortality when the birds were challenged with *E. coli* either 1 or 3 days after pre-treatment with bacteriophage (Huff *et al.*, 2002b). Although these data were variable we do believe that there were enough positive results to suggest that prophylactic use of bacteriophages in juvenile chickens may be possible and further research is needed to better establish the prophylactic efficacy of bacteriophage in colibacillosis.

The ability of bacteriophage to treat colibacillosis was investigated by a thoracic airsac challenge with *E. coli* (10^4 CFU) followed by bacteriophage treatment either as an aerosol spray or intra-muscular injection (Huff *et al.*, 2003a). Birds were challenged with *E. coli* at 7 days of age and then either immediately, 24, or 48 h post *E. coli* challenge treated with either an aerosol spray of bacteriophage or an intramuscular (i.m.) injection of bacteriophage at 10^9 PFU of bacteriophage per ml. These studies were concluded 2 weeks post *E. coli* challenge when the birds were three weeks of age. There was very little efficacy of bacteriophage to treat colibacillosis when administered as an aerosol spray. All bacteriophage treatments administered i.m. significantly reduced mortality compared to the *E. coli* challenged and untreated controls. In this experimental model colibacillosis quickly becomes a systemic infection, and few birds had detectable levels of bacteriophage in blood when bacteriophage were administered as an aerosol spray, while high systemic titres of bacteriophage were achieved in all the birds administered bacteriophage i.m., which is thought to explain the difference in efficacy of bacteriophage administration. Similar results were obtained in additional research where multiple i.m. bacteriophage injections were evaluated, with multiple injections more efficacious than a single injection (Huff *et al.*, 2003b).

Treatment efficacy of bacteriophage to treat *E. coli* in poultry was also demonstrated by Barrow *et al.* (1998). In these studies bacteriophage administered i.m. simultaneously at different titres was able to significantly reduce mortality induced by either an intramuscular or intracranial *E. coli* challenge. Bacteriophage titres as low as 10^4 PFU significantly decreased mortality caused by an intramuscular challenge with *E. coli*, while titres of 10^8 PFU significantly reduced mortality in birds intracranially challenged with *E. coli* at 3 weeks of age, and 10^6 PFU significantly reduced mortality in birds challenged with *E. coli* at hatching.

Oliveira *et al.* (2010) has published research on the field application of bacteriophage to reduce the impact of colibacillosis in commercial poultry production facilities. In a laboratory trial, administration of 10^9 PFU of bacteriophage orally and by spray immediately after challenging the birds with *E. coli* significantly decreased mortality, morbidity, and pathology scores. In actual field studies, 11 commercial poultry flocks were identified as having an outbreak of colibacillosis. Bacteriophage were administered to the flocks in the drinking water and by spray. One week post bacteriophage treatment mortality in 5 of the 11 flocks was considered controlled being less than 0.5%. By the third week mortality in 10 of the 11 flocks was considered controlled. Although there were no comparisons made to untreated commercial flocks, this work demonstrated that a practical application of bacteriophage in a commercial poultry facility could be used to decrease mortality caused by colibacillosis.

Studies comparing the efficacy of bacteriophage to the antibiotic enrofloxacin and the efficacy of combining bacteriophage and enrofloxacin to treat colibacillosis were conducted (Huff *et al.*, 2004). The birds were challenged with *E. coli* at 7 days of age and treated with either 50 parts per million (ppm) enrofoxacin in the drinking water for 7 days post challenge, treated with a single i.m. injection of 10^9 PFU of bacteriophage at 7 days of age, or the combination of bacteriophage and enrofloxacin treatment. Both enrofloxacin and bacteriophage therapy significantly decreased mortality from 68% in the *E. coli*-positive control to 3% and 15%, respectively, and the enrofloxacin treatment significantly decreased mortality compared to the bacteriophage treatment. There was complete protection of the birds when enfloxacin and bacteriophage were combined and this synergistic interaction was significant. These data suggest that there may be benefits from combining antibiotic and bacteriophage therapy, which could allow the use of lower doses of antibiotics. A comparative study on the efficacy of bacteriophage with antibiotics was also conducted by Xie *et al.* (2005) and found that bacteriophage therapy of *E. coli* challenged poultry was more effective than the treatments with the antibiotic chloromycetin.

Although another chapter in this book is devoted to bacteriophage applications in food safety, there is so much research on foodborne pathogens in poultry that a brief overview seems appropriate. There have been mixed results with trying to reduce *Salmonella* in the intestinal tract of birds with reductions varying from no observed efficacy to reductions of 5 logs (Fiorentin *et al.*, 2005; Atterbury *et al.*, 2007; Andreatti Filho *et al.*, 2007; Borie *et al.*, 2008; Hurley *et al.*, 2008). Toro *et al.* (2005) and Borie *et al.* (2009) investigated the efficacy of combining bacteriophage treatment with competitive exclusion to reduce *Salmonella enteritidis* infection in poultry. These researchers demonstrated reductions in the incidence of birds found positive for *Salmonella* and intestinal colonization with both the competitive exclusion and bacteriophage treatment, and the efficacy was enhanced when the two approaches were combined. Higgins *et al.* (2005) demonstrated that bacteriophage rinses of poultry carcasses could reduce the levels of *Salmonella* on the skin. The ability of bacteriophage to prevent and treat *Campylobacter* colonization of the intestinal tract has had similar variability as seen with *Salmonella* with results varying between a 0.5 and 5 log reduction in recovery of *Campylobacter* from the intestinal tract (Loc Carrillo *et al.*, 2005; Wagenaar *et al.*, 2005; El-Shibiny *et al.*, 2009). As with *Salmonella* there have been more consistent results obtained with bacteriophage rinses of carcasses to reduce carcass contamination with *Campylobacter* (Atterbury *et al.*, 2003; Goode *et al.*, 2003), with a 2 to 3 log reduction when levels of bacteria were high, to elimination of contamination when bacteria levels were low. Zhang *et al.* (2013) reported that bacteriophage treatment of ready to eat chicken surfaces contaminated with *Shigella* could reduce and even eliminate the *Shigella* contamination. There is also some very interesting work on the incidence of *Campylobacter* colonization of poultry flocks and the presence of bacteriophage in these flocks (Connerton *et al.*, 2004; Atterbury *et al.*, 2005; El-Shibiny *et al.*, 2005). One of the most encouraging developments in the regulation of bacteriophage products is the FDA approval of a bacteriophage product by the Baltimore, Maryland based company, Intralytix, to be used to reduce *Listeria monocytogenes* contamination of further processed meats (USFDA, 2007). Since then, three other phage preparations have been approved by FDA for different food applications (Sulakvelidze *et al.*, 2013).

Bacteriophage therapy in swine production

Smith and Huggins (1983) conducted research to investigate the efficacy of bacteriophage to treat severe *E. coli* induced diarrhoea in piglets. The piglets were challenged orally with *E. coli* that was followed by the oral administration of two bacteriophages. One of the bacteriophage used was selected based on its activity to the challenge culture and the additional bacteriophage used had no activity to the challenge culture, but had activity to isolates of the challenge strain of *E. coli* that were resistant to the activity of the first bacteriophage. Mortality was prevented in the piglets that were administered both phages, and the severity of diarrhoea was decreased. Treatment of piglets with only the bacteriophage with activity to the challenge culture was also effective in reducing the severity of diarrhoea.

A series of trials were conducted by Jamalludeen *et al.* (2009) to determine the efficacy of bacteriophages to both prevent and treat post-weaning diarrhoea in pigs. Six bacteriophage isolates were individually evaluated, and when these bacteriophage were orally administered immediately after an oral *E. coli* challenge, all six bacteriophage decreased mean weight changes and composite diarrhoea scores, and three of the bacteriophages decreased the duration of diarrhoea and mean diarrhoea scores. In a subsequent trial the pigs were pretreated with the antibiotic florfenicol and an antacid, sodium bicarbonate, prior to challenging the pigs with *E. coli*. A mixture of three bacteriophages were administered immediately after the *E. coli* challenge and were effective in reducing the duration of diarrhoea and severity of diarrhoea. In the treatment research two bacteriophages were administered three times after the onset of diarrhoea at 6-hour intervals. The bacteriophage treatment improved weight gain, and reduced the duration and severity of diarrhoea.

There has not been as much research on the application of bacteriophage to reduce contamination of pork products as there have been in poultry. However, Holck and Berg (2009) have been able to demonstrate that the application of a protective culture of *Lactobacillus sakei* in combination with bacteriophage could reduce the levels and outgrowth of *L. monocytogenes* in cooked hams. Wall *et al.* (2010) demonstrated that the *Salmonella* colonization could be reduced by 2 to 3 logs when bacteriophage were administered simultaneously with *Salmonella*, and that pigs administered bacteriophage prior to contact exposure with infected animals reduced *Salmonella* colonization. Callaway *et al.* (2011) also demonstrated that *Salmonella* in the cecum and rectum of swine could be reduced when they were challenged with *Salmonella* and treated with bacteriophage 24 and 48 h post challenge. Saez *et al.* (2011) found that microencapsulating bacteriophage improved the efficacy of the bacteriophage to reduce *Salmonella* colonization when bacteriophage were administered every day for 5 days prior to challenge with *Salmonella*.

Bacteriophage therapy in beef, dairy and cattle production

Smith *et al.* (1987) reported a number of studies on the efficacy of oral administration of bacteriophage to treat diarrhoea in calves orally challenged with *E. coli*. In an initial study all calves treated with low doses of bacteriophage (10^5 PFU) after the onset of diarrhoea recovered within 20 h after treatment, while challenged untreated calves developed severe diarrhoea. A trial was conducted to compare the efficacy of bacteriophage that differed in *in vitro* activity to the challenge *E. coli* strain to prevent diarrhoea in calves. The calves were orally administered the bacteriophage at 10^5 PFU immediately (10 min) after challenge with *E. coli*. Two bacteriophage with identical *in vitro* activity differed in their ability to prevent the onset of diarrhoea, one totally prevented diarrhoea while five out of six calves developed diarrhoea with the other bacteriophage; however, these five calves did recover. Bacteriophage shown to have less *in vitro* activity failed to prevent diarrhoea and all three calves died. Moreover, these researchers clearly demonstrated the importance of bacteriophage titre to the prevention of diarrhoea in calves. When calves were given bacteriophage 6 h prior to challenging the animals with *E. coli*, none of the calves given 10^5 PFU of bacteriophage developed diarrhoea and all the calves given 10^2 PFU of bacteriophage developed diarrhoea. Similar results were found when bacteriophage were administered 3 h prior to an *E. coli* challenge. When low titres of bacteriophage (10^2 PFU) were administered 10 min before, or 6 and 12 h after the *E. coli* challenge, no calves developed diarrhoea, but when treatment was delayed to 18 h post challenge diarrhoea developed. Bacteriophage administered twice daily either 4 h prior to, or 2, 4, 6, 8, 10 or 12 h post challenge with *E. coli* was effective at preventing diarrhoea up to 8 h post challenge with only mild diarrhoea observed when bacteriophage were administered 10 and 12 h post challenge. In additional work, calves were placed in a room previously occupied by calves challenged with *E. coli* and treated with bacteriophage. After being in the room for 3 h the calves were challenged with *E. coli* and did not develop diarrhoea, while calves untreated with bacteriophage developed severe diarrhoea. In a follow up study, the bedding in the rooms was sprayed with either 10^3 or 10^6 PFU per ml of bacteriophage and calves placed in these rooms were challenged with *E. coli* 3 h after placement. None of the calves developed diarrhoea, which was repeated with similar results using a different challenge strain of *E. coli* and bacteriophage. Even if calves were challenged with *E. coli* at 0, 6, or 12 h prior to being placed in bacteriophage sprayed rooms, protection from diarrhoea was observed in calves challenged 6 h prior to placement, and only two of the three calves challenged 12 h prior to placement developed diarrhoea. They also demonstrated that spraying the bedding with bacteriophage was more effective at preventing the onset of diarrhoea in calves than oral administration of bacteriophage.

Barrow *et al.* (1998) were able to protect calves from septicaemia when orally challenged with *E. coli* and treated with an intramuscular injection of bacteriophage. There is also a research effort to treat postpartum uterine diseases in cattle with bacteriophage (Bicalho *et al.*, 2010; Santos *et al.*, 2010). These researchers have reported on the isolation of *E. coli* from the uterus of postpartum dairy cows, and the isolation of bacteriophage from the dairy environment with activity to these pathogenic *E. coli*, however at this time no *in vivo* work has been reported.

Mastitis caused by *Staphylococcus aureus* is a significant problem in diary production, and has been an attractive target for bacteriophage therapy. The bacteriophages used to treat mastitis to date have not been shown to be effective in this application (Lerondelle and Poutrel, 1980; Gill *et al.*, 2006). The lack of efficacy of bacteriophage therapy of mastitis is thought to include immune interference, inhibition due to non-specific binding with milk proteins, and agglutination of *S. aureus* with fat globules (O'Flaherty *et al.*, 2005a; Gill *et al.*, 2006). However, efforts continue in an attempt to isolate bacteriophage that may have efficacy to prevent and treat mastitis (O'Flaherty *et al.*, 2005b; Garcia *et al.*, 2009a,b). There is also an effort to isolate genes coding for bacteriophage endolysins and express these genes in the mammary gland of cattle through transgenic transformation to provide indigenous expression of pathogen specific agents (Donovan *et al.*, 2006; Celia *et al.*, 2008). Schmelcher *et al.* (2012) demonstrated that two chimeric bacteriophage lysins could significantly reduce the numbers of *S. aureus* in a mouse model of mastitis, but were not as effective as lysostaphin. However, when one of these lysins was combined with lysostaphin, the numbers of *S. aureus* in this mouse model of mastitis were significantly reduced in a synergistic manner and may provide an approach to prevent and treat mastitis in dairy cattle.

In cattle, the emphasis on food-borne pathogens has focused on *E. coli* O157:H7. Although bacteriophage therapy resulting in complete elimination of *E. coli* O157:H7 has not been demonstrated, reduction of colonization of the intestinal tract has been shown to be possible (Sheng *et al.*, 2006; Raya *et al.*, 2006; Callaway *et al.*, 2008). Environmental studies in cattle feedlots suggest that bacteriophage to *E. coli* O157:H7 are prevalent, and that these bacteriophage may play a role in the prevention of colonization of cattle (Kudva *et al.*, 1999; Yoichi *et al.*, 2004; Callaway *et al.*, 2006; Oot *et al.*, 2007; Niu *et al.*, 2009).

Bacteriophage therapy in aquaculture

Aquaculture continues to be one of the fastest growing segments of animal production, and controlling diseases in these production systems is challenging. The efficacy of bacteriophage therapy to prevent and treat bacterial infections in these aquaculture systems has been explored. Ayu (*Plecoglossus altivelis*) is a popular cultured fish in Japan and one of its disease problems is known as bacterial haemorrhagic ascites disease. This disease is caused by *Pseudomonas plecoglossicida*, and is characterized by high mortality and bloody ascites. Park *et al.* (2000) reported on work to evaluate the efficacy of two bacteriophage they isolated to treat this disease. In their first trial the fish were orally challenged with *P. plecoglossicida* and immediately fed pellets impregnated with the bacteriophage. There was a significant decrease in mortality in the bacteriophage-treated fish from 65% in the untreated fish to 23% in the bacteriophage-treated fish. In a subsequent trial the fish were challenged with *P. plecoglossicida* and bacteriophage treatment was administered 1 or 24 h post challenge. Mortality in the untreated fish was 79%, and complete protection, based on mortality, was seen when the fish were treated 1 h post challenge, and a significant decrease in mortality (13%) was seen in the fish treated 24 h post challenge. In additional research (Park and Nakai, 2003) demonstrated that ayu challenged with *P. plecoglossicida* and either administered feed treated with bacteriophage designated PPpW-3, PPpW-4, or the combination of these two bacteriophage, significantly reduced mortality from 93% to 53%, 40% and 20%, respectively. In a field trial, fish naturally infected with *P. plecoglossicida* were fed bacteriophage-treated feed and mortality in this commercial facility was reduced approximately 5% per day.

Nakai *et al.* (1999) reported on research to treat a disease in the fish known as yellowtail (*Seriola seriolicida*) caused by *Lactococcus garvieae* with bacteriophage. When these fish were challenged with *L. garvieae* intraperitoneally (i.p.) and administered bacteriophage via i.p. injection, survival in the bacteriophage-treated fish was 90% compared to only a 40% survival rate in the untreated fish. When bacteriophage treatment was administered 0, 1 or 24 h post challenge with *L. garvieae*, survival was 100%, 80% and 50% compared with a 10% survival rate in the untreated fish. Studies were also conducted with fish challenged by anal intubation with *L. garvieae* followed by feeding the fish with bacteriophage-treated feed. Mortality was reduced from 65% in the untreated fish to 10% in the bacteriophage-treated fish.

Furunculosis is a significant disease in aquaculture caused by *Aeromonas salmonicida* that results in morbidity, mortality, weakened fish and skin ulcers. Imbeault *et al.* (2006) reported studies using bacteriophage to treat furunculosis in brook trout (*Salvelinus fontinalis*). The brook trout were exposed to *A. salmonicida* by inoculating the aquarium water with *A. salmonicida* followed by the addition of a bacteriophage to the water. Mortality was reduced from 100% in the challenged and untreated fish to 10% in the bacteriophage-treated fish. However, bacteriophage were not effective in treating furunculosis in Atlantic salmon (*Salmo salar*) when challenged i.p. with *A. salmonicida* followed by an i.p. treatment with bacteriophage (Verner-Jeffreys *et al.*, 2007). The authors suggested that the rapid clearance of bacteriophage from the peritoneal cavity may have reduced the treatment efficacy of the bacteriophage.

A significant disease problematic in shrimp hatcheries is luminous vibriosis caused by *Vibrio harveyi* which causes high mortality. Vinod *et al.* (2006) found in laboratory trials that when *V. harveyi* was introduced into the water with bacteriophage either simultaneously with the *V. harveyi* administration, or simultaneously and 24 h post challenge with *V. harveyi*, that survival of the shrimp (*Penaeus monodon*) larvae was only 25% in the challenged untreated group and 70 and 80% in the bacteriophage-treated groups, respectively. In additional studies in commercial hatcheries comparing the efficacy of the antibiotics oxytetracycline and kanamycin with bacteriophage therapy, bacteriophage therapy was clearly superior with only a 17% survival rate in the untreated controls, 40% with antibiotics, and 86% in the bacteriophage treatment. Additional research also demonstrated that bacteriophage increased shrimp larvae survival in commercial hatcheries naturally infected with *V. harveyi* compared to treatment with either oxytetracycline or kanamycin (Karunasagar *et al.*, 2007). Crothers-Stomps *et al.* (2010) have also reported the isolation of bacteriophage to *V. harveyi* with potential to be used to prevent and treat luminous vibriosis.

Matsuoka *et al.* (2007) reported that bacteriophage therapy was effective at reducing mortality in Japanese flounder (*Paralichthys olivaceus*) challenged with an i.p. inoculation of *Streptococcus iniae* followed by i.p. inoculation of bacteriophages either 1, 12, or 24 h post challenge. Wu *et al.* (1981) found that when the fish pathogen that causes eel's red-fin disease, *Aeromonas hydrophila*, was infected by bacteriophage prior to challenging loaches (*Misgurnus anguillicaudatus*) that infectivity and mortality was prevented. Moreover, bacteriophages have been isolated to the catfish pathogen *Edwardsiella ictaluri*, the causative agent of enteric septicaemia (Walakira *et al.*, 2008), and to the pathogen *Flavobacterium psychrophilum* (Stenholm *et al.*, 2008) that is the agent for two diseases of trout, rainbow trout fry syndrome and cold water disease. Castillo *et al.* (2012) also isolated bacteriophage to *Flavobacterium psychrophilum* and reported that mortality could be significantly reduced when fish were infected with *F. psychrophilum* containing bacteriophage. Columnaris disease has become a significant problem in aquaculture, with *Flavobacterium columnare* recognized as the etiologic agent. Prasad *et al.* (2011) challenged catfish with *F. columnare* i.m. then treated the fish with a bacteriophage administered i.m., in the water, or fed 24 h post challenge. Reductions in *F. columnare* from the sera, gills, liver, and kidney were found in all three methods of bacteriophage administration, with the i.m. bacteriophage treatment having the best efficacy of the three methods of bacteriophage administration. The bacteriophage treatments rescued the catfish with no mortality observed in the bacteriophage-treated catfish.

The application of bacteriophage to prevent food-borne contamination of salmon and catfish fillets with *Listeria monocytogenes* has demonstrated efficacy to reduce the levels of *L. monocytogenes* by as much as 3 logs (Soni *et al.*, 2010; Soni and Nannapaneni, 2010). There is also some interest in the use of bacteriophage to treat oysters contaminated with *Vibrio vulnificus* (Pelon *et al.*, 2005).

General discussion

This review of the literature on the use of bacteriophages to prevent and treat bacterial diseases in animal production clearly demonstrates the viability of bacteriophage therapy as an alternative to antibiotics. It is important to recognize that bacteriophage present an alternative to antibiotic therapy and are not a replacement for antibiotic therapy. There are some applications of bacteriophage therapy that are better than antibiotic therapy, but there are also some limitations to bacteriophage therapy. Unlike antibiotics, bacteriophage are self replicating and self limiting, they represent a natural approach to the control of bacterial diseases, they are safe having no known activity to animal or plant cells, they are plentiful in nature, they have diverse biological properties, and they can specifically target pathogens without effect on commensal bacteria. Bacteriophage therapy is especially important as a means to treat antibiotic resistant diseases. In addition, bacteriophage therapy combined with antibiotic therapy may improve the efficacy of both antibiotic and bacteriophage therapy. One concern with bacteriophage therapy is the development of resistance. However, any time that any pressure is placed on a bacterial population, the emergence of resistant subpopulations is expected. Bacteriophages have a real advantage over antibiotics with respect to combating bacterial resistance because bacteriophage can be isolated with activity to these resistant bacteria. This allows the development of a cocktail of bacteriophages with activity to the target pathogen as well as any resistant variants of the pathogen. In addition, there is research that demonstrates that resistant bacteria that emerge from a bacteriophage challenge have lost some pathogenicity (Smith and Huggins, 1983; Park *et al.*, 2000). Another concern with bacteriophage therapy is immune interference when the treatment of a bacterial disease requires multiple doses of bacteriophage over time. We have demonstrated that immune interference with bacteriophage therapy can occur with the systemic administration of bacteriophage to treat colibacillosis in poultry, with an approximately 40% decrease in efficacy in bacteriophage therapy when attempted in animals with prior exposure to a single bacteriophage (Huff *et al.*, 2010). Immune interference with bacteriophage therapy may be overcome with the isolation of bacteriophage that differ antigenically. The diversity represented by bacteriophage might make this possible. Probably one of the most significant problems in the development of bacteriophage therapeutics to treat animal production diseases is developing practical yet still effective methods of bacteriophage administration. In modern animal production there are generally thousands of animals housed in production facilities. If therapeutic agents cannot be administered in the drinking water or the feed they have little therapeutic efficacy, it is simply not practical to handle the animals to treat them with injections. However, oral administration of bacteriophage therapeutics presents many hurdles in the gastro-intestinal tract that include very low pH values, proteolytic enzymes, non-specific binding with digesta, non-specific binding with non-target bacteria, target bacteria residing in the crypts of the intestinal villi, and interference due to mucin, all of which can limit the efficacy of bacteriophage to treat enteric pathogens as well as prevent the translocation of bacteriophage from the intestine to treat systemic infections. These issues with oral administration of bacteriophage can be overcome through encapsulation, isolation of bacteriophage that can survive the digestive processes, and discovery of bacteriophage that translocate from either the respiratory or the intestinal tract to provide therapeutic systemic titres. It would also be expected that bacteriophage would have limited efficacy to treat bacterial diseases that are primarily intracellular.

In general, our experience has brought us to the conclusion that if sufficient titres of bacteriophage can be delivered to the site of a bacterial infection bacteriophage are effective in eliminating the infection. There is limited practical efficacy of injecting bacteriophage to prevent or treat bacterial diseases in animal production systems. However, there are some real opportunities to use bacteriophage to prevent and treat animal diseases in animal production systems that do not require the administration of bacteriophage by injection. One of these applications is augmenting animal production facilities with bacteriophage to lower the infectious dose of pathogens in these environments. It has always been a mystery why some animal facilities seem to continually have outbreaks of bacterial diseases and why some animals within facilities seem to be refractile to infection. D'Hérelle (1926) strongly believed that this observation could be explained by facilities and animals that had indigenous populations of bacteriophage that protected them from infection. The efficacy of augmenting animal facilities with bacteriophage has been shown by Smith *et al.* (1987), who demonstrated that spraying the bedding of calve pens with bacteriophage was more effective than oral treatment with bacteriophage in preventing *E. coli* induced diarrhoea, by Oliveira *et al.* (2010) who demonstrated that morbidity and mortality in commercial poultry facilities from colibacillosis could be reduced by spraying these facilities with bacteriophage and administration of bacteriophage in the drinking water, and by Vinod *et al.* (2006) who found that mortality due to luminous vibriosis in commercial shrimp hatcheries could be reduced with the addition of bacteriophage to the water. Another application of bacteriophage is the aerosol spray administration of bacteriophage to both prevent and treat respiratory tract infection. Respiratory tract infections are basically an extrinsic infection and antibiotic therapy has limited treatment efficacy, while an aerosol spray of bacteriophage can deliver high titres to the site of infection, and has been shown to prevent the onset of colibacillosis in poultry which starts as a respiratory tract infection prior to becoming a systemic infection (Huff *et al.*, 2002b). In the poultry industry *in ovo* injection and aerosol spray of newly hatched chicks are routine practices and both provide a practical means of administering bacteriophage to prevent a number of poultry diseases (Miller *et al.*, 2010).

In conclusion, it has been demonstrated since 1927 that bacteriophage can be used to prevent and treat bacterial diseases in animal production. There are no reports of any adverse effects of bacteriophage, regardless of how they are administered, in animals. Bacteriophage kill bacteria, represent a natural approach to limit the impact of diseases in animals, are safe, plentiful in nature, self replicating, self limiting, can be targeted to pathogens without affecting commensal bacteria, and have diverse biological properties. Although bacteriophages have great potential as an alternative to antibiotics in animal production there is a huge need for more research on how to best use this potential to control animal production diseases.

References

Andreatti Filho, R.L.A., Higgins, J.P., Higgins, S.E., Gaona, G., Wolfenden, A.D., Tellez, G., and Hargis, B.M. (2007). Ability of bacteriophages isolated from different sources to reduce *Salmonella enterica* serovar Enteritidis *in vitro* and *in vivo*. Poultry Sci. *86*, 1904–1909.

Atterbury, R.J., Connerton, P.L., Dodd, C.E.R., Rees, C.E.D., and Connerton, I.F. (2003). Application of host-specific bacteriophages to the surface of chicken skin leads to a reduction in recovery of *Campylobacter jejuni*. Appl. Environ. Microbiol. *69*, 6302–6306.

Atterbury, R.J., Dillon, E., Swift, C., Connerton, P.L., Frost, J.A., Dodd, C.E.R., Rees, C.E.D., and Connerton, I.F. (2005). Correlation of *Campylobacter* bacteriophage with reduced presence of hosts in broiler chicken ceca. Appl. Environ. Microbiol. *71*, 4885–4887.

Atterbury, R.J., Van Bergen, M.A.P., Ortiz, F., Lovell, M.A., Harris, J.A., De Boer, A., Wagenaar, J.A., Allen, V.M., and Barrow, P.A. (2007). Bacteriophage therapy to reduce *Salmonella* colonization of broiler chickens. Appl. Environ. Microbiol. 73, 4543–4549.

Bardina, C., Spricigo, D.A., Cortes, P., and Llagostera, M. (2012). Significance of the bacteriophage treatment schedule in reducing Salmonella colonization of poultry. Appl. Environ. Microbiol. *78*, 6600–6607.

Barnes, H.J., Nolan, L.K., and Vallancourt, J.P. (2008). Colibacillosis. In Diseases of Poultry, Y.M. Saif, ed. (Oxford, UK, Blackwell Publishing), pp. 691–732.

Barrow, P., Lovell, M., and Berchieri, Jr., A. (1998). Use of lytic bacteriophage for control of experimental *Escherichia coli* septicemia and meningitis in chickens and calves. Clin. Diagn. Lab. Immunol. *5*, 294–298.

Berchieri, A., Lovell, M.A., and Barrow, P.A. (1991). The activity in the chicken alimentary tract of bacteriophages lytic for *Salmonella typhimurium*. Res. Microbiol. *142*, 541–549.

Bicalho, R.C., Santos, T.M.A., Gilbert, R.O., Caixeta, L.S., Teixeira, L.M., Bicalho, M.L.S., and Machado, V.S. (2010). Susceptibility of *Escherichia coli* isolated from uteri of postpartum dairy cows to antibiotic and environmental bacteriophages. Part I: Isolation and lytic activity estimation of bacteriophages. J. Dairy Sci. *93*, 93–104.

Borie, C., Albala, I., Sanchez, P., Sanchez, M.L., Ramirez, S., Navarro, C., Morales, M.A., Retamales, J., and Robeson, J. (2008). Bacteriophage treatment reduces *Salmonella* colonization of infected chickens. Avian Dis. *52*, 64–67.

Borie, C., Sanchez, M.L., Navarro, C., Ramirez, S., Morales, M.A., Retamales, J., and Robeson, J. (2009). Aerosol spray treatment with bacteriophages and competive exclusion reduces *Salmonella* enteritidis infection in chickens. Avian Dis. 53, 250–254.

Callaway, T.R., Edrington, T.S., Brabban, A.D., Keen, J.E., Anderson, R.C., Rossman, M.L., Engler, M.J., Genovese, K.J., Gwartney, B.L., Reagan, J.O., *et al.* (2006). Fecal prevalence of *Escherichia coli O157, Salmonella, Listeria*, and bacteriophage infecting *E. coli* O157:H7 in feedlot cattle in the Southern Plains region of the United States. Foodborne Pathog. Dis. *3*, 234–244.

Callaway, T.R., Edrington, T.S., Brabban, A.D., Anderson, R.C., Rossman, M.L., Engler, M.J., Genovese, K.J., Keen, J.E., Looper, M.L., Kutter, E.M., *et al.* (2008). Bacteriophage isolated from feedlot cattle can reduce *Escherichia coli* O157:H7 populations in ruminant gastrointestinal tracts. Foodborne Pathog. Dis. *5*, 183–191.

Callaway, T.R., Edrington, T.S., Brabban, A., Kutter, B., Karriker, L., Stahl, C., Wagstrom, E., Anderson, R., Poole, T.L., Genovese, K., *et al.* (2011). Evaluation of phage treatment as a strategy to reduce *Salmonella* populations in growing swine. Foodborne Pathogens and Dis. *8*, 261–266.

Castillo, D., Higuera, G., Villa, M., Middelboe, M., Dalsgaard, I., Madsen, L., and Espejo, R.T. (2012). Diversity of *Flavobacterium psychrophilum* and the potential use of its phages for protection against bacterial cold water disease in salmonids. J. Fish Dis. 35, 193–201.

Celia, L.K., Nelson, D., and Kerr, D.E. (2008). Characterization of a bacteriophage lysin (Ply700) from *Streptococcus uberis*. Vet. Microbiol. *130*, 107–117.

Connerton, P.L., Carrillo, C.M.L., Swift, C., Dillon, E., Scott, A., Rees, C.E.D., Dodd, C.E.R., Frost, J., and Connerton, I.F. (2004). Longitudinal study of *Campylobacter jejuni* bacteriophages and their hosts from broiler chickens. Appl. Environ. Microbiol. *70*, 3877–3883.

Crothers-Stomps, C., Hoj, L., Bourne, D.G., Hall, M.R., and Owens, L. (2010). Isolation of lytic bacteriophage against *Vibrio harveyi*. J. Appl. Microbiol. *108*, 1744–1750.

Donovan, D.M., Lardeo, M., and Foster-Frey, J. (2006). Lysis of staphylococcal mastitis pathogens by bacteriophage phi11 endolysin. FEMS Microbiol. Lett. *265*, 133–139.

El-Shibiny, A., Connerton, P.L., and Connerton, I.F. (2005). Enumeration and diversity of *Campylobacters* and bacteriophages isolated during the rearing cycles of free-range and organic chickens. Appl. Environ. Microbiol. *71*, 1259–1266.

El-Shibiny, A., Scott, A., Timms, A., Metawea, Y., Connerton, P., and Connerton, I. (2009). Application of a group II *Campylobacter* bacteriophage to reduce strains of *Campylobacter jejuni* and *Campylobacter coli* colonizing broiler chickens. J. Food Prot. *72*, 733–740.

Fiorentin, L., Vieira, N.D., and Barioni Jr., W. (2005). Oral treatment with bacteriophages reduces the concentration of *Salmonella* enteritidis PT4 in caecal contents of broiler. Avian Pathol. *34*, 258–263.

Garcia, P., Madera, C., Martinez, B., Rodriguez, A., and Suarez, J.E. (2009a). Prevalence of bacteriophages infecting *Staphylococcus aureus* in dairy samples and their potential as biocontrol agents. J. Dairy Sci. *92*, 3019–3026.

Garcia, P., Martinez, B., Obeso, J.M., Lavigne, R., Lurz, R., and Rodriguez, A. (2009b). Functional genomic analysis of two *Staphylococcus aureus* phages isolated from the dairy environment. Appl. Environ. Microbiol. *75*, 7663–7673.

Gill, J.J., Pacan, J.C., Carson, M.E., Leslie, K.E., Griffiths, M.W., and Sabour, P.M. (2006). Efficacy and pharmacokinetics of bacteriophage therapy in treatment of subclinical *Staphylococcus aureus* mastitis in lactating dairy cattle. Antimicrob. Agents Chemother. *50*, 2912–2918.

Goode, D., Allen, V.M., and Barrow, P.A. (2003). Reduction of experimental *Salmonella* and *Campylobacter* contamination of chicken skin by application of lytic bacteriophage. Appl. Environ. Microbiol. *69*, 5032–5036.

Heller, E.D., and Drabkin, N. (1977). Some characteristics of pathogenic *E. coli* strains. Br. Vet. J. *133*, 572–578.

d'Hérelle, F. (1917). Sur un microbe invisible antagonists des *bacilles dysenteriques*. Comptes Rendus Acad. Sci. Paris *165*, 373–375.

d'Hérelle, F. (1926). The behavior of the bacteriophage in epidemics. 1. Avian typhosis. In The Bacteriophage and its Behavior (Baltimore, MD, The Williams & Wilkins Company), pp. 490–497.

Higgins, J.P., Higgins, S.E., Guenther, K., Huff, W.E., and Hargis, B.M. (2005). Use of a specific bacteriophage treatment to reduce *Salmonella* in poultry products. Poultry Sci. *84*, 1141–1145.

Holck, A., and Berg, J. (2009). Inhibition of *Listeria monocytogenes* in cooked ham by virulent bacteriophages and protective cultures. Appl. Environ. Microbiol. *75*, 6944–6946.

Huff, W.E., Huff, G.R., Rath, N.C., Balog, J.M., Xie, H., Moore, P.A., and Donoghue, A.M. (2002a). Prevention of *Escherichia coli* respiratory infection in broiler chickens with bacteriophage (SPR02). Poultry Sci. *81*, 437–441.

Huff, W.E., Huff, G.R., Rath, N.C., Balog, J.M., and Donoghue, A.M. (2002b). Prevention of *Escherichia coli* infection in broiler chickens with a bacteriophage aerosol spray. Poultry Sci. *81*, 1486–1491.

Huff, W.E., Huff, G.R., Rath, N.C., Balog, J.M., and Donoghue, A.M. (2003a). Evaluation of aerosol spray and intramuscular injection of bacteriophage to treat an *Escherichia coli* respiratory infection. Poultry Sci. *82*, 1108–1112.

Huff, W.E., Huff, G.R., Rath, N.C., Balog, J.M., and Donoghue, A.M. (2003b). Bacteriophage treatment of a severe *Escherichia coli* respiratory infection in broiler chickens. Avian Dis. *47*, 1399–1405.

Huff, W.E., Huff, G.R., Rath, N.C., Balog, J.M., and Donoghue, A.M. (2004). Therapeutic efficacy of bacteriophage and Baytril (enrofloxacin) individually and in combination to treat colibacillosis in broilers. Poultry Sci. *83*, 1944–1947.

Huff, W.E., Huff, G.R., Rath, N.C., and Donoghue, A.M. (2010). Immune interference of bacteriophage efficacy when treating colibacillosis in poultry. Poultry Sci. *89*, 805–900.

Hurley, A., Maurer, J.J., and Lee, M.D. (2008). Using bacteriophages to modulate *Salmonella* colonization of the chicken's gastrointestinal tract: Lessons learned from *in silico* and *in vivo* modeling. Avian Dis. *52*, 599–607.

Imbeault, S., Parent, S., Lagace, M., Uhland, C.F., and Blais, J. (2006). Using bacteriophages to prevent furunculosis caused by *Aeromonas salmonicida* in farmed brook trout. J. Aquat. Anim. Health. *18*, 203–214.

Jamalludeen, N., Johnson, R.P., Shewen, P.E., and Gyles, C.L. (2009). Evaluation of bacteriophage for prevention and treatment of diarrhea due to experimental enterotoxigenic *Escherichia coli* O149 infection of pigs Vet. Microbiol. *136*, 135–141.

Kaldhusdal, M., and Lovland, A.V. (2000). The economical impact of *Clostridium perfringens* is greater than anticipated. World Poult. *16*, 50–51.

Karunasagar, I., Shivu, M.M., Girisha, S.K., Krohne, G., and Karunasagar, I. (2007). Biocontrol of pathogens in shrimp hatcheries using bacteriophages. Aquaculture. *268*, 288–292.

Kudva, I.T., Jelacic, S., Tarr, P.I., Youderian, P., and Hovde, C.J. (1999). Biocontrol of *Escherichia coli* O157 with O157-specific bacteriophages. Appl. Environ. Microbiol. *65*, 3767–3773.

Lerondelle, C., and Poutrel, B. (1980). Bacteriophage treatment trials on staphylococcal udder infection in lactating cows. Ann. Rech. Vet. *11*, 421–426.

Lim, T.H., Lee, D.H., Lee, Y.N., Park, J.K., Youn, H.N., Kim, M.S., Lee, H.J., Yang, S.Y., Cho, Y.W., Lee, J.B., *et al.* (2011). Efficacy of bacteriophage therapy on horizontal transmission of *Salmonella gallinarum* on commercial layer chickens. Avian Dis. *55*, 435–438,

Lim, T.H., Kim, M.S., Lee, D.H., Lee, Y.N., Park, J.K., Youn, H.N., Lee, H.J., Yang, S.Y., Cho, Y.W., Lee, J.B., *et al.* (2012). Use of bacteriophage for biological control of *Salmonella enteritidis* infection in chicken. Res. Vet. Sci. *93*, 1173–1178.

Loc Carrillo, C., Atterbury, R.J., El-Shibiny, A., Connerton, P.L., Dillon, E., Scott, A., and Connerton, I.F. (2005). Bacteriophage therapy to reduce *Campylobacter jejuni* colonization of broiler chickens. Appl. Environ. Microbiol. *71*, 6554–6563.

Matsuoka, S., Hashizume, T., Kanzaki, H., Iwamoto, E., Park, S.C., Yoshida, T., and Nakai, T. (2007). Phage therapy against β-hemolytic streptococcicocis of Japanese flounder *Paralichthys olivaceus*. Fish Pathol. *42*, 181–189.

Miller, R.W., Skinner, J., Sulakvelidze, A., Mathis, G.F., and Hofacre, C.L. (2010). Bacteriophage therapy for control of necrotic enteritis of broiler chickens experimentally infected with *Clostridium perfringens*. Avian Dis. *54*, 33–40.

Nakai, T., Sugimoto, R., Park, K.H., Matsuoka, S., Mori, K., Nishioka, T., and Maruyama, K. (1999). Protective effects of bacteriophage on experimental *Lactococcus garvieae* infection in yellowtail. Dis. Aquat. Org. *37*, 33–41.

Niu, Y.D., McAllister, T.A., Xu, Y., Johnson, R.P., Stephens, T.P., and Stanford, K. (2009). Prevalence and impact of bacteriophages on the presence of *Escherichia coli* O157:H7 in feedlot cattle and their environment. Appl. Environ. Microbiol. *75*, 1271–1278.

O'Flaherty, S., Coffey, A., Meaney, W.J., Fitzgerald, G.F., and Ross, R.P. (2005a). Inhibition of bacteriophage K proliferation on *Staphylococcus aureus* in raw bovine milk. Let. Appl. Microbiol. *41*, 274–279.

O'Flaherty, S., Ross, R.P., Flynn, J., Meaney, W.J., Fitzgerald, G.F., and Coffey, A. (2005b). Isolation and characterization of two anti-staphylococcal bacteriophages specific for pathogenic *Staphylococcus aureus* associated with bovine infections. Lett. Appl. Microbiol. *41*, 482–486.

Oliveira, A., Sereno, R., and Azeredo, J. (2010). *In vivo* efficiency evaluation of a phage cocktail in controlling severe colibacillosis in confined conditions and experimental poultry houses. Vet. Microbiol. *146*, 303–308.

Opengart, K. (2008). Necrotic enteritis. In Diseases of Poultry, Y.M. Saif, ed. (Oxford, UK, Blackwell Publishing), pp. 872–879.

Oot, R.A., Raya, R.R., Callaway, T.R., Edrington, T.S., Kutter, E.M., and Brabban, A.D. (2007). Prevalence of *Escherichia coli* O157 and O157:H7-infecting bacteriophage in feedlot cattle feces. Lett. Appl. Microbiol. *45*, 445–453.

Park, S.C., and Nakai, T. (2003). Bacteriophage control of *Pseudomonas plecoglossicida* infection in ayu *Plecoglossus altivelis*. Dis. Aquat. Org. *53*, 33–39.

Park, S.C., Shimamura, I., Fukunaga, M., Mori, K., and Nakai, T. (2000). Isolation of bacteriophages specific to a fish pathogen, *Pseudomonas plecoglossicida*, as a candidate for disease control. Appl. Environ. Microbiol. *66*, 1416–1422.

Pelon, W., Luftig, R.B., and Johnston, K.H. (2005). *Vibrio vulnificus* load reduction in oysters after combined exposure to *Vibrio vulnificus*-specific bacteriophage and to an oyster extract component. J. Food Prot. *68*, 1188–1191.

Prasad, Y., Arpana, Kumar, D., and Sharma, A.K. (2011). Lytic bacteriophages specific to *Flavobacterium columnaris* rescue catfish, *Clarias batrachus* (Linn.) from columnaris disease. J. Environ. Biol. *32*, 161–168.

Raya, R.R., Varey, P., Oot, R.A., Dyen, M.R., Callaway, T.R., Edrington, T.S., Kutter, E.M., and Brabban, A.D. (2006). Isolation and characterization of a new T-even bacteriophage, CEV1, and determination of its potential to reduce *Escherichia coli* O157:H7 levels in sheep. Appl. Environ. Microbiol. *72*, 6405–6410.

Saez, A.C., Zhang, J., Rostagno, M.H., and Ebner, P.D. (2011). Direct feeding of microencapsulated bacteriophages to reduce *Salmonella* colonization in pigs. Foodborne Pathogens and Dis. *8*, 1269–1274.

Santos, T.M.A., Gilbert, R.O., Caixeta, L.S., Machado, V.S., Teixeira, L.M., and Bicalho, R.C. (2010). Susceptibility of *Escherichia coli* isolated from uteri of postpartum dairy cows to antibiotic and environmental bacteriophages. Part II: *in vitro* antimicrobial activity evaluation of a bacteriophage cocktail and several antibiotics. J. Dairy Sci. *93*, 105–114.

Schmelcher, M., Powell, A.M., Becker, S.C., Camp, M.J., and Donovan, D.M. (2012). Chimeric phage lysins act synergistically with lysostaphin to kill mastitis-causing *Staphylococcus aureus* in murine mammary glands. Appl. Environ. Microbiol. *78*, 2297–2305.

Sheng, H., Knecht, H.J., Kudva, I.T., and Hovde, J. (2006). Application of bacteriophage to control intestinal *Escherichia coli* O157:H7 levels in ruminants. Appl. Environ. Microbiol. *72*, 5359–5366.

Smith, H.W., and Huggins, M.B. (1983). Effectiveness of phages in treating experimental *Escherichia coli* diarrhoea in calves, piglets and lambs. J. Gen. Microbiol. *129*, 2659–2675.

Smith, H.W., Huggins, M.B., and Shaw, K.M. (1987). The control of experimental *Escherichia coli* diarrhoea in calves by means of bacteriophages. J. Gen. Microbiol. *133*, 1111–1126.

Sojka, W.J. (1965). *Escherichia coli* in Domestic Animals and Poultry (Farnham Royal, UK, Commonwealth Agricultural Bureau).

Soni, K.A., and Nanapaneni, R. (2009). Reduction of *Listeria monocytogenes* on the surface of fresh channel catfish fillets by bacteriophage Listex P100. Foodborne Pathog. Dis. *7*, 427–434.

Soni, K.A., and Nannapaneni, R. (2010). Bacteriophage significantly reduces *Listeria monocytogenes* on raw salmon fillet tissue. J. Food Prot. *73*, 32–38.

Stenholm, A.R., Dalsgaard, I., and Middelboe, M. (2008). Isolation and characterization of bacteriophages infecting the fish pathogen *Flavobacterium psychrophilum*. Appl. Environ. Microbiol. *74*, 4070–4078.

Sulakvelidze, A. (2013). Using lytic bacteriophages to eliminate or significantly reduce contamination of food by food-borne bacterial pathogens. J. Sci. Food Agric. 93, 3137–3146.

Sulakvelidze, A., Alavidze, Z., and Morris, J.G. (2001). Bacteriophage therapy. Antimicrob. Agents Chemother. *45*, 649–659.

Toro, H., Price, S.B., McKee, S., Hoerr, F.J., Krehling, J., Perdue, M., and Bauermeister, L. (2005). Use of bacteriophages in combination with competitive exclusion to reduce *Salmonella* from infected chickens. Avian Dis. *49*, 118–124.

Twort, F.W. (1915). An investigation on the nature of ultramicroscopic viruses. Lancet 2, 1241–1243.

USFDA. (2007). U.S. Food and Drug Administration Agency. Response Letter GRAS Notice No. 000218. Available at www.accessdata.fda.gov/scripts/fcn/gras_notices/701456A.PDF (accessed 10 August 2010)

Van Immerseel, F., De Buck, J., Pasmans, F., Huyghebaert, G., Haesebrouck, F., and Ducatelle, R. (2004). *Clostridium perfringens* in poultry: an emerging threat for animal and public health. Avian Pathol. 33, 537–549.

Verner-Jeffreys, D.W., Algoet, M., Pond, M.J., Virdee, H.K., Bagwell, N.J., and Roberts, E.G. (2007). Furunculosis in Atlantic salmon (*Salmo salar L.*) is not readily controlled by bacteriophage therapy. Aquaculture *270*, 475–484.

Vinod, M.G., Shivu, M.M., Umesha, K.R., Rajeeva, B.C., Krohne, G., Karunasagar, I., and Karunasagar, I. (2006). Isolation of *Vibrio harveyi* bacteriophage with a potential for biocontrol of luminous vibriosis in hatchery environments. Aquaculture 255, 117–124.

Wagenaar, J.A., Van Bergen, M.A.P., Mueller, M.A., Wassenaar, T.M., and Carlton, R.M. (2005). Phage therapy reduces *Campylobacter jejuni* colonization in broilers. Vet. Microbiol. *109*, 275–283.

Walakira, J.K., Carrias, A.A., Hossain, M.J., Jones, E., Terhune, J.S., and Liles, M.R. (2008). Identification and characterization of bacteriophages specific to the catfish pathogen, *Edwardsiella ictaluri*. J. Appl. Microbiol. *105*, 2133–2142.

Wall, S.K., Zhang, J., Rostagno, M.H., and Ebner, P.D. (2010). Phage therapy to reduce preprocessing *Salmonella* infections in market-weight swine. Appl. Environ. Microbiol. *76*, 48–53.

Wu, J.L., Lin, H.M., Jan, L., Hsu, Y.L., and Chang, L.H. (1981). Biological control of fish bacterial pathogen, *Aeromonas hydrophila*, by bacteriophage AH 1. Fish Pathol. *15*, 271–276.

Xie, H., Zhuang, X., Kong, J., Ma, G., and Zhang, H. (2005). Bacteriophage Esc-A is an efficient therapy for *Escherichia coli* 3–1 caused diarrhea in chickens. J. Gen. Appl. Microbiol. *51*, 159–163.

Yoichi, M., Morita, M., Mizoguchi, K., Fischer, C.R., Unno, H., and Tanji, Y. (2004). The criterion for selecting effective phage for *Escherichia coli* O157:H7 control. Biochem. Eng. J. *19*, 221–227.

Zhang, H., Wang, R., and Bao, H. (2013). Phage inactivation of food-borne *Shigella* on ready-to-eat spiced chicken. Poultry Sci. 92, 211–217.

The Use of Phages as Biocontrol Agents in Foods

9

Jan Borysowski and Andrzej Górski

Abstract

Bacteriophages have several features which make them novel antibacterial agents with which to prevent bacterial food-borne infections. Phages can be used to control the growth of bacteria both in food products and on food contact surfaces. In recent years, many studies have shown relatively high efficacy of phages against several major food-borne pathogens, especially *Listeria monocytogenes*, *Salmonella enterica*, *Escherichia coli* including *E. coli* O157:H7, *Campylobacter jejuni*, and *Staphylococcus aureus* in a variety of foods; phages were also successfully used against *Yersinia enterocolitica*, *Shigella* spp., *Bacillus cereus* and *Cronobacter* spp. In addition, some attempts were made to use phages to eliminate food spoilage bacteria. In most studies, the use of phages resulted in significant, log-fold reductions in the bacterial counts in foods; such reductions are known to substantially decrease a risk of food-borne infections. Main factors that determine the efficacy of bacteriophages as biocontrol agents include phage particles density in/on a food product, a level of bacterial contamination, the development of bacterial resistance to phages, and phage stability in different foods. Recent clearance by FDA of four bacteriophage preparations for food applications shows that bacteriophages are gradually gaining acceptance as a means of prevention of food-borne infections.

Introduction

Food-borne infections are considered a major public health threat (Batz *et al.*, 2012; O'Brien, 2012). For example, according to World Health Organization (WHO) estimates, in 2005 1.8 million people died from diarrhoeal diseases, most of which were caused by contaminated food and drinking water (Newell *et al.*, 2010). In the USA, food-borne pathogens cause 1,000 reported disease outbreaks, an estimated 48 million illnesses, 128,000 hospitalizations, and 3,000 deaths annually (Anonymous, 2011; Scallan *et al.*, 2011a,b). The importance of the problem is also reflected by the fact that food-borne infections cost the US economy $10–83 billion per year (Nyachuba, 2010). In spite of considerable efforts of scientific community, governmental agencies, and food industry to improve food safety, the incidence of food-borne illness has not substantially decreased over recent years; in fact, these infections are predicted to continue to be a major public health problem worldwide (Newell *et al.*, 2010; Nyachuba, 2010). Major factors that contribute to the prevalence of food-borne infections include large-scale production and wide distribution of foods, globalization of the food supply, inappropriate eating habits of consumers, genetic variability and emergence of new food-borne pathogens, as well as a growing number of high-risk consumers (Nyachuba, 2010). Main vehicles for the transmission of human pathogens are both foods of animal origin such as meat, poultry, eggs, and unpasteurized milk (DuPont, 2007), and fresh produce (Berger *et al.*, 2010). Infectious agents implicated in food-borne diseases include bacteria, viruses, parasites, and prions; of these, bacteria are the most frequent and well-studied group of aetiological agents of food-borne infections (Nyachuba, 2010; Newell *et al.*, 2010).

Standard methods employed at food processing

facilities to eliminate food-borne pathogens are based on mild heat treatments, high pressure processing, pulsed electric fields, intensive light pulses, chemical antibacterial agents, sanitizers, and disinfectants (Rajkovic *et al.*, 2010; Ahn *et al.*, 2013). However, these methods have significant drawbacks including inadequate efficacy, toxicity, corroding of food contact surfaces, and a negative influence on the environment, natural food microflora and the organoleptic properties of food (Sulakvelidze, 2013). This has created a need for the development of novel approaches to improve food safety that do not have the above-mentioned drawbacks. A growing number of studies have shown that bacteriophages may provide a basis for such approach (Hagens and Loessner, 2010; Coffey *et al.*, 2010; Mahony *et al.*, 2011; Goodridge and Bisha, 2011; Sillankorva *et al.*, 2012; Brovko *et al.*, 2012; Sulakvelidze, 2013).

Bacteriophages as biocontrol agents

Bacteriophages have several features which make them unique antibacterial agents with apparent potential for food applications. First, antibacterial range of phages is very narrow, which allows elimination of specific food-borne pathogens without killing bacteria from food and gut microflora (Guenther *et al.*, 2011; Bandara *et al.*, 2012). To obtain preparations with whole bacterial species coverage one can use broad host range phages (Carlton *et al.*, 2005) or employ phage cocktails, i.e. preparations containing several different bacteriophages (Leverentz *et al.*, 2003). The latter approach could be particularly useful in cases where antibacterial ranges of individual phages are too narrow to enable their practical applications (Greer and Dilts, 2002). In view of the presence in different food products of antibiotic-resistant bacteria which constitute substantial public health threat (Verraes *et al.*, 2013), an important feature of phages is their capacity to kill antibiotic-resistant bacterial strains (Górski *et al.*, 2009). Moreover, phages were shown to exert additive antibacterial effects with other antimicrobial agents and means of controlling food-borne bacteria including bacteriocins (Leverentz *et al.*, 2003), lauric arginate (LAE) and potassium lactate–sodium diacetate (PL-SD) mixture (Soni *et al.*, 2012), essential oils (Viazis *et al.*, 2011a), antagonistic bacteria (Ye *et al.*, 2010), hydrostatic pressure treatments (Tabla *et al.*, 2012), and quaternary ammonium compounds (Roy *et al.*, 1993). This indicates a possibility of inclusion of phages to 'hurdle technology' which is based on simultaneous use of different preservation methods to improve microbial food safety (Leistner, 2000). Furthermore, in view of their ability to multiply in bacterial cells, phages are self-replicating and self-limiting antibacterials (Smith and Huggins, 1982). Bacteriophages are also considered to be safe, and their oral administration to experimental animals and humans practically does not result in any serious side-effects (Carlton *et al.*, 2005; Międzybrodzki *et al.*, 2012; Sarker *et al.*, 2012; Kang *et al.*, 2013; McCallin *et al.*, 2013). Moreover, as particles omnipresent in the biosphere phages are regarded safe for environment; they are also not likely to adversely affect food contact surfaces and organoleptic properties of foods (Sulakvelidze, 2013). Given biofilm-forming ability of some food-borne bacteria, an important feature of at least some phages is their capacity to eliminate biofilms (Soni and Nannapaneni, 2010a; Siringan *et al.*, 2011). Furthermore, phages are relatively stable on different foods and retain their antibacterial activity in at least some food processing and storage conditions (Garcia *et al.*, 2009; Coffey *et al.*, 2011; Soni *et al.*, 2012).

The use of phages to prevent food-borne infections, termed phage biocontrol, involves three major approaches (Sulakvelidze, 2013). First, bacteriophages can be used to prevent or reduce colonization by pathogenic bacteria of live production animals; this approach is presented in Chapter 8 and will not be discussed here. The second approach is based on the use of phage preparations as biosanitizing agents to eliminate bacteria from food contact surfaces to prevent cross-contamination of foods (Viazis *et al.*, 2011b). Moreover, phages can be added directly to food products to control the growth of specific food-borne pathogens (Carter *et al.*, 2012; Guenther *et al.*, 2012). It is noteworthy here that some authors use the term 'biocontrol' only referring to the use of phages to reduce colonization by pathogenic bacteria of raw foods at the stage of industrial food processing (Garcia *et al.*, 2008; Sillankorva *et al.*, 2012). However, in this chapter, all phage-based methods of preventing food-borne infections will be referred to as 'biocontrol' (Hagens and Loessner, 2010; Goodridge and Bisha, 2011; Sulakvelidze, 2013).

A critical stage in all phage biocontrol approaches is a selection of specific bacteriophage(s) to be used in a phage preparation. General considerations on choosing correct phages to eliminate bacteria are presented in Chapter 2. In addition, desirable characteristics of phages intended for food applications were listed by Hagens and Loessner (2010). Such phages should have broad host range, be strictly lytic, propagated on non-pathogenic host, and lack an ability to transduce bacterial DNA. Their genomes ought to be sequenced and do not encode for proteins involved in pathogenicity or likely to induce any allergic reactions. Safety trials in experimental animals should show a lack of side-effects following oral administration of phage preparation. In addition, such phages should be formally approved for food applications by relevant regulatory agencies, be stable, and available in amount sufficient to produce preparations at commercial scale.

Standard form of bacteriophage preparations used as biocontrol agents is phage suspension that can be used as liquid or sprayed onto foods or food contact surfaces (Viazis *et al.*, 2011b; Carter *et al.*, 2011; Patel *et al.*, 2011; Bueno *et al.*, 2012). Moreover, foods can be immersed in buffer containing phage (Jassim *et al.*, 2012). It was also shown that phages can retain significant antibacterial activity after immobilization on inert surfaces (Anany *et al.*, 2011). While the main way of application of bacteriophage preparations seems to be their use as spray, a possibility of using phages in such different forms broadens the range of their potential applications in food industry. Insightful considerations on use of phage preparations in different forms to prevent food-borne infections were presented by Hagens and Loessner (2010).

By June 2013, one phage preparation (EcoShield™) has been cleared by FDA as a food contact substance and another one (ListShield™) has been cleared as a food additive (in addition, FDA permitted ListShield™ to be updated by inclusion of new phages provided that they meet specific requirements). ListShield™ has also been approved by the US Environmental Protection Agency (EPA) to reduce bacterial contamination of food processing plants. Moreover, two other preparations (Listex™ and SalmoFresh™) have been granted GRAS status by FDA. Regulatory issues associated with the use of these preparations were recently discussed in detail by Sulakvelidze (2013).

The use of phages as biocontrol agents against specific food-borne bacteria

Recent analysis lists 21 major bacterial food-borne pathogens (Scallan *et al.*, 2011a). However, the majority of studies on the use of phages as biocontrol agents focused on *Listeria monocytogenes*, *Salmonella enterica*, *Escherichia coli*, *Campylobacter jejuni* and *Staphylococcus aureus*. Except for *S. aureus*, all these bacterial species belong to the most prevalent food-borne bacteria.

Listeria monocytogenes

L. monocytogenes is the aetiological agent of listeriosis, whose clinical manifestations range from febrile gastroenteritis to severe invasive forms including meningitis, rhombencephalitis, sepsis, perinatal infections, as well as abortions. Listeriosis is most commonly associated with consumption of contaminated food including cheese, meats, and vegetables; transmission of *L. monocytogenes* by food is facilitated by its ability to withstand food-processing technologies and to multiply at low temperatures. Although listeriosis is relatively rare, it remains one of the leading causes of death among food-borne infections. In the last years the rate of listeriosis has increased in some European countries (Schlech, 2000; Allerberger and Wagner, 2010; Scallan *et al.*, 2011a; Anonymous, 2013).

The efficacy of phages in controlling the growth of *L. monocytogenes* has been extensively evaluated in different foods (Dykes and Moorhead, 2002; Leverentz *et al.*, 2003, 2004; Carlton *et al.*, 2005; Guenther *et al.*, 2009; Holck and Berg, 2009; Soni and Nannapaneni, 2010b; Soni *et al.*, 2010, 2012; Guenther *et al.*, 2011; Bigot *et al.*, 2011) and on materials used in food contact surfaces (Roy *et al.*, 1993; Hibma *et al.*, 1997; Soni and Nannapaneni, 2010a).

The vast majority of *Listeria* phages are temperate and have a narrow antibacterial range (Carlton *et al.*, 2005, and references therein); essentially, these are unsuitable for use as biocontrol agents. One of the very few known phages that are strictly lytic and have a broad antibacterial range within the genus *Listeria* is P100 (Carlton *et al.*, 2005). Preparation containing this phage (Listex™) has been granted GRAS status by FDA in 2006 (Sulakvelidze, 2013). Antibacterial effects of P100 were evaluated on cheese (Carlton *et al.*, 2005; Soni *et al.*, 2012), raw salmon fillet tissue (Soni and Nannapaneni, 2010b), and fresh channel catfish fillets (Soni *et al.*, 2010).

Carlton *et al.* (2005) investigated antibacterial effects of P100 on experimentally contaminated surface-ripened soft cheese. While repeated application of the phage at the concentration of 1.5×10^8 PFU/ml (corresponding to approximately 2×10^6 PFU/cm^2) led to 2–3 log reductions in the *L. monocytogenes* counts from an initial density of approximately 2×10^1 CFU/cm^2, the use of a higher phage concentration (3×10^9 PFU/ml, corresponding to approximately 6×10^7 PFU/cm^2) resulted in complete eradication of bacteria; in the latter case, eradication of *L. monocytogenes* was achieved following the application of both repeated and single dose of the phage. No phage-resistant bacteria were detected in cheeses treated with the lower phage dose.

In another study the antibacterial effects of P100 were compared with those of other GRAS antimicrobials – lauric arginate (LAE) and potassium lactate-sodium diacetate (PL-SD) mixture in queso fresco cheese stored at 4°C (Soni *et al.*, 2012). In the phage-treated pieces, after an initial reduction in the bacterial count from 3.5 log CFU/cm^2 to an undetectable level on day 1, the regrowth of *L. monocytogenes* was found, resulting in the bacterial count of ~4 log CFU/cm^2 on day 28. Likewise, in the LAE-treated pieces, the bacterial count increased after an initial reduction. Unlike P100 and LAE exerting listericidal activity, PL-SD, being a listeriostatic agent, did not reduce the *L. monocytogenes* count, but rather completely inhibited the growth of bacteria. The authors of the study also compared the efficacy of different combinations of the three studied antibacterial agents and found that it was highest following use of a combination of a listericidal agent (P100 or LAE) with a listeriostatic agent (PL-SD); the use of such a combination not only resulted in a rapid decrease in the bacterial count, but also prevented the growth of *L. monocytogenes* throughout the study period.

Antibacterial activity of P100 was also evaluated on raw salmon fillet tissue (Soni and Nannapaneni, 2010b). At the dose of 10^4 PFU/g P100 did not significantly reduce the *L. monocytogenes* population on salmon tissue samples inoculated with 4 log CFU/g of bacteria. Application of the doses of 10^5 and 10^6 PFU/g resulted in very small, though significant reductions in the bacterial counts, while the use of 10^7 and 10^8 PFU/g led to 2-log and 3.5-log CFU/g reductions, respectively. Antibacterial effects of P100 were also bacterial cell density-dependent, the reductions in the *L. monocytogenes* counts being 3, 2.5–2.9, and 1.9 log for samples inoculated with 4-, 3-, and 2-log CFU/g; these effects were comparable after 30 min and 2 h, as well as at 4 and 22°C. During the 10-day storage period of fillets at 4°C P100 reduced the bacterial count by 2.3 log compared with the control. High efficacy of this phage was also shown on fresh channel catfish fillets (Soni *et al.*, 2010).

Another broad host range lytic *L. monocytogenes* phage that was investigated as a potential biocontrol agent is A511 (Loessner and Busse, 1990). Guenther *et al.* (2009) evaluated the antibacterial effects of A511 on different ready-to eat (RTE) foods including hot dogs, mixed seafood, chocolate milk, cabbage, sliced cooked turkey breast, smoked salmon, mozzarella cheese brine, and iceberg lettuce. Food samples were spiked with *L. monocytogenes* (1×10^3 CFU/g), the phage was added at different doses corresponding to densities ranging between 3×10^6 PFU/g and 3×10^8 PFU/g, and samples were stored at 6°C for 6 days. In general, in liquid foods, the phage decreased the bacterial counts below the detection level within 2–3 days, while on solid foods the bacterial counts were reduced by 0.4 to 5.0 log units (45.7 to 100%). The highest efficiency of A511 was found on cabbage, where the *L.*

monocytogenes population was reduced to undetectable level at day 1, whereas relatively lower efficiency of bacterial elimination was observed on turkey breast and on mixed seafood. The phage was also effective at 20°C. Antibacterial effects of A511 were concentration dependent, the concentration necessary for optimal activity being not less than 10^8 PFU/g. It was also shown that the efficacy of A511 was comparable to that of P100.

In another study the potential of A511 for controlling the growth of *L. monocytogenes* was evaluated on soft ripened white mould and red-smear cheeses (Guenther *et al.*, 2011). The phage was applied in single or repeated doses of 3×10^8 or 1×10^9 PFU/cm^2 on the surface of cheeses inoculated with *L. monocytogenes* (10^1–10^3 CFU/cm^2), and the growth of bacteria was monitored for up to 22 days. Generally, all treatment protocols were effective, producing the reductions in the bacterial counts within the range 2 to >6 logs. The efficacy of A511 was higher at higher doses and lower levels of initial bacterial contamination. It was also shown that 30% of *L. monocytogenes* recovered from phage-treated cheeses displayed phage resistance.

Another phage that was used as a biocontrol agent against *L. monocytogenes* is LH7. Dykes and Moorhead (2002) compared antibacterial effects of LH7 with those of the lantibiotic nisin (a bacteriocin approved as a food preservative in many countries) in broth and two model food systems (PBS and vacuum-packaged fresh beef, both stored for 4 weeks at 4°C). In broth, antibacterial effects of nisin were superior to those of the phage, which had a very low or no activity depending on the experimental conditions, and a combination of nisin and the phage exerted a higher antibacterial activity than nisin alone. The results obtained in both model food systems were similar to one another. It was shown that while nisin (as well as a combination of nisin and the phage) displayed significant antibacterial activity against *L. monocytogenes*, the activity of the phage used alone was insignificant. Although the authors of the study suggested that the likely reasons for inadequate antibacterial activity of LH7 were the absence of $CaCl_2$ in the experimental milieu and suboptimal multiplicity of infection (MOI) values, this was not verified experimentally (according to Hagens and Loessner (2010) ineffectiveness of phage resulted from inadequate concentration of phage particles). This study also showed that a combination of nisin and the phage, unlike either agent used alone, exerted activity against bacteria at the stationary phase of growth.

Bigot *et al.* (2011) evaluated antibacterial effects of *L. monocytogenes* phage FWLLm1 on RTE chicken breast rolls at two different temperatures (30 and 5°C). At 30°C, the treatment with phages (2.5×10^7 PFU/cm^2) of rolls inoculated with *L. monocytogenes* (10^5 CFU/cm^2) resulted in a rapid, 2.5 $\log_{10}$ CFU/cm^2 reduction in the bacterial count followed by the regrowth of bacteria at a rate similar to that found in the untreated rolls. At 5°C and the initial bacterial density of 10^4 CFU/cm^2, the application of FWLLm1 (1.5×10^6 PFU/cm^2) led to a reduction in the *L. monocytogenes* count by 1.5 $\log_{10}$ CFU/cm^2 without bacterial regrowth. However, the highest efficacy of FWLLm1 was found at 5°C and a lower initial density of *L. monocytogenes* (10^2 CFU/cm^2), when the phage decreased the bacterial count to an undetectable level at day 11. Although this level was maintained until day 18, after 21 days a slight increase in the *L. monocytogenes* count was noted.

Apart from preparations containing single bacteriophages, also phage cocktails were investigated as biocontrol agents against *L. monocytogenes*. For example, Leverentz *et al.* (2003) evaluated the capacity of cocktails LM-103 and LMP-102 to eliminate *L. monocytogenes* from experimentally contaminated fresh-cut melons and apples stored at 10°C for 7 days (LMP-102 is currently known under the trade name ListShield™). In general, the antibacterial effects of phages were stronger on melons than on apples. On melons, phages reduced the *L. monocytogenes* counts by 2.0 to 4.6 log units compared to the control, while on apples the bacterial count was reduced by less than 0.4 log units. On melon slices phages were more effective than nisin. Application of a combination of a cocktail and nisin resulted in the reductions in the *L. monocytogenes* counts by 4.3–5.7 log units and by 2.3 log units over the control on melons and on apples, respectively. This combination was more effective than nisin alone on both melons and apples.

Another study of this group aimed to optimize timing and dosage of phage application (Leverentz *et al.*, 2004). LMP-102 at the concentrations ranging between 10^4 PFU/ml and 10^8 PFU/ml was applied to melon pieces at 1, 0.5, or 0 h before, or 0.5, 1, 2 or 4 h after contamination with *L. monocytogenes*. Although all phage applications reduced bacterial numbers, phages had highest efficacy when applied at the time of or up to 1 h before contamination with bacteria; this resulted in the reductions in the *L. monocytogenes* counts by up to 6.8 log units during storage of melon pieces at 10°C for 7 days. Higher phage concentrations were more effective in controlling the growth of bacteria; only the highest concentration of LMP-102 (10^8 PFU/ml) reduced *L. monocytogenes* population to undetectable level.

Holck and Berg (2009) showed that antibacterial effects of *L. monocytogenes* phages can be enhanced by bacterial protective cultures, especially over longer storage of foods. In particular, protective cultures might be used to prevent the regrowth of bacteria which are not eliminated by phages upon the treatment of a food product with a phage preparation. The efficacy of a combination of bacteriophages and a *Lactobacillus sakei* culture was shown on *L. monocytogenes*-inoculated sliced cooked ham.

Moreover, studies were conducted to evaluate potential of *L. monocytogenes* phages as biosanitizing agents (Roy *et al.*, 1993; Hibma *et al.*, 1997; Soni and Nannapaneni, 2010a). Roy *et al.* (1993) investigated the efficacy of three listerial phages in sanitizing stainless-steel (SS) and polypropylene (PP) surfaces contaminated with bacteria. The use of individual phages at the concentration of 3.5×10^8 PFU/ml resulted in about 3.4 log-unit reductions in the *L. monocytogenes* counts within 1 h on both SS and PP surfaces. When used in combination, the studied phages exerted synergistic antibacterial effects. The mixture of all three phages was about as efficient as a 20 ppm solution of the quaternary ammonium compound QUATAL, and a combination of the phage mixture and QUATAL was significantly more effective than either treatment used alone.

Hibma *et al.* (1997) employed phage breeding to obtain a phage specific for L-form of *L. monocytogenes*. This phage was shown both to inhibit L-form *L. monocytogenes* biofilm formation on stainless steel and to inactivate established biofilms; in the latter case, the phage was as effective as lactic acid (130 ppm).

Soni and Nannapaneni (2010a) showed high efficacy of P100 phage in eliminating *L. monocytogenes* biofilms from stainless steel coupons, the reductions in the counts of the five-strain mixture of *L. monocytogenes* being 3.5 and 5.4 log CFU/cm^2 for 1-week multilayered *L. monocytogenes* biofilm cells and cells firmly attached to the coupons for 2 days, respectively.

Salmonella spp.

Salmonella spp. is one of the most prevalent aetiological agents of food-borne infections (Newell *et al.*, 2010; Scallan *et al.*, 2011a). It is estimated that 1.0 million cases of gastroenteritis due to *Salmonella* spp. occur each year in the USA. Over 50% of reported cases of outbreaks of food-borne salmonellosis are caused by two serovars – *S. enterica* serovar Enteritidis and *S. enterica* serovar Typhimurium (Fatica and Schneider, 2011).

Many studies were conducted to evaluate the efficacy of *Salmonella* phages as biocontrol agents in different foods (Leverentz *et al.*, 2001; Modi *et al.*, 2001; Whichard *et al.*, 2003; Pao *et al.*, 2004; Bigwood *et al.*, 2008; Kocharunchitt *et al.*, 2009; Ye *et al.*, 2010; Guenther *et al.*, 2012). In addition, antibacterial effects of these phages were investigated on chicken and pig skin (Goode *et al.*, 2003; Hooton *et al.*, 2011; Kang *et al.*, 2013).

In the majority of studies on the use of *Salmonella* phages as biocontrol agents, antibacterial effects of these phages were evaluated on fresh produce. For example, Leverentz *et al.* (2001) examined the capacity of a phage cocktail SCPLX-1 to eliminate S. Enteritidis from experimentally contaminated fresh-cut melons and apples stored at different temperatures. While SCPLX-1 at the concentration of 2×10^8 PFU/ml significantly reduced bacterial populations on melon slices, its effects on *Salmonella* on apple slices were insignificant likely due to inactivation of phage particles by the acidic pH. On melon slices phages decreased the *Salmonella* counts by approximately 3.5 logs when slices were stored at 5°C and 10°C, and by approximately 2.5 logs when slices were stored at 20°C for 7 days. Antibacterial effects of SCPLX-1 on melon slices were stronger than the maximal effect reportedly obtained using hydrogen peroxide.

Pao *et al.* (2004) evaluated the capacity of two phages, Phage-A (capable of lysing S. Typhimurium and S. Enteritidis) and Phage-B (capable of lysing *Salmonella* Montevideo) to curb the growth of bacteria on sprouting broccoli and mustard seeds. Phages (MOI of 10^4–10^5) were added to water in which sprouting seeds contaminated with bacteria (10^2–10^3 CFU/g) were soaked (this protocol was supposed to mimic a stage in the sprout production during which seeds are soaked in water). Phage-A significantly suppressed the growth of bacteria on mustard and broccoli seeds inoculated with S. Typhimurium and S. Enteritidis within 24 h at 25°C. Furthermore, a mixture of Phage-A and Phage-B significantly reduced the bacterial counts on broccoli seeds inoculated simultaneously with S. Typhimurium, S. Enteritidis, and S. Montevideo. However, in another study phages failed to efficiently control the growth of *Salmonella* Oranienburg on alfalfa seeds, the reductions in the bacterial counts being only 1 $\log_{10}$ CFU/g (Kocharunchitt *et al.*, 2009). Inadequate antibacterial effects of phages did not result from the development of phage resistance, as all bacterial cells isolated from phage-treated seeds were sensitive to phages.

Ye *et al.* (2010) examined a possibility of using a phage cocktail in a combination with antagonistic bacteria (AB) to control the growth of *S. enterica* on sprouting mung beans. While both the cocktail and AB (an *Enterobacter asburiae* strain) reduced the *Salmonella* counts by ~3.5–5.5 log CFU/g, an additive antibacterial effect was found following the application of a combination of the cocktail with AB; in the latter case, viable *Salmonella* cells could be detected by enrichment only. The combination of phages and AB was equally efficacious at the range of temperatures between 20 and 30°C (standard temperatures at commercial sprout production), and it was also effective in controlling the growth of *Salmonella* on sprouting alfalfa seeds. However, the combination of phages and AB failed to prevent the growth of *Salmonella* Javiana on postharvest tomatoes (Ye *et al.*, 2009). Furthermore, although both phages and AB significantly reduced the *Salmonella* counts within the rhizosphere of tomatoes and in preharvest plants, no additive effects were found following the application of the combined treatment (Ye *et al.*, 2009).

Modi *et al.* (2001) used SJ2 phage as a biocontrol agent against S. Enteritidis during manufacture, ripening, and storage of Cheddar cheese produced from raw and pasteurized milk. Milks were inoculated to contain 10^4 CFU/ml of *Salmonella* and 10^8 PFU/ml of the phage; next the milks were processed into cheese using standard procedures. SJ2 significantly reduced the level of bacteria in cheeses during both 24-hour manufacture process and storage for 99 days at 8°C, in the latter case the reductions in the *Salmonella* counts being 2–3 log cycles compared to cheeses made from milk to which no phage was added. The phage was significantly more effective in controlling the growth of bacteria in cheese made from pasteurized milk compared with cheese made from raw milk. However, SJ2 failed to completely eradicate bacteria from cheese within 90 days.

Bigwood *et al.* (2008) investigated the influence of different experimental conditions on the capacity of phage P7 to lyse S. Typhimurium on cooked and raw beef. These conditions included two different temperatures (5°C and 24°C), low and high bacterial cell density (<100 CFU/cm^2 and 10^4 CFU/cm^2, respectively), as well as high and low MOI values (10^4 and 10^1, respectively). In general, antibacterial effects of the phage varied substantially depending on experimental conditions. These effects were weakest (and mostly insignificant) at 5°C and low bacterial cell density, as well as at 24°C, low bacterial cell density, and low MOI (however, at high bacterial cell density, the use of phage at low MOI resulted in significant reductions in the *Salmonella* counts). The greatest reductions in the bacterial counts (2 $\log_{10}$/cm^2 at 5°C and over 5.9 $\log_{10}$/cm^2 at 24°C) were found at high values of both bacterial cell density and MOI. No phage-resistant bacteria were isolated from phage-treated meat pieces.

Guenther *et al.* (2012) used a broad host range phage FOI-E2 to control the growth of S. Typhimurium in a variety of RTE foods including hot dogs, cooked and sliced turkey breast, mixed seafood, chocolate milk, and egg yolk. Samples of foods experimentally contaminated with *Salmonella* (10^3 CFU/g or ml) were treated with the phage (3×10^8 PFU/g or ml), and incubated at 8 or 15°C for up to 6 days. At 8°C phages decreased the bacterial counts to a level undetectable by direct plating after 1 day of incubation (however, when phage-treated food samples were subjected to enrichment protocol after 5 and 6 days, low levels of bacterial cells were detected in all samples). At 15°C the reductions in the bacterial counts were highest in chocolate milk and on turkey deli meat, then on hot dogs, and on mixed seafood. Overall, in all foods tested, a similar general pattern of changes of the bacterial count was observed: an initial decrease within 1–2 days followed by some bacterial regrowth in spite of the presence of infective phage particles. Possible reasons include the immobilization of phage particles on the food surface and matrix, as well as the development of phage resistance (phage-resistant bacterial cells were detected in phage-treated food samples).

Whichard *et al.* (2003) evaluated antibacterial effects of phage against S. Typhimurium on chicken frankfurters. 10-g samples of frankfurters were inoculated with 300 CFU of *Salmonella* and 5.25×10^6 PFU of the phage. The phage reduced the numbers of bacteria by 2 $\log_{10}$ units during 24 h of incubation at 22°C.

Moreover antibacterial effects of *Salmonella* phages were investigated on chicken and pig skin with a view to using phages to reduce bacterial contamination of animal carcasses (Goode *et al.*, 2003; Hooton *et al.*, 2011; Kang *et al.*, 2013). Goode *et al.* (2003) evaluated the capacity of phages 12, P22, and 29C to control the growth of S. Enteritidis on experimentally contaminated chicken skin. Phage 12 applied at a MOI of 1 on skin pieces inoculated with bacteria ($\log_{10}$ 3.91 CFU/cm^2) significantly reduced the bacterial count after both 24 and 48 h at 4°C. Antibacterial effects of phages P22 and 29C used at higher MOIs (between 100 and 1,000) were highly significant as early as after 30 min. When the level of bacterial contamination was low (between 10 and 10^2 CFU/cm^2 of skin) and phage 29C was used at a MOI of 10,000, the bacterial counts were reduced to undetectable levels within 30 min. Furthermore, application of this phage at a very high MOI (10^7) allowed killing of bacteria from two *Salmonella* strains which were resistant to the phage due to restriction, but to which the phage could attach. In another study phage used at a MOI of 5×10^3 reduced the S. Enteritidis count on chicken skin pieces from an initial value of 10^3–10^4 CFU/cm^2 by $\log_{10}$ 2.5 CFU/cm^2 within 7 days at 8°C (Kang *et al.*, 2013).

Hooton *et al.* (2011) applied a phage cocktail to reduce experimental S. Typhimurium contamination of pig skin. Skin pieces were inoculated with different doses of bacterial cells and phages to obtain MOI values ranging between 0.01 and 10,000, and stored for 96 h at 4°C. The efficacy of the cocktail was MOI dependent, MOI of 10 being the minimal value necessary for inducing significant reductions in the *Salmonella* counts. The highest efficacy of phages was found when the cocktail was applied at MOIs of 10 and 100, and the initial level of bacterial contamination was low (10^3 CFU/piece); in these cases phages reduced the bacterial counts to undetectable levels.

Escherichia coli

E. coli is one of the most frequent food-borne pathogens (Newell *et al.*, 2010; Scallan *et al.*, 2011a). In the context of food-borne infections, of particular importance is *E. coli* O157:H7, the most common type of Shiga toxin-producing *E. coli* (STEC) and the most common serotype causing haemolytic–uraemic syndrome (HUS). It is estimated that there are an average of 17 outbreaks 75,000 cases of illness, and 60 deaths due to *E. coli* O157:H7 each year in the USA. A major reservoir of this serotype is cattle. Although outbreaks are most frequently caused by consumption of undercooked chopped meat, recently they have been increasingly associated with contaminated vegetables and fruit juice (Bavaro *et al.*, 2012).

Several studies were conducted to evaluate a possibility of using coliphages to prevent food-borne infections due to *E. coli*. In these studies, antibacterial effects of phages were investigated in different foods (O'Flynn *et al.*, 2004; Abuladze *et al.*, 2008; Sharma *et al.*, 2009; Carter *et al.*, 2012; Jassim *et al.*, 2012; Tomat *et al.*, 2013), as well as on standard materials used in food contact surfaces (Sharma *et al.*, 2005; Viazis *et al.*, 2011b) and on food-processing equipment (Patel *et al.*, 2011).

One of the most well-studied bacteriophage preparations against *E. coli* is the cocktail ECP-100 (currently bearing the trade name EcoShield™) that contains three coliphages (Abuladze *et al.*, 2008; Sharma *et al.*, 2009; Carter *et al.*, 2012). Application of one dose of this cocktail (10^9 PFU/ml) significantly reduced experimental contamination by *E. coli* O157:H7 of tomato, spinach, broccoli, and ground beef during their storage at 10°C for 24 h, 120 h, and 168 h, the reductions in the bacterial counts being within the range between 94–100% compared to the control (Abuladze *et al.*, 2008). ECP-100 also significantly reduced the *E. coli* O157:H7 counts on fresh-cut cantaloupe pieces at 4 and 20°C and on lettuce pieces at 4°C (Sharma *et al.*, 2009). In another study, Carter *et al.* (2012) showed that one dose of EcoShield™ (3×10^6 PFU/g) applied to beef samples contaminated with *E. coli* (2×10^3 CFU/g) reduced the bacterial count by 94% after 5 min. of incubation at 4°C; similar reductions in the *E. coli* levels (by 94–98%) were also found for other incubation times (15–1440 min.). Importantly, the initial reduction in the *E. coli* count following the application of a single dose of EcoShield™ was maintained for at least 7 days of beef storage at 4°C. However, one dose of EcoShield™ failed to protect against recontamination when beef samples were recontaminated with *E. coli* 24 h after the application of phages. The lack of continued protection by EcoShield™ did not result from a reduction in the phage concentration, since the level of phages in beef samples after the application of one dose of EcoShield™ was not significantly reduced throughout the study period; it was also not due to the development of phage resistance, since no phage-insensitive bacterial cells were found in phage-treated beef samples. In addition, this study showed that EcoShield™ can significantly reduce the *E. coli* count on lettuce in a concentration-dependent manner.

Jassim *et al.* (2012) conducted a comprehensive study to evaluate the efficacy of phages in controlling the growth of *E. coli* on both naturally and artificially contaminated lettuce, cabbage, ground meat, and eggs. Two phage preparations were used in the study: the first contained CEH-162 coliphage and the other was a cocktail containing reportedly 140 coliphages. These were applied either for immersion of contaminated food samples (10^3 PFU/ml) or in the form of spray (10^5 PFU/ml). In the first set of experiments artificially contaminated lettuce samples (10^5, 10^7, or 10^9 CFU/200 g) were immersed in buffer containing CEH-162 and incubated for up to 48 h at 4°C, 24°C, or 37°C. In all experimental variants significant reductions in the *E. coli* counts were found; in particular, after 48 h a complete eradication of bacteria at 4°C and 24°C was observed regardless of the initial bacterial load. CEH-162 was also effective in reducing the level of bacteria in artificially inoculated ground meat (10^5 CFU/sample), especially at 4°C (complete eradication of bacteria was found after 48 h). At 24°C, lower but significant reductions in the *E. coli* counts were noted. The second set of experiments aimed to evaluate the efficacy of a cocktail containing reportedly 140 coliphages used for immersion of naturally contaminated eggs, lettuce, and cabbage samples (the initial level of *E. coli* in these experiments was at least 10^3 CFU/200 g). In general, phages were most effective in controlling the bacterial growth on eggs, then cabbage and lettuce. The highest efficacy of the cocktail was found at 4°C, then 24°C and 37°C.

O'Flynn *et al.* (2004) examined antibacterial effects of phages specific for *E. coli* O157:H7 on beef meat. When a phage cocktail composed of three phages (2×10^8 PFU) was pipetted onto nine pieces of meat inoculated with 2×10^2 bacterial cells each, no bacteria were detectable in seven pieces and in another two pieces the *E. coli* counts were less than 10 CFU/ml after 1 h, while in the control meat pieces the *E. coli* counts increased to $\sim 10^5$ CFU/ml.

Another study showed that coliphages used either individually or as a cocktail at a MOI of 2.3×10^5 can efficiently lyse *E. coli*, including *E. coli* O157:H7 in milk at 37°C; in the case of some bacterial strains the reductions in the *E. coli* count to undetectable levels within 1 h were found (Tomat *et al.*, 2013). While antibacterial effects of the phages against other *E. coli* strains were relatively weaker (reductions lower than 1 log CFU/ml after 24 h), this may have resulted from use of suboptimal MOI values (3.5×10^3).

It was also shown that a combination of coliphage and *trans*-cinnamaldehyde (a plant-derived essential oil) was more effective in controlling the growth of *E. coli* O157:H7 on lettuce and spinach leaves than either agent used alone. Higher efficacy of this combination was found at all temperatures (4, 8, 23 and 37°C) and all bacterial inoculum sizes (10^4, 10^5 and 10^6 CFU/ml; Viazis *et al.*, 2011a).

Moreover, there are studies to indicate that coliphages can be used as biosanitizing agents on standard materials used in food contact surfaces (Sharma *et al.*, 2005; Viazis *et al.*, 2011b) and on food processing equipment (Patel *et al.*, 2011). For example, Sharma *et al.* (2005) showed that coliphage KH1 can eliminate *E. coli* O157:H7 cells attached to stainless steel coupons with high efficiency comparable to that found in the case of planktonic bacteria. Initial populations of bacteria attached to coupons (2.6 log CFU/coupon) were reduced by 1.2 log CFU/coupon following the treatment with KH1 (7.7 log PFU/ml) for 4 days at 4°C. However, *E. coli* cells in biofilms formed on coupons were generally less susceptible to lysis by bacteriophage. Viazis *et al.* (2011b) investigated antibacterial effects of a phage cocktail BEC8 against *E. coli* O157:H7 on stainless steel chips (SSCs), ceramic tile chips (CTCs), and high density polyethylene chips (HDPECs). Chips were inoculated with 10^4, 10^5, or 10^6 CFU/chip, treated with the cocktail (10^6 PFU/chip), and incubated for 10 min, 1 h, or 24 h at 4, 12, 23, or 37°C. BEC8 significantly reduced the counts of both dried bacterial cells and bacteria in liquid cultures on SSCs, CTCs, and HDPECs, with the efficacy of phages increasing with increasing time of incubation, temperature, and MOI values. In the majority of experimental conditions the reductions in the *E. coli* counts by at least 1 log CFU were found, and in some cases complete eradication of bacteria was achieved. No phage-resistant bacterial cells were identified in this study following the exposure of *E. coli* to BEC8.

Moreover, Abuladze *et al.* (2008) reported high efficacy of the ECP-100 cocktail in reducing experimental contamination by *E. coli* O157:H7 of glass coverslips and gypsum boards (reductions in the bacterial counts within the range 85–99.99% depending on the phage dose).

In another study one dose of an *E. coli*-specific phage cocktail used in the form of spray (8 log PFU/ml) reduced the initial population of *E. coli* O157:H7 attached to spinach harvester blades (4.87 log CFU/blade) by 4.5 log CFU/blade within 2 h, while on the control blades, the *E. coli* counts were 4.4 log CFU/blade after 2 h (Patel *et al.*, 2011). The results of this study suggest that bacteriophages could be employed to eliminate bacteria from contaminated equipment used in harvesting which may be a source of contamination of fresh or fresh-cut produce.

Campylobacter spp.

Campylobacter spp., especially *C. jejuni*, is one of the most prevalent food-borne pathogens. The major reservoir of *Campylobacter* spp. are production animals in which bacteria asymptomatically colonize the gastrointestinal tract (Horrocks *et al.*, 2009). Campylobacteriosis, clinically manifested by gastroenteritis, is typically caused by consumption of contaminated raw or undercooked meats or by-products. A recent literature survey showed that *Campylobacter* contamination of retail poultry meats and by-products is prevalent worldwide. For example, approximately 71%, 69%, and 59% of analysed poultry meat samples in the US, UK, and Japan, respectively, were contaminated with *Campylobacter* (Suzuki and Yamamoto, 2009). However, in spite of high prevalence of campylobacteriosis, the number of studies on the use of *Campylobacter* phages as biocontrol agents at the post-harvest stage has been rather low (Atterbury *et al.*, 2003; Bigwood *et al.*, 2008; Orquera *et al.*, 2012).

Bigwood *et al.* (2008) evaluated the efficacy of *C. jejuni* phage Cj6 in lysing bacteria on cooked and raw beef at different experimental conditions. These conditions included two different temperatures (5°C and 24°C), low or high bacterial cell density (<100 CFU/cm^2 and 10^4 CFU/cm^2), as well as high or low MOI values (10^4 and 10^1). The efficiency of bacterial elimination was greatest at high values of both bacterial cell density and MOI, the reductions in the bacterial counts being 2.2 $\log_{10}/cm^2$ at 24°C and 1.5 $\log_{10}/cm^2$ after 24 h at 5°C. In experiments with either low bacterial cell density or low MOI antibacterial effects of Cj6 phage were substantially weaker.

In another study phage used at a MOI of 100 failed to reduce the *C. jejuni* and *Campylobacter coli* counts in raw chicken meat at 4°C; however, actual reason for the ineffectiveness of the phage was not identified by the authors of the study (Orquera *et al.*, 2012).

Atterbury *et al.* (2003) evaluated the capacity of ϕ2 phage to control the growth of *C. jejuni* on chicken skin. Sections of skin inoculated with different doses of *C. jejuni* (10^2, 10^4 or 10^7 PFU/section) and the phage (10^4, 10^5, or 10^7 PFU/section) were stored at 4°C or −20°C for 5 days. While the application of the lowest phage dose did not result in significant decreases in the bacterial counts, the use of the highest dose reduced the *C. jejuni* counts by $\log_{10}$ 1.1 to $\log_{10}$ 1.3 CFU and by $\log_{10}$ 2.3-$\log_{10}$ 2.5 CFU compared to the control in sections stored at 4 and −20°C, respectively. Because no increase in phage titre was found during the experiments, phages eliminated bacteria likely as a result of lysis from without (i.e. killing of bacteria without infecting and replicating in bacterial cells). No phage-resistant bacteria were identified in the study following the exposure of bacteria to phage.

Staphylococcus aureus

Food-borne illness due to *S. aureus* is caused especially by heat-stable enterotoxins that can be produced by staphylococci during their growth in different foods including meat and poultry products, eggs, dairy products, pastries, potatoes, and salads. Food-borne diseases caused by staphylococcal enterotoxins are estimated to affect over 240,000 persons in the US each year (Doyle *et al.*, 2012).

Several studies have been conducted to evaluate the efficacy of staphylococcal phages as biocontrol agents in different foods (Garcia *et al.*, 2007, 2009; Bueno *et al.*, 2012; Tabla *et al.*, 2012). Garcia *et al.* (2007) showed high effectiveness of phages Φ88 and Φ35 in controlling the growth of *S. aureus* during curd manufacturing processes. When the initial bacterial concentration was 10^6 CFU/ml, a mixture of both phages used at MOI of 250–350 completely eradicated bacteria within 4 h and 1 h of incubation during acid curd and enzymatic curd manufacturing, respectively. Likewise, in UHT whole milk the phages (MOI 100) reduced the *S. aureus* count from an initial value of 3×10^6 CFU/ml to undetectable level after 2 h of incubation at 37°C.

Garcia *et al.* (2009) evaluated antibacterial effects of a mixture of two staphylococcal phages (ΦH5 and ΦA72) in milk. The initial level of *S. aureus* contamination of milk was 10^2 CFU/ml, and the phages were applied at the concentration of 10^4–10^5 PFU/ml. It was shown that phages could lyse *S. aureus* at both 22 and 37°C, but not at 4°C. The efficacy of phages was highest in UHT and pasteurized whole-fat milk, relatively lower in semi-skimmed raw milk, and clearly decreased in whole-fat raw milk (however, in all kinds of milk the reductions in the bacterial counts were significant after 8–11 h). Antibacterial effects of phages were weaker in raw milks either due to the slower growth of *S. aureus* or because of virion inactivation being a result of aggregation mediated by some milk components (another reason for a decrease in antibacterial activity of phages in raw milk may be a reduction in phage adsorption due to staphylococcal agglutination; O'Flaherty *et al.*, 2005). Although in no experiment was the eradication of *S. aureus* achieved, the authors of the study suggested that the reductions in the bacterial counts would be sufficient to prevent the production of staphylococcal toxins to the level resulting in milk intoxication (however, this was not verified experimentally).

It was also shown that staphylococcal phages can exert synergistic antibacterial effects with high hydrostatic pressure (HHP) treatments in milk (Tabla *et al.*, 2012). Synergy between phages and HHP treatments was found at both levels of initial *S. aureus* milk contamination (1×10^4 and 1×10^6 CFU/ml). Thus the application of phages can enable one to employ milder hydrostatic pressure treatments to eliminate *S. aureus*. Another study evaluated a possibility of using staphylococcal phages in a combination with nisin (Martinez *et al.*, 2008). While during shorter incubation times in pasteurized milk phages exhibited synergistic antibacterial effects with nisin, nisin-adapted staphylococcal cells became partially resistant to phages possibly as a result of some modifications of lipoteichoic acids in the bacterial cell wall. These results indicate a need of detailed investigation of the interactions between bacteriophages and other antibacterial agents to be used in foods in combination with phage preparations.

Bueno *et al.* (2012) evaluated the effectiveness of a cocktail containing two staphylococcal bacteriophages in controlling the growth of *S. aureus* in fresh and hard-type cheeses. The initial concentration of *S. aureus* in pasteurized milks used for production of cheeses was approximately 10^6 CFU/ml, and phages were added at MOI of 1. In both kinds of cheese phages significantly reduced the bacterial counts to the extent sufficient to prevent the production of enterotoxins.

Other bacteria

Although the majority of studies on the use of phages as biocontrol agents have focused on *L. monocytogenes*, S. Enteritidis and S. Typhimurium, *E. coli*, *C. jejuni*, and *S. aureus*, there are also reports on successful application of phages against other major food-borne pathogens including *Yersinia enterocolitica* (Orquera *et al.*, 2012), *Shigella* spp. (Zhang *et al.*, 2013), *Bacillus cereus* (Bandara *et al.*, 2012), and *Cronobacter* spp. (Kim *et al.*, 2007; Zuber *et al.*, 2008). In addition, studies were conducted to evaluate antibacterial effects of phages against food spoilage bacteria (Ellis *et al.*, 1973; Greer, 1988; Greer and Dilts, 2002; Deasy *et al.*, 2011).

Yersinia enterocolitica

Y. enterocolitica is one of the most frequent food-borne pathogens (Scallan *et al.*, 2011a). The most common clinical form of food-borne illness due to *Y. enterocolitica* is gastroenteritis. The major reservoir of *Y. enterocolitica* are pigs, and pork products have been implicated in many cases of food-borne *Y. enterocolitica* infections, as well as outbreaks (Bolton *et al.*, 2013). In a recent study use of PY100 phage at a MOI of 10,000 reduced the *Y. enterocolitica* count in raw pork meat by 2 $\log_{10}$ after 24 h at 4°C (Orquera *et al.*, 2012).

Shigella spp.

Shigella spp. is the aetiological agent of shigellosis, an acute invasive enteric infection that is endemic in the majority of developing countries and occasionally can cause outbreaks in industrialized countries. Shigellosis is one of the most prevalent food-borne infections; it is estimated to affect over 160 million persons worldwide each year. *Shigella* spp. can be transmitted by a wide range of different foods including fruits, vegetables, low pH foods, prepared foods, and foods held in modified atmosphere or vacuum packaging (Warren *et al.*, 2006; Mokhtari *et al.*, 2012). Recent study showed high efficacy of phages in controlling the growth of *Shigella* spp. in RTE chicken (Zhang *et al.*, 2013). In chicken samples inoculated with a mixture of three *Shigella* species (*S. flexneri*, *S. dysenteriae* and *S. sonnei*; 3×10^4 CFU/g), a cocktail of three phages at a MOI of 10,000 reduced the bacterial count by up to 2 $\log_{10}$/g after 48 h, and after 72 h no bacteria could be detected in the treated samples. In chicken samples inoculated with one of the three species (10^4 CFU/g) phage at the same MOI significantly decreased the bacterial count within 72 h at 4°C.

Bacillus cereus

B. cereus can cause diarrhoeal syndrome following the ingestion of spores or vegetative cells that can be transmitted by a wide range of foods and produce toxins in the intestines, or emetic syndrome caused by preformed toxins present in food (Logan, 2012). Bandara *et al.* (2012) examined antibacterial effects of two *B. cereus* phages in fermented soybean paste. Higher efficacy of both phages was found upon the addition of divalent cations (Ca^{2+}, Mg^{2+} and Mn^{2+}) to the experimental milieu, which enabled the bacterial count to be reduced to an undetectable level within 24 h.

Cronobacter spp.

Cronobacter spp. (previously known as *Enterobacter sakazakii*) is an opportunistic bacterial pathogen that can cause necrotizing enterocolitis, septicaemia, and meningitis associated with a high mortality rate in neonates and infants. Although *Cronobacter* spp. has been isolated from a variety of foods, the main route of infection is the consumption of contaminated infant formula (Norberg *et al.*, 2012). So far, two studies have been conducted to evaluate the potential of phages for controlling the growth of *Cronobacter* spp. in infant formula (Kim *et al.*, 2007; Zuber *et al.*, 2008). In the first study, the initial bacterial concentration in the formula was 10^2 CFU/ml, and phages (ESP 1-3 and ESP 732-1) were used at the final concentration of 10^7, 10^8, or 10^9 PFU/ml. Both phages exerted significant antibacterial activity, their antibacterial effects being concentration and temperature dependent. The efficiency of bacterial lysis was higher at 24°C than at 37°C and 12°C; at 24°C the higher phage concentrations (10^8 and 10^9 PFU/ml) completely eradicated bacteria within 2h (Kim *et al.*, 2007). In the second study a cocktail of five phages (10^8 PFU/ml) prevented the outgrowth of 36 out of 40 *Cronobacter* spp. strains tested (initial bacterial concentration – 10^2 CFU/ml) during 4-hour incubation at 30°C in reconstituted infant formula (Zuber *et al.*, 2008).

Spoilage bacteria

While the main potential application of phages in food industry is prevention of food-borne infections, some attempts were also made to use phages to prevent food spoilage caused by bacteria (Ellis *et al.*, 1973; Greer, 1988, Greer and Dilts, 2002; Deasy *et al.*, 2011).

Ellis *et al.* (1973) showed that phage used at titres 0.1, 10, or 1000 PFU/ml could significantly reduce the *Pseudomonas* count in milk stored at 7°C only when the initial bacterial cells density was 100,000 CFU/ml, while at the initial level of bacterial contamination of 1000 CFU/ml antibacterial effects of phage were insignificant. Moreover, Greer (1988) evaluated potential of phage for controlling *Pseudomonas*-induced beef spoilage. While antibacterial activity of phage was not influenced by temperature within the range 1–10°C, it was significantly affected by the interactive effects of phage concentration and initial bacterial cells density. Significant biocontrol of beef spoilage was achieved only after crossing the threshold values of initial bacterial density and phage concentration (46 CFU/cm^2 and 9.7×10^5 PFU/cm^2, respectively). A maximum increase in the retail case life of beef steaks (from 3.4 to 6.4 days) required phage titre of 9.7×10^7 PFU/cm^2 and initial density of bacterial cells of 4.6×10^3 CFU/cm^2. Overall, in view of inadequate efficacy of phage at low bacterial cells densities the authors of the study concluded that potential of phage for biocontrol of beef spoilage is rather questionable.

Greer and Dilts (2002) evaluated the efficacy of phage A3 in controlling spoilage caused by *Brochothrix thermosphacta* on pork adipose tissue. Adipose tissue discs inoculated with *B. thermosphacta* (5.5 log CFU/cm^2) and the phage (5–6 log PFU/cm^2) were stored at 2 or 6°C. Under both storage conditions phages decreased bacterial numbers by about 2 log cycles within 1 to 2 days, resulting in suppression of offensive off-odour associated with the growth of bacteria and extension of the storage life of adipose tissue from 4 days in the untreated control samples to 8 days in phage-treated samples. However, after an initial reduction in bacterial numbers, the growth of bacteria in phage-treated adipose tissue discs continued at a rate comparable to that found in untreated discs. Resistance to phages developed rapidly, since 20% to 65% of bacteria isolated from adipose tissue discs coinoculated with bacteria and phages were resistant to phages, while no phage-resistant bacteria could be isolated from discs inoculated with bacteria only.

Another study (Deasy *et al.*, 2011) evaluated antibacterial effects of phages against *Lactobacillus brevis*, the major bacterial species implicated in beer spoilage. Beer samples were inoculated with 10^3, 10^4 or 10^5 CFU/ml of bacteria and the phage (10^8 PFU). At the lowest level of bacterial contamination, the phage completely eradicated bacteria within 24h of incubation at 30°C, while at the higher levels of contamination bacteria were completely eradicated after 48h. Importantly, no regrowth of *L. brevis* was found by 96h.

Main factors that determine the efficacy of phages as biocontrol agents

Bacterial cells density

One of the most important factors that determine the efficacy of phages as biocontrol agents is the level of bacterial contamination of food. In particular, phages will be able to efficiently eliminate bacteria only when the density of bacterial cells crosses a critical threshold, which is very important because in natural settings the level of bacterial contamination of food is usually very low (Hudson *et al.*, 2005; Greer, 2005; Hagens and Loessner, 2010). In the study by Greer (1988) the threshold value of bacterial cells density required for phage to induce significant antibacterial effects was 46 CFU/cm^2. Ellis *et al.* (1973) found that while phage multiplication in contaminated milk occurred both at high and low bacterial cells level (100,000 and 1000 CFU/ml, respectively), only in the former case was a decrease in the bacterial count in milk significant. Zuber *et al.* (2008) showed that at low initial levels of bacteria and phages in broth (10^2 CFU/ml and 10^4 PFU/ml, respectively), phage multiplication and lysis occurred only upon crossing by bacteria a threshold of 10^4–10^5 CFU/ml; however, even then phages were unable to completely eradicate bacteria. In another study, antibacterial effects of phages were weaker when the bacterial cell density was below 100 CFU/cm^2 compared with the level of contamination of 10^4 CFU/cm^2 (Bigwood *et al.*, 2008). Soni and Nannapaneni (2010a) showed that the effectiveness of phages was reduced with decreasing the density of bacterial cells within the range of 4–2 log CFU/g. However, Bigot *et al.* (2011) reported that phages were most effective when the density of bacterial cells was as low as 10^2 CFU/cm^2. It was also shown that for a low level of contamination, a significant reduction in the bacterial count (Bigwood *et al.*, 2008) or even bacterial eradication (Goode *et al.*, 2003; Kim *et al.*, 2007; Zuber *et al.*, 2008) can be achieved by using a high phage dose. Furthermore, in two other studies it was a combination of a low level of bacteria with a high phage dose that resulted in the maximal biocontrol efficacy (Hooton *et al.*, 2011; Guenther *et al.*, 2011). Thus, based on the results of the above studies, an important conclusion can be drawn that phage will likely be able to significantly reduce the bacterial count in foods even at low initial level of bacterial contamination provided that a sufficiently high number of phage particles are used.

Phage concentration

Several studies have consistently shown that the efficiency of antibacterial effects of bacteriophages is dependent on phage concentration in a food product (Leverentz *et al.*, 2004; Carlton *et al.*, 2005; Kim *et al.*, 2007; Zuber *et al.*, 2008; Guenther *et al.*, 2009, 2011; Soni and Nannapaneni, 2010b). For example, in the study by Soni and Nannapaneni (2010b) the efficacy of phage rose with increasing the concentration of phage particles from 10^4 to 10^8 PFU/g. Likewise, in another study phage concentration required for maximal effectiveness was at least 10^8 PFU/g (Guenther *et al.*, 2009). Kim *et al.* (2007) showed that the minimal phage concentration necessary for complete eradication of bacteria was 10^8 PFU/ml. Zuber *et al.* (2008) reported that phage at a concentration of 10^8 PFU/ml could eradicate bacteria in broth regardless of an initial bacterial cells density, while the effects obtained by using phage at a concentration of 10^4 PFU/ml were substantially weaker. Thus, it is a consistent finding that a phage concentration around 10^8 PFU/g or ml is required for optimal antibacterial effects.

The number of phage particles can also be expressed not as an absolute concentration of phages in a food product, but rather as a ratio of the number of phage particles to bacterial cells. This ratio is commonly expressed as MOI value. While some authors suggested that the PFU/CFU ratio is more relevant to the application of phages as biocontrol agents (Brovko *et al.*, 2012), it is the term MOI that is used much more frequently. Several studies have shown that antibacterial effects of phages are also MOI-dependent (Atterbury *et al.*, 2003; Bigwood *et al.*, 2008; Whichard *et al.*, 2003; Goode *et al.*, 2003; Hooton *et al.*, 2011). For example, Hooton *et al.* (2011) showed that a minimal MOI value required for significant antibacterial effects of phages was 10, while in the study by Whichard *et al.* (2003) phages were effective at MOI of at least 100.

Phage stability in foods

Phage stability was investigated in a wide range of foods including beef (Carter *et al.*, 2012), cheese (Soni *et al.*, 2012), milk (Garcia *et al.*, 2009), salmon (Soni and Nannapaneni, 2010b), fresh channel catfish fillets (Soni *et al.*, 2010), vegetables (Guenther *et al.*, 2009), as well as melons and apples (Leverentz *et al.*, 2001, 2003). In general, these studies showed that phages are fairly stable on different foods except for some vegetables (Guenther *et al.*, 2009) and apples (Leverentz *et al.*, 2001,2003). Rapid inactivation of phages on apples is likely caused by a low pH (Leverentz *et al.*, 2003), while their relatively lower stability on vegetables may result from inactivation of phage particles by organic acids or tannins, compounds known to adversely affect viruses (Guenther *et al.*, 2009). Recent study showed that phages retain significant antibacterial activity in broth at different experimental conditions relevant to the food chain environment including pH values ranging between 4 and 9, temperatures between 4 and 37°C, water activity values of 0.87 or 0.91 to 1.00, and NaCl concentrations of 1–2.5% (Coffey *et al.*, 2011). A more general discussion of the effects of different physicochemical factors including temperature, pH, salinity, and ions on phage stability was recently published by Jończyk *et al.* (2011).

Since bacteriophages are most likely to be used in combination with other antibacterial agents and means of prevention of food-borne infections, an important question is whether they can adversely affect phage particles. Unfortunately, relevant data are quite fragmentary. Soni *et al.* (2012) showed that neither LAE or PL-SD inactivated phage viability during co-incubation in broth. In the study by Tabla *et al.* (2012) high hydrostatic pressure treatments up to 400 MPa did not adversely affect phage viability. Roy *et al.* (1993) found that a quaternary ammonium compound at a concentration up to 50 ppm did not inactivate phage. However, most chemical sanitizers adversely affect phage particles, which needs to be taken into account when designing phage biocontrol protocols (Sulakvelidze, 2013).

Phage resistance

One of the main factors that can limit the efficacy of phages as biocontrol agents is the development of phage resistance (Hudson *et al.*, 2005; Greer, 2005; Hagens and Loessner, 2010). General mechanisms evolved by bacteria to prevent phage infections are discussed in detail in Chapter 4. In the context of food applications of phages of particular threat to long-term effectiveness of phage preparations would be situations in which resistant bacteria could prevail in environmental niches with high phage selective pressure; such bacteria could be a source of contamination of foods. On the other hand, the development of phage resistance in bacteria present on a food product that leaves a production facility shortly after phage treatment would not pose a high risk of reducing the efficacy of phages in this facility, especially when the level of bacterial contamination is low, as likely would be the case in natural settings (Hagens and Loessner, 2010). The use of bacteriophages at such 'epidemiological endpoints' in the food processing chain was proposed as a strategy of reducing a risk of development of bacterial resistance to phages (Sulakvelidze, 2013). Moreover, the use of phage cocktails rather than preparations containing single phages may lower a risk of development of phage resistance (Sulakvelidze, 2013). However, should resistance to phages develop, bacteriophage preparations could be updated by inclusion of new phages capable of killing resistant bacteria; as mentioned in the section 'Bacteriophages as biocontrol agents', FDA indeed permitted one of the approved cocktails to be updated.

Other factors

Several other factors need to be taken into consideration when discussing phage biocontrol. For example, antibacterial effects of phages are temperature-dependent, so the efficacy of biocontrol will vary depending on temperature (Leverentz *et al.*, 2001; Jassim *et al.*, 2012). In particular, an important question is whether phages can eliminate bacteria at temperatures non-permissive for their growth (Hudson *et al.*, 2005). Although some authors reported that phages significantly reduced the bacterial counts in foods stored at refrigeration temperatures (Leverentz *et al.*, 2001; Guenther *et al.*, 2009; Modi *et al.*, 2009; Soni *et al.*, 2010, 2012), in fact this may have resulted not from killing of bacteria during food storage, but rather from lysing them over bacterial counting procedure by phages which earlier only attached to bacterial cells (Hudson *et al.*, 2005). However, inability of phages to replicate at low temperatures does not preclude their use to control the growth of bacteria in foods that are kept at refrigeration temperatures. First, phages could lyse bacteria when a refrigerated food product begins to warm (Hudson *et al.*, 2005). Second, in some cases phages can kill bacteria as a result of lysis from without which can occur even at low temperatures (Atterbury *et al.*, 2003).

The efficacy of phage biocontrol also varies between different food products (Leverentz *et al.*, 2001, 2003; Guenther *et al.*, 2009, 2012; Jassim *et* al., 2012). Factors that may reduce antibacterial activity of phages in some foods include unfavourable pH (Leverentz *et al.*, 2003), the presence of compounds which adversely affect phage particles (Guenther *et al.*, 2009), non-specific interactions between phages and non-host bacteria or some particulate material present in a food product (Greer, 2005; Hudson *et al.*, 2005), as well as poorer accessibility of targeted bacteria and smaller volume of fluid necessary for phage diffusion (Hagens and Loessner, 2010).

Can phages adversely affect any properties of foods?

In the context of use of phages as biocontrol agents, an important question is whether they can adversely affect any properties of foods. This problem has been addressed by several studies, the majority of them focusing on the effects of phage preparations on cheese. These studies showed that phage application has no significant influence on macroscopic features of cheese (general appearance, colour, and texture), physicochemical characteristics (pH, fat, protein, salt, and moisture content), or the cheese microflora including bacteria, yeasts and moulds. Phages also did not seem to adversely affect the course of the ripening process (Modi *et al.*, 2001; Carlton *et al.*, 2005; Guenther *et al.*, 2011; Bueno *et al.*, 2012). Another study showed that phage application does not reduce the yield of mung bean sprouts and does not negatively influence the sprout microflora (Ye *et al.*, 2010). Thus available experimental data suggest that phage preparations should not adversely affect properties of foods. These results are not unexpected when one takes into account that phages are naturally present in large numbers in different foods (Hudson *et al.*, 2005, and references therein).

Concluding remarks

Bacteriophages have several features which make them novel antibacterials with which to combat food-borne pathogens. In particular, narrow antibacterial range of phages provides a unique opportunity to eliminate specific food-borne bacteria (regardless of their antibiotic-resistance) without adversely affecting bacteria from food microflora. The use of broad range bacteriophages or phage cocktails enables one to obtain preparations with whole bacterial species coverage.

Over the last decade the use of phage preparations to improve food safety has emerged as a major biomedical application of bacteriophages. So far, the majority of studies have focused on the use of phages against *L. monocytogenes*, S. Enteritidis and S. Typhimurium, *E. coli*, especially *E. coli* O157:H7, *C. jejuni*, and *S. aureus*. There are also reports on successful use of phages against other major food-borne pathogens, especially *Y. enterocolitica*, *Shigella* spp., *B. cereus*, and *Cronobacter* spp. In addition, some attempts were made to prevent bacteria-mediated food spoilage using phages. It can be expected that bacteriophages against practically any bacteria can be fairly easily isolated from different environmental samples or other sources. Thus, phage preparations can be developed to eliminate any major food-borne bacterial pathogen.

In general, most studies have shown relatively high efficacy of phage preparations in controlling the growth of bacteria in a variety of foods (as well as on standard materials used in food contact surfaces), the most frequent result of phage application being a significant, log-fold reduction in the bacterial count. While such reductions do not eliminate completely a risk of food-borne infection, they may substantially decrease it (Hagens and Loessner, 2010; Sulakvelidze, 2013). Furthermore, in some cases (e.g. in *S. aureus*) a significant decrease in the level of bacteria can reduce the risk of production of toxins implicated in food-borne intoxication (Garcia *et al.*, 2009; Bueno *et al.*, 2012). However, several factors need to be taken into account when interpreting the results of studies on the use of phages as biocontrol agents. First, experimental conditions (especially temperature and the host cells density) employed in some studies substantially differ from those that occur in natural settings (Modi *et al.*, 2001; Whichard *et al.*, 2003; Abuladze *et al.*, 2008). Second, some authors reported a lack of efficacy of phages in foods (Dykes and Moorhead, 2002; Kocharunchitt *et al.*, 2009; Ye *et al.*, 2009); unfortunately actual reasons for inadequate antibacterial effects of phages found in these studies have not been identified. Of particular concern is a lack of long-term protection of food after use of a single dose of a phage preparation (Modi *et al.*, 2001; Guenther *et al.*, 2011; Carter *et al.*, 2012). However, this problem could be solved by using repeated doses of bacteriophages (Carlton *et al.*, 2005) or phage preparations in a combination with other antibacterial agents, especially those that ensure long-lasting antimicrobial effects (Holck and Berg, 2009; Soni *et al.*, 2012).

A number of studies have shown that both preparations containing single phages (Carlton *et al.*, 2005; Guenther *et al.*, 2009; Soni and Nannapaneni, 2010b) and phage cocktails (Leverentz *et al.*, 2001, 2003; Abuladze *et al.*, 2008) can be successfully used to control the growth of pathogenic bacteria both in foods and on materials used in food contact surfaces. However, studies in which single phages were compared with cocktails consistently showed that these are cocktails that have higher efficacy (Roy *et al.*, 1993; Kudva *et al.*, 1999; Tomat *et al.*, 2013). The main advantages of cocktails are a broader antibacterial range and a lower risk of development of phage resistance (Chan *et al.*, 2013). Thus it appears that phage cocktails are the preferable form of preparations intended for food applications.

It was also shown that phages can exert additive effects with other antibacterial agents and means of antimicrobial therapy including antagonistic bacteria (Ye *et al.*, 2010; Holck and Berg, 2009), bacteriocins (Dykes and Moorhead, 2002; Leverentz *et al.*, 2003; Martinez *et al.*, 2008), LAE and PL–SD mixture (Soni *et al.*, 2012), essential oils (Viazis *et al.*, 2011a), quaternary ammonium compounds (Roy *et al.*, 1993), as well as with HHP (Tabla *et al.*, 2012). This is an additional argument for inclusion of phages to 'hurdle technology' based on simultaneous use of different methods of improving microbial food safety.

In the context of use of phages to prevent food-borne infections, it is also very important that phages do not seem to adversely affect any properties of food (Modi *et al.*, 2001; Carlton *et al.*, 2005; Guenther *et al.*, 2011; Bueno *et al.*, 2012). This distinguishes phages from some other antibacterial agents and technologies used to eliminate food-borne bacteria which can adversely affect organoleptic properties of foods.

Although the development of bacterial resistance to phages could decrease long-term efficacy of bacteriophage preparations, some strategies have been proposed to minimize this risk. In this context, an important development is permission granted by FDA to update one of the phage cocktails approved for food applications; such updating should substantially improve long-term effectiveness of phages.

Acknowledgements

This work was supported by funds from the Operational Program Innovative Economy, 2007–2013, Priority axis 1. Research and Development of Modern Technologies, Measure 1.4 Support for goal-oriented projects, project No. UDA-POIG.01.04.00-24-133/11-00 entitled 'The usefulness of specific bacteriophages against drug resistant bacterial strains in animal herds' and from project No. INNOTECH-K2/IN/7/181844/NCBR/13 entitled 'Microbiological stabilization of drinking water'.

References

Abuladze, T., Li, M., Menetrez, M.Y., Dean, T., Senecal, A., and Sulakvelidze, A. (2008). Bacteriophages reduce experimental contamination of hard surfaces, tomato, spinach, broccoli, and ground beef by *Escherichia coli* O157:H7. Appl. Environ. Microbiol. *74*, 6230–6238.

Ahn, D.U., Kim, I.S., and Lee, E.J. (2013). Irradiation and additive combinations on the pathogen reduction and quality of poultry meat. Poult. Sci. *92*, 534–545.

Allerberger, F., and Wagner, M. (2010). Listeriosis: a resurgent food-borne infection. Clin. Microbiol. Infect. *16*, 16–23.

Anany, H., Chen, W., Pelton, R., and Griffiths, M.W. (2011). Biocontrol of *Listeria monocytogenes* and *Escherichia coli* O157:H7 in meat by using phages immobilized on modified cellulose membranes. Appl. Environ. Microbiol. *77*, 6379–6387.

Anonymous. (2011). Vital signs: incidence and trends of infection with pathogens transmitted commonly through food – food-borne diseases active surveillance network, 10 U.S. sites, 1996–2010. MMWR Morb. Mortal. Wkly. Rep. *60*, 749–755.

Anonymous. (2013). Vital signs: listeria illnesses, deaths, and outbreaks – United States, 2009–2011. MMWR Morb. Mortal. Wkly. Rep. *62*, 448–452.

Atterbury, R.J., Connerton, P.L., Dodd, C.E.R., Rees, C.E.D., and Connerton, I.F. (2003). Application of host-specific bacteriophages to the surface of chicken skin leads to a reduction in recovery of *Campylobacter jejuni*. Appl. Environ. Microbiol. *69*, 6302–6306.

Bandara, N., Jo, J., Ryu, S., and Kim, K.-P. (2012). Bacteriophages BCP1-1 and BCP8-2 require divalent cations for efficient control of *Bacillus cereus* in fermented foods. Food Microbiol. *31*, 9–16.

Batz, M.B., Hoffmann, S., and Morris, J.G. Jr. (2012). Ranking the disease burden of 14 pathogens in food sources in the United States using attribution data from outbreak investigations and expert elicitation. J. Food Prot. *75*, 1278–1291.

Bavaro, M.F. (2012). *E. coli* O157:H7 and other toxigenic strains: the curse of global food distribution. Curr. Gastroenterol. Rep. *14*, 317–323.

Berger, C.N., Sodha, S.V., Shaw, R.K., Griffin, P.M., Pink, D., Hand, P., and Frankel, G. (2010). Fresh fruit and vegetables as vehicles for the transmission of human pathogens. Environ. Microbiol. *12*, 2385–2397.

Bigot, B., Lee, W.-J., McIntyre, Wilson, T., Hudson, J.A., Billington, C., and Heinemann, J.A. (2011). Control of *Listeria monocytogenes* growth in ready-to-eat poultry product using a bacteriophage. Food Microbiol. *28*, 1448–1452.

Bigwood, T., Hudson, J.A., Billington, C., Carey-Smith, G.V., and Heinemann, J.A. (2008). Phage inactivation of food-borne pathogens on cooked and raw meat. Food Microbiol. *25*, 400–406.

Bigwood, T., Hudson, J.A., and Billington, C. (2009). Influence of host and bacteriophage concentrations on the inactivation of food-borne pathogenic bacteria by two phages. FEMS Microbiol. Lett. *291*, 59–64.

Bolton, D.J., Ivory, C., and McDowell, D. (2013). Thermal inactivation of Yersinia enterocolitica in pork slaughter plant scald tank water. Meat Sci. *95*, 668–671.

Brovko, L.Y., Anany, H., and Griffiths, M.W. (2012). Bacteriophages for detection and control of bacterial pathogens in food and food-processing environment. Adv. Food Nutr. Res. *67*, 241–288.

Bueno, E., Garcia, P., Martinez, B., and Rodriguez, A. (2012). Phage inactivation of *Staphylococcus aureus* in fresh and hard-type cheeses. Int. J. Food Microbiol. *158*, 23–27.

Carlton, R.M., Noordman, W.H., Biswas, B., de Meester, E.D., and Loessner, M.J. (2005). Bacteriophage P100 for control of *Listeria monocytogenes* in foods: genome sequence, bioinformatics analyses, oral toxicity study, and application. Reg. Toxicol. Pharmacol. *43*, 301–312.

Carter, C.D., Parks, A., Abuladze, T., Li, M., Woolston, J., Magnone, J., Senecal, A., Kropinski, A.M., and Sulakvelidze, A. (2012). Bacteriophage cocktail significantly reduces *Escherichia coli* O157:H7 contamination of lettuce and beef, but does not protect against recontamination. Bacteriophage 2, 175–185.

Chan, B.K., Abedon, S.T., and Loc-Carrillo, C. (2013). Phage cocktails and the future of phage therapy. Future Microbiol. *8*, 769–783.

Coffey, B., Mills, S., Coffey, A., McAuliffe, O., and Ross, R.P. (2010). Phage and their lysins as biocontrol agents for food safety applications. Annu. Rev. Food. Sci. Technol. *1*, 449–468.

Coffey, B., Rivas, L., Duffy, G., Coffey, A., Ross, R.P., and McAuliffe, O. (2011). Assessment of *Escherichia coli* O157:H7-specific bacteriophages e11/2 and e4/1c in model broth and hide environments. Int. J. Food Microbiol. *147*, 188–194.

DuPont, H.L. (2007). The growing threat of food-borne bacterial enteropathogens of animal origin. Clin. Infect. Dis. *45*, 1353–1361.

Deasy, T., Mahony, J., Neve, H., Heller, K.J., van Sinderen, D. (2011). Isolation of a virulent Lactobacillus brevis phage and its application in the control of beer spoilage. J. Food Prot. *74*, 2157–2161.

Doyle, M.E., Hartmann, F.A., and Lee Wong, A.C. (2012). Methicillin-resistant staphylococci: implications for our food supply? Anim. Health Res. Rev. *13*, 157–180.

Dykes, G.A., and Moorhead, S.M. (2002). Combined antimicrobial effect of nisin and a listeriophage against *Listeria monocytogenes* in broth but not in buffer or on raw beef. Int. J. Food Microbiol. 73, 71–81.

Ellis, D.E., Whitman, P.A., and Marshall, R.T. (1973). Effects of homologous bacteriophage on growth of *Pseudomonas fragi* WY in milk. Appl. Microbiol. *25*, 24–25.

Fatica, M.K., and Schneider, K.R. (2011). Salmonella and produce: survival in the plant environment and implications in food safety. Virulence. *2*, 573–579.

Garcia, P., Madera, C., Martinez, B., and Rodriguez, A. (2007). Biocontrol of *Staphylococcus aureus* in curd manufacturing processes using bacteriophages. Int. Dairy J. *17*, 1232–1239.

García P., Martínez, B., Obeso, J.M., and Rodríguez, A. (2008). Bacteriophages and their application in food safety. Lett. Appl. Microbiol. *47*, 479–485.

García, P., Madera, C., Martínez, B., Rodríguez, A., and Evaristo Suárez, J. (2009). Prevalence of bacteriophages infecting *Staphylococcus aureus* in dairy samples and their potential as biocontrol agents. J. Dairy Sci. *92*, 3019–3026.

Goode, D., Allen, V.M., and Barrow, P.A. (2003). Reduction of experimental *Salmonella* and *Campylobacter* contamination of chicken skin by application of lytic bacteriophages. Appl. Environ. Microbiol. *69*, 5032–5036.

Goodridge, L.D., and Bisha, B. (2011). Phage-based biocontrol strategies to reduce food-borne pathogens in foods. Bacteriophage. *1*, 130–137.

Górski, A., Miedzybrodzki, R., Borysowski, J., Weber-Dabrowska, B., Lobocka, M., Fortuna, W., Letkiewicz, S., Zimecki, M., and Filby, G. (2009). Bacteriophage therapy for the treatment of infections. Curr. Opin. Investig. Drugs. *10*, 766–774.

Greer, G.G. (1988). Effects of phage concentration, bacterial density, and temperature on phage control of beef spoilage. J. Food Sci. 53, 1226–1227.

Greer, G.G., and Dilts, B.D. (2002). Control of *Brochothrix thermosphacta* spoilage of pork adipose tissue using bacteriophages. J. Food Prot. *65*, 861–863.

Guenther, S., and Loessner, M.J. (2011). Bacteriophage biocontrol of *Listeria monocytogenes* on soft ripened white mold and red-smear cheeses. Bacteriophage. *1*, 94–100.

Guenther, S., Huwyler, D., Richard, S., and Loessner, M.J. (2009). Virulent bacteriophage for efficient biocontrol of *Listeria monocytogenes* in ready-to-eat foods. Appl. Environ. Microbiol. 75, 93–100.

Guenther, S., Herzig, O., Fieseler, L., Klumpp, J., and Loessner, M.J. (2012). Biocontrol of *Salmonella*

Typhimurium in RTE foods with the virulent bacteriophage F01-E2. Int. J. Food Microbiol. *154*, 66–72.

Hagens, S., and Loessner, M.J. (2010). Bacteriophage for biocontrol of food-borne pathogens: calculations and considerations. Curr. Pharm. Biotechnol. *11*, 58–68.

Hibma, A.M., Jassim, S.A.A., and Griffiths, M.W. (1997). Infection and removal of L-forms of *Listeria monocytogenes* with bred bacteriophage. Int. J. Food Microbiol. *34*, 197–207.

Holck, A., and Berg, J. (2009). Inhibition of *Listeria monocytogenes* in cooked ham by virulent bacteriophages and protective cultures. Appl. Environ. Microbiol. *75*, 6944–6946.

Hooton, S.P., Atterbury, R.J., and Connerton, I.F. (2011). Application of a bacteriophage cocktail to reduce Salmonella Typhimurium U288 contamination on pig skin. Int. J. Food Microbiol. *151*, 157–163.

Horrocks, S.M., Anderson, R.C., Nisbet, D.J., and Ricke, S.C. (2009). Incidence and ecology of Campylobacter jejuni and coli in animals. Anaerobe. *15*, 18–25.

Hudson, J.A., Billington, C., Carey-Smith, G., and Greening, G. (2005). Bacteriophages as biocontrol agents in food. J. Food Prot. *68*, 426–437.

Jassim, S.A.A., Abdulamir, A.S., and Abu Bakar, F. (2012). Novel phage-based bio-processing of pathogenic *Escherichia coli* and its biofilms. World J. Microbiol. Biotechnol. *28*, 47–60.

Jończyk, E., Kłak, M., Międzybrodzki, R., and Górski, A. (2011). The influence of external factors on bacteriophages – review. Folia Microbiol. (Praha). *56*, 191–200.

Kang, H.W., Kim, J.W., Jung, T.S., and Woo, G.J. (2013). Wksl3, a new biocontrol agent for *Salmonella enterica* serovars enteritidis and typhimurium in foods: characterization, application, sequence analysis, and oral acute toxicity study. Appl. Environ. Microbiol. *79*, 1956–1968.

Kim, K.-P., Klumpp, J., and Loessner, M.J. (2007). *Enterobacter sakazakii* bacteriophages can prevent bacterial growth in reconstituted infant formula. Int. J. Food Microbiol. *115*, 195–203.

Kocharunchitt, C., Ross, T., and McNeil, D.L. (2009). Use of bacteriophages to control *Salmonella* associated with seed sprouts. Int. J. Food Microbiol. *128*, 453–459.

Kudva, I.T., Jelacic, S., Tarr, P.I., Youderian, P., and Hovde, C.J. (1999). Biocontrol of *Escherichia coli* O157 with O157-specific bacteriophages. Appl. Environ. Microbiol. *65*, 3767–3773.

Leistner, L. (2000). Basic aspects of food preservation by hurdle technology. Int. J. Food Microbiol. *55*, 181–186.

Leverentz, B., Conway, W.S., Alavidze, Z., Janisiewicz, W.J., Fuchs, Y., Camp, M.J., Chighladze, E., and Sulakvelidze, A. (2001). Examination of bacteriophage as a biocontrol method for *Salmonella* on fresh-cut fruit: a model study. J. Food. Prot. *64*, 1116–1121.

Leverentz, B., Conway, W.S., Camp, M.J., Janisiewicz, W.J., Abuladze, T., Yang, M., Saftner, R., and Sulakvelidze, A. (2003). Biocontrol of *Listeria monocytogenes* on fresh-cut produce by treatment with lytic bacteriophages and a bacteriocin. Appl. Environ. Microbiol. *69*, 4519–4526.

Leverentz, B., Conway, W.S., Janisiewicz, W and Camp, M.J. (2004). Optimizing concentration and timing of a phage spray application to reduce *Listeria monocytogenes* on honeydew melon tissue. J. Food Prot. *67*, 1682–1686.

Loessner, M.J., and Busse, M. (1990). Bacteriophage typing of *Listeria* species. Appl. Environ. Microbiol. *56*, 1912–1918.

Logan, N.A. (2012). Bacillus and relatives in food-borne illness. J. Appl. Microbiol. *112*, 417–429.

McCallin, S., Alam Sarker, S., Barretto, C., Sultana, S., Berger, B., Huq, S., Krause, L., Bibiloni, R., Schmitt, B., Reuteler, G., *et al.* (2013). Safety analysis of a Russian phage cocktail: From MetaGenomic analysis to oral application in healthy human subjects. Virology *443*, 187–196.

Mahony, J., McAuliffe, O., Ross, R.P., and van Sinderen, D. (2011). Bacteriophages as biocontrol agents of food pathogens. Curr. Opin. Biotechnol. *22*, 157–163.

Martinez, B., Obeso, J.M., Rodriguez, A., and Garcia, P. (2008). Nisin–bacteriophage crossresistance in *Staphylococcus aureus*. Int. J. Food Microbiol. *122*, 253–258.

Międzybrodzki, R., Borysowski, J., Weber-Dąbrowska, B., Fortuna, W., Letkiewicz, S., Szufnarowski, K., Pawełczyk, Z., Rogóż, P., Kłak, M., Wojtasik, E., *et al.* (2012). Clinical aspects of phage therapy. Adv. Virus Res. *83*, 73–121.

Modi, R., Hirvi, Y., and Griffiths, M.W. (2001). Effect of phage on survival of *Salmonella* Enteritidis during manufacture and storage of cheddar cheese made from raw and pasteurized milk. J. Food Prot. *64*, 927–933.

Mokhtari, W., Nsaibia, S., Majouri, D., Ben Hassen. A., Gharbi, A., and Aouni, M. (2012). Detection and characterization of Shigella species isolated from food and human stool samples in Nabeul, Tunisia, by molecular methods and culture techniques. J. Appl. Microbiol. *113*, 209–222.

Newell, D.G., Koopmans, M., Verhoef, L., Duizer, E., Aidara-Kane, A., Sprong, H., Opsteegh, M., Langelaar, M., Threfall, J., Scheutz, F., van der Giessen, J., and Kruse, H. (2010). Food-borne diseases – the challenges of 20 years ago still persist while new ones continue to emerge. Int. J. Food Microbiol. *139 Suppl 1*, S3–15.

Norberg, S., Stanton, C., Ross, R.P., Hill, C., Fitzgerald, G.F., and Cotter, P.D. (2012). Cronobacter spp. in powdered infant formula. J. Food Prot. *75*, 607–620.

Nyachuba, D.G. (2010). Food-borne illness: is it on the rise? Nutr. Rev. *68*, 257–269.

O'Brien, S.J. (2012). The public health impact of food-related illness. Curr. Opin. Infect. Dis. *25*, 537–545.

O'Flaherty, S., Coffey, A., Meaney, W.J., Fitzgerald, G.F., and Ross, R.P. (2005). Inhibition of bacteriophage K proliferation on *Staphylococcus aureus* in raw bovine milk. Lett. Appl. Microbiol. *41*, 274–279.

O'Flynn, G., Ross, R.P., Fitzgerald, G.F., and Coffey, A. (2004). Evaluation of a cocktail of three bacteriophages for biocontrol of *Escherichia coli* O157:H7. Appl. Environ. Microbiol. *70*, 3417–3424.

Orquera, S., Gölz, G., Hertwig, S., Hammerl, J., Sparborth, D., Joldic, A., and Alter, T. (2012). Control of Campylobacter spp. and Yersinia enterocolitica by virulent bacteriophages. J. Mol. Genet. Med. *6*, 273–278.

Pao, S., Randolph, S.P., Westbrook, E.W., and Shen, H. (2004). Use of bacteriophages to control *Salmonella* in experimentally contaminated sprout seeds. J. Food Sci. *69*, M127-M130.

Patel, J., Sharma, M., Millner, P., Calaway, T., and Singh, M. (2011). Inactivation of *Escherichia coli* O157:H7 attached to spinach harvester blade using bacteriophage. Food-borne Path. Dis. *8*, 541–546.

Rajkovic, A., Smigic, N., and Devlieghere, F. (2010). Contemporary strategies in combating microbial contamination in food chain. Int. J. Food Microbiol. *141 Suppl 1*, S29–42.

Roy, B., Ackermann, H.-W., Pandian, S., Picard, G., and Goulet, J. (1993). Biological inactivation of adhering *Listeria monocytogenes* by listeriaphages and a quaternary ammonium compound. Appl. Environ. Microbiol. *59*, 2914–2917.

Sarker, S.A., McCallin, S., Barretto, C., Berger, B., Pittet, A.C., Sultana, S., Krause, L., Huq, S., Bibiloni, R., Bruttin, A., Reuteler, G., and Brüssow, H. (2012). Oral T4-like phage cocktail application to healthy adult volunteers from Bangladesh. Virology *434*, 222–232.

Scallan, E., Hoekstra, R.M., Angulo, F.J., Tauxe, R.V., Widdowson, M.A., Roy, S.L., Jones, J.L., and Griffin, P.M. (2011a). Food-borne illness acquired in the United States-major pathogens. Emerg. Infect. Dis. *17*, 7–15.

Scallan, E., Griffin, P.M., Angulo, F.J., Tauxe, R.V., and Hoekstra, R.M. (2011b). Food-borne illness acquired in the United States-unspecified agents. Emerg. Infect. Dis. *17*, 16–22.

Schlech, W.F. 3rd. (2000). Food-borne listeriosis. Clin. Infect. Dis. *31*, 770–775.

Sharma, M., Ryu, J.-H., and Beuchat, L.R. (2005). Inactivation of *Escherichia coli* O157:H7 in biofilm on stainless steel by treatment with an alkaline cleaner and a bacteriophage. J. Appl. Microbiol. *99*, 449–459.

Sharma, M., Patel, J.R., Conway, W.S., Ferguson, S., and Sulakvelidze, A. (2009). Effectiveness of bacteriophages in reducing *Escherichia coli* O157:H7 on fresh-cut cantaloupes and lettuce. J. Food Prot. *72*, 1481–1485.

Sillankorva, S.M., Oliveira, H., and Azeredo, J. (2012). Bacteriophages and their role in food safety. Int. J. Microbiol. *2012*, 863945.

Siringan, P., Connerton, P.L., Payne, R.J., and Connerton, I.F. (2011). Bacteriophage-mediated dispersal of *Campylobacter jejuni* biofilms. Appl. Environ. Microbiol. *77*, 3320–3326.

Smith, H.W., and Huggins, M.B. (1982). Successful treatment of experimental *Escherichia coli* infections in mice using phage: its general superiority over antibiotics. J. Gen. Microbiol. *128*, 307–318.

Soni, K.A., and Nannapaneni, R. (2010a). Removal of *Listeria monocytogenes* biofilms with bacteriophage P100. J. Food Prot. *73*, 1519–1524.

Soni, K.A., and Nannapaneni, R. (2010b). Bacteriophage significantly reduces *Listeria monocytogenes* on raw salmon fillet tissue. J. Food Prot. *73*, 32–38.

Soni, K.A., Nannapaneni, R., and Hagens, S. (2010). Reduction of *Listeria monocytogenes* on the surface of fresh channel catfish fillets by bacteriophage Listex P100. Food-borne Pathog. Dis. *7*, 427–434.

Soni, K.A., Desai, M., Oladunjoye, A., Skrobot, F., and Nannapaneni, R. (2012). Reduction of *Listeria monocytogenes* in queso fresco cheese by a combination of listericidal and listeriostatic GRAS antimicrobials. Int. J. Food Microbiol. *155*, 82–88.

Sulakvelidze, A. (2013). Using lytic bacteriophages to prevent or significantly reduce contamination of food by food-borne bacterial pathogens. J. Sci. Food Agric. *93*, 3137–3146.

Suzuki, H., and Yamamoto, S. (2009). Campylobacter contamination in retail poultry meats and by-products in the world: a literature survey. J. Vet. Med. Sci. *71*, 255–261.

Tabla, R., Martinez, B., Rebollo, J.E., Gonzalez, J., Ramirez, M.R., Roa, I., Rodriguez, A., and Garcia, P. (2012). Bacteriophage performance against *Staphylococcus aureus* in milk is improved by high hydrostatic pressure treatments. Int. J. Food. Microbiol. *156*, 209–213.

Tomat, D., Mercanti, D., Balagué, C., and Quiberoni, A. (2013). Phage biocontrol of enteropathogenic and Shiga toxin-producing *Escherichia coli* during milk fermentation. Lett. Appl. Microbiol. *57*, 3–10.

Verraes, C., Van Boxstael, S., Van Meervenne, E., Van Coillie, E., Butaye, P., Catry, B., de Schaetzen, M.A., Van Huffel, X., Imberechts, H., Dierick, K., Daube, G., Saegerman, C., De Block, J., Dewulf, J., and Herman, L. (2013). Antimicrobial resistance in the food chain: a review. Int. J. Environ. Res. Public Health. *10*, 2643–2669.

Viazis, S., Akhtar, M., Feirtag, J., and Diez-Gonzalez, F. (2011a). Reduction of *Escherichia coli* O157:H7 viability on leafy green vegetables by treatment with a bacteriophage mixture and *trans*-cinnamaldehyde. Food Microbiol. *28*, 149–157.

Viazis, S., Akhtar, M., Feirtag, J., and Diez-Gonzalez, F. (2011b). Reduction of *Escherichia coli* O157:H7 viability on hard surfaces by treatment with a bacteriophage mixture. *145*, 37–42.

Warren, B.R., Parish, M.E., and Schneider, K.R. (2006) Shigella as a food-borne pathogen and current methods for detection in food. Crit. Rev. Food Sci. Nutr. *46*, 551–567.

Whichard, J.M., Sriranganathan, N., and Pierson, F.W. (2003). Suppression of *Salmonella* growth by wild-type and large-plaque variants of bacteriophage Felix O1 in liquid culture and on chicken frankfurters. J. Food Prot. *66*, 220–225.

Ye, J., Kostrzynska, M., Dunfield, K., and Warriner, K. (2009). Evaluation of a biocontrol preparation consisting of *Enterobacter asburiae* JX1 and a lytic bacteriophage cocktail to suppress the growth of *Salmonella* Javiana associated with tomatoes. J. Food Prot. *72*, 2284–2292.

Ye, J., Kostrzynska, M., Dunfield, K., and Warriner, K. (2010). Control of *Salmonella* on sprouting mung bean and alfalfa seeds by using a biocontrol preparation based on antagonistic bacteria and lytic bacteriophages. J. Food Prot. 73, 9–17.

Zhang, H., Wang, R., and Bao, H. (2013). Phage inactivation of food-borne Shigella on ready-to-eat spiced chicken. Poult. Sci. 92, 211–217.

Zuber, S., Boissin-Delaporte, C., Michot, L., Iversen, C., Diep, B., Brüssow, H., and Breeuwer, P. (2008). Decreasing *Enterobacter sakazakii* (*Cronobacter* spp.) food contamination level with bacteriophages: prospects and problems. Microbial Biotechnol. *1*, 532–543.

Phage Therapy: Experiments Using Animal Infection Models 10

Shigenobu Matsuzaki, Jumpei Uchiyama, Iyo Takemura-Uchiyama and Masanori Daibata

Abstract

Phage therapy is a method of controlling bacterial infections by bacteriophage (phage) administration. Many animal experiments have been performed to demonstrate the efficacy and safety of phage therapy. Bacterial species that are currently prevalent in clinical settings have been selected as the targets of experimental phage therapy. These include Gram-positive bacteria such as *Staphylococcus aureus* including methicillin-resistant strains (MRSA) and *Enterococcus* species including vancomycin-resistant strains (VRE), as well as Gram-negative bacteria such as *Pseudomonas aeruginosa* (and its relatives), *Escherichia coli*, *Klebsiella pneumoniae* and *Acinetobacter baumanii*. The infections selected have included meningitis, lung infections, wound infections, urinary tract infections, gastrointestinal infections, and septicaemia, all of which are clinically relevant. In most animal experiments, phage therapy has been demonstrated to be very effective and safe. Furthermore, phage therapy has been shown to be effective in immunosuppressed and diabetic animals, as well as in a murine model of chronic infection. Methods to overcome the problem of anti-phage antibodies and of phage therapy against intracellular bacteria such as *Mycobacterium* species have also been demonstrated. Moreover, genetically modified phages, which can kill the host bacteria without lysing them, have also been developed, and their efficacy has been demonstrated. This accumulated evidence encourages us to develop phage therapy systems for the future treatment of human infections.

Introduction

Since the middle of the 20th century when the first antibiotic, penicillin, was introduced to practical clinical use, humans have become able to control practically all bacterial infections. However, the recent rapid increase in multidrug-resistant bacteria is threatening current medical practice all over the world. Therefore, although the availability of antibiotics will likely not change in the future, effective alternatives need to be developed to solve the problem of multidrug-resistant pathogenic bacteria.

Bacteriophages (phages) are viruses that specifically infect and destroy bacteria. Because phages cannot infect eukaryotic cells including human cells, they have been considered to have potential as antibacterial agents. Today, phage therapy, which harnesses phage particles themselves and/or their products (e.g. endolysins) as bioagents to control bacterial infections, is one of the most promising alternatives to antibiotics (Alisky *et al.*, 1998; Merril *et al.*, 2003; Fischetti, 2010). Phage therapy was first used by d'Hérelle, one of the discoverers of phages, around 1920. However, the history of phage therapy is complex and quite different from that of other antibacterial agents such as antibiotics (Carlton, 1999; Ho, 2001; Summers, 2001). In the 1920s and 1930s phage therapy was widely used, but after the successful introduction of antibiotics to medical practice in the 1940s, it has had different histories in Eastern and Western countries. Its use has been continued until the present in the East (Georgia, Poland and Russia), but was abandoned in most Western countries in the 1940s. However, the increase in multidrug-resistant pathogenic bacteria has led to phage therapy being reevaluated in Western countries since the 1980s. This complex history means, curiously enough, that although phage therapy is a part of medicine in Georgia, Poland, and Russia, it is still in the development phase in the West. These different stages of development may cause difficulty in understanding phage therapy.

The use for over 80 years in Eastern countries of phage therapy against infectious bacterial diseases in humans has demonstrated the effectiveness and safety of this method (Alisky *et al.*, 1998; Weber-Dabrowska *et al.*, 2000; Kutateladze and Adamia, 2008; Miedzybrodzki *et al.*, 2012; Chanishvili *et al.*, 2012). However, for worldwide acceptance of phage therapy, experimental studies demonstrating its effectiveness and safety and detailed examination of phage kinetics *in vivo* are essential, as is also the case in the development of other antimicrobial agents such as antibiotics. Many experiments using animal models of bacterial infections have been performed as preclinical studies before clinical trials. These experiments have clearly shown the efficacy and safety of phage therapy. This chapter will provide an overview of studies on experimental phage therapy.

Two directions in experimental phage therapy: phage particles versus their products

There are essentially two forms of experimental phage therapy (i.e. therapy performed on animal models of infection). The first is based on virulent (lytic) phage particles themselves and the other involves the use of their proteins (e.g. endolysins) to control bacterial infection (Merril *et al.*, 2003; Fischetti, 2010). Phage therapy using virulent phages takes advantage of the whole phage life cycle from adsorption to the release of progeny phages. Usually, tailed phages (myoviruses, siphoviruses, or podoviruses) have been used for this method, although there are also non-tailed phages having filamentous, cubic, or pleomorphic morphology (Ackermann, 2006). In the life cycle of tailed phages, the median latent period, which is the time interval from adsorption to the release of progeny phages, and the median burst size, which is the number of progeny phages per infected cell, are 40–60 min and 50–100, respectively (Ackermann and Dubow, 1987). In contrast, most pathogenic bacteria multiply by binary fission at one doubling per 30–60 min. Therefore, even if the number of phages is very small compared with that of the host bacteria at the time of administration of a phage preparation, all cells will be infected by phages within a short time, and the whole population of bacteria will be disrupted. Therefore, if only relatively few phage particles can reach the infected tissues, even within internal organs of an experimental animal, they will proliferate in the bacterial cells, release many progeny phages, and rapidly decrease the number of bacteria. If bacteria have been eliminated by the phage before the time of irreversible commitment to death by bacterial infection, the animal will survive. As described below, this situation is presumed to actually occur *in vivo* in the treatment of experimental infections.

The second form of experimental phage therapy is based on phage-derived proteins such as endolysins (bacteriolytic enzymes produced at the final stage of phage proliferation) or tail-associated proteins (enzymes that degrade bacterial cell surface components such as the capsule or the peptidoglycan). Because the peptidoglycan layer is exposed on the outside of Gram-positive bacteria, the endolysin can degrade it even when added from outside the cell. Tail-associated endosialidase, tail-associated lytic enzyme, and tail spike protein have also been proposed as candidate molecules for experimental phage therapy (Mushtaq *et al.*, 2004; Rashel *et al.*, 2008; Waseh *et al.*, 2010). Tail-associated endosialidase or lytic enzyme partially degrades the capsule or the peptidoglycan layer, respectively, on the bacterial cell surface to allow the tail tip to contact the outer membrane and/or cytoplasmic membrane for DNA injection. Phage-associated endosialidase has been shown to protect mice from death caused by *E. coli* K1 (Mushtaq *et al.*, 2004). Tail spike protein of phage has also been shown to reduce colonization of *Salmonella* spp. (Waseh *et al.*, 2010).

In this chapter, experimental phage therapy using lytic phage will be described principally. Phage therapy using endolysins is discussed in detail in Chapter 14.

Bacterial species and infections targeted in experimental phage therapy

Although phage therapy can be potentially used to treat infections caused by all species of bacteria, researchers have selected their target bacterial species based on present incidence in clinical practice of infections due to different bacteria. Recently, Vincent *et al.* (2009) reviewed the prevalence of infections in intensive care units (ICUs) in several countries (Tables 10.1 and 10.2), which reflect the current worldwide status of infectious diseases. In ICU infections, *Staphylococcus aureus* including methicillin-resistant strains (MRSA) (30.7%), *Enterococcus* species including vancomycin-resistant strains (VRE) (10.9%), *Staphylococcus epidermidis* (10.8%), and *Streptococcus pneumoniae* (4.1%) were the most prevalent Gram-positive bacterial species. *Pseudomonas* species (19.9%), *Escherichia coli* (16.0%), *Klebsiella* species (12.7%), *Acinetobacter* species (8.8%), and *Enterobacter* species (7.0%) were the most prevalent Gram-negative species. The most prevalent sites of infection included the respiratory tract (63.5%), abdomen (19.6%), bloodstream (15.1%), renal/urinary tract (14.3%), skin (6.6%), and central nervous system (CNS) (2.9%).

Consistent with this current prevalence of bacterial infections as seen in clinical medicine,

Table 10.1 Phage therapy of infections caused by clinically relevant bacteria

Bacterial species	Frequency[1]	Papers reporting effectiveness of phage therapy in humans[2]	Studies of effectiveness of phage therapy in animal models of infection
Gram-positive bacteria			
S. aureus (including MRSA)	30.7%	(A), (K), (M), (W)	Matsuzaki *et al.* (2003), Wills *et al.* (2005), Capparelli *et al.* (2007), Zimecki *et al.* (2009, 2010) Hsieh *et al.* (2011), Gupta *et al.* (2011), Chhibber *et al.* (2013)
Enterococcus spp. (including VRE)	10.9%	(L), (M)	Biswas *et al.* (2002), Uchiyama *et al.* (2008b)
S. epidermidis	10.8%	–	–
S. pneumoniae	4.1%	–	–[3]
Gram-negative bacteria			
Pseudomonas spp.	19.9%	(A), (K), (M), (W)	Soothil (1992), McVay *et al.* (2007) Watanabe *et al.* (2007), Vinodkumar *et al.* (2008), Heo *et al.* (2009), Seed *et al.* (2009), Debarbieux *et al.* (2010), Morello *et al.* (2011), Alemayehu *et al.* (2012)
E. coli	16%	(A), (K), (M), (W)	Chibani-Chennoufi *et al.* (2004), Tanji *et al.* (2005), Sheng *et al.* (2006), Nishikawa *et al.* (2008), Denou *et al.* (2009), Pouillot *et al.* (2012), Maura *et al.* (2012)
Klebsiella spp.	12.7%	(A), (M), (W)	Chhibber *et al.* (2008), Kumari *et al.* (2009, 2010a,b, 2011)
Acinetobacter spp.	8.8%	–	Soothill (1992)
Enterobacter spp.	7.0%	–	–

This table includes selected studies of effectiveness of phage therapy against clinically relevant bacteria. The efficacy of phage therapy against these bacteria was shown both in animal models of infection and in humans. [1]Based on (Vincent *et al.*, 2009). The percentages show frequency of infections caused by individual bacterial species in intensive care units (ICUs). [2]Effectiveness of phage therapy in humans in infections caused by individual bacterial species was reported in one or more of the following articles: (A), Alisky *et al.* (1998); (K), Kutateladze *et al.* (2008); (L), Letkiewicz *et al.* (2009); (M), Miedzybrodzki *et al.* (2012); (W), Weber-Dabrowska *et al.* (2000). [3]While there are no reports on therapeutic use of pneumococcal phages in animal models of infections, some studies showed the effectiveness of endolysins in the treatment of pneumococcal infections in mice (Grandgirard *et al.*, 2008).

Table 10.2 Phage therapy of infections in different locations

Site of infection	Frequency[1]	Papers reporting effectiveness of phage therapy in humans	Studies of effectiveness of phage therapy in animal models of infection
Respiratory tract	63.5%	Lung infections[2,3,4]	Chhibber *et al.* (2008), Carmody *et al.* (2010), Debarbieux *et al.* (2010), Morello *et al.* (2011), Alemayehu *et al.* (2012)
Abdominal cavity	19.6%	Gastrointestinal infections[2,4]	Chibani-Chennoufi *et al.* (2004), Tanji *et al.* (2005), Shen *et al.* (2006), Weiss *et al.* (2009), Mai *et al.* (2010), Maura *et al.* (2012)
Blood stream	15.1%	Septicaemia[2,3,4]	Smith *et al.* (1982), Soothill (1992), Merril *et al.* (1996), Biswas *et al.* (2002), Matsuzaki *et al.* (2003), Capparelli *et al.* (2007), Uchiyama *et al.* (2008b), Vinodkumar *et al.* (2008)
Renal/urinary tract	14.3%	Urinary tract infections[3,4]	Nishikawa *et al.* (2008), Tóthová *et al.* (2011)
Skin	6.6%	Wound infections[2,3,4]	Cerveny *et al.* (2002), Wills *et al.* (2005), McVay *et al.* (2007), Capparelli *et al.* (2007), Kumari *et al.* (2009, 2010a,b, 2011), Chhibber *et al.* (2013)
Central nervous system	2.9%	Meningitis[3]	Dubos *et al.* (1942), Smith *et al.* (1982), Barrow *et al.* (1998), Pouillot *et al.* (2012)

This table includes selected studies of effectiveness of phage therapy in infections in different locations. The efficacy of phage therapy in these infections was shown both in animal models and in humans. [1]Based on (Vincent *et al.*, 2009). The percentages show frequency of individual infections in intensive care units (ICUs). [2]Alisky *et al.* (1998). [3]Weber-Dabrowska *et al.*(2000). [4]Kutateladze *et al.* (2008).

the infectious diseases caused by bacterial species such as *S. aureus, E. faecalis, E. faecium, S. pneumoniae, E. coli, P. aeruginosa, Klebsiella pneumoniae,* and *Acinetobacter baumanii* have been selected as main targets for experimental phage therapy performed on animal infection models (Table 10.1). In addition, several other bacterial pathogens were also used as target bacteria in experimental phage therapy; these include *Burkholderia cenocepacia* (Carmody *et al.*, 2010), *Cronobacter* spp. (Tóthová *et al.*, 2011), *Listeria monocytogenes* (Mai *et al.*, 2010), *Vibrio vulnificus* (Cerveny *et al.*, 2002), *Mycobacterium tuberculosis* (Broxmeyer *et al.*, 2002), *Salmonella enterica* (Capparelli *et al.*, 2010), and *Yersinia pestis* (Filippov *et al.*, 2012). Experimental phage therapy was performed on animal models of meningitis, lung infections, wound infections, urinary tract infections, gastrointestinal infections, and septicaemia. In most experiments using animal infection models, phage therapy has been demonstrated to be effective and safe, indicating strongly that phages could be effective against many bacterial infections of humans that currently pose serious therapeutic problems. In fact, the effectiveness of phage therapy against these infectious diseases in humans has been reported from Georgia, Poland, and Russia (Tables 10.1 and 10.2).

In this chapter, experimental phage therapy performed on animal infection models will be described for different kinds of infections.

Model animals used for phage therapy experiments

With the aim of future application of phage therapy to human infectious diseases, many treatment experiments using mammals such as mice, rats, and rabbits have been performed, with mice being the most frequently used model animals for phage therapy. The BALB/c mouse strain has been used most frequently; the next most frequently used strains are C57BL/6, C3H, and CD1 (ICR). Swiss Webster and CBA mice have also been used in a few experiments. BALB/c, C57BL/6, and C3H are inbred mouse strains (genetically homogenous), while CD1 (ICR) and Swiss Webster are closed-colony strains (genetically heterogeneous). Although the ages of mice used range from 1 to 12 weeks, 6- to 8-week-old mice have been used most frequently. Female mice have been used more frequently than male mice. Although the strains, age, and sex of mice used vary between experiments, this does not seem to be related to the bacterial species or target organ.

Recently, in addition to mammalian models, insect infection models using adult fruit flies (*Drosophila melanogaster*) and larvae of a honeycomb moth (*Galleria mellonella*) have also been reported (Heo *et al.*, 2009; Seed *et al.*, 2009). Administration of phages resulted in delayed death of the infected insects in these models. Our group is developing a system to evaluate phage therapy using the silkworm (*Bombyx mori*) (Takemura-Uchiyama *et al.*, 2013). Experiments using small insects may also be valuable for evaluating the efficacy of phage therapy, because they allow the investigators to avoid problems with animal ethics and restrictions on the number of experimental animals, and reduce the cost of the experiments.

Although in this chapter we will focus on murine models of infection, it is noteworthy that many studies have been performed to evaluate the efficacy of experimental phage therapy also in production animals. These were discussed in Chapter 8.

Pharmacokinetics of phages

In general, research into pharmacokinetics of phages in experimental animals is an essential preliminary stage of studies on their therapeutic use. In particular, this line of research provided important insights into the absorption of phages from the intestines following oral administration and the kinetics in the peripheral blood of systemically administered phages.

Absorption of phages from intestine to blood

In clinical medicine the main route of administration for the majority of drugs including antibiotics is the oral route. Therefore, an important problem in phage therapy is a capacity of phages to be absorbed from the gastrointestinal tract. In general, although phages administered orally to healthy mice reach the intestine and can multiply there, they do not seem to be absorbed into blood at substantial titre, at least not in healthy animals. For example, Denou *et al.* (2009) observed that in spite of the presence of T4 phages at high titre in the caecum and colon following oral phage administration, phages were not detected in the blood, liver, or spleen. However, because of both the low absorption efficiency into the blood of phages administered orally and their rapid clearance from the blood, such absorption may not have been detected. Indeed, there is one report showing the possibility of absorption into the blood of phages administered to mice by oral route, although with extremely low efficiency. In this study, Geier *et al.* (1973) found that phages administered intraperitoneally or subcutaneously rapidly penetrated into the blood. In addition, they also showed that a small number of phages administered orally to mice may be absorbed into the blood. However, the absorbed phages disappeared rapidly from the blood, because their number was extremely small. We also observed a similar phenomenon in our preliminary study of oral gavage administration of coliphage T4 and staphylococcal phage ΦMR11 to BALB/c mice (S. Matsuzaki *et al.*, unpublished data).

Thus, the efficiency of absorption of phages administered orally is rather low in healthy animals. However, there is one study to show that it may increase in animals with immunodeficiency. In this study, Watanabe *et al.* (2007) showed the effectiveness of phage therapy in *P. aeruginosa* gut sepsis using mice that were made immunodeficient by cyclophosphamide injection. Oral inoculation of *P. aeruginosa* D4 resulted in bacterial translocation from the intestine into the blood, followed by migration of bacteria to organs, and death of the mice about 2 weeks after inoculation. However, phage KPP10 administered orally at 2 days after bacterial inoculation was also absorbed into the blood, disseminated into spleen, liver, and mesenteric lymph nodes, and significantly decreased the bacterial number and the level of pro-inflammatory cytokines (TNF-α, IL-1, and IL-6) in the organs. Phage administration also significantly reduced mortality of mice (survival rate 66.7% for the phage-treated group versus 0% for the saline-treated control group). This was the first report clearly showing translocation of phages, although it was observed in immunosuppression and/or bacterial infection. This may explain the previously reported efficacy of orally administered phages in the treatment of bacterial infections in humans (Weber-Dabrowska *et al.*, 2000). It is possible that phage absorption, like bacterial translocation, may be promoted by several different factors including inflammatory reactions in the gut and immunodeficiencies (Górski *et al.*, 2006).

Kinetics in mouse blood of administered phages

The kinetics of administered phages in mouse blood seems to be independent of their size or shape. Phage ΦMR11 (*S. aureus* siphovirus, 43 kbp DNA), KPP10 (*P. aeruginosa* myovirus, 88 kbp DNA), ΦEF24C (*E. faecalis* myovirus, 143 kbp DNA), or KEP10 (*E. coli* myovirus, estimated 170 kbp DNA) were injected into the peritoneal cavity of mice at 10^{11} plaque-forming units (PFU), and active phages in the blood were counted by plaque count at 2, 4, 8, 12, 24, 48, and 72 h after injection (Uchiyama *et al.*, 2009). The number of phages rapidly decreased over 8 to 12 h and then gradually decreased over 3 days. The patterns of titre decrease for all phages were very similar and seemed to fit a two-compartment model. The initial phase of rapid decrease (reduction of 10^{-2} to 10^{-3} per 12 h) and the following phase of gradual decrease (reduction of 10^{-2} to 10^{-3} per 36–60 h) were presumed to be caused by distribution of the drug (phage) to organs and elimination of the drug (phage), respectively. These data suggest that the clearance pattern of phages in blood is not markedly influenced by their morphology (size, family, etc.). Classic studies showed that rapid clearance of phages from blood is mediated largely by cells of the reticulo-endothelial system of the spleen and liver (Inchley, 1969; Geier *et al.*, 1973). Another major factor involved in phage clearance is humoral immunity (Srivastava *et al.*, 2004). Since a relatively short half-life of phages is a very important problem in phage therapy, some experimental attempts were made to obtain phages with improved pharmacokinetic features (Merril *et al.*, 1996; see 'Sepsis caused by Gram-negative bacteria').

The patterns of decrease in phage titre were also examined in mice made neutropenic by injection of cyclophosphamide. Cyclophosphamide is toxic to T and B lymphocytes and causes severe transient leucopenia (Vogels *et al.*, 1994; Hemendinger and Bloom, 1996). However, the patterns of titre decrease in neutropenic mice did not differ from those in healthy mice, suggesting that the phages were not affected by innate immunity (Uchiyama *et al.*, 2009).

In contrast, as described below, in mice infected with bacteria, administered phages can replicate and are maintained at high concentration in organs until the host bacteria disappear (e.g. Dubos *et al.*, 1943; Smith and Huggins, 1982; Matsuzaki *et al.*, 2003).

Experimental phage therapy

Meningitis

Dubos and colleagues (Dubos *et al.*, 1943) showed the efficacy of phage therapy for bacterial meningitis using a mouse model. Preliminary experiments showed that the intracerebral injection of *Shigella dysenteriae* 2308 at 100 times LD_{50} (50% lethal dose) caused meningitis, and 95% of the injected mice died within 2 to 5 days. In this model, although a large number of bacteria were present in the brains of the mice, peritoneal washings were sterile, and bacteria were only occasionally detected in cardiac blood. The authors showed that phages administered intraperitoneally could multiply in the brains of the mice infected with *S. dysenteriae*, and could decrease the bacterial number. In another experiment, to test the therapeutic ability of phage administration, the mice were inoculated intracerebrally with a dose of 60–400 times LD50 of *S. dysenteriae*. The survival rate was only 3.6% in the control mice that received no treatment, but when active anti-Shiga phages (approximately 1×10^7 to 5×10^9 PFU) were administered intraperitoneally immediately after intracerebral injection of *S. dysenteriae*, the survival rate increased to 72%.

Forty years after the experiment of Dubos and his colleagues, Smith and Huggins (1982) opened the way for the modern studies of phage treatment in the West. At that time, it became impossible for clinicians to ignore the increase in antibiotic-resistant bacteria. Therefore, Smith and Huggins examined the possibility of phage therapy against

E. coli MW (O18ac: K1: H7 ColV$^+$) meningitis in mice. *E. coli* with K1-antigen is a major strain causing meningitis in human infants (Korhonen *et al.*, 1985). *E. coli* MW (5×10^2 colony-forming units (CFU)) inoculated intracerebrally proliferated in the brain. The number of bacteria reached $10^{9.7}$/g brain tissue at 24 h after inoculation and 75% of the mice died within 96 h. Consistent with the results of Dubos *et al.* (1943), the number of bacteria in the blood and spleen remained at a low level. Administration of phage R (3×10^8 PFU) into gastrocnemius muscle simultaneously with the bacterial inoculation reduced the bacterial number to $10^{4.6}$/g brain tissue at 24 h after injection of bacteria and phage, and all mice survived.

Recent study (Pouillot *et al.*, 2012) confirmed high efficacy of phages in the treatment of experimental meningitis. Extended-spectrum beta-lactamase (ESBL)-producing *E. coli* cells (2×10^2 or 2×10^6 CFU) were injected to rat pups intracerebrally, and phage EC200^{PP} (10^8 PFU) was injected intraperitoneally 1 h later. In both cases the phage decreased mortality from 100% in the control group to 0% at day 5 and significantly reduced the bacterial counts in brain tissues (in some animals, even those which were injected higher number of bacterial cells, EC200^{PP} completely eradicated bacteria from the infected tissues).

The above reports indicate that in animals with meningitis phages administered intraperitoneally or intramuscularly can rapidly penetrate to the brain through the bloodstream, and that phages can multiply and kill bacteria in the brain. The brain is isolated from the other organs by the blood–brain barrier that regulates the influx and efflux of water, nutrients, and drugs between blood and brain (Abbott *et al.*, 1996). However, in animals with meningitis caused by bacterial infection, phages can penetrate the damaged blood–brain barrier and kill the bacteria in the brain.

Lung infections

As shown in Table 10.2, the respiratory tract is the site with the highest incidence of infection. Therefore, the development of a phage therapy system for respiratory tract infections is especially important. The efficacy of phage therapy against bacterial lung infections was shown by five studies (Chhibber *et al.*, 2008; Carmody *et al.*, 2010; Debarbieux *et al.*, 2010; Morello *et al.*, 2011; Alemayehu *et al.*, 2012).

Chhibber *et al.* (2008) examined the efficacy of phage therapy against lung infection caused by *K. pneumoniae*. *K. pneumoniae* B5055 (5×10^6 CFU) were administered intranasally to mice. This dose caused infection in 100% of the mice without causing mortality. In this model, the titre of the inoculated bacteria increased to approximately 10^9 CFU/ml lung tissue homogenate at the third day after inoculation, and then gradually decreased. When phage SS (multiplicity of infection (MOI) of 200) was administered intraperitoneally simultaneously with inoculation of *K. pneumoniae*, the number of bacteria in lung tissue homogenate decreased by 8 $\log_{10}$ at the third day, and was reduced to undetectable level at day 5, while in the control group complete clearance of bacteria from lungs was found at day 10. Furthermore, administration of phages 3 h prior to bacterial infection showed a preventive effect against infection. However, phages were ineffective in this model when administered at 6 or 24 h after bacteria.

The efficacy of intraperitoneal administration of phages against lung infection was confirmed by Carmody *et al.* (2010), who used a non-lethal mouse model of *B. cenocepacia* lung infection. Mice were infected via tracheotomy with *B. cenocepacia* AU0728 (1×10^7 or 1×10^8 CFU), and were treated with either intranasal inhalation or intraperitoneal injection of phage BcepIL02 (MOI of 100) 24 h after infection. Intraperitoneal injection of the phage significantly reduced bacterial density, as well as macrophage inflammatory protein 2 (MIP-2) and tumour necrosis factor α (TNF-α) levels in lungs. The decrease in bacterial number and suppression of MIP-2 and TNF-α in mice treated with intranasal inhalation of the phage were less than those in mice treated by intraperitoneal administration of the phage, but the reason for this is not clear.

While in the above study (Carmody *et al.*, 2010) intranasal administration of phages was essentially ineffective, there are some reports showing that intranasal inhalation of phages can be effective in experimental models of bacterial lung infection (Debarbieux *et al.*, 2010; Morello *et al.*, 2011).

Debarbieux *et al.* (2010) described the efficacy of intranasal phage administration in a murine model of *P. aeruginosa* lung infection. Instillation of 10^7 of *P. aeruginosa* bioluminescent PAK strain, which allows identification of the location and amount of bacteria in each internal organ by luminescence measurement, caused 100% mortality within 2 days. However, when phage PAK-P1 (a myovirus) was intranasally administered at an MOI of >1 at 2 h after the bacterial instillation, the mice were completely protected from death caused by the infection. Furthermore, if the animals received the phage intranasally at 24 h before infection, 100% survived the infection, suggesting that the phage also has the potential to prevent *P. aeruginosa* lung infections.

Moreover, the same group reported that intranasal inhalation of phages is also effective for lung infection by a *P. aeruginosa* strain CHA (multidrug-resistant and mucoid) isolated from a cystic fibrosis patient (Morello *et al.*, 2011). Instillation of 3×10^6 of *P. aeruginosa* CHA strain to mice caused elevation of the number of bacteria (by at least two orders of magnitude) and the levels of cytokines IL-6, KC (pro-inflammatory markers) and lactate dehydrogenase (LDH, an organ toxicity marker) in the lungs resulting in 100% mortality rate of the mice within 2 days. However, intranasal administration of phage P3-CHA (3×10^7 PFU) 2 h after the bacterial instillation significantly reduced bacterial density and levels of IL-6, KC, and LDH in lungs, and protected the majority of mice from death. Phage P3-CHA is a myovirus distantly related to the above-mentioned phage PAK-P1 used in the study by Debarbieux *et al.* (2010). Furthermore, if the animals received the phage (3×10^8 PFU) intranasally 4 days before infection, 100% survived the bacterial infection, suggesting that phage P3-CHA, like PAK-P1, has the potential to prevent *P. aeruginosa* lung infections.

In another study, Alemayehu *et al.* (2012) evaluated the efficacy of a mixture of two phages specific for *P. aeruginosa* (ΦNH-4 and ΦMR299–2) in a murine model of lung infection. Mice received bacteria ($1–2.5 \times 10^7$ CFU) intranasally and phages at MOI of 10 were administered by the same route after 2 h. It was shown that phages reduced the bacterial counts in the lungs by at least 3–4 log units within 6 h. Since phages could also kill bacteria growing as biofilm on a cystic fibrosis bronchial epithelial cell line *in vitro*, the authors concluded that phage therapy could be effective in the treatment of chronic *P. aeruginosa* infections in cystic fibrosis patients which typically involve biofilms.

The finding that intraperitoneal administration of a phage, in addition to intranasal administration, is also very effective for the treatment of bacterial lung infection is important when considering the use of phage therapy for lung infection. The effectiveness of intraperitoneal and intranasal administration may differ depending on the mouse strain and/or the causative bacterial species or strains. Therefore, simultaneous phage administration through both intraperitoneal and intranasal routes may be needed for treatment of bacterial lung infections in some cases.

Wound infections

The potential of phage therapy against wound infection caused by *P. aeruginosa* was evaluated by McVay *et al.* (2007). Mice were challenged by subcutaneous injection of *P. aeruginosa* ($2–3 \times 10^2$ CFU) directly under the anterior end of an experimental burn wound. A cocktail of three phages (3×10^8 PFU in total) was administered to the infected mice by the intramuscular, subcutaneous, or intraperitoneal route. The authors reported that phages administered through the intraperitoneal route were delivered to the site of infection at a higher number, more rapidly, and more stably than by the other routes. The survival rate at 72 h post infection was 28%, 22%, and 88% following intramuscular, subcutaneous, and intraperitoneal phage administration, respectively, while it was only 6% in the untreated control mice.

The efficacy of phage therapy against burn wound infections caused by *K. pneumoniae* was examined in a series of studies (Kumari *et al.*, 2009, 2010a,b, 2011). In all four studies mice were challenged subcutaneously directly under the anterior end of an experimental burn wound with *K. pneumoniae* B5055 ($10^6–10^8$ CFU). In the first study (Kumari *et al.*, 2009) phages Kpn5, Kpn12, Kpn13, Kpn17, or Kpn22 (podoviruses; MOI of 1.0) were administered intraperitoneally immediately after the burn/bacterial challenge.

Administration of individual phages or a cocktail containing all five phages decreased the bacterial number in blood, lung, skin, and peritoneal fluid, and increased the survival rate from approximately 6% in the control group to 80–97% at 72 h. Of these phages, Kpn5 was most effective, and was effective even if the administration was delayed up to 18 h after the burn/bacterial challenge (Kumari *et al.*, 2009, 2010a). In addition, intraperitoneal administration of Kpn5 decreased the level of both pro-inflammatory and anti-inflammatory cytokines in sera and lungs (Kumari *et al.*, 2010a). Another study showed that a single topical administration of Kpn5 is more effective than multiple topical doses of either of two natural antibacterial products (honey and aloe vera gel; Kumari *et al.*, 2010b). Furthermore, when Kpn5 formulated in hydrogel was administered topically to the burn wound site at an MOI of 200, a significant reduction in mouse mortality was shown, and a single administration of the phage was more effective than multiple administrations of other antibacterial agents such as silver nitrate and gentamycin (Kumari *et al.*, 2011).

Cerveny *et al.* (2002) examined the potential use of phages against local and systemic disease caused by *Vibrio vulnificus* using a mouse model. This bacterium is an opportunistic pathogen of humans, causing septicaemia after ingestion of contaminated seafood and necrotizing fasciitis after contamination of wounds in people with risk factors such as genetic mutation, liver damage, immunosuppression, or diabetes (Linkous and Oliver, 1999). *V. vulnificus* MLT403 (10^6 CFU) was injected subcutaneously after intraperitoneal iron-dextran injection, which increases the sensitivity of mice to the bacterium. 14 to 18 h after injection of *V. vulnificus*, approximately 10^8 CFU/g lesion tissue and 10^4–10^5 CFU/g liver tissue were detected, and the lesion score was 3 (maximum). The presence of *V. vulnificus* in the lesion and liver tissues indicated local and systemic infection, respectively. When 10^8 PFU of phage CK-2 were intravenously injected to mice immediately after the bacterial inoculation, the CFU/g in lesion and liver tissue decreased to 10^5–10^6 and 10^2–10^3, respectively, and the lesion score was also lowered to 2.3. Although none of the untreated control mice survived, 100% of the treated group survived at 18 h. Of two other studied phages, 153-A5 and 153A-7, only the former was effective *in vivo*, while the latter failed to cure infection.

Wills *et al.* (2005) examined the effect of phage administration on *S. aureus*-induced abscess in a rabbit model. *S. aureus* 2698 (8×10^7 CFU) and phage LS2a (at 0, 6×10^5, 6×10^6, or 6×10^7 PFU) were injected subcutaneously at the same site resulting in a phage dose-dependent inhibition of abscess formation. The median of the abscess area was 120 mm^2, 80 mm^2, 64 mm^2, and 58 mm^2, respectively, in the four groups 4 days after injection of bacteria and/or phage. The bacterial number (median) in the abscess was 6.0×10^6, 6.6×10^4, 2.0×10^4 and 6×10^3 CFU, respectively. Administration of phage LS2a also showed a prophylactic effect against *S. aureus* 2698-induced abscess.

Suppression by phage administration of abscess formation caused by *S. aureus* was also reported by Capparelli *et al.* (2007). Mice were inoculated subcutaneously on both flanks with *S. aureus* A170 (10^8 CFU). Abscess formation was assessed by abscess number, CFU/abscess count, and weight of abscess. When phage M^{Sa} (10^9 PFU) was administered subcutaneously on both sides of the abdomen immediately after the bacterial inoculation, abscess formation was almost completely inhibited. When the phage was administered 4 days after bacterial infection to mice with well established abscesses, CFU/abscess count and abscess weight were significantly decreased compared with the control, although the phage failed to reduce the abscess number. Furthermore, when four daily doses of the phage were given starting at 4 day after bacterial infection, the CFU/abscess count and the weight of abscesses were decreased to a greater extent compared with mice which received a single dose of the phage.

In another study, Chhibber *et al.* (2013) evaluated the efficacy of MR-10 phage in the treatment of acute hindpaw foot infection due to MRSA in alloxan-induced murine model of diabetes. MRSA cells (10^4 CFU) were injected into the hindpaw, followed by injection of the phage (MOI of 100) 30 min later near the site of bacterial inoculation. It was shown that the efficacy of a single dose of phage was comparable to that of linezolid administered per os. However, the maximal effects in terms of reducing the bacterial count, myeloperoxidase activity in foot paw samples, and oedema and lesion scores, as well as hastening wound healing were found following the application of a combination of the phage and antibiotic.

These results indicate that phage therapy may be effective against wound infections caused by both Gram-negative and Gram-positive bacteria. Furthermore, phages administered by intraperitoneal, intravenous, or subcutaneous routes were effective against these infections.

Urinary tract infections

There are only two reports on phage therapy of urinary tract infection (UTI) in a mouse model. In the first study, Nishikawa *et al.* (2008) examined the efficacy of phage therapy against urosepsis caused by uropathogenic *E. coli*. When 5×10^9 cells of *E. coli* ECU5 were transurethrally injected into the bladder of mice, urosepsis developed and most infected mice died within 3 days. However, when phages T4 or KEP10 were administered intraperitoneally at an MOI of 60 immediately after bacterial inoculation, 90–100% of the infected mice survived. KEP10 is a phage morphologically similar to T4, but its host range against uropathogenic *E. coli* strains is broader than that of T4. In this experiment, a proportion of the phage particles administered intraperitoneally was probably secreted into urine. Both T4 and KEP10 are very stable in the urine of both mice and humans for at least 24 h. Therefore, the protective effect of phage administration was suggested to be the result of destruction of *E. coli* not only in mouse internal organs, but also in urine. Another study showed high efficacy of a mixture of two specific phages in a murine model of UTI due to *C. turicensis* (Tóthová *et al.*, 2011).

Gastrointestinal infections

One of the advantages of phage therapy is a possibility to kill only the target pathogenic bacteria, without any effects on the normal microflora. However, is this feasible in an environment such as the intestine, in which a great number of bacteria (often including the same species as the target) are present? For example, is it possible to kill only an invading pathogenic *E. coli* strain without killing the *E. coli* strains from the gut normal flora? There are some reports addressing this question.

Chibani-Chennoufi *et al.* (2004) examined the effect of oral administration of a mixture of T4-type phages on *E. coli* in the gut normal microflora in mice. Faeces samples were spread on Drigalski agar plates to distinguish *E. coli* from other bacteria. *E. coli* forms yellow colonies (lactose fermentation) on these plates. When mice drank water containing 0, 10^3, 10^5, or 10^7 phage/ml, the number of *E. coli* colonies on Drigalski plates was $10^{6.2 \pm 0.04}$, $10^{5.9 \pm 0.10}$, $10^{6.1 \pm 0.05}$, or $10^{5.7 \pm 0.06}$, respectively, indicating that phage administration had a very small and biologically insignificant effect on the *E. coli* strains in the gut microflora, though the majority of the strains from the microflora were sensitive to phages and could be lysed *in vitro*. This indicates that the *E. coli* strains in the intestine might be physically or physiologically shielded from phage infection. By contrast, *E. coli* strains freshly introduced to the intestines of mice could be lysed by phages *in vivo*. The inability of orally administered coliphages to lyse sensitive *E. coli* strains from the gut normal microflora was also described by Weiss *et al.* (2009). The authors also showed, using an *E. coli* strain having an ampicillin resistance marker to discriminate it from the *E. coli* in normal bacterial flora in the gut, that orally administered phage could lyse *E. coli* artificially colonizing the mouse intestine. These results indicate that therapeutic phages administrated orally can selectively remove *E. coli* strains that have recently invaded the intestine without adversely affecting *E. coli* strains of the gut normal microflora, although the mechanism of this shielding of residential *E. coli* strains is not known.

Phage therapy without disturbance of the gut normal microflora has also been reported in *L. monocytogenes* food-borne infection. *L. monocytogenes* is a food-borne pathogen and is translocated from the intestine to the blood. It causes perinatal infections (abortion, stillbirth, and neonatal sepsis and meningitis) and meningitis and encephalitis in adults (Schlech III, 1998). Mai *et al.* (2010) examined the efficacy of phage administration in a murine model of *L. monocytogenes* infection. Mice were challenged by oral gavage with a single dose (approximately

10^5 CFU) of *L. monocytogenes* Lm370. The challenged mice lost up to 10% of their body weight within 3 days, probably because of diarrhoea, and the bacteria were detected not only in faecal matter but also translocated to the spleen and liver. However, oral gavage with ListShield™ (10^5 PFU), which is a mixture of 6 *L. monocytogenes* phages, daily for 3 days before and after challenge suppressed diarrhoea and weight loss caused by *L. monocytogenes* infection. The bacterial numbers in faecal material, the spleen, and liver were also significantly reduced compared to the control group. Although oral gavage of a single dose of ampicillin (25 mg/g) at 15 min post challenge also decreased to comparable extent the bacterial numbers in faecal material, the spleen, and liver, it could not suppress the weight loss and diarrhoea, probably because of disturbance of the gut microflora. Thus, unlike chemotherapy, oral administration of phages can cure gut infection without disturbance of the gut microflora. Even bi-weekly treatment with ListShield™ of mice for 90 days did not adversely affect the gut microflora.

Moreover, there are two reports on elimination of enterohaemorrhagic *E. coli* (*E. coli* O157:H7) from mouse intestine by phage therapy. In the first study, Tanji *et al.* (2005) orally administered a phage cocktail (SP15–21–22) in buffer containing 0.25% $CaCO_3$ to mice that were orally administered *E. coli* O157:H7 cells 2 days earlier. They showed that although a single administration of the phages could not decrease the *E. coli* O157:H7 counts in the faeces, daily administration of phages completely eliminated the bacteria from the lower small intestine and colon and substantially reduced the bacterial counts in upper small intestine and caecum. Sheng *et al.* (2006) also showed that repeated oral administration of phages for 3 days could eliminate *E. coli* O157:H7 from faeces (and probably also from the intestine) in a murine model of intestinal *E. coli* carriage.

In another study performed on a murine model of intestinal *E. coli* colonization, Maura *et al.* (2012) showed that bacterial permissivity to infection by a cocktail of three lytic phages varies between different parts of the gut. While *E. coli* cells in the ileum were permissive to all three phages, bacteria in the faeces were permissive to only one phage. These results are very important because they indicate that a necessary condition of successful phage therapy of alimentary tract infections is a detailed characterization of the interactions between phages and their target bacteria in the gut.

Septicaemia caused by Gram-negative bacteria

Smith and Huggins (1982) carried out very detailed treatment experiments in an *E. coli* sepsis mouse model. They injected *E. coli* MW (3×10^7 CFU) into the gastrocnemius muscle, and then administered phage R (3×10^8 PFU) into the opposite gastrocnemius muscle, 8 h after bacterial injection. Although the survival rate was approximately 7% (2/30 mice) in the control group that received *E. coli* cell lysate, it increased to approximately 93% (28/30 mice) in the phage-treated group. These investigators also performed a detailed examination of the dynamics of *E. coli* and phage R in each organ, and showed that administration of the phage lowered the number of bacteria in the organs which was accompanied by phage multiplication, and led to the survival of mice. Furthermore, they showed that the effectiveness of only one dose of the phage was higher than that of eight doses of antibiotics (streptomycin, tetracycline, ampicillin, chloramphenicol, and trimethoprim/sulfafurazole); multiple administrations of the antibiotics led to survival rates in the range of 3–90%. In addition, this study revealed that phage can efficiently eliminate bacteria also following intravenous administration.

Soothil (1992) demonstrated the efficacy of phage therapy in *A. baumanii*- and *P. aeruginosa*-infected mice. When 9.5×10^7 CFU of *A. baumanii* AC54 was injected intraperitoneally into mice, 100% of the mice died. However, simultaneous administration of a very small dose (MOI of 10^{-6}) of phage BS46 increased the survival rate to 80%. However, for perfect protection of mice inoculated intraperitoneally by *P. aeruginosa* 3719 (1.5×10^8 CFU), simultaneous administration of phage BS24 at an MOI of approximately 2 was needed. These results indicate strongly that the effective phage dose varies considerably depending on the bacterial and/or phage species. Efficacy of phage therapy against sepsis caused by *P. aeruginosa* was confirmed by Wang *et al.* (2006a), Watanabe *et al.*

2007; see the section 'Absorption of phages from intestine to blood'), Vinodkumar *et al.* (2008), and Heo *et al.* (2009). Other studies showed high effectiveness of phages in the treatment of sepsis due to *K. pneumoniae* (Vinodkumar *et al.*, 2005; Gu *et al.*, 2012), *S. enterica* (Capparelli *et al.*, 2010), and *E. coli* (Wang *et al.*, 2006b; Pouillot *et al.*, 2012).

Merril *et al.* (1996) devised an ingenious method to obtain a persistently circulating mutant phage. In general, as described above, the number of phages in mouse blood decreases rapidly after systemic administration largely due to removal by cells of the reticulo-endothelial system of the spleen and liver (see 'Kinetics in mouse blood of administered phages'). Merril *et al.* (1996) isolated a persistently circulating mutant of λcI phage by repeated intraperitoneal injection of phage into mouse followed by its *in vitro* amplification. The mutant phage had improved therapeutic ability in a murine model of bacteraemia. Interestingly, the isoelectric point of its major head protein, gpE, was changed. Although the reason why this mutation caused persistent circulation of the phage is unknown, this may suggest a general method to obtain mutant phages with improved pharmacokinetic features and therapeutic ability. As described below (see 'Septicaemia caused by Gram-positive bacteria', below), Capparelli *et al.* (2007) successfully isolated a persistently circulating mutant of an *S. aureus* phage by a similar method. Thus such methods could be used to obtain similar mutant phages regardless of the phage species.

Septicaemia caused by Gram-positive bacteria

Biswas *et al.* (2002) showed the efficacy of phage therapy against sepsis caused by vancomycin-resistant *E. faecium* (VRE). Injection of 10^9 cells of *E. faecium* VRE strain CRMEN 44 into the peritoneal cavity of mice induced bacteraemia, and 100% of them died within 48 h. However, a single intraperitoneal administration of phage ENB6 (myovirus) at the dose of 3×10^8 PFU 45 min after the bacterial challenge rescued 100% of the mice. A higher dose (3×10^9 PFU) could reduce also the morbidity of the mice and rescued 100% of them when administered at 5 h after bacterial challenge. Furthermore, the administration of a high dose of phage even 18 h after bacterial challenge, could rescue about 60% of the mice. The group also showed that multiple injections of the phage did not cause adverse effects such as anaphylactic reactions or changes in core body temperature.

Uchiyama *et al.* (2008b) examined the efficacy of phage therapy in *E. faecalis*-infected mice. They injected two *E. faecalis* strains, EF14 and VRE2 (a vancomycin-resistant strain), which had different sensitivities to phage ΦEF24C (myovirus), into the peritoneal cavity of mice. The efficiencies of plating of phage ΦEF24C on *E. faecalis* EF14 and VRE2 were 1 and 0.032, respectively (Uchiyama *et al.*, 2008a). Intraperitoneal inoculation of EF14 (1×10^{10} cells) and VRE2 (4.2×10^9 cells) caused sepsis and resulted in 100% mortality rate of mice within 2 days. However, in mice injected with EF14, intraperitoneal administration of ΦEF24C at an MOI more than 0.01 at 20 min after the bacterial injection saved 100% of the mice. The administration of the phage at an MOI more than 0.1 rescued 100% of mice infected with VRE2, and at an MOI of 0.01 could save 80% of the mice. These results indicate that the efficient propagation of ΦEF24C *in vivo* and its therapeutic efficacy do not correlate significantly with the *in vitro* sensitivity of the host to this phage. The study also showed no anaphylactic reactions after multiple injections of the phage (seven doses at 4-day intervals), consistent with the report of Biswas *et al.* (2002) described above. In addition, this group also identified the endolysin (ORF9) of EF24C, and described its potential application to controlling *E. faecalis* infection (Uchiyama *et al.*, 2011).

Matsuzaki *et al.* (2003) observed the life-saving effect of phage administration in mice infected with *S. aureus*, including MRSA strains. When 8×10^8 cells of *S. aureus* SA37 were injected intraperitoneally, more than 80% of the mice died within 24 h. However, intraperitoneal administration of ΦMR11 (siphovirus) at MOI more than 0.1 immediately after the bacterial inoculation showed a significant protective effect, and administration of phages at MOI more than 1 saved almost 100% of the mice. The administration of a high dose of phage (MOI of 200) showed protective effects even when the phage was administered 1 h after the bacterial challenge. The administered phages were stably maintained at high levels in the blood as long as the bacteria were present, and their level decreased rapidly with the disappearance of the bacteria. Intraperitoneal administration of another phage ΦMR25 (siphovirus) also showed similar life-saving effect in mice infected with *S. aureus* SA37 (Hoshiba *et al.*, 2010).

In another study, this group purified an endolysin (MV-L) derived from ΦMR11, and showed that its intranasal administration reduced the number of MRSA colonizing the nasal mucosa of mice (Rashel *et al.*, 2007). Furthermore, intraperitoneal administration of MV-L within 30 min after a lethal inoculation of *S. aureus* also rescued 100% of the mice (Rashel *et al.*, 2007). They also identified a tail-associated lysin (Gp61), and described its potential application to controlling *S. aureus* infection (Rashel *et al.*, 2008).

Capparelli *et al.* (2007) carried out more detailed analysis of phage therapy against *S. aureus* infection in mice. First, they successfully isolated a persistently circulating mutant phage M^{Sa} from wild-type phage W^{Sa} using a similar method to that of Merril *et al.* (1996). Interestingly, the mutant phage M^{Sa} was not neutralized by circulating antibodies against M^{Sa}. This finding is extremely important because it demonstrates the method to overcome the problem of anti-phage antibodies in phage therapy. Intravenous injection of *S. aureus* A170 (10^8 CFU) killed 100% of mice within 4 days. However, when M^{Sa} (10^9 PFU) was administered intravenously immediately after the bacterial challenge, the bacterial number in organs (kidney, heart, spleen, and lung) of infected mice decreased dramatically and 97% of them survived. It was also shown that phage can multiply *in vivo* only when bacterial density in the peripheral blood crosses a threshold of approximately 10^4 CFU/ml. Importantly, some experiments showed very high efficacy of M^{Sa} in treating infection caused by MRSA. The authors also examined the possibility of sterilization of mice chronically infected with *S. aureus* by phage administration. When *S. aureus* A170 (5×10^6 CFU) was inoculated intravenously, bacterial cells were maintained at high levels in the blood even 20 days after bacterial inoculation without killing the host. However, when phage M^{Sa} (10^9 PFU) was administered 10 days after the bacterial inoculation, *S. aureus* A170 was cleared from blood, kidney, spleen, and heart within 10 days. This result indicates that the treatment of chronic *S. aureus* infection by phage administration is possible. This group also reported the efficacy of phage M^{Sa} in the treatment of abscesses caused by *S. aureus* (Capparelli *et al.*, 2007; see 'Wound infections') and its lytic activity against the bacteria in macrophages (Capparelli *et al.*, 2007; see 'Potential use of phage therapy against intracellular bacteria'). More recently, Gupta and Prasad (2011) and Hsieh *et al.* (2011) also showed the efficacy of phage therapy in murine *S. aureus* sepsis models.

Sunagar *et al.* (2010) examined the efficacy of phage therapy against *S. aureus* bacteraemia in mice with streptozotocin-induced diabetes. Streptozotocin-treated mice exhibited increased blood glucose levels (>250 mg/ml), glycosuria, ruffled fur, and excessive drinking of water. Intraperitoneal inoculation of 2×10^8 CFU of *S. aureus* RCS21 resulted in 80–100% mortality rate in both non-diabetic and diabetic mice within 2 days. However, intraperitoneal administration of GRCS phage (a myovirus) at 30 min after the bacterial challenge resulted in 90% and 100% survival rates in diabetic and non-diabetic mice, respectively. These lifesaving effects were greater than those of single or multiple intraperitoneal injections of oxacillin (20 mg/kg body weight), and the numbers of bacteria in the blood of phage-treated mice were also lower than those in oxacillin-treated mice in both non-diabetic and diabetic groups. Although survival after delayed phage treatment (4–20 h after the bacterial challenge) was slightly lower in diabetic mice than in non-diabetic mice, the above results indicate the efficacy of phage therapy against *S. aureus* infection in diabetic mice (and probably in diabetic human patients).

Zimecki *et al.* (2009) examined the prophylactic effect of phage administration in mice immunocompromised by cyclophosphamide injection. Four days after intraperitoneal administration of cyclophosphamide, 5×10^6 cells of *S. aureus* L were injected into mice intravenously. Phage A5/L (1×10^6) was administered intraperitoneally 30 min before bacterial inoculation. The prophylactic administration of the phage significantly decreased the bacterial numbers in the spleen, kidney, and liver, and lowered the levels of pro-inflammatory cytokines (IL-6 and TNF-α) in sera of the mice. In addition, the administration of phage resulted in accelerated renewal of immune cells of both myelocytic and lymphocytic lineages depleted by cyclophosphamide. These results suggest that phage administration may be useful for prophylaxis against *S. aureus* infections in immunocompromised hosts.

This group conducted another study to examine the prophylactic potential of phage administration in situations mimicking *S. aureus* infection during bone marrow transplantation (Zimecki et al., 2010). Mice were subjected sequentially to chemotherapy-induced immunosuppression, bone marrow injection, and infection with *S. aureus* L. After 4 oral doses of busulfan (a myeloablative agent), and two intraperitoneal injections of cyclophosphamide (an immunosuppressive agent), syngeneic bone marrow cells (2.5×10^6) were injected into the lateral tail vein of mice. Four days later, 1×10^7 CFU of *S. aureus* L were injected intravenously. In prophylaxis experiments, phage A5 (1×10^6) was administered intraperitoneally 30 min before the bacterial injection. The prophylactic administration of phage A5 caused a significant decrease in the number of bacteria in the liver and spleen compared with the control, and saved 72% of the mice from death, while the survival rate in the control group was only 8.2%. Administration of the phage also increased leucocyte and neutrophil counts in the peripheral blood and the number of cells of the myelocytic cell lineage in the bone marrow. These data indicate that phage administration is potentially beneficial for prophylaxis of bacterial infections during preparation for bone marrow transplantation.

Other models of infection

Animal models of infection used in experimental phage therapy most frequently were discussed above. However, the efficacy of phages was also evaluated in several other models. For example, Hung *et al.* (2011) reported successful use of phages in a murine model of *K. pneumoniae*-induced liver abscesses with bacteraemia. Fukuda *et al.* (2012) showed high effectiveness of phage formulated in eye drops in a murine model of keratitis due to *P. aeruginosa*. In addition, Filippov *et al.* (2012) reported some efficacy of phage in the treatment of experimental bubonic plague in mice.

Potential use of phage therapy against intracellular bacteria

Currently, tuberculosis caused by multidrug-resistant *Mycobacterium tuberculosis* is a serious clinical problem (Kant *et al.*, 2010). Unfortunately, in general, phage therapy is believed to be unsuitable for treating infections caused by facultative intracellular pathogenic bacteria such as *Salmonella, Brucella, Yersinia* and *Mycobacterium* (Barrow and Soothill, 1997). However, to solve this problem, Broxmeyer *et al.* (2002, 2004) designed an ingenious method of introducing bacteriophage into eukaryotic cells by avirulent bacteria transiently infected with the phage to kill intracellular pathogenic bacteria. Specifically, they examined the possibility of lysis by phage TM4 of *M. tuberculosis* and *M. avium* in the mouse peritoneal macrophage cell line RAW 264.7. A monolayer of RAW 264.7 cells was infected with *M. tuberculosis* or *M. avium*, and then exposed to *M. smegmatis*, phage TM4, or *M. smegmatis* carrying phage TM4. *M. smegmatis* is a low-pathogenicity and fast-growing *Mycobacterium* species which was used as a vehicle for mycobacteriophage. The number of *M. tuberculosis* or *M. avium* increased with time in macrophages treated with *M. smegmatis* alone or phage TM4 alone and in the control macrophages which were not subjected to any treatment. However, the bacterial number in macrophages was decreased following their exposure to *M. smegmatis* carrying phage TM4, showing that phage-containing *M. smegmatis* was internalized by macrophages and that the phage could then infect and lyse *M. tuberculosis* and *M. avium* within the macrophages. Peng *et al.* (2006) confirmed this observation in experiments in which *M. tuberculosis*-infected mouse macrophages were exposed to *M. smegmatis* carrying phage D29.

A phenomenon similar to the above was also observed in experiments conducted by Capparelli (2007) who examined the lytic activity of phage M^{Sa} against *S. aureus* A170 in mouse peritoneal macrophages. Mouse macrophages (10^5 cells/well) were infected with *S. aureus* A170 (10^4 CFU) for 1 h. At 3 h, after gentamycin treatment for 1 h to eliminate viable extracellular bacteria, the macrophages were treated with phage M^{Sa} (10^6 PFU) alone or *S. aureus* A170 cells (10^4 CFU) that had adsorbed phage M^{Sa}. The number of viable intramacrophage *S. aureus* cells after treatment with phage M^{Sa} alone for 48 h was almost the same as in the control cells which were not subjected to any treatment. However, treatment with *S. aureus* that had adsorbed the phage decreased the number of *S. aureus* in the macrophages to about 1/100 of the number of staphylococci found in the control cells. These data strongly suggest that *S. aureus* cells within macrophages can be destroyed by a phage if the phage is transferred into the macrophage.

These results suggest the possibility of phage therapy against tuberculosis caused by multidrug-resistant *M. tuberculosis*, which is a serious clinical problem and possibly against some other intracellular bacteria.

Phage therapy using genetically modified phages

In phage therapy against infections caused by Gram-negative bacteria, the rapid release of lipopolysaccharide (LPS) from the bacterial outer membrane following cell lysis poses the risk of endotoxic shock in the treated patients. Therefore, some researchers developed phage therapy methods involving the use of genetically modified phages that can kill bacteria but do not have lytic activity. This approach is based on either filamentous phage used as a delivery vehicle for genes encoding antibacterial toxins (Westwater *et al.*, 2003; Hagens *et al.*, 2004), or on engineered lysis-deficient virulent or temperate phage (Matsuda *et al.*, 2005; Paul *et al.*, 2011). In general, all these studies showed the efficacy of such phages in murine models of infection. In addition, in two studies engineered non-lytic phages were shown to reduce mortality of mice to a greater extent compared with unmodified virulent phages (Hagens *et al.*, 2004; Matsuda *et al.*, 2005); in the other two studies (Westwater *et al.*, 2003; Paul *et al.*, 2011), similar comparisons were not performed. Higher survival rates obtained using non-lytic phages resulted most likely from a reduction in the release of endotoxin from killed bacteria (Hagens *et al.*, 2004). Genetic engineering was also used to modify antibacterial range of phage. To that end, Cao *et al.* (2000) constructed M13 phage expressing a single-chain variable fragment (ScFv), which binds the *Helicobacter pylori* surface antigen. Mice were inoculated orally with *H. pylori* strain 244, which is a good colonizer of the mouse stomach. After deacidification of the stomach by oral administration of omeprazole, the mice were inoculated with *H. pylori* alone or *H. pylori* plus the M13 phage expressing ScFv. It was shown that bacterial number per cm^2 of the mucosa was about five times lower in phage-treated mice than in untreated mice at day 10 after administration of bacteria and phage.

Lu and Collins (2009) constructed a filamentous phage, M13, carrying the *lexA* gene, which codes a repressor (LexA) of the SOS regulon implicated in reducing the efficacy of antibiotic therapy, and used it as an adjuvant of antibiotics. Infection of *E. coli* with this phage improved the efficacy of aminoglycosides, β-lactams, and quinolones as a result of suppression of the SOS network by the LexA repressor. Intravenous administration of the phage with one of the antibiotics at 1 h after 8.8×10^7 CFU of *E. coli* were injected intraperitoneally into mice resulted in improved survival compared with mice which were administered only antibiotic.

Safety of phage therapy

The safety of the phage therapy, with the effectiveness, is an important concern. However, to the best of our knowledge, there are no reports on serious side-effects in animals following systemic administration of phages. Our group has also confirmed repeatedly that intraperitoneal administration of a large number of phages (10^{11} to 10^{12} PFU) to mice did not affect their behaviour or survival rate (Matsuzaki *et al.*, 2003; Uchiyama *et al.*, 2008; Nishikawa *et al.*, 2008; Hoshiba *et al.*, 2010). Furthermore, oral administration of phages was also shown not to exert any adverse effects such as abnormal histological changes or morbidity, or to increase mortality in rats (Carlton *et al.*, 2005). Likewise, the administration of one dose of phage per os to mice did not result in any clinical side-effects (Kang *et al.*, 2013). In addition, repeated intraperitoneal administrations of phage to mice have shown no signs of adverse immunological reaction such as anaphylactic shock (Biswas *et al.*, 2002; Uchiyama *et al.*, 2010). These results strongly suggest the safety of phage therapy, at least in terms of a lack of short-term side-effects following the administration of phage preparations.

Conclusions

Phage therapy against human infectious diseases has a long history: about 80 years in Georgia, Poland and Russia. Many physicians and researchers in these countries have demonstrated the efficacy and safety of phage therapy for human infectious diseases. They have reported the effectiveness of phage therapy against a number of infections including meningitis, lung infections, wound infections, urinary tract infections, gastrointestinal infections, and septicaemia caused by *S. aureus*, *Enterococcus* species, *E. coli*, *Pseudomonas* species, or *Klebsiella* species (Tables 10.1 and 10.2). Although phage therapy remains at a developmental or experimental stage in other countries, most of the animal experiments conducted in those countries essentially seem to confirm the results of researchers and physicians in Georgia, Poland, and Russia.

The animal experiments have confirmed the effectiveness of phage therapy against the diseases that are problematic in humans (Tables 10.1 and 10.2). Furthermore, phage therapy has also been shown to be effective in immunosuppressed and diabetic mice, as well as in a murine model of chronic infection. These results indicate that phage therapy has the potential to treat most of the currently prevalent bacterial infections in humans.

Furthermore, methods to overcome the problems of anti-phage antibodies and of phage therapy against intracellular bacteria such as *M. tuberculosis* have been reported. Namely, it is possible to isolate spontaneous mutant phages insensitive to antibodies against natural unmodified phage particles. Bacteria in macrophages can be killed by phages that are introduced into the cells with the bacteria, indicating that phage therapy against tuberculosis may be possible (Broxmeyer *et al.*, 2002).

Genetically modified phages have also been developed to reduce the release of LPS from bacterial outer membranes, for expression of growth-inhibitory proteins, or for increasing antibiotic sensitivity of bacteria, and their efficacy has been demonstrated.

In the future, more detailed experiments using animal infection models will further confirm the efficacy and safety of phage therapy. Together with further accumulation of evidence of the usefulness of phage therapy in animal experiments, it is necessary to advance research into the practical use of phage therapy in humans. At minimum, the construction of a therapeutic phage bank in which all members are adequately characterized, effective, and safe, together with the establishment of preparation, preservation, and administration methods for phages may be necessary before human trials. Such work has already started (Wright *et al.*, 2009; Merabishvili *et al.*, 2009).

Acknowledgements

This study was supported by the System Glycobiology Center, the Center of Biomembrane Functions Controlling Biological Systems, and the Center for Innovative and Translational Medicine, Kochi University.

References

Abbott, N.J., and Romero, I.A. (1996). Transporting therapeutics across the blood–brain barrier. Mol. Med. Today 3, 106–113.

Ackermann, H.-W. (2006). Classification of bacteriophages. In The Bacteriophages, R. Calendar, ed. (New York, USA, Oxford University Press), pp. 8–16.

Ackermann, H.-W., and Dubow, M.S. (1987). Viruses of prokaryotes. Volume 1. General properties of bacteriophages (Florida, USA, CRC Press).

Alemayehu, D., Casey, P.G., McAuliffe, O., Guinane, C.M., Martin, J.G., Shanahan, F., Coffey, A., Ross, R.P., and Hill, C. (2012). Bacteriophages ΦMR299-2 and ΦNH-4 can eliminate *Pseudomonas aeruginosa* in the murine lung and on cystic fibrosis lung airway cells. MBio. 3, e00029-12.

Alisky, J., Iczkowski, K., Rapoport, A., and Troitsky, N. (1998). Bacteriophages show promise as antimicrobial agents. J. Infect. *36*, 5–15.

Barrow, P.A., and Soothill, J.S. (1997). Bacteriophage therapy and prophylaxis: rediscovery and renewed assessment of potential. Trends Microbiol. 5, 268–271.

Biswas, B., Adhya, S., Washart, P., Paul, B., Trostel, A.N., Powell, B., Carlton, R., and Merril, C.R. (2002). Bacteriophage therapy rescues mice bacteremic from a clinical isolate of vancomycin-resistant *Enterococcus faecium*. Infect. Immun. *70*, 204–210.

Broxmeyer, L. (2004). Bacteriophages: antibacterials with a future? Med. Hypotheses. *62*, 889–893.

Broxmeyer, L., Sosnowska, D., Miltner, E., Chacón, O., Wagner, D., McGarvey, J., Barletta, R.G., and Bermudez, L.E. (2002). Killing of *Mycobacterium avium* and *Mycobacterium tuberculosis* by a mycobacteriophage delivered by a nonvirulent mycobacterium: a model for phage therapy of intracellular bacterial pathogens. J. Infect. Dis. *186*,1155–60.

Cao, J., Sun, Y., Berglindh, T., Mellgård, B., Li, Z., Mårdh, B., and Mårdh, S. (2000). *Helicobacter pylori*-antigen-binding fragments expressed on the filamentous M13 phage prevent bacterial growth. Biochim. Biophys. Acta. *1474*, 107–113.

Capparelli, R., Parlato, M., Borriello, G., Salvatore, P., and Iannelli, D. (2007). Experimental phage therapy against *Staphylococcus aureus* in mice. Antimicrob. Agents Chemother. *51*, 2765–2773.

Capparelli, R., Nocerino, N., Iannaccone, M., Ercolini, D., Parlato, M., Chiara, M., and Iannelli, D. (2010). Bacteriophage therapy of *Salmonella* enterica: a fresh appraisal of bacteriophage therapy. J. Infect. Dis. *201*, 52–61.

Carlton R.M. (1999). Phage therapy: past history and future prospects. Arch. Immunol. Ther. Exp. (Warsz). *47*, 267–274.

Carlton, R.M., Noordman, W.H., Biswas, B., de Meester, E.D., and Loessner, M.J. (2005). Bacteriophage P100 for control of *Listeria monocytogenes* in foods: genome sequence, bioinformatics analyses, oral toxicity study, and application, Reg. Toxicol. Pharmacol. *43*, 301–312.

Carmody, L.A., Gill, J.J., Summer, E.J., Sajjan, U.S., Gonzalez, C.F., Young, R.F., and LiPuma, J.J. (2010). Efficacy of bacteriophage therapy in a model of *Burkholderia cenocepacia* pulmonary infection. J. Infect. Dis. *201*, 264–271.

Cerveny, K.E., DePaola, A., Duckworth, D.H., and Gulig, P.A. (2002). Phage therapy of local and systemic disease caused by *Vibrio vulnificus* in iron-dextran-treated mice. Infect. Immun. *70*, 6251–6262.

Chanishvili, N. (2012). Phage therapy – history from Twort and d'Hérelle through Soviet experience to current approaches. Adv. Virus Res. *83*, 3–40.

Chhibber, S., Kaur, S., and Kumari, S. (2008). Therapeutic potential of bacteriophage in treating *Klebsiella pneumoniae* B5055-mediated lobar pneumonia in mice. J. Med. Microbiol. *57*, 1508–1513.

Chhibber, S., Kaur, T., and Kaur, S. (2013). Co-therapy using lytic bacteriophage and linezolid: effective treatment in eliminating methicillin resistant *Staphylococcus aureus* (MRSA) from diabetic foot infections. PLoS One *8*, e56022.

Chibani-Chennoufi, S., Sidoti, J., Bruttin, A., Kutter, E., Sarker, S., and Brüssow, H. (2004). *In vitro* and *in vivo* bacteriolytic activities of *Escherichia coli* phages: implications for phage therapy. *48*, 2558–2569.

Debarbieux, L., Leduc, D., Maura, D., Morello, E., Criscuolo, A., Grossi, O., Balloy, V., and Touqui, L. (2010). Bacteriophages can treat and prevent *Pseudomonas aeruginosa* lung infections. J. Infect. Dis. *201*, 1096–1104.

Denou, E., Bruttin, A., Barretto, C., Ngom-Bru, C., Brüssow, H., and Zuber, S. (2009). T4 phages against *Escherichia coli* diarrhea: potential and problems. Virology *388*, 21–30.

Dubos, R.J., Straus, J.H., and Pierce, C. (1943). The multiplication of bacteriophage *in vivo* and its protective effect against an experimental infection with *Shigella dysenteriae*. J. Exp. Med. *78*, 161–168.

Filippov, A.A., Sergueev, K.V., He, Y., Huang, X.Z., Gnade, B.T., Mueller, A.J., Fernandez-Prada, C.M., and Nikolich, M.P. (2012). Bacteriophage therapy of experimental bubonic plague in mice. Adv Exp Med Biol. *954*, 337–348.

Fischetti, V.A. (2010). Bacteriophage endolysins: a novel anti-infective to control Gram-positive pathogens. Int. J. Med. Microbiol. *300*, 357–362.

Fukuda, K., Ishida, W., Uchiyama, J., Rashel, M., Kato, S., Morita, T., Muraoka, A., Sumi, T., Matsuzaki, S., Daibata, M., *et al.* (2012). *Pseudomonas aeruginosa* keratitis in mice: effects of topical bacteriophage KPP12 administration. PLoS One *7*, e47742.

Geier, M.R., Trigg, M.E., and Merril, C.R. (1973). Fate of bacteriophage lambda in non-immune germ-free mice. Nature *246*, 221–223.

Górski, A., Wazna, E., Dabrowska, B.W., Dabrowska, K., Switała-Jeleń, K., and Miedzybrodzki, R. (2006). Bacteriophage translocation. FEMS Immunol. Med. Microbiol. *46*, 313–319.

Grandgirard, D., Loeffler, J.M., Fischetti, V.A., and Leib, S.L. (2008). Phage lytic enzyme Cpl-1 for antibacterial therapy in experimental pneumococcal meningitis. J. Infect. Dis. *197*, 1519–1522.

Gu, J., Liu, X., Li, Y., Han, W., Lei, L., Yang, Y., Zhao, H., Gao, Y., Song, J., Lu, R., *et al.* (2012). A method for generation phage cocktail with great therapeutic potential. PLoS One *7*, e31698.

Gupta, R., and Prasad, Y. (2011). Efficacy of polyvalent bacteriophage P-27/HP to control multidrug resistant *Staphylococcus aureus* associated with human infections. Curr. Microbiol. *62*, 255–260.

Hagens, S., Habel, A., von Ahsen, U., von Gabain, A., and Bläsi, U. (2004). Therapy of experimental pseudomonas infections with a nonreplicating genetically modified phage. Antimicrob. Agents Chemother. *48*, 3817–3822.

Hemendinger, R.A., and Bloom, S.E. (1996). Selective mitomycin C and cyclophosphamide induction of apoptosis in differentiating B lymphocytes compared to T lymphocytes in vivo. Immunopharmacology *35*, 71–82.

Heo, Y.J., Lee, Y.R., Jung, H.H., Lee, J., Ko, G., and Cho, Y.H. (2009). Antibacterial efficacy of phages against *Pseudomonas aeruginosa* infections in mice and *Drosophila melanogaster*. Antimicrob. Agents Chemother. *53*, 2469–2474.

Ho, K. (2001). Bacteriophage therapy for bacterial infections. Rekindling a memory from the pre-antibiotics era. Perspect. Biol. Med. *44*, 1–16.

Hoshiba, H., Uchiyama, J., Kato, S., Ujihara, T., Muraoka, A., Daibata, M., Wakiguchi, H., and Matsuzaki, S.

(2010). Isolation and characterization of a novel *Staphylococcus aureus* bacteriophage, ΦMR25, and its therapeutic potential. Arch. Virol. *155*, 545–552.

Hsieh, S.-E., Lo, H.-H., Chen, S.-T., Lee, M.-C., and Tseng, Y.-H. (2011). Wide host range and strong lytic activity of *Staphylococcus aureus* lytic phage Stau2. Appl. Environ. Microbiol. *77*, 756–761.

Hung, C.H., Kuo, C.F., Wang, C.H., Wu, C.M., and Tsao, N. (2011). Experimental phage therapy in treating *Klebsiella pneumoniae*-mediated liver abscesses and bacteremia in mice. Antimicrob. Agents Chemother. *55*, 1358–1365.

Inchley, C.J. (1969). The actvity of mouse Kupffer cells following intravenous injection of T4 bacteriophage. Clin. Exp. Immunol. *5*, 173–187.

Kang, H.W., Kim, J.W., Jung, T.S., and Woo, G.J. (2013). wksl3, a new biocontrol agent for *Salmonella enterica* serovars Enteritidis and Typhimurium in foods: characterization, application, sequence analysis, and oral acute toxicity study. Appl. Environ. Microbiol. *79*, 1956–1968.

Kant, S., Maurya, A.K., Kushwaha, R.A., Nag, V.L., and Prasad, R. (2010). Multi-drug resistant tuberculosis: an iatrogenic problem. Biosci. Trends. *4*, 48–55.

Korhonen, T.K., Valtonen, M.V., Parkkinen, J., Väisänen-Rhen, V., Finne, J., Orskov, F., Orskov, I., Svenson, S.B., and Mäkelä, P.H. (1985). Serotypes, hemolysin production, and receptor recognition of *Escherichia coli* strains associated with neonatal sepsis and meningitis. Infect. Immun. *48*, 486–491.

Kumari, S., Harjai, K., and Chhibber, S. (2009). Efficacy of bacteriophage treatment in murine burn wound infection induced by *Klebsiella pneumoniae*. J. Microbiol. Biotechnol. *19*, 622–628.

Kumari, S., Harjai, K., and Chhibber, S. (2010a). Evidence to support the therapeutic potential of bacteriophage Kpn5 in burn wound infection caused by *Klebsiella pneumoniae* in BALB/c mice. J. Microbiol. Biotechnol. *20*, 935–941.

Kumari, S., Harjai, K., and Chhibber, S. (2010b). Topical treatment of *Klebsiella pneumoniae* B5055 induced burn wound infection in mice using natural products. J. Infect. Dev. Ctries. *4*, 367–377.

Kumari, S., Harjai, K., and Chhibber, S. (2011). Bacteriophage versus antimicrobial agents for the treatment of murine burn wound infection caused by *Klebsiella pneumoniae* B5055. J. Med. Microbiol. *60*, 205–210.

Kutateladze, M., and Adamia, R. (2008). Phage therapy experience at the Eliava Institute. Med. Mal. Infect. *38*, 426–430.

Letkiewicz, S., Międzybrodzki, R., Fortuna, W., Weber-Dąbrowska, B., and Górski, A. (2009). Eradication of *Enterococcus faecalis* by phage therapy in chronic bacterial prostatitis – case report. Folia Microbiol. *54*, 457–461.

Linkous, D.A., and Oliver, J.D. (1999). Pathogenesis of *Vibrio vulnificus*. FEMS Microbiol. Lett. *174*, 207–214.

Lu, T.K., and Collins, J.J. (2009). Engineered bacteriophage targeting gene networks as adjuvants for antibiotic therapy. Proc. Natl. Acad. Sci. U.S.A. *106*, 4629–4634.

McVay, C.S., Velásquez, M., and Fralick, J.A. (2007). Phage therapy of *Pseudomonas aeruginosa* infection in a mouse burn wound model. Antimicrob. Agents Chemother. *51*, 1934–1938.

Mai, V., Ukhanova, M., Visone, L., Abuladze, T., and Sulakvelidze, A. (2010). Bacteriophage administration reduces the concentration of *Listeria monocytogenes* in the gastrointestinal tract and its translocation to spleen and liver in experimentally infected mice. Int. J. Microbiol. *2010*, 624234.

Matsuda, T., Freeman, T.A., Hilbert, D.W., Duff, M., Fuortes, M., Stapleton, P.P., and Daly, J.M. (2005). Lysis-deficient bacteriophage therapy decreases endotoxin and inflammatory mediator release and improves survival in a murine peritonitis model. Surgery *137*, 639–646.

Matsuzaki, S., Yasuda, M., Nishikawa, H., Kuroda, M., Ujihara, T., Shuin, T., Shen, Y., Jin, Z., Fujimoto, S., Nasimuzzaman, M.D., *et al.* (2003). Experimental protection of mice against lethal *Staphylococcus aureus* infection by novel bacteriophage ΦMR11. J. Infect. Dis. *187*, 613–624.

Maura, D., Galtier, M., Le Bouguénec, C., and Debarbieux, L. (2012). Virulent bacteriophages can target O104:H4 enteroaggregative *Escherichia coli* in the mouse intestine. Antimicrob. Agents Chemother. *56*, 6235–6242.

Merabishvili, M., Pirnay, J-P., Verbeken, G., Chanishvili, N., Tediashvili, M., Lashkhi, N., Glonti, T., Krylov, V., Mast, J., Van Parys, L., *et al.* (2009). Quality-controlled small-scale production of a well-defined bacteriophage cocktail for use in human clinical trials. PLoS One *4*, e4944.

Merril, C.R., Biswas, B., Carlton, R., Jensen, N.C., Creed, G.J., Zullo, S., and Adhya, S. (1996). Long-circulating bacteriophage as antibacterial agents. Proc. Natl. Acad. Sci. *93*, 3188–3192.

Merril, C.R., Scholl, D., and Adhya, S.L. (2003). The prospect for bacteriophage therapy in Western medicine. Nat. Rev. Drug Discov. *2*, 489–497.

Międzybrodzki, R., Borysowski, J., Weber-Dąbrowska, B., Fortuna, W., Letkiewicz, S., Szufnarowski, K., Pawełczyk, Z., Rogóż, P., Kłak, M., Wojtasik, E., and Górski, A. (2012). Clinical aspects of phage therapy. Adv. Virus Res. *83*, 73–121.

Morello, E., Saussereau, E., Maura, D., Huerre, M., Touqui, L., and Debarbieux, L. (2011). Pulmonary bacteriophage therapy on *Pseudomonas aeruginosa* cystic fibrosis strains: first steps towards treatment and prevention. PLOS ONE *6*, e16963.

Mushtaq, N., Redpath, M.B., Luzio, J.P., and Taylor, P.W. (2004). Prevention and cure of systemic *Escherichia coli* K1 infection by modification of the bacterial phenotype. Antimicrob. Agents Chemother. *48*, 1503–1508.

Nishikawa, H., Yasuda, M., Uchiyama, J., Rashel, M., Maeda, Y., Takemura, I., Sugihara, S., Ujihara, T., Shimizu, Y., Shuin, T., and Matsuzaki, S. (2008). T-even-related bacteriophages as candidates for treatment of *Escherichia coli* urinary tract infections. Arch. Virol. *153*, 507–515.

Paul, V.D., Sundarrajan, S., Rajagopalan, S.S., Hariharan, S., Kempashanaiah, N., Padmanabhan, S., Sriram, B., and Ramachandran, J. (2011). Lysis-deficient phages as novel therapeutic agents for controlling bacterial infection. BMC Microbiol. *11*, 195.

Peng, L., Chen, B.W., Luo, Y.A., and Wang, G.Z. (2006). Effect of mycobacteriophage to intracellular mycobacteria *in vitro*. Chin. Med. J. *119*, 692–695.

Pouillot, F., Chomton, M., Blois, H., Courroux, C., Noelig, J., Bidet, P., Bingen, E., and Bonacorsi, S. (2012). Efficacy of bacteriophage therapy in experimental sepsis and meningitis caused by a clone O25b:H4-ST131 *Escherichia coli* strain producing CTX-M-15. Antimicrob. Agents Chemother. *56*, 3568–3575.

Rashel, M., Uchiyama, J., Ujihara, T., Uehara, Y., Kuramoto, S., Sugihara, S., Yagyu, K., Muraoka, A., Sugai, M., Hiramatsu, K., *et al.* (2007). Efficient elimination of multidrug-resistant *Staphylococcus aureus* by cloned lysin derived from bacteriophage ΦMR11. J. Infect. Dis. *196*, 1237–1247.

Rashel, M., Uchiyama, J., Takemura, I., Hoshiba, H., Ujihara, T., Takatsuji, H., Honke, K., and Matsuzaki, S. (2008). Tail-associated structural protein gp61 of *Staphylococcus aureus* phage ΦMR11 has bifunctional lytic activity. FEMS Microbiol. Lett. *284*, 9–16.

Schlech III, W.F. (1998). Infections caused by *Listeria monocytogenes* and *Erysipelothrix rhusiopathiae*. In Microbiology and Microbial Infections, Vol. 3, W.J. Hausler Jr and M. Sussman, eds. (New York, USA, Oxford University Press), pp. 593–600.

Seed, K.D., and Dennis, J.J. (2009). Experimental bacteriophage therapy increases survival of *Galleria mellonella* larvae infected with clinically relevant strains of the *Burkholderia cepacia* complex. Antimicrob. Agents Chemother. *53*, 2205–2208.

Sheng, H., Knecht, H.J., Kudva, I.T., and Hovde, C.J. (2006). Application of bacteriophages to control intestinal *Escherichia coli* O157:H7 levels in ruminants. Appl. Environ. Microbiol. *72*, 5359–5366.

Smith, H.W., and Huggins, M.B. (1982). Successful treatment of experimental *Escherichia coli* infections in mice using phage: its general superiority over antibiotics. J. Gen. Microbiol. *128*, 307–318.

Soothill, J.S. (1992). Treatment of experimental infections of mice with bacteriophages. J. Med. Microbiol. *37*, 258–261.

Srivastava, A.S., Kaido, T., and Carrier, E. (2004). Immunological factors that affect the *in vivo* fate of T7 phage in the mouse. J. Virol. Methods, *115*, 99–104.

Summers, W.C. (2001). Bacteriophage therapy. Annu. Rev. Microbiol. *55*, 437–451.

Sunagar, R., Patil, S.A., and Chandrakanth, R.K. (2010). Bacteriophage therapy for *Staphylococcus aureus* bacteremia in streptozotocin-induced diabetic mice. Res. Microbiol. *161*, 854–860.

Takemura-Uchiyama, I., Uchiyama, J., Kato, S., Inoue, T., Ujihara, T., Ohara, N., Daibata, M., and Matsuzaki, S. (2013). Evaluating efficacy of bacteriophage therapy against *Staphylococcus aureus* infections using a silkworm larval infection model. FEMS Microbiol. Lett. *347*, 52–60.

Tanji, Y., Shimada, T., Fukudomi, H., Miyanaga, K., Nakai, Y., and Unno, H. (2005). Therapeutic use of phage cocktail for controlling *Escherichia coli* O157:H7 in gastrointestinal tract of mice. J. Biosci. Bioeng. *100*, 280–287.

Tóthová, L., Celec, P., Bábíčková, J., Gajdošová, J., Al-Alami, H., Kamodyova, N., Drahovská, H., Liptáková, A., Turňa, J., and Hodosy, J. (2011). Phage therapy of Cronobacter-induced urinary tract infection in mice. Med. Sci. Monit. *17*, BR173–178.

Uchiyama, J., Rashel, M., Maeda, Y., Takemura, I., Sugihara, S., Akechi, K., Muraoka, A., Wakiguchi, H., and Matsuzaki, S. (2008a). Isolation and characterization of a novel *Enterococcus faecalis* bacteriophage ΦEF24C as a therapeutic candidate. FEMS Microbiol. Lett. *278*, 200–206.

Uchiyama, J., Rashel, M., Takemura, I., Wakiguchi, H., and Matsuzaki, S. (2008b). *In silico* and *in vivo* evaluation of bacteriophage ΦEF24C, a candidate for treatment of *Enterococcus faecalis* infections. Appl. Environ. Microbiol. *74*, 4149–4163.

Uchiyama, J., Maeda, Y., Takemura, I., Chess-Williams, R., Wakiguchi, H., and Matsuzaki, S. (2009). Blood kinetics of four intraperitoneally administered therapeutic candidate bacteriophages in healthy and neutropenic mice. Microbiol. Immunol. *53*, 301–304.

Uchiyama, J., Takemura, I., Hayashi, I., Matsuzaki, S., Satoh, M., Ujihara, T., Murakami, M., Imajoh, M., Sugai, M., and Daibata, M. (2011). Characterization of lytic enzyme open reading frame 9 (ORF9) derived from *Enterococcus faecalis* bacteriophage ΦEF24C. Appl. Environ. Microbiol. *77*, 580–585.

Vincent, J.L., Rello, J., Marshall, J., Silva, E., Anzueto, A., Martin, C.D., Moreno, R., Lipman, J., Gomersall, C., Sakr, Y., Reinhart, K.; EPIC II Group of Investigators. (2009). International study of the prevalence and outcomes of infection in intensive care units. JAMA *302*, 2323–2329.

Vinodkumar, C.S., Neelagund, Y.F., and Kalsurmath, S. (2005). Bacteriophage in the treatment of experimental septicemic mice from a clinical isolate of multidrug resistant *Klebsiella pneumoniae*. J. Commun. Dis. *37*, 18–29.

Vinodkumar, C.S., Kalsurmath, S., and Neelagund, Y.F. (2008). Utility of lytic bacteriophage in the treatment of multidrug-resistant *Pseudomonas aeruginosa* septicemia in mice. Indian J. Pathol. Microbiol. *51*, 360–366.

Vogels, M.T., Hermsen, C.C., Huys, H.L., Eling, W.M., and van der Meer, J.W. (1994). Roles of tumor necrosis factor, platelet-activating factor and arachidonic acid metabolites in interleukin 1-induced resistance to infection in neutropenic mice. Infect. Immun. *62*, 2065–2070.

Wang, J., Hu, B., Xu, M., Yan, Q., Liu, S., Zhu, X., Sun, Z., Reed, E., Ding, L., Gong, J., *et al.* (2006a). Use of bacteriophage in the treatment of experimental animal bacteremia from imipenem-resistant *Pseudomonas aeruginosa*. Int. J. Mol. Med. *17*, 309–317.

Wang, J., Hu, B., Xu, M., Yan, Q., Liu, S., Zhu, X., Sun, Z., Tao, D., Ding, L., Reed, E., *et al.* (2006b). Therapeutic effectiveness of bacteriophages in the rescue of mice with extended spectrum beta-lactamase-producing *Escherichia coli* bacteremia. Int. J. Mol. Med. *17*, 347–355.

Waseh, S., Hanifi-Moghaddam, P., Coleman, R., Masotti, M., Ryan, S., Foss, M., MacKenzie, R., Henry, M., Szymanski, C.M., and Tanha, J. (2010). Orally administered P22 phage tailspike protein reduces salmonella colonization in chickens: prospects of a novel therapy against bacterial infections. PLoS One *5*, e13904.

Watanabe, R., Matsumoto, T., Sano, G., Ishii, Y., Tateda, K., Sumiyama, Y., Uchiyama, J., Sakurai, S., Matsuzaki, S., Imai, S., and Yamaguchi, K. (2007). Efficacy of bacteriophage therapy against gut-derived sepsis caused by *Pseudomonas aeruginosa* in mice. Antimicrob. Agents Chemother. *51*, 446–452.

Weber-Dabrowska, B., Mulczyk, M., and Górski, A. (2000). Bacteriophage therapy of bacterial infections: an update of our institute's experience. *48*, 547–551.

Weiss, M., Denou, E., Bruttin, A., Serra-Moreno, R., Dillmann, M.L., and Brüssow, H. (2009). In vivo replication of T4 and T7 bacteriophages in germ-free mice colonized with *Escherichia coli*. Virology *393*, 16–23.

Westwater, C., Kasman, L.M., Schofield, D.A., Werner, P.A., Dolan, J.W., Schmidt, M.G., and Norris, J.S. (2003). Use of genetically engineered phage to deliver antimicrobial agents to bacteria: an alternative therapy for treatment of bacterial infections. Antimicrob. Agents Chemother. *47*, 1301–1307.

Wills, Q.F., Kerrigan, C., and Soothill, J.S. (2005). Experimental bacteriophage protection against *Staphylococcus aureus* abscesses in a rabbit model. Antimicrob. Agents Chemother. *49*, 1220–1221.

Wright, A., Hawkins, C.H., Anggård, E.E., and Harper, D.R. (2009). A controlled clinical trial of a therapeutic bacteriophage preparation in chronic otitis due to antibiotic-resistant *Pseudomonas aeruginosa*; a preliminary report of efficacy. Clin. Otolaryngol. *34*, 349–357.

Zimecki, M., Artym, J., Kocieba, M., Weber-Dabrowska, B., Borysowski, J., and Górski, A. (2009). Effects of prophylactic administration of bacteriophages to immunosuppressed mice infected with *Staphylococcus aureus*. BMC Microbiol. *9*, 169.

Zimecki, M., Artym, J., Kocieba, M., Weber-Dabrowska, B., Borysowski, J., and Górski, A. (2010). Prophylactic effect of bacteriophages on mice subjected to chemotherapy-induced immunosuppression and bone marrow transplant upon infection with *Staphylococcus aureus*. Med. Microbiol. Immunol. *199*, 71–79.

Clinical Phage Therapy

11

Elizabeth Kutter, Jan Borysowski, Ryszard Międzybrodzki, Andrzej Górski, Beata Weber-Dąbrowska, Mzia Kutateladze, Zemphira Alavidze, Marina Goderdzishvili and Revaz Adamia†

Abstract

Since the first therapeutic use of phages in humans in 1919, a large number of studies have been conducted to evaluate the safety and the efficacy of phage therapy in a wide range of bacterial infections, and phage therapy has become a well-accepted part of clinical practice in some parts of the world. Elsewhere, interest in phage therapy has been growing strongly in parallel with the emerging threats of antibiotic resistance worldwide and the paucity of new approaches, but progress has been hampered by the lack of the rigorous double-blind clinical trials now widely required for governmental approval and the challenges in getting funding for such studies. Earlier therapeutic phage applications are important for further development of clinical phage therapy in multiple ways. First, their results strongly suggest the safety of bacteriophage preparations; no serious side-effects have been reported despite their use by hundreds of thousands of people, suggesting that any such effects are almost certainly rare and/or subtle. Secondly, they may help to select specific diseases and approaches to be best targeted by future more tightly controlled trials of phage therapy. Thirdly, more in-depth exploration of the strongest earlier work and of recent advances in the field should help spur private, governmental and academic investment in further research. The main goal of this chapter is to discuss in detail the recent controlled trials of bacteriophage preparations along with the clinical experiences of the two major centres of phage therapy, the Ludwik Hirszfeld Institute of Immunology and Experimental Therapy, Wrocław, Poland, and the George Eliava Institute of Bacteriophages, Microbiology and Virology, Tbilisi, Georgia. The positive results of hitherto conducted studies, as well as dramatic increase in the prevalence of antibiotic-resistant bacteria, warrant concerted further explorations of clinical phage therapy as one element in our arsenal.

Introduction

Concern continues to grow worldwide about the very serious problem of antibiotic resistance and about the need for new and creative approaches to address the crisis (Giamarellou, 2010; Nathan, 2012; Calfee, 2012). The World Health Organization (WHO) has been particularly actively involved, as reflected in many pronouncements in recent years and in the book *The Evolving Threat Of Antimicrobial Resistance – Options for Action*, launched on 8 March 2012 (http://whqlibdoc.who.int/publications/2012/9789241503181_eng.pdf). This book is the result of a collaboration between WHO scientists and policy makers and more than 50 experts in the field from around the world. However, there has been far too little discussion of a potential role for phage therapy in combating the problem. This is beginning to change. For example, in May, 2011, when Jim Watson convened a Banbury conference involving 28 international experts on the topic from industrial, academic and governmental bodies to explore possible steps forward, phage therapy was included in the discussions (Antibiotic Resistance: Past, Present, Future, Cold Spring Harbour, 15–18 May; Bush *et al.*, 2011). There is significant evidence and experience supporting the efficacy of phage in a variety of clinical infections, including those caused by antibiotic-resistant bacteria, and extensive data on that work are becoming increasingly available (Sulakvelidze and Kutter, 2005; Kutter *et al.*, 2010; Abedon *et al.*, 2011; Burrowes *et al.*, 2011; Chanishvili, 2012b; Międzybrodzki *et al.*, 2012). However, wider expansion of phage therapy has been largely stymied by the lack of double-blind randomized controlled trials (RCTs) and the difficulties in getting the enormous financial support needed for such trials from either industrial or governmental sources. In recent years, several more detailed small-scale RCTs of phage preparations have been conducted that support the safety of phage therapy as reported in numerous earlier less well documented trials (Bruttin and Brüssow, 2005; Wright *et al.*, 2009; Rhoads *et al.*, 2009; Sarker *et al.*, 2012; Brüssow, 2012; McCallin *et al.*, 2013). We believe that the largely positive results of the great number of applications through the years and the recent RCTs, along with substantial progress of knowledge about phage biology, make phage therapy a potentially attractive option to combat multidrug-resistant bacteria in clinical medicine. The earlier studies, in spite of apparent methodological shortcomings, strongly suggest the safety of phage therapy and help to define the specific applications of phage preparations which could best be evaluated by further controlled trials. Academic–corporate–governmental collaborations and key mergers of companies involved in phage therapy research appear to finally be laying the groundwork for more extensive clinical trials that build on this experience.

The main objectives of this chapter are to explore in depth the recent clinical trials of phage preparations and the very extensive clinical experiences of the two major long-time centres of phage therapy research and implementation (the Institute of Immunology and Experimental Therapy (IIET), Wrocław, Poland, and the Eliava Institute, Tbilisi, Georgia), and to look at work under way that draws on combined expertise from many sources.

Randomized controlled trials of phage therapy

The first trial of phage therapy in humans was carried out in 1919 by Félix d'Hérelle, the co-discoverer of bacteriophages (Abedon *et al.*, 2011). However, d'Hérelle delayed the publication of his results pending further studies of the nature of phage, and the first report on therapeutic use of phage in humans was published by Bruynoghe and Maisin (1921). Since then, a great number of studies have been conducted to evaluate the efficacy and safety of phage therapy and much clinical experience has been accumulated (for detailed reviews, see Alisky *et al.*, 1998; Summers, 2001; Sulakvelidze *et al.*, 2001; Sulakvelidze and Kutter, 2005; Górski *et al.*, 2009; Kutter *et al.*, 2010; Abedon *et al.*, 2011; Harper *et al.*, 2011; Burrowes *et al.*, 2011; Chanishvili, 2012b). However, these earlier studies did not appear to meet the current rigorous standards for clinical trials, and many key data were unavailable. Only recently have the first double-blind, extensively controlled trials of phage therapy started.

The first known phase I clinical trial meeting current rigorous standards was carried out in Britain by Exponential Biotherapies, Inc. under Richard Carlton, MD in 2000, and involved a single, largely sequenced phage targeting vancomycin-resistant *Enterococcus* (VRE) (R. Carlton, personal communication, 21 December 2012). The trial, approved and supervised by the institutional review board, is particularly noteworthy for having tested an intravenous application for safety and for pharmacokinetics in the bloodstream. On days 1, 3 and 9 of the trial, 12 healthy volunteers were each injected i.v. with 5×10^6 phage particles, an amount designed to give the same phage concentration as that of VRE found in VRE septicaemia – about 10^3 per ml. The only adverse event observed was a transient rash on one volunteer's arm, with no sequelae. After the first injection, the expected level of phage was observed. However, after the day-3 injection, the level was an order of magnitude lower, and only a few score phage per ml were detected after the day-9 injection. It thus appeared that the volunteers' immune systems were specifically activated by the initial doses to produce antibodies that inactivated later doses, even though the phage used had been selected by their proprietary method to be long-circulating. This made it unlikely that it could successfully treat VRE septicaemia, which often results from ongoing seeding into the bloodstream from VRE-colonized tissues, such as the skin or the gastrointestinal tract. Multiple doses of an anti-infective agent would then be needed. Under the circumstances, they felt they could not raise sufficient funds for trials of other approaches of administration, and they did not proceed to the planned double-blind phase Ib/IIa trial, which would have involved larger numbers and actual patients, allowing them to also collect initial efficacy data. Instead, the company switched to other therapeutic targets, and never published this phase Ia trial.

By July 2013 the results of five RCTs of phage preparations have been published in the peer-reviewed literature (Bruttin and Brüssow, 2005; Wright *et al.*, 2009; Rhoads *et al.*, 2009; Sarker *et al.*, 2012; McCallin *et al.*, 2013). At least two other trials are now under way: the very extensive Bangladeshi phase II/III trial of the Nestlé infant-diarrhoea product (www.clinicaltrials.gov/ct2/show/NCT00937274) and a series of compassionate-use studies under the Helsinki protocol at the Military Hospital in Brussels (Merabishvili *et al.*, 2012). A pilot clinical study was also conducted to evaluate the safety of a phage preparation in burn wound patients, with the methodology of phage characterization, preparation and administration discussed in great detail (Merabishvili *et al.*, 2009). We here explore those trials in some depth, consider some of the challenges that have delayed the implementation of more such formal double-blind trials, and then explore some of the complementing broad data from the widespread clinical application of phage therapy over the last 90 years, largely in Poland and Georgia, particularly as related to issues of safety.

The first of these formal Western RCTs was conducted at the Nestlé Research Center in Lausanne, Switzerland (Bruttin and Brüssow, 2005). It was a double-blinded randomized placebo-controlled phase I trial whose main objective was to evaluate the safety of oral administration of phage. To that end, 15 healthy adult volunteers received 150 ml of mineral water containing a low dose of T4 phage (10^3 PFU/ml), a higher dose of the phage (10^5 PFU/ml), or placebo three times per day for 2 consecutive days. There was no formal separate control group in this trial – each subject served as an internal control and received all treatments in a random order (repeated measures design). This study showed no significant side-effects following the administration of the phage preparation. The levels of markers for liver toxicity were also measured; no increases in serum alanine aminotransferase or aspartate aminotransferase were observed. Although mild adverse events occurred for 5 volunteers over the course of the trial, these did not seem to be related to phage administration and no-one required any particular treatment. The authors also showed that oral administration of T4 did not disturb the balance of the *E. coli* gut population. Not even the higher dose reduced the total *E. coli* faecal counts. However, pre-trial 12 of 168 random colonies (7%) from the 15 volunteers were found to be susceptible to T4, while afterwards, only 1 of 116 random colonies was susceptible (0.9%). This apparent ability of phage to greatly reduce strains targeted by the specific phage used without shifting the overall species balance is encouraging, emphasizing the potential usefulness of phage in maintaining non-pathogenic natural intestinal microflora while targeting pathogens.

Three main facts should be considered when interpreting and evaluating the results of the study.

First, although the upper total phage dose of 1.5×10^7 PFU employed by the investigators falls within the range of the standard doses of about 10^6 to 10^8 PFU reportedly administered to patients at both major centres of phage therapy, the IIET and the Eliava Institute (Górski *et al.*, 2009; Chanishvili, 2012a), it was given for a shorter time.

Second, there was no indication here that T4 phage particles were absorbed from the intestine to the blood, since neither T4 phage nor anti-phage antibodies were detected in the peripheral blood of the subjects. However, it should be noted that no anti-T4 antibodies could be expected within the time frame of this study.

Third, T4, due to its narrow range on the common serious pathogenic strains of *E. coli*, is not likely to be useful in phage therapy. Nevertheless, it was used in this trial because it has been characterized at the genetic, molecular, and physiological levels in such great detail (see Miller *et al.*, 2003) and because a large fraction of the therapeutic phages targeting serious *E. coli* pathogens are rather close relatives of T4 (in particular, all of the phages used in their later phase II trial in Bangladesh are T4-related, see below). In this context it is also noteworthy that T4-like phages are known to have highly varied and overlapping host ranges, as well as to be able to recognize different receptors on different hosts. For example, on *E. coli* B, T4 itself uses B's unique lipopolysaccharide as its receptor, while on K12 strains it primarily targets outer-membrane protein OmpC. T4 itself hits 28/72 of the members of the *E. coli* collection of reference (ECOR), which is a higher fraction than that achieved by two-thirds of a tested set of 49 T4-like phages from around the world, with the highest hitting 47 strains, the lowest only four strains, and only seven targeting over 36 strains (Kutter, 2009). Every ECOR strain was hit by at least one of this set of phages.

The choice of T4 for this initial human safety trial becomes clearer when put into context. This particular trial is but one small step in the long, complex and expensive process of developing and testing a targeted commercial phage therapeutic by H. Brüssow's group, who had long been working on the very serious problem of 3rd-world infant diarrhoea due to *E. coli*. A number of the broadest-spectrum T4-like phages from the study cited above (Kutter, 2009) were tested against bacteria collected from infant diarrheal patients in Dakka, Bangladesh; some worked, but their efficiency of plating was not high enough to seem promising. Therefore, new phages were isolated from the stools of those young Bangladeshi patients (Chibani-Chennoufi *et al.*, 2004). Two different *E. coli* strains were chosen for the selection: a widespread enteropathogenic *E. coli* (EPEC) strain, O127:K63, and the common lab strain K803. Using the EPEC strain, phages were isolated from 18 out of 120 stool samples; all of these were *Siphoviridae*, from three different families, and none of them hit more than a few of their *E. coli* strains isolated from the Bangladeshi infant stool samples. In contrast, when K803 was used for the selection, all isolated phages but one were T4-like. Many of these had very broad host ranges on 10 common Bangladeshi strains and on a large set of EPEC and enterohaemorrhagic *E. coli* (EHEC) strains reflecting 21 different O antigens, 10 K antigens and 10 H antigens. Cross-hybridization patterns showed that only 12 out of 31 phages tested were T-even *sensu stricto*, while 19 represented more distant members of the T4 family in different genera.

Building on the successes of these preliminary trials, both the search for the most effective phages for applications in Bangladesh and the animal studies were expanded very substantially (Denou *et al.*, 2009). Six sequenced phages, representing four of the five known subgroups of T4-like coliphages, were tested against a London panel of 25 EPEC and enterotoxigenic *E. coli* (ETEC) strains with known O, K and H serotypes that were associated with infant diarrhoea in the 1970s. RB49 lysed 8 out of the 25 strains, T4 lysed two strains, and the other phages lysed up to four strains; a cocktail of all 6 phages lysed 13 strains (52%). They then obtained 46 epidemiologically independent pathogenic *E. coli* patient isolates from the International Centre for Diarrheal Disease Research in Dhaka, Bangladesh (ICDDR,B), including 15 EHEC strains, 15 ETEC strains, 10 enteroaggregative *E. coli* (EAggEC) strains, three enteroinvasive *E. coli* (EIEC) strains and three verotoxin-producing *E. coli* (VTEC) strains. The above cocktail showed only 18% coverage against these recent pathogenic *E. coli*. They then tested their entire collection of about 140 T4-like phages from around the world against these strains and found a set of 10 phages giving 52% and 16 giving two-thirds coverage. They also evaluated the safety of oral administration of phages, their effects on the gut microbiota, and investigated the gut survival of phages from 3 different subgroups, measuring concentrations in different segments of the gut in mice, with or without their first having been inoculated with 10^{10} phage-sensitive *E. coli* K12 by intragastric tube (Denou *et al.*, 2009).

This series of studies makes clearer the necessity and complexities of using complex cocktails to deal with many kinds of enteric infections, and the challenges facing efforts to develop appropriate phage preparations that fit the Western paradigm for such applications. It should be noted that the Georgian Intestiphage (see 'Phage therapy in Georgia', below) contains hundreds of different phages that together target over 28 bacterial species, including phages targeting gut-derived *Pseudomonas* and *Staphylococcus* strains in addition to a wide range of bacteria usually classified as 'enteric'. The Nestlé infant diarrhoea studies have provided the opportunity to at least narrow the spectrum to focusing on a defined and widely problematic group of *E. coli*, but that range is still very broad.

The next step of H. Brüssow's group was testing the genetic safety of their entire collection of 99 T4-like phages and the clinical risk of a subset of them in an oral cocktail (Sarker *et al.*, 2012). The sequencing and analysis of the whole set of DNA revealed no genes posing either a direct threat for clinical phage application or elements that might contribute to the sort of gene transfer sometimes mediated by temperate phages. The biological safety of a cocktail of nine phages selected from the set on the basis of ability to lyse relevant *E. coli* diarrheal strains was examined in a randomized, double-blind, placebo-controlled phase I trial, conducted at ICDDR,B. Fifteen healthy adult volunteers were subjected to the experimental design similar to that employed in the study of Bruttin and Brüssow (2005), but using higher phage doses: the low dose used was 3×10^7 PFU and the high dose was 3×10^9 PFU of total phage, 100-fold greater than the previous higher T4 dose. Clinical observation as well as thorough analysis of the changes in the wide panel of biochemical and haematological parameters gave no indication that phage application could cause any significant adverse events even when phages were applied at the higher dose. There was no significant difference in the frequency of phage detection in faeces between the periods when the subjects were treated with placebo (28% of positive samples) and low phage dose (30%), but for the period with a high phage dose this parameter was significantly higher (64%). In addition, two small safety trials of phage preparations were performed in healthy children in Bangladesh (Brüssow, 2012).

Similar metagenomic analysis and clinical safety testing were also carried out for a commercial therapeutic coliphage preparation called ColiProteus made by the Russian pharmaceutical company Microgen (McCallin *et al.*, 2013) which is being used as a third arm in the Bangladeshi clinical studies. The metagenomic analysis of this cocktail documented its expected high degree of complexity and its total dependence on lytic phages, with no signs of elements potentially related to pathogenicity. A substantial fraction of the phages were related to T4, including members of the T4D, RB69, and RB49 subgroups but not the JS98 subgroup first observed in the new T4-related species that was fairly prominent among the phages isolated from the stools of infant diarrhoeal patients in Bangladesh. Other *Myoviridae* were also seen: some related to the recently described 140-kb coliphage rv5 and others to the 86-kb *Salmonella* phage Felix 01. However, the most abundant phages were *Podoviridae*, particularly 40-kb ones related to T7, as well as some related to 73-kb N4. Only one siphovirus was seen – an unclassified phage of 46-kb. It will be very interesting to see whether there are significant differences between the efficacy of these two quite different sorts of cocktails. In addition, a safety trial of the cocktail was performed on 15 healthy individuals (five adults and 10 children). Like the trials by Bruttin and Brüssow (2005) and by Sarker (2012), this trial did not involve a separate control group; rather, each subject received a higher phage dose (at least 1.4×10^8 PFU), a tenfold lower phage dose and placebo three times daily for 2 days in a random order thus serving as his/her own control. Safety of phage application was evaluated based on a panel of clinical, clinical chemistry, and haematological parameters. Overall, this trial showed that oral administration of the cocktail was safe. While some laboratory abnormalities were observed in some subjects, these were not associated with phage administration. Likewise, reported minor adverse events were not attributable to administration of the higher phage dose.

Phase II of the trial by Brüssow's group started in August, 2009 in the main diarrhoeal disease hospital in Bangladesh. It is a major double blind, placebo-controlled study evaluating the safety and efficacy of an orally administered cocktail of 9 of these T4-like phages that were isolated from infant diarrheal patients in treating diarrhoea due to ETEC and EPEC in children aged 6–24 months. (www.clinicaltrials.gov/ct2/show/NCT00937274) In addition to the study group treated with the Nestlé cocktail of T4-like phages and the control group that receives placebo, the study uses, as a third group, children treated with the cocktail ColiProteus. The plan is to enrol 450 children between 6 and 24 months old with non-cholera, non-rotavirus, non-invasive diarrhoea. Phage preparations are being administered for up to 5 days at a dose of 10^6 PFU/ml in the standard rehydration fluid each child is receiving. The main outcome measures for the study include duration of diarrhoea, daily and cumulative stool output, volume of oral rehydration solution intake, stool frequency, time to recovery, phage excretion levels and weight gain. The current estimated completion date for this very extensive phase of the trial is October 2014.

The fact that a major corporation like Nestlé has invested in such a major and thorough trial, has been involved in publishing such in-depth safety studies and has convinced the regulatory bodies of the safety of carrying out this trial in infants is considered very important and encouraging for the field. The already-published clinical trials have been on a much smaller scale.

The first published modern RCT to evaluate not only the safety but also the efficacy of phage therapy was conducted at the UCL Ear Institute and Royal National Throat, Nose, and Ear Hospital in London by the British company Biophage (Wright *et al.*, 2009). The scientists involved in developing Biophage included James Soothill, who had conducted earlier studies of phage therapy against dog-ear infections and other phage therapy related explorations (Soothill *et al.*, 2004, Hawkins *et al.*, 2010). In this phase I/II study, 24 patients with long-term chronic otitis due to antibiotic-resistant *P. aeruginosa* were administered a single topical dose of Biophage-PA (containing six *P. aeruginosa*-specific phages) or a placebo. The trial showed a significant clinical improvement and a decrease in *P. aeruginosa* counts in ear swabs in patients who received phages compared with the control group. Moreover, no treatment-related adverse events occurred in patients who were administered Biophage-PA. The planned follow-up extensive phase 2 trial with a larger population and expanded protocol was delayed due to financial challenges. However, the merger of Biophage with the US-based Targeted Genetics to form a new company, AmpliPhi Biosciences, and its further merger with Australia-based Special Phage Services, which has conducted phage work for nearly a decade, has now given them the crucial resources, the range of expertise of various kinds and the added US and broadly international base and credibility to move ahead on this clinical trial and several others involving other pathogens in the near future.

The first published FDA-approved clinical trial of phage therapy was a self-funded physician-initiated phase I study performed in 2007 and 2008 at the Southwest Regional Wound Care Center in Lubbock, Texas, USA (Rhoads *et al.*, 2009). It was undertaken as a result of extensive successful application, on a compassionate use basis, of Pyophage cocktails brought in from the Eliava Institute in Tbilisi. However, the FDA approval required that this first safety trial use a well-defined small set of well-characterized phages. Therefore, they arranged with Intralytics Inc. in Baltimore to produce a special cocktail, WPP-201, containing eight sequenced phages: two targeting *S. aureus*, five targeting *P. aeruginosa* and one targeting *E. coli.* This was a prospective, controlled double-blind study whose main objective was to evaluate the safety of local administration of this phage preparation in patients with chronic venous leg ulcers, with or without clinical signs of infection. Once a week for 12 weeks, either saline or WPP-201 was given topically to wounds of 42 patients (of whom 39 completed the trial), along with other standard treatments designed particularly to deal with biofilms. The study showed that topical administration of WPP-201 does not result in any adverse events, nor does it adversely affect wound healing. Although without outside funding the range of parameters studied was somewhat disappointing and efficacy could not be established, the lack of safety problems was clearly documented, and they continue to make some use of phage on a compassionate-use basis in their practice though they do not have resources to carry out formal phase II trials.

Another clinical study of phage therapy was conducted at the Burn Centre of the Queen Astrid Military Hospital in Brussels, Belgium (Merabishvili *et al.*, 2009; Kutter *et al.*, 2010; Pirnay *et al.*, 2012). The trial was cleared by a leading Belgian Medical Ethical Committee and its main objective was to evaluate the effects of a phage cocktail called BFC-1 in patients with burn wound infections due to antibiotic-resistant strains of *S. aureus* and *P. aeruginosa*. The cocktail contains two *P. aeruginosa* phages (14/1 and PNM) and one *S. aureus* phage (ISP) at a concentration of 10^9 PFU/ml each. Phages were suspended in sterile 0.9% NaCl solution and were administered to patients locally (sprayed directly on one part of an infected burn wound while its untreated distant part served as control). Ten applications of BFC-1 were done in 9 patients; no adverse events or changes in clinical or laboratory parameters related to the phage use were found (the follow-up period was 3 weeks). This study, also done in non-commercial fashion with little outside funding, is noteworthy for the breadth of the collaboration among Belgian, Georgian and Russian scientists and physicians and for the care that was taken in documenting the preparation of the cocktail used (Merabishvili *et al.*, 2009). Currently they are conducting a placebo controlled multicentre clinical trial focused on nasal/throat decontamination of *S. aureus* as well as *P. aeruginosa* in intensive care patients (Merabishvili, 2012). Forty patients are intended to be enrolled into this study. The group continues to do a commendable job of explicitly exploring the murky challenges of the regulatory issues involved (Verbeken *et al.*, 2012). This topic, and the challenges it brings to developing and carrying out the badly needed formal clinical trials of phage therapy, are discussed in Chapter 12. In addition, the importance of regulated clinical trials for the development of phage therapeutics was recently discussed by Parracho *et al.* (2012).

Earlier studies of phage therapy in the West

Increasingly, we are becoming aware of very extensive research and clinical phage work that was carried out in France for much of the last century and in the US in the middle part of that century. However, since these topics have recently been quite extensively reviewed (Kutter *et al.*, 2010; Abedon *et al.*, 2011), they will be mentioned here only briefly. In France, phage therapy was used on a larger scale until the early 1990s. Phages have been used in the treatment of a wide range of infections including orthopaedic infections, alimentary tract infections, skin infections, surgical infections, urinary tract infections, respiratory tract infections, otolaryngological infections, and septicaemia. The results of phage therapy conducted in France were published in a number of papers, documenting substantial success in most cases (Abedon *et al.*, 2011, and references therein). Since the mid-1990s, when phage preparations stopped being produced in France, phage therapy has been practiced by a few physicians who obtain phage preparations from Georgia and Russia. In the US, an interest in phage therapy developed in the 1920s and the 1930s. However, this interest largely dwindled following the publication of two series of unfavourable reviews of most phage therapy by Eaton and Bayne-Jones, and by Krueger and Scribner in 1934 and 1941, respectively. One wonders what the results would have been had they included the French literature, where d'Hérelle maintained a strong influence in phage production and implementation and the reviews were generally positive, and not limited themselves to English-language papers. Quite extensive, very successful clinical work continued only for *S. aureus* infections and for typhoid fever well into the 1940s (Abedon *et al.*, 2011 and references therein).

Phage therapy in Poland

Early trials of phage therapy in Poland

Poland is one of the very few countries to have extensive experiences in the use of phage therapy in humans, and it is the one that has had the strongest focus on tracking applications, collecting data and publishing the work. To the best of our knowledge, the first Polish report on therapeutic use of phage was published in 1923 in the journal 'Military Doctor' (Kalinowski and Czyż, 1923). Shortly thereafter, in 1927, Jasieński from the Surgery Clinic of the Jagiellonian University in Cracow published the results of another study of the efficacy of phage therapy in patients with different suppurative staphylococcal infections (Jasieński, 1927).

Following these early studies, phage therapy was applied at various centres around Poland. For example, phages were used in civilians during the Second World War (Lityński, 1950), and the tradition of publishing the results of clinical studies was continued. In another paper, Witoszka and Strumiłło (1963) reported the treatment of 50 patients with surgical site infections due to coagulase-positive staphylococcal infection. A multivalent staphylococcal phage cocktail prepared by the Phage Laboratory of the Warsaw Epidemiological Station was administered locally (2–3 ml) as moist dressings or spray once a day. No more than three doses of phages were applied. In 35 patients (70%), the treatment was effective as evaluated by negative wound swabs, subsiding of the local inflammation, and normal wound healing.

Phage therapy supervised by the Ludwik Hirszfeld Institute of Immunology and Experimental Therapy in Wrocław

The development of research on phages and phage therapy in Poland was greatly strengthened in 1952 when the Polish Academy of Sciences established the Institute of Immunology and Experimental Therapy (IIET) in Wrocław under the direction of professor Ludwik Hirszfeld, a renowned Polish microbiologist and immunologist who had used bacteriophages in epidemiologic studies of *Salmonella typhi* infection (Hirszfeld, 1948). An important line of research at the IIET was phage typing of *Shigella* bacilli, including *Shigella flexneri* and *Shigella sonnei* (Ślopek *et al.*, 1972; Ślopek *et al.*, 1973). The IIET played a key role in the development of clinical phage therapy in Poland by producing phage preparations that were subsequently distributed among different hospitals (largely located in Silesia) and by coordinating and assessing the efficacy of phage therapy performed there. Detailed results from the 1980s were reported in a number of papers that contain clinical data on the treatment of about 2000 patients – papers that are of central importance to the English-language literature on clinical phage therapy. Here, we discuss the results presented in three of these articles (Ślopek *et al.*, 1987; Weber-Dąbrowska *et al.*, 2000, 2003).

The first article (Ślopek *et al.*, 1987) presents an overall analysis of the results of phage therapy which were discussed in detail in a series of earlier reports (Ślopek *et al.*, 1983a,b, 1984, 1985a–c). It summarizes the results of virtually all of the phage therapy performed between 1981 and 1986 in 550 patients (52.4% men and 47.6% women), with infections caused by *Staphylococci, Klebsiella, Escherichia, Proteus,* and *Pseudomonas.* These infections and diseases included alimentary tract infections, septicaemia, blepharitis, conjunctivitis, otitis media, meningitis, varicose veins with ulcer and inflammation, pericarditis, infections of the respiratory tract (including bronchopneumonia), gingivitis, peritonitis, abdominal abscesses, genitourinary tract infections, furunculosis, diseases of the connective tissue and lymphatic vessels, decubitus ulcers, arthritis, myositis, osteomyelitis of the long bones, wound infections, osteitis of the long bones after fracture, and fistulas. Almost all patients (94.2%) had been previously treated with antibiotics without success. The age of the patients ranged between 1 week and 86 years. Polyinfections were seen in 32.4% of the cases. Appropriate individual phages for each patient were selected from the extensive Institute phage library, which was constantly being expanded and refined, and the sensitivity of the bacteria to the phage used was monitored throughout the treatment; the bacteriophages were changed if their bacteria developed resistance. In 72.4% of the patients phage preparations were applied as a stand-alone therapy, while 27.6% were administered phages along with antibiotics. Phage preparations were administered orally three times daily in a volume of 10 ml before meals, after neutralization of the gastric acid. In addition, phages were administered locally (as moist formulations, inserted into the peritoneal and pleural cavity, and as ear, nose, or eye drops). Overall, phage therapy was assessed as effective in 92.4% of patients. The percentages of the positive results of the treatment were 94.1% and 88.8% in patients with monoinfections and polyinfections, respectively (the difference between the two groups was statistically significant, especially in the case of infections caused by Gram-negative bacteria). Moreover, treatment with phages alone was significantly more effective than parallel use of phages and antibiotics. The efficacy of phage therapy ranged between 75% and 100% depending on the kind of infection. The highest efficacy (100%) was found in patients with infections of the alimentary tract, pericarditis, and furunculosis, while phages were least effective in patients with varicose veins with ulcer and inflammation (a 75% efficacy rate).

The second paper discusses the results of phage therapy performed in the years 1987–1999 (Weber-Dąbrowska *et al.*, 2000). The analysis included 1307 patients with a wide range of suppurative antibiotic-resistant infections, including septicaemia, otitis media, meningitis, varicose ulcers, bronchitis, laryngitis, rhinitis, bronchopneumonia, empyema, pleuritis, peritonitis, urinary tract infections, furunculosis, decubitus with infections, arthritis and myositis, osteomyelitis, osteitis after bone fractures, infections of burns, postoperative infections, fistulas, sinusitis, and mastitis. Etiologic agents of these infections included *E. coli*, *Klebsiella*, *Proteus*, *Enterobacter*, *Pseudomonas*, and *S. aureus*; in 30.8% of patients polyinfections were observed. Before administering phage preparations to patients, the sensitivity of the etiologic agent to available phages was evaluated in each patient. Phage preparations were administered to patients *per os* three times daily for 1–12 weeks (32 days on average). In some patients, phage preparations were also administered locally. Overall, positive results were reported in 85.9% of patients. The highest efficacy was noted in patients with meningitis and furunculosis (100%). Phage therapy also proved to be very effective in septicaemia, otitis media, peritonitis, arthritis and myositis, osteomyelitis, burn wound infections, mastitis, and fistulas. On the other hand, in 10.4% of the patients only transient improvement was found, and in 3.8% of the patients phage preparations were ineffective.

The third paper reports the results of phage therapy in 94 patients with sepsis caused by antibiotic-resistant strains of different bacterial species (Weber-Dąbrowska *et al.*, 2003). In 33 of these patients, infection was caused by a single bacterial species (*S. aureus*, *P. aeruginosa*, *E. coli*, or *K. pneumoniae*), while 61 had polyinfections caused by some combination of the above-mentioned bacteria as well as *Proteus mirabilis*, *Morganella morganii*, and *Enterobacter*. In these patients sepsis was a complication of postoperative wound infections (36 cases), tissue injury (18 cases), phlegmonous cellulitis (10 cases), urinary tract infections (five cases), peritonitis (five cases), or other kinds of infections. Phage preparations were administered to patients orally three times daily. In 71 patients phage preparations were used alongside antibiotics, and 23 patients received only phages. The median time of the treatment was 29 days. The authors reported that phages were effective in 80 patients (85.1%), while in 14 patients (14.9%) phage therapy was ineffective. An important finding of the study was that there was no statistically significant difference in the efficacy of bacteriophages compared with the treatment involving both phages and antibiotics.

These three major Polish papers largely reported a high level of efficacy for phage therapy. However, these studies have significant shortcomings, including a lack of control groups and of precise efficacy criteria (they simply relied on the fact that virtually none of the patients had responded to other treatments and thus, in effect, served as their own controls). Furthermore, these were generally studies of hospitalized acute cases being managed by hospital physicians, who made the reports back to the IIET; therefore, no follow-up information was available to the Institute after the patients left the hospital, although they were being monitored closely and constantly while in the hospital. These shortcomings clearly limit the scientific value of those reports. However, for several reasons we believe that these results are still potentially very useful in assessing those areas where phage therapy is likely to be most useful and successful, as are the data from the most recent Polish paper on clinical phage therapy (Międzybrodzki *et al.*, 2012; see also 'The Phage Therapy Unit at the IIET in Wrocław', below). First, it has been emphasized that observational studies can also provide valuable information on the efficacy of new medicines (Vandenbroucke, 2004). Secondly, it should be noted that the data summarized in the first paper (Ślopek *et al.*, 1987) were discussed in much more detail in a series of papers that include an exploration of just what was going on for each unsuccessful patient; for example, some were already so sick that they died before very much could be done (Ślopek *et al.*, 1983a,b, 1984, 1985a–c). In addition to these detailed reports, a number of other papers were published on phage therapy in Poland. Some of the most important data from selected articles are

Table 11.1 Summary of the selected reports presenting the results of phage therapy conducted in Poland after 1950. This table does not include the data from studies which are discussed in detail in this chapter (Ślopek *et al.*, 1987; Weber-Dąbrowska *et al.*, 2000, 2003; Miedzybrodzki *et al.*, 2012)

Kind of infection	Aetiological factor	Number of patients	Route of administration of phage preparation	Therapeutic efficacy	Reference
Carrier state of *Shigella sonnei*	*S. sonnei*	21	Oral	Eradication of the carrier state in 100% of carriers	Mulczyk and Ślopek (1971)
Puerperal mastitis	*S. aureus* *S. albus* *E. coli*	6	Oral and topical	Marked local improvement in all patients	Cisło *et al.* (1988)
Different infections including sinusitis and osteomyelitis of the jaws	*S. aureus* *P. aeruginosa* *E. coli* *Proteus*	16	Oral and topical	Very high efficacy in two patients, marked improvement in nine patients	Kaczkowski *et al.* (1990)
Different infections including mastitis, osteomyelitis, skin and subcutaneous tissue infections	*S. aureus* *E. coli* *P. aeruginosa* *Proteus* *Streptococcus*	62	Oral and topical	Complete remission in 59 patients, marked improvement in three patients	Dworski *et al.* (1990)
Urinary tract infections	*E. coli* *Pseudomonas* *Klebsiella* *Enterobacter* *S. aureus*	15	Oral	About 30% efficacy rate	Boratyńska *et al.* (1994)
Recurrent subphrenic and subhepatic abscess with the jejunal fistula after stomach resection	*E. coli*	1	Oral and topical	Complete remission	Kwarciński *et al.* (1994)
Sinusitis	*S. aureus* *S. epidermidis* *S. saprophyticus* *E. coli* *Pseudomonas*	32	Oral and topical	Complete remission in 24 patients, marked improvement in six patients	Weber-Dąbrowska *et al.* (1996)
Otitis media	*S. aureus* *S. epidermidis* *S. saprophyticus* *Proteus* *Klebsiella* *Pseudomonas*	22	Oral and topical	Complete remission in 19 patients, partial remission in three patients	Weber-Dąbrowska *et al.* (1997)
Cerebrospinal meningitis	*K. pneumoniae*	1	Oral	Complete remission	Strój *et al.* (1999)
Different infections in cancer patients	*S. aureus* *P. aeruginosa* *K. pneumoniae* *K. oxytoca* *E. coli*	20	Oral and topical	100% efficacy rate	Weber-Dąbrowska *et al.* (2000)
MRSA colonization of the intestines	*S. aureus* (MRSA)	1	Oral	Eradication of the colonization	Leszczyński *et al.* (2006)
Chronic bacterial prostatitis	*E. faecalis*	3	Intrarectal	Eradication of bacteria and marked improvement of the clinical state in all of patients	Letkiewicz *et al.* (2009)

Kind of infection	Aetiological factor	Number of patients	Route of administration of phage preparation	Therapeutic efficacy	Reference
Chronic bacterial prostatitis	*E. faecalis* *E. coli* *K. pneumoniae* *P. aeruginosa* *S. haemoliticus*	22	Intrarectal, oral, and/or topical (on the glans penis)	Eradication of the pathogen in 50% of cases	Letkiewicz *et al.* (2010)
Wound infection in a patient with acute leukaemia	*P. aeruginosa*	1	Oral and topical	Eradication of bacteria from the wound and marked improvement of the general state	Trelińska *et al.* (2010)
Laryngeal ulcer in a patient with acute leukaemia	*P. aeruginosa*	1	Oral	Healing of the ulcer and marked improvement of the general state	Trelińska *et al.* (2010)

presented in Table 11.1. This table includes studies involving smaller groups of patients and case reports.

In our opinion, regardless of the actual extent of efficacy of treatment, earlier Polish studies discussed in this section appear to confirm the relative lack of significant side-effects and strongly suggest the safety of phage therapy. Notably, even though no formal prolonged post-treatment follow-up of those patients was available, the IIET never received any claims at all related to possible side-effects or other harm that could be directly related to a patient's phage treatment. Therefore, we believe that the major input of these studies is the confirmation of the safety of phage therapy, at least as applied externally. It should be noted, however, that these studies say nothing about the level of risk that might be associated with intravenous applications or use of purified highly concentrated stocks, neither of which has been used in Poland.

The Phage Therapy Unit at the IIET in Wrocław

Background

In 2005, after Poland joined the European Union, the IIET took another major step forward in its exploration and support of phage therapy, building on the many years of collaborative experience described above. The Institute has opened its own outpatient Phage Therapy Unit (PTU; http://surfer.iitd.pan.wroc.pl/pl/OTF) to support long-term treatment of longstanding chronic infections, complementing its ongoing work supplying phage and support services for physicians in regional hospitals.

This is the first dedicated centre of phage therapy in the European Union, equipped with its own specially trained physicians and test facilities. The Unit began operating within the protocol 'Experimental phage therapy of drug-resistant bacterial infections, including MRSA infections', which was approved by the bioethics commission and registered also at the ClinicalTrials.gov, although the effort is not a clinical trial (ID: NCT00945087). It performs phage therapy under the rules of a *therapeutic experiment* on the basis of the respective Polish regulations (which include the laws regulating the physician's profession, pharmacological law, and regulations of the Minister of Health) as well as, more generally, the Declaration of Helsinki (Międzybrodzki *et al.*, 2007). Here, a *therapeutic experiment* is defined to occur when 'a physician introduces new or only partially tested diagnostic, therapeutic, or prophylactic methods for the direct benefit of the person being treated'. Such therapeutic experiments may be implemented only when available treatment has failed. These complex issues are discussed in some detail in Chapter 12.

According to the current protocol, the PTU accepts only ambulatory adult patients over 18 years old for whom no other effective therapy is available (e.g. targeted antibiotic therapy is ineffective, the infection is caused by multidrug resistant bacteria or the use of the potential targeted drug(s) is contraindicated). Patients must sign a special informed consent form. The Unit deals with very long-term cases involving chronic bacterial infections of the skin, subcutaneous tissues, bones and joints, fistulas, wounds and bedsores, urinary and reproductive tracts, digestive tract, middle ear, sinuses, tonsils, and upper as well as lower respiratory tract. The presence of bacteria in the site of infection must be confirmed by microbiological culture at the Institute and the bacteria must be sensitive to at least one phage from the Bacteriophage Collection of the IIET (see below).

At this point, the protocol excludes pregnant or breastfeeding women, some cases with malabsorption syndrome, allergy to food or animal protein, advanced hepatic insufficiency, and patients with allergy to components of the phage preparation. It also excludes patients where surgery is required or recommended, though patients can be accepted for whom the active bacterial infection is a contraindication to the surgery or there is a need for non-invasive phage treatment before surgery because of a high risk that the infection would otherwise persist after surgery. The qualifying physician may also disqualify a subject when, in his opinion, the patient's health condition precludes conducting this experimental therapy.

For each prospective patient, extensive lab tests and phage typing of their bacteria are carried out and specific lytic phage preparation(s) confirmed to be active against the pathogenic bacterial strain are selected for the treatment. This requires demonstrating complete lysis of the tested strain by the proposed phage, followed by testing of a dilution series to see individual plaques and be certain that the preparation has high enough activity. As discussed above, the IIET Phage Collection was established in 1948 by Professor L. Hirszfeld, who was then working with *Salmonella* phages, and was extended over the years through collaboration with other laboratories, mainly in Poland and other parts of Europe. Over the last decade, the collection was enriched with new phages isolated by the Bacteriophage Laboratory of the IIET from environmental samples, including crude and purified city sewage, inland and sea waters, and filtrates from biological samples according to methods developed by the IIET (Weber-Dabrowska *et al.*, 2007). Currently, the collection consists of 524 bacteriophages, which are specific to *E. coli* (121 phages, 77 of them routinely tested), *Klebsiella* (95 phages), *Enterococcus* (73), *Enterobacter* (48), *Shigella* (39) *Citrobacter* (38), *Pseudomonas* (37), *Salmonella* (32), *Stenotrophomonas* (18), *Serratia* (17), *Proteus* (17), *Morganella* (14), *Staphylococcus* (7), *Acinetobacter* (5), and *Burkholderia* (2). The majority of phages examined under the electron microscope were *Myoviridae* and *Siphoviridae*, with only a few *Podoviridae.* Since 2006, most of the phage preparations used for the phage therapy have been produced by BIOMED SA in Kraków, Poland, as sterile lysates under Good Manufacturing Practice (GMP) conditions. Importantly, the majority of *S. aureus*, *P. aeruginosa*, and *E. faecalis* strains from patients diagnosed at the PTU have been sensitive to phages from the collection (90%, 72%, and 77% of the strains, respectively).

The therapeutic bacterial lysates are prepared according to a modified version of Ślopek's method (Ślopek *et al.*, 1983a; Zimecki *et al.*, 2003; Letkiewicz *et al.*, 2009). Briefly, phages and the appropriate live bacteria are added to peptone broth and incubated at 37°C until complete lysis occurs (approximately 3–6 h). After overnight storage at 4°C, the suspension is filtered through an 0.22-μm Millipore filter. The phage titres in these preparations range between 10^6 and 10^9 PFU/ml. They are administered orally, rectally or topically as wet compresses as well as by gargling, irrigation of the abscess cavity, fistula or vagina, nose or eardrops, aerosol inhalation, or sitz baths (Górski *et al.*, 2009). For oral application, 10–20 ml of the bacteriophage formulation is administered three times daily, at least 30 min before meals. The phage is protected from gastric-juice inactivation by drinking 10 ml of dihydroxyaluminium sodium carbonate oral suspension up to 20 min before the phage preparation. Rectally, 10–20 ml of the preparation is administered twice daily. Topically, phages are applied once or twice daily. When two or more types of phage need to be used, they may be taken alternately.

As indicated above, a major difference between the treatment documented here (see 'Retrospective analysis of phage therapy results at the Phage Therapy Unit, January 2008 to December 2010', below) and the earlier documented set of applications (presented in the section 'Phage therapy supervised by the Ludwik Hirszfeld Institute of Immunology and Experimental Therapy in Wrocław') is that these patients are not hospitalized and do not require or implement surgery. Their infections are not life threatening, although they have often long had debilitating consequences despite all efforts at treatment. As an Experimental Protocol, state insurance does not cover phage treatment; the patients generally have to pay for their own initial screening visit, interim visits, phage preparations and laboratory tests. The Unit has its own set of physicians from various specialties, experienced in phage therapy, available daily to see the patients, who generally come for control visits every 2–4 weeks during the treatment. Patients can also call at any time during business hours with questions and concerns, and are also given contact numbers that can be used at any time for perceived emergencies. The standard course of treatment generally lasts 6–12 weeks. It may be even more prolonged in cases with good ongoing clinical response to therapy or when sensitivity of the targeted pathogen to the initial phage has changed. To help support protocol compliance and provide relevant data for the in-depth analysis, each patient has a diary to fill out each day.

Retrospective analysis of phage therapy results at the Phage Therapy Unit, January 2008 to December 2010

The PTU therapeutic protocol is undergoing periodic revision with approval of the bioethics committee. Some major changes were made in early 2008, including adding rectal and inhalation modes of administration and adding new medical specialists. Thus, the latest analysis of the therapeutic results focuses on the more recent current work. The results of this analysis were discussed in detail in a recent paper (Międzybrodzki *et al.*, 2012); here, we will present the most important data from the paper. Between January 2008 and December 2010, 157 patients with a wide range of very long-standing antibiotic-resistant bacterial infections were accepted for experimental phage therapy in the PTU. In 153 cases (68 women and 85 men), it was possible to evaluate the results of the treatment. The treated conditions included orthopaedic infections (37 cases), soft tissue infections (31 cases), urinary tract infections and/or chronic bacterial prostatitis in men (29 cases), urinary and/or vaginal infections in women (22 cases), respiratory tract infections (24 cases) and skin infections (10 cases). The major etiologic agents included *Staphylococcus* (51% of the cases), *Enterococcus* (11.1%), *E. coli* (11.1%), *Pseudomonas* (9.8%), and a few cases of *Klebsiella, Citrobacter, Enterobacter, Proteus,* and *Salmonella*; 20% of the patients had polyinfections. The patients were generally referred to the PTU by other physicians and had an initial application interview at the Institute, followed by a thorough work-up as described above.

The selected phage preparations were administered to patients orally, topically, and intrarectally. The majority of the patients (71.9%) were only treated with phage preparations, directed against one or more pathogens. The others also used antibiotics (41 cases), disinfectants (eight cases) and/or herbs and supplements (for four cases of urinary tract infections) at some point during the phage application or in the breaks between the therapeutic cycles. These were introduced due to lack of an active phage against one or more of the pathogens in patients with polyinfections (five cases), the development of a different kind of infection not subjected to the experimental procedure (seven cases), superinfection by a different bacterium during phage treatment of an existing infection (five cases), exacerbation of the symptoms during the breaks between the therapeutic cycles (two cases) or continuation of antibiotic therapy started due to exacerbation of symptoms during the process of qualification for the phage treatment (nine cases) or during phage use (16 cases). The median cumulative treatment duration (i.e. the sum of days when the phage preparation was used) was 55 days (with relevant data available for 149 patients). However, it was very individually diversified depending on the nature of the infection, treatment results, route(s) of phage administration and incidence of adverse reactions. The minimum duration was 3 days, the maximum duration for a single course of phage therapy was 12 weeks, and the maximum cumulative duration overall was 328 days, discussed in detail below. Interruptions in a single course of treatment were permitted for up to 4 weeks for justified reasons, such as the need for preparing new phages for the treatment. The cases involving very extensive use and repeated treatment cycles provide particularly important data for evaluating phage safety.

As mentioned above, the patients treated at PTU differ from those discussed in the earlier three major Polish papers (Ślopek *et al.*, 1987; Weber-Dąbrowska *et al.*, 2000, 2003) in that their infections are neither acute nor life threatening and do not require hospitalization or surgery, but they have been intransigent for a very long time. Here, we will focus on a few of the specific cohorts where the groups were large enough to give informative results. In all cases, the efficacy was evaluated based on the results of bacterial cultures and specific diagnostic tests, assessment of the intensity of infection symptoms, and the opinions of consulting medical specialists, during and following the period of treatment.

The highest percentage of positive outcomes was found in the cohort of 29 men dealing with urogenital problems – diagnosed just with chronic bacterial prostatitis (13) or a urinary tract infection (10), or with both (six men). Overall, there was a 48.3% good response rate, and 11 of the patients showed full recovery and/or pathogen eradication. Of the five subjects with *E. coli* as the sole pathogen, one each showed full eradication, a good clinical result, or no response, while two others showed transient clinical improvement. The results were most clear cut when enterococcus was the sole pathogen – seven patients showed pathogen eradication while five showed no response, and there were no intermediate results. Six others showed enterococcus combined with *Staphylococcus* (two cases), *Klebsiella*, *E. coli*, *Citrobacter* or *Pseudomonas*. Here, full eradication was observed only in one prostatitis patient infected with both enterococcus and *Pseudomonas*, as well as two men with urinary tract infections, one of them treated with *Enterobacter* phage, the other with phage against *Pseudomonas*. In general, rectal application seemed particularly efficacious here. Moreover, earlier observations of Letkiewicz *et al.* (2010) showed that phage treatment of chronic bacterial prostatitis also resulted in significant improvement in maximum urinary flow rate, decrease in prostate volume and leucocyte count in expressed prostatic fluid.

More exploration is clearly needed in trying to deal with urogenital infections in women, 10 of which were caused by *E. coli* alone, seven more by *E. coli* plus enterococcus, and five by other Gram-negative and Gram-positive pathogens. Only three of 22 patients experienced pathogen eradication and full recovery, and those three all differed in aetiology and in method of phage application. Some clinical improvement was seen for five more; none were classified in between with 'good clinical result'.

The largest group of patients (37 cases) were being treated for orthopaedic infections, of which 34 were staphylococcal, two were caused by *Pseudomonas*, and one involved both staph and *Enterobacter*. The major obvious variable was the route of administration of the phage. Those who received phage only topically (25 patients, including 16 cases with osteomyelitis, four cases with joint infection, four cases with prosthetic joint infection, and one case with osteomyelitis/joint infection, who were using phage preparations for irrigation of the fistula and/or as wet compresses on the external orifice of the fistula) had a much lower success rate (only 28% of good responses), compared with 83% for those who received them only orally or both orally and topically (12 patients, including five cases with osteomyelitis, four cases with prosthetic joint infection, one case with joint infection, one case with osteomyelitis/joint infection, and one case with discitis). The two groups were quite similar in terms of age, gender, the fraction of staph monoinfections, and cumulative time of treatment (53 and 70 days, respectively). It thus appears that orally administered phage are remarkably efficacious in such infections.

The longest treatment (328 days of cumulative

phage use) involved an 82 year old woman with *P. aeruginosa* –infected chronic leg ulcers due to chronic venous insufficiency, who used phages only topically. A good response was observed during the first course of phage application, but her ulcers were only partially healed. Since her condition was not further improved by standard treatment, including hyperbaric therapy, done during a break between phage therapy courses (> 4 months according to protocol) and the previous phage treatment was safe and well tolerated, she was admitted to a second phage-treatment course. Her response to phages was again a partial healing of the ulcers accompanied by reduction of pain and improvement in walking.

General lessons for phage therapy from these studies

The analysis carried out here has revealed important findings regarding the safety of phage therapy and the development of resistance to the therapeutic phages (Międzybrodzki *et al.*, 2012). First of all, the effects of phage preparations on organ and system function were extensively analysed by comparing haematological, urinary and biochemical serum parameters over multiple periods of cumulative phage use (3–6, 7–20, 21–48 and 49–84 days) with the results of tests done before phage therapy was initiated. There were no significant changes in haematocrit, haemoglobin, leucocytes, lymphocytes, monocytes, neutrophils, eosinophils or basophils as well as total protein, bilirubin, alanine transaminase, alkaline phosphatase or amylase. Only minor but significant transient decreases in mean platelet count and gamma-glutamyl transpeptidase as well as a small transient increase in mean aspartate transaminase and blood sugar were observed. There was no significant influence of phage administration on renal function; in only a few cases, there were small, mainly transient and clinically insignificant fluctuations in various urinary parameters.

General evaluation of the reactions to the phage preparations indicated that they were well tolerated by 77.8% of patients (with no complaints about ways of using the preparation and no actual ailments, whether or not adverse reactions were observed). In 18.3% of the cases, minor reactions or complaints such as a bad taste (five reports from 39 patients who received phages orally) were noted. The most common were short-lived minor responses to local administration reported by 24 of the 141 patients (18%) who received phage intrarectally or topically. The other frequent reactions included a rise in body temperature (6.5% of cases) and superinfections (4.6% of cases). Three patients (7.9%) who applied preparations orally reported nausea. Adverse reactions which led to terminating the treatment were observed in only 3.9% of the subjects. These included a fever after two days of combined oral and topical treatment, a local pain, an aggravation of atopic dermatitis in the region of the skin where the phage preparation was applied, a superinfection resulting in oedema and pain in the wound, and limb oedema following phage topical use. The most serious case involved acute groin and leg swelling following oral phage administration, suggestive of a possible venous thrombosis, in a 77 year old diabetic man with a chronic right hip infection following implantation of a prosthesis which then had had to be removed. He had tolerated a first course of topical staph phage very well. The oedema appeared 7 h after the first oral dose at the beginning of a second, now oral, course of phage therapy, so the phage therapy was stopped, even though there were also other risk factors involved.

Moreover, in 53 cases involving *S. aureus*, 14 each involving *E. coli* and *E. faecalis*, and 11 involving *Pseudomonas*, they carefully examined changes in the phage typing profile as well as the development of bacterial resistance *in vitro* to the therapeutic phage. Resistance developed to only nine of the staph phages (17%), but to four of the *Pseudomonas* phages (36%), 6 of the *E. faecalis* phages (43%) and a whopping 86% of the *E. coli* phages, requiring frequent changes of phages there. Far more developed changes in their phage typing protocol: 70% of the *S. aureus*-infected patients, 10/11 of those infected with *P. aeruginosa*, and in all cases infected with *E. coli* and *E. faecalis*. Of most concern, development of resistance to all strains in the extensive IIET collection was observed for 28.6% of the *E. coli* cases, 27.3% of *P. aeruginosa* cases, 21.4% for *E. faecalis*, but just 7.4% of the staph cases. These results support the widely held perception that it is highly preferable to use well-constructed phage cocktails rather than individual phages, and the IIET is starting to implement use of multiple phages.

An earlier retrospective analysis of some parameters of patients subjected to phage therapy was carried out between January 2006 and October 2007 (Międzybrodzki *et al.*, 2009). This included 37 patients with osteomyelitis, prosthetic joint infections, skin and soft tissue infections, or lower respiratory tract infections caused largely by *S. aureus*. They were treated according to the experimental protocol with oral and/or topical phage. Interesting changes of C-reactive protein (CRP) serum concentration and white blood cell count (WBC) were observed during treatment. These inflammatory markers, although stable after 5–8 days of phage administration, significantly decreased after 9–32 days of treatment (CRP: from 23.3 mg/l at baseline to 16.1 mg/l, $n = 26$, and WBC: from 7.8×10^3 cells/mm^3 at baseline to 7.1×10^3 cells/mm^3). These findings suggested that phage treatment could diminish the inflammatory reaction triggered by bacterial infection. This may result not only from direct antibacterial activity of the phages but also from immunomodulating effects of phages, as has been reported during immunological monitoring of the PTU patients (Górski *et al.*, 2012).

Over the last few decades, most of the Polish phage use had been relatively local. It is encouraging to see the increasingly broad acceptance and implementation of phage therapy in Poland, in response to the better-defined path for its introduction, easier availability of GMP preparations, broader knowledge base and ever-growing concern about the crisis in antibiotic resistance. The Polish standards now explicitly require the approval of a bioethics committee for phage therapy to be conducted by an individual physician or health centre, with focus on the concept of 'therapeutic experiments', as discussed above. So far, 13 of the regional Polish bioethics committees have approved the use of phage therapy as needed locally, and the committee for Lower Silesia has extended that approval to all regional hospitals intending to use IIET phage treatment, requiring only that they report the initiation of treatment under the special IIET protocol, supervised by the IIET (Międzybrodzki *et al.*, 2012). Furthermore, the mayor of Wroclaw approved a small pilot project and payment of phage therapy costs for citizens of the city with leg ulcers, which was a very substantial help for the IIET PTU and its patients as well as, ultimately, for the increasing usefulness of their detailed research and analysis in the context of their phage therapy work.

Phage therapy in Georgia and Russia

Phage therapy in Russia

The most extensive work on phage therapy was done in the former Soviet Union, where clinical use began back in the 1920s, research remained strong even after the advent of antibiotics and phage continue to be accepted in main-line medicine. Phage cocktails of various sorts are a standard part of medical practice in the Republic of Georgia and are available on the web as well as in some clinics, hospitals and pharmacies in Russia.

Phage therapy was initiated in the Soviet Union by George Eliava, head of the Microbiology Institute in Tbilisi, Georgia. Eliava spent 1918–21 at the Pasteur Institute in Paris and was drawn to Félix d'Hérelle's work there, beginning with d'Hérelle's use of phage to treat chickens with fowl typhoid caused by *Salmonella gallinarum*. The two formed close collaborative ties, and d'Hérelle worked with Eliava to expand the Tbilisi Microbiology Institute to include bacteriophage research by 1923. Interest in phage therapy quickly spread to a number of centres in the Soviet Union, as has been extensively reviewed (Alisky *et al.*, 1998; Sulakvelidze *et al.*, 2001; Summers, 2001; Chanishvili *et al.*, 2001; Sulakvelidze and Kutter, 2005; Häusler, 2006; Kutateladze and Adamia, 2008, 2010; Kutter *et al.*, 2010; Abedon *et al.*, 2011; Chanishvili, 2012b). However, strong political, linguistic and cultural barriers long blocked awareness in the West of the extent and depth of that work.

A more complete window into early Soviet phage research was opened by the detailed analysis, funded by the British Ministry of Defence, of the material in the Eliava Institute library – books, journal articles, meeting reports, theses and documentation for the official approval of phage cocktails (Chanishvili, 2012a,b). In depth animal studies are discussed, starting with a 1924 meeting report by Krestnikova on phage targeting dysentery and extending through the 1960s with various isolated gut loop experiments. There are 25 pages detailing extensive human gastrointestinal phage trials, beginning with phage isolated against *Shigella Shiga-Kruse* in 1929 in the Ukraine. It also includes extensive work with other dysentery strains and *Salmonella* serovars, *E. coli* and *Proteus*, and an equally in depth exploration of prophylactic uses of phage. Civilian surgical applications began in the early 1930s using phage cocktails manufactured in Tbilisi and in Moscow. During the Finnish Campaign in 1938–39, phage cocktails were used very extensively and successfully both as emergency wound treatment by the mobile sanitary brigades and in the front-line hospitals, as discussed in great detail in a 1941 book by Tsulukidze, available in the Eliava library and extensively summarized by Chanishvili *et al.* (2001) and Chanishvili (2012a,b).

One of the top Soviet priorities was development of a low-cost, stable, high-titre anti-dysentery phage cocktail in tablet form, as described by Sulakvelidze and Kutter (2005). Finally, Sergienko (1945) developed a successful approach involving (1) growing phage lysates on solid medium (1–2% agar) on big trays, (2) exposing them to chloroform fumes for 18 h and (3) mixing the harvested phage-containing layer with starch and horse serum and making it into tablets, which were stable and had high titres. No toxic effect was seen in animals even when the tablets were dissolved and injected. These tablets, produced by the Alma-Ata branch of the Central Institute of Immunology and Microbiology, were used for controlled clinical trials involving 13,913 people in the treatment group and 12,690 untreated controls. Tablets were administered at 5-day intervals. Only 0.17% of the treated group developed dysentery, versus 1.5% in the control group. In children under age 10 (21% of the total), only 0.3% of those treated got dysentery vs. 5.2% of the control (Sergienko, 1945). Such tablets were in full-scale production in a number of centres in 1943 and continued to be produced and used in large quantities throughout the Soviet era and beyond. There were also major phage research and production centres in Russian cities such as Ufa, where work continued on through the break-up of the Soviet Union.

Eventually, the Soviet-era production centres were all taken over by the Russian pharmaceutical giant Microgen, which now makes many phage-related products available online and in some pharmacies as well as supplying them to hospitals and clinics. Their production and testing now appear to be covered with a mantle of corporate secrecy, with no new therapy papers out, even in Russian. However, Microgen phage cocktails are being used in the above-mentioned Nestlé Bangladesh infant diarrhoea clinical trials as a control treatment for Nestlé's own cocktail of T4-related phages, and Russian colleagues refer to ongoing clinical use there of phage therapy in such areas as burn and wound treatment. Start-up companies like MicroWorld Ltd, Moscow, have also been developing various new phage therapy products such as their recently approved phage preparation called Phagoderm, a gel with virulent phages against 14 pathogens associated with festering and inflammatory problems of skin and soft tissue, as reported by Zurabov *et al.* (2012). Russian phage biologists are also still publishing much interesting basic research, some of it directly applicable to phage therapy (Letarov *et al.*, 2010).

Phage therapy in Georgia

The Republic of Georgia: the Eliava Institute

In 1924, George Eliava founded what is now the thriving Eliava Institute of Bacteriophages, Microbiology and Virology in Tbilisi with the aid of Félix d'Hérelle and extensive financial support from the Paris Pasteur Institute. Together, Eliava and d'Hérelle dreamed of making this the world centre of therapeutic phage research and application, and drew in key political support leading to construction of their current park-like expansive facilities (Häusler, 2006). Despite Eliava's January 1937 arrest and subsequent execution, Elena Makashvili and others who had been trained by Eliava and d'Hérelle persevered, aided by their ongoing contacts with French phage therapists and by the broadening military interest in phage therapy.

By 1938, the Microbiology Institute in Tbilisi had become the Institute of Microbiology, Epidemiology and Bacteriophages of the All-Union Ministry of Health. Eliava scientists isolated and extensively characterized phages against clinical isolates from all corners of the Soviet Union, using them to develop a number of liquid phage preparations. After extensive testing and documentation, the approved products were made by the Eliava Institute for therapy, prophylaxis and diagnostics and used throughout the whole Soviet Union. According to the regulations developed by the Ministry of Healthcare of the USSR, new phage components active against those new clinical isolates were added to commercial phage cocktails at least once every six months. All of this work also led to very large collections of bacterial pathogens and phages targeting them in various labs at the Institute. Much of the basic research, animal experimentation and human application carried out during that period has been made much more accessible through the very extensive literature review carried out by Nina Chanishvili and her colleagues (2012a,b), focusing particularly on phage therapy in wound infections and surgery, ophthalmology, prevention and treatment of enteric infections and septicaemia, development of vaccines and the potential applicability of phage against agents of bioterrorism.

Since the early days of the Institute, its two most widely used cocktails have been Intestiphage and Pyophage, both of which evolved from the original work of Félix d'Hérelle. They are generic names, with various spellings, originally coined in Paris for these particularly useful formulations; licensed versions of both cocktails also are still produced commercially in Russia by Microgen and in Georgia by Biochimpharm, a company which recently grew out of part of the privatized old Eliava production facility. Since these various versions have evolved separately in recent years, they vary somewhat in antibacterial ranges and thus their relative efficacy may be dependent on each patient's specific aetiological agent(s) of infection and can be checked in the diagnostic laboratory. Intestiphage was initially produced at the Pasteur Institute in Paris to deal with dysentery and other dysbioses and was first used in Georgia in 1934. It has been updated regularly by Eliava scientists to deal with current problematic strains, and phages targeting many additional enteric bacteria have been added to the cocktail. Eliava Intestiphage now contains phages which target most pathogenic strains of about 28 different bacterial species in all, including gut-derived *Staphylococcus* and *Pseudomonas* as well as a very wide range of more common enteric pathogens. The other cocktail, Pyophage, directed against purulent-septic infections, consists of phages against *S. aureus*, *P. aeruginosa*, *E. coli*, and several *Streptococcus* and *Proteus* species which are often found in wounds, diabetic ulcers, infectious complications of burns, osteomyelitis, and other infections.

Phage preparations targeting *S. aureus* have always been of particular interest to Eliava scientists, reflecting its key roles in purulent wounds and hospital-acquired infections as well as the particularly high efficacy of such phages. In the 1970s, Eliava scientists developed a highly purified Intravenous Staph Phage preparation, intended for the treatment of acute and chronic sepsis. After animal trials and tests on 20 volunteer patients showed no side-effects, permission was granted on 11 April 1979 to carry out clinical trials on a limited basis at three medical centres in Georgia and three in Moscow, involving 250 patients. The results were discussed in detail by Chanishvili (2012a,b), drawing particularly on the very extensive compendium 'Results of Clinical Trials' (1983) from the Eliava Institute records. Further trials were conducted on 653 patients in two additional hospitals in Tbilisi and two in Moscow (Kutateladze and Adamia, 2008). A substantial fraction of the first batch of the preparation was commandeered by the Soviet Ministry of Health to treat a severe outbreak of sepsis involving hundreds of premature infants in a Moscow hospital. Altogether, 149 children (99 in Moscow and 50 in Tbilisi, 106 of them under 1 month of age) were treated with the intravenous phage, mostly in a combination with antibiotics, while 98 children were in a control group and received only antibiotics. The infants receiving the phage recovered much more rapidly in terms of both physical symptoms and laboratory indices. Only one of the phage-treated infants died (0.67%), while 8 infants from the control group died (8.2%). This highly purified staphylococcal phage preparation was also used to treat peritonitis, osteomyelitis, mastitis, purulent arthritis, lung abscesses, chronic pneumonia and bronchitis, septic staphylodermia, purulent cysts and chronic pelvic inflammatory disease in adults. In the 1980s, clinical trials of intravenous and intra-arterial applications were performed in the Scientific-Research Institute of Clinical and Experimental Surgery of Georgia, Georgian Centre of Sepsis, Central Clinical Hospital, Scientific Centre of Traumatology and Orthopedics, and Scientific-Research Institute of Human Generative Functions. The route of administration of the phage depended on the nature and location of the infection. Intravenous and intra-arterial administration of the preparation was used in cases of traumatic osteomyelitis. For lung disease, the preparation was administered intravenously as well as by direct bronchoscopic introduction or intrathoracal injection into the site of the infection. These trials were performed on 653 patients in all (age 15–75; 355 men, 298 women). The studied group consisted of 345 patients, 130 of whom were treated only with phage preparations, while 215 were administered phages and antibiotics. The control group included 308 patients who received no phages. The rates of total recoveries in the group treated only with phages, the group which received both phages and antibiotics, and the control group were 41.3%, 77.5%, and 11%, respectively.

Phage Sb-1, which is a main component of this preparation, still reveals a broad spectrum of lytic activity *in vitro* against freshly isolated, genetically different clinical samples (including MRSA) from clinics in a wide range of different geographical areas. Analysis of the sequenced phage genome showed no integrase or potentially harmful genes such as those that might contribute to bacterial virulence. This phage is now again being prepared in a defined minimal medium and highly purified. It is legally approved for external applications, available in pharmacies and being successfully used for patient treatment (Kvachadze *et al.*, 2011).

By the 1980s, the Institute had expanded to over 1200 people, ¾ of them working in large production facilities where up to 2 tons of phage products were made weekly, with 80% being purchased by the Soviet military. They also developed improved vaccines (anti-rabies, anti-anthrax, Brucella, smallpox) and sera (for treatment of diphtheria, tetanus, gangrene, scarlet fever), sets of typing phages for identification and characterization of pathogenic strains of *Salmonella* and *Shigella*, probiotic cocktails, and therapeutic enzymes like hyaluronidase, all strongly supported and extensively utilized by the Soviet Ministry of Health. The phage typing schemes they developed for *Salmonella typhimurium* and *Shigella sonnei* were approved by the WHO. They also isolated the first bacteriophages specific to *Brucella* spp., now considered by WHO to be reference phages for identification of *Brucella*. Different labs in the scientific section of the Institute worked on developing new and updated forms of therapeutic phage cocktails and getting them approved, carrying out electron microscope and molecular characterization of some of the phages, optimizing probiotic compositions, and isolating and characterizing phages targeting anaerobic bacteria.

Close collaborations between Eliava scientists, clinics and hospitals

A key aspect of the Georgian work to optimize phage therapy has always been the ongoing close collaboration of Tbilisi clinics and hospitals with phage producers. Eliava scientists worked closely with the Surgery Department at the Central Republican Hospital (Dr G. Gvasalia and Academician N. Kipshidze), the Sepsis Centre (Academician V. Bochorishvili), Gudushauri Hospital, Burn Centre, Infectious Diseases Hospital and many others. They carried out monitoring every 6–12 months of the patterns of pathological microbial flora taken from infected wounds, other infectious sources and the hospital environment. Both hospital and non-hospital bacterial strains were constantly used to adapt the commercial Eliava phage cocktails via serial passage, and the adapted phage preparation, produced in large quantity using sets of carefully selected bacterial strains, was then employed clinically.

The scientists updating and upgrading the cocktails worked closely with key physicians not only in Tbilisi but throughout the Soviet Union. They often developed special versions of Pyophage for particular hospitals to cover pathogenic bacterial strains that were particularly problematic there, as well as isolating and preparing new phages targeting bacteria for specific patients whose chronic infections were not sensitive to the available commercial phage cocktails and Eliava phage banks. The best of these 'eigenphages' – phages isolated specifically for particular patients – eventually were characterized and added to Pyophage or Intestiphage at the next updating of those cocktails.

Phage targeting of nosocomial infections

From 1975 to 1990, the Eliava Institute lab primarily charged with updating the commercial preparations collaborated closely with leading medical organizations in Moscow, Kazan, St. Petersburg, Kemerovo and Tolyatti, as well as in Tbilisi and Batumi, to explore use of phage preparations in problem areas of the hospital environment as well as in the patients. Bacterial strains from four general areas were characterized in terms of the distribution of various pathogens: traumatology (emergency rooms), burn centres, general and purulent surgery, and urology. As detailed in the internal documents of the Institute and various meeting presentations, they confirmed that a very broad range of bacteria and fungi were present in traumatology, while the major problems involved Gram-positive bacteria in purulent surgery and Gram-negative pathogens in urology, and they tailored the Pyophage cocktails used for sanitation accordingly, with much success in reducing problems with nosocomial infections.

For example, at Kazan's Scientific-Research Institute of Traumatology and Orthopedics, a study was carried out involving 400 patients, using Pyophage especially tailored to the conditions there for sanitation of the inpatient ward as well as for treating the patients. At first, phages were administered to 50 patients whose bacteria were antibiotic resistant, which resulted in a rapid improvement in the patients' general condition, reduction of the extent of purulent discharge, and finally elimination of the pathogen. This initiated phage cocktail use during surgery. The remaining 350 patients were randomly divided between phage and undefined standard treatment. Patient recovery was 90% in the former group, and 63% in the latter, leading to routine use of phage in that department before and during surgery, which resulted in a reduction in complications and shortened hospital stays. The results and methodological recommendations were compiled into a paper entitled 'Application of new combined Pyobacteriophage against wound infections in traumatic-orthopaedic patients' drawn up in 1987, and sent as a directive by the Soviet Ministry of Health to all major traumatology and orthopaedics hospitals in the Soviet Union. It is still available at the Eliava Institute library but was not otherwise published.

The burn unit in St. Petersburg's hospital No. 27 was having major problems with *Acinetobacter calcoaceticus*, found to be infecting 68.5% of their patients, so a special Pyophage cocktail augmented with *A. calcoaceticus* phages was prepared at the Eliava Institute and used in aerosol as well as liquid form twice daily by patients (Alavidze, unpublished results). The patient rooms, operating theatre, dressing room and room for physical therapy were treated using aerosol as well. Within a month, only 32% of 303 patient samples showed any *Acinetobacter*, and within 3 months the level was down to 2–3%, allowing plastic surgery to be performed much earlier and patients to leave hospital far more quickly (Iafaev *et al.*, 1998).

Similar studies involving phages targeting *Pseudomonas* species causing problems in hospitals were carried out in the Tbilisi Children's Infectious Disease Hospital for the treatment and prophylaxis of intestinal infections, in Saratov's Allergy Centre, Kazan's Traumatology unit, in the Tolyatti Automobile Factory Hospital Surgical Department, and in St. Petersburg's Urological Centre (see Zueva *et al.*, 1985). In addition to being distributed to the relevant medical facilities throughout the Soviet Union, the results of these trials were presented orally at medical conferences, for example: (1) 'Application of adapted Pseudomonas bacteriophage against nosocomial infections' by T.G. Chanishvili, A.N. Orlov, I.N. Okolov, L.P. Zueva, P.X. Iafaev, and Z.I. Alavidze, presented at the VI Ukraine Microbiology and Epidemiology Conference in Kharkov, 1988, and (2) 'Epidemiological overview of prevention of purulent infections caused by Pseudomonas with bacteriophages in burn centres' by Z.I. Alavidze and I.N. Okolov, presented at the Conference at the Gabrichevski Institute of Epidemiology and Microbiology in Moscow, 1988.

Unfortunately, contact between the Eliava Institute and these various Russian experimental centres was lost and resources for such studies were no longer available after the Soviet Union broke apart in 1991.

Phage therapy in Georgia since independence

When Georgia left the Soviet Union, the Eliava Institute lost all of its long-term support from the Soviet Ministry of Health and lucrative sales to the Soviet Army. The large production facilities were closed, and those facilities were soon transferred to commercial investors in the wave of privatization that swept the Soviet Union. The long-standing research arm of the Institute began supplying the country with therapeutic phage preparations. A diagnostic laboratory and pharmacy was established on the Institute grounds. Leading scientists carried out the analyses and the income helped keep the labs going while the new bacterial strains they collected helped keep their collections up to date. The lab that had been making the experimental versions of the commercial phages in 30-litre batches for the hospitals used those facilities to produce Pyophage and Intestiphage for Georgian patients and physicians.

The relationship between top Georgian surgeons and these Eliava scientists became even closer in 1992–1993, during the battles in the Abkhazia region of Georgia. Dr Guram Gvasalia, their strongest collaborator, headed the physicians treating the soldiers, and the standard phage preparations were specially adapted and routinely used for the prophylaxis and treatment of gunshot wound infections throughout the war (Alavidze *et al.*, 2007; Gvasalia *et al.*, 2010). Samples from wounds were sent to the Eliava laboratory almost daily. Investigations of 462 of the earliest wounded soldiers identified 18 different microbes with the potential to cause purulent wound infections in soft-tissue wounds in the first 4–6 h, but the severe infections that soon developed involved primarily five of them: *S. aureus* (24.9%), *E. coli* (24.7%), *Proteus* (21.6%), *Enterococci* (21.4%) and *Klebsiella* (7.4%). Resistant bacterial strains were used to isolate new phages from sewage water, and after their quick adaptation, concentration and proliferation, a number of them were added to the commercial Pyophage cocktail for battlefield purposes. Of the 612 bacterial strains originally isolated and tested at the start of the war, only 55% were sensitive to Pyophage. The host range of Pyophage on the wartime strains quickly reached 95% after the addition of these newly isolated phages. These special Pyophage cocktails were prepared in ampules and also in aerosol form. In this way, the troops were provided with a specific phage preparation, based on the most frequently occurring causative agents isolated from purulent gunshot wounds. The preparations could be used immediately after injury, since each soldier carried the aerosol form of Pyophage in his individual pack to spray around the wounded area for disinfection on the battlefield. Thus, the phage preparations were being used both as therapeutics in already-complicated wounds and for prevention. However, this prophylactic treatment of gunshot wound infections was not meant to be an alternative to surgical manipulations such as debridement, but rather increased the length of the 'golden' period for still achieving the most effective surgical management. Such gunshot wound prophylaxis was found to reduce the risk of a wound becoming purulent from 14% to 4% and to reduce recovery time markedly (Gvasalia *et al.*, 2010).

Sanitation of medical care units using bacteriophages (phagosanitation) may be a cheap and effective measure. The use of phage cocktails for cleaning of hospital rooms (floor, walls, water basin, beds, end tables, etc.), and operating theatres as well as for hand washing by medical personnel has been carried out since 1991, when a systematic study of the effects of phagosanitation on the environmental microflora was conducted at Tbilisi Central Hospital Departments of Surgery and Critical Medicine. An analysis of 732 bacteriologic samples taken from the environment at various times showed a large decrease in the detection of *Pseudomonas*, *Proteus*, *E. coli*, and *Staphylococcus* following the use of phage preparations (Table 11.2; Gvasalia *et al.*, 2010). After 6 months of phagosanitation the frequency

Table 11.2 Results of phagosanitation carried out in Tbilisi Central Hospital Departments of Surgery and Critical Medicine

	Frequency of occurrence of bacteria in tested samples			
Analysed bacterial strains	Baseline values	After first phagosanitation	After 2 months	After 6 months
P. aeruginosa	7.2%	3.6%	1.2%	0.3%
Proteus	11.2%	6.3%	3.2%	1.8%
E. coli	11.4%	6.5%	3.3%	1.9%
Staphylococcus	13.6%	8.2%	3.4%	0.9%

of isolation of bacteria was reduced by 6 to 24 times depending on the analysed genus or species of bacteria.

Due to the challenging economic conditions of the last 20 years, these were the last major detailed phage therapy studies in Georgia, and the results were only published rather recently (Alavidze *et al.*, 2007; Gvasalia *et al.*, 2010).

International support of phage research and development in Georgia

Through much of the 1990s, the Eliava Institute and much of the whole health care system were in survival mode, along with the rest of Georgia. However gradually, Georgia began reviving, with substantial help from Western nations, while the Western world became more aware of phage therapy, especially the Georgian work. Major factors which contributed to this included a 1995 article on phage therapy and the Eliava Institute in Discover Magazine, followed by hour-long documentaries by the British Broadcasting Company (BBC) and the Canadian Broadcasting Company (CBC). Additionally, support from NATO, as well as the US Defense Threat Reduction Agency (DTRA), the International Science and Technology Centers (ISTC), the Civilian Research and Development Foundation (CRDF), Science and Technology Centers of the Ukraine (STCU), UK Global Threat Reduction Programme (GTRP) and European Union programs helped resurrect research efforts. The Eliava Institute was a major beneficiary of grants for their work with phage for defence against potential agents of bioterrorism and for therapeutic applications (Chanishvili *et al.* 2001; Kutter *et al.*, 2010; Abedon *et al.*, 2011).

Even during the challenging period of the 1990s, a major innovation came out of the focus on treating severe wound and burn infections, coupled with new collaborations between research groups from Georgia and other countries. This innovation was a novel phage preparation called PhagoBioDerm, a biocomposite based on Pyophage embedded in a non-toxic biodegradable polymer which is characterized by prolonged release and does not require repeated application (Tsitlanadze and Khosroashvili, 1996). As a result of polymer biodegradation by α-chymotrypsin and trypsin, phages targeting *Staphylococcus, Streptococcus, Proteus, E. coli* and *Pseudomonas* are gradually released from the film with zero-order kinetics, ideal for releasing phage in a well-targeted and controlled fashion. In 1998, the required pre-clinical animal and other trials of PhagoBioDerm conducted under the 'rules for preclinical evaluation of safety of pharmacological media (GLP) RD G4-126-91, Moscow, 1992' (lead scientist Professor L. Tuskia; project general director T. Kshutashvili) were successfully completed at the Tbilisi Medical University Experimental and Medical Institute (Alavidze *et al.*, 1998) and published in a 131-page official institutional report approved by Professor G. Dumbadze. Markoishvili *et al.* (2002) reported the use of PhagoBioDerm in treating post-operative complications, infected bite wounds, eczema, bed sores, and infected venous ulcers in patient volunteers who had failed to respond to conventional therapy; the wounds or ulcers healed completely in 67 of 96 cases (70%) where follow-up data were available. PhagoBioDerm also was successfully used to treat severe *S. aureus*-infected radiation burns in two patients who were not responding to antibiotics or other treatments (Stone, 2002; Jikia *et al.*, 2005).

Supported by a 2004 STCU grant, an upgraded biocomposite film, 'Mycoliz', was created by the Eliava Institute using Pyophage enriched with phages specific for *Klebsiella*, *Enterococcus* and *Acinetobacter*, Pimafucin (an antifungal agent), and trypsin. Incorporation of phages targeting these additional pathogens into the Pyophage significantly increased its range against the most problematic purulent bacteria. Infections caused by conditionally pathogenic fungi are a growing problem for contemporary clinical therapy. Pimafucin was chosen because it is less toxic than other available antifungal remedies and it does not inactivate phage. The biocomposite film thus formulated has a wide spectrum of antimicrobial activity (89.3% of 941 tested bacterial and fungal strains) and provides effective controllable phage release (Alavidze, 2004; Goderdzishvili *et al.*, 2009). This highly active preparation has complementary bacteriophage and antifungal components for dealing with mixed fungal and bacterial infections, often eliminating the need for separate antibiotic use.

Training physicians and tailoring commercial phage cocktails to local needs

One major key to the ongoing success of phage therapy in Georgia is the close collaboration between Eliava scientists and local medical facilities, including producing special batches of their commercial phages with added components as needed. Furthermore, when the standard cocktails are not effective enough, active individual phages can often be selected from the Institute collections. Various labs of the Institute work intensely on the regular updating of the classic commercial cocktails to deal with the most prevalent current pathogenic bacterial strains, improving methods of phage delivery and developing new formulations. This can include adding phages targeting such pathogens as *Enterococci* and *Klebsiella*, increasingly recognized as causing serious, hard to treat antibiotic-resistant infections, as exemplified by the development of the Mycoliz biocomposite film and the battlefield work during the 1991–92 conflict.

While many Georgian physicians were experienced in incorporating phage therapy into their practice during Soviet times, Western training and medical practices became more prevalent and antibiotics more readily available. Relatively few of the hospitals and clinics still used phage as a standard part of medical practice, and the intricate skills of using phage in preventing and treating infection in complex wounds was in danger of being lost. Thus, the Eliava Institute began periodically providing lectures and training sessions for local and regional physicians. In addition, in 2005 the Medical University Department of Infectious Surgery started offering a year-long 'Surgical Infections and Phage Therapy' training program for young Georgian surgeons, headed by Dr Guram Gvasalia and combining extensive supervised clinical experience in phage-assisted wound treatment, with targeted in-depth academic work in clinical microbiology, pathology and immunology, phage biology, and surgical infections. As well as improving patient care, this will facilitate collection of data needed for publishable studies of phage therapy and, eventually, for more formal clinical trials with such long-term chronic applications as infected diabetic ulcers. Several small new hospitals now have the resources to carry out complex phage therapy studies, especially if outside funding can be obtained.

Dr Teona Danelia, a graduate of this program, now heads the in-patient surgery department at the David Kodua Center (DKC) Hospital. Dr Danelia's doctoral dissertation (Danelia, 2006) on phage therapy as applied to maxillofacial wounds is a good example of the current level of work possible there. Her studies involved 394 patients with maxillofacial infections (suppurative wounds, local abscesses, maxillary sinusitis, necrotic cellulitis, osteomyelitis of the jaw) and 184 patients with fresh street wounds, ranging from accidents to gunshots, which were treated prophylactically to prevent infection; 118 of the former group of patients and 96 of the latter served as controls, given the same treatment but without the inclusion of phage. This work is especially challenging, since maxillofacial infections easily spread to become fatal CNS infections. Phage therapy was used in combination with the removal of necrotic tissues, opening of blind wound pouches, and lavage with 4% sodium bicarbonate solution. For severe maxillofacial injuries PhageBioDerm powder was sprinkled in the wound just before radical surgery to decrease the risk of infection and/or cavities were drained and phage introduced via thin catheters three or four times daily. Initially, only 64% of the nosocomial opportunistic pathogens were sensitive to the commercial Pyophage and 72% to the laboratory bank of phages, but by adapting the phage preparation to these particular nosocomial strains, the level of sensitivity was increased to 88% (Danelia, 2006). This prophylactic use of Pyophage reduced the risk of suppuration 4-fold and sped recovery. In addition, she reported that oral phage sanitation played a significant role in preventing infections in intensive care patients, since medical devices in the oral cavity provide excellent conditions for propagation and dissemination of oral pathogenic flora.

New therapeutic directions at the Eliava Institute

Today, the Eliava Institute continues research on the use of bacteriophages against a wide variety of pathogenic and opportunistic bacteria, including *Klebsiella* and *Acinetobacter* as well as on the traditional components of Intestiphage and Pyophage. Bacterial strains are received from the medical centres and hospitals of various countries and virulent phages with broad host range are selected from a wide variety of sources and characterized in detail. Work is systematically carried out on the development or updating of therapeutic phage preparations to be produced by Eliava Biopreparations and marketed, usually as sterile lysates after approval of the appropriate legal documents and testing of the properties of each batch in the Georgian federal quality-control laboratory. For example, scientists of the Eliava Institute have continued to work on adaptation and isolation of new staph phages on freshly isolated clinical collections of *S. aureus*, including MRSA, obtained from various countries. Recently, Eliava Biopreparations obtained a marketing license for local use of a new highly purified preparation of their sequenced and otherwise extensively characterized staphylococcal phage, Sb-1 (Kvachadze *et al.*, 2011). This phage, a main component of the intravenous phage preparation, developed and applied clinically in the 1980s, was described above (see the section 'The Republic of Georgia: the Eliava Institute'). This commercial staphylococcal phage preparation, available in local pharmacies, is still approved only for local use, mainly in monoinfections caused by *S. aureus*.

Eliava scientists are also working together with different foreign centres to better analyse their phages, including sequencing some of them, and to develop new phage applications (Kutateladze and Adamia, 2008, 2010; Kvachadze *et al.*, 2011; Ceyssens *et al.*, 2011; Kusradze *et al.*, 2011; Karumidze *et al.*, 2012; Vandersteegen *et al.*, 2011). They also collaborate with international partners to supply phage and support services for clinical work in other countries (see Merabishvili *et al.*, 2009; Khawaldeh *et al.*, 2011).

Phage formulations produced by the Eliava Institute are still used as a standard part of medical practice in Georgia, particularly in burns and in surgical practice, both prophylactically and where infection is already present. Physicians are building on the clinical trials and general experience from earlier decades. As far as actual formal clinical trials go, cystic fibrosis and prostatitis/urinary tract infections are the only situations in recent years where they have managed to find funding for their own small-scale clinical trials – and little of that work, discussed below, has yet been published. Institute scientists are currently working together with the physicians at the Eliava Phage Therapy Center on 'guidelines for phage therapy' to be approved by the Ministry of Healthcare of Georgia, drawing also on the long years of experience there, as made more accessible by Chanishvili (2012a,b). This document will provide a cohesive set of methods and guidelines for physicians interested in using phage for the treatment of various infectious complications.

New therapeutic directions in phage therapy in Georgia include, among others, the treatment of cystic fibrosis (CF), which is the leading lethal genetic disorder among northern Europeans but was only recently recognized as an important medical problem in Georgia; until recently, it had led to undiagnosed failure to thrive and early death from a combination of causes. Several researchers at the Eliava have been isolating and characterizing phages that encode and carry enzymes capable of degrading the biofilm glycocalyx of *P. aeruginosa* CF strains, enhancing phagocytic uptake of bacteria, and reducing the viscoelasticity of CF sputum (Glonti *et al.*, 2010). Others at the Eliava Institute are now carrying out trials of phage therapy with the goal of improving quality of life in this population of patients (Kutateladze and Adamia, 2010).

All children brought to the new Cystic Fibrosis National Center in Tbilisi from 2007 to 2009 with symptoms suggestive of CF were diagnosed using a sweat conductivity analyser. Where CF was thus confirmed, they were followed in terms of weight for height ratio, general state of health, lung function, and detailed analysis of lung infections. In some cases, genetic testing identified the specific mutation. *S. aureus* and *P. aeruginosa* were, as expected, the main causative agents of their lung infections. All bacterial isolates were tested against the commercial Eliava Pyophage, Intestiphage and *S. aureus* phage Sb-1 and several recently isolated *Pseudomonas* phages; virtually all were lysed by at least one of these phage preparations. The genetic relatedness of the bacterial isolates was examined by pulsed-field gel electrophoresis (PFGE). In only a few cases, multiple genetically different *S. aureus* strains were detected in the same patient. In some cases, genetically similar strains were isolated at different times from patients who had had no contact with each other.

Thirteen of these Georgian children, aged 2–8 years, were enrolled in a new Eliava CF phage treatment programme from 2007 to 2009, and are still receiving phage treatment. Their lung infections were mainly caused by *S. aureus* and *P. aeruginosa*. Most had sweat tests from 110 to 140 mmol/l and complaints of tiredness, while some had dark-coloured phlegm and tachycardia. The bacterial concentration in clinical specimens was determined before and after phage treatment. The phages were diluted into isotonic saline and generally administered over the course of an hour once to twice a day for 5 days using the nebulizer they also used for administration of other medications, with several interludes between blocks of treatment. The results were carefully monitored, with sputum samples analysed just before and after each treatment block. The patients also continued using the standard therapeutic protocol, including anti-mucus medications, enzymes, vitamins and courses of conventional antibiotics as needed. In all cases, the general condition of the patients improved with the use of phage inhalation as needed: they gained in weight, their frequencies of exacerbation decreased approximately 2-fold, and their oxygen saturation levels increased significantly (Kvachadze *et al.*, 2011).

The first patient, a 7-year-old girl in whom CF was diagnosed in 2002, provides a good case study (Kvachadze *et al.*, 2011). She is homozygous for the mutation 1677 delTA and had a chloride concentration of 128–135 mmol/l. She presented with a chronic colonization of the lungs with *P. aeruginosa* and *S. aureus*, both unresponsive for several years to broad spectrum antibiotics. Pyophage was administered daily by nebulizer for 9 days at a time, at 4- to 6-week intervals. Prior to administration of the first dose of phage, the bacterial titre in her sputum was 10^7 CFU/ml for *S. aureus* and 8×10^6 CFU/ml for *P. aeruginosa.* After the first phage application, the titre of *P. aeruginosa* rapidly decreased to 7×10^3 CFU/ml, but the *S. aureus* titre dropped only to 10^6 CFU/ml, reflecting the weak susceptibility of this strain to Pyophage observed also *in vitro*. During a month when no phage therapy was administered, the *P. aeruginosa* titre remained low: 7×10^4 CFU/ml.

This augmented Pyophage cocktail was applied daily by nebulizer for five days. The *S. aureus* level then decreased markedly and remained at 10^3–10^5 CFU/ml throughout a medication-free month; for *P. aeruginosa*, it was reduced to 10^1–10^2 CFU/ml during this period. No adverse effects were seen after application either of Pyophage or of the Sb-1 phage. Three to four hours after phage application, phage were detected in the sputum at about 10^2 PFU/ml, but by the second day no phage were observed. The sputum level of *S. aureus* showed a steady decrease over 2 months and then slipped below the detection level. For several months, no *P. aeruginosa* was detected, but then the concentration briefly increased to10^3–3×10^4 CFU/ml three times over the following 12 months. There was no significant change in the chest X-ray or blood tests, but the total periods of antibiotic use decreased by 50%, and the general health of the patient was improved. After the 5th round of phage application, the level of phage-neutralizing antibodies in the patient's serum was explored to determine whether phage inhalation stimulated antibody formation (M. Kutateladze and N. Balarjishvili, unpublished data). Anti-staphylococcal phage antibodies did appear, but their concentration was low ($K=4.5^{-1}$). In spite of the multiple rounds of treatment, no antibodies were detected against the *P. aeruginosa* phages. The patient continues to be monitored and treated as needed.

The Cystic Fibrosis National Center has now also started working with adult CF patients from abroad. The first was a 25 year old male (M. Kutateladze and T. Topuria, unpublished data). He was diagnosed with CF at the age of 15 months and is heterozygous for the ΔF508 CF transmembrane conductance regulator gene mutation. His oxygen saturation before phage treatment was 90%, with a bacterial titre in sputum of 6×10^6 CFU/ml for *S. aureus* and 10^6 CFU/ml for *P. aeruginosa*. Pyophage showed high lytic activity on these strains *in vitro*. Phage treatment was applied daily by nebulizer for 10 days. The titre of *S. aureus* decreased to10^2 CFU/ml, and *P. aeruginosa* could no longer be detected. The patient then returned home and took phage in his nebulizer once daily for 5 days every month for 6 months. A year later, when he returned to Tbilisi for further treatment, his oxygen saturation was 94–96%. Bacteriological analyses of the sputum and throat showed 10^6 CFU/ml of *P. aeruginosa* (hem+), 6×10^6 CFU/ml of *S. aureus* (hem–), 10^5 CFU/ml of *S. pyogenes* (hem+), and 3×10^5 CFU/ml of *Enterococcus*. No phage-neutralizing antibodies were detected against any of the respective phages. After 10 days of Pyophage treatment in Tbilisi, the titre dropped to 4×10^2 CFU/ml for *S. aureus*, but was still 5×10^5 CFU/ml for *P. aeruginosa*, 10^5 CFU/ml *for S. pyogenes* (hem-) and 5×10^6 CFU/ml for *Enterococcus*. A month later, the concentration was 10^2 PFU/ml *for S. aureus* and *P. aeruginosa*, but *S. pyogenes* (hem–) was up at 5×10^6 PFU/ml. A year after the previous analyses, the *P. aeruginosa* was $<10^3$ CFU/ml, but the concentration of *S. aureus* (hem–) was back to 5×10^6 CFU/ml and *S. pyogenes* (hem–) was 10^6 CFU/ml. Pyophage was again sent to this patient for further treatment; he now applies phage continually, with 2–3 month interruption periods. The patient's health status has continued to improve, he has started physical exercise, and he has even gotten a job.

In the general opinion of the Eliava scientists, a major advantage of phage therapy for CF has been a decrease in antibiotic use by 50%. They also observed that the general health of all patients is significantly better including improved psycho-emotional status, recovered appetite, decreased frequency of illnesses, long remissions, weight gains, and increases in oxygen saturation.

Phages have also been used to treat men with chronic bacterial prostatitis and urogenital infections. Researchers at the Eliava Institute have isolated strains of the aetiological agents of infection as well as specific phages against them which were used for preparation of a new commercial phage cocktail called UroPhage (Goderdzishvili *et al.*, 2013).

Before the beginning of the clinical studies, samples taken from 314 patients were examined (250 patients with chronic bacterial prostatitis and 40 with acute or chronic urethritis, along with sperm samples from an additional 24 patients). Bacteria identified in their urine, sperm and prostate fluid included 185 *E. coli* isolates, 112 *Staphylococcus* isolates, 67 *P. aeruginosa* isolates, 55 *Proteus* isolates, 53 *Enterococcus* isolates, 50 *Klebsiella* isolates, and 38 *Streptococcus* isolates. The strains found to be most antibiotic resistant included *Klebsiella*, *Enterococcus*, and *P. aeruginosa*. Only 147 of the 560 strains were sensitive to the current commercial Intestiphage. Pyophage does not contain phages targeting *Klebsiella* or *Enterococcus*; only 109 of the other 437 strains were sensitive to the Pyophage. The generally observed low sensitivities of bacteria from prostatitis and urinary tract infections to the Eliava Pyophage and Intestiphage presumably reflect the fact that only bacterial strains isolated from purulent wounds and the alimentary tract, respectively, had been used in their formulation and production. Therefore, many new phages were isolated against resistant strains of the above bacteria, followed by studies of their host range, stability, virulence, cross-reactivity and other relevant properties. Their incorporation into Urophage resulted in its high activity against 86.2% of this set of 560 clinical strains.

An open set of clinical trials involving 118 patients with chronic bacterial prostatitis caused by various combinations of *Staphylococcus*, *Pseudomonas*, *Klebsiella*, *Enterococcus*, *Proteus*,

Streptococcus spp. and *E. coli* was carried out in the Urology Department at the Central Clinical Hospital in Tbilisi to evaluate the safety and efficacy of Urophage. All of the patients had previously been treated unsuccessfully in the same department with targeted antibiotics and anti-inflammatory medications. The patients were divided into seven sets, each composed of individuals who all had the same combination of microbes, with control and various treatment groups in each set. All patients of each set exhibited similar characteristic symptoms of frequent urination, false sense of need to urinate, burning pain while urinating, increased leucocytes in prostate fluid, and sexual difficulty. It was observed by the physicians conducting the study that the combination of phage and antibiotics succeeded in lowering the bacterial concentration below the detection level, while the antibiotic alone (where bacteria were still sensitive to some antibiotic) or the phage alone left some residual bacterial cells in prostatic fluid that required further treatment beyond the 19-day treatment period and 10-day follow-up. In general, the symptoms were substantially lowered over this period. In conclusion, conditions have been established for carrying out full-scale initial clinical trials by adding phage to the standard protocols for treating prostatitis and urinary tract infections; such trials are currently being designed in Georgia, but funding is still a problem.

Concluding remarks

Since the first clinical use of bacteriophage by d'Hérelle in 1919, the attitude of the scientific and medical community towards phage therapy has undergone considerable evolution which ultimately has shaped its current status. Although in the first half of the 20th century phage therapy was used in a number of European countries as well as in the USA, India and elsewhere, it had mixed success for a variety of reasons and was almost completely forgotten in most of the Western World after antibiotics became widely available. However, it has continued to be employed on a relatively large scale in some eastern countries, especially Poland, Georgia, and Russia. As a result, phage therapy paradoxically started to be regarded with suspicion as an 'Eastern' treatment which was not covered by Western regulations and therefore not worthy of interest (Stone, 2002).

The extensive Nestlé study of phage targeting infant diarrhoea emphasizes the possibilities of phage therapy, but also the challenges and costs of requiring standard western double-blind clinical trials of fully characterized products prior to commercialization. It is very important that this is being done at least once, but how many lives will be lost if this depth of study is required to bring each phage product to market, and is that justifiable at this point in time? As Brüssow (2012) discusses, there are many barriers to the private sector carrying out this sort of very expensive studies and commercializing the most urgently needed products. What is needed is a public health approach, with public/private partnerships and the minimization, not maximization, of the use of such new products to prolong their usefulness. It is not accidental that phage therapy was so widely developed in parts of the world where public health was a collaborative effort, not so dependent on economic forces.

Despite the lack of studies meeting current Western RCT standards, hundreds of thousands of people were treated using phage therapy in Georgia and Russia (where a few phage preparations are registered for standard use), or Poland (where the highly individualized approach is based on the rules of experimental therapy, as defined by Declaration of Helsinki, but does not lend itself to double-blind trials). Thus, their increasingly available extensive data can and should substantially contribute to exploration of the areas of potential applications where the knowledge is sufficient to warrant moving forward, at least on a compassionate-use basis, as is happening in some places. We believe that these studies help supply direction to appropriate detailed clinical trials of phage therapy and also provide strong arguments for the general safety of phage therapy as applied topically or orally, at a level and breadth that will never be reachable through specific phase I trials of any achievable size, though the latter may well help point out some specific potential considerations if very well designed.

Particularly relevant safety studies are seen in the recent in-depth work of the IIET, published by Międzybrodzki *et al.* (2012). As in the Nestlé

Bangladesh studies, people are used as their own controls, though not in blinded fashion in the Polish therapeutic applications. For example, the effects of various individual therapeutic phages on such factors as blood morphology including white blood cell subsets, a variety of liver enzymes, serum proteins, blood sugar and renal function were compared between multiple extended periods of cumulative use of therapeutic phages and the results of tests before initiation of phage therapy. There were a few complaints about taste and 18% had short-lived minor local responses to topical application. While resistance developed rarely in patients treated with *S. aureus* phages, it was found in 43% of *Pseudomonas* phage cases and 86% of patients who were administered coliphages, requiring frequent phage changes; worse yet, 29% of the *E. coli* cases and 27% of the *P. aeruginosa* developed resistance to all of the IIET cognate phages. These results strongly support the broad perception that phage therapy is safe and that well-constructed phage cocktails are highly preferable to individual phages for most therapeutic applications, and the IIET is starting to implement use of multiple phages. This further emphasizes the necessity of accommodating such cocktails in regulatory pathways and clinical trials, along with reasonable rules for updating the cocktails on a regular basis. For reviews on phage cocktails, see (Chan and Abedon, 2012) and (Chan *et al.*, 2013).

Complex cocktails, such as those used in Georgia, make possible large-scale commercial production for widespread application, the safety-testing and licensing of the cocktails by the state, simplified testing of the product against the bacteria in question and double-blind studies of efficacy. The metagenomic testing of the Russian and Nestlé cocktails provides the blueprint for developing extensive understanding of the make-up and absence of potentially pathogenic components of the Pyophage and Intestiphage cocktails believed to be so efficacious, and using that understanding to support use of those cocktails and/or to explicitly design and monitor new cocktails with a fairly similar distribution of components but complementary targets. It is clear that the specific hosts used in developing particular cocktails are important. For example, the staph and pseudomonas phages used in Intestiphage and Pyophage were isolated against bacteria from the gut and from purulent infections, respectively. Intestiphage and Pyophage are excellent for problems in those areas, but showed mixed success at treating prostatitis and urinary tract infections, even though Intestiphage includes phages targeting all of the bacterial species commonly found in such infections. With the support of the ISTC, the Eliava developed a new cocktail, using 560 bacterial isolates from 314 urological patients, 250 of them with chronic bacterial prostatitis – *E. coli*, *Staphylococcus*, *P. aeruginosa*, *Proteus*, *Enterococcus*, *Klebsiella* and *Streptococcus* for phage isolation and characterization. They created a new Urophage cocktail that hit 86% of these strains rather than the 28% hit by Intestiphage and appeared in small-scale clinical trials to be much more successful for prostatitis and other UTIs; it is now on the path towards Georgian approval.

Clearly, there is much more to successful phage therapy than just the particular product, especially in such cases as the treatment of complex wounds. The Georgian clinical data are often extensive enough to use in designing appropriate efficacy studies even if not sufficient on its own for approval in the Western world, and there is a strong pool of physicians there with the knowledge and experience to effectively use phage clinically. The data related to military phage use in the 1992 fighting in Abkhazia points to the great usefulness of carefully monitoring the subset of initial bacteria actually leading to long-term complications, rapidly isolating phage against those pathogens, and adding them to the approved cocktail in cases where speed of response and treatment is of the essence.

When discussing phages as an alternative to antibiotics one has to consider not only such advantages as different mode of action, high specificity, and self-replication in the infection site and challenges such as the potential development of phage resistance, immunogenicity, varied bioavailability and complex pharmacokinetics. We are also restricted by the current regulations which get in the way of exploiting the full potential of phages or their registration for therapeutic use (as described in Chapter 12). Hopefully ongoing and future RCTs and deeper studies in the context of extensive current use will not only confirm the safety and efficacy of phage therapy, but also will outline a way to overcome regulatory difficulties and stimulate interest in phage therapy development, while our increasing knowledge of phage biology helps us choose the most efficacious and safe phages.

Acknowledgements

This work was supported by the Phagebiotics Research Foundation and, for the Polish portions, by funds from the Operational Program Innovative Economy, 2007–2013, Priority axis 1. Research and Development of Modern Technologies, Measure 1.3 Support for R&D projects for entrepreneurs carried out by scientific entities, Submeasure 1.3.1 Development projects as project No. POIG 01.03.01-02-003/08 entitled "Optimization of the production and characterization of bacteriophage preparations for therapeutic use".

References

Abedon, S.T., Kuhl, S.J., Blasdel, B.G., and Kutter, E.M. (2011). Phage treatment of human infections. Bacteriophage *1*, 66–85.

Alavidze, Z., Meipariani, A., Kvatadze, N., Katsarava, R., Tsitlanadze, G., Nadirasvili, N., Dgebuadze, Sh., Dumbadze, G., Tuskia, E., and Goderdzishvili, M. (1998). Preclinical studies of a novel therapeutic wound dressing containing bacteriophages, 'Phagobioderm'. Proceedings of the 12th Evergreen International Phage Biology Meeting, p27.

Alavidze, Z., Meiphariani, A.L., Dzidzishvili, L., Chkonia, I., Goderdzishvili, M., Kvatadze, N., Jgenti, D., Makhatadze, N., Zhgenti, D., and Dolidze, N. (2004). Application of a new Combined Preparation –'Mycoliz' in Gynecology. Bull. Georg. Acad. Sci. *170*, 379–382.

Alavidze, Z., Meiphariani, A., Dzidzishvili, L., Chkonia, I., Goderdzishvili, M., Kvatadze, N., Jgenti, D., Makhatadze, N., Gudumidze, N., and Gvasalia, G. (2007). Treatment of the complicated forms of inflammatory wounds by bacteriophages. Med. J. Georg. 2,123–127.

Alisky, J., Iczkowski, K., Rapoport, A., and Troitsky, N. (1998). Bacteriophages show promise as antimicrobial agents. J. Infect. *36*, 5–15.

Boratyńska, M., Szewczyk, Z., and Weber-Dąbrowska, B. (1994). Clinical assessment of bacteriophage use in the treatment of urinary tract infections. Post. Med. Klin. Dośw. 3, 7–11 [article in Polish].

Brüssow H. (2012). What is needed for phage therapy to become a reality in Western medicine? Virology *434*, 138–142.

Bruttin, A., and Brüssow, H. (2005). Human volunteers receiving *Escherichia coli* phage T4 orally: a safety test of phage therapy. Antimicrob. Agents. Chemother. *49*, 2874–2878.

Bruynoghe, R., and Maisin, J. (1921). Essais de thérapeutique au moyen du bactériophage du Staphylocoque. J. Compt. Rend. Soc. Biol. *85*, 1120-1121.

Burrowes, B., Harper, D.R., Anderson, J., McConville, M., and Enright, M.C. (2011). Bacteriophage therapy: potential uses in the control of antibiotic-resistant pathogens. Expert Rev. Anti. Infect. Ther. *9*, 775–785.

Bush, K., Courvalin, P., Dantas, G., Davies, J., Eisenstein, B., Huovinen, P., Jacoby, G.A., Kishony, R., Kreiswirth, B.N., Kutter, E., *et al.* (2011). Tackling antibiotic resistance. Nat. Rev. Microbiol. *9*, 894–896.

Calfee, D.P. (2012). Methicillin-resistant *Staphylococcus aureus* and vancomycin-resistant enterococci, and other Gram-positives in healthcare. Curr. Opin. Infect. Dis. *25*, 385–394.

Ceyssens, P.J., Glonti, T., Kropinski, N.M., Lavigne, R., Chanishvili, N., Kulakov, L., Lashkhi, N., Tediashvili, M., and Merabishvili, M. (2011). Phenotypic and genotypic variations within a single bacteriophage species. Virol. J. *8*, 134.

Chan, B.K., and Abedon, S.T. (2012). Phage therapy pharmacology: phage cocktails. Adv. Appl. Microbiol. *78*, 1–23.

Chan, B.K., Abedon, S.T., and Loc-Carrillo, C. (2013). Phage cocktails and the future of phage therapy. Future Microbiol. *8*, 769–783.

Chanishvili, N. (2012a). A Literature Review of the Practical Application of Bacteriophage Research. N. Chanishvili, ed. (Hauppauge, NY, Nova Science).

Chanishvili, N. (2012b). Phage therapy-history from Twort and d'Hérelle through Soviet experience to current approaches. Adv. Virus Res. *83*, 3–40.

Chanishvili, N., Chanishvili, T., Tediashvili, M., and Barrow, P.A. (2001). Phages and their application against drug-resistant bacteria. J. Chem. Technol. Biotechnol. *76*, 689–699.

Chibani-Chennoufi, S., Sidoti, J., Bruttin, A., Dillmann, M.L., Kutter, E., Qadri, F., Sarker, S.A., and Brüssow, H. (2004). Isolation of *Escherichia coli* bacteriophages from the stool of pediatric diarrhea patients in Bangladesh. J. Bacteriol. *186*, 8287–8294.

Cisło, M., Popiela, A., Blok, K., Weber-Dąbrowska, B., and Dąbrowski, M. (1988). Bacteriophage therapy in pospartum mastitis. Ginekol. Pol. *59*, 160–163 [article in Polish].

Danelia, T. (2006). Options for the Optimization of Prophylaxis and Treatment of Surgical Infections in the Maxillofacial Surgery. Dissertation for the Degree of Candidate of Medical Sciences (Tbilisi, Georgia).

Denou, E., Bruttin, A., Barretto, C., Ngom-Bru, C., Brüssow, H., and Zuber, S. (2009). T4 phages against *Escherichia coli* diarrhea: Potential and problems. Virology *388*, 21–30.

Dworski, M., Pelzer, M., Chruściel, A., Markiewicz, Z., Weber-Dąbrowska, B., and Dąbrowski, M. (1990). Bacteriophage therapy in selected clinical cases. Prace Karkonoskiego Towarzystwa Naukowego *57*, 221–227 [article in Polish].

Giamarellou, H. (2010). Multidrug-resistant Gram-negative bacteria: how to treat and for how long. Int. J. Antimicrob. Agents. *36*, S50-S54-Supplement2.

Glonti, T., Chanishvili, N., and Taylor, P.W. (2010). Bacteriophage-derived enzyme that depolymerizes the alginic acid capsule associated with cystic fibrosis isolates of *Pseudomonas aeruginosa*. J. Appl. Microbiol. *108*, 695–702.

Goderdzishvili, M., Alavidze, Z., Meipariani, A., Chkonia, I., Dzizdzishvili, L., Kvatadze, N., Jgenti, D., Rigvava, S., Dvalidze, T., Kusradze, I., *et al.* (2009). Creation and study of antimicrobial-antifungal composite against mixed infection using biodegradable polymer as a matrix. Proc. Georg. Acad. Sci., Biological series B 7, 57–65.

Goderdzishvili M., Meipariani, A., Chkonia, I., Dzizdzishvili, M., Kvatadze, N., Jgenti, D., Makhatadze, N., Rigvava, S., Karumidze, N., Kusradze, I., *et al.* (2013). Development and initial testing of a phage cocktail to treat chronic bacterial prostatitis. Microbiol. Biotechnolog. *In press*.

Górski, A., Międzybrodzki, R., Borysowski, J., Weber-Dabrowska, B., Lobocka, M., Fortuna, W., Letkiewicz, S., Zimecki, M., and Filby, G. (2009). Bacteriophage therapy for the treatment of infections. Curr. Opin. Investig. Drugs *10*, 766–774.

Górski, A., Międzybrodzki, R., Borysowski, J., Dąbrowska, K., Wierzbicki, P., Ohams, M., Korczak-Kowalska, G., Olszowska-Zaremba, N., Łusiak-Szelachowska, M., Kłak, M., *et al.* (2012). Phage as a modulator of immune responses: practical implications for phage therapy. Adv. Virus Res. *83*, 41–71.

Gvasalia, G., Kutter, E., Alavidze, Z., and Danelia, T. (2010). Phage therapy of surgical infections. (Thirty years' experience and issues to be solved). Exp. Clin. Med. *57*, 9–14.

Harper, D.R., Anderson, J., and Enright, M.C. (2011). Phage therapy: delivering on the promise. Ther. Deliv. *2*, 935–947.

Hawkins, C., Harper, D., Burch, D., Anggård, E., and Soothill, J. (2010). Topical treatment of *Pseudomonas aeruginosa* otitis of dogs with a bacteriophage mixture: a before/after clinical trial. Vet. Microbiol. *146*, 309–313.

Häusler, T. (2006). Viruses vs. Superbugs: A Solution to the Antibiotics Crisis? (London, New York, Macmillan).

Hirszfeld, L. (1948). The struggle of the visible against invisible world. Sprawozdania Wrocławskiego Towarzystwa Naukowego 3, 3–18 [article in Polish].

Ilfaev, P.X., Zueva, L.P., Lubimova, A.M., Aclanov, B.M., and Okolov, I.N. (1998). Prospects for application of bacteriophages for therapeutic and preventive purposes. In Infection Control in Medical-Prophylactic Units. B.M. Taits and L.P. Zueva, eds. (Saint Peterburg, State Medical Academy), pp.55–61 [chapter in Russian].

Jasieński, J. (1927). Attempts of application of bacteriophage therapy in surgery. Polska Gazeta Lekarska 4, 67–73 [article in Polish].

Jikia, D., Chkhaidze, N., Imedashvili, E., Mgaloblishvili, I., Tsitlanadze, G., Katsarava, R., Glenn Morris, J. Jr, and Sulakvelidze, A. (2005). The use of a novel biodegradable preparation capable of the sustained release of bacteriophages and ciprofloxacin, in the complex treatment of multidrug-resistant *Staphylococcus aureus*-infected local radiation injuries caused by exposure to Sr90. Clin. Exp. Dermatol. *30*, 23–26.

Kaczkowski, H., Weber-Dąbrowska, B., Dąbrowski, M., Zdrojewicz, Z., and Cwioro, F. (1990). Use of bacteriophages in the treatment of chronic bacterial diseases. Wiad. Lek. *43*, 136–141 [article in Polish].

Kalinowski, W., and Czyż, J. (1923). The report on the epidemic of dysentery in the year 1922. Lekarz Wojskowy *4*, 286–293 [article in Polish].

Karumidze, N., Thomas, J.A., Kvatadze, N., Goderdzishvili, M., Hakala, K., Weintraub, S.T., Alavidze, Z., and Hardies, S.C. (2012). Characterisation of lytic *P. aeruginosa* bacteriophages via biological properties and genomic sequences. Appl. Microbiol. Biotechnol. *94*, 1609–1617.

Khawaldeh, A., Morales, S., Dillon, B., Alavidze, Z., Ginn, A.N., Thomas, L., Chapman, S.J., Dublanchet, A., Smithyman, A., and Iredell, J.R. (2011). Bacteriophage therapy for refractory *Pseudomonas aeruginosa* urinary tract infection. J. Med. Microbiol. *60 (Pt 11)*, 1697–1700.

Krylov, V.N. (2001). Phagotherapy in terms of bacteriophage genetics: hopes, perspectives, limitations. Genetika *37*, 869–887.

Kuchment A. (2012). The Forgotten Cure: The Past and Future of Phage Therapy (New York, Copernicus Books).

Kutateladze, M., and Adamia, R. (2008). Phage therapy experience at the Eliava Institute. Med. Mal. Infect. *38*, 426–430.

Kutateladze, M., and Adamia, R. (2010). Bacteriophages as potential new therapeutics to replace or supplement antibiotics. Trends Biotechnol. *28*, 591–595.

Kutter, E. (2009). Phage host range and efficiency of plating. Methods Mol. Biol. *501*, 141–149.

Kutter, E., De Vos, D., Gvasalia, G., Alavidze, Z., Gogokhia, L., Kuhl, S., and Abedon, S.T. (2010). Phage therapy in clinical practice: treatment of human infections. Curr. Pharm. Biotechnol. *11*, 69–86.

Kusradze, I., Seydina, M., Diene, Goderdzishvili, M., Raoult, D., and Rolain, J.M. (2011). Molecular detection of OXA-carbapenemase genes in multidrug-resistant *Acinetobacter baumannii* isolates from Iraq and Georgia. Int. J. Antimicrob. Agents *38*, 164–168.

Kvachadze, L., Balarjishvili, M., Meskhi, T., Tevdoradze, E., Skhirtladze, N., Pataridze, T., Adamia, R., Topuria, T., Kutter, E., Rohde, C., and Kutateladze, M. (2011). Evaluation of lytic activity of staphylococcal bacteriophage Sb-1 against freshly isolated clinical pathogens. Microbiol. Biotechnol. *4*, 643–650.

Kwarciński, W., Lazarkiewicz, B., Weber-Dąbrowska, B., Rudnicki, J., Kamiński, K., and Sciebura, M. (1994). Bacteriophage therapy in the treatment of recurrent subphrenic and subhepatic abscess with jejunal fistula after stomach resection. Pol. Tyg. Lek. *49*, 535 [article in Polish].

Leszczyński, P., Weber-Dąbrowska, B., Kohutnicka, M., Łuczak, M., Górecki, A., and Górski, A. (2006).

Successful eradication of methicillin-resistant *Staphylococcus aureus* (MRSA) intestinal carrier status in a healthcare worker – case report. Folia Microbiol. *51*, 236–238.

Letarov, A.V., Golomidova, A.K., and Tarasyan, K.K. (2010). Ecological Basis for Rational Phage Therapy. Acta Naturae 2, 60–72.

Letkiewicz, S., Międzybrodzki, R., Fortuna, W., Weber-Dąbrowska, B., and Górski, A. (2009). Eradication of *Enterococcus faecalis* by phage therapy in chronic prostatitis – case report. Folia Microbiol.*54*, 457–461.

Letkiewicz, S., Międzybrodzki, R., Kłak, M., Bubak, B., Jończyk, E., Weber-Dąbrowska B., and Górski, A. (2010). The perspectives of the application of phage therapy in chronic bacterial prostatitis. FEMS Immunol. Med. Microbiol. *60*, 99–112.

Lityński, M. (1950). Treatment of infection of *E. coli* using specific bacteriophages. Przegl. Lek. *6*, 13–19 [article in Polish].

McCallin S., Alam Sarker S., Barretto, C., Sultana, S., Berger, B., Huq, S., Krause, L., Bibiloni, R., Schmitt, B., Reuteler, G., *et al.* (2013). Safety analysis of a Russian phage cocktail: From MetaGenomic analysis to oral application in healthy human subjects. Virology *443*, 187–196.

Markoishvili, K., Tsitianadze, G., Katsarava, R., Morris, J.G., Jr., and Sulakvelidze, A. (2002). A novel sustained release matrix based on biodegradable poly(ester amide)s and impregnated with bacteriophages and an antibiotic shows promise in management of infected venous stasis ulcers and other poorly healing wounds. Int. J. Dermatol. *41*, 453–458.

Merabishvili, M. (2012). The phage therapy experience of the Brussels burn wound centre [abstract]. In EuroPhages 2012: Bacteriophage in Medicine, Food and Biotechnology; Conference Handbook, p 21.

Merabishvili, M., Pirnay, J.P., Verbeken, G., Chanishvili, N., Tediashvili, M., Lashkhi, N., Glonti, T., Krylov, V., Mast, J., Van Parys, L., *et al.* (2009). Quality-controlled small-scaleproduction of a well-defined bacteriophage cocktail for use in human clinical trials. PLoS One *4*, e4944.

Międzybrodzki, R., Fortuna, W., Weber-Dąbrowska, B., and Górski, A. (2007). Phage therapy of staphylococcal infections (including MRSA) may be less expensive than antibiotic treatment. Post. Hig. Med. Dośw. *61*, 461–465.

Międzybrodzki, R., Fortuna, W., Weber-Dąbrowska, B., and Górski, A. (2009). A retrospective analysis of changes in inflammatory markers in patients treated with bacterial viruses. Clin. Exp. Med. *9*, 303–312.

Międzybrodzki, R., Borysowski, J., Weber-Dąbrowska, B., Fortuna, W., Letkiewicz, S., Szufnarowski, K., Pawełczyk, Z., Rogóż, P., Kłak, M., Wojtasik, E., *et al.* (2012). Clinical aspects of phage therapy. Adv. Virus Res. *83*, 73–121.

Miller, E.S., Kutter, E., Mosig, G., Arisaka, F., Kunisawa, T., and Rüger, W. (2003). Bacteriophage T4 genome. Microbiol. Mol. Biol. Rev. *67*, 86–156.

Mulczyk, M., and Ślopek, S. (1971). Role of specific bacteriophages in combating Shigella sonnei carrier state. Przegl. Epidemiol. *25*, 475–476 [article in Polish].

Nathan, C. (2012). Fresh approaches to anti-infective therapies. Sci. Transl. Med. *4*, 140sr2.

Parracho, H.M., Burrowes, B.H., Enright, M.C., McConville, M.L., and Harper, D.R. (2012). The role of regulated clinical trials in the development of bacteriophage therapeutics. J. Mol. Genet. Med. *6*, 279–286.

Rhoads, D.D., Wolcott, R.D., Kuskowski, M.A., Wolcott, B.M., Ward, L.S., and Sulakvelidze, A. (2009). Bacteriophage therapy of venous leg ulcers in humans: results of a phase I safety trial. J. Wound Care *18*, 237–238, 240–243.

Sarker, S.A., McCallin, S., Barretto, C., Berger, B., Pittet, A.C., Sultana, S., Krause, L., Huq, S., Bibiloni, R., Bruttin, A., Reuteler, G., and Brüssow, H. (2012). Oral T4-like phage cocktail application to healthy adult volunteers from Bangladesh. Virology *434*, 222–232.

Sergienko, F.E. (1945). Dry bacteriophages, their preparation and use. In Microbiology and Epidemiology, B. Babsky, I.G. Kochergin, and V.V. Parin, eds. (Medical Publications Ltd, London).

Soothill, J., Hawkins, C., Anggård, E., and Harper, D. (2004). Therapeutic use of bacteriophages. Lancet Infect. Dis. *4*, 544–555.

Stone, R. (2002). Bacteriophage therapy. Stalin's forgotten cure. Science *298*, 728–731.

Strój, L., Weber-Dąbrowska, B., Partyka, K., Mulczyk, M., and Wójcik, M. (1999). Successful treatment with bacteriophage in purulent cerebrospinal meningitis in a newborn. Neurol. Neurochir. Pol. *33*, 693–698 [article in Polish].

Sulakvelidze, A., and Kutter, E. (2005). Bacteriophage therapy in humans. In Bacteriophages: Biology and Applications, E. Kutter and A. Sulakvelidze, eds. (Boca Raton, FL, CRC Press), pp. 381–436.

Sulakvelidze, A., Alavidze, Z., and Morris, J.G. Jr. (2001). Bacteriophage therapy. Antimicrob. Agents Chemother. *45*, 649–659.

Summers, W.C. (2001). Bacteriophage therapy. Annu. Rev. Microbiol. *55*, 437–451.

Ślopek, S., Durlakowa, I., Kucharewicz-Krukowska, A., Krzywy, T., Ślopek, A., and Weber, B. (1972). Phage typing of *Shigella flexneri*. Arch. Immunol. Ther. Exp. *20*, 1–60.

Ślopek, S., Durlakowa, I., Kucharewicz-Krukowska, A., Krzywy, T., Ślopek, A., and Weber, B. (1973). Phage typing of *Shigella sonnei*. Arch. Immunol. Ther. Exp. *21*, 1–161.

Ślopek, S., Durlakowa, I., Weber-Dąbrowska, B., Kucharewicz-Krukowska, A., Dąbrowski, M., and Bisikiewicz, R. (1983a). Results of bacteriophage treatment of suppurative bacterial infections. I. General evaluation of the results. Arch. Immunol. Ther. Exp. *31*, 267–291.

Ślopek, S., Durlakowa, I., Weber-Dąbrowska, B., Kucharewicz-Krukowska, A., Dąbrowski, M., and Bisikiewicz, R. (1983b). Results of bacteriophage treatment of suppurative bacterial infections. II. Detailed evaluation of the results. Arch. Immunol. Ther. Exp. *31*, 293–327.

Ślopek, S., Durlakowa, I., Weber-Dąbrowska, B., Dąbrowski, M., and Kucharewicz-Krukowska, A. (1984). Results of bacteriophage treatment of suppurative bacterial infections. III. Detailed evaluation of the results obtained in further 150 cases. Arch. Immunol. Ther. Exp. 32, 317–335.

Ślopek, S., Kucharewicz-Krukowska, A., Weber-Dąbrowska, B., and Dąbrowski, M. (1985a). Results of bacteriophage treatment of suppurative bacterial infections. IV. Evaluation of the results obtained in 370 cases. Arch. Immunol. Ther. Exp. 33, 219–240.

Ślopek, S., Kucharewicz-Krukowska, A., Weber-Dąbrowska, B., and Dąbrowski, M. (1985b) Results of bacteriophage treatment of suppurative bacterial infections. V. Evaluation of the results obtained in children. Arch. Immunol. Ther. Exp. 33, 241–259.

Ślopek, S., Kucharewicz-Krukowska, A., Weber-Dąbrowska, B., and Dąbrowski, M. (1985c). Results of bacteriophage treatment of suppurative bacterial infections. VI. Analysis of treatment of suppurative staphylococcal infections. Arch. Immunol. Ther. Exp. 33, 261–273.

Ślopek, S., Weber-Dąbrowska, B., Dąbrowski, M., and Kucharewicz-Krukowska A. (1987). Results of bacteriophage treatment of suppurative bacterial infections in the years 1981–1986. Arch. Immunol. Ther. Ex. 35, 569–583.

Zelińska, J., Stolarska, M., Zalewska-Szewczyk, B., Kuzański, W., and Młynarski, W. (2010). The use of bacteriophage therapy in *Pseudomonas aeruginosa* infection in patients with acute leukemias – two cases report. Onkol Pol. 13, 50–53.

Tsitalanadze, G., and Khosroashvili, T. (1960). Amino acid based bioanalogous polymers. Some biological studies of regular poly(ester amide)s and bioactive components based on them [abstract]. Proceedings of the International Symposium on Biodegradable Materials, Hamburg, Germany, p. 122.

Vandenbroucke, J.P. (2004). When are observational studies as credible as randomized trials? Lancet 363, 1728–1731.

Vandersteegen, K., Mattheus, W., Ceyssens, P.J., Bilocq, F., De Vos, D., Pirnay, J.P., Noben, J.P., Merabishvili, M., Lipinska, U., Hermans, K., *et al.* (2011). Microbiological and molecular assessment of bacteriophage ISP for the control of *Staphylococcus aureus*. PLoS One 6, e24418.

Verbeken, G., Pirnay, J.P., De Vos D., Jennes, S., Zizi, M., Lavigne, R., Casteels, M., and Huys, I. (2012). Optimizing the European regulatory framework for sustainable bacteriophage therapy in human medicine. Arch. Immunol. Ther. Exp. 60,161–172.

Weber-Dąbrowska, B., Koźmińska, J., Mulczyk, M., and Kaczkowski, H. (1996). The use of bacteriophages in the treatment of chronic suppurative sinusitis. Post. Med. Klin. Dośw. 5, 291–293 [article in Polish].

Weber-Dąbrowska, B., Przestalski, M., and Mulczyk, M. (1997). The use of bacteriophages in the treatment of chronic suppurative otitis. Otolaryngologia Polska 51, 773–775 [article in Polish].

Weber-Dąbrowska, B., Mulczyk, M., and Górski, A. (2000). Bacteriophage therapy of bacterial infections: an update of our Institute's experience. Arch. Immunol. Ther. Exp. 48, 547–551.

Weber-Dąbrowska, B., Mulczyk, M., and Górski, A. (2001). Bacteriophage therapy for infections in cancer patients. Clin. Appl. Immunol. Rev. 1, 131–134.

Weber-Dąbrowska, B., Mulczyk, M., Górski, A., Boratyński, J., Łusiak-Szelachowska, M., and Syper, D. (2002). Methods of polyvalent bacteriophage preparation for the treatment of bacterial infections. US Patent No. US 7232564 B2 filed: 18 July 2002. Issued on 19 June 2007.

Weber-Dąbrowska, B., Mulczyk, M., and Górski, A. (2003). Bacteriophages as an efficient therapy for antibiotic-resistant septicemia in man. Transplant. Proc. 35, 1385–1386.

Witoszka, M., and Strumillo, B. (1963). Bacteriophage therapy of infected surgical wounds. Pol. Tyg. Lek. 18, 430–432 [article in Polish].

Wright, A., Hawkins, C.H., Anggård, E.E., and Harper, D.R. (2009). A controlled clinical trial of a therapeutic bacteriophage preparation in chronic otitis due to antibiotic-resistant *Pseudomonas aeruginosa*; a preliminary report of efficacy. Clin. Otolaryngol. 34, 349–357.

Zimecki, M., Weber-Dąbrowska, B., Łusiak-Szelachowska, M., Mulczyk, M., Boratyński, J., Syper, D., and Górski, A. (2003). Bacteriophages provide regulatory signals in mitogen- induced murine splenocyte proliferation. Cell. Mol. Biol. Lett. 8, 700–711.

Zueva, L.P., Diatkov, M.N., and Okolov, I.N. (1985). Epidemiology of Pseudomonas infection. Antibiot. Med. Biotekhnol. 30, 308–313 [article in Russian].

Zurabov, A., Yu., Zhilenkov, E.L., Popova, V.M., and Popov, V.G. (2012). Phage research carried out by 'MicroWorld' company. Bacteriophages and probiotics – alternatives to antibiotics. July 1–4, 2012, Tbilisi, Georgia, Abstract book, p. 2.

Reintroducing Phage Therapy in Modern Medicine: The Regulatory and Intellectual Property Hurdles

12

Daniel De Vos, Gilbert Verbeken, Carl Ceulemans, Isabelle Huys and Jean-Paul Pirnay

Abstract

Antibiotic resistance is a life-threatening problem worldwide and the industrial pipeline is dry. Other therapeutic options are needed and one of them is 'phage therapy'. Bacteriophages, phages in short, have proven to be effective in combating (multidrug-resistant) bacterial infections. However, legal obstacles and intellectual property rights are impeding the implementation of phage therapy in modern medicine and triggering ethical discussions. Worldwide, medicinal product regulations are directed towards standardized marketing authorization for 'classical' medicinal products. But phage are of a different nature than antibiotics on which most of our current regulation is based. Phage therapy is not covered by a specific regulatory pathway. Exceptions defined under the medicinal products legislation do not include the idea of phage therapy. Another hurdle is the Intellectual Property issue. Owning patents is essential in our current industrial economic model. But natural phages are evolving biological lifelike entities and thus difficult to cover by patents. In the future the adapted legal framework should allow the coexistence of a 'sur-mesure' pathway beside a 'prêt-à-porter' road. Taking into account the sustainability concept, all relevant safety measures and quality production controls, the 'sur-mesure' pathway should enable the use of the most fruitful and efficiency based phage therapeutic approach at regional or hospital level.

Introduction

Bacteria increasingly evolve to outsmart the available antibiotics. Yet the antibiotic pipeline is running dry while new life threatening bacterial strains are emerging continuously at an ever increasing rate (Levy and Marshal, 2004; Kumarasamy *et al.*, 2010; Brzuskiewicz *et al.*, 2011; Bush *et al.*, 2011; Cooper and Shlaes, 2011). Bacteria that are resistant to all commercially available antibiotics, so-called superbugs, are emerging worldwide. The emergence and evolution of antibiotic resistance is complex and multifactorial requiring a challenging and multidisciplinary approach if we want to control it. Indeed, this still not yet fully understood, biological phenomenon of drug resistance is a typical emergent characteristic of a dynamic, highly complex, and self-organizing system that evolves at the edge of chaos (Martinez and Baquero, 2002; Baquero *et al.*, 2003). In this setting, some laboratories and a handful of small pharmaceutical companies are (re)turning to (bacterio)phage therapy (Thiel, 2004). Phage are natural viruses that specifically infect bacteria. They are considered to be the most abundant and ubiquitous lifelike entities on Earth and co-evolve with their hosts, the bacteria (Bergh *et al.*, 1989; Fuhrman, 1999; Hendrix, 2002; Bamford, 2003). Up to 50% of bacterial mortality is thought to be due to bacteriophage predation (Wommack and Colwell, 2000). Lytic phage bind to receptors on the bacterial cell surface, inject their genetic material, use the bacterium's biochemical reproductive machinery to replicate and subsequently destroy (lyse) the bacterium, irrespective of its resistance to antibiotics, releasing the newly formed phages (virons) to seek out new hosts. It really is an evolving self amplifying antibacterial drug obviously different than a classic static chemical drug.

In 1919, d'Hérelle used phage to treat dysentery in Paris, in what was probably the first attempt to use phage therapeutically. D'Hérelle eventually developed a commercial laboratory in Paris that produced phage preparations against various bacterial infections, which were marketed by what later became the large French company L'Oreal (Sulakvelidze, 2001). In the 1930s, therapeutic phages were also marketed in the United States by major pharmaceutical companies including Eli Lilly, Squibb & Sons (today Bristol-Meyers Squibb) and the Swan-Meyers division of Abbott Laboratories. Scientific controversies and the advent of antibiotics, however, relegated phage therapy to complete obscurity in most of the Western world. Therefore, the current 'knowledge' of the therapeutic effect of phage is mainly based on theoretical grounds, basic laboratory observations, animal models and decades of empirical medical experience, accumulated mainly in Eastern Europe and the former Soviet Union with activities centered at the Eliava Institute of Bacteriophages, Microbiology, and Virology (EIBMV) in Tbilisi, Georgia, several institutes in Russia, and the Ludwik Hirszfeld Institute of Immunology and Experimental Therapy in Wrocław, Poland (d'Hérelle, 1917, 1919; Sulakvelidze *et al.*, 2001; Biswas *et al.*, 2002; Wills *et al.*, 2005; Brüssow, 2005; McVay *et al.*, 2007; Chanishvili, 2009, 2012; Gorski *et al.*, 2009; Kutateladze and Adamia, 2010; Kutter *et al.*, 2010; Maura and Debarbieux, 2011). According to most supporters, phage therapy has been proven safe through the massive application of lytic bacteriophages in humans, mostly in former USSR states and Poland (Dabrowska *et al.*, 2003; Brüssow, 2005; Chanishvili, 2009), and in the Western world in animal studies (re-)initiated by UK researchers in the 1980s and reboosted by others two decades later (Smith *et al.*, 1982. Biswas *et al.*, 2002; Chibani-Chennoufi *et al.*, 2004; Wills *et al.*, 2005; Marza *et al.*, 2006; McVay *et al.*, 2007; Debarbieux *et al.*, 2010) and safety trials in healthy volunteers and patients (Brüttin and Brüssow, 2005; Rhoads *et al.*, 2009; Sarker *et al.*, 2012; McCallin *et al.*, 2013).

This chapter discusses the problems encountered when trying to reintroduce 'old school' phage therapy in our modern liberal and precautionary society.

Bacteriophage: a sustainable, evolving and self-amplifying drug

A virus is a natural biological entity, consisting of a molecular assemblage of nucleic acids (the genome) surrounded by proteins, that behaves as a genetic replicative parasite and co-evolves with its host. However, this co-evolutionary aspect, essential for sustainable phage therapy, has been practically under-estimated and consequently underutilized. Indeed, the dynamic bacteria–phage interactions (co-evolution) give rise, *in vitro* and *in vivo*, to an 'arms race', consisting of the repeated origin and fixation of new phage virulence and bacterial host defence traits (Buckling and Rainey, 2002; Faruque *et al.*, 2005a,b).

Are these natural biological 'lifelike' entities living? At first sight, viruses are 'not alive' since they are acellular and lack metabolism. On the other hand, they replicate and evolve, in a Darwinian sense, which is a typical trait of living systems. These characteristics, however, only emerge once the virion (the extra cellular viral particle or phage) has transferred its genome to an organismal cellular environment, *in casu* a bacterial cell. Research has shown that viruses, including phage, play a fundamental role in the emergence and evolution of cellular (organismal) life.

New ideas on the definition of life and the tree of life are emerging with strong debate among scientists (Ward, 2005; Cleland, 2007; Raoult and Forterre, 2008; Brüssow, 2009; Benner, 2010; Tirard *et al.*, 2010). Viruses are more and more seen as essential elements in the origin and organization of life itself (Villarreal, 2005; Villarreal and Witzany, 2010). The fundamental transition of the RNA world into the DNA world, as well as the emergence of placental organisms was probably mediated by viruses (Mi *et al.*, 2000; Forterre, 2001). Mimivirus research brought Raoult and Forterre (2008) to conceive a new tree of life model including viruses as 'capsid encoding organisms' versus 'ribosomal encoding organisms'.

In view of these fundamental scientific developments on the nature of phage and the empirical evidence of their therapeutic usefulness, it is clear that therapeutic phage are very different from classical (chemical molecular) drugs such as antibiotics (Chanishvili, 2009; Kutateladze and

Adamia, 2010). And this challenges our actual regulatory frame which is not well adapted for phage documentation in order to setup clear and well designed standard clinical trials which are required to prove phage therapeutic effectiveness in different clinical settings as well as for the clinical optimization of this therapeutic approach (Verbeken *et al.*, 2007, 2012; Gilmore, 2012; Parracho *et al.*, 2012; Brüssow, 2012).

In fact, natural lytic phages are biological entities playing an important role in maintaining equilibrium in bacterial populations of ecological environments including man. Hence, we should not see them as a conventional chemically stable drug, but more as an interactive and evolving phage/bacterium couplet.

A phage is in fact a nucleic acid and protein based, self-amplifying and evolving medicine, exhibiting particular pharmacokinetics (Payne *et al.*, 2000; Payne and Jansen, 2003; Levin and Bull, 2004). Their action depends mainly on the susceptibility and the concentration of the targeted bacteria, the emergence of phage resistance as well as the physicochemical and immunological conditions present at the site of infection, while also the route of phage delivery, the galenic formulation and phage concentration beside the frequency and time of administration will be of importance for a successful phage therapy (Payne *et al.*, 2000; Debarbieux *et al.*, 2010; Ryan *et al.*, 2011, Morello *et al.*, 2011; Hall *et al.*, 2012).

This interactive and evolving phage/bacterium couplet, seen as a medicine, requires a new way of thinking that fits well in the emerging field of Darwinian medicine. But exactly this is less compatible with current drug development and marketing models. Indeed more than 150 years after Darwin's first publication of 'The Origin of Species' in 1859, our society as a whole, and not even the biomedical sciences as a whole, have still not well grasped and surely not yet fully integrated the idea and consequences of evolution, the cornerstone of biology, which is a basic science on its own for medicine (Darwin, 1968; Ewald, 1994; Corbellini, 2008; Williams, 2010). We agree with Shanks and Pyles that 'it is important for the public, as consumers of medical services, and for medical practitioners themselves to have a greater appreciation of the medical implications of evolutionary biology' as well as of the science of biology itself (Mayr, 2004; Shanks and Pyles, 2007). A worldview that is too deterministic and mechanistic hampers the development of the evolutionary medicine's approach as well as the consequently required adaptations of our regulatory setting and our current non-sustainable biopharmaceutical industry model. The current focus on the 'how' questions (mechanistic explanations) with immediate effects should be equilibrated with the 'why' questions (evolutionary explanations) that typically are resulting in longer term views, based on evolutionary insights (Mayr, 2004; Shanks and Pyles, 2007; Williams, 2010; Valenti, 2012).

Ethical considerations

It might be a good idea to consider some ethical issues.

Does a patient have a moral right of access to an unapproved medical treatment? Reflecting on such a question requires us to identify some of the ethical principles that seem to underlie such discussions. One of the basic principles in the realm of biomedical ethics is that of non-maleficence (Beauchamp and Childress, 2009): a minimal moral duty of any health care professional is to abstain from inflicting harm on his or her patients. There exists a general moral and legal consensus that no patient should be subjected to experimental – and therefore unapproved – medical treatments, based on this fundamental moral principle. It is necessary to conduct rigorous testing, followed by the approval by official public health agencies, in order to make sure that a proposed treatment has the beneficial health effects claimed; while at the same time it is demonstrated not to cause unacceptable health risks.

Yet is a patient's right absolute? Can they be subjected to a not (yet) approved treatment under certain conditions? If we assume that this is the case, then we need to find out under what conditions this right can justifiably be overridden. First of all, there has to be a very good reason for putting aside a patient's moral protection against being subjected to possible hazardous medical treatments. The moral weight of whatever it is we want to achieve with this therapy has to be sufficiently important – for example, saving the life of a patient may constitute a viable ethical motive. Secondly, we need to make sure that all the moral agents involved are motivated by ethically proper intentions. If indeed a patient's life is at stake, our intention for using an unlicensed treatment has to be about improving the patient's health condition, and not about commercial, research or cost-reducing benefits. Based on the results of ongoing experimental trials, we must also have sufficient reason to believe that the treatment under investigation will indeed produce the beneficial health effects it claims to have. What is more, there must be a good prospect that the probable health benefits will outweigh the risks of subjecting the patient to the treatment.

Another criterion is that the unapproved treatment needs to be a last resort. All existing treatments must have been tried with little or no success. Finally, a decision to subject a patient to a not yet approved treatment needs to be made in respect with the patient's right to autonomy (informed consent). Notice here, however, that a patient giving his informed consent to be subjected to an unapproved treatment does not automatically lead to the waiving of his right *not* to be subjected to that therapy. A patient's moral protection against being subjected to unapproved treatment does not simply disappear because he wants it too and gives his consent as there may be ethical issues above the level of the patient that have to be considered by medical practitioners and others. Besides a patient's informed consent, all the other criteria just cited (good reason, moral intention, reasonable chance of success, proportionality, and last resort) need to be satisfied before that patient's right not to be subjected to an unapproved treatment, may justifiably be set aside.

But what if all these conditions are satisfied? Does it automatically follow from this that a patient has a *right* to access to an unapproved treatment? Well, not necessarily. If it can be shown that all these safeguards have been met, it becomes morally permissible to subject a patient to that therapy. It does not follow, however, that there exists a *duty* to provide a patient with that therapy. It might be said that the right not to be subjected to an unapproved treatment corresponds with a negative duty (duty not to harm), whereas the right to have access to an unapproved treatment corresponds with a positive duty (duty not to allow a harm to happen). Do those responsible for the development of a new treatment have such a positive duty, once it has been established that they are no longer constrained by the negative duty not to harm? Given the fact that most of those involved in developing new treatments are private companies wanting to maximize their economic profits as much as they can, it is certainly not easy to see how one could argue that such positive duty exists. A possible defence for such a duty could be based on the principle that anyone engaged in the realm of public health – and this includes the pharmaceutical companies – implicitly accepts some moral responsibilities towards society, such as making available experimental treatments in sufficiently large quantities when called for. Such a 'public responsibility'-argument is of course only one example of a possible defence, and would require a lot more development than we can provide here, but that the problem of ethical conflicts in public health research and practice exists is well known. Specifically in the area of drug resistance, as well as in the research for new antibacterial products and/or treatment approaches, ethical issues emerge at the global public health policy level, in relation with the industry, and at the individual (patient/doctor) treatment level (Aiello *et al.*, 2006; Selgelid, 2007; Leibovici *et al.*, 2012).

The Declaration of Helsinki

In 1964, the World Medical Association (WMA) elaborated a set of ethical principles for the medical community regarding human experimentation, the Declaration of Helsinki. Today, it is widely regarded as the cornerstone document of human research ethics (Bošnjak, 2001; Tyebkhan, 2003). Although it is not a legally binding instrument in international law, it draws its authority from the degree to which it has been codified in, or influenced, national or regional legislation and regulations (Human and Fluss, 2001).

In paragraph 35 of the Declaration of Helsinki ([35] of Declaration of Helsinki (2008: Sixth revision, 59th Meeting, Seoul) specifically states:

> In the treatment of a patient, where proven interventions do not exist or have been ineffective, the physician, after seeking expert advice, with informed consent from the patient or a legally authorized representative, may use an unproven intervention if in the physician's judgement it offers hope of saving life, re-establishing health or alleviating suffering. Where possible, this intervention should be made the object of research, designed to evaluate its safety and efficacy. In all cases, new information should be recorded and, where appropriate, made publicly available.

During the latter part of the 20th century, phage therapy was largely used in Poland and the former Soviet Union. One of the major centres of phage therapy is the L. Hirszfeld Institute of Immunology and Experimental Therapy, where since the 1970s specific phages have been used for the treatment of patients with suppurative bacterial infections, in whom a routine antibiotic therapy failed. The results obtained so far show that phage therapy is safe and highly effective (the majority of patients were cured) as reported by Slopek and Weber-Dabrowska (Slopek *et al.*, 1987; Weber-Dabrowska *et al.*, 2000).

Today, Poland is a European Union (EU) Member State in which this therapy is still possible. Phage therapy is considered an 'experimental treatment' under the Declaration of Helsinki and the national adapted Act of 5 December 1996 on the Medical Profession (Polish Law Gazette, 2011, No. 277 item 1634). On this basis the L. Hirszfeld Institute of Immunology and Experimental Therapy (an institute of the Polish Academy of Sciences) offers phage therapy to treat patients infected with drug-resistant bacteria (Miedzybrodzki *et al.*, 2012).

More recently in June 2005, the Ethical Committee of the Medical Academy in Wroclaw authorized a study entitled *Experimental Phage Therapy in Antibiotic-Resistant Bacterial Infection, Including MRSA Infection*. The results of experimental phage therapy (conducted at the Phage Therapy Unit at the Hirszfeld Institute) were discussed in a recent publication (Miedzybrodzki *et al.*, 2012).

Routine therapeutic application and marketing

The paragraph 35 of the Declaration of Helsinki is only applicable when therapeutic methods do not exist or have been ineffective. So, strictly speaking, the Declaration of Helsinki can only justify the application of phage in critical patients with untreatable MDR bacterial infections. Fortunately, today this is only the case for a small number of patients. As a consequence we feel that the Declaration of Helsinki is not the appropriate tool to – once and for all – evaluate the efficacy of phage therapy, which requires (according to modern medicine) prospective, double blind, placebo-controlled clinical trials, in order to eventually use phage therapy as a routine antibacterial treatment.

Medicinal product regulation

A medicinal product is defined by the European regulation as 'any substance presented for treating or preventing disease in human beings' (European Parliament, 2003). With 'substance' is meant 'any matter, irrespective of origin or intrinsic nature of that matter'. The current definition of a medicinal product is so broad that it covers a wide array of products, including foods, herbs, nutrients, micro-organisms, whole animals and even water. Consequently, leeches and fly larvae (maggots) are today indeed classified as medicinal products. Although, as previously mentioned, phage are included in the tree of life, they are definitely of a different nature than the above-mentioned 'organisms'. Nevertheless, if we strictly follow the medicinal product definition, therapeutic natural phage are undeniably medicinal products.

Based on the medicinal product regulation, they would in all probability be considered biological medicinal products, in analogy with leeches and maggots. According to some (commercial) players in the field this positioning of phage as biologicals was adopted for a transitional period of a few years. But we were not able to find any official document or report attesting this transitional phage status. At the European level, a 2008 EMEA/ECDC Technical Report 'Bacterial Challenge: Time to React' mentions phage as 'therapeutic agents', no more.

In our experience it is difficult to document phage as if they were classical medicinal products. In addition, given the specific nature of phage (i.e. natural bacterial co-evolving biological entities), we feel that it is irrational to develop and market them as if they were classical medicines (i.e. static chemical substances). In practice it is possible to produce (*ad hoc*) effective phage against current problematic bacterial infections (e.g. EHEC, NDM-1 carrying Gram-negatives, MRSA etc.) within days to weeks (Merabishvili *et al.*, 2012). If the classical medicinal product development and licensing pathways were being followed, selected organism-specific phages would only become available several years later. As a consequence, the flexibility of phage therapy would be lost (Pirnay *et al.*, 2010, 2012). Eventually, rapid updating and licensing procedures could be adopted (e.g. as in the case of influenza vaccines), but these would only reduce the postponement of use to, at best, several months. Recently Merabishvili and colleagues showed that it is possible to respond quickly on an emergent outbreak of for example the epidemic enteroaggregative Shiga toxin/verotoxin-producing *Escherichia coli* (EAggEC STEC/VTEC) strain O104:H4 since the isolation, selection and characterization of a candidate therapeutic bacteriophage could be done in the timeframe of days (Merabishvili *et al.*, 2012). Therefore, a specific regulatory framework for phage therapy should be considered (Verbeken *et al.*, 2007, 2012).

In the advanced therapy medicinal product (ATMP) regulation there is a gene therapy section that allows the use of genetically modified viruses targeting eukaryotic cells. This section, however, is not applicable to natural phage, which are not genetically modified and do not target eukaryotic cells. Indeed, no phage-related nucleic acid sequence could be found in our genome, in contrast with the huge amount of retroviral remnants present in our human core genome. Up to 8–10% of the human genome consists of human endogenous retroviral sequences (HERV's) (Lander *et al.*, 2001).

Furthermore, it is impossible for a phage to interfere or multiply in an eukaryotic cell system since it requires the specific prokaryotic cell wall receptors and biochemical machinery for its attachment and replication (e.g. prokaryotic polymerases). Some phage-related polymerase gene sequences were, however, identified in human mitochondrial DNA. It is common knowledge that mitochondria originated from Rickettsia-like ancestor bacteria that started a symbiotic relationship with prototype eukaryotic cells. A similar event occurred in plants and gave rise to the chloroplast. This evolutionary process dates back from the endosymbiotic era, the time when the evolutionary split occurred between the pro- and eukaryotes. This remnant of phage DNA was likely introduced in the eukaryotic cell during the bacterial phase of the mitochondrion, the actual energy production unit of our cells. Recent work also suggests that even the eukaryotic nucleus is a viral import (Bell *et al.*, 2001). Very recently Jeremy Barr and colleagues published a paper showing a new kind of an, until now, unrecognized active mucosal protection system based on a symbiotic relationship between phage and metazoan hosts, including humans. The study showed the permanent and host-mucus dependent presence of an increased lytic phage density in the epithelial mucus layers protecting the host against invading bacteria (Barr *et al.*, 2013). All this taking into consideration shows that phage, and phage therapy, is intrinsically safe for us, human beings constituted of eukaryotic cells with whom the phage will not be able to interact.

Notwithstanding the unadapted regulatory status of phage therapy in the Western world (Verbeken *et al.*, 2007, 2012) some studies were recently conducted or are ongoing beside some sporadic therapeutic applications (Kutter *et al.*, 2010). These recent studies in conjunction with earlier empiric experiences show at least significant potentialities and basic patients safety.

As in burn centres all over the world, also the clinicians of the burn centre of the Queen Astrid Military Hospital in Brussels are increasingly confronted with multi-drug resistant (MDR) bacteria causing virtually untreatable infections (Pirnay *et al.*, 2003a). A leading Belgian Ethical Committee authorized a pilot clinical trial in which a Good Manufacturing Production (GMP)-like produced phage cocktail (Merabishvili *et al.*, 2009) was applied on burn wounds infected with MRSA and/or MDR *Pseudomonas aeruginosa*. In the preparation phase of this pilot clinical trial researchers of the Brussels burn wound centre were confronted with the false perception of viruses as 'enemies of life', an observation that was previously expressed by Villarreal (2005). This 'fear', for example, resulted in a ten times too high study insurance fee and an unwarranted request to notify the National Bio-safety Council, which normally is only entitled to rule on the release of genetically modified organisms and pathogenic micro-organisms in the environment. In addition, this cocktail was also used to successfully treat a critical patient, amongst others, suffering since months from a pelvic osteomyelitis infection accompanied with frequent septic episodes, after a car crash and subsequent surgical intervention, with MRSA and MDR *P. aeruginosa* under the umbrella of the Declaration of Helsinki.

In the UK a small phage therapy company conducted a phase I/II clinical trial in chronic otitis approved, on a national level, by the UK Medicines and Healthcare products Regulatory Agency (MHRA) and the Central Office for Research Ethics Committees (COREC). A preliminary report of efficacy was published (Wright *et al.*, 2009).

In France therapeutic made-to-order phage preparations from the Institut Pasteur of Paris and Lyon were available for medical use till the beginning of the nineties (Dublanchet, 2009; Kutter *et al.*, 2010). Although phage therapy was not retained in the current French Medicinal Product Regulation, a French MD, Alain Dublanchet, still applies phage therapy to treat patients infected with MDR bacterial infections. He obtains his phage products from pharmacies in Russia and/or Georgia where they are legally and commercially available (Personal communication of Dublanchet).

In the United States a Food and Drug Administration (FDA)-approved phase I clinical trial was performed (Rhoads *et al.*, 2009). A recent publication from an international group describes in a case report the application of therapeutic bacteriophages for refractory *P. aeruginosa* urinary tract infection in a 67-year-old woman that underwent extensive intra-abdominal resections and irradiation for adenocarcinoma. This treatment was approved by the Western Sydney Human Research Ethics Committee on a compassionate use basis and after patient's informed consent was obtained (Khawaldeh *et al.*, 2011).

Other potential paths

Several alternative application paths for phage can be taken into consideration.

Magistral or extemporaneous preparations

At first sight, phage preparations could be produced and dispensed as Magistral or Extemporaneous preparations by a (hospital) pharmacist. The dispenser (pharmacist) stands between the prescriber and the patient and only a very intimate acquaintance with the characters and doses of medicines will enable him to successfully perform his duty to each. The successful performance of this medication has to be preceded by knowledge of the physical and chemical characters of its (active) components and it is debatable whether this is currently the case for phage. Furthermore, in most countries the raw materials of Magistral or Extemporaneous preparations need to be licensed in/or accompanied by an adequate certificate of analysis. Also the traceability has to be guaranteed in a way which conforms to current regulations.

Compassionate use

'Compassionate use' is a method of providing experimental therapeutics (investigational drugs) prior to final approval for use in humans. This procedure is used with seriously ill individuals who have no other treatment options. This rule only applies to non-registered drugs that are already in a clinical-study phase and have proven potential. Phage preparations that are currently tested in approved clinical trials and have shown efficacy could thus theoretically, and under certain conditions, be used to treat patients with life threatening MDR bacterial infections. Recently, an official question regarding the compassionate use, eventually under or in combination with the French 'statut d'Authorisation Temporaire d'Utilisation' (ATU), of therapeutic bacteriophages against life threatening resistant bacterial infections in patients with functional or vital bad prognosis was asked by two French MDs, Larché and Lenoir from the non-profit organization PhagEspoirs to the French competent authorities. The answer is still pending. In Australia however phage was recently successfully applied under a compassionate use basis (Khawaldeh *et al.*, 2011).

GRAS and QPS

Antibiotics and other antibacterial (preservation) products play an important role in the food and agro-biological industry. More than half of the antibiotic consumption worldwide is for non-human medical use, often as animal growth promoter/enhancer. The routine and widespread (over) use of antibiotics in the agro-biological industry, which are also used in human medicine, is no longer generally accepted since it was shown to accelerate the development of antibiotic-resistant strains of bacteria. Several studies showed the presence of antibiotic-resistant strains and molecular antibiotic gene cassettes in the inanimate environment (Kümmerer and Henninger, 2003; Pirnay *et al.*, 2005; Allen *et al.*, 2010). Although recent surveys and specific studies show the widespread presence of antibiotic resistance genes in the environment, even in remote pristine environments, we do not agree to quickly change the paradigm that 'the overuse and misuse of antibiotics (OMUAB) is at the origin of the emergence of the increased antibiotic resistant bacteria' (Selgelid, 2007; Bertoloni *et al.*, 2009; Allen *et al.*, 2010; D'Costa *et al.*, 2011; Rolain *et al.*, 2012). The effect of OMUAB might indeed not be specifically at the origin of the *resistance genes* as such, but it is apparently clearly related to their *mobilization* from the so-called resistome (D'Costa *et al.*, 2006). This phenomenon is currently enhanced as a consequence of the worldwide massive use of antibiotics especially in the agrobio industry. We actually have to realize that in fact our knowledge of the natural role of antibiotics, as secondary metabolites, is incomplete or too low. They can have a signalling function(s), similar to cytokines in eukaryotes, in their natural ecological setting. Clearly, a better use of antibiotics requires more integrated and fundamental research. Research results through metagenomic studies will surely help and bring new useful insights (Schmieder and Edwards, 2012). Realizing this, it is not surprising, and even a good initiative of the food and agro-biological industry, to also try to exploit phage–host interactions for biotechnological applications (Shapiro and Kushmaro, 2011; Mahony *et al.*, 2011) in order to avoid the use of antibiotics or at least to minimize it or use it more appropriately.

Some companies decided to penetrate the food market first in order to accustom the public and the regulatory authorities to bacteriophages, which should facilitate future clinical trials whilst already generating revenues. In the USA, the Food and Drug Administration (FDA) approved the use of the first phage preparation for food safety applications in April 2006. Since then, three other preparations have been approved. FDA classified two of these preparations under the Generally Recognized As Safe (GRAS) product regulation, one was cleared as a food additive, and another one as a food contact substance (Burdock and Carabin, 2004; FDA, 2006; Sulakvelidze, 2013).

In 2009, the European Food Safety Agency (EFSA) issued a product regulation similar to GRAS, the Qualified Presumption of Safety (QPS) regulation. In addition, the EFSA maintains a list of QPS micro-organisms that can be intentionally added to food or feed. Phage are not included on the list, but similar to the GRAS regulation, they can be authorized as antimicrobial agents in the industrial food production after assessment on a case-by-case basis. The review process pays specific attention to the characterization of the phage at the genomic and proteomic level, which should ascertain that the phage are exclusively lytic and are not carrying potential toxins and/or virulence factors that are potentially transmissible in the natural ecological environment. As such two different cocktails of phages active against *L. monocytogenes* were approved and are commercially in use and available, Intralytix's ListShield and Micreos' Listex.

Phage probiotics?

Phage are extremely abundant and ubiquitous. They are present in environments as diverse as seawater, drinking water, activated sludge, food and cosmetics, and inhabit our bodies, especially the digestive tract, outnumbering our own bodies eukaryotic cell number. In the food industry phage play an important role not only in 'starter culture mixes' but also as 'in-process agents of concern', especially in all kind of fermentation products where they can have huge negative impact on bacterial fermentation based production processes (Shapiro and Kushmaro, 2011). Unknowingly we constantly consume phage with our drinking water and food (e.g. yoghurt, cheese, salami).

An equilibrated food intake is the basis for health. Yoghurts containing probiotics such as bifidobacteria, also available on the European market, are claimed to have a positive influence on our health through the restoration of the intestinal flora. This means, in fact, that there is a commercial claim that these yoghurts generate positive therapeutic effects. However, at the regulatory level, the therapeutic effects for this type of product are not perceived, since otherwise such yoghurts would be considered medicinal products according to European regulation. Fermented-milk drinks containing living *Lactobacillus* spp. such as *Lactobacillus casei shirota* or *Lactobacillus casei immunitas*, but also their respective phage, are also present on the European market.

According to the Japanese Foods for Specified Health Use (FOSHU) regulation, the following therapeutic effects of probiotics can be claimed: 'regulation of the gastrointestinal condition', 'reduction of harmful bacteria' and 'suitable for therapeutic use against acute diarrhoea' (Sanders and Huis in't Veld, 1999; Berman *et al.*, 2006). These products did not receive the medicinal product status, either.

In The Netherlands, some milk drinks are claimed to 'improve bowel habits' in subjects who are susceptible to constipation and to 'support a well-balanced gut microbiome' through an increase in the number of *Lactobacilli*. These claims were validated by a committee of independent scientists on specific request of The Netherlands Nutrition Centre and a major milk-drink manufacturer. Despite these valorized claims, these products were not classified as a medicinal product, either (Agarwal *et al.*, 2002).

Apparently, for food products, therapeutic claims did not have to be backed up by clinical (safety) trials or, in fact, by any study whatsoever. And what to think about other, more 'classical' nutritional supplements such as various vitamins or minerals like calcium, iron or magnesium amongst others, who in fact are considered as medicinal products. This shows that although a medicinal product was defined there are several exceptions that enable to use the 'same substance' as medicinal product or non-medicinal product. However, questions have increasingly been raised regarding the status of these 'grey-zone' products. New health claim regulations in Europe, and a tightening regulatory environment in the US make it harder to make therapeutic claims. Several claims are unfavourably assessed because they were not supported by any relevant studies in humans. Today, such studies are central to the establishment of a cause and effect relationship between the food and/or substance concerned and the beneficial health effect claimed. But even with some good scientific evidence 'probiotics as therapeutics' are actually facing approval problems as described by Reid (2011).

Phage could eventually be considered as food additives with therapeutic claims based for example on the restoration of the commensal flora of the gastrointestinal tract, by eliminating pathogens which could promote the growth of commensals, but the health claim regulation and licensing pathways for food-additives are, again, not compatible with the flexibility and sustainability we pursue for phage therapy.

Intellectual property rights

As mentioned earlier the rise of multi-resistant bacterial infections shows the current limits of common antibiotics. In the bio-medical community, there is currently a renewed interest in phage therapy, but (bio)pharmaceutical companies seem generally not interested to push it forward. The main reason for this is that, apart from the lack of an appropriate regulatory framework, it is hard to get patents for phage and their therapeutic applications, hence no way to secure a return on investment costs (Thiel, 2004). The fact that intellectual property (IP) rights or patents give problems in the biomedical field is increasingly recognized (Aiello, 2006; Selgelid, 2007, Taubman, 2008; Gold *et al.*, 2009; Kapczynski, 2009; Van Overwalle, 2009, 2010; Kesselheim, 2010).

In Europe, the four essential pre-conditions governing the patentability of inventions under the European Patent Convention (EPC) are laid down in Art. 52(1) EPC, which reads: 'European patents shall be granted for any *inventions*, in all fields of technology, provided that they are new, involve an inventive step and are susceptible of industrial application'. Thus, in a first step, in order to be patentable, there needs to be an invention. Thereafter, that invention has to fulfil the patentability requirements of novelty, inventive step and industrial applicability. The concept of 'invention' as such is not defined in the EPC, but the Implementing Regulations to the EPC do specify that the invention must have a technical character (Rule 29(1)), that is related to a technical field (Rule 27(1)(a)) and concerned with a technical problem (Rule 27(1)(c)). It is clear from these rules that 'technicality' is a key precondition for qualification as a patentable invention in Europe.

Article 52(2) EPC lists exclusions which should not be regarded as inventions, if claimed 'as such' (Article 52(3) EPC), because they are abstract in nature (*discoveries*) or non-technical in nature (scientific theories or methods for performing mental acts).

Thus in Europe, whether natural phage and cocktails of natural phage are to be regarded as *inventions* eligible for patent protection, or whether they are *discoveries* or principles of nature and thus excluded from patentability, depends on the technical character related to the claimed subject matter. There must be something more than mere disclosure of a natural phenomenon. Statutory, any 'biological material isolated from its natural environment or produced by means of a technical process, may be the subject of an invention even if it previously occurred in nature' (Art. 3 Biotechnology Directive 98/44/EC). Thus, in principle, a naturally occurring phage might become patentable as soon as some human intervention is needed to isolate the phage from its natural environment by any technical means as long as the phage can be properly characterized by either the process by which it is obtained (product-by-process claim), by its structure or by other means. Even so, cocktails of naturally occurring phage might be patentable as a result of a technical contribution.

In the US patent law Code (U. C), general requirements for patentability are listed in the 35 USC §101. That article defines that subject matter may be patentable, provided it or its improvement (1) belongs to one of four distinct classes, namely a process, machine, manufacture, or composition of matter, that is (2) new and (3) useful and (4) non-obvious (§103). A precondition for patentability as to its 'technical character' does not seem to be explicitly present in US patent law. Even though the US does not have a statutory counterpart to Article 52(2) EPC, exceptions to patentability are established by case law and certain categories are also excluded from patentability, such as products of nature, laws of nature and/or natural phenomena, and abstract ideas or basic human knowledge or thought. But what exactly means 'abstract ideas or basic human knowledge or thought'?

The US case law is clear on the aspect of non-naturally occurring organisms.

Most notably, the Supreme Court addressed the question in the 1980 landmark case *Diamond v. Chakrabarty*, 447 U.S. 303 considering a live, human-made genetically modified bacterium as patentable subject matter, because the newly formed bacterium had markedly different characteristics from any found in nature. According to the Court, 'a mere purification of known materials does not result in a patentable product, unless the product obtained has properties and characteristics which were different in kind from those of the known product rather than in degree'. The *Chakrabarty* case has been viewed as mandating the patentability of non-naturally occurring, non-human multicellular organisms such as transgenic animals, genetic materials, and purified biologically produced compounds such as enzymes.

For naturally occurring organisms, the US case law has also been developed in a specific way.

In 1948 the Supreme Court of the US considered a patent case known and recorded as Funk Brothers Seed Co. V. Kalo Inoculant Co. 333 U.S. 127 related to a mixture of several naturally occurring species of nitrogen-fixing bacteria. The Court held that the qualities of these bacteria are the work of nature and hence these qualities are not patentable.

However, other cases in the US Court of Customs and Patent Appeals (CCPA, which was renamed in 1982 as the Court of Appeals for the Federal Circuits) such as the 1977 *In re Application of Bergy*, 563 F.2d 1031 (CCPA) and the 1979 *In re Kratz*,592 F.2d 1169 (CCPA) have provided some support for the patentability of various biological materials, including cells, proteins, and organisms if isolated or purified from their natural environment or pre-existing material. On the other hand, recently, the ruling in the US Court for the Southern District of New York by Judge Robert Sweet in a case between the *Association of Molecular Pathology (AMP)* and the *US Patent and Trademark Office (USPTO)* invalidated seven patents claiming genes and genetic diagnostic methods held by Myriad Genetics (Huys *et al.*, 2009; Akst, 2010; Cho, 2010; Huys *et al.*, 2011). Very recently however a new US court decision reversed Myriad gene patent ruling for *BRCA1* and *BRCA2* (Brower, 2011).

Thus the discussion is still ongoing and shows that the topic is hotly debated and not yet set.

Although related to naturally occurring genes, this decision restarted the discussion about the patentability for genes, but also for other types of naturally occurring biological (lifelike) entities, such as phage, and organisms. A similar situation is found in another medical therapeutic situation, namely the Faecal Microbiota Transplantation (FMT) intervention for diseases such as pseudomembranous colitis due to *Clostridium difficile* and inflammatory bowel disease (IBD) including Crohn's syndrome. Indeed bacterio-transplantation therapy faces analogous hurdles of IP and regulatory classification as phage therapy before being recognized and introduced in routine medical practices (McKenna, 2011; Landy *et al.*, 2011).

It is a very thin line that separates *inventions* from *discoveries* under both US law and European law. Therefore, it is very useful to investigate the second very important patentability requirement, namely novelty, which is of high relevance for naturally occurring phage. An invention can be patented only if it is new. An invention is considered to be new if it does not form part of the state of the art. European case law established that a natural substance which has been isolated for the first time and which had no previously recognised existence, does not lack novelty because it has always been present in nature as was the case for the Howard Florey Institute's Application on Relaxin OJEPO 1995, 388 (V 0008/94). In other words, this means that for a phage or the cocktail of phages as claimed in a patent, it should never have been isolated or produced before. Scientific literature with respect to phage as natural source to treat human bacterial infections exists and is increasing as mentioned in the introduction. In addition some clinical studies and reports on humans, using phage and performed in the Eastern part of Europe, have recently been translated into English (Chanishvili, 2009). Therefore, many phage and their uses have been disclosed over the past decades to almost a century. European law allows the patentability of known substances (as phage might be) if claimed for use in a medical method, provided that such use is new, meaning that such use may not be comprised in the state of the art (first medical use claim). We have to keep in mind that for novelty, the US applies a different 'state of the art' as the EU. In the US, the invention is considered as being part of the state of the art (hence not patentable) if it was known or used by others only in the US itself, or if it is patented or described in a printed publication in the US or outside the US. Thus in case such isolated phage are known in another country, it can still be patentable in the US.

In the past, patents on purified natural products of many kinds have been granted. For instance, in 1873, Louis Pasteur was granted US141072 patent for 'Yeast, free from organic germs of disease, as an article of manufacture'. And in 1903 Takamine obtained US730176 patent for adrenaline purified from gland tissue. More recently, in the field of phage applications, several patents for phage used in the food sector were granted, such as US7507571 (food additive) owned by Intralytix, Inc., claiming 'an isolated phage of a phage strain selected from a [specific] group, [somewhere] deposited under a [specific] accession number, together with variants thereof, wherein said variants retain the phenotypic characteristics of said deposited phage and wherein said phage, and variants thereof, have lytic activity against *Listeria monocytogenes* strains'. More important for therapeutic use is the US patent 7459272 of Intralytix, Inc., claiming 'a method for reducing the risk of bacterial infection or sepsis in a person colonized with pathogenic bacteria comprising treating the colonized person with a pharmaceutical composition containing phage of one or more strains which produce lytic infections in said pathogenic bacteria'. In 2001, a European patent application (EP1250143 A2) was filed claiming 'a method for reducing the risk of bacterial infection or sepsis in a susceptible patient by treating the susceptible patient with a pharmaceutical composition containing phage of one or more strains which produce lytic infections in pathogenic bacteria'. Such claim was considered not to be novel. The applicant tried to rescue the claim by using a second medical use wording, however, in 2004, this application was deemed to be withdrawn. Only recently 'a method for production of compositions of phage with a specific titre and total yield' was claimed in the US by Phage Biopharm LLC (US7588929). No European counterpart has been published yet. Two other interesting patents are the US patent 7758856 and US patent 7807149, both of Biocontrol Limited and granted in 2010. They claim 'a composition for treating a bacterial biofilm', as well as 'a method for treating a biofilm infection' and 'a bacteriophage containing therapeutic agents'. Those 'patents' however, when analysed in depth do not patent phage as such, but a whole process from which phage is a part. They also are very broad in definition. One might ask what the real value of such a patent is in practical terms? A similar patent, covering also the use of phage for the treatment of biofilms, has been granted in Europe under the EPI code EP1587520 B1 and is owned by the Health Protection Agency (GB).

In summary, diverging views across Europe and the US exist on patenting biological material. Aside from the requirements of novelty, inventive step and industrial applicability (that are the same for Europe and the US), in order to be patentable in Europe, some technical intervention is needed to isolate the phage from its natural environment and the phage needs to be properly characterized. But this 'technical intervention' has been basically known since d'Hérelle while the fact that the phage 'needs to be well characterized' seems obvious and technically not a big problem (Merabishvili *et al.*, 2009). In the US, the phage, as claimed in a patent, needs to have markedly different characteristics from any found in nature. However, if we talk about naturally occurring, exclusively lytic phage, which are our object of concern here, then they remain as they are found in nature. It is only when working with genetically modified phage that we can agree with the US statement. The use of 'manipulated or engineered phage' has certainly applications, which could be patented, but, considering the current concerns about potential risks for public health and the environment which may arise from genetically modified organisms (GMOs), they are not likely to be given more easily a market licensing approval in the near future even if a smoother regulatory pathway was constructed. Phage-derived products (e.g. cell wall-degrading enzymes such as endolysins) can and probably will be licensed and marketed within a few years. They may also select resistance, but presumably at a lower rate than antibiotics. Of course, these phage-derived products lack the capacity of self-replication and adaptation in the infectious site.

In this chapter we focus on natural lytic phage just because of their natural intrinsic bacterial co-evolutionary aspect allowing a sustainable antibacterial treatment approach that is suitable for quick, flexible, and potentially cheaper therapeutic applications.

Patents claiming genes or natural entities like plants and, for instance, phage seem difficult or even (and perhaps correctly) 'impossible'. Even 'inventing around' as Van Overwalle (2010) writes is not easy and therefore Taubman (2009) thinks that 'technology specific interventions' are required. From an industrial point of view this is difficult to accept, but this is mainly because we are still fixed on the classical thinking paths. As biologists we should know that nothing is stable and as a consequence our 'laws' and 'attitudes' should adapt and co-evolve.

At a concrete level, however, research groups are trying to propose constructive solutions such as the establishment of 'genetic pools', 'clearing houses', 'patent pool', 'alternative licensing models' amongst others (Taubman, 2009; Van Overwalle, 2009). They are trying to work out the organization of several types of interface institutions such as clearing houses in order to facilitate the access to patents and technologies by research exemptions, the creation of patent pools or 'collaborative patent systems' for affordable redistribution among interested developers. This matter was recently well reviewed by Van Zimmeren *et al.* (2011).

The renewed interest in natural phage as therapeutic agents might trigger scientists and entrepreneurs' creativity in defining the contours of appropriate patent claims for phage. Such new ideas on patentability should not be based on the existing classical model but on a broader 'new' philosophy in relation to sustainable economic and industrial development. Thinkers such as Ricardo Petrella (and his colleagues from the Group of Lisbon) and Jeffrey D. Sachs are leading this potential paradigm shift (The Group of Lisbon, 1995; Sachs, 2008). Petrella claims that today 'being competitive' is no longer a tool for increased development but an aim on itself. This increasingly implies, that the possession of patents, often as a strategic weapon, is thought to be more important (in a short-term perspective) than owning a truly functional innovative technology. This kind of attitude tends to block the development of new biomedical approaches. The patent tragedy is indeed exemplified by the millions of AIDS victims that die while treatment drugs exist and raises deep questions about global intellectual property rights and Western Ethics. How can the benefits of a global patent system that provides incentives for innovation and continuous development be combined with an assurance that the targeted people (rich and poor) gain access to the medical care that they need and have right to? As mentioned earlier the patentability of phage or phage-related procedures is key, but thorny, for the development of the potential of this rediscovered tool, which is urgently needed in an increasingly antibiotic resistant world (Levy and Marshal, 2004; Thiel, 2004; Kümmerer and Henninger, 2003; Kümmerer, 2004; Pirnay *et al.*, 2005; Kumarasamy, 2010; Abraham, 2011; Cooper and Shlaes, 2011). In our opinion, changes to the patents, patentability and restrictive licensing, common in the current pharmaceutical environment are essential for our global medical future. This need for change, after an analysis of the multifactorial aspects of antibacterial resistance, with a focus on our current Western societal profitable attitudes brings us to a 'medicine™' situation which we, amongst others, discussed already previously (Pirnay *et al.*, 2003b). This 'medicine™' situation is the consequence of the current industrialization of medicine, the 'medical industry', which neglects the fact that biological systems do not behave and consequently do not respond to markets as expected for typical mechanical behaving systems. Medicine, basically an applied biological science, based on biology a fundamental science on its own, is governed by inherent different rules than classic mechanics, and consequently should be considered as a common public good.

Along this line of thinking, the Group of Lisbon, led by Ricardo Petrella, proposes an evolution to world cooperative governance, which is based on a global contract that requires that each decision must be linked to the fact that each person should have access to all basic livelihoods, including health access, which actually is often blocked by our out-dated economic model. This is also the result of the analysis of Selgelid in his well documented article 'Ethics and drug resistance' who concludes that 'public goods warrant special treatment among which governmental intervention/funding will play a key role' (Selgelid, 2007). In this context, a 'tailor-made' approach as proposed and discussed by Pirnay and colleagues for phage therapy could eventually be developed under the umbrella of, for example, the WHO (Pirnay *et al.*, 2010; Pirnay *et al.*, 2012). The WHO recognizes the importance of the worldwide antibiotic resistance issues and is discussing new incentives to push the pharmaceutical industry in alliance with governments to launch new antimicrobial drug research and development projects (http://www.who.int/drugresistance/en/). Specific working groups were created such as the WHO Advisory Group on Integrated Surveillance of Antimicrobial Resistance (AGISAR) and the Intergovernmental Task Force on Antimicrobial Resistance (ITFAR). The 2011 World Health Day (7 April 2011) was specifically dedicated to the antimicrobial resistance issue. The strong eye catching slogan was as follows: '*Antimicrobial resistance: no action today, no cure tomorrow*' (http://www.who.int/world-health-day/2011/en/). We are therefore surprised that, although an awareness of the problem exists, 'phage therapy' is, as far as we are aware, still not mentioned by this organization as one of the potential options in the fight against bacterial infections.

Conclusions

Having analysed several major aspects (negatively) influencing the reintroduction of phage therapy as a potential antibacterial treatment we have to realise that Ethics, Science/Technology and Economy are intertwined and form a field of tension which can result in conflicting, although eventually constructive, situations.

Medicinal product regulation and marketing models evolved to predominantly accommodate stable, precautionary and profitable drugs with sufficient IP protection. Unfortunately, sustainable phage therapy seems to lack strong IP protection, which hampers its re-introduction in practical medicine today and justifiably raises ethical discussions.

In this respect we can ask ourselves 'is our current drug development and marketing model still the best suited and ethically consistent?' We live in a globalizing society with a growing awareness of evolution and sustainability. This should also trickle down into the field of drug development. Are the current overregulation and IP issues ethically tolerable in a time when increasing numbers of patients die as a consequence of antibiotic resistant infections?

The current (European) regulatory setting allows for eventual sporadic clinical trials under the responsibility and supervision of leading Medical Ethical Committees and/or borderline applications under the umbrella of the Declaration of Helsinki. Some strategies, for example the strategy adopted in Poland are possible but they will not enable the further optimal development of phage therapy in clinical practice at large. This is due to the difficulty of setting up well designed large scale and/or multi-centric (European) studies in order to prove the efficacy of this approach. Also will there be a need for studies which could optimize this therapeutic antibacterial approach in the different medical application fields (especially for such aspects as for example the most appropriate galenic formulation, the application frequency, the optimal phage concentration amongst other essential clinical parameters). There is a clear need for an adapted European regulation tailored to phage therapy, allowing the justified and broad therapeutic use of phage as an alternative or complement to antibiotics. In order to make full and safe use of phage therapy the frame has to allow, beside a 'prêt-à-porter' development a real flexible near the patient 'sur-mesure' approach, the latter being intrinsically the better suited of both models.

We support a long-term solution, such as the creation of a specific section for phage therapy under the actual Advanced Therapy Medicinal Product (ATMP) Regulation (EC) No 1394/2007, amending Directive 2001/83/EC and Regulation (EC) No 726/2004. This Regulation presently includes products for Gene Therapy, and for Somatic Cell Therapy and Tissue Engineered Products. We propose the creation of a fourth section, named Bacteriophage Therapy Preparations, which could be divided into two distinct sections: 'tailor-made use in a hospital environment' (via a specific hospital exemption regulation) versus 'industrially produced for market placement' (Verbeken *et al.*, 2007; Pirnay *et al.*, 2010). Amending Commission Directive 2003/63/EC could also be considered. Placing Bacteriophage Therapy Preparations under or next to the current sections foreseen for Biological Medicinal Products or Particular Medicinal Products could also be a possible approach.

A major hurdle is the lack of societal realization that the classic economic, ethical, medical and scientific ways of thinking have to be adapted to our natural environment, which is a world in constant movement and evolution. Patent related issues should be seen from a more societal impact perspective and not only from the pure straight-line economic point of view, currently at the forefront. Our ethical views also have to be re-evaluated if we prevent patients' access to therapy, which has at least empirically proven to be efficacious, under the umbrella of 'patient's safety'. The role of fundamental scientists and medical practitioners is also to show new insights and to advocate for some change, especially when the current societal views, often too reductionist and based upon short term considerations, are blocking a further promising development.

This view of the future of 'Phage Therapy' was described in a primary opinion paper by Pirnay and colleagues (Pirnay *et al.*, 2010) and further discussed in two recent review articles (Pirnay *et al.*, 2012; Verbeken *et al.*, 2012). During a recent EMBO sponsored meeting on 'Viruses of microbes' organized in Brussels at the Royal Military Academy a whole workshop and debate session was dedicated to the issue (Brüssow, 2012). Phage therapy and antibiotic therapy, as well as the two proposed developmental models, should not be mutually exclusive as also discussed by Maura and Debarbieux (Maura and Debarbieux, 2011). At the contrary some studies even argue for a synergistic effect of phages and antibiotics as well as a limitation of the emergence of antibiotic resistance as a result of using phages (Comeau *et al.*, 2007; Zhang and Buckling, 2011; Ryan *et al.*, 2012; Kirby, 2012).

On 14 February 2011 the question 'What is the European View on the status of Phage as antibacterial agent,' was officially asked by the Belgian Christian Democrat Ivo Belet and his French parliamentarian colleague from the Socialist fraction Catherine Trautmann to the European Commission (http://www.europarl.europa.eu/sides/getDoc.do?pubRef=-//EP//TEXT+WQ+E-2011-001144+0+DOC+XML+V0//EN&language=CS). This shows that reflections on the idea of reintroducing phage therapy diffuses out of the world of biomedical science and the pharmaceutical industry to the political arena as a topic for public discussion in a time of antibiotic shortages. The official answer, received on 29 March 2011 was formulated as follows: 'The Commission considers that the existing regulatory framework is adequate for bacteriophage therapy without the need for an extra set of documentation for bacteriophage therapy' (http://www.europarl.europa.eu/sides/getAllAnswers.do?reference=E-2011–001144&language=CS). Thus the official answer was, as expected, that phage is covered by the actual European medicinal product regulation. This official statement was our starting point for opening discussions, as described, at the level of the European Medicines Agency and its specifically dedicated Innovation Task Force (ITF) unit. Those discussions are still going on and will hopefully result in an adequate change enabling the implementation of phage therapy in clinical practice.

Acknowledgements

We would like to thank Mr William Anderson for reviewing the text for language. We would also like to thank Jérôme Larché, MD, MSc and Head of the Intensive Care Unit, Hospital of Narbonne, France. As president of PhagEspoirs, a non-profit organization aiming at promoting research on phage, he reviewed and commented actively this document. We also would like to thank Geert Laire, MD, Surgeon General and Chief of the Belgian Medical Defence Component, Pierre Neirinckx, MD, Colonel and Director of the Queen Astrid Military Hospital and Serge Jennes, MD, Lt. Colonel and Head of the Burn Wound Centre of the Queen Astrid Military Hospital for their continuous support to our phage therapy project. And last but not least we would also like to thank Valérie De Vos, for the multiple reflections and discussions on traditional medical systems versus our current medical industry, based on her African cultural anthropological background and her crisis management approach in conflicting fields of globalization.

Web resources

- WHO. Policy Package to Combat Antimicrobial Resistance. http://www.who.int/bulletin/volumes/89/5/11-088435/en/ (accessed July 2013).
- WMA Declaration of Helsinki–Ethical Principles for Medical Research Involving Human Subjects. http://www.wma.net/en/30publications/10policies/b3/index.html.pdf?print-media-type&footer-right=[page][toPage (accessed July 2013).

References

Abedon, S.T., Kuhl, S.J., Blasdel, B.G., and Kutter, E.M. (2011). Phage treatments of human infections. Bacteriophage. *1*, 66–85.

Agarwal, K.N., and Bhasin, S.K. (2002). Feasibility studies to control acute diarrhoea in children by feeding fermented milk preparations Actimel and Indian Dahi. Eur. J. Clin. Nutr. *56(S4)*, 556–559.

Aiello, A.E., King, N.B., and Foxman, B. (2006). Ethical conflicts in public health research and practice. Am. J. Public Health. *96*, 1910–1914.

Akst, J. (2010). Key cancer patents killed. http://www.the-scientist.com/blog/display/57265/ (accessed19/10/10), part of The Scientist.http://www.the-scientist.com/(accessed on 19/10/10).

Allen, H.K., Donato, J., Wang, H.H., Cloud-Hansen, K.A., Davies, J., and Handelsman, J. (2010). Call of the wild: antibiotic resistance genes in natural environments. Nat. Rev. Microbiol. *8*, 251–259.

Association for Molecular Pathology. (2010). U.S. patent and trademark office, Southern District of New York (29 March 2010).

Bamford, D.H. (2003). Do viruses form lineages across different domains of life? Res. Microbiol. *154*, 231–236.

Baquero, F., Coque, T.M., and Canton, R. (2003). Antibiotics, complexity and evolution. ASM News. *69*, 547–552.

Barr, J.J., Auro, R., Furlan, M., Whiteson, K.L., Erb, M.L., Pogliaéno, J., Stotland, A., Wolkowicz, R., Cutting, A.S., Doran, K.S., *et al.* (2013). Bacteriophage adhering to mucus provide a non-host-derived immunity. Proc. Natl. Acad. Sci. U.S.A. *110*, 10771–10776.

Bartoloni, A., Pallechi, L., Rodriguez, H., Fernandez, C., Mantella, A., Bartalesi, F., Strohmeyer, M., Kristiansson, C., Gotuzzo, E., Paradisi, F., *et al.* (2009). Antibiotic resistance in a very remote Amazonas community. Int. J. Antimicrob. Agents. 33, 125–129.

Beauchamp, T.L., and Childress, J.F. (2009). Principles of biomedical ethics (New York/Oxford, Oxford University Press).

Bell, P.J.L. (2001). Viral eukaryogenesis: was the ancestor of the nucleus a complex DNA virus? J. Mol. Evol. 53, 251–256.

Benner, S.A. (2010). Defining life. Astrobiology *10*, 1021–1030.

Bergh, O., Borsheim, K.Y., Bratbak, G., and Heldal, M. (1989). High abundance of viruses found in aquatic environments. Nature *340*, 476–468.

Berman, S.H., Eichelsdoerfer, P., Yim, D., Elmer, G.W., and Wenner, C.A. (2006). Daily ingestion of a nutritional probiotic supplement enhances innate immune function in healthy adults. Nutr. Res. *26*, 454–459.

Biswas, B., Adhya, S., Washart, P., Paul, B., Trostel, A.N., Powell, B., Carlton, R., and Merril, C.R. (2002). Bacteriophage therapy rescues mice bacteremic from a clinical isolate of vancomycin-resistant *Enterococcus faecium*. Infect. Immun. *70*, 204–210.

Bošnjak, S. (2001). The Declaration of Helsinki – the cornerstone of research ethics. Arch. Oncol. *9*, 179–184.

Brewer, V. (2012). US court reverses gene patent ruling for *BRCA1* and *BRC2*. Lancet Oncol. *12*, 835.

Brüssow, H. (2005). Phage therapy: the *Escherichia coli* experience. *151*, 2133–2140.

Brüssow, H. (2009). The not so universal tree of life or the place of viruses in the living world. Phil. Trans. R. Soc. B *364*, 2263–2274.

Brüsssow, H. (2012). What is needed for phage therapy to become a reality in Western medicine? *434*, 138–142.

Bruttin, A., and Brüssow, H. (2005). Human volunteers receiving *Escherichia coli* phage T4 orally: a safety test of phage therapy. Antimicrob. Agents and Chemother. 49, 2874–2878.

Brzuszkiewicz, E., Thurmer, A., Schuldes, J., Leinbach, A., Lieregang, H., Meyer, F.D., Boelter, J., Petersen, H., Gottschalk, G., and Daniel, R. (2011). Genome sequence analyses of two isolates from the recent *Eschericha coli* outbreak in Germany reveal the emergence of a new pathogen type Entero-Aggregative-Haemorhagic-*Escherichia coli* (EAHEC). Arch. Microbiol. *193*, 883–891.

Buckling, A., and Rainey, P.B. (2002). Antagonistic coevolution between a bacterium and a bacteriophage. Proc. Biol. Sci. *269*, 931–936.

Burdock, G.A., and Carabin, I.G. (2004). Generally recognized as safe (GRAS): history and description. Toxicol. Lett. *150*, 3–18.

Bush, K., Courvalin, P., Dantas, G., Davies, J., Eisenstein, B., Huovinen, P., Jacoby, G.A., Kishony, R., Kreiswirth, B.N., Kutter, E., *et al.* (2011). Tackling antibiotic resistance. Nat. Rev. Microbiol. *9*, 894–896.

Chanishvili, N. (2009). A Literature Review of the Practical Application of Bacteriophage Research (Tbilisi, Georgia, UK HPA, Eliava Institute of Bacteriophage, Microbiology and Virology).

Chanishvili, N. (2012). Phage therapy – history from Twort and d'Hérelle through Soviet experience to current approaches. Adv. Virus Res. *83*, 3–40.

Cherniak, E.P. (2010). Bugs as drugs, Part 1: Insects. The 'New' alternative Medicine for the 21st century. Altern. Med. Rev. *15*, 124–135.

Chibani-Chenouffi, S., Sidoti, J., Brüttin, A., Kutter, E., Sarker, S., and Brüssow, H. (2004). *In vitro* and *in vivo* bacteriolytic activities of *Escherichia coli* phages: implications for phage therapy. Antimicrob. Agents Chemother. *48*, 2558–2569.

Cho, M. (2010). Patently unpatentable: implications of the Myriad court decision on genetic diagnostics. Trends Biotechnol. *28*, 548–551.

Cleland, C.E. (2007). Epistemological issues in the study of microbial life: alternative terran biospheres? Stud. Hist. Phil. Biol. & Biomed. Sci. *38*, 847–861.

Comeau, A.M., Tétart, F., Trojet, S.N., Prère, M.F., and Krish, H.M. (2007). Phage-antibiotic synergy (PAS): beta-lactam and quinolone antibiotics stimulate virulent phage growth. PLoS One *2*, e799.

Cooper, M.A., and Shlaes, D. (2011). Fix the antibiotics pipeline. Nature *472*, 32.

Corbellini, G. (2008). Evolution based medicine (EBM): an epistemological framework for thinking of, and dealing with the so called « crisis of medicine ». J. Hist. Med. *20*, 115–139.

Darwin, C. (1968). The Origin of Species. London, UK, Penguin Classics, Penguin Group.

D'Costa, V.M., King, C.E., Kalan, L., Morar, M., Sung, W.W.L., Schwarz, C., Froese, D., Zazula, G., Calmels, F., Debruyne, R., Golding, G.B., Poinar, H.N., and Wright, G.D. (2011). Antibiotic resistance is ancient. Nature *477*, 457–461.

D'Costa, V.M., McGrann, K.M., Hughes, D.W., and Wright, G.D. (2006). Sampling the antibiotic resistome. Science *311*, 374–377.

Dabrowska, K., Switala-Jelen, K., Opolski, A., Weber-Dabrowska, B., and Gorski, A. (2003). Bacteriophage penetration in vertebrates. J. Appl. Microbiol. *98*, 7–13.

Debarbieux, L., Leduc, D., Maura, D., Morello, E., Criscuolo, A., Grossi, O., Balloy, V., and Touqui, L. (2010). Bacteriophages can treat and prevent *Pseudomonas aeruginosa* lung infections. J. Infect. Dis. *201*, 1096–1104.

Dublanchet, A. (2009). Des virus pour combattre les infections. Favre, Lausanne, Switzerland.

European Parliament. (2003). Commission Directive 2003/63/EC of 25 June 2003 amending Directive 2001/83/EC of the European Parliament and the council of the Community code relating to medicinal products for human use. Official Journal of the European Union (translated into Belgian Law by Royal Decree of 04.03.2004, published in the Belgian Official Journal on 10.03.2004).

European Parliament. (2004). Directive 2001/83/EC of the European Parliament and of the Council of 6 November 2001 on the Community code relating to medicinal products for human use. Official journal of the European Communities (translated into Belgian Law by Royal Decree of 04.03.2004, published in the Belgian Official Journal on 10.03.2004).

Faruque, S.M., Naser, I.B., Islam, M.J., Faruque, A.S.G., Ghosh, A.N., Nair, G.B., Sack, D.A., and Mekalanos, J.J. (2005a). Seasonal epidemics of cholera inversely correlate with the prevalence of environmental cholera phages. Proc. Natl. Acad. Sci. U.S.A. *102*, 1701–1707.

Faruque, S.M., Islam, M.J., Ahmad, Q.S., Faruque, A.S.G., Sack, D.A., Nair, G.B., and Mekalanos, J.J. (2005b). Self-limiting nature of seasonal cholera epidemics: role of host-mediated amplification of phage. Proc. Natl. Acad. Sci. U.S.A. *102*, 6119–6124.

FDA (2006). FDA 21 CFR Part 172: Food Additives Permitted for Direct Addition to Food for Human Consumption; Bacteriophage Preparation. Federal Register, Rules and Regulations No. 160. *71*, 47729–47732.

Forterre, P. (2001). Genomics and early cellular evolution, the origin of the DNA world. Life Sci. *324*, 1067–1076.

Forterre, P. (2006). The origin of viruses and their possible role in major evolutionary transitions. Virus Res. *117*, 5–16.

Fuhrman, J.A. (1999). Marine viruses and their biogeochemical and ecological effects. Nature *399*, 541–548.

Gilmore, B.F. (2012). Bacteriophages as anti-infective agents: recent developments and regulatory challenges. Expert. Rev. Anti Infect. Ther. *10*, 533–535.

Gold, E.R., Kaplan, W., Orbinski, J., Harland-Logan, S., and N-Marandi, S. (2009). Are patents impeding medical care and innovation? PLoS Medicine 7, e1000208.

Gorski, A., Miedzybrodzki, R., Borysowski, J., Weber-Dabrowska, B., Lobocka, M., Fortuna, W., Letkiewiecz, S., Zimecki, M., and Filby, G. (2009). Bacteriophage therapy for the treatment of infections. Curr. Opin. Investig. Drugs *10*, 766–774.

Hall, A.R., De Vos, D., Friman, V.P., Pirnay, J.P., and Buckling, A. (2012). Effects of sequential and simultaneous applications of bacteriophages on populations of *Pseudomonas aeruginosa in vitro* and in Wax Moth larvae. Appl. Environment. Microbiol. *78*, 5646–5652.

Hendrix, R.W. (2002). Bacteriophages: evolution of the majority. Theor. Popul. Biol. *61*, 471–480.

d'Hérelle, F. (1917). Sur un microbe invisible antagoniste des bactéries dysentériques. C.R. Acad. Sci. *165*, 373–375.

d'Hérelle, F. (1919). Sur le rôle du microbe bacteriophage dans la typhose aviaire. C.R. Acad. Sci. *169*, 932–934.

Human, D., and Fluss, S.S. (2001). The World Medical Association's Declaration of Helsinki: historical and contemporary perspectives. Available at: http://www.wma.net

Huys, I., Berthels, N., Matthijs, G., and Van Overwalle, G. (2009). Legal uncertainty in area of genetic diagnostic testing. Nat. Biotechnol. *27*, 903.

Huys, I., Van Overwalle, G., and Matthijs, G. (2011). Gene and genetic diagnostic method patent claims: a comparison under current European and US patent law. Eur. J. Hum. Genet. *19*, 1104–1107.

Kapczynski, A. (2009). Innovation policy for a New Era. J. Law Med. Ethics *37*, 264–268.

Kesselheim, A.S. (2010). Using market-exclusivity incentives to promote pharmaceutical innovation. N. Engl. J. Med. *363*, 1855–1862.

Khawaldeh, A., Morales, S., Dillon, B., Alavidze, Z., Ginn, A.N., Thomas, L., Chapman, S.J., Dublanchet, A., Smithyman, A., and Iredell, J.R. (2011). Bacteriophage therapy for refractory *Pseudomonas aeruginosa* urinary tract infection. J. Med. Microbiol. *60*, 1697–1700.

Kirby, A.E. (2012). Synergistic action of gentamicin and bacteriophage in a continous culture population of *Staphylococcus aureus*. PLoS One 7, e51017.

Kumarasamy, K., Toleman, M.A., Walsh, T.R., Bagaria, J., Butt, F., Balakrishnan, R., Chaudhary, U., Doumith, M., Giske, C.G., Irfan, S., *et al.* (2010). Emergence of a new antibiotic resistance mechanism in India, Pakistan, and the UK: a molecular, biological, and epidemiological study. Lancet Infect. Dis. *10*, 597–602.

Kümmerer, K. (2004). Resistance in the environment. J. Antimicrob. Chemother. *54*, 311–320.

Kümmerer, K., and Henninger, A. (2003). Promoting resistance by the emission of antibiotics from hospitals and households into effluent. Clin. Microbiol. Infect. *9*, 1203–1214.

Kutateladze, M., and Adamia, R. (2010). Bacteriophages as potential new therapeutics to replace or supplement antibiotics. Trends Biotechnol. *28*, 591–595.

Kutter, E., De Vos, D., Gvasalia, G., Alavidze, Z., Gogokhia, L., Kuhl, S., and Abedon, S.T. (2010). Phage therapy in clinical practice: treatment of human infections. Curr. Pharm. Biotechnol. *11*, 69–86.

Lander, E.S., Linton, L.M., Birren, *et al.* (the international human genome consortium) (2001). Initial sequencing and analysis of the human genome. Nature *409*, 860–921.

Landy, J., Al-Hassi, H.O., McLaughlin, S.D., Walker, A.W., Ciclitira, P.J., Nicholls, R.J., Clark, S.K., and Hart, A.L. (2011). Review article: faecal transplantation therapy for gastrointestinal disease. Aliment. Pharmacol. Ther. *34*, 409–415.

Leibovici, L., Paul, M., and Ezra, O. (2012). Ethical dilemmas in antibiotic treatment. J. Antimicrob. Chemother. *67*, 12–16.

Levy, S.B., and Marshal, B. (2004). Antibacterial resistance worldwide: causes, challenges and responses. Nat. Med. *10*, S122–S129.

McCallin, S., Alam Sarker, S., Barretto, C., Sultana, S., Berger, B., Huq, S., Krause, L., Bibiloni, R., Schmitt, B., Reuteler, G., *et al.* (2013). Safety analysis of a Russian phage cocktail: From MetaGenomic analysis to oral application in healthy human subjects. Virology *443*, 187–196.

McKenna, M. (2011). Swapping germs: should fecal transplants become routine for debilitating diarrhea? Sci. Am. *305*, 34.

McVay, C.S., Velasquez, M., and Fralick, J.A. (2007). Phage therapy of *Pseudomonas aeruginosa* infection in a mouse burn wound model. Antimicrob. Agents Chemother. *51*, 1934–1938.

Mahony, J., McAuliffe, O., Ross, R.P., and van Sinderen, D. (2011). Bacteriophages as biocontrol agents of food pathogens. Curr. Opin. Biotechnol. *22*, 157–163.

Martinez, J.L., and Baquero, F. (2002). Interactions among strategies associated with bacterial infection: pathogenicity, epidemicity, and antibiotic resistance. Clin. Microbiol. Rev. *15*, 647–679.

Marza, J.A., Soothill, J.S., Boydell, P., and Collyns, T.A. (2006). Multiplication of therapeutically administered bacteriophages in *Pseudomonas aeruginosa* infected patients. Burns *32*, 644–646.

Maura, D., and Debarbieux, L. (2011). Bacteriophages as twenty-first century antibacterial tools for food and medicine. Appl. Microbiol. Biotechnol. *90*, 851–859.

Mayr, E. (2004). What makes biology unique? Considerations on the autonomy of a discipline. Cambridge University Press.

Merabishvili, M., Pirnay, J.-P., Verbeken, G., Chanishvili, N., Tediashvili, M., Lashkhi, N., Gionti, T., Krylov, V., Mast, J., Van Parijs, L., *et al.* (2009). Quality-controlled small-scale production of a well-defined bacteriophage cocktail for use in human clinical trials. PLoS One *4*, e4944.

Merabishvili, M., De Vos, D., Verbeken, G., Kropinski, A.M., Vandenheuvel, D., Lavigne, R., Wattiau, P., Mast, J., Ragimbeau, C., Mossong, J., *et al.* (2012). Selection and characterization of a candidate therapeutic bacteriophage that lyses the *Escherichia coli* O104:H4 strain from the 2011 outbreak in Germany. PLoS One, *7*, e52709.

Mi, S., Lee, X., Li, X.-P., Veldman, G., Finnerty, H., Racie, L., LaVilie, E., Tang, X.-Y., Edouard, P., Howes, S., *et al.* (2000). Syncitin is a captive retroviral envelope protein involved in human placental morphogenesis. Nature *403*, 785–789.

Międzybrodzki, R., Fortuna, W., Weber-Dąbrowska, B., and Górski, A. (2007). Phage therapy of staphylococcal infections (including MRSA) may be less expensive than antibiotic treatment. Post. Med. Hig. Dośw. *61*, 461–465.

Medzybrodzki, R., Borysowski, J., Weber-Dabrowska, B., Fortuna, W., Letkiewicz, S., Szufnarowski, K., Pawelczyk, Z., Rogoz, P., Klak, M., Wojtasik, E., *et al.* (2012). Clinical aspects of phage therapy. Adv. Virus Res. *83*, 73–119.

Morello, E., Saussereau, E., Maura, D., Huerre, M., Touqui, L., and Debarbieux, L. (2011). Pulmonary bacteriophage therapy on *Pseudomonas aeruginosa* cystic fibrosis strains: first steps towards treatment and prevention. PLoS One *6*, e16963.

Parracho, H.M., Burrowes, B.H., Enright, M.C., McConville, M.L., and Harper, D.R. (2012). The role of regulated clinical trials in the development of bacteriophage therapeutics. *6*, 279–286.

Pirnay, J.-P., De Vos, D., Cochez, C., Bilocq, F., Pirson, J., Struelens, M., Duinslaeger, L., Cornelis, P., Zizi, M., and Vanderkelen, A. (2003a). Molecular epidemiology of *Pseudomonas aeruginosa* colonization in a burn unit: persistence of a multidrug-resistant clone and a silver sulfadiazine-resistant clone. J. Clin. Microbiol. *41*, 1192–1202.

Pirnay, J.-P., De Vos, D., Zizi, M., and Heyman, P. (2003b). No easy way to exterminate 'superbugs' at the dawn of the third millennium. Exp. Rev. Anti-Infectiv. Ther. *1*, 523–525.

Pirnay, J.-P., Mathijs, S., Colak, H., Chablain, P., Bilocq, F., Van Eldere, J., De Vos, D., Zizi, M., Triest, L., and Cornelis, P. (2005). Global *Pseudomonas aeruginosa* biodiversity as reflected in a Belgian river. Env. Microbiol. *7*, 969–980.

Pirnay, J.-P., De Vos, D., Verbeken, G., Merabishvili, M., Chanishvili, N., Vaneechoutte, M., Zizi, M., Laire, G., Lavigne, R., Huys, I., *et al.* (2010). The Phage therapy paradigm: prêt-à-porter or sur-mesure? Pharm. Res. *28*, 934–937.

Pirnay, J.-P., Verbeken, G., Rose, T., Jennes, S., Zizi, M., Huys, I., Lavigne, R., Merabishvili, M., Vaneechoutte, M., Buckling, A., and De Vos, D. (2012). Introducing yesterday's phage therapy in today's medicine. Future Virol. *7*, 379–390.

Raoult, D., and Forterre, P. (2008). Redefining viruses: lessons from the mimivirus. Nat. Rev. Microbiol. *6*, 315–319.

Reid, G. (2011). Quo vadis-EFSA? Benef. Microb. *2*, 177–179.

Rhoads, D.D., Wolcott, R.D., Kuskowski, M.A., Wolcott, B.M., Ward, L.S., and Sulakvelidze, A. (2009). Bacteriophage therapy of venous leg ulcers in humans: results of a phase I safety trial. J. Wound Care *18*, 237–243.

Rolain, J.M., Canton, R., and Cornaglia, G. (2012). Emergence of antibiotic resistance: need for a new paradigm. Clin. Microbiol. Infect. *18*, 615–616.

Ryan, E.M., Gorman, S.P., Donnelly, R.F., and Gilmore, B.F. (2011). Recent advances in bacteriophage therapy: how delivery routes, formulation, concentration and timing influence the success of phage therapy. J Pharm. Pharmacol. *63*, 1253–1264.

Ryan, E.M., Alkawareek, M.Y., Donnelly, R.F., and Gilmore, B.F. (2012). Synergistic phage–antibiotic combinations for the control of *Escherichia coli* biofilms in vitro. FEMS Immunol. Med. Microbiol. *65*, 395–398.

Sachs, J.D. (2008). Common Wealth: Economics for a crowded Planet. Penguin, London, UK.

Sanders, M.E., and Huis in 't Veld, J. (1999). Bringing a probiotic-containing functional food to the market:

microbiological, product, regulatory and labeling issues. Antonie van Leeuwenhoek. *76*, 293–315.

Sarker, S.A., McCallin, S., Barretto, C., Berger, B., Pittet, A.C., Sultana, S., Krause, L., Huq, S., Bibiloni, R., Bruttin, A., *et al.* (2012). Oral T4-like phage cocktail application to healthy adult volunteers from Bangladesh. Virology *434*, 222–232.

Schmieder, R., and Edwards, R. (2012). Insights into antibiotic resistance through metagenomic approaches. Future Microbiol. *7*, 73–89.

Selgelid, M.J. (2007). Ethics and drug resistance. Bioethics. *21*, 218–229.

Shanks, N., and Pyles, R.A. (2007). Evolution and medicine: the long reach of 'Dr. Darwin'. Philos. Ethics Humanit. Med. *2*, 4.

Shapiro, O.H., and Kushmaro, A. (2011). Bacteriophage ecology in environmental biotechnology processes. Curr. Opin. Biotechnol. *22*, 1–7.

Slopek, S., Weber-Dabrowska, B., Dabrowski, M., and Kucharewicz-Krukowska, A. (1987). Results of bacteriophage treatment of suppurative bacterial infections in the years 1981–1986. Arch. Immunol. Ther. Exp. *35*, 569–583.

Smith, H.W., and Huggins, M.B. (1982). Successful treatment of experimental *Escherichia coli* infections in mice using phage: its general superiority over antibiotics. J. Gen. Microbiol. *128*, 307–318.

Sulakvelidze, A. (2013). Using lytic bacteriophages to eliminate or significantly reduce contamination of food by food-borne bacterial pathogens. J. Sci. Food Agric. *93*, 3137–3146.

Sulakvelidze, A., Alavidze, Z., and Morris, J.G. Jr. (2001). Bacteriophage therapy. Antimicrob. Agents Chemother. *45*, 649–659.

Taubman, A. (2008). The international patent system and biomedical research: reconciling aspiration, policy and practice. AAPS J. *10*, 526–536.

Thiel, K. (2004). Old dogma, new tricks – 21st. century phage therapy. Nat. Biotechnol. *22*, 31–36.

Tirard, S., Morange, M., and Lazcano, A. (2010). The definition of life: a brief history of an elusive scientific endeavour. Astrobiology *10*, 1003–1009.

Tyebkhan, G. (2003). Declaration of Helsinki: the ethical cornerstone of human clinical research. Indian J. Dermatol. Venerol. Leprol. *69*, 245–247.

The Group of Lisbon. (1995). Limits to Competition. The MIT Press, Cambridge, MA, USA.

Valenti, R. (2012). Time, evolution and physical reductionism. EMBO Rep. *13*, 181–185.

Van Overwalle, G. (2009). Gene Patents and Collaborative Licensing Models, Patent Pools, Clearing Houses, Open Source Models and Liability Regimes (Cambridge: Cambridge University Press).

Van Overwalle, G. (2010). Turning patent swords into shares. Science *230*, 1630–1631.

Van Zimmeren, E., Vanneste, S., Matthijs, G., Vanhaverbeke, W., and Van Overwalle, G. (2011). Patent pools and clearing houses in the life sciences. *29*, 569–576.

Verbeken, G., De Vos, D., Vaneechoutte, M., Merabishvili, M., Zizi, M., and Pirnay, J.-P. (2007). European regulatory conundrum of phage therapy. Future Microbiol. *2*, 485–491.

Verbeken, G., Pirnay, J.-P., De Vos, D., Jennes, S., Zizi, M., Lavigne, R., Casteels, M., and Huys, I. (2012). Optimizing the European regulatory framework for sustainable bacteriophage therapy in human medicine. Arch. Immunol. Ther. Exp. *60*, 161–172.

Villarreal, L.P. (2005). Overall issues of viruses and host evolution. In Viruses and the Evolution of Life, L.P. Villareal, ed. (Washington, DC, ASM press), pp. 1–28.

Villarreal, L.P., and Witzany, G. (2010). Viruses are essential agents within the roots and stem of the tree of life. J. Theor. Biol. *262*, 698–710.

Ward, P. (2005). Life As We Do Not Know It (New York, USA, Viking Penguin).

Weber-Dabrowska, B., Mulczyk, M., and Gorski, A. (2000). Bacteriophage therapy of bacterial infections: an update of our institute's experience. Arch. Immunol. Ther. Exp. *48*, 547–551.

Williams, P.D. (2010). Darwinian interventions: taming pathogens through evolutionary ecology. Trends Parasitol. *26*, 83–92.

Wills, Q.F., Kerrigan, C., and Soothill, J.S. (2005). Experimental bacteriophage protection against *Staphylococcus aureus* abscess in a rabbit model. Antimicrob. Agents. Chemother. *49*, 1220–1221.

Wommack, K.E., and Colwell, R.R. (2000). Virioplankton: viruses in aquatic ecosystems. Microbiol. Mol. Biol. Rev. *64*, 69–114.

World Medical Association. (2008). Declaration of Helsinki. Ethical Principles for Medical Research Involving Human Subjects. http://www.wma.net/en/30publications/10policies/b3/index.html

Wright, A., Hawkins, C.H., Anggärd, E.E., and Harper, D.R. (2009). A controlled clinical trial of a therapeutic bacteriophage preparation in chronic otitis due to antibiotic resistant *Pseudomonas aeruginosa*: a preliminary report of efficacy. Clin. Otolaryngol. *34*, 349–357.

Zhang, Q.G., and Buckling, A. (2011). Phages limit the evolution of bacterial antibiotic resistance in experimental microcosms. Evol. Appl. *5*, 572–582.

The Use of Bacteriophages and Bacteriophage-derived Enzymes for Clinically Relevant Biofilm Control

13

Sanna Sillankorva and Joana Azeredo

Abstract

Biofilm formation occurs spontaneously on both inert and living systems and is an important bacterial survival strategy. Biofilms are often associated with several chronic and acute infections such as wound infections, vaginitis, upper respiratory tract infections, otitis media, and endocarditis among others. There is an inherent tolerance of cells embedded in biofilms towards antibiotics and therefore, alternative agents, such as bacteriophages have been the focus of interest. This is an overview of the phage–biofilm interaction research carried out, starting with a general introduction to biofilm structure and composition, and to clinically relevant biofilms and their inherent tolerance to antimicrobials. Then, the use of phages and phage derived enzymes as alternative biofilm control agents will be discussed in detail based on the results from a number of *in vitro* experiments found in literature. Although focus is almost exclusively given to the results of biofilm control and biofilm prevention experiments using phages and phage-derived enzymes, other relevant topics, such as the diffusion of phages through the matrix, and the effect of prophages, will also be addressed.

Introduction

It has long been understood that the development of complex structured communities or biofilms occurs spontaneously on both inert and living systems and is an important bacterial survival strategy. Although the existence of biofilms has been documented for almost a century, studies of planktonic cells, from their isolation and identification to control have dominated medical microbiology. The clinical relevance of biofilms was recognized only after finding that they are involved in persistent infections (Potera, 1999). Now, it is well known that biofilms are often associated with chronic and acute infections such as wound infections, vaginitis, upper respiratory tract infections, otitis media, and endocarditis among others. Biofilm-related infections are thought to affect more patients than heart disease and cancer combined (Balaban, 2008). Biofilms provide an immobilized but dynamic environment in which the bacterial cells are able to attain homeostasis (Sutherland, 2001b).

While many infectious biofilms are composed of a single bacterial species, the possibility of multispecies infectious biofilms must also be considered. In these, synergistic interactions between the bacteria may lead to enhanced biofilm formation and to increased resistance to antimicrobial agents (Burmolle *et al.*, 2006). The pathogens involved in biofilm-associated infections show typically reduced *in vitro* susceptibility to antimicrobial agents and host defence mechanisms (Donlan and Costerton, 2002), especially in interspecies biofilms where several interactions occur between the individual species. This decreases the efficacy of the antimicrobial agents and leads to serious problems particularly regarding to reduced microbial eradication efficiency (Budhani and Struthers, 1998; Cowan *et al.*, 2000; Erb *et al.*, 1997; Skillman *et al.*, 1999).

Bacteriophages (phages) are host-specific bacterial viruses. This specificity makes them ideal candidates for antimicrobial therapy and alternative to antibiotics for the treatment of biofilm related infections since they do not interfere with the normal microflora of the environment where they are to be applied. Phages have also a self-replicating nature at the infection site as long as the host bacterium is present and, orally and intraperitoneally administered phages have been recovered for instance from the blood, liver, kidneys, spleen, lungs, muscles, brain (Weber-Dabrowska *et al.*, 1987; Nishikawa *et al.*, 2008). Furthermore, several studies have shown that phages can effectively infect and lyse cells present in single and polymicrobial species biofilms which strongly supports the concept of phage therapy as an alternative or complementary strategy to prevent and control biofilm-related infections (reviewed in Azeredo and Sutherland, 2008; Donlan, 2009). Also, the first studies on the use of exopolymeric matrix and cell wall degrading enzymes of phage origin have produced encouraging results for biofilm control.

Biofilm structure and composition

Direct observations of microbial biofilms have been reported since the first microscopes were built. For instance, the Dutch researcher Antonie van Leeuwenhoek (1632–1723) observed this type of life form when examining the 'scuff' from his teeth using a self-constructed microscope capable of enlargements of up to 500 times. Since the description and naming of these sessile communities back in 1978 (Costerton *et al.*, 1978), biofilm science and biofilm engineering have been active fields of study. Today, it is well established that the majority of microbes persist attached to surfaces within a structured biofilm ecosystem and not as free-floating organisms.

The definition of biofilm has changed considerably since the first direct observations. The most recent definition states that a biofilm is a three-dimensional microbial structure consisting of a multicellular community composed of prokaryotic and/or eukaryotic cells embedded in a matrix composed, at least partially, of material synthesized by the sessile cells in the community (Costerton, 2007). Biofilm development stages include reversible and irreversible attachment of cells, followed by their proliferation into small clusters and excretion of extracellular polymeric substances and attachment of other microbes such as different species of bacteria, fungi or protozoa (Davey and O'Toole, 2000; O'Toole *et al.*, 2000) (Fig. 13.1). Direct structural examination of biofilms has shown that, in terms of volume, only 15% of the biofilms consist of cells (De Beer *et al.*, 1994b). The remaining 85%, which is the biofilm matrix, consists of a range of macromolecular products secreted by replicating adherent bacteria, of which insoluble gelatinous exopolymers are the major macromolecular constituent; the matrix also consists of water (90–97%), enzymes, proteins, bacteriocins, and low mass solutes and nucleic acids which are released through cell lysis (Sutherland, 2001a, 2004). The function of the matrix is to provide structural stability and protection to the biofilm. Within biofilms, microbial cells mainly form matrix–enclosed clusters between ramifying water channels (Donlan and Costerton, 2002) that carry bulk fluid into the community by convective flow (De Beer *et al.*, 1994b), facilitating also the oxygen transport from the surrounding environment to the cell clusters (De Beer *et al.*, 1994c).

Microorganisms adopt the biofilm mode of growth as a survival strategy to resist against environmental stresses. In fact, there are several advantages of living in a biofilm community, including increased availability of nutrients for growth; increased binding of water molecules, reducing the possibility of dehydration; proximity to progeny and other bacteria, facilitating plasmid transfer (Lappin-Scott and Costerton, 1995) and protection against antimicrobials. Therefore, one of the major problems associated with biofilms is their inherent tolerance to host defences and antibiotic therapies, which are effective against free-floating cells (Bourion and Cerf, 1996; Chavant *et al.*, 2004; Costerton *et al.*, 1999; Frank *et al.*, 2003; Mattila *et al.*, 1990; Knowles *et al.*, 2005; Kumar and Anand, 1998; Norwood and Gilmour, 2000; Reisner *et al.*, 2006; Ryu and Beuchat, 2005).

Biofilms are structurally and metabolically heterogeneous as established by direct measurement of several parameters by the use of microelectrodes. Furthermore, aerobic and anaerobic processes occur simultaneously in different parts of the multicellular community. Biofilm formation is governed by a number of biological, physical and chemical processes whose relative contribution

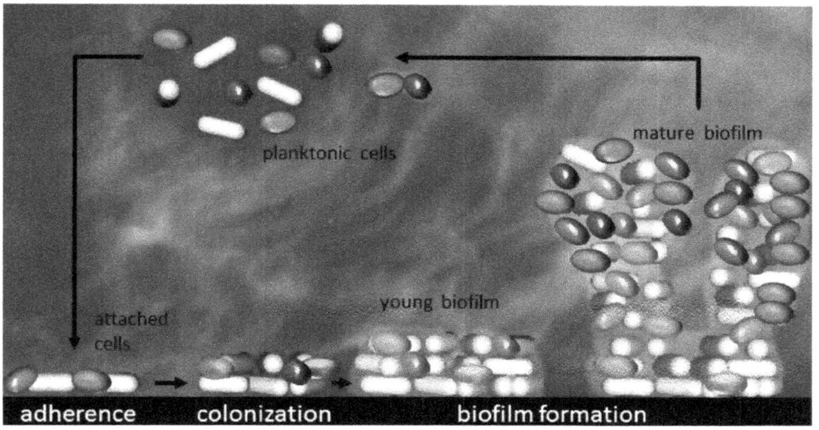

Figure 13.1 Biofilm formation steps. First bacteria adhere to conditioned substratum surface, then colonisation starts with production of extracellular metabolites, the young biofilms proliferate and in the mature biofilms portions of cells slough off the main biofilm structure. A colour version of this figure is available in the plate section at the back of the book.

changes throughout biofilm development and depends on established environmental and hydrodynamic conditions (Bryers, 2000). It is also well established that the biofilm communities are regulated by intercellular signals which were first described in 1998 (Davies *et al.*, 1998).

In nature, many species of bacteria, fungi, protozoa and algae form biofilms. However, bacteria have received the most attention and it is estimated that 99% of the bacteria present in natural and pathogenic ecosystems can grow in these metabolically integrated communities. The cell densities commonly found in a biofilm range from 10^5 to 10^8 CFU/cm^2 depending on the bacterial species, the available nutrients, and the underlying substratum corresponding to a thickness in the range of tens to hundreds of µm (Jackson *et al.*, 2001; Sillankorva *et al.*, 2010; Stoodley *et al.*, 2011).

Generally, microbial cells remain viable and metabolically active, although this depends on their position within the biofilm. Moreover, cells can change their phenotype and even the position in the community in order to reach a scenario of better survival possibility. In monospecies and polymicrobial biofilms, the position of individual bacterial cells does not occur randomly. It appears that some undetermined species-specific rules and certain classes of pili can dictate cell positioning. The volume and extent of biofilms are fully dependent on the nutritional and environmental factors present (pH, temperature, nature of biotic and abiotic surface, etc.) and the attachment and biofilm formation by microorganisms is influenced also by signals from adjacent cooperative and competitive bacterial species that affect cell positioning and the activity in the community (Costerton *et al.*, 1987; Lappin-Scott and Costerton, 1995).

Clinically relevant biofilms

Biofilms can occur within an animal host where they can potentially lead to severe and recurrent infection (Donlan and Costerton, 2002) From a medical perspective, both commensal and pathogenic microorganisms form biofilm consortia that are associated with the epithelial or endothelial tissue; embedded in the lung, intestinal or vaginal mucus layer; attached to the teeth or medical implant surfaces, or formed intracellularly. It has been hypothesized that biofilms may be present in more than 60% of microbial infections (Lewis, 2001). For instance, in the sputum and in postmortem lungs of cystic fibrosis patients, biofilms of *Pseudomonas aeruginosa* were observed to consist of cells surrounded by very large expanses of matrix material (Lam *et al.*, 1980). There is also evidence of independent mucosal bacterial communities in the human digestive tract

(Croucher *et al.*, 1983; Macfarlane *et al.*, 2004). The causing agents of osteomyelitis grew well in biofilm communities both in patients as well as in lab animals (Lambe *et al.*, 1991). In general, a number of pathologies in humans, such as endocarditis, urinary tract infections, burn infections, chronic otitis media, chronic bacterial prostatitis, respiratory infections in cystic fibrosis patients and periodontitis have been clearly associated with biofilm related infections. There is a variety of bacterial species identified as causative pathogens of biofilm-associated infections in humans (Tables 13.1 and 13.2).

Single species microbial biofilms can be found in infections of heart valves, catheters and medical prosthetic devices. However, other infections, such as the ones found in different regions of the gastrointestinal tract are typically formed by multispecies biofilm consortia (Macfarlane and Dillon, 2007).

The prevalence of biofilm pathogens depends on the type of infection (Costerton *et al.*, 1999b). Moreover, the biofilm composition varies immensely and while in some infections biofilms can be composed almost exclusively of bacterial cells and matrix material, in others the microbial

Table 13.1 Examples of infections involving biofilms formed on biomedical devices

Biomedical device	Common bacterial species
Arteriovenous shunts	Staphylococci
Cerebrospinal fluid shunts	Staphylococci
Contact lenses	*P. aeruginosa*, Gram-positive cocci
Endovascular catheter	Staphylococci
Intrauterine devices (IUDs)	*Actinomyces israelii* and others
Mechanical heart valves	Staphylococci
Orthopaedic prosthesis	Staphylococci
Peritoneal dialysis catheters	Various species
Sutures	Staphylococci
Urinary catheter	*E. coli* and other Gram-negative rods
Vascular grafts	Gram-positive cocci
Mechanical ventilator	Gram-negative rods
Voice prostheses	Staphylococci, *P. aeruginosa*

Table 13.2 Examples of infections involving biofilms formed on human tissues

Infection	Common causative agents
Bacterial prostatitis	*E. coli* and other Gram-negative bacteria
Biliary tract infection	Enteric bacteria
Chronic tonsillitis	Various species
Cystic fibrosis pneumonia	*P. aeruginosa*, *Burkholderia cepacia*
Dental caries	Acidogenic Gram-positive cocci (*Streptococcus* sp.)
Endocarditis	Viridans group streptococci, staphylococci
Infectious kidney stones	Gram-negative rods
Musculoskeletal infections	Gram-positive cocci
Necrotizing fasciitis	Group A streptococci
Osteomyelitis	Various species
Otitis media	Non-typeable *Haemophilus influenzae*
Periodontitis	Gram-negative anaerobic oral bacteria

consortia can also contain fragmented platelets and host molecules (e.g. fibrin) resulting in the bacteria being widely dispersed within the biofilm (Costerton *et al.*, 2003).

Despite the sophistication of both indwelling and subcutaneous biomedical implants, all medical devices or tissue engineering constructs are susceptible to microbial colonization and infection (Bryers and Ratner, 2004; Castelli *et al.*, 2006); some examples are shown in Table 13.1. One of the consequences of biofilm formation on these devices is the constitution of a microbial reservoir that often leads to symptomatic but non-culturable inflammation, recurrent or persistent infection and spread of infectious emboli. For instance, indwelling catheters are a major route by which healthcare-associated bloodstream infections can develop. The human body reacts to the presence of a biomedical device by coating the device with a film consisting of proteins and glycoproteins, such as fibronectin, vitronectin, fibrinogen, albumin, and immunoglobulins. Upon insertion of the biomedical device, regardless of its material (e.g. glass, silicone, ceramic, titanium, polyether urethane), the surface chemistry is instantly altered by macromolecule adsorption (Rosenberg, 2006; Strevett and Chen, 2003; Verran and Whitehead, 2005). After bacterial cells are transported to the substratum–liquid interface by a combination of transport mechanisms (i.e. diffusion, convection, sedimentation, and motility) (Bryers, 2000), cells will be only reversibly attached to a surface, remaining susceptible to antibiotics (Marshall, 1992). Bacterial cells adhere by non-specific mechanisms such as physico-chemical attraction forces and/or specific receptor-ligand adhesion mechanisms (An and Friedman, 1998; Busscher *et al.*, 1992). After a short period, through a process termed *quorum sensing* (Bjarnsholt and Givskov, 2007; Camilli and Bassler, 2006; Pritchard, 2006), adherent cells up-regulate the secretion of certain cell signal molecules that induce community-wide phenotypic responses such as: secretion of extracellular polymers (Hall-Stoodley and Stoodley, 2005) and up-regulation of virulence factors (Branda *et al.*, 2005; Otto, 2006). Biofilms continue to mature by cell division and recruitment of other bacterial species or mammalian cells and their fragments (e.g. platelets) (Brunstedt *et al.*, 1995) growing in thickness. At this stage, biofilms show maximum tolerance to antibiotics. Clusters of biofilm containing bacterial cells mixed with platelets can detach from the matured biofilm (Sauer *et al.*, 2004) and be carried downstream (a potentially life-threatening event known as a thromboembolism) (Wenzel, 2007).

Antibiotic tolerance of biofilm

In vivo antibiotic efficacy has been predicted by evaluating the minimal inhibitory concentration (MIC) and the minimal bactericidal concentration (MBC) *in vitro*. According to the standard procedures (NCCLS) these determinations are made using planktonic bacteria in the exponential phase of growth, therefore it can be said that MIC and MBC only predict antibiotic efficacy against rapidly dividing bacteria in acute infections, such as septicaemia. However, the properties of biofilm growing bacteria are different from those of planktonic bacteria which has significant diagnostic and therapeutic consequences. Biofilm bacteria are able to tolerate significantly higher levels of antibiotics than planktonic bacteria, antibiotic resistance is reportedly up to 1000-fold greater in biofilm bacterial cells. However, reliable methods to compare the antibiotic sensitivities of planktonic bacteria to cells in biofilms are lacking (Ceri *et al.*, 1999). A study in which the same cell density was used to compare the susceptibility of planktonic *versus* biofilm bacteria to antibiotics reported a 10 fold higher resistance in biofilm bacteria (Cerca *et al.*, 2005). Therefore, very often, the antibiotic concentration needed to eradicate the biofilm is above the peak serum concentration of the antibiotic (Monzón *et al.*, 2002), rendering it ineffective in treating biofilm infections. The clinical relevance of this phenomenon is underscored by the occurrence of biofilm-related infections that are tolerant to antibiotic therapy (Arciola *et al.*, 2002; Monzón *et al.*, 2002). It has even been observed that in the case of *P. aeruginosa* biofilms, aminoglycosides induced the Lon protease which is involved in biofilm formation (Marr *et al.*, 2007).

The mechanisms of biofilm tolerance to antibiotherapy are still not completely understood. It should be emphasized that the current understanding of the clinical relevance of biofilm-associated infections is mainly based on indirect evidence derived from laboratory findings, as well as on clinical observations regarding foreign body infections, which are presumed to be biofilm-associated. The success of an antimicrobial agent is dependent upon its ability to inactivate and remove biofilm organisms. Several explanations for biofilm tolerance focus on the diffusion limitation of the antibiotics through the biofilm matrix since positively charged antimicrobials can bind to the negatively charged biofilm matrix, which prevents the agents from reaching the cells within a colony (De Beer *et al.*, 1994a; Hoyle *et al.*, 1992; Suci *et al.*, 1994). Indeed, certain antibacterials show delayed or reduced penetration into biofilms. However, other antibacterials may adequately diffuse into biofilms through water channels (Donlan and Costerton, 2002; Sutherland, 2001b) but still cannot kill the cells in the biofilm (Anderl *et al.*, 2000; Walters *et al.*, 2003) and thus it is not well established whether they can adequately penetrate into microbial cell clusters (De Beer *et al.*, 1994c; Sutherland, 2001b; Lacqua *et al.*, 2006).

Other factors that may account for the increased antibiotic tolerance of sessile communities include: slow growth of the cells; stationary phase of bacterial growth; specialized persister cells which can lead to cycles of biofilm re-growth after the antibiotherapy (Lewis, 2008).

Due to the failures of antibiotherapy against bacterial biofilms and the emergence of antibiotic resistance of pathogenic bacteria and their prominent threat to public health, there is today a renewed interest in alternative therapeutic modalities to treat biofilm associated infections. Possible alternatives to synthetic antibiotics include probiotics (Hong *et al.*, 2005), human antimicrobial peptides (defensins, cathelicidins, and histatins) (De Smet and Contreras, 2005), and bacteriophage and bacteriophage lytic enzymes (Fischetti, 2008; Kutateladze and Adamia, 2010).

Biofilm–phage interactions

The interactions between phage and biofilms are a rather complex process (Sutherland *et al.*, 2004). Theoretically, a biofilm should be rapidly infected because cells are closer to each other which can enhance phage–host interaction, when compared to the less accessible bacteria of planktonic cultures (Wiggins and Alexander, 1985). On the other hand, the structure and composition of the biofilm as well as the physiology of the biofilm cells may impose some limitations on phage infection. In fact, recent findings that bacterial strains tolerant to phage infection increase their biofilm formation ability, have suggested that the biofilm phenotype might be an additional strategy of bacteria to escape phage infection (Lacqua *et al.*, 2006). Nevertheless, there are several studies showing that phages can indeed infect biofilms *in vitro* (Ahiwale *et al.*, 2011; Cerca *et al.*, 2007; Chibeu *et al.*, 2012; Doolittle *et al.*, 1995; Hughes *et al.*, 1998a; Kelly *et al.*, 2012; Pires *et al.*, 2011; Sillankorva *et al.*, 2004b, 2010; Tait *et al.*, 2002; Thawal *et al.*, 2012; Yele *et al.*, 2012). These studies show that in natural environments phages are equally able to act on their hosts when they are in a biofilm form. Based on experimental data one can hypothesize that the general mechanism of a virulent phage infection of biofilm, in a very simplistic model, includes four stages (Fig. 13.2):

1 Transport of the phage particles through the biofilm matrix (by diffusion or convection mechanisms). One of the first biofilm–phage interaction studies performed by Doolittle *et al.* (1996b) showed, using fluorescent labelled phage particles, that progeny phage propagate radially through a biofilm.
2 Settlement and/or attachment of phages onto bacterial cells embedded in the biofilm matrix, followed by adsorption and phage replication inside host cells.
3 Release of phage progeny to planktonic and biofilm phases, through host cell lysis and infection of neighbourhood biofilm cells resulting in biofilm biomass reduction.
4 Detachment of biofilm portions and phages into the planktonic phase (Fig. 13.2).

Under suitable conditions, the presence of bacterial cells in a biofilm does not prevent access to and infection by phage particles. However, the outcome of phage adsorption and entry to the

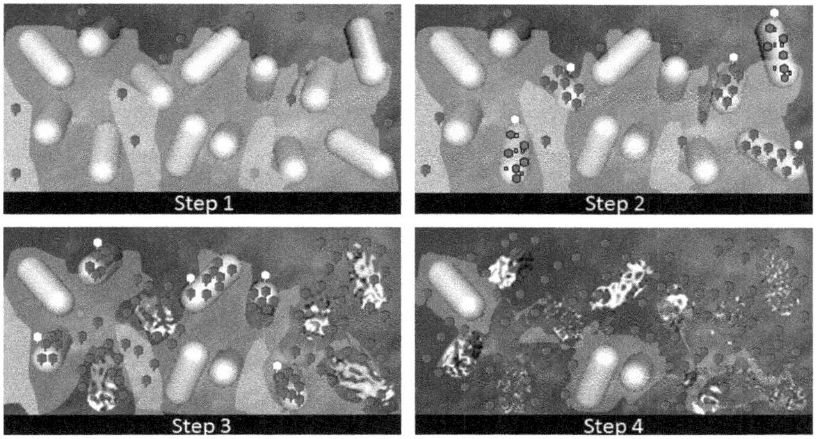

Figure 13.2 Model of biofilm infection by phage. Step 1: phage transport through the biofilm matrix and water channels through diffusion and/or convection; Step 2: Settlement of phages and phage cycle progression in the biofilm host cell; Step 3: Release of phage progeny through host cell lysis and infection of neighbourhood biofilm cells causing biomass reduction; Step 4: Detachment of biofilm portions and phages into the planktonic phase and phage cycle progression in planktonic cells. A colour version of this figure is available in the plate section at the back of the book.

host bacterium depends on a number of different factors, such as:

The biofilm matrix

The polymeric matrix in which the microbial cells are embedded consists of various polymers in addition to exopolysaccharides (Flemming *et al.*, 2007). This poses a problem for the adsorption of phages to their hosts and thus, the phage must be able to penetrate the matrix either by diffusion or with the aid of enzymes associated with the viral particles (Hughes *et al.*, 1998a) and destroy enough amount of the matrix to release the embedded bacteria and make them vulnerable to the phage.

According to Sutherland *et al.* (2004), many biofilms possess an open architecture with water-filled channels, which allows phage access to the inner biofilm layers. The application of confocal laser scanning microscopy and similar techniques to study biofilms altered the perception of biofilm structure and function (Lawrence *et al.*, 1991) demonstrating the micro-heterogeneity of biofilms with variable distribution of cells, matrix and fluid-filled channels and pores (Wood *et al.*, 2000). Hu and colleagues (Hu *et al.*, 2010) used a simple two-chamber diffusion method to study the diffusion properties of phages and concluded that the diffusion coefficient was dependent on agarose concentration. Thus, phage effectiveness will depend not only on the extent to which the matrix is composed of polysaccharides but also the way in which they interact with all the other polymers and ions in the matrix.

Doolittle *et al.* (1996) demonstrated that while T4 phage particles moved through *E. coli* biofilms, the access of phage E79 to cells of *P. aeruginosa* in deeper layers of the biofilms was somewhat restricted as this phage only adsorbed to cells on the outer surface of the biofilm. More recently, the research group of Fontaine-Aupart has published two papers on the use of fluorescence correlation spectroscopy (FCS) under two-photon excitation to characterize the diffusion of phages in biofilms of *Stenotrophomonas maltophilia* and to study the reactivity of phages when they come into contact with the embedded bacteria (Lacroix-Gueu *et al.*, 2005; Briandet *et al.*, 2008). Both studies show that phages penetrate and diffuse inside the matrix of mucoid biofilms and suggest that, like phages, other pathogenic viruses must also benefit from the biofilms' protective characteristics towards antimicrobials and that the biofilms which harbour pathogenic viruses are a probable cause of persistent infections (Lacroix-Gueu *et al.*, 2005; Briandet *et al.*, 2008).

To overcome the presence of the biofilm matrix, some phages produce polysaccharide depolymerases. These enzymes degrade the exopolymeric matrix, allowing phages to penetrate through the exopolysaccharide layers in order to reach the cell surface of their hosts. On plating out phages possessing depolymerase activity, the plaques obtained from single infected bacterial cells appear surrounded by haloes of varying size. The haloes represent susceptible bacteria from which the associated EPS has been removed by excess soluble enzyme released on lysis of each infected bacterium. Polysaccharide depolymerases (Adams and Park, 1956; Eklund and Wyss, 1962; Mare and Smit, 1969) can either be produced during phage replication and released into the matrix after cell lysis (Bartell *et al.*, 1966; Kimura and Itoh, 2003) or be associated with the phages' tail spikes (Eklund and Wyss, 1962; Mare and Smit, 1969). In general, depolymerase activity results in a reduction of bacterial slime viscosity and a release of monomeric sugars from the substrate (Bartell *et al.*, 1966). For instance, in 1956, Adams and Park (1956) showed that certain *K. pneumoniae* phages produced a depolymerase able to target the bacterial capsular polysaccharide, allowing the phage to penetrate the capsule and reach the host cell surface. Sutherland (1967) has also described a phage-induced depolymerase specific for the exopolysaccharide and capsular polysaccharides of *K. aerogenes* strains. Bartell *et al.* (1968) isolated the first *P. aeruginosa* depolymerase, active against polymers of a few strains and found that the activity resided in a 180 kDa component most likely a particulate component of the phage tail spike assemblage consisting of either three or six enzyme subunits (Sutherland, 1977). In literature, there are several other studies also reporting the positive effects of phage-associated depolymerases on the adsorption of phages to bacterial species protected by a slime layer (Bartell *et al.*, 1966; Kimura and Itoh, 2003; Mare and Smit, 1969; Reese and *et al.* 1974). For instance, Hanlon and colleagues (2001) have shown, *in vitro*, that *P. aeruginosa* phages diffused through alginate and that phage-free suspensions decreased the viscosity and the molecular weight of purified alginate suggesting that these phages almost certainly carry alginate-degrading enzymes. Wong *et al.* (2000) have also identified a range of alginate lyases of *P. aeruginosa* that differ significantly in their mode of polysaccharide degradation, substrate specificity, molecular mass, pH and cation requirements. These authors also suggest that the enzymes most likely to be used therapeutically would be those that have co-evolved with the bacterial hosts. The therapeutic use of phages possessing polysaccharide depolymerases could prove beneficial to treat biofilm-related infections, such as those caused by *P. aeruginosa* where cells produce copious amounts of O-acetylated alginic acids (Wong *et al.*, 2000) which mediate attachment to epithelium (e.g. tracheal) and enable bacteria to grow as a genetically heterogeneous and antibiotic refractory biofilm community (Atabek and Camesano, 2007; Costerton *et al.*, 1999; Mah *et al.*, 2003). The polysaccharide depolymerase of the *Enterobacter* (now *Pantoea*) *agglomerans* phage SF153b was shown to be highly active under different mono and dual-species conditions which suggests that phages gain access to the receptors on the cell wall and infect the bacteria even in mixed consortia (Tait *et al.*, 2002). EPS-degrading activity has recently been reported in two *Staphylococcus epidermidis* phages (Gutiérrez *et al.*, 2012) carrying a pectin lyase-like domain in the tail fibre and in the C-terminal part of Gp19 of *P. putida* phages AF and φ15 (Cornelissen *et al.*, 2011, 2012).

The potential of phages to degrade the biofilm matrix by depolymerases may prove superior to chemotherapy for the treatment of biofilm related infections. As shown in the examples described above, phages can lyse biofilm cells and also cause a disintegration of the biofilm (Sillankorva *et al.*, 2004b; Tait *et al.*, 2002) and this disruption of the biofilm structure would, in principle, enable the host immune system to resolve the infection and even restore the efficacy of antibiotics.

Some studies have also shown that the biofilm exopolymers could protect phages from being cleared and inactivated by the immune system. For instance, Landy *et al.* (1958) showed that a variety of polysaccharides, including polysaccharides produced by bacteria, blocked the phage-neutralizing activity of human serum. However, the biofilm matrix is also a reservoir of proteolytic enzymes as well as endoglucanases which can lead to bacteriophage inactivation (Azeredo and Sutherland, 2008). The phage morphology can also be a factor that plays an important role in phage diffusion through the matrix. For example, using fluorescence correlation spectroscopy under two-photon excitation Lacroix-Gueu *et al.* (2005) showed that the diffusion of phage C2 (infecting *Lactococcus lactis* subsp. *lactis*) in *Stenotrophomonas maltophilia* biofilms was slowed due to the long tail of the phage.

Another important property of the biofilm matrix that may influence phage action is its charge and hydrophobicity (Sutherland *et al.*, 2004). In a study of coliphage interaction with both hydrophobic and hydrophilic membranes, it was observed that a critical factor in the retention of the phage was its isoelectric point (van Voorthuizen *et al.*, 2001). That study also showed that retention of viral particles on hydrophilic membranes was dominated by hydrophobic interactions and further increased with high ionic strength (van Voorthuizen *et al.*, 2001).

Physiological state of the host bacteria

The success of phage replication in a biofilm is affected by the reduced metabolic activity of biofilm cells, especially those in the inner layers oxygen and nutrient deprived. Experiments comparing phage infection of planktonic and biofilm cells have shown that planktonic cells are more rapidly lysed and the rate of biomass reduction is dependent on the physiological state of the cells rather than on the diffusion barrier imposed by the biofilm matrix (Cerca *et al.*, 2007; Sillankorva *et al.*, 2004a). Nevertheless, despite the rate of cell lyses upon phage attack being slower in biofilms than in planktonic conditions, equivalent biomass reductions have been observed (Cerca *et al.*, 2007; Sillankorva *et al.*, 2004a). Phage infection and life cycle generally strongly depend on the growth stage of the host bacterium (Los *et al.*, 2007; Sillankorva *et al.*, 2004b). Phage infection is considerably faster in rapidly growing bacteria which poses a challenge for the treatment of slowly growing cells in biofilms. The production of phages inside stationary cells commonly results in a smaller burst size (Hadas *et al.*, 1997; Sillankorva *et al.*, 2004b). The reduced lysis of cells near or at the stationary phase can also be due to the reduction of phage adsorption sites and thickening of the cell wall. The reduction of phage adsorption sites has also been ascribed to adaptation of bacteria for example to differing nutrient conditions (Spanakis and Horne, 1987). Nevertheless, it has also been suggested that with the sloughing off of dead cells from biofilms, potential new viral receptor sites may become available (Sutherland *et al.*, 2004). On the other hand, studies have shown that adsorption of phages to stationary growth phase cells is lower than to exponentially growing cells (Hadas *et al.*, 1997). Other possible causes of less effective lysis of stationary cells may be the cell wall thickening and even an increased number of abortive infections. As early as 1958, Wilkinson (Wilkinson, 1958) showed that phage resistance in bacteria correlated with the thickness of their capsule. In his studies, he observed that the thickness of the capsule was greater than the length of the phage tail which prevented the injection of the phage DNA into the cytoplasm of the bacteria.

Furthermore, the presence of dead cells can also be considered a difficulty for the progression of the phage infection, as they provide potential viral receptor sites. In general, older biofilms containing more dead and stationary phase cells are more difficult to kill (Sillankorva *et al.*, 2004a). Nevertheless, recent work has shown a successful phage lysis of matured biofilms of *P. fluorescens* by phage phiIBB-PF7A (Fig. 13.3), a T7-like phage which caused nearly a 5 $\log_{10}$ reduction of viable cells in 7-day-old biofilms (Sillankorva *et al.*, 2008). Also persister cells have been demonstrated to be susceptible to bacteriophage action (Pearl *et al.*, 2008). These fairly recent successful experiments demonstrate another advantage of some phages – their ability to infect stationary cells – over antibiotics which fail to kill both planktonic and biofilm cells at the stationary phase (Spoering and Lewis, 2001).

Bacterial resistance and co-existence of the hosts with their phages

Some studies have explored the role of biofilms in the development of phage resistance. After exposure of biofilms to phages, depending on the exposure period, the emergence of subpopulations

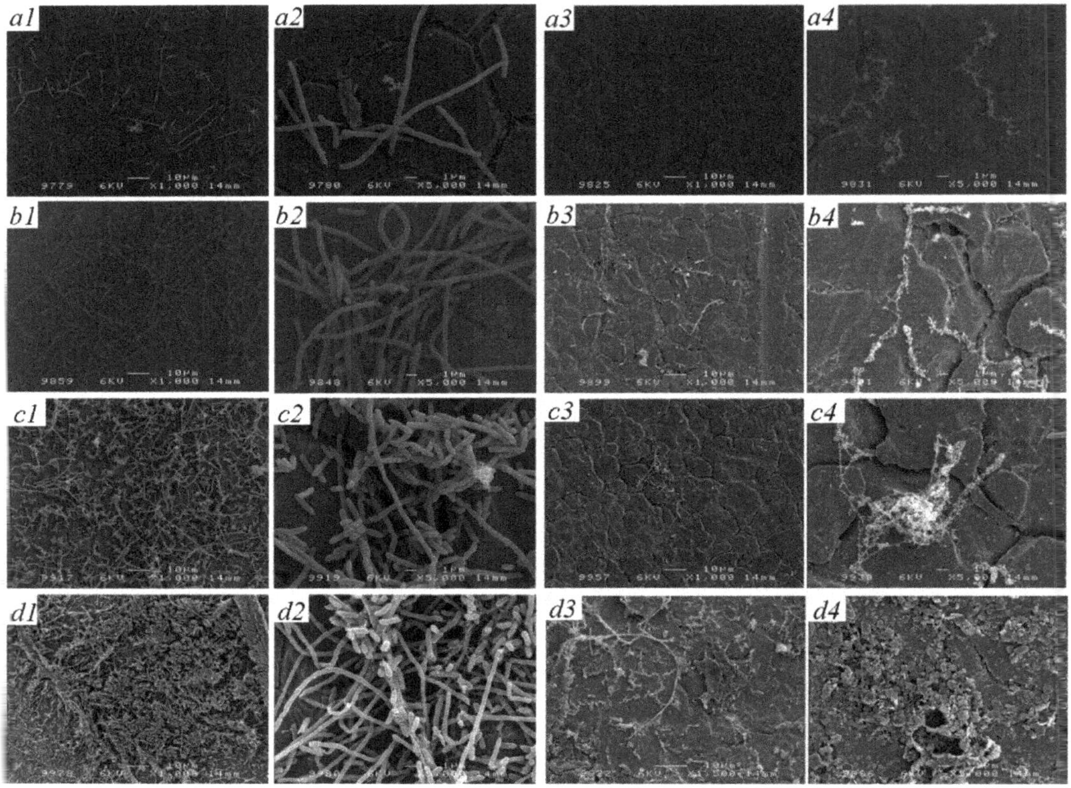

Figure 13.3 Successful *P. fluorescens* biofilm removal from stainless steel surfaces by a T7-like phage (phiIBB-PF7A) observed by field emission scanning electron microscopy (FSEM). (a) 1-day, (b) 3-day, (c) 5-day and (d) 7-day old biofilms; 1 and 2 are biofilms before phage treatment (magnification 1000× and 5000×, respectively); 3 and 4 are biofilms after 4 h of phage treatment (magnification 1000× and 5000×, respectively). a2, b2, c2 and d2 have been previously published (Sillankorva *et al.*, 2008).

with a stable phenotype and resistant to lysis is common. Lacqua *et al.* (2006) showed that the subpopulations developed after exposure of *E. coli* biofilms to two different lytic phages (OP7061 and OP10081) exhibited increased clumping in liquid media, enhanced biofilm formation and produced fimbria-like extracellular structures. All these characteristics, present in the biofilm subpopulations, were absent in the parent strain suggesting that growth as a biofilm might be a mechanism of phage resistance. It has been suggested that the use of phage cocktails can reduce bacterial resistance (Fu *et al.*, 2010). These cocktails may be developed to exhibit a broad spectrum to cover the range of the varying host mutants that arise from the exposure to phages. Application of cocktails reduces the emergence of phage-resistant phenotypes and also increases the efficacy of phages for mono and dual species biofilm control.

In a study describing the interaction of phages with dual species biofilms, the treatment of biofilms with phages and polysaccharide depolymerases was unsuccessful. These authors observed that the structural heterogeneity of the biofilm produced pockets of unattainable susceptible bacteria (Tait *et al.*, 2002).

Besides the biological mechanisms already described, many other factors can affect phage reproduction inside the biofilm cells, including host cell density, availability of nutrients, and temperature, among others (De Paepe and Taddei, 2006; Moebus, 1996; Sillankorva *et al.*, 2004b). These also have a tremendous impact on the efficacy of phages for biofilm control.

It has also been shown that bacteria and phage

can co-exist stably in biofilms. Such co-existence occurs not only after exposure of single-species biofilms to phages, but also in more complex biofilm structures (Tait *et al.*, 2002).

Resch *et al.* (2005) did not focus their studies on phages; however, in their works with lysogenic *S. aureus* planktonic cultures and biofilms, they found that both types of cells occasionally released phages resembling the *S. aureus* phages φ11 and possibly φ13. Furthermore, while phages in planktonic phases were degraded by secreted proteases, phages in biofilms were detected over a long period. Also, Webb *et al.* (2004) observed that the phenotypic variation in *P. aeruginosa* biofilms was a response to Pf1-like filamentous phage, which increased the number of small colony variants, exhibiting enhanced attachment and accelerated biofilm development.

In vitro efficacy studies of phages and phage-derived enzymes

Although it is a known fact that bacterial lifestyle on surfaces adopts the form of biofilms, there is a limited number of studies on phage–bacteria interactions performed with sessile communities compared to their planktonic counterparts. Since phage discovery, in the beginning of the 20th century, one of the first phage–biofilm interaction studies was reported only in 1995 by Doolittle *et al.* (1995) (Table 13.3). In this study, *E. coli* biofilms developed on the surface of polyvinylchloride using a modified Robbins device, were targeted with T4 phage which caused an overall 6 $\log_{10}$ reduction of biofilm viable counts (from the initial 10^8 CFU/ml to 10^2 CFU/ml) after 5 h of treatment with a low multiplicity of infection (MOI calculated to be about 0.6–0.7). Since then, several similar studies have been performed to evaluate the effects of both *Podoviridae* (T7-like) and *Myoviridae* (T4-like) phages on *E. coli* biofilms, but none of them reported such a great reduction in viable cell counts. Table 13.3 summarizes the results of several *in vitro* studies on the efficacy of phages against mono and dual-species biofilms. The majority of these phage–biofilm interaction studies have been performed with *E. coli* biofilms, and the most commonly used phage genus has been the T7-like phage genus, which besides *E. coli* has also been used to control *Pseudomonas* sp., and *K. pneumoniae* (Table 13.3). The choice of using T7-like phages for biofilm control experiments may be due to the fact that T4-like phages are not so effective in lysing stationary cells (Los *et al.*, 2003; Chan *et al.*, 2007) and therefore their efficacy is limited to young biofilms. Furthermore, T7 phages aside from their high rate of development compared to T4 phages are also interesting for their small genome, which decreases the chance of transferring extended fragments of chromosomal DNA with pathogenicity islands to host strains (Krylov, 2001).

In the first studies on mixed biofilms performed by Tait *et al.* (2002), dual species biofilms of *Enterobacter* (now *Pantoea*) *agglomerans* Ent and 5920 were formed and infected with a cocktail of three phages. The cocktail was highly effective and caused a 6 $\log_{10}$ reduction of *E. cloace* 5920. Since then, only one study of phage interaction with dual species biofilms has been performed (Sillankorva *et al.*, 2010). These authors used a phage cocktail against biofilms consisting of Gram positive and Gram negative hosts and observed a significant reduction in the viable counts of biofilm and planktonic cells which had been released from the biofilms. Furthermore, when challenging the mixed consortia with a single phage for the less predominant bacteria, these authors observed an increase of the number of non-susceptible host in the planktonic phase showing that this phage besides killing the susceptible host, could also cause a partial destruction of the biofilms.

The first studies on preventing bacterial adhesion and consequent formation of biofilms were performed by Curtin and Donlan (2006). These authors pretreated hydrogel-coated silicone catheters with phages and observed a 4.47 $\log_{10}$ reduction in the mean viable count per cm^2 of *S. epidermidis* after 24 h compared to biofilms formed on untreated surfaces. The same strategy was adopted by Fu *et al.* (2010) against *P. aeruginosa*. The authors demonstrated that catheter surfaces pretreated with phage M4 or a cocktail of phages had less 4 $\log_{10}$ CFU/cm^2 and 3 $\log_{10}$ CFU/cm^2 after 24 h with phage M4 and 48 h with the phage cocktail respectively.

Apart from unmodified virulent phages, also

Table 13.3 Studies of *in vitro* effects of phages on mono- and dual-species biofilms

Bacteria	Phage	Experimental approach	Results	Reference
Single species biofilms				
E. coli 3000XIII	T4D	28 h biofilms treated with 10^9 or 10^{10} PFU per ml of phage for 30 min to 8 h	6 log_{10} reduction in biofilm viable count within 5 h	Doolittle *et al.* (1995)
E. agglomerans	SF53b	24 h biofilms (10^6 CFU per cm^2) treated with 10^{10} PFU per ml of phage for 15 min which was then washed out from the system.	3 log_{10} reduction in biofilm viable count in 5 h (chemostat and batch)	Hughes *et al.* (1998b)
E. coli K12	T4	72 h biofilms treated with phage at a MOI of 10 or 100 for 30 min	3 log_{10} reduction in biofilm viable count. Biofilm regrowth after 3 h. A MOI of 100 resulted in greater change in the biofilm structure.	Corbin *et al.* (2001)
P. aeruginosa NCIMB10548	F116 (*Podoviridae*)	Biofilms of different ages treated with phage at a MOI of 100 or 1000 for 24 h	A MOI of 100 caused 1 log_{10} reduction in biofilm viable count; a MOI of 1000 caused 2 log_{10} reduction	Hanlon *et al.* (2001)
P. fluorescens ATCC 27663	phiS1	5 day biofilms treated with 10^9 PFU per ml of phage for 0–200 min at different temperatures	Up to 84% reduction in biofilm biomass after 200 min, with greatest reduction at 26°C	Sillankorva *et al.* (2004a)
E. coli O157:H7	KH1	72 h biofilms treated with 10^7 PFU per ml of phage for 1–4 days	1.2 log_{10} reduction in biofilm viable count	Sharma *et al.* (2005)
S. epidermidis (several strains)	456	Pre-treatment of catheter surfaces with phages	4.5 log_{10} and 2.3 log_{10} reduction in catheter colonization when pre-treated with phages and divalent cations and phages alone, respectively	Curtin and Donlan (2006)
E. coli TG1	T7 and T7 DspB	24 h biofilms treated with 10^1–10^5 PFU per ml of phage for 0–50 h	Greater than 3 log_{10} reduction in biofilm viable count over 48 h. T7 phage engineered with dispersin B gene was most effective	Lu and Collins (2007)
S. epidermidis (several strains)	K (*Myoviridae*)	24 h biofilms treated with 10^8 PFU per ml of phage for 24 h	Significant reduction in biofilm biomass of several strains assessed by crystal violet staining	Cerca *et al.* (2007)
P. fluorescens	philBB-PF7A (T7-like)	1 up to 7 day old biofilms formed under different growth conditions treated for 4 h with phage (10^7 PFU per ml)	1 log_{10} up to 5 log_{10} reduction of viable cell counts after 4 h depending on the biofilm age and phages infection conditions	Sillankorva *et al.* (2008)
K. pneumoniae B5055	KPO1K2 (T7-like)	12 h biofilms subjected to antibiotic (ciprofloxacin) (10 μg/l) and phage (MOI of 1) treatment, alone or in combination	Antibiotic caused 1 log_{10} reduction of viable cell counts after 3 h, phages alone reduced viable cell counts by 6 log_{10} and combined therapy resulted in a 5 log_{10} reduction of viable cell counts	Verma *et al.* (2009)
K. pneumoniae B5055	Unidentified	1- to 8-day-old biofilms treated with a MOI of 0.01 and amoxicillin	256 μg per ml of amoxicillin reduced planktonic cultures and 1-day-old biofilms by 4 and 1 log_{10}, and was ineffective against 8 day-old biofilms. 512 μg per ml of amoxicillin reduced by 3 log_{10} the viable cell counts and phages alone (MOI = 0.01) reduced by 3.5 log_{10} the number of viable cell counts in 8 day old biofilms	Bedi *et al.* (2009)

E. coli	M13mp18	24 h biofilms treated with 10^1–10^5 PFU per ml of phage for 0–50 h	The engineered bacteriophage together with ofloxacin enhanced the killing of biofilm cells by about 1.5 and 2 $\log_{10}$ compared with the non-engineered phage plus ofloxacin and ofloxacin alone, respectively	Lu and Collins (2009)
E. coli ATCC 11303 and *P. mirabilis* 13 HER1094	T4 for *E. coli* and unidentified phage for *P. mirabilis*	10^6 CFU of a single species biofilm populations treated with 200 ml of bacteriophage at 10^6 PFU per ml, 10^3 PFU per ml or 10 PFU per ml	Up to 4 $\log_{10}$ reduction of viable cell counts in *E. coli* biofilm (at the highest phage titre used) and 1 $\log_{10}$ reduction of viable cells in *P. mirabilis* biofilm	Carson *et al.* (2010)
K. pneumoniae B5055	KPO1K2 (T7-like)	1- up to 7-day-old biofilms subjected to antibiotic (ciprofloxacin) (10 µg per l), phage (MOI of 1) and phage depolymerase (phiNDP) treatment, alone or in combination	1 $\log_{10}$ up to 6 $\log_{10}$ reduction of viable cell counts depending on the biofilm age. A synergistic effect of ciprofloxacin and phage was observed in 3-day-old biofilms	Verma *et al.* (2010)
P. aeruginosa	M4	Pretreatment of hydrogel-coated catheters for 2h with 10^{10} PFU per ml of phage or phage cocktail	M4 phage pretreated surfaces had less 4.03 $\log_{10}$ CFU cm^2 after 24 h than untreated surfaces, and phage cocktail pretreated surfaces had less 3 $\log_{10}$ CFU cm^2 after 48 h	Fu *et al.* (2010)
S. aureus	cell-wall degrading phage enzyme (SAL-2)	5, 10, 15, and 20 µg of SAL-2 per well tested against biofilms (48 h). The SAL-2 incubation period was 2 h.	SAL-2 (1 µg per ml) showed lytic activity against *S. aureus*, including MRSA strains	Son *et al.* (2010)
P. aeruginosa	BVPaP-3	Prevention of biofilm formation using phage at a MOI of 0.001 and 24 h control of biofilms (18 h old) with phages at MOI of 200	Phage at a MOI of 0.001 prevented biofilm formation by a clinical strain. The dispersal of clinical strains, evaluated by determining the percentage of log reduction, varied with the strain used but in all cases a significant disruption was observed	Ahiwale *et al.* (2011)
P. aeruginosa	philBB-PAA2, philBB-PAP21	Control of 24 h and 48 h biofilms using phage at a MOI of 1.0	Phage efficacy towards biofilms was phage and host dependent. Phage philBB-PAA2 reduced gradually the viable cells by 3 $\log_{10}$ by 24 h of treatment with no emergence of resistant strains. Maximum viable cell reduction was observed after 6 h of treatment with phage philBB-PAP21 but at 24 h resistant phenotypes emerged.	Pires *et al.* (2011)
S. aureus	SAP-26	2 up to 24 h treatment with phage (10^8 PFU) alone or combined with antibiotic (rifampicin) (0.6 mg/l) of biofilms (24 h)	Phage alone killed 28% of the biofilm bacteria. Combined treatment caused a strong biofilm removal effect killing 65% of the viable cells.	Rahman *et al.* (2011)
S. aureus	Phage K and 6 modified derivatives	Control of 24 h up to 72 h biofilms and prevention of biofilm formation using a mixture of phages (unknown concentration or MOI)	Biofilm biomass removal significantly greater after 72 h than after 24–48 h	Kelly *et al.* (2012)
A. baumannii	AB7-IBB2	Prevention using phage at a MOI of 1 and control of biofilms	Phage inhibited by 50% and 60% using a MOI of 0.1 and 10. Phage reduced up to 80% of viable cells from biofilm.	Thawal *et al.* (2012)
A. baumannii	AB7-IBB1	Biofilm inhibition on abiotic surface and on human HK293 cell line using 10^1 to 10^7 PFU of phage per well and biofilm control	Phage inhibited by 40 and more than 50% using a MOI of 0.1 and 100. Phage inhibited by 50% biofilm formation on HK293. Phage removed by 75% preformed biofilms	Yele *et al.* (2012)

Table 13.3 (Continued)

Bacteria	Phage	Experimental approach	Results	Reference
E. coli	vB_EcoP_ACG-C91, vB_EcoM_ACG-C40, vB_EcoS_ACG-M12	Control of established biofilms using phage solutions of 10^5, 10^7 and 10^9 PFU per ml	All phages reduced within 2–12 h of incubation the number of viable cells of strain Can91. The reduction was not dose-dependent as low phage concentrations were as effective as high concentrations	Chibeu *et al.* (2012)
S. aureus (MRSA) *P. aeruginosa*		48 rats were divided into 4 groups. 12 rats received phage and antibiotic (20 mg/kg/day of teicoplanin for MRSA; 120 mg/kg/day imipenem + cilastatin treatment + 25 mg/kg/day for *P. aeruginosa*). 12 rats received phage alone (local injection of 10^7 PFU of phage per 0.1 ml suspension) on 3 consecutive days. Treatments during 15 days	MRSA biofilm was absent only in the antibiotic + bacteriophage subgroup. In *P. aeruginosa* biofilm the number of viable cells after combined antibiotic-phage treatment was significantly lower than the corresponding values in the other subgroups	Yilmaz *et al.* (2013)
S. aureus	Romulus, Remus, ISP	Control of biofilms (24 h) with 10^7 and 10^9 PFU of phage per peg	10^9 of ISP, Romulus and Remus phage per peg caused a 37.8%, 34.4% and 60.4% biofilm biomass destruction	Vandersteegen *et al.* (2013)
Dual species biofilms				
E. agglomerans and *E. cloacae* 5920	11229, ϕ*Ent*, ϕ1.15, Blackburn, Philipstown	16-h single (10^8 CFU/cm^2) and dual species (10^9 CFU/cm^2) biofilms treated with single phages and a phage cocktail using a MOI of 0.001 up to 0.1	Combinations of three phages completely eradicated single *E. cloacae* biofilms and decreased by 6 log_{10} *E. cloacae* 5920 in a mixed biofilm	Tait *et al.* (2002)
P. fluorescens + *S. lentus*	philBB-PF7A (T7-like) for *P. fluorescens* and phiBB-SL58B (unidentified *Podoviridae*) for *S. lentus*	24 and 72 h biofilms formed under different steering and mode of growth (batch and fed-batch)	1.5 log_{10} up to 4.2 log_{10} reduction of viable cell counts after 4 h depending on the biofilm age and phages infection conditions	Sillankorva *et al.* (2010)
P. aeruginosa	ΦE2005-A	Urinary catheters pretreated with benign *E. coli* HU2117 and/or phage and incubated up to 72 h in human urine	*P. aeruginosa* adherence was 4 log_{10} units lower when pretreated with *E. coli* plus phage compared to the control after 24h and 3 log_{10} units lower after 72 h. Phage alone did not generate significant decrease in bacterial adherence	Liao *et al.* (2012)

CFU, colony-forming units; PFU, plaque-forming units.

genetically engineered phages could be used to eliminate biofilms. Lu and Collins (2009) genetically engineered phage M13mp18 specific for *E. coli* and showed that this engineered phage was able to suppress the SOS network and enhanced killing by quinolones against both planktonic cells and biofilm cells *in vitro*.

Recently, the application of phage lysins to control biofilms of Gram positive bacteria has also been investigated. Sass and Bierbaum (2007) used two recombinant endolysins (φ11 and φ12) and demonstrated that φ11, that seems to be a multifunctional hydrolase, hydrolysed heat-killed staphylococci and staphylococcal biofilms (except *S. epidermidis* O-47 biofilms) while φ12 was unable to exert efficient hydrolytic activity on staphylococcal murein and caused aggregation of the cells. Furthermore, the effect of φ11 on *S. aureus* NCTC8325 was comparable to or slightly lower than that of lysostaphin, an endopeptidase from *S. simulans* that has antimicrobial activity against *S. aureus*. Son *et al.* (2010) have also studied the antibacterial and biofilm removal activity of phage SAP-2 (φ29-like phage) and its cell wall degrading enzyme (SAL-2) and reported that SAL-2 had a broader spectrum of activity than phage SAP-2 in its ability to remove the *S. aureus* biofilms. As a conclusion of their work, the authors also suggest that this enzyme could potentially be used to treat and prevent *S. aureus* biofilm related infections alone or in combination with other enzymes with anti-*S. aureus* activity.

Overall, the different published results are difficult to compare because of the varying conditions of biofilm formation and phage infection. For example, in some experiments phage infections were performed after scrapping the biofilms from the surfaces, while in others phages were applied directly on the surfaces, where biofilms were formed. Furthermore, in many reported experiments, it is also not clear what type of phage was used in the assays since it is even not known to what family of phages they belong.

Potential application of phages to control biofilms

Several approaches have been suggested to apply phages as biocontrol agents of bacterial biofilms. The first ones were proposed in the 1990s in two US patents in which the inventors suggested different methods to deliver a phage cocktail against *S. sanguinis* and other pathogens involved in dental plaque formation by incorporating phages into dental tablets, toothpaste, chewing gum or mouth wash solutions (US patents 4891210 and 4957686). Treatment of biofilm infections causing skin ulcers by incorporating phages in wound dressings is another approach currently used in some Eastern European countries and US (Kutateladze and Adamia, 2008). Moreover, a pioneering clinical trial for the treatment of burn wound infections is being carried out at the Burn Centre of the Queen Astrid Military Hospital in Brussels, Belgium, using a highly purified and fully defined bacteriophage cocktail (BFC-1) (Merabishvili *et al.*, 2009).

The impregnation of the surface of a medical device with a phage solution is another strategy suggested by Curtin and Donlan to control infectious biofilms associated with the use of indwelling devices (Donlan and Curtin, 2006). These authors showed that incorporating a coagulase-negative staphylococci phage into the hydrogel coating of a catheter can significantly reduce the attachment and biofilm formation by *S. epidermidis* (the phage host). The same strategy was later followed by Donlan's team to control *P. aeruginosa* biofilms formed on catheters (Fu *et al.*, 2010). A similar approach was recently suggested by Son *et al.* (Son *et al.*, 2010; Yoon *et al.*, 2009). The authors used phages SAP-1, SAP-2 and lytic proteins derived from those phages to soak and spray catheters as a way to prevent *S. aureus* biofilm formation.

All these approaches are based on the direct application of phage particles onto the infectious biofilm. However, if phages were able to pass through the gut barrier reaching the peripheral lymph and blood, invasive administration methods would be avoided. This would be of particular importance for the treatment of infections caused by biofilms formed on internal implants such as pacemakers and internal prostheses. Górski *et al.* (2006) suggested that phages can translocate by a process similar to the well-known bacterial translocation. This hypothesis is supported by the results of several studies in which phage particles were detected in human body (e.g. saliva, urine, faeces and human sera) (Dabrowska *et al.*, 2005). Bruttin and Brussow (2005) were, however, unable to detect T4 coliphages in the blood of volunteers following oral administration. According to Górski *et al.* (2006) the phage translocation may be determined by the phage type.

Despite several studies showing phage efficacy against mono and multi-species biofilms there are only few commercial preparations of phages and phage-derived enzymes. The most well-known commercial product is PhagoBioDerm®, a biodegradable polymer wound dressing composed of ciprofloxacin, benzocaine, chymotrypsin, bicarbonate, and six lytic phages. These phages have shown activity against *S. aureus*, *P. aeruginosa*, *E. coli*, *Streptococcus* species, and *Proteus* species. Another phage cocktail under evaluation is WPP-201, a mixture of eight lytic phages that possess the same spectrum of activity as PhagoBioDerm®.

Another example is a preparation consisting of 6 phages – (Biophage-PA) against *P. aeruginosa* that was used in a clinical trial involving 24 patients suffering from chronic otitis. This product reduced significantly the counts of *P. aeruginosa* in the phage-treated group. Furthermore, no treatment related adverse event was reported (Wright *et al.*, 2009).

Future trends

An analysis of the different biofilm–phage interaction studies performed so far shows that it is extremely difficult to compare them and take conclusions about phages performance in a biofilm environment. This is mostly due to the different approaches employed to form biofilms and to perform the infections assays. Biofilms are formed on a variety of devices under different hydrodynamic and feed conditions that greatly influence biofilm structure and cell physiology which, as already discussed, influences phage–biofilm interactions. Infection assays have been carried out on established biofilms attached to surfaces or biofilms scrapped from the surfaces, in which the original structure might have been lost. However, as stated throughout this chapter phage interacts differently with biofilm cells compared to their planktonic counterparts and any conclusion regarding phage performance based on assays involving suspended cells cannot be extrapolated to biofilms.

In order to deepen the knowledge about the interactions between phages and their hosts occurring in biofilms, the following questions briefly presented below should be addressed.

For instance, it is well known that phage infection of planktonic cells depends on the physiological state of bacteria, therefore studies of phage infection of the different biofilm populations (active cells, transient and dormant cells or persister cells) should be performed as their response to the phages will likely be different. Although a few studies have shown that long periods of exposure of bacteria to phages induces the growth of resistant phenotypes both in suspension and biofilm, it is still not completely clear if this phenomenon is enhanced in biofilm compared to planktonic conditions. Furthermore, while planktonic cells that survived phage infection showed transient resistance and could revert to be phage susceptible (O'Flynn *et al.*, 2004), there are no studies on resistance/susceptibility of the surviving biofilm cells.

Further research providing evidence that endolysins and phage depolymerases are efficient in biofilms should be carried out since these are reportedly important bacterial control agents. Also, it is still controversial whether antibiotics can enhance phage efficacy in biofilms, thus studies should be performed to address the effect of combinatory therapy in well defined biofilm models.

It is also necessary to increase the number of species used for the *in vitro* biofilm formation and phage infection experiments since, today, studies on interactions of phages with multispecies biofilms are only limited to dual-species biofilms overlooking the complex biofilm communities present in some infections. Furthermore, only biofilms formed in *in vitro* conditions have been targeted so far, thus it is important to expand phage–biofilm interaction studies to real biofilms removed from biofilm-related infection sites. Also, focus should be given to *in vivo* studies, performed on animal models of infection, of phage efficacy against biofilm infections and to investigate the interactions between phages and the host immune system in the presence of biofilms. Moreover, it is important to develop efficient ways of phage delivery considering the stability of phage preparations in for example wound dressings or catheters coatings.

References

Adams, M.H., and Park, B. (1956). An enzyme produced by a phage-host cell system II. The properties of the polysaccharide depolymerase. Virology 2, 719–736.

Ahiwale, S., Tamboli, N., Thorat, K., Kulkarni, R., Ackermann, H., and Kapadnis, B. (2011). In Vitro Management of Hospital *Pseudomonas aeruginosa* Biofilm Using Indigenous T7-Like Lytic Phage. Curr. Microbiol. *62*, 335–340.

An, Y.H., and Friedman, R.J. (1998). Concise review of mechanisms of bacterial adhesion to biomaterial surfaces. J. Biomed. Mat. Res. *43*, 338–348.

Anderl, J., Franklin, M., and Stewart, P. (2000). Role of antibiotic penetration limitation in *Klebsiella pneumoniae* biofilm resistance to ampicillin and ciprofloxacin. Antimicrob. Agents Chemother. *44*, 1818–1824.

Arciola, C.R., Campoccia, D., Gamberini, S., Cervellati, M., Donati, E., and Montanaro, L. (2002). Detection of slime production by means of an optimised Congo red agar plate test based on a colourimetric scale in *Staphylococcus epidermidis* clinical isolates genotyped for ica locus. Biomaterials *23*, 4233–4239.

Atabek, A., and Camesano, T. (2007). Atomic force microscopy study of the effect of lipopolysaccharides and extracellular polymers on adhesion of *Pseudomonas aeruginosa*. J. Bacteriol. *189*, 8503–8509.

Azeredo, J., and Sutherland, I.W. (2008). The use of phages for the removal of infectious biofilms. Curr. Pharm. Biotechnol. *9*, 261–266.

Balaban, N. (2008). Preface. In Control of biofilm infections by signal manipulations, Springer-Verlag).

Bartell, P., Orr, T., and Lam, G. (1966). Polysaccharide depolymerase associated with bacteriophage infection. J. Bacteriol. *92*, 56–62.

Bartell, P., Lam, G., and Orr, T. (1968). Purification and properties of polysaccharide depolymerase associated with phage-infected *Pseudomonas aeruginosa*. J. Biol. Chem. *243*, 2077–2080.

Bedi, M.S., Verma, V., and Chhibber, S. (2009). Amoxicillin and specific bacteriophage can be used together for eradication of biofilm of *Klebsiella pneumoniae* B5055. World J. Microbiol. Biotechnol. *25*, 1145–1151.

Bjarnsholt, T., and Givskov, M. (2007). Quorum-sensing blockade as a strategy for enhancing host defences against bacterial pathogens. Phil. Trans. Royal Soc. B-Biol. Sci. *362*, 1213–1222.

Bourion, F., and Cerf, O. (1996). Disinfection efficacy against pure-culture and mixed-population biofilms of *Listeria innocua* and *Pseudomonas aeruginosa* on stainless steel, teflon(r) and rubber. Sci. Aliments *16*, 151–166.

Branda, S.S., Vik, A., Friedman, L., and Kolter, R. (2005). Biofilms: the matrix revisited. Trends Microbiol. *13*, 20–26.

Briandet, R., Lacroix-Gueu, P., Renault, M., Lecart, S., Meylheuc, T., Bidnenko, E., Steenkeste, K., Bellon-Fontaine, M.N., and Fontaine-Aupart, M.P. (2008). Fluorescence correlation spectroscopy to study diffusion and reaction of bacteriophages inside biofilms. Appl. Environ. Microbiol. *74*, 2135–2143.

Brunstedt, M.R., Sapatnekar, S., Rubin, K.R., Kieswetter, K.M., Ziats, N.P., Merritt, K., and Anderson, J.M. (1995). Bacteria blood material interactions. I. Injected and preseeded slime-forming *Staphylococcus epidermidis* in flowing blood with biomaterials. J. Biomed. Mater. Res. *29*, 455–466.

Bryers, J.D. (2000). Biofilms., J.D. Bryers, ed. (New York, Wiley Interscience).

Bryers, J.D., and Ratner, B.D. (2004). Bioinspired implant materials befuddle bacteria. ASM News *70*, 232–237.

Budhani, R.K., and Struthers, J.K. (1998). Interaction of *Streptococcus pneumoniae* and *Moraxella catarrhalis*: Investigation of the indirect pathogenic role of beta-lactamase-producing moraxellae by use of a continuous-culture biofilm system. Antimicrob. Agents Chemother. *42*, 2521–2526.

Bruttin, A., and Brüssow, H. (2005). Human volunteers receiving *Escherichia coli* phage T4 orally: a safety test of phage therapy. Antimicrob. Agents Chemother. *49*, 2874–2878.

Burmolle, M., Webb, J.S., Rao, D., Hansen, L.H., Sorensen, S.J., and Kjelleberg, S. (2006). Enhanced biofilm formation and increased resistance to antimicrobial agents and bacterial invasion are caused by synergistic interactions in multispecies biofilms. Appl. Environ. Microbiol. *72*, 3916–3923.

Busscher, H.J., Cowan, M.M., and Vandermei, H.C. (1992). On the relative importance of specific and nonspecific approaches to oral microbial adhesion. FEMS Microbiol. Rev. *88*, 199–209.

Camilli, A., and Bassler, B.L. (2006). Bacterial small-molecule signaling pathways. Science *311*, 1113–1116.

Carson, L., Gorman, S., and Gilmore, B. (2010). The use of lytic bacteriophages in the prevention and eradication of biofilms of *Proteus mirabilis* and *Escherichia coli*. FEMS Immunol. Med. Microbiol. *59*, 447–455.

Castelli, P., Caronno, R., Ferrarese, S., Mantovani, V., Piffaretti, G., Tozzi, M., Lomazzi, C., Rivolta, N., and Sala, A. (2006). New trends in prosthesis infection in cardiovascular surgery. Surg. Infect. *7(Suppl 2)*, s45-s47.

Cerca, N., Martins, S., Cerca, F., Jefferson, K.K., Pier, G.B., Oliveira, R., and Azeredo, J. (2005). Comparative assessment of antibiotic susceptibility of coagulase-negative staphylococci in biofilm versus planktonic culture as assessed by bacterial enumeration or rapid XTT colorimetry. J. Antimicrob. Chemother. *56*, 331–336.

Cerca, N., Oliveira, R., and Azeredo, J. (2007). Susceptibility of *Staphylococcus epidermidis* planktonic cells and biofilms to the lytic action of staphylococcus bacteriophage K. Lett. Appl. Microbiol. *45*, 313–317.

Ceri, H., Olson, M., Stremick, C., Read, R., Morck, D., and Buret, A. (1999). The calgary biofilm device: new technology for rapid determination of antibiotic susceptibilities of bacterial biofilms. J. Clin. Microbiol. *37*, 1771–1776.

Chan, T., Lee, K., and Roulin, P. (2007). *Escherichia coli* biofilm development and bacteriophage T4 susceptibility. J. Exp. Microbiol. Immunol *11*, 73–80.

Chavant, P., Gaillard-Martinie, B., and Hebraud, M. (2004). Antimicrobial effects of sanitizers against planktonic and sessile *Listeria monocytogenes* cells according to the growth phase. FEMS Microbiol. Lett. *236*, 241–248.

Chibeu, A., Lingohr, E.J., Masson, L., Manges, A., Harel, J., Ackermann, H.W., Kropinski, A., and Boerlin, P. (2012). Bacteriophages with the ability to degrade uropathogenic *Escherichia coli* biofilms. Viruses *4*, 471–487.

Corbin, B.D., McLean, R.J.C., and Aron, G.M. (2001). Bacteriophage T4 multiplication in a glucose-limited *Escherichia coli* biofilm. Can. J. Microbiol. *47*, 680–684.

Cornelissen, A., Ceyssens, P.J., T'Syen, J., Van Praet, H., Noben, J.P., Shaburova, O.V., Krylov, V.N., Volckaert, G., and Lavigne, R. (2011). The T7-related Pseudomonas putida phage φ15 displays virion-associated biofilm degradation properties. PLoS One *6*, e18597.

Cornelissen, A., Ceyssens, P.J., Krylov, V.N., Noben, J.P., Volckaert, G., and Lavigne, R. (2012). Identification of EPS-degrading activity within the tail spikes of the novel *Pseudomonas putida* phage AF. Virology *434*, 251–256.

Costerton, J.W. (2007). The biofilm primer. (Berlin Heidelberg, Springer-Verlag).

Costerton, J.W., Geesey, G.G., and Cheng, G. (1978). How bacteria stick. Sci. Am. *238*, 86–95.

Costerton, J.W., Cheng, K.J., Geesey, G.G., Ladd, T.I., Nickel, J.C., Dasgupta, M., and Marrie, T.J. (1987). Bacterial biofilms in nature and disease. Annu. Rev. Microbiol. *41*, 435–464.

Costerton, J.W., Stewart, P.S., and Greenberg, E.P. (1999). Bacterial biofilms: a common cause of persistent infections. Science *284*, 1318–1322.

Cowan, S.E., Gilbert, E., Liepmann, D., and Keasling, J.D. (2000). Commensal interactions in a dual-species biofilm exposed to mixed organic compounds. Appl. Environ. Microbiol. *66*, 4481–4485.

Croucher, S.C., Houston, A.P., Bayliss, C.E., and Turner, R.J. (1983). Bacterial populations associated with different regions of the human colon wall. Appl. Environ. Microbiol. *45*, 1025–1033.

Curtin, J.J., and Donlan, R.M. (2006). Using bacteriophages to reduce formation of catheter-associated biofilms by *Staphylococcus epidermidis*. Antimicrob. Agents Chemother. *50*, 1268–1275.

Dabrowska, K., Switala-Jelen, K., Opolski, A., Weber-Dabrowska, B., and Gorski, A. (2005). Bacteriophage penetration in vertebrates. J. Appl. Microbiol. *98*, 7–13.

Davies, D.G., Parsek, M., Pearson, J., Iglewski, B., Costerton, J.W., and Greenberg, E.P. (1998). The involvement of cell-to-cell signals in the development of a bacterial biofilm. Science *280*, 295–298.

De Beer, D., Srinivasan, R., and Stewart, P.S. (1994a). Direct measurement of chlorine penetration into biofilms during disinfection. Appl. Environ. Microbiol. *60*, 4339–4344.

De Beer, D., Stoodley, P., and Lewandowski, Z. (1994b). Liquid flow in heterogeneous biofilms. Biotechnol. Bioeng. *44*, 636–641.

De Beer, D., Stoodley, P., Roe, F., and Lewandowski, Z. (1994c). Effects of Biofilm Structures on Oxygen Distribution and Mass-Transport. Biotechnol. Bioeng. *43*, 1131–1138.

De Paepe, M., and Taddei, F. (2006). Viruses' life history: towards a mechanistic basis of a trade-off between survival and reproduction among phages. PLoS Biol *4*, 193.

De Smet, K., and Contreras, R. (2005). Human antimicrobial peptides: Defensins, cathelicidins and histatins. Biotechnol. Lett. *27*, 1337–1347.

Donlan, R.M. (2009). Preventing biofilms of clinically relevant organisms using bacteriophage. Trends Microbiol. *17*, 66–72.

Donlan, R.M., and Costerton, J.W. (2002). Biofilms: Survival mechanisms of clinically relevant microorganisms. Clin. Microbiol. Rev. *15*, 167–193.

Donlan, R.M., and Curtin, J.J. (2006). Inhibition of biofilm formation using bacteriophage. PCT/US2005/044498[WO/2006/063176].

Doolittle, M.M., Cooney, J.J., and Caldwell, D.E. (1995). Lytic infection of *Escherichia coli* biofilms by bacteriophage T4. Can. J. Microbiol. *41*, 12–18.

Doolittle, M., Conney, J., and Caldwell, D. (1996). Tracing the interaction of bacteriophage with bacterial biofilms using fluorescent and chromogenic probes. J. Ind. Microbiol. *16*, 331–341.

Eklund, C., and Wyss, O. (1962). Enzyme associated with bacteriophage infection. J. Bacteriol. *84*, 1209–1215.

Erb, T., Christen, P., Kern, C., Hampl, K.F., and Frei, F.J. (1997). Children breathing spontaneously via a laryngeal mask airway: Comparison of sevoflurane and halothane. Anesthesiology *87*, A1043.

Fischetti, V. (2008). Bacteriophage lysins as effective antibacterials. Curr. Opin. Microbiol. *11*, 393–400.

Flemming, H.-C., Neu, T., and Wozniak, D. (2007). The EPS Matrix: The 'House of Biofilm Cells'. J. Bacteriol. *189*, 7945–7947.

Frank, J.F., Ehlers, J., and Wicker, L. (2003). Removal of *Listeria monocytogenes* and poultry soil-containing biofilms using chemical cleaning and sanitizing agents under static conditions. Food Prot. Trends *23* 654–663.

Fu, W., Forster, T., Mayer, O., Curtin, J.J., Lehman, S. and Donlan, R.M. (2010). Bacteriophage cocktail for the prevention of biofilm formation by *Pseudomonas aeruginosa* on catheters in an *in vitro* model system. Antimicrob. Agents Chemother. *54*, 397–404.

Górski, A., Wazna, E., Dabrowska, B.W., Dabrowska, K., Switala-Jelen, K., and Miedzybrodzki, R. (2006). Bacteriophage translocation. FEMS Immunol. Med. Microbiol. *46*, 313–319.

Gutiérrez, D., Martínez, B., Rodríguez, A., and García, P. (2012). Genomic characterization of two *Staphylococcus epidermidis* bacteriophages with anti-biofilm potential. BMC Genomics *13*, 228.

Hadas, H., Einav, M., Fishov, I., and Zaritsky, A. (1997). Bacteriophage T4 development depends on the physiology of its host *Escherichia coli*. Microbiology *143*, 179–185.

Hall-Stoodley, L., and Stoodley, P. (2005). Biofilm formation and dispersal and the transmission of human pathogens. Trends Microbiol. *13*, 7–10.

Hanlon, G.W., Denyer, S.P., Olliff, C.J., and Ibrahim, L.J. (2001). Reduction in exopolysaccharide viscosity as an aid to bacteriophage penetration through *Pseudomonas aeruginosa* biofilms. Appl. Environ. Microbiol. *67*, 2746–2753.

Hong, H., Ducle, H., and Cutting, S. (2005). The use of bacterial spore formers as probiotics. FEMS Microbiol. Rev. *29*, 813–835.

Hoyle, B.D., Alcantara, J., and Costerton, J.W. (1992). *Pseudomonas aeruginosa* biofilm as a diffusion barrier to piperacillin. Antimicrob. Agents Chemother. *36*, 2054–2056.

Hu, J., Miyanaga, K., and Tanji, Y. (2010). Diffusion properties of bacteriophages through agarose gel membrane. Biotechnol. Prog. *26*, 1213–1221.

Hughes, K.A., Sutherland, I., Clark, J., and Jones, M.V. (1998a). Bacteriophage and associated polysaccharide depolymerases – novel tools for study of bacterial biofilms. J. Appl. Microbiol. *85*, 583–590.

Hughes, K.A., Sutherland, I.W., and Jones, M.V. (1998b). Biofilm susceptibility to bacteriophage attack: the role of phage-borne polysaccharide depolymerase. Microbiology *144*, 3039–3047.

Jackson, G., Beyenal, H., Rees, W., and Lewandowski, Z. (2001). Growing reproducible biofilms with respect to structure and viable cell counts. J. Microbiol. Methods *47*, 1–10.

Kelly, D., McAuliffe, O., Ross, R.P., and Coffey, A. (2012). Prevention of *Staphylococcus aureus* biofilm formation and reduction in established biofilm density using a combination of phage K and modified derivatives. Lett. Appl. Microbiol. *54*, 286–291.

Kimura, K., and Itoh, Y. (2003). Characterization of poly-γ-glutamate hydrolase encoded by a bacteriophage genome: possible role in phage infection of *Bacillus subtilis* encapsulated with poly- γ -glutamate. Appl. Environ. Microbiol. *69*, 2491–2497.

Knowles, J.R., Roller, S., Murray, D.B., and Naidu, A.S. (2005). Antimicrobial action of carvacrol at different stages of dual-species biofilm development by *Staphylococcus aureus* and *Salmonella enterica* serovar Typhimurium. Appl. Environ. Microbiol. *71*, 797–803.

Krylov, V.N. (2001). Phage therapy in terms of bacteriophage genetics: Hopes, prospects, safety, limitations. Russ. J. Gen. *37*, 715–730.

Kumar, C.G., and Anand, S.K. (1998). Significance of microbial biofilms in food industry: a review. Int. J. Food Microbiol. *42*, 9–27.

Kutateladze, M., and Adamia, R. (2010). Bacteriophages as potential new therapeutics to replace or supplement antibiotics. Trends Biotechnol. *28*, 591–595.

Kutter, E., De Vos, D., Gvasalia, G., Alavidze, Z., Gogokhia, L., Kuhl, S., and Abedon, S.T. (2010). Phage Therapy in Clinical Practice: Treatment of Human Infections. Curr. Pharmaceut. Biotechnol. *11*, 69–86.

Lacqua, A., Wanner, O., Colangelo, T., Martiotti, M., and Landini, P. (2006). Emergence of biofilm-forming subpopulations upon exposure of *Escherichia coli* to environmental bacteriophages. Appl. Environ. Microbiol. *72*, 956–959.

Lacroix-Gueu, P., Briandet, R., Lévêque-Fort, S., Bellon-Fontaine, M.-N., and Fontaine-Aupart, M.P. (2005). *In situ* measurements of viral particles diffusion inside mucoid biofilms. C. R. Biol. *328*, 1965–1072.

Lam, J., Chan, R., Lam, K., and Costerton, J.W. (1980). Production of mucoid microcolonies by *Pseudomonas aeruginosa* within infected lungs in cystic fibrosis. Infect. Immun. *28*, 546–556.

Lambe, D.J., Ferguson, K., Mayberry-Carson, K., Tober-Meyer, B., and Costerton, J.W. (1991). Foreign-body-associated experimental osteomyelitis induced with *Bacteroides fragilis* and *Staphylococcus epidermidis* in rabbits. Clin. Orthop. *266*, 285–294.

Landy, M., Trapani, R., and Shear, M. (1958). Interference by endotoxic polysaccharides with inactivation of coliphage by human serum. Br. J. Exp. Pathol. *39*, 618–630.

Lappin-Scott, H.M., and Costerton, J.W. (1995). Microbial biofilms (Cambridge, UK, Cambridge University Press).

Lawrence, J., Korber, D., Hoyle, B.D., Costerton, J.W., and Caldwell, D.E. (1991). Optical sectioning of microbial biofilms. J. Bacteriol. *173*, 6558–6567.

Lewis, K. (2001). Riddle of biofilm resistance. Antimicrob. Agents Chemother. *45*, 999–1007.

Lewis, K. (2008). Multidrug tolerance of biofilms and persister cells. In Bacterial biofilms., T. Romeo, ed. (Berlin, Springer-Verlag).

Liao, K.S., Lehman, S.M., Tweardy, D.J., Donlan, R.M., and Trautner, B.W. (2012). Bacteriophages are synergistic with bacterial interference for the prevention of *Pseudomonas aeruginosa* biofilm formation on urinary catheters. J. Appl. Microbiol. *113*, 1530–1539.

Los, M., Wegrzyn, G., and Neubauer, P. (2003). A role for bacteriophage T4 rI gene function in the control of phage development during pseudolysogeny and in slowly growing host cells. Res. Microbiol. *154*, 547–552.

Los, M., Golec, P., Los, J., Weglewska-Jurkiewicz, A., Czyz, A., Wegrzyn, A., Wegrzyn, G., and Neubauer, P. (2007). Effective inhibition of lytic development of bacteriophages lambda, P1 and T4 by starvation of their host, *Escherichia coli*. BMC Biotechnol. *26*, 7–13.

Lu, T.K., and Collins, J.J. (2007). Dispersing biofilms with engineered enzymatic bacteriophage. Proc. Natl. Acad. Sci. U.S.A. *104*, 11197–11202.

Lu, T.K., and Collins, J. (2009). Engineered bacteriophage targeting gene networks as adjuvants for antibiotic therapy. Proc. Natl. Acad. Sci. U.S.*A. 106*, 4629–4634.

Macfarlane, S., and Dillon, J.F. (2007). Microbial biofilms in the human gastrointestinal tract. J. Appl. Microbiol. *102*, 1187–1196.

Macfarlane, S., Furrie, E., Cummings, J.H., and Macfarlane, G.T. (2004). Chemotaxonomic analysis of bacterial populations colonizing the rectal mucosa in patients with ulcerative colitis. Clin. Infect. Dis. 38, 1690–1699.

Mah, T., Pitts, B., Pellock, P., Walker, G., Stewart, P.S., and O'Toole, G.A. (2003). A genetic basis for *Pseudomonas aeruginosa* biofilm antibiotic resistance. Nature *426*, 306–310.

Mare, I., and Smit, J. (1969). A capsule-depolymerizing enzyme from *Alcaligenes faecalis* infected with bacteriophage A6. J. Gen. Virol. *5*, 551–552.

Marr, A.K., Overhage, J., Bains, M., and Hancock, R.E.W. (2007). The Lon protease of *Pseudomonas aeruginosa* is induced by aminoglycosides and is involved in biofilm formation and motility. Microbiology *153*, 474–482.

Marshall, K. (1992). Biofilms: an overview of bacterial adhesion, activity, and control at surfaces. ASM News *58*, 207.

Mattila, T., Manninen, M., and Kylasiurola, A.L. (1990). Effect of cleaning-in-place disinfectants on wild bacterial strains isolated from a milking line. J. Dairy Res. *57*, 33–39.

Merabishvili, M., Pirnay, J., Verbeken, G., Chanishvili, N., Tediashvili, M., Lashkhi, N., Glonti, T., Krylov, V., Mast, J., Van Parys, L., *et al.* (2009). Quality-controlled small-scale production of a well-defined bacteriophage cocktail for use in human clinical trials. PLoS One *4*, e4922.

Moebus, K. (1996). Marine bacteriophage reproduction under nutrient-limited growth of host bacteria. I. Investigations with six phage-host systems. Mar. Ecol. Prog. Ser. *144*, 1–12.

Monzón, M., Oteiza, C., Leiva, J., Lamata, M., and Amorena, B. (2002). Biofilm testing of Staphylococcus epidermidis clinical isolates: low performance of vancomycin in relation to other antibiotics. Diagn. Microbiol. Infect. Dis. *44*, 319–324.

Nishikawa, H., Yasuda, M., Uchiyama, J., Rashel, M., Maeda, Y., Takemura, I., Sugihara, S., Ujihara, T., Shimizu, Y., Shuin, T., and Matsuzaki, S. (2008). T-even-related bacteriophages as candidates for treatment of *Escherichia coli* urinary tract infections. Arch. Virol. *153*, 507–515.

Norwood, D.E., and Gilmour, A. (2000). The growth and resistance to sodium hypochlorite of *Listeria monocytogenes* in a steady-state multispecies biofilm. J. Appl. Microbiol. *88*, 512–520.

O'Flynn, G., Ross, R.P., Fitzgerald, G.F., and Coffey, A. (2004). Evaluation of a cocktail of three bacteriophages for biocontrol of *Escherichia coli* O157:H7. Appl. Environ. Microbiol. *70*, 3417–3424.

O'Toole, G.A., Kaplan, H., and Kolter, R. (2000). Biofilm formation as microbial development. Annu. Rev. Microbiol. *54*, 49–79.

Otto, M. (2006). Bacterial evasion of antimicrobial peptides by biofilm formation. Antimicrob. Pept. Hum. Dis. *306*, 251–258.

Pearl, S., Gabay, C., Kishony, R., Oppenheim, A., and Balaban, N. (2008). Nongenetic individuality in the host–phage interaction. PLoS Biol *6*, e120.

Pires, D., Sillankorva, S., Faustino, A., and Azeredo, J. (2011). Use of newly isolated phages for control of *Pseudomonas aeruginosa* PAO1 and ATCC 10145 biofilms. Res. Microbiol. *162*, 798–806.

Potera, C. (1999). Microbiology – forging a link between biofilms and disease. Science *283*, 1837.

Pritchard, D.I. (2006). Immune modulation by *Pseudomonas aeruginosa* quorum-sensing signal molecules. Int. J. Med. Microbiol. *296*, 111–116.

Rahman, M., Kim, S., Kim, S.M., Seol, S.Y., and Kim, J. (2011). Characterization of induced *Staphylococcus aureus* bacteriophage SAP-26 and its anti-biofilm activity with rifampicin. Biofouling *27*, 1087–1093.

Reese, J., Dimitracopoulos, G., and Bartell, P.F. (1974). Factors influencing the adsorption of bacteriophage 2 to cells of *Pseudomonas aeruginosa*. J. Virol. *13*, 22–27.

Reid, G., Howard, J., and Bing, S. (2001). Can bacterial interference prevent infection? Trends Microbiol. *9*, 424–428.

Reisner, A., Krogfelt, K.A., Klein, B.M., Zechner, E.L., and Molin, S. (2006). *In vitro* biofilm formation of commensal and pathogenic *Escherichia coli* strains: impact of environmental and genetic factors. J. Bacteriol. *188*, 3572–3581.

Resch, A., Fehrenbacher, B., Eisele, K., Schaller, M., and Gotz, F. (2005). Phage release from biofilm and planktonic *Staphylococcus aureus* cells. FEMS Microbiol. Lett. *252*, 89–96.

Rosenberg, M. (2006). Microbial adhesion to hydrocarbons: twenty-five years of doing MATH. FEMS Microbiol. Lett. *262*, 129–134.

Ryu, J.H., and Beuchat, L.R. (2005). Biofilm formation by *Escherichia coli* O157:H7 on stainless steel: effect of exopolysaccharide and Curli production on its resistance to chlorine. Appl. Environ. Microbiol. *71*, 247–254.

Sass, P., and Bierbaum, G. (2007). Lytic activity of recombinant bacteriophage phi 11 and phi 12 endolysins on whole cells and biofilms of *Staphylococcus aureus*. Appl. Environ. Microbiol. *73*, 347–352.

Sauer, K., Cullen, M.C., Rickard, A.H., Zeef, L.A.H., Davies, D.G., and Gilbert, P. (2004). Characterization of nutrient-induced dispersion in *Pseudomonas aeruginosa* PAO1 biofilm. J. Bacteriol. *186*, 7312–7326.

Sharma, M., Ryu, J.H., and Beuchat, L.R. (2005). Inactivation of *Escherichia coli* O157:H7 in biofilm on stainless steel by treatment with an alkaline cleaner and a bacteriophage. J. Appl. Microbiol. *99*, 449–459.

Sillankorva, S., Oliveira, R., Vieira, M.J., Sutherland, I.W., and Azeredo, J. (2004a). Bacteriophage phiS1 infection of *Pseudomonas fluorescens* planktonic cells versus biofilms. Biofouling *20*, 133–138.

Sillankorva, S., Sutherland, I., Vieira, M.J., Oliveira, R., and Azeredo, J. (2004b). *Pseudomonas fluorescens* infection by bacteriophage phiS1 – the influence of temperature, host growth phase and media. FEMS Microbiol. Lett. *241*, 13–20.

Sillankorva, S., Neubauer, P., and Azeredo, J. (2008). *Pseudomonas fluorescens* biofilms subjected to phage philBB-PF7A. BMC Biotechnol. *8*, 79.

Sillankorva, S., Neubauer, P., and Azeredo, J. (2010). Phage control of dual species biofilms of *Pseudomonas fluorescens* and *Staphylococcus lentus*. Biofouling *26*, 567–575.

Skillman, L.C., Sutherland, I.W., and Jones, M.V. (1999). The role of exopolysaccharides in dual species biofilm development. J. Appl. Microbiol. *85*, 13S-18S.

Son, J.-S., Lee, S.-J., Jun, S., Yoon, S., Kang, S., Paik, H., Kang, J., and Choi, Y.-J. (2010). Antibacterial and biofilm removal activity of a podoviridae *Staphylococcus aureus* bacteriophage SAP-2 and a derived recombinant cell-wall-degrading enzyme. Appl. Microbiol. Biotechnol. *86*, 1439–1449.

Spanakis, E., and Horne, M. (1987). Co-adaptation of *Escherichia coli* and coliphage lambda vir in continuous culture. J. Gen. Microbiol. *133*, 353–360.

Spoering, A.L., and Lewis, K. (2001). Biofilms and planktonic cells of *Pseudomonas aeruginosa* have similar resistance to killing by antimicrobials. J. Bacteriol. *183*, 6746–6751.

Stoodley, P., Wilson, S., Hall-Stoodley, L., Boyle, J., Lappin-Scott, H.M., and Costerton, J.W. (2011). Growth and detachment of cell clusters from mature mixed-species biofilms. Appl. Environ. Microbiol. *67*, 5608–5613.

Strevett, K.A., and Chen, G. (2003). Microbial surface thermodynamics and applications. Res. Microbiol. *154*, 329–335.

Suci, P.A., Mittelman, M.W., Yu, F.P., and Geesey, G.G. (1994). Investigation of ciprofloxacin penetration into *Pseudomonas aeruginosa* biofilms. Antimicrob. Agents Chemother. *38*, 2125–2133.

Sutherland, I. (1967). Phage-induced fucosidases hydrolyzing the exopolysaccharide of *Klebsiella aerogenes* Type 54. Biochem. J. *104*, 278–285.

Sutherland, I.W. (1977). Enzymes acting on bacterial surface carbohydrates. In Surface Carbohydrates of the Bacterial Cell, I.W. Sutherland, ed. (New York, Academic Press), pp. 209–245.

Sutherland, I.W. (2001a). Biofilm exopolysaccharides: a strong and sticky framework. Microbiology-Uk *147*, 3–9.

Sutherland, I.W. (2001b). The biofilm matrix – an immobilized but dynamic microbial environment. Trends Microbiol. *9*, 222–227.

Sutherland, I.W., Hughes, K.A., Skillman, L.C., and Tait, K. (2004). The interaction of phage and biofilms. FEMS Microbiol. Lett. *232*, 1–6.

Tait, K., Skillman, L.C., and Sutherland, I.W. (2002). The efficacy of bacteriophage as a method of biofilm eradication. Biofouling *18*, 305–311.

Thawal, N.D., Yele, A.B., Sahu, P.K., and Chopade, B.A. (2012). Effect of a novel podophage AB7-IBB2 on *Acinetobacter baumannii* biofilm. Curr. Microbiol. *65*, 66–77.

Vandersteegen, K., Kropinski, A.M., Mash, J.H.E., Noben, J.P., Hermans, K., and Lavigne, R. (2013). Romulus and Remus, two phage isolates representing a distinct clade within the Twortlikevirus genus display suitable properties for phage therapy applications. J. Virol. 87, 3237–3247.

Verma, V., Harjai, K., and Chhibber, S. (2009). Restricting ciprofloxacin-induced resistant variant formation in biofilm of *Klebsiella pneumoniae* B5055 by complementary bacteriophage treatment. J. Antimicrob. Chemother. *64*, 1212–1218.

Verma, V., Harjai, K., and Chhibber, S. (2010). Structural changes induced by a lytic bacteriophage make ciprofloxacin effective against older biofilm of *Klebsiella pneumoniae*. Biofouling *26*, 729–737.

Verran, J., and Whitehead, K. (2005). Factors affecting microbial adhesion to stainless steel and other materials used in medical devices. Int. J. Art. Org. *28*, 1138–1145.

van Voorthuizen, E., Ashbolt, N., and Schäfer, A. (2001). Role of hydrophobic and electrostatic interactions for initial enteric virus retention by MF membranes. J. Membr. Sci. *194*, 69–79.

Walters, M.C., Roe, F., Bugnicourt, A., Franklin, M.J., and Stewart, P.S. (2003). Contributions of antibiotic penetration, oxygen limitation, and low metabolic activity to tolerance of *Pseudomonas aeruginosa* biofilms to ciprofloxacin and tobramycin. Antimicrob. Agents Chemother. *47*, 317–323.

Webb, J.S., Lau, M., and Kjelleberg, S. (2004). Bacteriophage and phenotypic variation in *Pseudomonas aeruginosa* biofilm development. J. Bacteriol. *186*, 8066–8073.

Weber-Dabrowska, B., Dabrowski, M., and Slopek, S. (1987). Studies on bacteriophage penetration in patients subjected to phage therapy. Arch. Immunol. Ther. Exp. *35*, 563–568.

Wenzel, R.P. (2007). Health care-associated infections: Major issues in the early years of the 21st century. Clin. Infect. Dis. *45*, 85–88.

Wiggins, B., and Alexander, M. (1985). Appl. Environ. Microbiol. *49*, 19–23.

Wilkinson, J. (1958). The extracellular polysaccharides of bacteria. Bacteriol. Rev. *22*, 46–73.

Wood, S., Kirkham, J., Marsh, P., Shore, R., Nattress, B., and Robinson, C. (2000). Architecture of intact natural human plaque biofilms studied by confocal laser scanning microscopy. J. Dent. Res. *79*, 21–27.

Wright, A., Hawkins, C., Anggard, E., and Harper, D. (2009). A controlled clinical trial of a therapeutic bacteriophage preparation in chronic otitis due to antibiotic resistant *Pseudomonas aeruginosa*: a preliminary report of efficacy. Clin. Otolaryngol. *34*, 349–357.

Yilmaz, C., Colak, M., Yilmaz, B.C., Ersoz, G., Kutateladze, M., and Gozlugol, M. (2013). Bacteriophage Therapy in Implant-Related Infections: An Experimental Study. J. Bone Joint Surg. Am. *95*, 117–125.

Yele, A.B., Thawal, N.D., Sahu, P.K., and Chopade, B.A. (2012). Novel lytic bacteriophage AB7-IBB1 of Acinetobacter baumannii: isolation, characterization and its effect on biofilm. Arch. Virol. *157*, 1441–1450.

Yoon, S., Choi, Y., Lee, S.J., Son, J., Jun, S., and Kang, S. (2009). Bacteriophage or lytic protein derived from the bacteriophage which effective for treatment of *Staphylococcus aureus* biofilm. PCT/KR2008/005434[WO/2009/035303].

Using What Phage Have Evolved to Kill Pathogenic Bacteria

14

Vincent A. Fischetti

Abstract

Bacteriophage (or phage) endolysins (or lysins) are highly evolved enzymes produced to cleave essential bonds in the bacterial cell wall peptidoglycan for phage progeny release. Small quantities of purified recombinant lysin added externally to Gram-positive bacteria cause immediate lysis resulting in log-fold death of the target bacterium. Lysins have now been used successfully in a variety of animal models to control pathogenic antibiotic resistant bacteria found on mucosal surfaces and in infected tissues. The advantages over antibiotics are their specificity for the pathogen even antibiotic resistant organisms without disturbing the normal flora, the low chance of bacterial resistance, and their ability to kill colonizing pathogens on mucosal surfaces, a capacity previously unavailable. Lysins therefore, may be a much-needed anti-infective (or enzybiotic) in an age of mounting antibiotic resistance.

Background

Viruses that specifically infect bacteria are called bacteriophage. After replicating inside its bacterial host the phage is faced with a problem, it needs to efficiently exit the bacterium to disseminate its progeny phage to begin a new cycle. To solve this, double-stranded DNA phages have evolved a lytic system to weaken the bacterial cell wall resulting in hypotonic bacterial lysis (Catalão *et al.*, 2013; Oliveira *et al.*, 2013). Phage lytic enzymes or lysins are highly efficient molecules that have been refined over millions of years of evolution for this very purpose. These enzymes target the integrity of the cell wall, and are designed to attack one of the five major bonds in the peptidoglycan. With few exceptions (Loessner *et al.*, 1997; Frias *et al.*, 2013), lysins do not have signal sequences, so they are not translocated through the cytoplasmic membrane to attack their substrate in the peptidoglycan, this movement is tightly controlled by a second phage gene product in the lytic system, the holin (Wang *et al.*, 2000). During phage development in the infected bacterium, lysin accumulates in the cytoplasm in anticipation of phage maturation. At a genetically specified time, holin molecules are inserted in the cytoplasmic membrane forming patches, ultimately resulting in generalized membrane disruption (Wang *et al.*, 2003; White *et al.*, 2011), allowing the cytoplasmic lysin to access the peptidoglycan, thereby causing immediate cell lysis and the release of progeny phage (Wang *et al.*, 2000). In contrast to large DNA phage, small RNA and DNA phages use a different release strategy. They call upon a phage-encoded protein to interfere with bacterial host enzymes responsible for peptidoglycan biosynthesis (Bernhardt *et al.*, 2001; Young *et al.*, 2000) resulting in misassembled cell walls and ultimate lysis. Scientists have been aware of the lytic activity of phage for nearly a century, and while whole phage have been used to control infection (Matsuzaki *et al.*, 2005; Chanishvili *et al.*, 2012), not until recently have lytic enzymes been exploited for bacterial control *in vivo* (Loeffler *et al.*, 2003; Nelson *et al.*, 2001; Schuch *et al.*, 2002). One of the main reasons that such an approach is now even being considered is the sharp increase in antibiotic resistance among pathogenic bacteria (Hawser, 2012). Current data indicate that lysins work only with Gram-positive bacteria, since they are able to make direct contact with the cell wall carbohydrates and peptidoglycan when added externally, whereas the outer membrane of Gram-negative bacteria prevents this interaction. This chapter will outline the remarkable potency these enzymes have in killing bacteria both *in vitro* and *in vivo*. Therapeutic potential of lysins has also been discussed in several reviews published over recent years (Fischetti, 2005, 2008, 2010, 2011; Borysowski *et al.*, 2006; Hermoso *et al.*, 2007; Donovan, 2007; Fenton *et al.*, 2010; Borysowski *et al.*, 2011; Schmelcher *et al.*, 2012; Nelson *et al.*, 2012).

Most human infections (viral or bacterial) begin at a mucous membrane site (upper and lower respiratory, intestinal, urogenital, and ocular). Furthermore, these human mucous membranes are the reservoir (and sometimes the only reservoir) for many pathogenic bacteria found in the environment (i.e. pneumococci, staphylococci, streptococci, *Haemophilus*) some of which are resistant to current antibiotics. In most instances, it is this mucosal reservoir that is the focus of infection in the population (Coello *et al.*, 1994; de Lencastre *et al.*, 1999; Eiff *et al.*, 2001). To date, except for polysporin and mupirocin ointments, which are the most widely used topically, there are no anti-infectives that are designed to control colonizing pathogenic bacteria on mucous membranes (Hudson and Hay, 1976; O'Grady *et al.*, 2009; Mehta *et al.*, 2013); we usually first wait for infection to occur before treating. Because of the fear of increasing the resistance problem, antibiotics are not indicated to control the carrier state of disease bacteria. It is acknowledged however, that by reducing or eliminating this human reservoir of pathogens in the community and controlled environments (i.e. hospitals and nursing homes), the incidence of disease will be markedly reduced (Eiff *et al.*, 2001; Hudson and Hay, 1976). Towards this goal, lysins have been developed to prevent infection by safely and specifically destroying disease bacteria on mucous membranes. For example, based on extensive animal results, enzymes specific for *S. pneumoniae*, *S. pyogenes*, and *S. aureus* may be used nasally and orally to control these organisms in the community as well as in nursing homes and hospitals to prevent or markedly reduce serious infections caused by these bacteria.

Like antibiotics, which are used by bacteria to control the organisms around them in the environment, phage lysins are the culmination of millions of years of development by the bacteriophage in their association with bacteria. Specific lysins have now been identified and purified that are able to kill specific Gram-positive bacteria seconds after contact (Loeffler *et al.*, 2001; Nelson *et al.*, 2001). For example, nanogram quantities of lysin could reduce 10^7 *S. pyogenes* by >6-logs seconds after enzyme addition. No known biological compounds, except chemical agents, kill bacteria this quickly. Because of their highly effective activity against bacteria for the control of disease, the term 'enzybiotics' was coined (Nelson *et al.*, 2001) to describe these novel anti-infectives.

Structural characteristics of lysins

Lysins from DNA-phage that infect Gram-positive bacteria are generally between 25–40 kDa in size except the PlyC lysin from the C1 phage for streptococci, which is 114 kDa. PlyC is unique because it is composed of two separate gene products, PlyCA and PlyCB. Based on biochemical, biophysical, and X-ray crystallographic studies, the catalytically active PlyC holoenzyme is composed of eight PlyCB subunits forming the binding domain. PlyCA is composed of two catalytic domains (Nelson *et al.*, 2006; McGowan *et al.*, 2012). A feature of all other Gram-positive phage lysins is their two-domain structure (Fig. 14.1) With rare exceptions (Diaz *et al.*, 1990; Garcia *et al.*, 1990), the N-terminal domain contains the catalytic activity of the enzyme. This activity may be either an endo-β-N-acetylglucosaminidase or N-acetylmuramidase (lysozymes), both of which act on the sugar moiety of the bacterial wall peptidoglycan, an endopeptidase which cleaves the peptide moiety, or an N-acetylmuramoyl-L-alanine amidase (or amidase), which hydrolyses the amide bond connecting the glycan strand and peptide moieties (Loessner, 2005; Young, 1992). The endopeptidase cleaves the cross-bridge. Recently an enzyme with γ-D-glutaminyl-L-lysine endopeptidase activity has also been reported (Pritchard *et al.*, 2007). In some cases, particularly staphylococcal lysins and a lysin from group B

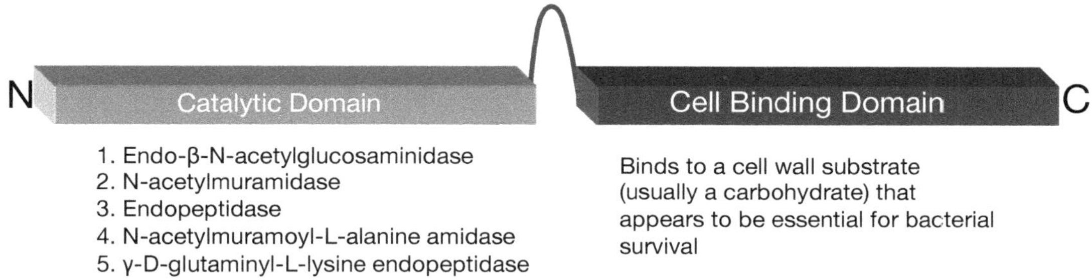

Figure 14.1 Basic structure of phage lysins. In general, lysins range between 25 kDa to 40 kDa in size and have a domain structure. The N-terminal domain is invariably the catalytic domain which cleaves one of the five major bonds in the peptidoglycan, and the C-terminal domain binds to a carbohydrate determinant in the cell wall. A colour version of this figure is available in the plate section at the back of the book.

streptococci, two and perhaps even three different catalytic domains may be linked to a single binding domain (Cheng *et al.*, 2005; Navarre *et al.*, 1999). The C-terminal cell binding domain (termed the CBD domain) on the other hand binds to a specific substrate (usually carbohydrate) found in the cell wall of the host bacterium (Garcia *et al.*, 1988; Lopez *et al.*, 1997, 1992; Low *et al.*, 2011). Efficient cleavage requires that the binding domain bind to its cell wall substrate, offering some degree of specificity to the enzyme since these substrates are only found in enzyme-sensitive bacteria.

The first complete crystal structure for the free and choline-bound states of the Cpl-1 lytic enzyme which lyses pneumococci has been published (Hermoso *et al.*, 2003). As suspected, the data suggest that choline recognition by the choline-binding domain of Cpl-1 may allow the catalytic domain to be properly oriented for efficient cleavage. An interesting feature of this lysin is its hairpin conformation suggesting that the two domains interact with each other prior to the interaction of the binding domain with its substrate in the bacterial cell wall (in this case choline). A second lysin PlyL encoded by a lysogen in the *Bacillus anthracis* genome and a third Ply21 from *B. cereus* phage TP21 have similar characteristics (Low *et al.*, 2005). Their data suggest that the structure of PlyL and Ply21 are also in a hairpin conformation where the C-terminal domain blocks the activity of the catalytic domain through intramolecular interactions that are reversed when the C-terminal domain binds to its substrate in the cell wall freeing the catalytic domain to cleave its substrate. This suggests that these lysins are autoregulated so that they are not able to act on substrates in the cytoplasm that are being synthesized for cell wall assembly, but function when they are outside the cell after interacting with their wall substrate.

When the sequences between lytic enzymes of the same enzyme class are compared, high sequence homology is seen within the N-terminal catalytic region and very little homology within the C-terminal cell-binding region. It seemed counterintuitive that the phage would design a lysin that was uniquely lethal for its host organism, however as we learned more about how these enzymes function, a possible reason for this specificity became apparent. Because of their specificity, enzymes that spilled out after cell lysis had a good chance of killing potential bacterial hosts in the vicinity of the released phage progeny thus competing for phage survival. To prevent this, we believe that the Gram-positive lysins have evolved binding domains that bind to their cell wall substrate at a high affinity (Loessner *et al.*, 2002) to limit the release of free enzyme. This is not the case for lysins produced by Gram-negative phage. Since spilled lysin after lysis is unable to penetrate through the outer membrane to cleave the peptidoglycan, nearly all lysins from Gram-negative phage did not evolve binding domains.

Because of their domain structure, it seemed plausible that different enzyme domains could be swapped resulting in lysins with different bacterial and catalytic specificities. This was actually accomplished by early detailed studies of Garcia and colleagues (Garcia *et al.*, 1990; Weiss *et al.*,

1999), in which the catalytic domains of lytic enzymes from *S. pneumoniae* phage could be swapped resulting in a new enzyme having the same binding domain for pneumococci, but able to cleave a different bond in the peptidoglycan. In addition, DNA mutagenesis has been used to create lysins with higher antibacterial activity (Cheng and Fischetti, 2006). This capacity allows for enormous potential in creating designer enzymes with high specificity and equally high cleavage potential. In recent years this idea has been capitalized and lysins have been engineered to achieve certain characteristics not present in the native lysin (Daniel *et al.*, 2010; Donovan *et al.*, 2006; Low *et al.*, 2011; Mayer *et al.*, 2011; Resch *et al.*, 2011a; Lukacik *et al.*, 2012; Heselpoth and Nelson, 2012).

Mechanism of action

When examined by thin section electron microscopy, it seems obvious that lysins exert their lethal effects by forming holes in the cell wall through peptidoglycan digestion. The high internal pressure of Gram-positive bacterial cells (roughly 15–25 atmospheres) is controlled by the highly cross-linked cell wall. Any disruption in the wall's integrity will result in the extrusion of the cytoplasmic membrane and ultimate hypotonic lysis (Fig. 14.2). Catalytically, a single enzyme molecule should be sufficient to cleave an adequate number of bonds to kill an organism; however, it is uncertain at this time whether this theoretical limit is possible. The reason comes from the work of Loessner (Loessner *et al.*, 2002), showing that a listeria phage enzyme had a binding affinity approaching that of an IgG molecule for its wall substrate, suggesting that phage enzymes, like cellulases (Jervis *et al.*, 1997), are one-use enzymes, likely requiring several molecules attacking a local region to sufficiently weaken the cell wall.

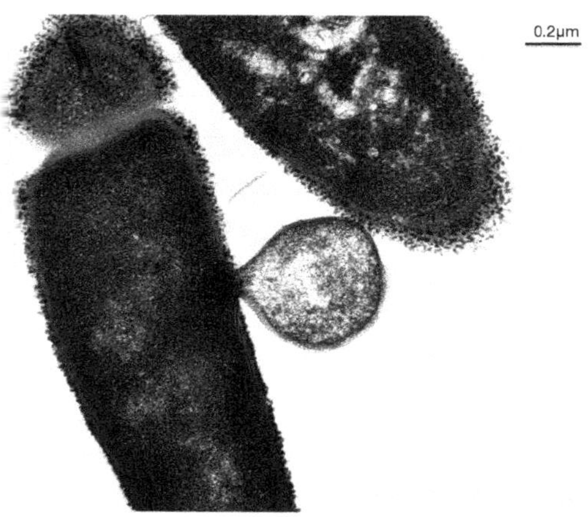

Figure 14.2 Electron microscopy of lysin-treated bacilli. One minute after treatment of *B. cereus* with lysin, membrane extrusion is observed prior to lysis and ultimate death of the bacterium.

Lysin efficacy

In general lysins only kill the species (or subspecies) of bacteria from which they were produced. For instance, enzymes produced from streptococcal phage kill certain streptococci, and enzymes produced by pneumococcal phage kill pneumococci (Loeffler *et al.*, 2001; Nelson *et al.*, 2001). Specifically, a lysin from the C1 streptococcal phage (PlyC) will kill group C streptococci, as well as groups A and E streptococci, the bovine pathogen *S. uberis* and the horse pathogen, *S. equi*, with essentially no effect on streptococci normally found in the oral cavity of humans and other Gram-positive bacteria. Similar results are seen with a pneumococcal-specific lysin, however in this case, the enzyme was also tested against strains of penicillin-resistant pneumococci and the killing efficiency was the same (Djurkovic *et al.*, 2005; Loeffler *et al.*, 2001). Unlike antibiotics, which are usually broad spectrum and kill many different bacteria found in the human body (some of which are beneficial) lysins may be identified which kill only the disease organism with little to no effect on the normal human bacterial flora. One of the most specific lysin reported is the lysin for *B. anthracis* (PlyG), this enzyme only kills *B anthracis* and some rare *B. cereus* strains (Schuch *et al.*, 2002). Another highly specific lysin is a chimeric lysin for staphylococci called ClyS (Daniel *et al.*, 2010; Pastagia *et al.*, 2011). Because this enzyme is an endopeptidase that cleaves the peptidoglycan cross-bridge, and only staphylococci have polyglycine in their cross bridge, this enzyme was shown to have lytic activity on all staphylococci and no other species of bacteria tested (Daniel *et al.*, 2010). In some cases however, phage enzymes may be identified with broad lytic activity. For example, an enterococcal phage lysin PlyV12 has recently been reported to not only kill enterococci but also a number of other Gram-positive pathogens such as *S. pyogenes*, group B streptococci and *Staphylococcus aureus*, making it one of the broadest acting lysins identified (Yoong *et al.*, 2004). However, its activity for these other pathogens was somewhat lower than for enterococci. Thus, the specificity of lysis may be either the result of the cell wall substrate bound by the lysin's binding domain or the unique cross-bridge for that bacterial species. Another lysin with broad lytic activity is PlySs2 from a *Streptococcus suis* phage that can lyse a number of bacterial species including *S. aureus*, *S. epidermidis*, *S. pyogenes*, and *S. pneumoniae* (Gilmer *et al.*, 2013).

Lysin synergy with other lysins and antibiotics

Several lysins have been identified from pneumococcal bacteriophage which are classified into two groups: amidases, and lysozymes. Exposure of pneumococci to either of these enzymes leads to efficient lysis. Both classes of enzymes have very different N-terminal catalytic domains but share a similar C-terminal choline-binding domain. When these enzymes were tested to determine whether their simultaneous use is competitive or synergistic the results clearly showed that they are synergistic (Loeffler and Fischetti, 2003). *In vivo*, the combination of two lysins with different peptidoglycan specificities was found to be more effective in protecting against disease than each of the single enzymes (Jado *et al.*, 2003; Loeffler and Fischetti, 2003). Thus, in addition to more effective killing, the application of two different lysins may significantly retard the emergence of enzyme-resistant mutants. Lysins were also shown to exert synergistic antibacterial effects with phage virion-associated lytic enzymes (Rodríguez-Rubio *et al.*, 2012).

When the pneumococcal lysin Cpl-1 was used in combination with certain antibiotics a similar synergistic effect was seen. Cpl-1 and gentamicin were found to be increasingly synergistic in killing pneumococci with a decreasing penicillin MIC, while Cpl-1 and penicillin showed synergy against an extremely penicillin-resistant strain (Djurkovic *et al.*, 2005). Synergy was also observed with staphylococcal-specific lysins and antibiotics with MRSA both *in vitro* (Fig. 14.3) (Daniel *et al.*, 2010) as well as *in vivo* (Daniel *et al.*, 2010; Rashel *et al.*, 2007). A possible reason for this effect may be partially explained by the finding that the cell wall of MRSA is less cross-linked than MSSA (Trotonda *et al.*, 2009), allowing small quantities of lysin to have a more dramatic effect on cell wall integrity. Thus, the right combination of enzyme and antibiotic could help in the control of antibiotic resistant bacteria as well as reinstate the use of certain antibiotics for which resistance has been established.

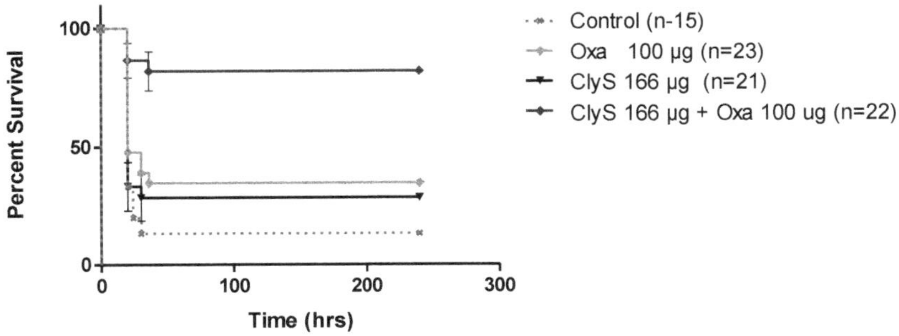

Figure 14.3 Synergistic effects of ClyS and oxacillin protected mice from MRSA septicaemia-induced death. Mice were intraperitoneally injected with ~5x10^5 CFU of MRSA strain MW2 in 5% mucin. Three hours post infection, mice received an i.p. injection of a suboptimal concentration of ClyS (166 µg) or 20 mM phosphate buffer along with an IM injection of oxacillin (100 µg) or saline control. Mice were monitored for survival for 10 days and the results of 5 independent experiments were combined and plotted in a Kaplan–Meier survival curve. A colour version of this figure is available in the plate section at the back of the book.

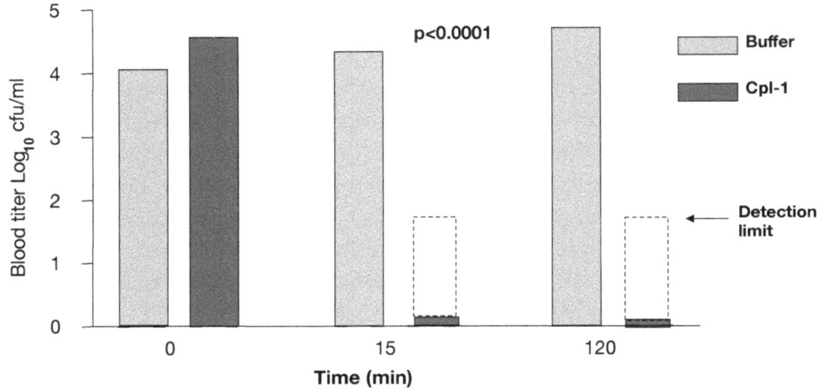

Figure 14.4 Survival of septic mice after treatment with Cpl-1. Mice pretreated IV with 10^8 pneumococci were bled 18 h later to determine the quantity of bacteria in the blood. Mice were then treated IV with a single 2 mg dose of Cpl-1 and groups were euthanized 15 and 120 min after treatment to determine remaining pneumococci. A single dose reduced the bacterial burden below detectable limit. A colour version of this figure is available in the plate section at the back of the book.

Animal models of infection

Animal models of mucosal colonization were used to test the capacity of lysins to kill organisms on these surfaces; perhaps the most important use for these enzymes. An oral colonization model was developed for *S. pyogenes* (Nelson *et al.*, 2001), a nasal model for pneumococci (Loeffler *et al.*, 2001), and a vaginal model for group B streptococci (Cheng *et al.*, 2005). In all three cases, when the animals were colonized with their respective bacteria and treated with a single dose of lysin, specific for the colonizing organism, these organisms were reduced by several logs (and in some cases below the detection limit of the assay) when tested again two to four hours after lysin treatment. These results lend support to the idea that such enzymes may be used in specific high-risk populations to control the reservoir of pathogenic bacteria and thus control disease. A perfect example is the prevention of secondary infections after influenza. Recent studies reveal that 50–90% of deaths resulting from influenza are due to a secondary infection caused predominantly by *S. pneumoniae, S. aureus*, and *S. pyogenes* in that order (Brundage and Shanks, 2007; Morens *et al.*, 2009). Reducing the carriage of these mucosal pathogens during flu season would have a significant impact on disease and death.

Similar to other proteins delivered intravenously to animals and humans, lysins have a short half-life ($T_{1/2}$ = 15–20 min) (Loeffler *et al.*, 2003). However, the action of lysins for bacteria is so rapid, that this may be sufficient time to observe a therapeutic effect (Jado *et al.*, 2003; Loeffler *et al.*, 2003) (Fig. 14.4). Mice intravenously infected with type 14 *S. pneumoniae* and treated 1 h later with a single bolus of 2.0 mg of Cpl-1 survived through the 48 h endpoint, whereas the median survival time of buffer-treated mice was only 25 h, and only 20% survival at 48 h. Blood and organ cultures of the euthanized surviving mice showed that only one Cpl-1-treated animal was totally free of infection at 48 h, suggesting that multiple enzyme doses or a constant infusion of enzyme would be required to eliminate the organisms completely in this application. Similar results were obtained when animals were infected and treated intraperitoneally with lysin (Jado *et al.*, 2003; Rashel *et al.*, 2007). Because of lysin's short half-life, it may be necessary to modify the lysins with polyethylene glycol (PEG) or the Fc region of IgG, to extend the residence time *in vivo* to several hours (Walsh *et al.*, 2003; Resch *et al.*, 2011b). However, in the latter study, even the smallest PEG blocked the activity of the lysin. A decrease in plasma clearance can also be achieved by dimerization of a lysin molecule (Resch *et al.*, 2011a). In recent studies, phage lysins have also been shown to be successful in the treatment of meningitis by adding the lysin directly to the brain intrathecally (Grandgirard *et al.*, 2008), endocarditis by delivering the lysin intravenously by constant IV infusion (Entenza *et al.*, 2005), and pneumonia by administration via inhalation (Doehn *et al.*, 2013) or IV bolus (Witzenrath *et al.*, 2009). Some of these applications would also benefit from modified long-acting lysins.

A significant lysin with respect to infection control is one directed to *Staphylococcus aureus* (Clyne *et al.*, 1992; O'Flaherty *et al.*, 2005; Rashel *et al.*, 2007; Sass and Bierbaum, 2007; Sonstein *et al.*, 1971; Borysowski *et al.*, 2011; Fenton *et al.*, 2011; Abaev *et al.*, 2013; Mishra *et al.*, 2013; Rodríguez-Rubio *et al.*, 2013; Mao *et al.*, 2013). However in most cases these enzymes show low activity or are difficult to produce in large quantities. In one recent publication (Rashel *et al.*, 2007), a staphylococcal enzyme was described that could be easily produced recombinantly and had a significant lethal effect on MRSA both *in vitro* and in a mouse model. In their animal experiments the authors showed that the enzyme may be used to decolonize staphylococci from the nose of the mice as well as protect the animals from an intraperitoneal challenge with MRSA. However, in the latter experiments, the best protection was observed if the lysin was added up to 30 min after the MRSA. Very similar results were published recently using a lysin termed LysGH15 (Gu *et al.*, 2011). In another publication, a chimera was produced linking the catalytic domain of the Twort phage lysin with the binding domain of a PhiNM3 lysin (Daniel *et al.*, 2010). This chimera had eliminated many of the bad features of native staphylococcal phage lysins in its activity and production. The enzyme was also shown to be more effective than mupirocin in a murine model of *S. aureus* skin colonization (Pastagia *et al.*, 2011).

The crucial challenge for lysins would be to determine whether they are able to cure an established infection. To approach this, a mouse pneumonia model was developed in which mice were transnasally infected with pneumococci and treated with Cpl-1 by repeated intraperitoneal injections after infection was established (Witzenrath *et al.*, 2009). From a variety of clinical measurements, as well as morphologic changes in the lungs, it was shown that at 24 h mice suffered from severe pneumonia. When treatment was initiated at 24 h and every 12 h thereafter, 100% of the mice survived otherwise fatal pneumonia and showed rapid recovery. Cpl-1 dramatically reduced pulmonary bacterial counts, and prevented bacteraemia.

Another potential application of lysins is the elimination of bacteria from foods in order to prevent food-borne infections, as recently reported for lysin against *Listeria monocytogenes* that could efficiently kill bacteria in soya milk (Zhang *et al.*, 2012).

Bacterial resistance to lysins

Though several attempts have been made, thus far bacteria that are resistant to the lytic action of native lysins have not been reported. In experiments similar to those that would reveal antibiotic resistance, lysin resistance has not been found. For example, exposure of bacteria grown on agar plates in the presence of low concentrations of lysin did not lead to the recovery of resistant strains even after over 40 cycles. Organisms in colonies isolated at the periphery of a clear lytic zone created by a 10 μl drop of diluted lysin on a lawn of bacteria always resulted in enzyme sensitive bacteria. Enzyme resistant bacteria could also not be identified after >10 cycles of bacterial exposure to low concentrations of lysin (from 5–20 units) in liquid culture (Loeffler *et al.*, 2001; Schuch *et al.*, 2002). These results may be explained for example by the fact that the cell wall receptor for the pneumococcal lysin Cpl-1 is choline (Garcia *et al.*, 1983), a molecule that is essential for pneumococcal viability and the receptor for PlyG, the lysin specific for *B. anthracis*, is also essential for viability (Schuch *et al.*, 2013). It is possible that lack of resistance is linked to the evolution of the binding domain to prevent lysin spill during lysis. Since lysin spill would compromise the survival of the phage progeny, lysin binding domains evolved to bind to cell wall substrates that the bacteria could not change easily, ultimately targeting essential wall components. Because of this, resistance would be a very rare event, rarer than the frequency of antibiotic resistance.

Identifying and isolating new lysins

There are a few ways in which lysins may be identified. The first and simplest is to identify a phage, shotgun clone its DNA and identify lytic activity by overlaying the plated clones with the phage-sensitive bacterium. In this case you usually have a lysin for the organism or species that the phage infects. A more general way of isolating lysins, in order to understand the diversity in this class of enzymes in the environment, is through functional metagenomic analysis (Schmitz *et al.*, 2008). This technique uses random environmental phage populations processed for metagenomic analysis. The twist here is to add an amplification step and an expression step to express and produce the products of the isolated lysin genes. This approach has the potential of identifying novel lysins with powerful biotechnological value. Another approach, which combines the general and the specific approach mentioned above, is to exploit the lysogens in a host genome. This approach, termed multigenomics, identifies the lysin genes in the lysogens within many strains of the same species. In this case the DNA from tens to hundreds of strains of the same species is processed as in the metagenomic analysis, except here the enzymes are from the variety of lysogens in the single species (Schmitz *et al.*, 2011). Interestingly, a recent study showed that the binding domains of lysins can also be used to identify novel targets for antibiotic development (Schuch *et al.*, 2013).

Concluding remarks

Lysins are a new reagent to control bacterial pathogens, particularly those found on the human mucosal surface. For the first time we may be able to specifically kill pathogens on mucous membranes without affecting the surrounding normal flora thus reducing a significant pathogen reservoir in the population. Since this capability has not been previously available, its acceptance may not be immediate. Nevertheless, like vaccines, we should be striving to develop methods to prevent rather than treat infection. Whenever there is a need to kill bacteria, and contact can be made with the organism, lysins may be freely utilized. Such enzymes will be of direct benefit in environments where antibiotic resistant Gram-positive pathogens are a serious problem, such as hospitals, day care centres and nursing homes. The lysins isolated thus far are remarkably heat stable (up to 60°C) and are relatively easy to produce in a purified state and in large quantities, making them amenable to these applications. The challenge for the future is to use this basic strategy and improve upon it, as was the case for second and third generation antibiotics. Protein engineering, domain swapping and gene shuffling all could lead to better lytic enzymes to control bacterial pathogens in a variety of environments. Since it is estimated that there are 10^{31} phage on earth, the potential to identify new lytic enzymes as well as those that kill Gram-negative bacteria is enormous. Perhaps some day phage lytic enzymes will be an essential component in our armamentarium against pathogenic bacteria.

Acknowledgements

I acknowledge the members of my laboratory who are responsible for much of the phage lysin work, Qi Chang, Mattias Collin, Anu Daniel, Sherry Kan, Jutta Loeffler, Daniel Nelson, Chad Euler, Jonathan Schmitz, Raymond Schuch, and Pauline Yoong, and with the excellent technical assistance of Peter Chahales, Adam Pelzek, Rachel Shively, Mary Windels, and Shiwei Zhu. I am indebted to my collaborators Stephen Leib, Jon McCullers, Philippe Moreillon, and Martin Witzenrath for their excellent work with the lysins in their model systems. Supported by DARPA and USPHS Grants AI057472 and AI11822.

References

Abaev, I., Foster-Frey, J., Korobova, O., Shishkova, N., Kiseleva, N., Kopylov, P., Pryamchuk, S., Schmelcher, M., Becker, S.C., and Donovan, D.M. (2013). Staphylococcal phage 2638A endolysin is lytic for *Staphylococcus aureus* and harbours an inter-lytic-domain secondary translational start site. Appl. Microbiol. Biotechnol. *97*, 3449–3456.

Bernhardt, T.G., Wang, I.N., Struck, D.K., and Young, R. (2001). A protein antibiotic in the phage Q-beta virion: diversity in lysis targets. Science *292*, 2326–2329.

Borysowski, J., Weber-Dabrowska, B., and Górski, A. (2006). Bacteriophage endolysins as a novel class of antibacterial agents. Exp. Biol. Med. (Maywood). *231*, 366–377.

Borysowski, J., Lobocka, M., Międzybrodzki, R., Weber-Dabrowska, B., and Górski, A. (2011). Potential of bacteriophages and their lysins in the treatment of MRSA: current status and future perspectives. BioDrugs. *25*, 347–355.

Brundage, J.F., and Shanks, G.D. (2007). What really happened during the 1918 influenza pandemic? The importance of bacterial secondary infections. J. Infect. Dis. *196*, 1717–1718.

Catalão, M.J., Gil, F., Moniz-Pereira, J., São-José, C., and Pimentel, M. (2013). Diversity in bacterial lysis systems: bacteriophages show the way. FEMS Microbiol. Rev. *37*, 554–571.

Chanishvili, N. (2012). Phage therapy – history from Twort and d'Hérelle through Soviet experience to current approaches. Adv. Virus Res. *83*, 3–40.

Cheng, Q., and Fischetti, V.A. (2007). Mutagenesis of a bacteriophage lytic enzyme PlyGBS significantly increases its lytic activity. Appl. Microbiol. Biotechnol. *74*, 1284–1291.

Cheng, Q., Nelson, D., Zhu, S., and Fischetti, V.A. (2005). Removal of group B streptococci colonizing the vagina and oropharynx of mice with a bacteriophage lytic enzyme. Antimicrob. Agents Chemother. *49*, 111–117.

Clyne, M., Birkbeck, T.H., and Arbuthnott, J.P. (1992). Characterization of staphylococcal λ-lysin. J. Gen. Microbiol. *138*, 923–930.

Coello, R., Jimenez, J., Garcia, M., Arroyo, P., Minguez, D., Fernandez, C., Cruzet, F., and Gaspar, C. (1994). Prospective study of infection, colonization and carriage of methicillin-resistant *Staphylococcus aureus* in an outbreak affecting 990 patients. Eur. J. Clin. Microbiol. Infect. Dis. *13*, 74–81.

Daniel, A., Euler, C., Collin, M., Chahales, P., Gorelick, K.J., and Fischetti, V.A. (2010). Synergism between a novel chimeric lysin and oxacillin protects against infection by methicillin-resistant *Staphylococcus aureus*. Antimicrob. Agents Chemother. *54*, 1603–1612.

Diaz, E., Lopez, R., and Garcia, J.L. (1990). Chimeric phage-bacterial enzymes: a clue to the modular evolution of genes. Proc. Natl. Acad. Sci. U.S.A. *87*, 8125–8129.

Djurkovic, S., Loeffler, J.M., and Fischetti, V.A. (2005). Synergistic killing of *Streptococcus pneumoniae* with the bacteriophage lytic enzyme Cpl-1 and penicillin or gentamicin depends on the level of penicillin resistance. Antimicrob. Agents Chemother. *49*, 1225–1228.

Doehn, J.M., Fischer, K., Reppe, K., Gutbier, B., Tschernig, T., Hocke, A.C., Fischetti, V.A., Löffler, J., Suttorp, N., Hippenstiel, S., and Witzenrath, M. (2013). Delivery of the endolysin Cpl-1 by inhalation rescues mice with fatal pneumococcal pneumonia. J. Antimicrob. Chemother. *68*, 2111–2117.

Donovan D.M. (2007). Bacteriophage and peptidoglycan degrading enzymes with antimicrobial applications. Recent Pat. Biotechnol. *1*, 113–122.

Donovan, D.M., Dong, S., Garrett, W., Rousseau, G.M., Moineau, S., and Pritchard, D.G. (2006). Peptidoglycan hydrolase fusions maintain their parental specificities. Appl. Environ. Microbiol. *72*, 2988–2996.

Eiff, C.V., Becker, K., Machka, K., Stammer, H., and Peters, G. (2001). Nasal carriage as a source of *Staphylococcus aureus* bacteremia. New Eng. J. Med. *344*, 11–16.

Entenza, J.M., Loeffler, J.M., Grandgirard, D., Fischetti, V.A., and Moreillon, P. (2005). Therapeutic effects of bacteriophage Cpl-1 lysin against *Streptococcus pneumoniae* endocarditis in rats. Antimicrob. Agents Chemother. *49*, 4789–4792.

Fenton, M., Ross, P., McAuliffe, O., O'Mahony, J., and Coffey, A. (2010). Recombinant bacteriophage lysins as antibacterials. Bioeng. Bugs. *1*, 9–16.

Fenton, M., Ross, R.P., McAuliffe, O., O'Mahony, J., and Coffey, A. (2011). Characterization of the staphylococcal bacteriophage lysin CHAP(K). J. Appl. Microbiol. *111*, 1025–1035.

Fischetti, V.A. (2005). Bacteriophage lytic enzymes: novel anti-infectives. Trends Microbiol. *13*, 491–496.

Fischetti, V.A. (2008). Bacteriophage lysins as effective antibacterials. Curr. Opin. Microbiol. *11*, 393–400.

Fischetti, V.A. (2010). Bacteriophage endolysins: a novel anti-infective to control Gram-positive pathogens. Int. J. Med. Microbiol. *300*, 357–362.

Fischetti, V.A. (2011). Exploiting what phage have evolved to control gram-positive pathogens. Bacteriophage. *1*, 188–194.

Frias, M.J., Melo-Cristino, J., and Ramirez, M. (2013). Export of the pneumococcal phage SV1 lysin requires choline-containing teichoic acids and is holin-independent. Mol. Microbiol. *87*, 430–445.

Garcia, E., Garcia, J.L., Arraras, A., Sanchez-Puelles, J.M., and Lopez, R. (1988). Molecular evolution of lytic enzymes of *Streptococcus pneumoniae* and its bacteriophages. Proc. Natl. Acad. Sci. U.S.A. *85*, 914–918.

Garcia, P., Garcia, E., Ronda, C., Tomasz, A., and Lopez, R. (1983). Inhibition of lysis by antibody against phage-associated lysin and requirement of choline residues in the cell wall for progeny phage release in *Streptococcus pneumoniae*. Curr. Microbiol. *8*, 137–140.

Garcia, P., Garcia, J.L., Garcia, E., Sanchez-Puelles, J.M., and Lopez, R. (1990). Modular organization of the lytic enzymes of *Streptococcus pneumoniae* and its bacteriophages. Gene *86*, 81–88.

Gilmer, D.B., Schmitz, J.E., Euler, C.W., and Fischetti, V.A. (2013). Novel bacteriophage lysin with broad lytic activity protects against mixed infection by Streptococcus pyogenes and methicillin-resistant *Staphylococcus aureus*. Antimicrob. Agents Chemother. *57*, 2743–2750.

Grandgirard, D., Loeffler, J.M., Fischetti, V.A., and Leib, S.L. (2008). Phage lytic enzyme Cpl-1 for antibacterial therapy in experimental pneumococcal meningitis. J. Infect. Dis. *197*, 1519–1522.

Gu, J., Xu, W., Lei, L., Huang, J., Feng, X., Sun, C., Du, C., Zuo, J., Li, Y., and Dia, Y. (2011). LysGH15, a novel bacteriophage lysin, protects a murine bacteremia model efficiently against lethal methicillin-resistant *Staphylococcus aureus* infection. J. Clin. Microbiol. *49*, 111–117.

Hawser, S. (2012). Surveillance programmes and antibiotic resistance: worldwide and regional monitoring of antibiotic resistance trends. Handb. Exp. Pharmacol. *211*, 31–43.

Hermoso, J.A., Monterroso, B., Albert, A., Galan, B., Ahrazem, O., Garcia, P., Martinez-Ripoli, M., Garcia, J.L., and Menendez, M. (2003). Structural basis for selective recognition of pneumococcal cell wall by modular endolysin from phage Cp-1. Structure *11*, 1239–1249.

Hermoso, J.A., García, J.L., and García, P. (2007). Taking aim on bacterial pathogens: from phage therapy to enzybiotics. Curr. Opin. Microbiol. *10*, 461–472.

Heselpoth, R.D., and Nelson, D.C. (2012). A new screening method for the directed evolution of thermostable bacteriolytic enzymes. J. Vis. Exp. *69*, pii: 4216.

Hudson, L., and Hay, F.C. (1976). Practical Immunology (Oxford, UK, Blackwell Scientific Publications).

Jado, I., Lopez, R., Garcia, E., Fenoll, A., Casal, J., and Garcia, P. (2003). Phage lytic enzymes as therapy for antiobiotic-resistant Streptococcus pneumoniae infection in a murine sepsis model. J. Antimicrob. Chemother. *52*, 967 – 973.

Jervis, E.J., Haynes, C.A., and Kilburn, D.G. (1997). Surface diffusion of cellulases and their isolated binding domains on cellulose. J. Biol. Chem. *272*, 24016–24023.

de Lencastre, H., Kristinsson, K.G., Brito-Avo, A., Sanches, I.S., Sa-Leao, R., Saldanha, J., Sigvaldadottir, E., Karlsson, S., Oliveira, D., Mato, R., *et al.* (1999). Carriage of respiratory tract pathogens and molecular epidemiology of *Streptococcus pneumoniae* colonization in healthy children attending day care centers in Lisbon, Portugal. Microb. Drug Resist. *5*, 19–29.

Loeffler, J.M., and Fischetti, V.A. (2003). Synergistic lethal effect of a combination of phage lytic enzymes with different activities on penicillin-sensitive and -resistant *Streptococcus pneumoniae* strains. Antimicrob. Agents Chemother. *47*, 375–377.

Loeffler, J.M., Nelson, D., and Fischetti, V.A. (2001). Rapid killing of *Streptococcus pneumoniae* with a bacteriophage cell wall hydrolase. Science *294*, 2170–2172.

Loeffler, J.M., Djurkovic, S., and Fischetti, V.A. (2003). Phage lytic enzyme Cpl-1 as a novel antimicrobial for pneumococcal bacteremia. Infect. Immun. *71*, 6199–6204.

Loessner, M.J. (2005). Bacteriophage endolysins- current state of research and applications. Curr. Op. Microbiol. *8*, 480–487.

Loessner, M.J., Maier, S.K., Daubek-Puza, H., Wendlinger, G., and Scherer, S. (1997). Three Bacillus cereus bacteriophage endolysins are unrelated but reveal high homology to cell wall hydrolases from different bacilli. J. Bacteriol. *179*, 2845–2851.

Loessner, M.J., Kramer, K., Ebel, F., and Scherer, S. (2002). C-terminal domains of *Listeria monocytogenes* bacteriophage murein hydrolases determine specific recognition and high-affinity binding to bacterial cell wall carbohydrates. Mol. Microbiol. *44*, 335–349.

Lopez, R., Garcia, J.L., Garcia, E., Ronda, C., and Garcia, P. (1992). Structural analysis and biological significance of the cell wall lytic enzymes of Streptococcus pneumoniae and its bacteriophage. FEMS Microbiol. Lett. *79*, 439–447.

Lopez, R., Garcia, E., Garcia, P., and Garcia, J.L. (1997). The pneumococcal cell wall degrading enzymes: a modular design to create new lysins? Microb. Drug Resist. 3, 199–211.

Low, L.Y., Yang, C., Perego, M., Osterman, A., and Liddington, R.C. (2005). Structure and lytic activity of a *Bacillus anthracis* prophage endolysin. J. Biol. Chem. *280*, 35433–35439.

Low, L.Y., Yang, C., Perego, M., Osterman, A., and Liddington, R. (2011). Role of net charge on catalytic domain and influence of cell wall binding domain on bactericidal activity, specificity, and host range of phage lysins. J. Biol. Chem. *286*, 34391–34403.

Lukacik, P., Barnard, T.J., and Buchanan, S.K. (2012). Using a bacteriocin structure to engineer a phage lysin that targets Yersinia pestis. Biochem. Soc. Trans. *40*, 1503–1506.

McGowan, S., Buckle, A.M., Mitchell, M.S., Hoopes, J.T., Gallagher, D.T., Heselpoth, R.D., Shen, Y., Reboul, C.F., Law, R.H., Fischetti, V.A., *et al.* (2012). X-ray crystal structure of the streptococcal specific phage lysin PlyC. Proc. Natl. Acad. Sci. U.S.A. *109*, 12752–12757.

Mao, J., Schmelcher, M., Harty, W.J., Foster-Frey, J., and Donovan, D.M. (2013). Chimeric Ply187 endolysin kills *Staphylococcus aureus* more effectively than the parental enzyme. FEMS Microbiol. Lett. *342*, 30–36

Matsuzaki, S., Rashel, M., Uchiyama, J., Sakurai, S., Ujihara, T., Kuroda, M., Ikeuchi, M., Tani, T., Fujieda, M., Wakiguchi, H., *et al.* (2005). Bacteriophage therapy: a revitalized therapy against bacterial infectious diseases. J. Infect. Chemother. *11*, 211–219.

Mayer, M.J., Garefalaki, V., Spoerl, R., Narbad, A., and Meijers, R. (2011). Structure-based modification of a *Clostridium difficile*-targeting endolysin affects activity and host range. J. Bacteriol. *193*, 5477–5486.

Mehta, M.S., Hacek, D.M., Kufner, B.A., Price, C., and Peterson, L.R. (2013). Dose-ranging study to assess the application of intranasal 2% mupirocin calcium ointment to eradicate *Staphylococcus aureus* nasal colonization. Surg. Infect. (Larchmt). *14*, 69–72.

Mishra, A.K., Rawat, M., Viswas, K.N., Abhishek, Kumar, S., and Reddy, M. (2013). Expression and lytic efficacy assessment of the *Staphylococcus aureus* phage SA4 lysin gene. J. Vet. Sci. *14*, 37–43.

Morens, D.M., Taubenberger, J.K., Folkers, G.K., and Fauci, A.S. (2009). An historical antecedent of modern guidelines for community pandemic influenza mitigation. Public Health Rep. *124*, 22–25.

Navarre, W.W., Ton-That, H., Faull, K.F., and Schneewind, O. (1999). Multiple enzymatic activities of the murein hydrolase from staphylococcal phage phi11. Identification of a D-alanyl-glycine endopeptidase activity. J. Biol. Chem. *274*, 15847–15856.

Nelson, D., Loomis, L., and Fischetti, V.A. (2001). Prevention and elimination of upper respiratory colonization of mice by group A streptococci by using a bacteriophage lytic enzyme. Proc. Natl. Acad. Sci. U.S.A. *98*, 4107–4112.

Nelson, D., Schuch, R., Chahales, P., Zhu, S., and Fischetti, V.A. (2006). PlyC: a multimeric bacteriophage lysin. Proc. Natl. Acad. Sci. U.S.A. *103*, 10765–10770.

Nelson, D.C., Schmelcher, M., Rodriguez-Rubio, L., Klumpp, J., Pritchard, D.G., Dong, S., and Donovan, D.M. (2012). Endolysins as antimicrobials. Adv. Virus Res. *83*, 299–365.

O'Flaherty, S., Coffey, A., Meaney, W., Fitzgerald, G.F., and Ross, R.P. (2005). The recombinant phage lysin LysK has a broad spectrum of lytic activity against clinically relevant staphylococci, including methicillin-resistant *Staphylococcus aureus*. J. Bacteriol. *187*, 7161–7164.

O'Grady, S., Hirji, Z., Pejcic-Karapetrovic, B., Fung, S., Dedier, H., Takata-Shewchuk, J., Zhang, K., and Conly, J. (2009). A double-blind, randomized, controlled trial of topical polysporin triple compound versus topical mupirocin for the eradication of colonization with methicillin-resistant *Staphylococcus aureus* in a complex continuing care population. Can. J. Infect. Dis. Med. Microbiol. *20*, e49–55.

Oliveira, H., Melo, L.D., Santos, S.B., Nóbrega, F.L., Ferreira, E.C., Cerca, N., Azeredo, J., and Kluskens, L.D. (2013). Molecular aspects and comparative genomics of bacteriophage endolysins. J. Virol. *87*, 4558–4570.

Pastagia, M., Euler, C., Chahales, P., Fuentes-Duculan, J., Krueger, J.G., and Fischetti, V.A. (2011). A novel chimeric lysin shows superiority to mupirocin for skin decolonization of methicillin-resistant and -sensitive *Staphylococcus aureus* strains. Antimicrob. Agents Chemother. *55*, 738–744.

Pritchard, D.G., Dong, S., Kirk, M.C., Cartee, R.T., and Baker, J.R. (2007). LambdaSa1 and LambdaSa2 prophage lysins of Streptococcus agalactiae. Appl. Environ. Microbiol. 73, 7150–7154.

Rashel, M., Uchiyama, J., Ujihara, T., Uehara, Y., Kuramoto, S., Sugihara, S., Yagyu, K., Muraoka, A., Sugai, M., Hiramatsu, K., *et al.* (2007). Efficient elimination of multidrug-resistant *Staphylococcus aureus* by cloned lysin derived from bacteriophage phi MR11. J. Infect. Dis. *196*, 1237–1247.

Resch, G., Moreillon, P., and Fischetti, V.A. (2011a). A stable phage lysin (Cpl-1) dimer with increased antipneumococcal activity and decreased plasma clearance. Int. J. Antimicrob. Agents. *38*, 516–521.

Resch, G., Moreillon, P., and Fischetti, V.A. (2011b). PEGylating a bacteriophage endolysin inhibits its bactericidal activity. AMB Express *1*, 29.

Rodríguez-Rubio, L., Martínez, B., Rodríguez, A., Donovan, D.M., and García, P. (2012). Enhanced staphylolytic activity of the *Staphylococcus aureus* bacteriophage vB_SauS-phiIPLA88 HydH5 virion-associated peptidoglycan hydrolase: fusions, deletions, and synergy with LysH5. Appl. Environ. Microbiol. *78*, 2241–2248.

Rodríguez-Rubio, L., Martínez, B., Rodríguez, A., Donovan, D.M., Götz, F., and García, P. (2013). The phage lytic proteins from the *Staphylococcus aureus* bacteriophage vB_SauS-phiIPLA88 display multiple active catalytic domains and do not trigger staphylococcal resistance. PloS One *8*, e64671.

Sass, P., and Bierbaum, G. (2007). Lytic activity of recombinant bacteriophage phi11 and phi12 endolysins on whole cells and biofilms of *Staphylococcus aureus*. Appl. Environ. Microbiol. 73, 347–352.

Schmelcher, M., Donovan, D.M., and Loessner, M.J. (2012). Bacteriophage endolysins as novel antimicrobials. Future Microbiol. 7, 1147–1171.

Schmitz, J.E., Daniel, A., Collin, M., Schuch, R., and Fischetti, V.A. (2008). Rapid DNA library construction for functional genomic and metagenomic screening. Appl. Environ. Microbiol. *74*, 1649–1652.

Schmitz, J.E., Ossiprandi, M.C., Rumah, K.R., and Fischetti, V.A. (2011). Lytic enzyme discovery through multigenomic sequence analysis in Clostridium perfringens. Appl. Microbiol. Biotechnol. *89*, 1783–1795.

Schuch, R., Nelson, D., and Fischetti, V.A. (2002). A bacteriolytic agent that detects and kills *Bacillus anthracis*. Science *418*, 884–889.

Schuch, R., Pelzek, A.J., Raz, A., Euler, C.W., Ryan, P.A., Winer, B.Y., Farnsworth, A., Bhaskaran, S.S., Stebbins, C.E., Xu, Y., *et al.* (2013). Use of a bacteriophage lysin to identify a novel target for antimicrobial development. PloS One *8*, e60754.

Sonstein, S.A., Hammel, J.M., and Bondi, A. (1971). Staphylococcal bacteriophage-associated lysin: a lytic agent active against *Staphylococcus aureus*. J. Bacteriol. *107*, 499–504.

Trotonda, M.P., Xiong, Y.Q., Memmi, G., Bayer, A.S., and Cheung, A.L. (2009). Role of mgrA and sarA in methicillin-resistant *Staphylococcus aureus* autolysis and resistance to cell wall-active antibiotics. J. Infect. Dis. *199*, 209–218.

Walsh, S., Shah, A., and Mond, J. (2003). Improved pharmacokinetics and reduced antibody reactivity of lysostaphin conjugated to polyethylene glycol. Antimicrob. Agents Chemother. *47*, 554–558.

Wang, I.N., Smith, D.L., and Young, R. (2000). Holins: the protein clocks of bacteriophage infections. Annu. Rev. Microbiol. *54*, 799–825.

Wang, I.N., Deaton, J., and Young, R. (2003). Sizing the holin lesion with an endolysin–beta-galactosidase fusion. J. Bacteriol. *185*, 779–787.

Weiss, K., Lavardiere, M., Lovgren, M., Delorme, J., Poirier, L., and Beliveau, C. (1999). Group A streptococcus carriage among close contacts of patients with invasive infections. Am. J. Epidemiol. *149*, 863–868.

White, R., Chiba, S., Pang, T., Dewey, J.S., Savva, C.G., Holzenburg, A., Pogliano, K., and Young, R. (2011). Holin triggering in real time. Proc. Natl. Acad. Sci. U.S.A. *108*, 798–803.

Witzenrath, M., Schmeck, B., Doehn, J.M., Tschernig, T., Zahlten, J., Loeffler, J.M., Zemlin, M., Muller, H., Gutbier, B., Schütte, H., *et al.* (2009). Systemic use of the endolysin Cpl-1 rescues mice with fatal pneumococcal pneumonia. Crit. Care Med. *37*, 642–649.

Yoong, P., Nelson, D., Schuch, R., and Fischetti, V.A. (2004). Identification of a broadly active phage lytic enzyme with lethal activity against antibiotic-resistant Enterococcus faecalis and Enterococcus faecium. J. Bacteriol. *186*, 4808–4812.

Young, R. (1992). Bacteriophage lysis: mechanism and regulation. Microbiol. Rev. *56*, 430–481.

Young, R., Wang, I.N., and Roof, W.D. (2000). Phages will out: strategies of host cell lysis. Trends Microbiol. *8*, 120–128.

Zhang, H., Bao, H., Billington, C., Hudson, J.A., and Wang, R. (2012). Isolation and lytic activity of the Listeria bacteriophage endolysin LysZ5 against *Listeria monocytogenes* in soya milk. Food Microbiol. 3, 133–136.

Genetically Engineered Phage as Antimicrobials and Biodetectors

15

Salim Manoharadas and Udo Bläsi

Abstract

With the advent of antibiotics the use of bacteriophage as antimicrobial agents has been abandoned in the western world. However, the increasing prevalence of multi-drug resistant bacterial pathogens has resulted in a resurgence of research efforts to use phage as antimicrobials. A side-effect of many antibiotics as well as of phage therapy with lytic phage is the release of cell wall components, e.g. endotoxins of Gram-negative bacteria, which mediate the general pathological aspects of septicaemia. In the last decade several strategies based on genetically engineered lysis-deficient phage have been devised with the aim to avoid disintegration of the cell envelope but to kill the bacterial target. These studies indicated that killing-proficient but lysis-defective recombinant phage can be exploited as efficient antimicrobials with reduced side-effects. Moreover, genetically engineered phage can be used to augment the antimicrobial efficacy of antibiotics and to reduce bacterial biofilms. Apart from these potential medical applications, modified phages have been used to detect bacterial pathogens in foodstuff. Here, we provide a review of these studies and briefly discuss the prospects of genetically modified phage in medicine and industry.

Introduction

The therapeutic potential of lytic bacteriophage to control bacterial pathogens is long known, and phage therapy is still in use in some central and eastern European countries. In contrast, the emergence of effective antibiotics with a broad host range together with their facile administration resulted in a negligence of the potential of therapeutic phage in the western world. However, the emergence of multi-drug resistant bacterial pathogens has spawned renewed interest in phage therapy. For instance, ~2 billion people worldwide are colonized with *Staphylococcus aureus*, out of which approximately 53 million (2.7%) are thought to carry MRSA (*m*ethicillin-*r*esistant *S. aureus*) strains. MRSA strains are resistant to all conventional anti-staphylococcal beta-lactam antibiotics. The glycopeptide antibiotics vancomycin and teicoplanin served for many years as the last resort to treat MRSA infections (Schentag *et al.*, 1998; Segarra-Newnham and Church, 2012). However, several of the newly discovered MRSA strains show resistance to both antibiotics. These MRSA strains are known as VISA (*v*ancomycin *i*ntermediate-resistant *S. aureus*) (Sieradzki and Tomasz, 1997; Schito, 2006). *S. aureus* isolates with decreased susceptibility to new antibiotics, including linezolid and daptomycin, were also reported (Sader *et al.*, 2013). In addition to *S. aureus* there are several other emerging multidrug-resistant pathogens that cause a serious menace in medicine. To name a few: *Acinetobacter baumanii, Clostridium difficile, Enterococcus faecalis, Klebsiella pneumoniae, Pseudomonas aeruginosa* and *Escherichia coli* (Woodford *et al.*, 2011; Chancey *et al.*, 2012).

Bacteriophage can offer advantages over antibiotics. They are natural entities, can be produced at a reasonable cost, are highly specific for the bacterial target, do not harm beneficial commensals, and they vanish once the bacterial target is eliminated (Kutateladze and Adamia, 2010; Burrowes *et al.*, 2011). The number of scientific reports on the treatment of experimental animal infections with intact phage particles or phage-derived proteins either solely or in combination with antibiotics has steadily increased in the last decade. For instance, phage therapy was used to treat respiratory infections caused by *E. coli* in broiler chickens. The intramuscular injection of bacteriophage protected chicken from infections when administered at 0 h, 24 h or 48 h after challenge with *E. coli* (Huff *et al.*, 2003). In another study, phage were successfully used to treat fish infections caused by *Pseudomonas plecoglossicida* (Park and Nakai, 2003). The efficacy of bacteriophage therapy was also assessed in mice with acute lung infections caused by a bioluminescent *P. aeruginosa* strain. The phage-treated mice showed a reduced pathogen load in the chest within 6 h, and were completely devoid of the pathogen after 24 h (Debarbieux *et al.*, 2010). In addition, successful curative and preventative treatment of lung infections in mice caused by *P. aeruginosa* strain CHA was also achieved using bacteriophage P3-CHA (Morello *et al.*, 2011).

The efficacy and safety of phage therapy in the treatment of a wide range of human infections has been demonstrated in many uncontrolled studies conducted largely in Poland, Georgia, and Russia (Kutadeladze and Adamia, 2010; Abedon *et al.*, 2011; Chanishvili *et al.*, 2012; Miedzybrodzki *et al.*, 2012). Recently, the first controlled clinical trials were conducted (Wright *et al.*, 2009; Harper *et al.*, 2011; Sarker *et al.*, 2012). They seem to confirm the results of earlier uncontrolled studies.

Host cell lysis by double-stranded DNA phage is commonly achieved by a phage encoded holin in conjunction with a muralytic enzyme (endolysin). In general, the holin induces a non-specific lesion in the cytoplasmic membrane, which allows the endolysin to gain access to and to degrade the peptidoglycan, leading to disruption of the cell envelope (Young, 1992; Catalão *et al.* 2013). In contrast to Gram-negative bacteria, endolysins encoded by phage of Gram-positive bacteria can act as exolysins, i.e. by degrading the peptidoglycan from without (Web resource 1 and 2). Purified phage endolysins have been used as antimicrobials against Streptococci in mice (Nelson *et al.*, 2001; Loeffler *et al.*, 2001; Gilmer *et al.*, 2013), and have been proven effective against other Gram-positive pathogens including *S. aureus* (Fenton *et al.*, 2013; Gilmer *et al.*, 2013), *Enterococcus faecalis* and *E. faecium* (Yoong *et al.*, 2004), *Clostridium perfringens* (Zimmer *et al.*, 2002; Simmons *et al.*, 2010), group B *Streptococcus* (Cheng *et al.*, 2005) and *Bacillus anthracis* (Schuch *et al.*, 2002). The term enzybiotic has been coined to classify these antimicrobials (Nelson *et al.*, 2001; Hermoso *et al.*, 2007; Fischetti, 2010). Moreover, recombinant endolysins have been engineered to improve the solubility of the proteins (Daniel *et al.*, 2010; Fenton *et al.*, 2010). For instance, a chimeric enzybiotic consisting of the N-terminal catalytic domain of the P16 endolysin and the C-terminal cell wall binding domain of the virion protein P17 of *S. aureus* phage P68 was constructed. The recombinant protein was soluble and exerted antimicrobial activity against *S. aureus* strains including MRSA when used alone or in combination with antibiotics (Manoharadas *et al.*, 2009). A chimeric enzybiotic (ClyS) encompassing the N-terminal domain of the endolysin of the *S. aureus* phage Twort and the C-terminal cell wall-targeting domain of the endolysin of *S. aureus* phage phiNM3, displayed antimicrobial activity against MRSA and VISA strains. ClyS also displayed synergistic antimicrobial activity with both vancomycin and oxacillin *in vitro*, thereby reducing the effective concentration of the drug (Daniel *et al.*, 2010). Similarly, a filamentous phage was shown to augment the antimicrobial efficacy of antibiotics, i.e. in the presence of the phage the effective dose of antibiotics could be reduced (Hagens *et al.*, 2006).

The US Food and Drug Administration (FDA) has issued GRAS (Generally Recognized As Safe) status for two phage preparations (Listex™ and SalmoFresh™). Moreover, one phage preparation (EcoShield™) has been approved as a food contact substance, and another one (ListShield™) as a food additive (Sulakvelidze *et al.*, 2013). This provided a major boost towards the application of bacteriophage in food industry (Web Resource 3; Guenther *et al.*, 2009; Sillankorva *et al.*, 2012; Sulakvelidze, 2013). In order to achieve efficient biocontrol by virulent phage a broad host range is desired (O'Flaherty *et al.*, 2005; Bielke *et al.*, 2007). Alternatively, a cocktail or mixture of different phage can be used in food processing. A cocktail containing three different phage was used to treat beef contaminated with *E. coli* O157:H7. After storage, no viable *E. coli* cells were observed in the majority of the samples (O'Flynn *et al.*, 2004).

Genetically engineered phage

As mentioned above, phage infection generally culminates in lysis of the host, and consequently mediates the release of cell-wall components, e.g. endotoxin release from Gram-negative bacteria. Orally administered phage have been detected in the blood of patients (Slopek *et al.*, 1983; Weber-Dabrowska *et al.*, 1987), which would induce endotoxin release upon phage-mediated lysis of Gram-negative pathogens, which in turn may lead to septic shock (Matsuda *et al.*, 2005; Paul *et al.*, 2011). In addition, as phages are replicating entities, the application of a defined dose is difficult, which could jeopardize the endeavour to obtain certification for phage as pharmaceuticals. Genetically engineered phage (Fig. 15.1) that effectively kill the host but do not cause lysis may alleviate these concerns without compromising efficacy. Moreover, phage can be genetically modified to extend their host range, to augment the activity of antibiotics or to reduce biofilms (Moradpour and Ghasemian, 2011).

Lysis-deficient phage

It has been shown that a recombinant filamentous phage encoding lethal but non-lytic proteins can efficiently kill *E. coli* with minimal release of endotoxin when compared to a strictly lytic phage (Hagens and Bläsi, 2003). The filamentous M13 phage (Fig. 15.1) was genetically engineered to encode either the phage lambda holin (λS105) or the *Bgl*II restriction endonuclease (R) with the reasoning that upon infection with the M13-λS105 and the M13R phage, the holin will cause a lethal lesion in the inner membrane and the *Bgl*II enzymes will lead to breaks in the chromosomal DNA, respectively. Two hours following infection with M13-λS105 at a MOI (*m*ultiplicity *o*f *i*nfection) of 10, 99% of the cells were killed. However, growth of the bacterial cells resumed after 2 hours, albeit slower than in the control. This was attributed to the emergence of *E. coli* mutants lacking F-pili, which are required by M13 to adsorb to the cell surface. The infection of *E. coli* cells with the M13R phage led likewise to a rapid decline in viable cell counts. When compared with M13-λS105, the killing efficiency was increased in a manner comparable with the performance of the lytic λcI- phage. Although 99.9% of the cells were killed, the number of *E. coli* cells remained constant over a 6-hour period following M13R infection. In contrast, the number of structurally intact cells mirrored the number of viable cells upon infection with phage λcI^-. As anticipated, the endotoxin levels released after infection with M13-λS105, M13R and λcI^- differed. After 1 h following infection with λcI^-, a 18-fold increase of the endotoxin levels were noticed as compared to a ~ 2-fold increase with either M13S105 or M13R (Hagens and Bläsi, 2003).

In a follow up study a non-replicating, non-lytic bacteriophage was engineered to combat an experimental *P. aeruginosa* infection (Hagens *et al.*, 2004). An export protein gene of the *P. aeruginosa* filamentous phage Pf3 was replaced with the gene encoding the *Bgl*II restriction endonuclease. This rendered the Pf3 variant (Pf3R) non-replicative, and concomitantly prevented the release of Pf3R from the target cell. The recombinant phage was propagated on a host strain expressing the *bglII* methylase gene and the Pf3 export function. The Pf3R phage efficiently killed *P. aeruginosa* while the structural integrity of the cells was maintained and endotoxin release was kept to a minimum. The release of endotoxin after Pf3R infection was approximately 10-fold lower when compared with the lytic *P. aeruginosa* phage Pt1. Pf3R treatment and treatment with the replicating lytic phage Pt1 of *P. aeruginosa* infections of mice resulted in a comparable survival rate upon challenge with a minimal lethal dose of 3. However, when compared with the lytic phage, the survival rate after phage therapy with Pf3R was significantly higher upon challenge with a minimal lethal dose of 5. After 48 h, the Pf3R and Pt1 phage-treated mice showed a survival rate of 73% and 20%, respectively. When compared to Pf3R treated mice, increased TNF-α and IL-6 levels were detected in serum samples of mice treated with the lytic phage Pt1. Thus, the increased survival rate

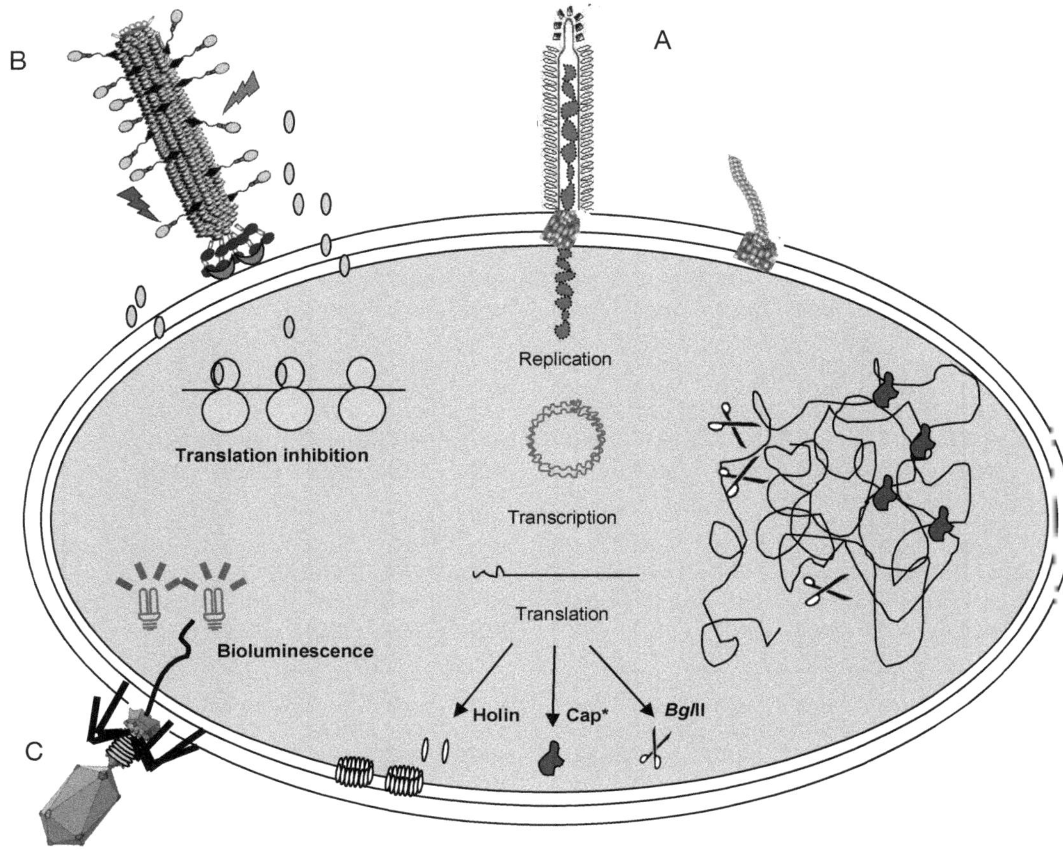

Figure 15.1 Host cell killing by genetically engineered phage. (A) Recombinant phage encoding lethal functions. M13 variants encoding the λ holin (kills the host cell by generation of lesions in the inner membrane), Cap* (binds chromosomal DNA non-specifically), or *Bgl*II (cleaves host DNA) are used to kill target bacteria without lysis. (B) Recombinant phage with engineered specificity for the bacterial target. M13 phage are used as a drug carrying platform. Upon targeted binding to a specific pathogen, the attached drug (chlorampenicol; yellow oval) is released and causes cell death. (C) Recombinant phage exploited as biodetectors. Phage encoding luciferase or GFP are employed to detect pathogens. For more details and references see text. A colour version of this figure is available in the plate section at the back of the book.

of Pf3R treated mice appeared to result from a reduced endotoxin release (Hagens *et al.*, 2004).

In a conceptually similar study, a M13 phagemid was used to deliver the genes encoding the toxic *E. coli* proteins Gef and ChpBK (Westwater *et al.*, 2003). By forming a pore in the inner membrane Gef dissipates the membrane potential, which leads to cell killing (Gerdes *et al.*, 1986). Unlike Gef, ChpBK acts as a sequence-specific endoribonuclease, which preferentially cleaves single-stranded RNAs (Zhang *et al.*, 2005). The *gef* and *chpBK* genes were cloned into a phagemid and the respective transformants were infected with the helper phage R408. This procedure allowed preferential packaging of the phagemid DNA, resulting in M13 phage variants encoding the lethal functions. Infection of *E. coli* ER2738 with either M13-Gef or M13-ChpBK phages resulted in a strong reduction of viable cells. The efficacy of these recombinant phages was also studied in a mouse infection model. A single intraperitoneal injection of 10^8 CFU (Colony Forming Units) of *E. coli* ER2738 was used to induce a transient infection in immunocompromised mice. When compared to the control, a significant reduction (>90%) in blood-circulating bacteria was observed 5 hours after treatment with either recombinant phage (Westwater *et al.*, 2003).

Catabolite activator protein (CAP) is a DNA-binding protein involved in carbon utilization (de Crombrugghe *et al.*, 1984). CAP in complex with cyclic-AMP (cAMP) binds to DNA target sites at or near many promoters in the *E. coli* chromosome and stimulates transcription (Musso *et al.*, 1977). A modified CAP variant with a replacement of Glu181→Gln displays unspecific DNA binding, and over-production of this protein is lethal (Lopata *et al.*, 1997). Using the same strategy as described above, Moradpour *et al.* (2009) cloned the mutant *cap** gene into the pBluescript II SK (+) vector to generate a M13 variant encoding the lethal function (Fig. 15.1). When compared to the uninfected control, a ~ 650-fold reduction in CFU was observed 60 min post infection of the pathogenic *E. coli* O157:H7 with the M13-*cap** phage. As *E. coli* O157:H7 is sometimes found in dairy products including milk and yoghurt (Kumar *et al.*, 2013), the efficacy of the M13-*cap** phage in reducing viable *E. coli* O157:H7 was tested in milk. Sterilized milk was inoculated with 10^7 bacterial cells and the M13-*cap** phage was added at a MOI of 100. After 120 min, the viable bacterial cell count in milk was reduced by 6 orders of magnitude (Moradpour *et al.*, 2009).

Bactericidal quinolone antibiotics such as ofloxacin induce hydroxyl radical formation that leads to damage of cellular DNA, proteins and lipids (Kohanski *et al.*, 2007; Cheng *et al.*, 2013). Consequently, the SOS response of the cell is induced, which results in repair of the damaged DNA (Lewin *et al.*, 1989; Miller *et al.*, 2004; Janion, 2008). The efficacy of such bactericidal antibiotics can be enhanced by disabling the SOS response. An M13 phage was engineered to over-express the *lexA3* gene, encoding a repressor of the SOS response (Lu and Collins, 2009). When compared to ofloxacin treated cells, a 4.5 log fold drop in CFU was observed when *E. coli* cells were additionally infected with the M13-*lexA3* phage. The M13-*lexA3* phage was also proficient in killing *E. coli* strains resistant to quinolone antibiotics. *E. coli* RFS289, which harbours a mutation that confers resistance against quinolone antibiotics, was susceptible to a combination therapy with ofloxacin and the modified phage (Lu and Collins, 2009). In addition, the efficacy of the combination therapy was tested in mice. Intraperitoneal injections of mice with 8.8×10^7 CFU/mouse of *E. coli* EMG2 killed the mice within 2 days. In contrast, 80% of the mice that received a combination therapy with M13-*lexA3* phage (10^9 phage/mouse) and ofloxacin (0.2 mg/kg) after 1 h of infection survived, whereas only 50% and 20% treated with unmodified phage/ofloxacin and ofloxacin alone survived, respectively (Lu and Collins, 2009). Moreover, the M13-*lexA3* phage augmented the antibacterial efficacy of other classes of antibiotics, especially aminoglycosides and β-lactams, reduced the development of antibiotic-resistance upon exposure of bacteria to subinhibitory concentrations of antibiotics and decreased the number of persister cells (Lu and Collins, 2009).

These authors further designed recombinant M13 variants equipped with genes, the products of which target gene networks other than the SOS response. The M13 variants carried the *soxR*, the *csrA* and the *ompF* gene, repectively. *SoxR* and *soxS* form part of the *soxRS* operon, which controls a cellular response to oxidative stress (Hidalgo, 1997; Chiang and Schellhorn, 2012). CsrA regulates glycogen synthesis and acts as a repressor of biofilm formation (Jackson *et al.*, 2002; Romeo *et al.*, 2013), whereas OmpF is a porin through which quinolones enter the bacterial cell (Hirai *et al.*, 1986; Mahendran *et al.*, 2010). In all cases a combination of the respective recombinant M13 phage and ofloxacin was more efficient in killing *E. coli* than the sole application of either the recombinant phage or the antibiotic.

Although the studies described above proved the efficacy of recombinant phage *in vitro* as well as in *in vivo* infection models, genetically modified phage are commonly believed to complicate the endeavour to obtain a certification as pharmaceuticals. An alternative to recombinant phage encoding toxic functions are endolysin-deficient phage. These phage retain the ability to kill by means of their holin function but do not lyse the host cell (Rietsch and Bläsi, 1993; Paul *et al.*, 2011). *Vice versa*, holin-deficient phage are non-lytic but can kill the host by means of phage-induced DNases and/or by interference with the bacterial metabolism. In one study a T4 phage (LyD) carrying an amber mutation in the holin gene *t* was tested for its efficacy in murine peritonitis. The mice were injected intraperitoneally with 1×10^{10} CFU of *E. coli*. Thirty minutes after infection, the first and second group was treated with 1×10^{11} LyD and T4 phage, respectively. The third group was treated with the β-lactam antibiotic LMOX at a final concentration of 25 mg/kg body weight. The untreated mice died after 20 h following the challenge. However, 81% of the mice treated with a single dose of 1×10^{11} LyD phage were rescued. Interestingly, the group treated with T4 wild-type phage and LMOX had a survival rate of only 52% and 33%, respectively (Matsuda *et al.*, 2005). The total endotoxin levels in peritoneal lavage fluid were significantly lower in the LyD phage-treated groups than in the other two groups, again showing that reduced endotoxin release correlated with an increased survival.

Genetically engineered phage with an extended host range

The specificity of phage is attributed to receptor binding proteins usually present on tail fibres. For instance, the specificity of phage T4 towards *E. coli* B is determined by gene *37*, encoding a 1026 amino-acid distal tail fibre protein (Oliver and Crowther, 1981; Bartual *et al.*, 2010). A mutation resulting in duplications of a small domain of the P37 tail fibre adhesin expanded the host range of phage T4 (Tétart *et al.*, 1996). In several studies phage were genetically modified to extend their host range.

In a study by Cao *et al.* (2000), the filamentous phage M13 was engineered to present the variable fragments of IgG (ScFv) against *Helicobacter pylori* on the surface. The recombinant phage could efficiently bind to spiral and coccoid forms of *H. pylori*. In addition, the ScFv-phage was effective in reducing the viable counts of *H. pylori in vitro* and *in vivo*. The recombinant phage was further able to selectively reduce *H. pylori* cells in a mixed population of bacteria, consisting of *S. aureus*, *Streptococcus* and *E. coli* strains. Moreover, a ScFv-phage/*H. pylori* mixture exhibited a somewhat reduced efficiency to colonize mice when compared with *H. pylori* alone (Cao *et al.*, 2000).

The filamentous phage fd was engineered to infect bacteria bearing both F and N pili. Phage fd is a F-pilus specific *E. coli* phage (Marvin *et al.*, 2006). The minor coat protein pIII is found on the tip of the virion and accounts for its host specificity. Binding of protein pIII to the F-pilus results in pilus retraction and thus in surface location of the phage. The translocation of phage DNA through the cell envelope requires *tolQRA* operon functions (Sun and Webster, 1987). In addition to binding to F-pili, pIII also interacts with the inner membrane protein TolA. Following TolA–pIII interaction, the major coat protein pVIII is inserted into the inner membrane and the phage DNA is translocated into the cytoplasm of the host cell (Click and Webster, 1998). The specificity for N-pili by phage fd was achieved through the creation of a fusion between genes *3* of phage fd and IKe, the latter of which encodes a receptor specific for N-pili. The recombinant fd phage effectively infected *E. coli* strains bearing either F or N pili (Marzari *et al.*, 1997).

Similarly, phage fd was modified to infect *Vibrio cholerae*. The lysogenic phage CTXΦ infects *V. cholerae* by binding to the toxin-co-regulated pilus (TCP) (Davis and Waldor, 2003). The product of CTXΦ OrfU ($pIII^{CTX}$) is homologous to the pIII protein of phage fd. *OrfU* was abutted with fd gene *3* and the fusion protein OrfU-pIII was shown to be exposed on the surface of the recombinant fd-$pIII^{CTX}$ phage. Phage fd-$pIII^{CTX}$ was effective in transducing both TCP^+ (1.1×10^{-4}) and TCP^- (3.3×10^{-5}) *V. cholerae* strains without affecting the effectiveness of infecting the host *E. coli* F^- TG1. Infection of the TCP- *V. cholerae* strain by phage fd-$pIII^{CTX}$ was attributed to the fact that different segments of $pIII^{CTX}$ mediate distinct steps in phage infection. Phage fd-$pIII^{CTX}$ encompasses aa 15–138 of $pIII^{CTX}$, which was sufficient to support infection of *V. cholerae* in a TolA-dependent, TCP-independent manner (Heilpern and Waldor, 2003).

Genetic recombination has also been used to switch the host range of a lytic phage. The genes *37/38* of *E. coli* phage T2 and IP008 encode long tail fibre proteins that recognize surface receptors on the bacterial cell (Riede *et al.*, 1987). Phage T2 and IP008 were able to infect ~7% and 33% of environmental *E. coli* strains, respectively (Mahichi *et al.*, 2009). The host range of phage T2 was successfully extended by exchanging genes *37/38* with the orthologous gene pair of phage IP008 (Mahichi *et al.*, 2009). The recombinant T2 phage acquired specificity towards *E. coli* strains that are not infected by wild-type T2 (Mahichi *et al.*, 2009).

As a consequence of co-evolution of bacteria and phage, phage-resistant bacteria have emerged and continue to emerge. One way of overcoming the dilemma of phage-resistant bacterial pathogens is the establishment of phage banks, from which virulent phage can be selected against any pathogen. In a recent pilot study, Pouillot *et al.* (2010) have created a genetically engineered T4-phage bank by means of nested PCR technology termed TAPE (*T*argeted *A*ccelerated *P*rotein *E*volution) and homologous recombination. Random mutations were created in the variable regions of the T4 genes *37/38*, encoding the tail fibre proteins. Following selection of the mutant alleles, they were recombined into the T4 *hoc* gene encoding the highly antigenic outer capsid protein. To permit recombination, *E. coli* DK8 was engineered to reversibly interrupt the lytic cycle of phage T4 by overexpressing a mutant Rho protein gene. The recombinant T4 particles were capable of infecting and lysing bacteria that are evolutionarily close (*Yersinia ruckeri* ATCC 29908) and distant (*Pseudomonas aeruginosa* ATCC 47053) from the phage T4 *E. coli* K12 host (Pouillot *et al.*, 2010).

Recombinant phage as vehicles for the delivery of drugs

Filamentous phage have further been engineered to deliver antimicrobials to bacteria other than their natural host. In the study by Yacoby *et al.* (2006), chloramphenicol was conjugated to M13 phage with the aim to specifically deliver the drug to *S. aureus* SH100 (Fig. 15.1). The authors used a phage display library and selected a M13 variant phage (A12C) that bound to *S. aureus* with high specificity, which resulted from a N-terminal lysine-free peptide displayed on M13 protein pVIII. In another approach, the sequence encoding the IgG-binding ZZ domain of *S. aureus* protein A was cloned into the phagemid vector fUSE5. This resulted in the polyvalent display of the ZZ domain on all copies of pIII of the recombinant phage (fUSE5-ZZ). Phage fUSE5-ZZ was further equipped with *S. aureus* targeting antibodies that were bound to the P3 linked ZZ domain (Yacoby *et al.*, 2006).

Exploitation of genetically engineered phage to reduce bacterial biofilms

Bacterial biofilms are cellular communities that adhere to an inert or living surface. They can resist treatment with antibiotics or the immune system (Costerton *et al.*, 1999; Costerton *et al.*, 2003; Römling and Balsalobre, 2012). The heterogeneous structure of biofilms is formed by hydrated EPS (*e*xtracellular *p*olymeric *s*ubstance) matrix generally composed of several polysaccharides, proteins, nucleic acids and lipids (Xavier *et al.*, 2005). Given the antibiotic resistance of bacterial biofilms and recent experimental hints that antibiotics can even induce biofilm formation (Hoffman *et al.*, 2005; Kaplan, 2011), there is an intense need for novel antimicrobials that disperse biofilms and eradicate biofilm communities. Disruption of biofilms can be achieved by enzymatic degradation of EPS components (Kaplan, 2009). For instance, the enzymatic degradation of the cell-bound EPS polysaccharide, adhesin (polymeric β-1,6-N-acetyl-D-glucosamine) by *Actinobacillus actinomycetemcomitans* dispersin B (DspB) has been shown to disrupt biofilms of several bacterial species including *S. aureus* and *E. coli* (Itoh *et al.*, 2005). However, while exogenous enzymes may perturb the biofilm mesh they do not necessarily kill the embedded bacteria. Another possibility to disrupt biofilms is the use of bacteriophage (Donlan, 2009; Brüssow, 2013). It has been reported that phage T4 can infect and replicate within *E. coli* biofilms and disrupt the biofilm by lysing the embedded bacteria (Doolittle *et al.*, 1995, 1996). In addition, bacteriophage have been shown to diffuse through alginate gels and to penetrate the *P. aeruginosa* EPS matrix (Hanlon *et al.*, 2001).

With the aim to engineer a phage that could effectively disperse *E. coli* biofilms, phage T7 was genetically modified to encode the *A. actinomycetemcomitans* DspB protein (Lu and Collins, 2007). Upon T7-mediated host cell lysis, the DspB protein was anticipated to be released and to degrade the EPS. The engineered T7 phage was tested on F-plasmid bearing *E. coli* TG1 biofilms (Ghigo, 2001; Da Re *et al.*, 2007). However, phage T7 was unable to infect the F-plasmid bearing *E. coli* strain (García and Molineux, 1995). To counter this problem, the *dspB* harbouring T7 phage was further modified by inserting gene *1.2* from phage T3 into the T7-*dspB* genome. The resulting phage infected *E. coli* TG1, and in addition showed an increased host range. *In vitro* studies using T7-*dspB-1.2* were promising, as the phage effectively reduced *E. coli* TG1 biofilms by an A_{600} factor of 10.3 (Lu and Collins, 2007). These authors also showed that the observed decrease of biofilms is accompanied by a reduction of viability of the biofilm embedded cells. These studies hold promise that phage harbouring EPS-degrading functions can be designed for other biofilm forming pathogens, like *P. aeruginosa*, which causes severe problems in patients afflicted with cystic fibrosis.

Use of genetically engineered phage as biodetectors

Salmonella spp., *Campylobacter* spp., *E. coli* and *L. monocytogenes* are prevalent bacterial pathogens that impose the greatest burden of food-borne illnesses world-wide. The fatality rate in listeriosis reaches 20 to 30% with higher numbers in immunocompromised individuals (Newell *et al.*, 2010). The rapid detection of food-contaminating pathogens is a primary task for prevention of food-borne diseases. The currently used detection methods for food-borne pathogens have several drawbacks; they are time-consuming and laborious, costly and sometimes lack sensitivity and specificity. In particular, the detection of a small number of bacteria in clinic, environment or food requires the time-consuming initial cultivation to increase the cell count. This time might be considerably shortened by using phage-based bioreporters.

Genetically engineered phage have been successfully used to rapidly detect food contaminants. Loessner *et al.* (1997) used phage A511, which infects 95% of all *L. monocytogenes* strains including serovars 1/2 and 4 that are implicated in human listeriosis. Phage A511 was genetically modified to direct synthesis of luciferase upon infection of the host (Fig. 15.1). The *Vibrio harveyi* derived chimeric luciferase gene was inserted downstream of the major capsid protein gene (*cps*). Upon addition of the oxidizable aldehyde substrate, luciferase activity, indicating successful infection of *L. monocytogenes* by the engineered A511 phage, was monitored in a luminometer. As few as 500–1000 cells per ml were detected within 2 h. Moreover, the engineered phage permitted detection of *L. monocytogenes* in a variety of foodstuffs including meat, cheese, milk and chocolate pudding upon deliberate contamination with the pathogen (Loessner *et al.*, 1997).

In a related study a T4*e*⁻ phage, harbouring an amber mutation in the endolysin gene, was engineered to produce a fusion protein composed of GFP (*g*reen *f*luorescent *p*rotein) and the phage small outer capsid protein Soc. The GFP-Soc protein did not impair phage adsorption to *E. coli*, and the recombinant endolysin-deficient T4*e*-(*gfp-soc*) phage did not provoke lysis. As monitored by epifluorescence microscopy, the recombinant phage specifically detected *E. coli* K12 in a mixed population. When compared to the lysis-proficient T4 phage harbouring the *gfp–soc* fusion gene, the fluorescence increased over time when the T4*e*-(*gfp-soc*) phage was used as a bioreporter (Tanji *et al.*, 2004).

Furthermore, 'fluoromycobacteriophage' were employed to detect *Mycobaterium tuberculosis*. A number of diagnostic tools have been developed to detect the presence of slow growing *M. tuberculosis*, including sputum smear microscopy. However, this method requires a minimum of 10 000 bacteria per ml of sputum (Watterson and Drobniewski, 2000). Mycobacteriophage TM4 was genetically engineered to express *gfp* or *Zsyellow* (fluorescent reporter genes) upon infection of the target bacterium. The rapid detection of as few as 100 cells of phage infected 'fluorescent protein synthesizing' cells was possible by either fluorescent microscopy or flow cytometry. Another advantage of using GFP as a reporter is its ability to withstand fixation of cells by paraformaldehyde, which in turn confers biosafety while analysing hazardous strains like XDR-TB (Piuri *et al.*, 2009). In addition to specific detection, phage based assays can also provide a means to assess antibiotic susceptibility. A modified phage expressing the gene encoding an enhanced GFP (EGFP-Phage) was used to detect isoniazid, rifampicin and streptomycin resistance in 155 strains of *M. tuberculosis* with a specificity of 94% (Rondón *et al.*, 2011).

In vivo biotinylation of phage T7 followed by conjugation of the T7 phage to streptavidin-coated quantum dots was performed to develop a related bioreporter. Quantum dots (QDs) are fluorescent colloidal semiconductor crystals of a few nanometers in diameter. They exhibit broad absorption spectra. The presence of a few atomic layers in the outer shell increases the quantum yield and further increases the photostability of the fluorescent probes. Bacteriophage T7 was genetically engineered such that a 15 aa biotinylation peptide was attached to the major capsid protein gp10A (> 400 copies per phage head). Following infection of the host bacterium with the recombinant phage, all progeny virions produced are anticipated to be biotinylated by the host cell biotin-ligase protein machinery (BLP-BirA in *E. coli*). The presence of biotinylated phage was detected by conjugation to streptavidin-linked QDs. The phage biodetector was able to specifically sense the *E. coli* host in quantities as low as 10 CFU/ml within a mixed population of *P. aeruginosa, Vibrio cholerae, Salmonella, Yersinia pseudotuberculosis* and *Bacillus subtilis,* all of which are not susceptible to bacteriophage T7. This method of bacterial detection appears to be particularly suited for slow growing bacteria or highly infectious agents like *B. anthracis* (Edgar *et al.*, 2006).

Conclusion

The described studies with recombinant phage hold promise that phage antimicrobials can be specifically tailored for a pathogen of interest. An inherent benefit of targeted killing of pathogens by recombinant phage may not only be the reduction of side-effects observed with wild-type phage or antibiotics, e.g. inflammatory mediator release caused by lysis of bacteria, but the preservation of beneficial gut commensals that are vital for immune development and function. A major hurdle in using recombinant phage as antimicrobials in systemic infections will be presumably imposed by regulatory agencies. To further develop the avenue of 'recombinant phage therapy' an increased number of successful studies using experimental infection models will be necessary to reduce concerns. In contrast to the exploitation of recombinant phage, phage derived products such as enzybiotics or biofilm degrading enzymes may encounter less obstacles towards certification. Apart from medical applications, we are confident that phage will find acceptance as biocontrol agents and biodetectors in the food industry.

Future trends

The rapid progress in understanding complex cellular machines, regulatory events and gene networks in bacterial pathogens at the molecular level will continue to reveal many 'Achilles heels' in bacterial pathogens. Specific targeting and delivery of chemical or proteinaceous agents to distinct pathogens by recombinant phage could provide a means to exploit this knowledge for therapy. Fast high throughput sequencing on one hand together with the possibility to synthesize whole genomes on the other will open unprecedented possibilities to design phage-based antimicrobials with a high selectivity and potency to kill the bacterial target. In addition, *de novo* assembly of phage genomes may entail the creation of phage with an extended host range. Similarly, phage derived enzybiotics can be further developed towards a broader spectrum of pathogens. Moreover, future developments may include optimized synthesis and activity of phage-encoded EPS degrading enzymes as well as exploitation of different activities that target distinct components of the EPS such as carbohydrates, proteins or lipids.

Web resources

1 http://www.rockefeller.edu/vaf/media/bac1a.html
2 http://www.rockefeller.edu/vaf/media/bac2a.html
3 http://www.accessdata.fda.gov/scripts/fcn/gras_notices/701456A.PDF

References

Abedon, S.T., Kuhl, S.J., Blasdel, B.G., and Kutter, E.M. (2011). Phage treatment of human infections. Bacteriophage. *1*, 66–85.

Bartual, S.G., Garcia-Doval, C., Alonso, J., Schoehn, G., and van Raaij, M.J. (2010). Two-chaperone assisted soluble expression and purification of the bacteriophage T4 long tail fibre protein gp37. Protein Expr. Purif. *70*, 116–121.

Bielke, L.R., Higgins, S.E., Donoghue, A.M., Donoghue, D.J., Hargis, B.M., and Tellez, G. (2007). Use of wide-host-range bacteriophages to reduce *Salmonella* on poultry products. Int. J. Poult. Sci. *6*, 754–757.

Brüssow, H. (2013). Bacteriophage–host interaction: from splendid isolation into a messy reality. Curr. Opin. Microbiol. *16*, 500–506.

Burrowes, B., Harper, D.R., Anderson, J., McConville, M., and Enright, M.C. (2011). Bacteriophage therapy: potential uses in the control of antibiotic-resistant pathogens. Expert Rev. Anti. Infect. Ther. *9*, 775–785.

Cao, J., Sun, Y., Berglindh, T., Mellgård, B., Li, Z., Mårdh, B., and Mårdh, S. (2000). *Helicobacter pylori*-antigen-binding fragments expressed on the filamentous M13 phage prevent bacterial growth. Biochim. Biophys. Acta. *6*, 107–113.

Catalão, M.J., Gil, F., Moniz-Pereira, J., São-José, C., and Pimentel, M. (2013). Diversity in bacterial lysis systems: bacteriophages show the way. FEMS Microbiol Rev. *37*, 554–571.

Chancey, S.T., Zähner, D., and Stephens, D.S. (2012). Acquired inducible antimicrobial resistance in Gram-positive bacteria. Future Microbiol. *7*, 959–978.

Chanishvili, N. (2012). Phage therapy – history from Twort and d'Hérelle through Soviet experience to current approaches. Adv. Virus Res. *8*, 33–40.

Cheng, Q., Nelson, D., Zhu, S., and Fischetti, V.A. (2005). Removal of group B *Streptococci* colonizing the vagina and oropharynx of mice with a bacteriophage lytic enzyme. Antimicrob. Agents Chemother. *49*, 111–117.

Cheng, G., Hao, H., Dai, M., Liu, Z., and Yuan, Z. (2013). Antibacterial action of quinolones: From target to network. Eur. J. Med. Chem. *66*, 555–562.

Chiang, S.M., and Schellhorn, H.E. (2012). Regulators of oxidative stress response genes in *Escherichia coli* and their functional conservation in bacteria. Arch. Biochem. Biophys. *525*, 161–169.

Click, E.M., and Webster, R.E. (1998). The TolQRA proteins are required for membrane insertion of the major capsid protein of the filamentous phage f1 during infection. J. Bacteriol. *180*, 1723–1728.

Costerton, J.W., Stewart, P.S., and Greenberg, E.P. (1999). Bacterial biofilms: a common cause of persistent infections. Science *284*, 1318.

Costerton, W., Veeh, R., Shirtliff, M., Pasmore, M., Post, C., and Ehrlich, G. (2003). The application of biofilm science to the study and control of chronic bacterial infections. J. Clin. Invest. *112*, 1466–1477.

de Crombrugghe, B., Busby, S., and Buc, H. (1984). Cyclic AMP receptor protein: role in transcription activation. Science *224*, 831–838.

Daniel, A., Euler, C., Collin, M., Chahales, P., Gorelick, K.J., and Fischetti, V.A. (2010). Synergism between a novel chimeric lysin and oxacillin protects against infection by methicillin-resistant *Staphylococcus aureus*. Antimicrob. Agents Chemother. *54*, 1603–1612.

Da Re, S., Le Quéré, B., Ghigo, J.M., and Beloin, C. (2007). Tight modulation of *Escherichia coli* bacterial biofilm formation through controlled expression of adhesion factors. Appl. Environ. Microbiol. *73*, 3391–3403.

Davis, B.M., and Waldor, M.K. (2003). Filamentous phages linked to virulence of Vibrio cholerae. Curr. Opin. Microbiol. *6*, 35–42.

Debarbieux, L., Leduc, D., Maura, D., Morello, E., Criscuolo, A., Grossi, O., Balloy, V., and Touqui, L. (2010). Bacteriophages can treat and prevent *Pseudomonas aeruginosa* lung infections. J. Infect. Dis. *201*, 1096–1104.

Donlan, R.M. (2009). Preventing biofilms of clinically relevant organisms using bacteriophage. Trends. Microbiol. *17*, 66–72.

Doolittle, M.M., Cooney, J.J., and Caldwell, D.E. (1995). Lytic infection of *Escherichia coli* biofilms by bacteriophage T4. Can. J. Microbiol. *41*, 12–18.

Doolittle, M.M., Cooney, J.J., and Caldwell, D.E. (1996). Tracing the interaction of bacteriophage with bacterial biofilms using fluorescent and chromogenic probes. J. Ind. Microbiol. *16*, 331–341.

Edgar, R., McKinstry, M., Hwang, J., Oppenheim, A.B., Fekete, R.A., Giulian, G., Merril, C., Nagashima, K., and Adhya, S. (2006). High-sensitivity bacterial detection using biotin-tagged phage and quantum-dot nanocomplexes. Proc. Natl. Acad. Sci. U.S.A. *103*, 4841–4845.

Fenton, M., Casey, P.G., Hill, C., Gahan, C.G., Ross, R.P., McAuliffe, O., O'Mahony, J., Maher, F., and Coffey, A. (2010). The truncated phage lysin CHAP(K) eliminates *Staphylococcus aureus* in the nares of mice. Bioeng. Bugs. *1*, 404–407.

Fenton, M., Keary, R., McAuliffe, O., Ross, R.P., O'Mahony, J., and Coffey, A. (2013). Bacteriophage-derived peptidase CHAP(K) eliminates and prevents staphylococcal biofilms. Int. J. Microbiol. *2013*, 625341.

Fischetti, V.A. (2010). Bacteriophage endolysins: a novel anti-infective to control Gram-positive pathogens. Int. J. Med. Microbiol. *300*, 357–362.

García, L.R., and Molineux, I.J. (1995). Incomplete entry of bacteriophage T7 DNA into F plasmid-containing *Escherichia coli*. J. Bacteriol. *177*, 4077–4083.

Gerdes, K., Bech, F.W., Jørgensen, S.T., Løbner-Olesen, A., Rasmussen, P.B., Atlung, T., Boe, L., Karlstrom, O., Molin, S., and von Meyenburg, K. (1986). Mechanism of postsegregational killing by the hok gene product of the parB system of plasmid R1 and its homology with the relF gene product of the *E. coli* relB operon. EMBO J. *5*, 2023–2039.

Ghigo, J.M. (2001). Natural conjugative plasmids induce bacterial biofilm development. Nature *412*, 442–445.

Gilmer, D.B., Schmitz, J.E., Euler, C.W., and Fischetti, V.A. (2013). Novel bacteriophage lysin with broad lytic activity protects against mixed infection by *Streptococcus pyogenes* and methicillin-resistant *Staphylococcus aureus*. Antimicrob. Agents Chemother. *57*, 2743–2450.

Guenther, S., Huwyler, D., Richard, S., and Loessner, M.J. (2009). Virulent bacteriophage for efficient biocontrol of *Listeria monocytogenes* in ready-to-eat foods. Appl. Environ. Microbiol. *75*, 93–100.

Hagens, S., and Bläsi, U. (2003). Genetically modified filamentous phage as bactericidal agents: a pilot study. Lett. Appl. Microbiol. *37*, 318–323.

Hagens, S., Habel, A., von Ahsen, U., von Gabain, A., and Bläsi, U. (2004). Therapy of experimental *Pseudomonas* infections with a nonreplicating genetically modified phage. Antimicrob. Agents Chemother. *48*, 3817–3822.

Hagens, S., Habel, A., and Bläsi, U. (2006). Augmentation of the antimicrobial efficacy of antibiotics by filamentous phage. Microb. Drug Resist. *12*, 164–168.

Hanlon, G.W., Denyer, S.P., Olliff, C.J., and Ibrahim, L.J. (2001). Reduction in exopolysaccharide viscosity as an aid to bacteriophage penetration through *Pseudomonas aeruginosa* biofilms. Appl. Environ. Microbiol. *67*, 2746–2753.

Harper, D.R., Anderson, J., and Enright, M.C. (2011). Phage therapy: delivering on the promise. Ther. Deliv. *2*, 935–947.

Heilpern, A.J., and Waldor, M.K. (2003). pIIICTX, a predicted CTXphi minor coat protein, can expand the host range of coliphage fd to include *Vibrio cholerae*. J. Bacteriol. *185*, 1037–1044.

Hermoso, J.A., García, J.L., and García, P. (2007). Taking aim on bacterial pathogens: from phage therapy to enzybiotics. Curr. Opin. Microbiol. *10*, 461–472.

Hidalgo, E., Ding, H., and Demple, B. (1997). Redox signal transduction: Mutations shifting [2Fe–2S] centers of the SoxR sensor-regulator to the oxidized form. Cell. *88*, 121–129.

Hirai, K., Aoyama, H., Irikura, T., Iyobe, S., and Mitsuhashi, S. (1986). Differences in susceptibility to quinolones of outer membrane mutants of *Salmonella typhimurium* and *Escherichia coli*. Antimicrob. Agents Chemother. *29*, 535–538.

Hoffman, L.R., D'Argenio, D.A., MacCoss, M.J., Zhang, Z., Jones, R.A., and Miller, S.I. (2005). Aminoglycoside antibiotics induce bacterial biofilm formation. Nature *436*, 1171–1175.

Huff, W.E., Huff, G.R., Rath, N.C., Balog, J.M., and Donoghue, A.M. (2003). Evaluation of aerosol spray and intramuscular injection of bacteriophage to treat an *Escherichia coli* respiratory infection. Poult. Sci. *82*, 1108–1112.

Itoh, Y., Wang, X., Hinnebusch, B.J., Preston, J.F. 3rd, and Romeo, T. (2005). Depolymerization of beta-1,6-N-acetyl-D-glucosamine disrupts the integrity of diverse bacterial biofilms. J. Bacteriol. *187*, 382–387.

Jackson, D.W., Suzuki, K., Oakford, L., Simecka, J.W., Hart, M.E., and Romeo, T. (2002). Biofilm formation and dispersal under the influence of the global regulator CsrA of *Escherichia coli*. J. Bacteriol. *184*, 290–301.

Janion, C. (2008). Inducible SOS response system of DNA repair and mutagenesis in *Escherichia coli*. Int. J. Biol. Sci. *4*, 338–344.

Kaplan, J.B. (2009). Therapeutic potential of biofilm-dispersing enzymes. Int. J. Artif. Organs. *32*, 545–554.

Kaplan, J.B. (2011). Antibiotic-induced biofilm formation. Int. J. Artif. Organs. *34*, 737–751.

Kohanski, M.A., Dwyer, D.J., Hayete, B., Lawrence, C.A., and Collins, J.J. (2007). A common mechanism of cellular death induced by bactericidal antibiotics. Cell. *130*, 797–810.

Kumar, A., Grover, S., and Kumar Batish, V. (2013). Application of multiplex PCR assay based on uidR and fliCH7 genes for detection of *Escherichia coli* O157:H7 in milk. J. Gen. Appl. Microbiol. *59*, 11–19.

Kutateladze, M., and Adamia, R. (2010). Bacteriophages as potential new therapeutics to replace or supplement antibiotics. Trends Biotechnol. *28*, 591–595.

Lewin, C.S., Howard, B.M., Ratcliffe, N.T., and Smith, J.T. (1989). 4-Quinolones and the SOS response. J. Med. Microbiol. *29*,139–144.

Loeffler, J.M., Nelson, D., and Fischetti, V.A. (2001). Rapid killing of *Streptococcus pneumoniae* with a bacteriophage cell wall hydrolase. Science *294*, 2170–2172.

Loessner, M.J., Rudolf, M., and Scherer, S. (1997). Evaluation of luciferase reporter bacteriophage A511::luxAB for detection of *Listeria monocytogenes* in contaminated foods. Appl. Environ. Microbiol. *63*, 2961–2965.

Lopata, M., Schlieper, D., von Wilcken-Bergmann, B., and Müller-Hill, B. (1997). A lethal mutant of the catabolite gene activator protein CAP of *Escherichia coli*. Biol. Chem. *378*, 1153–1162.

Lu, T.K., and Collins, J.J. (2007). Dispersing biofilms with engineered enzymatic bacteriophage. Proc. Natl. Acad. Sci. U.S.A. *104*, 11197–11202.

Lu, T.K., and Collins, J.J. (2009). Engineered bacteriophage targeting gene networks as adjuvants for antibiotic therapy. Proc. Natl. Acad. Sci. U.S.A. *106*, 4629–4634.

Mahendran, K.R., Hajjar, E., Mach, T., Lovelle, M., Kumar, A., Sousa, I., Spiga, E., Weingart, H., Gameiro, P., Winterhalter, M., *et al.* (2010). Molecular basis of enrofloxacin translocation through OmpF, an outer membrane channel of *Escherichia coli* – when binding does not imply translocation. J. Phys. Chem. B. *114*, 5170–5179.

Mahichi, F., Synnott, A.J., Yamamichi, K., Osada, T., and Tanji, Y. (2009). Site-specific recombination of T2 phage using IP008 long tail fiber genes provides a targeted method for expanding host range while retaining lytic activity. FEMS Microbiol. Lett. *295*, 211–217.

Manoharadas, S., Witte, A., and Bläsi, U. (2009). Antimicrobial activity of a chimeric enzybiotic towards *Staphylococcus aureus*. J. Biotechnol. *139*, 118–123.

Marvin, D.A., Welsh, L.C., Symmons, M.F., Scott, W.R., and Straus, S.K. (2006). Molecular structure of fd (f1, M13) filamentous bacteriophage refined with respect to X-ray fibre diffraction and solid-state NMR data supports specific models of phage assembly at the bacterial membrane. J. Mol. Biol. *355*, 294–309.

Marzari, R., Sblattero, D., Righi, M., and Bradbury, A. (1997). Extending filamentous phage host range by the grafting of a heterologous receptor binding domain. Gene. *185*, 27–33.

Matsuda, T., Freeman, T.A., Hilbert, D.W., Duff, M., Fuortes, M., Stapleton, P.P., and Daly, J.M. (2005). Lysis-deficient bacteriophage therapy decreases endotoxin and inflammatory mediator release and improves survival in a murine peritonitis model. Surgery. *137*, 639–646.

Miller, C., Thomsen, L.E., Gaggero, C., Mosseri, R., Ingmer, H., and Cohen, S.N. (2004). SOS response induction by beta-lactams and bacterial defense against antibiotic lethality. Science *305*, 1629–1631.

Moradpour, Z., and Ghasemian, A. (2011). Modified phages: novel antimicrobial agents to combat infectious diseases. Biotechnol Adv. *29*, 732–738.

Moradpour, Z., Sepehrizadeh, Z., Rahbarizadeh, F., Ghasemian, A., Yazdi, M.T., and Shahverdi, A.R. (2009). Genetically engineered phage harbouring the lethal catabolite gene activator protein gene with an inducer-independent promoter for biocontrol of *Escherichia coli*. FEMS Microbiol. Lett. *296*, 67–71.

Morello, E., Saussereau, E., Maura, D., Huerre, M., Touqui, L., and Debarbieux, L. (2011). Pulmonary bacteriophage therapy on *Pseudomonas aeruginosa* cystic fibrosis strains: first steps towards treatment and prevention. PloS One *15*, e16963.

Musso, R.E., Di Lauro, R., Adhya, S., and de Crombrugghe, B. (1977). Dual control for transcription of the galactose operon by cyclic AMP and its receptor protein at two interspersed promoters. Cell. *12*, 847–854.

Nelson, D., Loomis, L., and Fischetti, V.A. (2001). Prevention and elimination of upper respiratory colonization of mice by group A streptococci by using a bacteriophage lytic enzyme. Proc. Natl. Acad. Sci. U.S.A. *98*, 4107–4112.

Newell, D.G., Koopmans, M., Verhoef, L., Duizer, E., Aidara-Kane, A., Sprong, H., Opsteegh, M., Langelaar, M., Threfall, J., Scheutz, F., *et al.* (2010). Food-borne diseases – the challenges of 20 years ago still persist while new ones continue to emerge. Int. J. Food Microbiol. *139*, 3–15.

O'Flaherty, S., Ross, R.P., Meaney, W., Fitzgerald, G.F., Elbreki, M.F., and Coffey, A. (2005). Potential of the polyvalent anti-*Staphylococcus* bacteriophage K for control of antibiotic resistant staphylococci from hospitals. Appl. Environ. Microbiol. *71*, 1836–1842.

O'Flynn, G., Ross, R.P., Fitzgerald, G.F., and Coffey, A. (2004). Evaluation of a cocktail of three bacteriophages for biocontrol of *Escherichia coli* O157:H7. Appl. Environ. Microbiol. *70*, 3417–3424.

Oliver, D.B., and Crowther, R.A. (1981). DNA sequence of the tail fibre genes 36 and 37 of bacteriophage T4. J. Mol. Biol. 153, 545–568.

Park, S.C., and Nakai, T. (2003). Bacteriophage control of *Pseudomonas plecoglossicida* infection in ayu *Plecoglossus altivelis*. Dis. Aquat. Organ. *53*, 33–39.

Paul, V.D., Sundarrajan, S., Rajagopalan, S.S., Hariharan, S., Kempashanaiah, N., Padmanabhan, S., Sriram, B., and Ramachandran, J. (2011). Lysis-deficient phages as novel therapeutic agents for controlling bacterial infection. BMC Microbiol. *11*, 195.

Piuri, M., Jacobs, W.R., Jr, and Hatfull, G.F. (2009). Fluoromycobacteriophages for rapid, specific, and sensitive antibiotic susceptibility testing of *Mycobacterium tuberculosis*. PloS One *4*, e4870.

Pouillot, F., Blois, H., and Iris, F. (2010). Genetically engineered virulent phage banks in the detection and control of emergent pathogenic bacteria. Biosecur. Bioterror. *8*, 155–169.

Riede, I., Drexler, K., Schwarz, H., and Henning, U. (1987). T-even-type bacteriophages use an adhesin for recognition of cellular receptors. J. Mol. Biol. *194*, 23–30.

Rietsch, A., and Bläsi, U. (1993). Non-specific hole formation in the *Escherichia coli* inner membrane by lambda S proteins is independent of cellular secY and secA functions and of the proportion of membrane acidic phospholipids. FEMS Microbiol. Lett. *107*, 101–105.

Romeo, T., Vakulskas, C.A., and Babitzke, P. (2013). Post-transcriptional regulation on a global scale: form and function of Csr/Rsm systems. Environ Microbiol. *15*, 313–324.

Römling, U., and Balsalobre, C. (2012). Biofilm infections, their resilience to therapy and innovative treatment strategies. J. Intern. Med. *272*, 541–561.

Rondón, L., Piuri, M., Jacobs, W.R. Jr, de Waard, J., Hatfull, G.F., and Takiff, H.E. (2011). Evaluation of fluoromycobacteriophages for detecting drug resistance in *Mycobacterium tuberculosis*. J. Clin. Microbiol. *49*, 1838–1842.

Sader, H.S., Flamm, R.K., and Jones, R.N. (2013). Antimicrobial activity of ceftaroline tested against staphylococci with reduced susceptibility to linezolid, daptomycin, or vancomycin from U.S. hospitals, 2008 to 2011. Antimicrob. Agents Chemother. *57*, 3178–3181.

Sarker, S.A., McCallin, S., Barretto, C., Berger, B., Pittet, A.C., Sultana, S., Krause, L., Huq, S., Bibiloni, R., Bruttin, A., *et al.* (2012). Oral T4-like phage cocktail application to healthy adult volunteers from Bangladesh. Virology *434*, 222–232.

Schentag, J.J., Hyatt, J.M., Carr, J.R., Paladino, J.A., Birmingham, M.C., Zimmer, G.S., and Cumbo, T.J. (1998). Genesis of methicillin-resistant *Staphylococcus aureus* (MRSA), how treatment of MRSA infections has selected for vancomycin-resistant *Enterococcus faecium*, and the importance of antibiotic management and infection control. Clin. Infect. Dis. *26*, 1204–1214.

Schito, G.C. (2006). The importance of the development of antibiotic resistance in *Staphylococcus aureus*. Clin. Microbiol. Infect. *12* Suppl 1, 3–8.

Schuch, R., Nelson, D., and Fischetti, V.A. (2002). A bacteriolytic agent that detects and kills *Bacillus anthracis*. Nature *418*, 884–889.

Segarra-Newnham, M., and Church, T.J. (2012). Pharmacotherapy for methicillin-resistant *Staphylococcus aureus* nosocomial pneumonia. Ann. Pharmacother. *46*, 1678–1687.

Sieradzki, K., and Tomasz, A. (1997). Inhibition of cell wall turnover and autolysis by vancomycin in a highly vancomycin-resistant mutant of *Staphylococcus aureus*. J. Bacteriol. *179*, 2557–2566.

Sillankorva, S.M., Oliveira, H., and Azeredo, J. (2012). Bacteriophages and their role in food safety. Int. J. Microbiol. *2012*, 863945.

Simmons, M., Donovan, D.M., Siragusa, G.R., and Seal, B.S. (2010). Recombinant expression of two bacteriophage proteins that lyse Clostridium perfringens and share identical sequences in the C-terminal cell wall binding domain of the molecules but are dissimilar in their N-terminal active domains. J. Agric. Food Chem. *58*, 10330–10337.

Slopek, S., Durlakowa, I., Weber-Dabrowska, B., Kucharewicz-Krukowska, A., Dabrowski, M., and Bisikiewicz, R. (1983). Results of bacteriophage treatment of suppurative bacterial infections. II. Detailed evaluation of the results. Arch. Immunol. Ther. Exp. (Warsz). *31*, 293–327.

Sulakvelidze, A. (2013). Using lytic bacteriophages to prevent or significantly reduce contamination of food by food-borne bacterial pathogens. J. Sci. Food Agric. *93*, 3137–3146.

Sun, T.P., and Webster, R.E. (1987). Nucleotide sequence of a gene cluster involved in entry of E colicins and single-stranded DNA of infecting filamentous bacteriophages into *Escherichia coli*. J. Bacteriol. *169*, 2667–2674.

Tanji, Y., Furukawa, C., Na, S.H., Hijikata, T., Miyanaga, K., and Unno, H. (2004). *Escherichia coli* detection by GFP-labeled lysozyme-inactivated T4 bacteriophage. J. Biotechnol. *114*, 11–20.

Tétart, F., Repoila, F., Monod, C., and Krisch, H.M. (1996). Bacteriophage T4 host range is expanded by duplications of a small domain of the tail fiber adhesin. J. Mol. Biol. *258*, 726–731.

Watterson, S.A., and Drobniewski, F.A. (2000). Modern laboratory diagnosis of mycobacterial infections. J. Clin. Pathol. *53*, 727–732.

Weber-Dabrowska, B., Dabrowski, M., and Slopek, S. (1987). Studies on bacteriophage penetration in patients subjected to phage therapy. Arch. Immunol. Ther. Exp. *35*, 563–568.

Westwater, C., Kasman, L.M., Schofield, D.A., Werner, P.A., Dolan, J.W., Schmidt, M.G., and Norris, J.S. (2003). Use of genetically engineered phage to deliver antimicrobial agents to bacteria: an alternative therapy for treatment of bacterial infections. Antimicrob. Agents Chemother. *47*, 1301–1307.

Woodford, N., Turton, J.F., and Livermore, D.M. (2011). Multiresistant Gram-negative bacteria: the role of high-risk clones in the dissemination of antibiotic resistance. FEMS Microbiol. Rev. *35*, 736–755.

Wright, A., Hawkins, C.H., Anggård, E.E., and Harper, D.R. (2009). A controlled clinical trial of a therapeutic bacteriophage preparation in chronic otitis due to antibiotic resistant *Pseudomonas aeruginosa;* a preliminary report of efficacy. Clin Otolaryngol. *34*, 349–357.

Xavier, J.B., Picioreanu, C., Rani, S.A., van Loosdrecht, M.C., and Stewart, P.S. (2005). Biofilm-control strategies based on enzymic disruption of the extracellular polymeric substance matrix-a modelling study. Microbiology *151*, 3817–3832.

Yacoby, I., Shamis, M., Bar, H., Shabat, D., and Benhar, I. (2006). Targeting antibacterial agents by using drug-carrying filamentous bacteriophages. Antimicrob. Agents Chemother. *50*, 2087–2097.

Yoong, P., Schuch, R., Nelson, D., and Fischetti, V.A. (2004). Identification of a broadly active phage lytic enzyme with lethal activity against antibiotic-resistant *Enterococcus faecalis* and *Enterococcus faecium*. J. Bacteriol. *186*, 4808–4812.

Young, R. (1992). Bacteriophage lysis: mechanism and regulation. Microbiol. Rev. *56*, 430–481.

Zhang, Y., Zhu, L., Zhang, J., and Inouye, M. (2005). Characterization of ChpBK, an mRNA interferase from *Escherichia coli*. J. Biol. Chem. *280*, 26080–26088.

Zimmer, M., Vukov, N., Scherer, S., and Loessner, M.J. (2002). The murein hydrolase of the bacteriophage phi3626 dual lysis system is active against all tested *Clostridium perfringens* strains. Appl. Environ. Microbiol. *68*, 5311–5317.

Engineered Filamentous Bacteriophages as Targeted Anti-bacterial Drug-carrying Nanomedicines

16

Lilach Vaks and Itai Benhar

Abstract

The increasing development of bacterial resistance to traditional antibiotics forces the scientists to develop new antimicrobial approaches. In traditional as well as newly developed antibiotics, the drug itself provides the target specificity, thus excluding potent but non-selective drugs from possible use. The conjugation of a toxic drug to a targeted carrier replaces the drug selectivity by the targeting moiety of the carrier thus maximizing the drug bio-availability to diseased tissues and minimizing the healthy tissues exposure. Bacteriophages (phages) possess unique characteristics such as modular nanometric structure, high specificity to host bacteria, an ability to display foreign proteins on the phage surface and varying levels of tolerance to chemical modification which suggest they may be perfect carriers for targeting and eradication of bacterial pathogens. The natural ability of host-specific phages to infect and lyse their host bacteria in animal disease models was already demonstrated in early 1940s. However, major challenges such as limited host specificity, uncontrolled replication and high immunogenicity remained mostly unaddressed. The application of phage carriers loaded with non-selective antibacterial drugs provides for drug accumulation at high concentration vicinal to the target pathogen and its effective killing with minimal effect on the organism infected by the pathogen. The targeting moiety is provided by either natural host-specific recognition or by genetic modification of the phage to display pathogen-specific peptide or antibody on phage surface. Such approaches were demonstrated mostly with filamentous phages, but also with tailed phages. The drug conjugation affects the phage *in vivo* properties and decreases its immunogenicity.

In conclusion, the unique natural properties of the phage carrier and its universal compatibility for drug conjugation and for display of target-specific proteins and peptides give the phage nanoparticle an exclusive advantage among the presently existing anti bacterial drug-delivery systems.

Introduction

The increasing development of bacterial resistance to antibiotics makes the development of new antimicrobial approaches an urgent necessity (Taylor, 2013). The historical observations show that whenever a new antibiotic, either a broader-spectrum form of an existing antibiotic, or even a new class of antibiotic is introduced into widespread use in people or livestock, clinically significant resistance soon emerges (Walsh, 2003). It is vital that there should be absolutely no cessation in the search for new antimicrobial agents (Boucher *et al.*, 2009; Lewis, 2013). While the chase after the identification of the new microbial targets is hard and unpredictable, the adoption of naturally existing biological systems for bacteria treatment may provide highly effective antimicrobial solutions. The ability of phage to infect and lyse the specific bacteria was demonstrated almost a century ago (Summers, 2012). Its potential for therapeutic application had been recognized and 'traditional' phage therapy was invented, based on the infection of the pathogenic bacteria by their specific phage, as described in other chapters of this book. This chapter will describe a new direction in phage-based antibacterial treatment:

targeted drug-carrying phages (mostly filamentous but also tailed phages). In this approach, a non-infectious phage particle serves as a high-capacity carrier for an otherwise non-potent or non-selective antibiotic (or another antibacterial agent) that is specifically targeted to the pathogen and provides upon controlled release at the target site the drug action vicinal the bacteria (Vaks and Benhar, 2011a).

The biology of Ff phage

The Ff class of the filamentous coliphages consists of M13, f1, and fd. The genomes of these three phages show 98% DNA sequence identity and consequently the protein sequence of their gene products are practically the same (Beck and Zink, 1981; Hill and Petersen, 1982; van Wezenbeek *et al.*, 1980; Rakonjac *et al.*, 2011). These *Escherichia coli* male-specific phages form particles of nanometric dimensions (~ 1 μm in length but less than 10 nm in diameter); hence we refer to them as phage nanoparticles. The genome of the filamentous phage is composed of a single-stranded circular DNA molecule of approximately 6400 bases. Phage ssDNA is converted to double-stranded replicative form once inside the infected cell. The genome encodes 11 distinct proteins that can be divided into three functional subgroups: proteins involved in replication (pII, pV, pX), proteins required for membrane-associated assembly (pI, pIV, pXI) and capsid proteins (pIII, pVI, pVII, pVIII and pIX) (Russel and Model, 1989). During the release step from the host bacteria, the phage genome is encapsulated in a proteinaceous capsid to form a long, thin, cylindrical nanoparticle (Glucksman *et al.*, 1992) (Fig. 16.1).

The protein coat contains five different proteins. The phage coat is primarily assembled from the 50 amino acid major coat protein pVIII (also referred to as g8p or p8). Several thousand copies of pVIII cover the length of the phage. For a wild type M13 particle, ~2700 copies of pVIII construct its protein coat. The pVIII monomers are packed quite tightly, as only three residues are accessible for protease cleavage (Terry *et al.*, 1997). The C-terminal 10–13 residues of pVIII form the inner wall of the phage coat. This region possesses four positively charged lysine residues

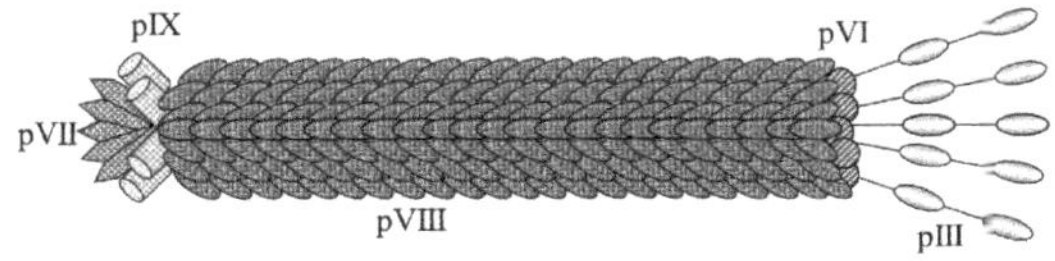

Figure 16.1 The schematic structure of filamentous phage. A colour version of this figure is available in the plate section at the back of the book.

that interact with the sugar phosphate backbone of the DNA present in the particle with the bases pointed inwards (Greenwood *et al.*, 1991; Marvin, 1998; Marvin *et al.*, 1994). The N-terminal portion of pVIII is present on the outside of the phage and is broadly used for polyvalent peptide display that will be described in following sections. The coat's dimensions, determined by the number of pVIII copies, adjusts according to the size of the single-stranded genome it packages. It was demonstrated that when the size of the phage genome was reduced from 6.4 kb to packaging signal-containing 221 bases, then the number of pVIII copies is decreased to about 95 (Specthrie *et al.*, 1992). Thus, it is possible to artificially modulate the length of the virion by manipulating the size of the packaged DNA.

The four minor coat proteins are present at about five copies per particle. The pVII and pIX minor coat proteins (33 and 32 amino acid residues, respectively) cap the end of the particle that emerges first from the host. These proteins initiate phage packaging, and in the absence of either protein, no particle is formed. The pIII minor coat protein (406-amino acid residues) and pVI minor coat protein (112-amino acid residues) cap the opposite end, both of which are needed in order for the phage to detach from the cell membrane. All five coat proteins contribute to the structural integrity and stability of the phage particle, while pIII is also necessary for host cell recognition and infection. Consequently, pIII, the largest and most complex of the coat proteins, contains three distinct domains; pIII-N1, pIII-N2 and pIII-CT separated by two long, presumably flexible glycine-rich linkers (Barbas *et al.*, 2001; Stengele *et al.*, 1990). The N1 domain is responsible for the injection of the phage genome into the bacterial cytoplasm during the infection process, the N2 domain is required for F pilus attachment and the C-terminus CT domain interacts with a pVIII coat protein, mediating the termination of assembly and release of phage from the cell (Barbas *et al.*, 2001; Crissman and Smith, 1984; Kremser and Rasched, 1994). All phage proteins are synthesized simultaneously, although diverse mechanisms ensure that each is produced in an appropriate rate.

Phage display

Phage display was pioneered in 1985 by George Smith, who postulated (and demonstrated) that if a DNA sequence encoding a peptide or protein will be fused in-frame with a gene encoding a phage coat protein, the result will be that the encoded peptide will be displayed on the surface of the phage particle (Smith, 1985). Phage display is a very powerful technique to obtain libraries containing millions (or even billions) of different peptides or proteins. Phage display has been used for affinity screening of combinatorial peptide libraries to identify ligands for peptide receptors, define epitopes for monoclonal antibodies, select enzyme substrates (Kay *et al.*, 1996), and screen cloned antibody repertoires. One of the most successful applications of phage display has been isolation of monoclonal antibodies, and fragments thereof, using large phage antibody libraries (Tohidkia *et al.*, 2012; Benhar, 2001; Hoogenboom *et al.*, 1998; Jespers *et al.*, 1994).

All five coat proteins have been used to display proteins and peptides. For protein display, pIII protein has been the preferred fusion protein because of its low copy number which is suited for affinity selection (Bradbury and Marks, 2004; Garrard *et al.*, 1991) and its relative tolerance for large insertions (McCafferty *et al.*, 1991). The major coat protein, pVIII has been commonly used for short (6–8 residues) peptide display. The display of larger polypeptides on every copy of pVIII would damage the efficient packaging of virion and significantly reduce the number of propagated phages (Iannolo *et al.*, 1995).

The three other minor coat proteins (pVI, pVII and pIX) are much less popular for polypeptide display, mainly because of small size and low surface exposure. However, since pIII is responsible for interaction with the host bacteria receptors during infection, pIII modification may result in reduced infectivity. This gave rise to alternative display systems utilizing other minor coat proteins. The pVI display system was first reported by the Stanssens group (Jespers *et al.*, 1995) for isolation of protease inhibitor from *Ancylostoma caninum*, followed by establishment of a set of pVI display vectors used for display and selection of cDNA libraries (Hufton *et al.*, 1999). The pVII coat protein was used for multivalent scFv display, while the phage infectivity was not compromised (Kwasnikowski *et al.*, 2005). The pIX protein fusion had been demonstrated for scFv (Gao *et al.*, 2002; Gao *et al.*, 1999) as well as Fab display (Tornetta *et al.*, 2010). The fusion of heavy and light chain variable regions to pVII and pIX and display on the same phage virion (Gao *et al.*, 1999) gave the opportunity for combinatorial display that can be applied in antibody phage display libraries. The combination of two and more display systems enables the parallel selection of multiple proteins or binding molecules on a single phage particle, so called 'multi-specific phages'.

Monovalent versus polyvalent display

The filamentous phage coat proteins pIII and pVIII are commonly used for display of foreign proteins and peptides. A variety of vector systems that were adapted for display via these proteins can be generally divided into three main categories:

1. Type 3 or 8 display system: the polypeptide is genetically fused to pIII or pVIII and displayed on every copy of the phage coat protein. The accepted polyvalent display permits avidity and is highly practical for selecting binders with low monovalent binding affinity but with higher functional affinity when the antigen is multivalent (O'Connell *et al.*, 2002).
2. Type 33 or 88 hybrid system: the phage genome is modified to express the displayed fusion protein as well as wild type protein copies. The resulting phage will display both the wild type and fusion proteins (Sidhu, 2001). The wild type protein is displayed in greater preference than the polypeptide-fused

copy in hybrid systems, providing monovalent (for pIII-fused proteins) or relatively low display (for pVIII-fused proteins) per single phage particle. The monovalent display allows the efficient selection of high affinity binders.

3 Type 3+3 or 8+8 display technology: phagemid-based system for hybrid display. Phagemids are small plasmid vectors which carry pIII or pVIII-fused protein for display. In addition, phagemids contain the Ff origin of replication that allows their rolling circle replication, yielding single-stranded copies that can be packaged into the phage particle. This is facilitated by co-infection of a phagemid carrying *E. coli* with helper phage that provides the auxiliary functions facilitating the production of single stranded phagemid DNA, its assembly and encapsulation (Sidhu, 2001). Similar to previous hybrid system, the resulting phage particles will carry fused as well as unfused coat proteins, providing a monovalent display.

Monovalent display is highly practiced in screening of peptide and antibody libraries. This technique is based on pure affinity and allows identifying the tightest antigen binders. The polyvalent display provides the avidity and the strongest binding of polypeptide-displaying phage to its target.

Targeting of the phage to specific bacteria

To be applied for target cell-specific drug delivery, the phage should be equipped with a target-specifying entity. The following sections will discuss the different approaches to ensure the specific targeting of the phage to the target cells. While the adoption of phage natural ability to recognize and infect its host bacteria gave rise for 'traditional' phage therapy, as described in other chapters of this book, drug carrying phages are largely directed against non-natural hosts. In a few cases, the natural ability of the phage is harnessed for targeting to the natural host, but still infection of the host is not involved (section 6.1, below). While such targeting is limited to a single target bacterium, the targeting mediated by peptide or antibody displayed on the phage particle enabled development of targeted drug-carrying phages that can be used against different bacterial pathogens (described in 'Antibody-based targeting' and 'Peptide-based targeting', below).

Host-specific phages

The 'traditional' phage therapy involves the infection and lysis of pathogenic bacteria with a specific lytic phage (Burrowes *et al.*, 2011). The infection is initiated by the adsorption of the phage to the host cell provided by a very specific interaction between the specific phage protein and the host receptor, as in the case of T4 phage gp37 tail protein and *E. coli* OmpC, OmpF or LPS molecules (Goldberg *et al.*, 1994; Tetart *et al.*, 1998; Wood *et al.*, 1994). An additional example is the J protein of phage lambda providing the specific binding to LamB membrane protein of *E. coli* K-12 bacteria (Randall-Hazelbauer and Schwartz, 1973). The natural phage host specificity is highly refined and each phage will only attack one species or in some cases a single strain of bacteria (Hanlon, 2007). However, the high specificity that is extremely desirable in particular therapy fields is identified as a limitation in phage-based therapy. The pathogenic bacteria have to be identified before an appropriate phage can be selected for therapy (Housby and Mann, 2009). A possible solution to that problem may be the complex treatment including the mixture of several phages of different host specificities (so called 'phage cocktail') that may provide 'broad-spectrum' protection (Tanji *et al.*, 2005; Chan *et al.*, 2013). Another option is the identification of naturally wide range phages, such as *staphylococcal* phages (Gupta and Prasad, 2011; Hsieh *et al.*, 2011) or *Salmonella* phages (Bielke *et al.*, 2007; Marti *et al.*, 2013) that are able to recognize a number of bacterial strains or even more than one bacterial genus.

It's important to mention that the 'traditional' phage therapy requires the infection of pathogen by its phage, while the anti-bacterial activity of targeted drug-carrying phages (will be described in 'Drug-carrying filamentous phages', below) is achieved by tight binding to the bacteria and do not involve infection. This fact may redefine the host range spectrum of particular phage that possibly binds the bacterial target but does not infect it. Such phages would be useless in 'traditional' phage therapy, but may be valuable for drug-carrying applications.

Antibody-based targeting

Antibody display on the filamentous phage coat proteins is widely used for isolation of monoclonal antibodies from phage antibody libraries, as described above. The modification of phage to display antibody targeted against a bacterial surface component guarantees its highly specific recognition and binding. Unlike the 'traditional' phage therapy that is based on hardly changeable natural narrow host recognition systems, the antibody- and peptide-based targeting (will be discussed in 'Peptide-based targeting', below) depend exclusively on antibody (or peptide) displayed on the phage surface. Produced against the strain-specific bacterial protein, the antibody will provide the specific targeting of the particular strain. On the other hand, the display of antibody that is able to bind the specie-defining protein will result in specific targeting to the all bacterial strains within the specie. In other words, it is feasible to use antibodies directed against bacterial cell wall of membrane components that are common to many potential target bacteria. Thus, the antibacterial targeting range can be controlled choosing an appropriate antibody–target pair. It is important to emphasize that the pathogenic bacteria are targeted through the surface exposed bacterial components, therefore the correct identification of the bacterial antigens will offer the useful information about the estimated targeting efficiency, range, specificity and selectivity. The binding of targeted *E. coli* filamentous phage to non-host bacteria does not involve infection or direct phage mediated influence on the pathogen. The therapeutic effect is mediated by drug loaded on the phage surface, which is released specifically near the target (as discussed in the following sections). As the chemical drug conjugation may damage the binding affinity of the displayed antibody, approaches that provide protection of antibody binding sites from such damage have to be evaluated. In the following section we will describe the two possible systems for production of targeted, relatively chemistry-tolerant filamentous phages.

Antibody-based ZZ domain-mediated targeting

ZZ domain, the tandem repeated, mutated domain B derived from *Staphylococcus aureus* protein A, is known for its ability to bind the antibody constant region Fc (Nilsson *et al.*, 1987). The display of ZZ domain on filamentous phages and incubation with HRP-conjugated rabbit IgG confirmed that fusion proteins had IgG-binding abilities (Djojonegoro *et al.*, 1994; Kushwaha *et al.*, 1994). ZZ displaying phages were shown to interact with immobilized human IgG and its various subclasses. In the above studies, monovalent display of ZZ domains was proposed, providing a phage population where ~90% of the particles do not display the desired molecule on their coat (Berdichevsky *et al.*, 1999). The polyvalent display of ZZ domain as a pIII coat protein fusion guaranteed the display on all pIII copies of all phages (Yacoby *et al.*, 2006, 2007). Further complexation of ZZ-displaying phage with anti-bacterial antibodies generated tight binding between phage and its targeting moiety (Fig. 16.2A). The fact that targeting depends on antibody exclusively, turns the ZZ-mediated system into a universal system for targeting any bacteria against which antibody can be obtained.

Antibody-based biotin–avidin-mediated targeting

The non-covalent interaction between biotin/streptavidin and biotin/avidin represents one of the strongest and the most specific interactions amongst biological molecules ($K_D = 10^{-14}$ to 10^{-15} M). Biotin labelling of drugs, proteins and viruses has traditionally been performed *in vitro* by chemical methods where an activated biotin derivative is conjugated to protein surface residues (commonly lysine residues) or carbohydrate moieties (Bayer and Wilchek, 1990; Diamandis and Christopoulos, 1991; Ohno *et al.*, 1996; Smith *et al.*, 1999). However, these methods result in random and heterogeneous modification, which can lead to the inactivation of biological function, cross-linking or aggregation after mixing with streptavidin or avidin. Antibody biotinylation by

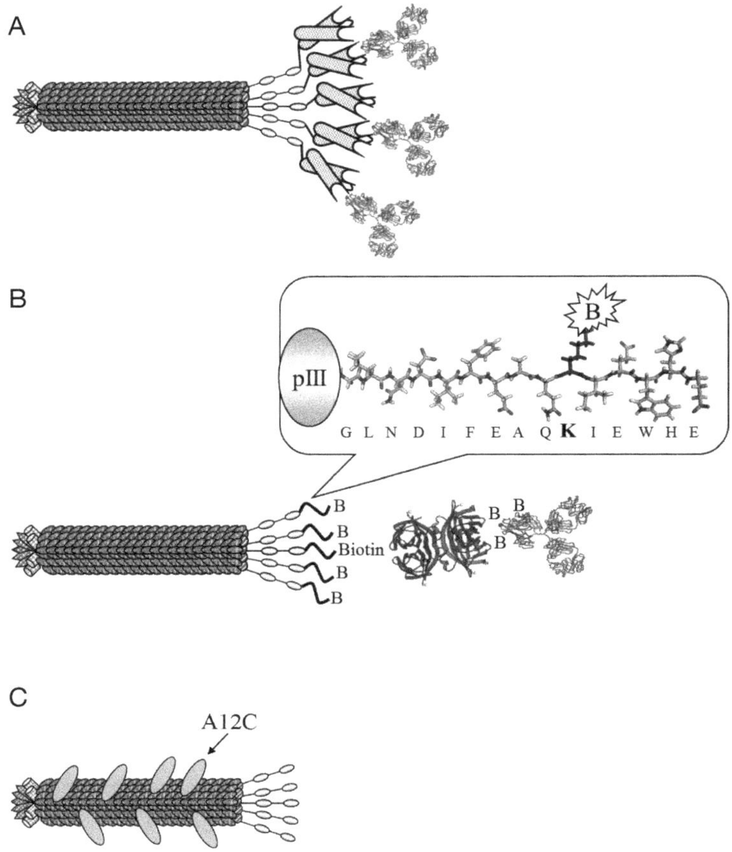

Figure 16.2 Schematic representation of antibody-based targeting of filamentous phage. (A) The anti-bacterial antibody provides the specific targeting via the ZZ domain displayed as a pIII phage coat protein fusion. (B) Biotin-displayed phages undergo complexation with biotinylated anti-bacterial antibody via a tetrameric avidin molecule. (C) A12C peptide displaying phage. A colour version of this figure is available in the plate section at the back of the book.

chemical methods generally leads to the production of heterogeneous conjugates. Furthermore, biotinylation of the residues in the binding site of antibodies can damage their binding properties (Saviranta *et al.*, 1998).

An alternative approach was first demonstrated by Cronan (1990). It was based on the biotin ligase BirA enzyme activity and its unique ability to catalyse the covalent ligation of biotin to a precisely defined lysine residue of the desired protein. The functional interaction between biotin ligases and their protein substrates showed a very high degree of conservation throughout evolution, since biotinylation occurred even with enzymes and substrates from widely divergent species (Chapman-Smith and Cronan, 1999).

A 15-residue peptide (GLNDIFEAQKIEWHE, Biotin AviTag, Avidity, LCC) (Beckett *et al.*, 1999) that can be very efficiently biotinylated was used for specific biotinylation of fusion protein in *E. coli*, insect and mammalian cells (de Boer *et al.*, 2003; Smith *et al.*, 1998; Tirat *et al.*, 2006; Viens *et al.*, 2004; Warren *et al.*, 2005; Wu *et al.*, 2002; Yang *et al.*, 2004). Utilizing this small peptide *in vivo* biotinylation has also been performed on the yeast surface (Parthasarathy *et al.*, 2005). Such highly specific biotinylation permits attachment of the protein to avidin/streptavidin derivatives, which in their turn are conjugated to other chemicals, surfaces or beads. Avidin/streptavidin derivatives are tetrameric and bind four biotin molecules, thus providing a straightforward method for production of stable complexes of biotinylated particles.

In vivo biotinylation of phages was previously described by Scholle *et al.* (2006). In that work biotinylation was used for immobilization of biotinylated phage library on a streptavidin-coated surface followed by controlled release by examined protease. The filamentous phage displaying AviTag peptide on pIII coat protein underwent efficient (~50%) biotinylation when co-expressed with BirA enzyme in *E. coli* bacteria (Scholle *et al.*, 2006).

The commercially available BirA-expressing *E. coli* strains (such as ABV101 carrying IPTG-inducible pBirAcm by Avidity, LLC, USA) can be used for phage propagation providing efficient biotinylation (our unpublished results). Phage biotinylation efficiency can be detected by ELISA using streptavidin-HRP or quantified by magnetic streptavidin beads assay. Moreover, biotin molecule is tolerant to amine- and carboxyl-chemistries which is exceptionally useful for the following drug conjugation step.

The further complexation of biotinylated phage and biotinylated antibody via avidin is carried out in two sequential steps. In order to provide phage-avidin-antibody binding and prevent a formation of phage-avidin or antibody-avidin aggregates, the exact molar concentrations of each component should be carefully adjusted (Fig. 16.2B). In the first step, phages are complexed with avidin molecules, while during the second step the biotinylated antibodies are added. The method's advantage is the possibility to use any available biotinylated (anti-bacterial) antibody as well as a polyclonal mixture. Polyclonal antibodies provide binding to a variety of antigenic epitopes and are on average less sensitive to biotin labelling chemistry. The author's group has recently produced monoclonal antibodies in the IgG format in *E. coli* (Hakim and Benhar, 2009; Luria *et al.*, 2012) including IgGs to which the BirA tag was fused at the C-terminus of the Fc. Such antibodies undergo efficient site-specific *in vivo* biotinylation with no risk of being damaged as is the case with chemical biotinylation (unpublished results).

Peptide-based targeting

Display of antibacterial peptide on phage coat proteins can serve as a perfect alternative to antibody-based targeting. As was the case with antibody targeting, the recognition of pathogen relies on peptide binding ability while the natural host specificity of the phage is irrelevant.

Peptides are usually expressed as pVIII major coat protein fusions and show an oligo- or polyvalent display in all display systems ('Monovalent vs. polyvalent display', above). While the binding affinity of peptide is usually lower than that of antibody, the polyvalent display provides an avidity that results in tight binding of the displaying phage to the target. Target binding peptides are usually isolated by phage display, which when successful yields many binders for each target. This makes it possible to select phages that display peptide sequences lacking amino acid residues that may be vulnerable to damage by the subsequent drug conjugation chemistry (like Lys, Glu or Asp) and are less likely to be damaged by chemical modifications than larger proteins.

A case in point is the anti-*S. aureus* A12C peptide (VHMVAGPGREPT) that was obtained from a 12-mer combinatorial type 88 phage-peptide display library (Enshell-Seijffers *et al.*, 2001) and applied for further work (Fig. 16.2C) (Yacoby *et al.*, 2006). The peptide was presented in a few tens of copies on the phage surface and was not influenced by the drug-conjugation step.

Drug-carrying filamentous phages

An efficient (targeted) drug-carrying agent should contain the following three components: a targeting moiety, a drug carrier and a cytotoxic drug. Filamentous phages and some tailed phages possess the unique spatial structure that may be applied as a nanometric, modular, high capacity drug carrying platform. The phage coat proteins can be genetically and chemically modified and used for drug conjugation. This approach includes the use of toxic and non-specific as well as non-potent drugs that are otherwise rarely used for systemic treatment, loading on the phage carrier and targeting to the pathogenic bacteria. The targeting will be provided either by phage natural host specificity (described in 'Host-specific drug-carrying phage', below) or antibacterial peptide or antibody display (described in 'Peptide- and antibody-targeted drug-carrying phage', below).

The phage carrier is a possible alternative to liposomes and other drug-loaded nano-platforms (Devalapally *et al.*, 2007; Mattheolabakis *et al.*, 2013). The advantages of phage drug carrier are its natural stability based on tight protein–protein interactions within the phage particle, extremely high drug payload, specific targeting and tolerance to chemical modifications.

Host-specific drug-carrying phage

The natural ability of phage to bind its host bacteria provides the specific targeting of the particle to an individual pathogen. The research carried out by the Wilson group (Embleton *et al.*, 2005) was based on *Staphylococcus aureus*-specific phage that was chemically modified to deliver tin(IV) chlorine e6 (SnCe6) to *S. aureus* bacteria. Photosensitizer SnCe6 is a light-activatable compound, which generates singlet oxygen and free radicals when exposed to light of a suitable wavelength (Wainwright, 1998). Previous studies demonstrated that the delivery of SnCe6 fused to non-specific IgG (targeting *S. aureus* protein A) and to anti-*S. aureus* IgG (targeting *S. aureus* capsule protein) resulted in the specific and efficient bacteria killing in comparison to free photosensitizer (Embleton *et al.*, 2002, 2004). However, bacteria had shown different sensitivity levels at different growth phases: much less susceptibility at the lag phase (for protein A targeting) and the stationary phase (for capsule targeting).

The replacement of immunoglobulin by the phage host recognition mechanism provided the system with targeting selectivity and drug delivery efficacy (Fig. 16.3A). The drug conjugation involved the linkage of the exposed phage proteins amine groups to the carboxyl side chains of the drug. The drug-carrying phages damaged the *S. aureus* bacteria and had no effect on the viability of human epithelial cells. In addition, phage particle enabled high-capacity loading of the drug and its concentration at the bacteria cell wall, thus

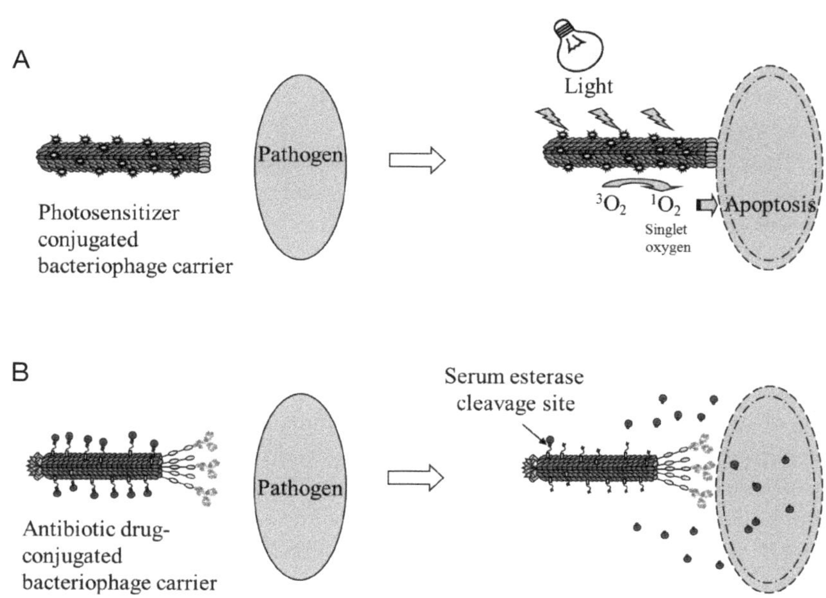

Figure 16.3 Mechanism of action of targeted drug-carrying phages. (A) Photosensitizer-conjugated phage tightly binds its host bacteria. Exposure to light at suitable wavelength generates singlet oxygen and free radicals release, providing the apoptosis of targeted pathogen. (B) The targeting moiety of drug-carrying phage provides the specific binding to selected bacteria. The drug linked to phage surface by serum esterase sensitive linker is released nearby the pathogen. The toxic effect is received by creating the high local drug concentration. A colour version of this figure is available in the plate section at the back of the book.

achieving more efficient killing. The conjugate demonstrated the more bactericidal effect than the unconjugated SnCe6. Furthermore, the killing was no more growth phase-dependent.

It was later proved that drug delivery ability of phage and its effect on the targeted bacteria did not require actual infection of the pathogen. It was suggested that phage/drug conjugate only needs to bind the bacterial cell to induce its efficient killing (Hope *et al.*, 2009). However, this approach still requires the isolation of a specific phage for each individual pathogen or a number of pathogens within the same species.

Peptide- and antibody-targeted drug-carrying phage

The replacement of native host-specific recognition ability of phage by peptide or antibody display (described in ('Antibody-based targeting' and 'Peptide-based targeting', above) provides the specific binding to antibody-defined target. The author's group developed a universal, modular solution for applying phages as a targeted drug-carrying platform for eradication of pathogenic bacteria (Yacoby *et al.*, 2006, 2007; Vaks and Benhar, 2011a,b). The platform was based on genetic and chemical modifications of filamentous phage whose natural host specificity is not relevant to its target specificity. The targeting moiety was provided by three alternative display methods: (1) anti-*S. aureus* A12C peptide ('Peptide-based targeting', above), (2) antibody-based ZZ domain mediated ('Antibody-based ZZ domain mediated targeting', above) or (3) antibody-based biotin-avidin mediated ('Antibody based biotin–avidin-mediated targeting', above) display on the phage surface. The drug was linked to the phages via chemical conjugation through a labile linker, subject to controlled release. The conjugated drug, unlike the photosensitizer used in the previous work (Embleton *et al.*, 2005), does not exert cytotoxic activity and may be considered as a prodrug activated upon its dissociation from the phage. The drug used in these studies was the bacteriostatic antibiotic chloramphenicol which is rarely used systemically due to its haemotoxicity (Turton *et al.*, 2002). Prior to conjugation chloramphenicol molecule was chemically modified to contain a labile ester bond-containing linker and a N-hydroxysuccinimide (NHS) reactive group, resulting in chloramphenicol prodrug (Yacoby *et al.*, 2006). Ester linker provided the controlled release by serum esterases, while NHS group was necessary for drug conjugation to the carrier (Fig. 16.3B).

The first generation of phage carriers: direct conjugation of chloramphenicol prodrug to phages

In our initial study, the bacterium *S. aureus* was used as the target and the phages were conferred with target specificity by a *S. aureus*-specific peptide selected from a peptide phage display library. The phage library was affinity-selected on *S. aureus* bacteria and specific *S. aureus* binders were identified. Of those, a *S. aureus*-specific peptide sequence that contained no lysine residues was selected, to minimize the risk of damaging target binding upon subsequent drug conjugation. NHS-containing chloramphenicol prodrug can be easily conjugated to amine groups (lysine residues) on phage coat proteins in aqueous solution. The phage major coat protein pVIII possesses two exposed amine groups; the alpha amino group of N-terminus and the epsilon amino group of lysine 8. Four C-terminal lysine residues are buried within the phage particle and are not accessible to conjugation. Thus up to two chloramphenicol molecules can theoretically be conjugated to each pVIII monomer. However, it was demonstrated that the high hydrophobicity of the drug resulted in solubility reduction of phage/chloramphenicol conjugate at high concentrations of the linked drug (Yacoby *et al.*, 2006). The mild conjugation (~3000 molecules per phage particle) to A12C-targeted phage retarded *S. aureus* growth as efficiently as a 20 times higher concentration of free drug, while antibody targeted anti-*S. aureus* phages resulted in one log efficiency shift of bacteria growth inhibition in comparison to free chloramphenicol. The possible explanations for this partial growth retardation were low drug-carrying capacity, relatively slow drug release and chemistry-sensitive targeting moiety.

The second generation of phage carriers: conjugation of chloramphenicol to phages through aminoglycosides

The direct conjugation of a hydrophobic drug to the phage carrier limited its actual capacity to maximum 3000 chloramphenicol molecules per phage ('The first generation of phage carriers: direct conjugation of chloramphenicol prodrug to phages', above). Therefore, neomycin was used as a hydrophilic aminoglycoside linker between the drug and the carrier. Applying highly water-soluble aminoglycosides as linkers, enhanced the complex solubility and enabled the conjugation of over 40,000 chloramphenicol molecules per phage without compromising the phage integrity (Yacoby *et al.*, 2007). The drug conjugation procedure included the preparation of chloramphenicol–neomycin adduct and its further conjugation to the phage surface by 1-ethyl-3-(3-dimethylaminopropyl) carbodiimide (EDC) chemistry. The conjugate showed about 20,000 times a higher drug potency in comparison to free chloramphenicol and demonstrated potent growth retardation of *S. aureus* bacteria *in vitro*. In addition, the targeted drug-carrying phage enabled the growth inhibition of *Streptococcus pyogenes* and *E. coli O78* pathogenic bacteria, each when targeted by a microbe-specific antibody. The targeting moiety in this study (the displayed ZZ domain) was still affected by drug conjugation chemistry, suggesting that better binding ability may be achieved by using the biotin-avidin mediated antibody display.

In vivo characterization of anti-*S. aureus* Neo-CAM-carrying phages

Previous studies on phage therapy had demonstrated a few important characteristics of *in vivo* phage-based treatment. Here we relate some of the key issues to observations that were made by the authors during *in vivo* studies using chloramphenicol drug-carrying phages targeted against *S. aureus* bacteria.

1 Topical and oral administration of phages to animals and humans was found safe and non-toxic (Carlton *et al.*, 2005; Międzybrodzki *et al.*, 2012; Sarker *et al.*, 2012; Kang *et al.*, 2013). Moreover, four phage preparations have been approved by FDA for food applications (Sulakvelidze, 2013). However, crude phage lysates may contain bacterial toxins and lipopolysaccharides (LPS), thus have to be highly purified before the use (Bogovazova *et al.*, 1992). While the phage carrier is a harmless entity, the drug-carrying phage loaded with toxic non-selective compound may potentially cause serious side-effects. The toxicity of chloramphenicol-conjugated phages (conjugation through neomycin aminoglycoside as described in 'The second generation of phage carriers: conjugation of chloramphenicol to phages through aminoglycosides', above), was evaluated in BALB/c mice that were injected with up to 10^{11} phages. No toxic effect could be observed (mice were monitored for weight loss or attenuation of behaviour). These observations suggest that the total concentration of conjugated drug is too low to cause a harmful effect and also the drug loses its toxic properties as a result of the conjugation (Vaks and Benhar, 2011b).

2 Phages are highly immunogenic causing a production of anti-phage antibodies upon injection into animals (Srivastava *et al.*, 2004). In fact, phages have been used as vaccine carriers and vaccination with peptide-displaying phage raised a high antibody titre against the displayed molecule and more so, against the phage itself (Grabowska *et al.*, 2000). The immunogenicity is probably caused by the protein (VIII) repeats on its surface and a large size of a phage particle or by the repetitive arrangement of coat proteins in other phages. In drug-carrying phages, drug conjugation by EDC chemistry which targets the free carboxyl groups on phage coat proteins (described in 'The second generation of phage carriers: conjugation of chloramphenicol to phages through aminoglycosides', above) was found to extremely influence phage physical and biological features. For example, EDC-treated phages could no longer be recognized by commercial anti-phage antibodies, could

no longer be precipitated by PEG/NaCl, and the majority of the EDC cross-linked phages lost their infectivity. The author's group established that chloramphenicol-neomycin carrying phages were far less immunogenic in comparison to unconjugated phages (Vaks *et al.*, 2011b). While in a prime-boost experiment the repeated administration of unmodified phages into BALB/c mice by intraperitoneal or intravenous route resulted in a very high titre of serum anti-phage antibodies (titre up to 150,000), mice injected with chloramphenicol-neomycin-carrying phages developed a much lower serum anti-phage antibody titre (up to 50,000). It was hypothesized that the reduced immunogenicity was due to the hydrophilic neomycin, used for chloramphenicol conjugation. Neomycin provided a 'dense sugar coat' for phage particle, shielding it from the immune response and reducing the immune response against the modified phages. It was also observed that chloramphenicol-neomycin carrying phages had an extended half-life in the circulation similarly to PEGylated particles reported by other authors (Vaks and Benhar, 2011b; Marik *et al.*, 2007; Romberg *et al.*, 2007; Weissig *et al.*, 2000).

3 In classic phage therapy phages infect bacteria and increase their titre by *in vivo* replication (Smith and Huggins, 1982). As long as susceptible bacteria are present, phage replication cannot be controlled (Chibani-Chennoufi *et al.*, 2004). The drug-carrying phage system that is based on *E. coli* specific phage may theoretically infect *E. coli* bacteria of the patients' natural flora. However, as was mentioned above, drug conjugated phages lose their infectivity and cannot infect host bacteria. Thus, the exposure of the treated individual is only to the administered dose of phages.

4 Filamentous (and other) phages are rapidly removed from the circulation 5 to 15 min following intravenous injection (Zou *et al.*, 2004). Additional research implies that the half life of filamentous phages in mice serum is about 4 h (Molenaar *et al.*, 2002). The mechanism of phage removal is mostly uptake by phagocytic cells of the reticuloendothelial system (RES). Possible solutions could be selection of long-circulating phage mutants (as was shown for phage lambda) capable of escaping or even repelling the host defence system (Merril *et al.*, 1996) and modifying the phage coat with polyethylene glycol (PEG) to 'shield it' from host defence surveillance mechanisms (Marik *et al.*, 2007; Weissig *et al.*, 2000).

5 Therapeutic efficacy of targeted drug carrying phages has not been reported. However, the author's group found in a preliminary challenge assay that BALB/c mice challenged with a lethal dose of *S. aureus* bacteria can be protected by co-administration of *S. aureus* targeted neomycin-chloramphenicol-carrying phages (unpublished results).

Phage-based treatment was previously suggested as highly problematic for *in vivo* application due to its immunogenicity and uncontrolled replication in the patient. It was demonstrated that drug-carrying phages lost their infectivity upon the drug conjugation, thus were not able to infect the pathogen while providing exceptionally the specific targeting. The immunogenicity concern may be removed by 'shielding' the phage surface by sugar-like linkers, the modification that may as well influence the kinetic properties of the system.

Future trends

This chapter described the promising properties of phage therapy generally and drug-carrying phage system in particular. The nanometric dimensions of the phage particle and the modular assemblage of its coat turn it into a novel and versatile drug-delivery platform. The vast diversity of phage world and the tight evolution of phage with its bacterial host enable the perfect targeting of the particle to specifically identified bacteria. On the other hand, the targeting may be augmented for one phage to bind any selected bacterial target cell by peptide or antibody display on phage coat proteins. The outstanding targeting moiety and the high drug-carrying capacity of the phage provide the effectiveness and specificity to non-potent or non-selective drugs.

Future accurate studies on pharmacokinetics and bio-distribution of drug-carrying phages will provide for better understanding of its therapeutic properties. The common concern of phage application as therapeutics is based on its high immunogenicity caused by pVIII protein repeats and by the fact that upon intravenous administration phages are rapidly removed from the circulation by RES cells. However, it was reported that the targeted drug-conjugated phages generated much lower anti-phage antibody titre than unconjugated particles in vaccinated mice. The kinetics and distribution of targeted drug-carrying phages in an animal model of infection has not been studied in sufficient detail, but we believe that the lower immunogenicity may result in prolonged circulation half-life and increase the drug accumulation in the infected area.

The replacement of anti-bacterial targeting moiety by anti-fungal or anti-mammalian cell antibodies will create the phage-based nanometric system for specific targeting of pathogenic fungi (unpublished results) as well as cancer cells (Bar *et al.*, 2008). The application of suitable drug and the appropriate drug release mechanism will provide the specific killing of the selected target.

The unique natural properties of the phage carrier and its universal compatibility for drug conjugation and targeting issue display give the phage nanoparticle an exclusive advantage among the present anti bacterial nanomedicines.

References

Bar, H., Yacoby, I., and Benhar, I. (2008). Killing cancer cells by targeted drug-carrying phage nanomedicines. BMC Biotechnol. *8*, 37.

Barbas, F.C., Burton, D.R., Scott, J.K., and Silverman, D.J. (2001). Phage Display, A Laboratory Manual (N.Y., Cold Spring Harbor Laboratory Press).

Bayer, E.A., and Wilchek, M. (1990). Protein Biotinylation. Method. Enzymol. *184*, 138–160.

Beck, E., and Zink, B. (1981). Nucleotide sequence and genome organisation of filamentous bacteriophages fl and fd. Gene *16*, 35–58.

Beckett, D., Kovaleva, E., and Schatz, P.J. (1999). A minimal peptide substrate in biotin holoenzyme synthetase-catalyzed biotinylation. Protein Sci. *8*, 921–929.

Benhar, I. (2001). Biotechnological applications of phage and cell display. Biotechnol. Adv. *19*, 1–33.

Berdichevsky, Y., Ben-Zeev, E., Lamed, R., and Benhar, I. (1999). Phage display of a cellulose binding domain from Clostridium thermocellum and its application as a tool for antibody engineering. J. Immunol. Meth. *228*, 151–162.

Bielke, L., Higgins, S., Donoghue, A., Donoghue, D., and Hargis, B.M. (2007). Salmonella host range of bacteriophages that infect multiple genera. Poultry Sci. *86*, 2536–2540.

de Boer, E., Rodriguez, P., Bonte, E., Krijgsveld, J., Katsantoni, E., Heck, A., Grosveld, F., and Strouboulis, J. (2003). Efficient biotinylation and single-step purification of tagged transcription factors in mammalian cells and transgenic mice. Proc. Natl. Acad. Sci. U.S.A. *100*, 7480–7485.

Bogovazova, G.G., Voroshilova, N.N., Bondarenko, V.M., Gorbatkova, G.A., Afanasyeva, E.V., Kazakova, T.B., Smirnov, V.D., Mamleeva, A.G., Glukharev, Y.A., Erastova, E.I., *et al.* (1992). Immunobiological Properties and Therapeutic Effectiveness of Klebsiella Bacteriophage Preparations. Zhurnal Mikrobiologii Epidemiologii I Immunobiologii, 30–33.

Boucher, H.W., Talbot, G.H., Bradley, J.S., Edwards, J.E., Gilbert, D., Rice, L.B., Scheld, M., Spellberg, B., and Bartlett, J. (2009). Bad bugs, no drugs: no ESKAPE! An update from the Infectious Diseases Society of America. Clin. Infect. Dis. *48*, 1–12.

Bradbury, A.R., and Marks, J.D. (2004). Antibodies from phage antibody libraries. J. Immunol. Methods *290*, 29–49.

Burrowes, B., Harper, D.R., Anderson, J., McConville, M., and Enright, M.C. (2011). Bacteriophage therapy: potential uses in the control of antibiotic-resistant pathogens. Expert Rev. Anti. Infect. Ther. *9*, 775–785.

Carlton, R.M., Noordman, W.H., Biswas, B., de Meester, E.D., and Loessner, M.J. (2005). Bacteriophage P100 for control of *Listeria monocytogenes* in foods: Genome sequence, bioinformatic analyses, oral toxicity study, and application. Regul. Toxicol. Pharmacol. *43*, 301–312.

Chan, B.K., Abedon, S.T., and Loc-Carrillo, C. (2013). Phage cocktails and the future of phage therapy. Future Microbiol. *8*, 769–783.

Chapman-Smith, A., and Cronan, J.E. (1999). In vivo enzymatic protein biotinylation. Biomol. Eng. *16*, 119–125.

Chibani-Chennoufi, S., Sidoti, J., Bruttin, A., Kutter, E., Sarker, S., and Brüssow, H. (2004). In vitro and *in vivo* bacteriolytic activities of *Escherichia coli* phages: Implications for phage therapy. Antimicrob. Agents Chemother. *48*, 2558–2569.

Crissman, J.W., and Smith, G.P. (1984). Gene-III protein of filamentous phages: evidence for a carboxyl-terminal domain with a role in morphogenesis. Virology *132*, 445–455.

Cronan, J.E. (1990). Biotination of Proteins In vivo – a Posttranslational Modification to Label, Purify, and Study Proteins. J. Biol. Chem. *265*, 10327–10333.

Devalapally, H., Chakilam, A., and Amiji, M.M. (2007). Role of nanotechnology in pharmaceutical product development. J. Pharm. Sci. *96*, 2547–2565.

Diamandis, E.P., and Christopoulos, T.K. (1991). The Biotin (Strept)Avidin System – Principles and Applications in Biotechnology. Clin. Chem. *37*, 625–636.

Djojonegoro, B.M., Benedik, M.J., and Willson, R.C. (1994). Bacteriophage surface display of an immunoglobulin-binding domain of *Staphylococcus aureus* protein A. Biotechnology (N Y) *12*, 169–172.

Embleton, M.L., Nair, S.P., Cookson, B.D., and Wilson, M. (2002). Selective lethal photosensitization of methicillin-resistant *Staphylococcus aureus* using an IgG-tin (IV) chlorin e6 conjugate. J. Antimicrob. Chemother. *50*, 857–864.

Embleton, M.L., Nair, S.P., Cookson, B.D., and Wilson, M. (2004). Antibody-directed photodynamic therapy of methicillin resistant *Staphylococcus aureus*. Microb. Drug Resist. *10*, 92–97.

Embleton, M.L., Nair, S.P., Heywood, W., Menon, D.C., Cookson, B.D., and Wilson, M. (2005). Development of a novel targeting system for lethal photosensitization of antibiotic-resistant strains of *Staphylococcus aureus*. Antimicrob. Agents Chemother. *49*, 3690–3696.

Enshell-Seijffers, D., Smelyanski, L., and Gershoni, J.M. (2001). The rational design of a 'type 88' genetically stable peptide display vector in the filamentous bacteriophage fd. Nucleic Acids Res. *29*, E50–50.

Gao, C., Mao, S., Lo, C.H., Wirsching, P., Lerner, R.A., and Janda, K.D. (1999). Making artificial antibodies: a format for phage display of combinatorial heterodimeric arrays. Proc. Natl. Acad. Sci. U.S.A. *96*, 6025–6030.

Gao, C., Mao, S., Kaufmann, G., Wirsching, P., Lerner, R.A., and Janda, K.D. (2002). A method for the generation of combinatorial antibody libraries using pIX phage display. Proc. Natl. Acad. Sci. U.S.A. *99*, 12612–12616.

Garrard, L.J., Yang, M., O'Connell, M.P., Kelley, R.F., and Henner, D.J. (1991). Fab assembly and enrichment in a monovalent phage display system. Biotechnology (N Y) *9*, 1373–1377.

Glucksman, M.J., Bhattacharjee, S., and Makowski, L. (1992). Three-dimensional structure of a cloning vector. X-ray diffraction studies of filamentous bacteriophage M13 at 7 A resolution. J. Mol. Biol. *226*, 455–470.

Goldberg, E., Grinius, L., and Letellier, L. (1994). Recognition, attachment, and injection. In Molecular biology of bacteriophage T4 (American Society for Microbiology (ASM) {a}), pp. 347–356.

Grabowska, A.M., Jennings, R., Laing, P., Darsley, M., Jameson, C.L., Swift, L., and Irving, W.L. (2000). Immunisation with phage displaying peptides representing single epitopes of the glycoprotein G can give rise to partial protective immunity to HSV-2. Virology *269*, 47–53.

Greenwood, J., Willis, A.E., and Perham, R.N. (1991). Multiple display of foreign peptides on a filamentous bacteriophage. Peptides from Plasmodium falciparum circumsporozoite protein as antigens. J. Mol. Biol. *220*, 821–827.

Gupta, R., and Prasad, Y. (2011). Efficacy of polyvalent bacteriophage P-27/HP to control multidrug resistant *Staphylococcus aureus* associated with human infections. Curr. Microbiol. *62*, 255–260.

Hakim, R., and Benhar, I. (2009). 'Inclonals': IgGs and IgG–enzyme fusion proteins produced in an *E. coli* expression-refolding system. mAbs *1*, 281–287.

Hanlon, G.W. (2007). Bacteriophages: an appraisal of their role in the treatment of bacterial infections. Int. J. Antimicrob. Agents *30*, 118–128.

Hill, D.F., and Petersen, G.B. (1982). Nucleotide sequence of bacteriophage f1 DNA. J. Virol. *44*, 32–46.

Hoogenboom, H.R., de Bruine, A.P., Hufton, S.E., Hoet, R.M., Arends, J.W., and Roovers, R.C. (1998). Antibody phage display technology and its applications. Immunotechnology *4*, 1–20.

Hope, C.K., Packer, S., Wilson, M., and Nair, S.P. (2009). The inability of a bacteriophage to infect *Staphylococcus aureus* does not prevent it from specifically delivering a photosensitizer to the bacterium enabling its lethal photosensitization. J. Antimicrob. Chemother. *64*, 59–61.

Housby, J.N., and Mann, N.H. (2009). Phage therapy. Drug Discov. Today *14*, 536–540.

Hsieh, S.E., Lo, H.H., Chen, S.T., Lee, M.C., and Tseng, Y.H. (2011). Wide host range and strong lytic activity of *Staphylococcus aureus* lytic phage Stau2. Appl. Environ. Microbiol. *77*, 756–761.

Hufton, S.E., Moerkerk, P.T., Meulemans, E.V., de Bruine, A., Arends, J.W., and Hoogenboom, H.R. (1999). Phage display of cDNA repertoires: the pVI display system and its applications for the selection of immunogenic ligands. J. Immunol. Methods *231*, 39–51.

Iannolo, G., Minenkova, O., Petruzzelli, R., and Cesareni, G. (1995). Modifying filamentous phage capsid: limits in the size of the major capsid protein. J. Mol. Biol. *248*, 835–844.

Jespers, L.S., Roberts, A., Mahler, S.M., Winter, G., and Hoogenboom, H.R. (1994). Guiding the selection of human antibodies from phage display repertoires to a single epitope of an antigen. Biotechnology (N Y) *12*, 899–903.

Jespers, L.S., Messens, J.H., De Keyser, A., Eeckhout, D., Van den Brande, I., Gansemans, Y.G., Lauwereys, M.J., Vlasuk, G.P., and Stanssens, P.E. (1995). Surface expression and ligand-based selection of cDNAs fused to filamentous phage gene VI. Biotechnology (N Y) *13*, 378–382.

Kang, H.W., Kim, J.W., Jung, T.S., and Woo, G.J. (2013). wksl3, a new biocontrol agent for Salmonella enterica serovars enteritidis and typhimurium in foods: characterization, application, sequence analysis, and oral acute toxicity study. Appl. Environ. Microbiol. *79*, 1956–1968.

Kay, B., Winter, J., and McCafferty, J. (1996). Phage Display of Peptides and Proteins: A Laboratory Manual (Academic Press).

Kremser, A., and Rasched, I. (1994). The adsorption protein of filamentous phage fd: assignment of its disulfide bridges and identification of the domain incorporated in the coat. Biochemistry 33, 13954–13958.

Kushwaha, A., Chowdhury, P.S., Arora, K., Abrol, S., and Chaudhary, V.K. (1994). Construction and characterization of M13 bacteriophages displaying functional

IgG-binding domains of staphylococcal protein A. Gene *151*, 45–51.

Kwasnikowski, P., Kristensen, P., and Markiewicz, W.T. (2005). Multivalent display system on filamentous bacteriophage pVII minor coat protein. J. Immunol. Methods *307*, 135–143.

Lewis, K. (2013). Platforms for antibiotic discovery. Nat. Rev. Drug Discov. *12*, 371–387.

Luria, Y., Raichlin, D., and Benhar, I. (2012). Fluorescent IgG fusion proteins made in *E. coli*. MAbs. *4*, 373–384.

McCafferty, J., Jackson, R.H., and Chiswell, D.J. (1991). Phage-enzymes: expression and affinity chromatography of functional alkaline phosphatase on the surface of bacteriophage. Protein Eng. *4*, 955–961.

Marik, J., Tartis, M.S., Zhang, H., Fung, J.Y., Kheirolomoom, A., Sutcliffe, J.L., and Ferrara, K.W. (2007). Long-circulating liposomes radiolabeled with [F-18] fluorodipalmitin ([F-18]FDP). Nucl. Med. Biol. *34*, 165–171.

Marti, R., Zurfluh, K., Hagens, S., Pianezzi, J., Klumpp, J., and Loessner, M.J. (2013). Long tail fibres of the novel broad-host-range T-even bacteriophage S16 specifically recognize Salmonella OmpC. Mol. Microbiol. *87*, 818–834.

Marvin, D.A. (1998). Filamentous phage structure, infection and assembly. Curr. Opin. Struct. Biol. *8*, 150–158.

Marvin, D.A., Hale, R.D., Nave, C., and Helmer-Citterich, M. (1994). Molecular models and structural comparisons of native and mutant class I filamentous bacteriophages Ff (fd, f1, M13), If1 and IKe. J. Mol. Biol. *235*, 260–286.

Mattheolabakis, G., Rigas, B., and Constantinides, P.P. (2012). Nanodelivery strategies in cancer chemotherapy: biological rationale and pharmaceutical perspectives. Nanomedicine (Lond). *7*, 1577–1590.

Merril, C.R., Biswas, B., Carlton, R., Jensen, N.C., Creed, G.J., Zullo, S., and Adhya, S. (1996). Long-circulating bacteriophage as antibacterial agents. Proc. Natl. Acad. Sci. (USA) *93*, 3188–3192.

Miedzybrodzki, R., Borysowski, J., Weber-Dąbrowska, B., Fortuna, W., Letkiewicz, S., Szufnarowski, K., Pawełczyk, Z., Rogóż, P., Klak, M., Wojtasik, E., and Górski, A. (2012). Clinical aspects of phage therapy. Adv. Virus Res. *83*, 73–121.

Molenaar, T.J.M., Michon, I., de Haas, S.A.M., van Berkel, T.J.C., Kuiper, J., and Biessen, E.A.L. (2002). Uptake and processing of modified bacteriophage M13 in mice: Implications for phage display. Virology *293*, 182–191.

Nilsson, B., Moks, T., Jansson, B., Abrahmsen, L., Elmblad, A., Holmgren, E., Henrichson, C., Jones, T.A., and Uhlen, M. (1987). A synthetic IgG-binding domain based on staphylococcal protein A. Protein Eng. *1*, 107–113.

O'Connell, D., Becerril, B., Roy-Burman, A., Daws, M., and Marks, J.D. (2002). Phage versus phagemid libraries for generation of human monoclonal antibodies. J. Mol. Biol. *321*, 49–56.

Ohno, K., Levin, B., and Meruelo, D. (1996). Cell-specific, multidrug delivery system using streptavidin protein A fusion protein. Biochem. Mol. Med. *58*, 227–233.

Parthasarathy, R., Bajaj, J., and Boder, E.T. (2005). An immobilized biotin ligase: Surface display of *Escherichia coli* BirA on Saccharomyces cerevisiae. Biotechnol. Prog. *21*, 1627–1631.

Rakonjac, J., Bennett, N.J., Spagnuolo, J., Gagic, D., and Russel, M. (2011). Filamentous bacteriophage: biology, phage display and nanotechnology applications. Curr. Issues Mol. Biol. *13*, 51–76.

Randall-Hazelbauer, L., and Schwartz, M. (1973). Isolation of the bacteriophage lambda receptor from *Escherichia coli*. J. Bacteriol. *116*, 1436–1446.

Romberg, B., Oussoren, C., Snel, C.J., Hennink, W.E., and Storm, G. (2007). Effect of liposome characteristics and dose on the pharmacokinetics of liposomes coated with poly(amino acid)s. Pharm. Res. *24*, 2394–2401.

Russel, M., and Model, P. (1989). Genetic analysis of the filamentous bacteriophage packaging signal and of the proteins that interact with it. J. Virol. *63*, 3284–3295.

Sarker, S.A., McCallin, S., Barretto, C., Berger, B., Pittet, A.C., Sultana, S., Krause, L., Huq, S., Bibiloni, R., Bruttin, A., *et al.* (2012). Oral T4-like phage cocktail application to healthy adult volunteers from Bangladesh. Virology *434*, 222–232.

Saviranta, P., Haavisto, T., Rappu, P., Karp, M., and Lovgren, T. (1998). In vitro enzymatic biotinylation of recombinant fab fragments through a peptide acceptor tail. Bioconjug. Chem. *9*, 725–735.

Scholle, M.D., Kriplani, U., Pabon, A., Sishtla, K., Glucksman, M.J., and Kay, B.K. (2006). Mapping protease substrates by using a biotinylated phage substrate library. Chembiochem. *7*, 834–838.

Sidhu, S.S. (2001). Engineering M13 for phage display. Biomolecular Eng. *18*, 57–63.

Skurnik, M., Pajunen, M., and Kiljunen, S. (2007). Biotechnological challenges of phage therapy. Biotechnol. Lett. *29*, 995–1003.

Smith, G.P. (1985). Filamentous fusion phage: novel expression vectors that display cloned antigens on the virion surface. Science *228*, 1315–1317.

Smith, H.W., and Huggins, M.B. (1982). Successful treatment of experimental *Escherichia coli* infections in mice using phage: its general superiority over antibiotics. J. Gen. Microbiol. *128*, 307–318.

Smith, J.S., Keller, J.R., Lohrey, N.C., McCauslin, C.S., Ortiz, M., Cowan, K., and Spence, S.E. (1999). Redirected infection of directly biotinylated recombinant adenovirus vectors through cell surface receptors and antigens. Proc. Natl. Acad. Sci. U.S.A. *96*, 8855–8860.

Smith, P.A., Tripp, B.C., DiBlasio-Smith, E.A., Lu, Z.J., LaVallie, E.R., and McCoy, J.M. (1998). A plasmid expression system for quantitative *in vivo* biotinylation of thioredoxin fusion proteins in *Escherichia coli*. Nucl. Acids Res. *26*, 1414–1420.

Specthrie, L., Bullitt, E., Horiuchi, K., Model, P., Russel, M., and Makowski, L. (1992). Construction of a microphage variant of filamentous bacteriophage. J. Mol. Biol. *228*, 720–724.

Srivastava, A.S., Kaido, T., and Carrier, E. (2004). Immunological factors that affect the *in vivo* fate of T7 phage in the mouse. J. Virol. Methods. *115*, 99–104.

Stengele, I., Bross, P., Garces, X., Giray, J., and Rasched, I. (1990). Dissection of functional domains in phage fd adsorption protein. Discrimination between attachment and penetration sites. J. Mol. Biol. *212*, 143–149.

Sulakvelidze, A. (2013). Using lytic bacteriophages to prevent or significantly reduce contamination of food by food-borne bacterial pathogens. J. Sci. Food. Agric. *93*, 3137–3146.

Summers, W.C. (2012). The strange history of phage therapy. Bacteriophage. *2*, 130–133.

Tanji, Y., Shimada, T., Fukudomi, H., Miyanaga, K., Nakai, Y., and Unno, H. (2005). Therapeutic use of phage cocktail for controlling *Escherichia coli* O157:H7 in gastrointestinal tract of mice. J. Biosci. Bioeng. *100*, 280–287.

Taylor, P.W. (2013). Alternative natural sources for a new generation of antibacterial agents. Int. J. Antimicrob. Agents. *42*, 195–201.

Terry, T.D., Malik, P., and Perham, R.N. (1997). Accessibility of peptides displayed on filamentous bacteriophage virions: susceptibility to proteinases. Biol. Chem. *378*, 523–530.

Tetart, F., Desplats, C., and Krisch, H.M. (1998). Genome plasticity in the distal tail fiber locus of the T-even bacteriophage: Recombination between conserved motifs swaps adhesin specificity. J. Mol. Biol. *282*, 543–556.

Tirat, A., Freuler, F., Stettler, T., Mayr, L.M., and Leder, L. (2006). Evaluation of two novel tag-based labelling technologies for site-specific modification of proteins. Int. J. Biol. Macromol. *39*, 66–76.

Tohidkia, M.R., Barar, J., Asadi, F., and Omidi, Y. (2012). Molecular considerations for development of phage antibody libraries. J. Drug Target. *20*, 195–208.

Tornetta, M., Baker, S., Whitaker, B., Lu, J., Chen, Q., Pisors, E., Shi, L., Luo, J., Sweet, R., and Tsui, P. (2010). Antibody Fab display and selection through fusion to the pIX coat protein of filamentous phage. J. Immunol. Methods *360*, 39–46.

Turton, J.A., Andrews, C.M., Havard, A.C., and Williams, T.C. (2002). Studies on the haemotoxicity of chloramphenicol succinate in the Dunkin Hartley guinea pig. Int. J. Exp. Pathol. *83*, 225–238.

Vaks, L., and Benhar, I. (2011a). Antibacterial application of engineered bacteriophage nanomedicines: antibody-targeted, chloramphenicol prodrug loaded bacteriophages for inhibiting the growth of *Staphylococcus aureus* bacteria. Methods Mol. Biol. *726*, 187–206.

Vaks, L., and Benhar, I. (2011b). In vivo characteristics of targeted drug-carrying filamentous bacteriophage nanomedicines. J. Nanobiotechnol. *9*, 58.

Viens, A., Mechold, U., Lehrmann, H., Harel-Bellan, A., and Ogryzko, V. (2004). Use of protein biotinylation *in vivo* for chromatin immunoprecipitation. Anal. Biochem. *325*, 68–76.

Wainwright, M. (1998). Photodynamic antimicrobial chemotherapy (PACT). J. Antimicrob. Chemother. *42*, 13–28.

Walsh, C. (2003). Antibiotics: actions, origins, resistance (Washington, DC, ASM Press).

Warren, D.J., Bjerner, J., Paus, E., Bormer, O.P., and Nustad, K. (2005). Use of an *in vivo* biotinylated single-chain antibody as capture reagent in an immunometric assay to decrease the incidence of interference from heterophilic antibodies. Clin. Chem. *51*, 830–838.

Weissig, V., Babich, J., and Torchilin, V. (2000). Long-circulating gadolinium-loaded liposomes: potential use for magnetic resonance imaging of the blood pool. Colloid Surf. B-Biointer. *18*, 293–299.

van Wezenbeek, P.M., Hulsebos, T.J., and Schoenmakers, J.G. (1980). Nucleotide sequence of the filamentous bacteriophage M13 DNA genome: comparison with phage fd. Gene *11*, 129–148.

Wood, W.B., Eiserling, F.A., and Crowther, R.A. (1994). Long tail fibers: Genes, proteins, structure, and assembly. In Molecular biology of bacteriophage T4 (American Society for Microbiology (ASM) {a}), pp. 282–290.

Wu, S.C., Yeung, J.C., Hwang, P.M., and Wong, S.L. (2002). Design, production, and characterization of an engineered biotin ligase (BirA) and its application for affinity purification of staphylokinase produced from *Bacillus subtilis* via secretion. Protein Expr. Purif. *24*, 357–365.

Yacoby, I., Shamis, M., Bar, H., Shabat, D., and Benhar, I. (2006). Targeting antibacterial agents by using drug-carrying filamentous bacteriophages. Antimicrob. Agents Chemother. *50*, 2087–2097.

Yacoby, I., Bar, H., and Benhar, I. (2007). Targeted drug-carrying bacteriophages as antibacterial nanomedicines. Antimicrob. Agents Chemother. *51*, 2156–2163.

Yang, J.B., Jaramillo, A., Shi, R.L., Kwok, W.W., and Mohanakumar, T. (2004). In vivo biotinylation of the major histocompatibility complex (MHC) class II/peptide complex by coexpression of B irA enzyme for the generation of MHC class II/tetramers. Hum. Immunol. *65*, 692–699.

Zou, J., Dickerson, M.T., Owen, N.K., Landon, L.A., and Deutscher, S.L. (2004). Biodistribution of filamentous phage peptide libraries in mice. Mol. Biol. Rep. *31*, 121–129.

Index

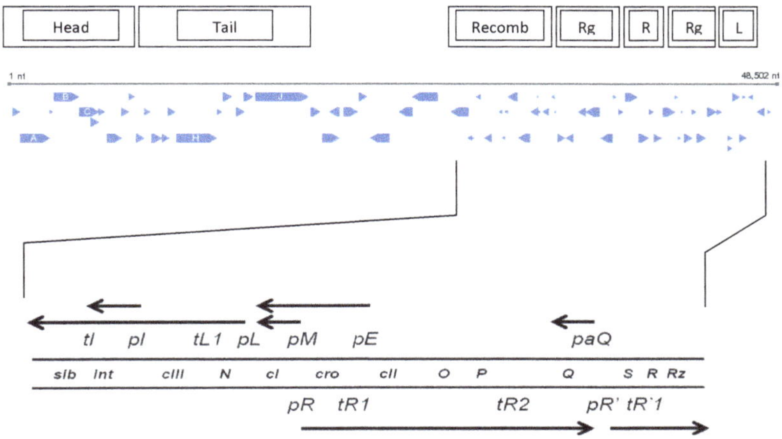

Figure 1.3 Map of the bacteriophage λ genome, showing the position of ORFs (open reading frames) the direction of transcription of particular genes (middle panel), and clusters of genes coding for proteins with specific functions (indicated above). Functions of gene products are abbreviated as follows: Head, assembly of the virion head; Tail, assembly of the virion tail; Recomb, DNA recombination; Rg, regulation of gene expression; R, phage DNA replication; L, lysis of host cell. A map of the right arm of the genome, not drawn to scale, indicates major transcription promoters (*p*) and transcription terminators (*t*) is shown below. Arrows indicate the direction of transcription. The whole genome sequence of phage λ may be found at (http://www.ncbi.nlm.nih.gov/genomes/genlist.cgi?taxid=10239&type=6&name=Phages) in the NCBI database.

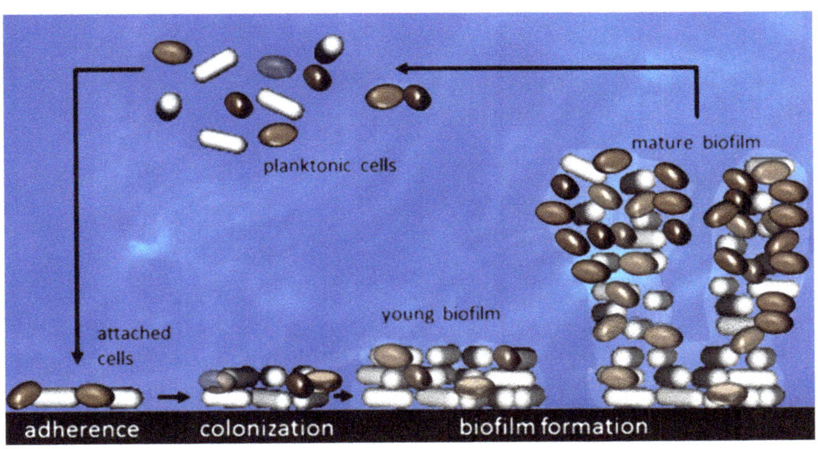

Figure 13.1 Biofilm formation steps. First bacteria adhere to conditioned substratum surface, then colonisation starts with production of extracellular metabolites, the young biofilms proliferate and in the mature biofilms portions of cells slough off the main biofilm structure.

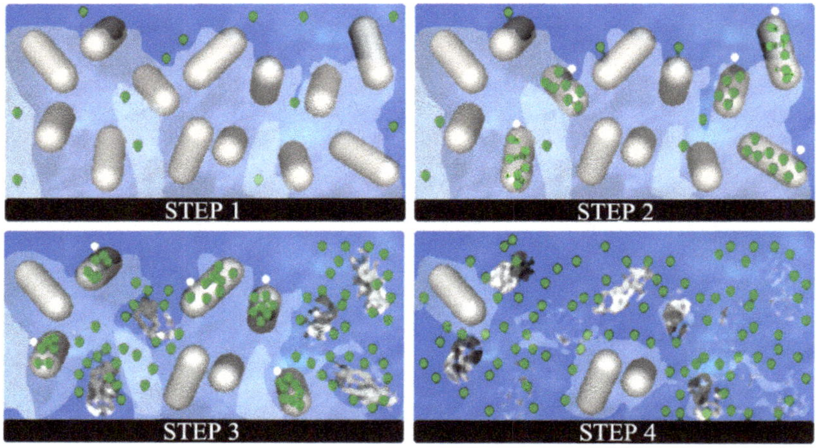

Figure 13.2 Model of biofilm infection by phage. Step 1: phage transport through the biofilm matrix and water channels through diffusion and/or convection; Step 2: Settlement of phages and phage cycle progression in the biofilm host cell; Step 3: Release of phage progeny through host cell lysis and infection of neighbourhood biofilm cells causing biomass reduction; Step 4: Detachment of biofilm portions and phages into the planktonic phase and phage cycle progression in planktonic cells.

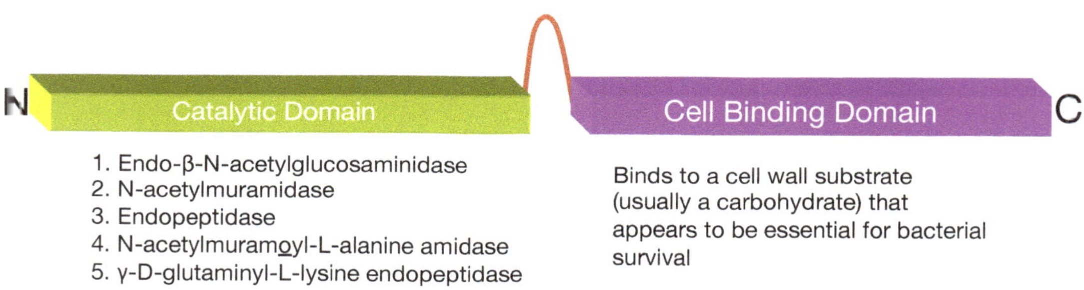

Figure 14.1 Basic structure of phage lysins. In general, lysins range between 25 kDa to 40 kDa in size and have a domain structure. The N-terminal domain is invariably the catalytic domain which cleaves one of the five major bonds in the peptidoglycan, and the C-terminal domain binds to a carbohydrate determinant in the cell wall.

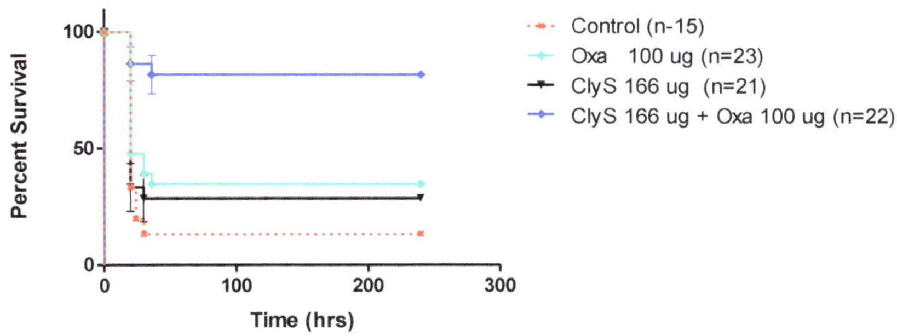

Figure 14.3 Synergistic effects of ClyS and oxacillin protected mice from MRSA septicaemia-induced death. Mice were intraperitoneally injected with ~5x10^{5} CFU of MRSA strain MW2 in 5% mucin. Three hours post infection, mice received an i.p. injection of a suboptimal concentration of ClyS (166μg) or 20mM phosphate buffer along with an IM injection of oxacillin (100μg) or saline control. Mice were monitored for survival for 10 days and the results of 5 independent experiments were combined and plotted in a Kaplan–Meier survival curve.

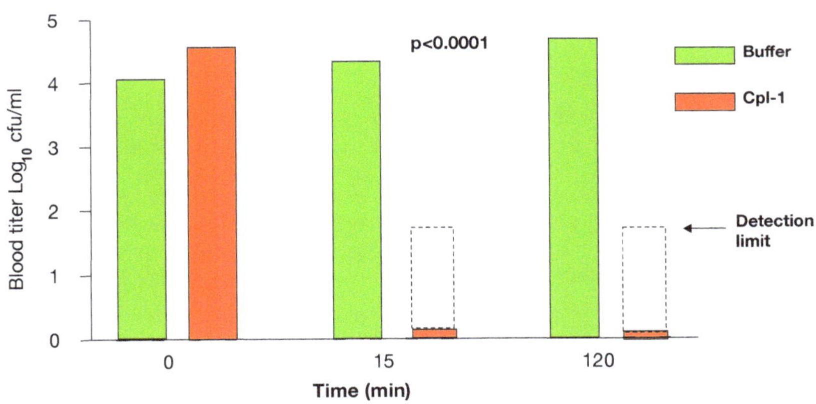

Figure 14.4 Survival of septic mice after treatment with Cpl-1. Mice pretreated IV with 10^{8} pneumococci were bled 18h later to determine the quantity of bacteria in the blood. Mice were then treated IV with a single 2mg dose of Cpl-1 and groups were euthanized 15 and 120 min after treatment to determine remaining pneumococci. A single dose reduced the bacterial burden below detectible limit.

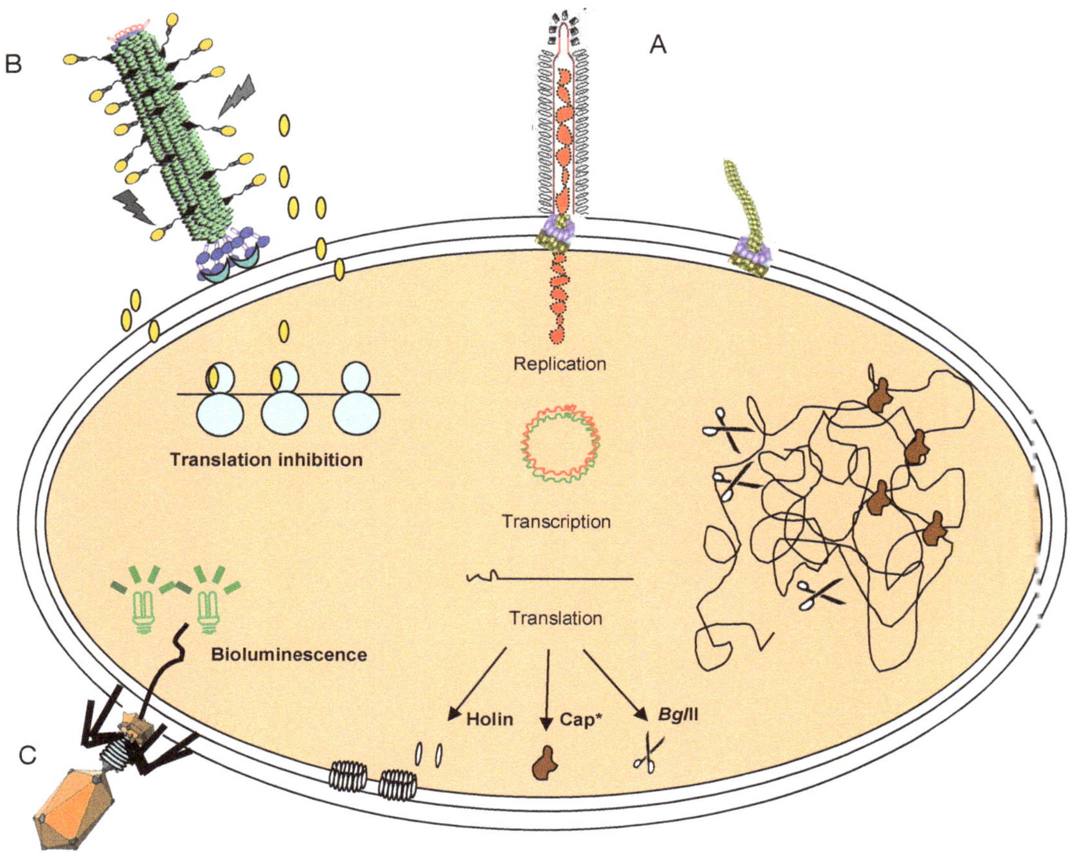

Figure 15.1 Host cell killing by genetically engineered phage. (A) Recombinant phage encoding lethal functions. M13 variants encoding the λ holin (kills the host cell by generation of lesions in the inner membrane), Cap* (binds chromosomal DNA non-specifically), or *Bgl*II (cleaves host DNA) are used to kill target bacteria without lysis. (B) Recombinant phage with engineered specificity for the bacterial target. M13 phage are used as a drug carrying platform. Upon targeted binding to a specific pathogen, the attached drug (chlorampenicol; yellow oval) is released and causes cell death. (C) Recombinant phage exploited as biodetectors. Phage encoding luciferase or GFP are employed to detect pathogens. For more details and references see text.

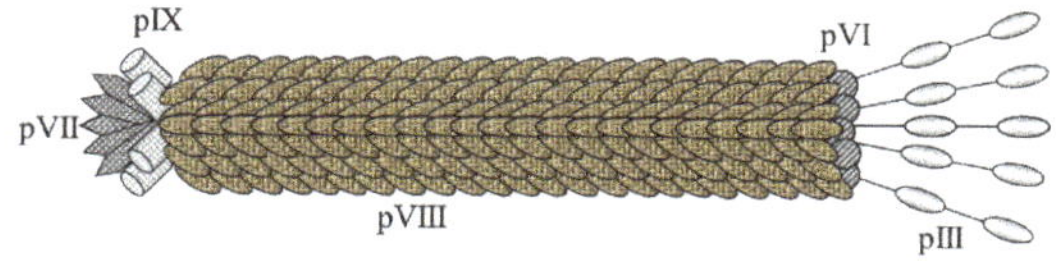

Figure 16.1 The schematic structure of filamentous phage.

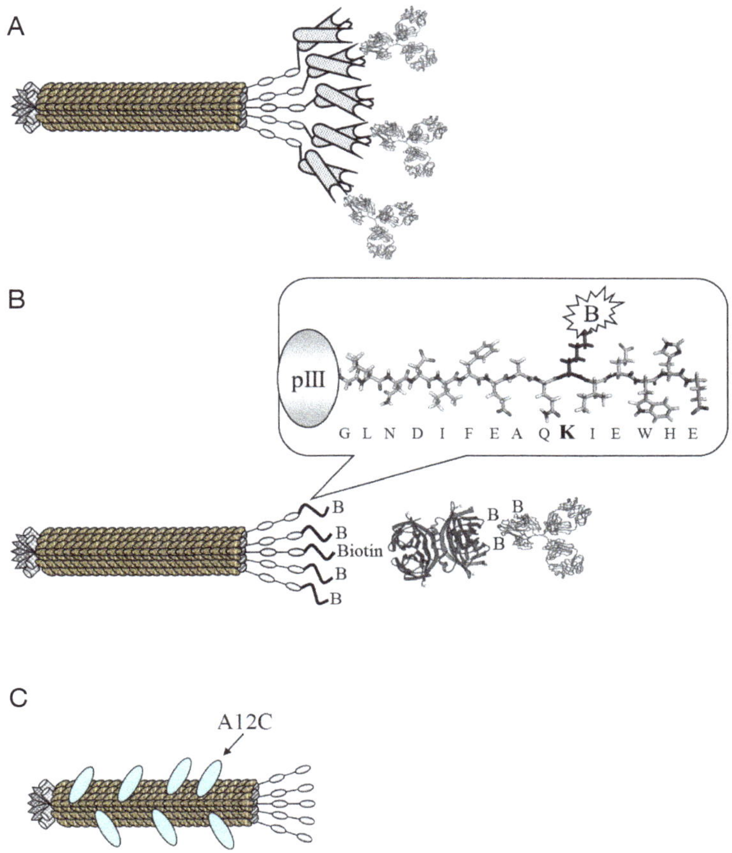

Figure 16.2 Schematic representation of antibody-based targeting of filamentous phage. (A) The anti-bacterial antibody provides the specific targeting via the ZZ domain displayed as a pIII phage coat protein fusion. (B) Biotin-displayed phages undergo complexation with biotinylated anti-bacterial antibody via a tetrameric avidin molecule. (C) A12C peptide displaying phage.

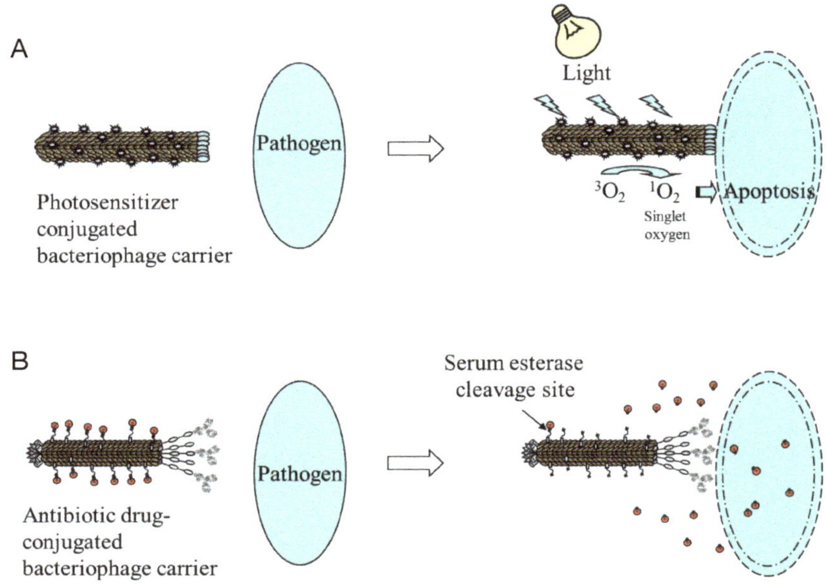

Figure 16.3 Mechanism of action of targeted drug-carrying phages. (A) Photosensitizer-conjugated phage tightly binds its host bacteria. Exposure to light at suitable wavelength generates singlet oxygen and free radicals release, providing the apoptosis of targeted pathogen. (B) The targeting moiety of drug-carrying phage provides the specific binding to selected bacteria. The drug linked to phage surface by serum esterase sensitive linker is released nearby the pathogen. The toxic effect is received by creating the high local drug concentration.

* 9 7 8 1 9 0 8 2 3 0 4 0 9 *